Lecture Notes in Computer Science 14760

Advanced Research in Computing and Software Science

Subline of Lecture Notes in Computer Science

More information about this series at https://link.springer.com/bookseries/558

Daniel Kráľ · Martin Milanič
Editors

Graph-Theoretic Concepts in Computer Science

50th International Workshop, WG 2024
Gozd Martuljek, Slovenia, June 19–21, 2024
Revised Selected Papers

Springer

Editors
Daniel Král'
Masaryk University
Brno, Czech Republic

Martin Milanič
University of Primorska
Koper, Slovenia

ISSN 0302-9743 ISSN 1611-3349 (electronic)
Lecture Notes in Computer Science
ISBN 978-3-031-75408-1 ISBN 978-3-031-75409-8 (eBook)
https://doi.org/10.1007/978-3-031-75409-8

This Springer imprint is published by the registered company Springer Nature Switzerland AG
The registered company address is: Gewerbestrasse 11, 6330 Cham, Switzerland

Preface

This LNCS volume is formed by the proceedings of the 50th International Workshop on Graph-Theoretic Concepts in Computer Science (WG), which was held in Gozd Martuljek, Slovenia, in the period June 19–21, 2024. The WG workshops have a long tradition going back to 1975: the previous workshops were held in Austria (twice), the Czech Republic (twice), France (three times), Germany (25 times), Greece, Israel, Italy, the Netherlands (five times), Norway, Poland, Slovakia, Spain, Switzerland (twice), Turkey, and the UK (twice). This year's WG workshop, which was organized jointly by the Faculty of Mathematics, Natural Sciences and Information Technologies and the Andrej Marušič Institute of the University of Primorska, the Slovenian Discrete and Applied Mathematics Society, and the Institute of Mathematics, Physics and Mechanics, was the first held in Slovenia. The Organizing Committee, chaired by Nina Chiarelli from the University of Primorska, prepared a very full social program for 53 participants from 16 countries (Austria, Canada, Czech Republic, France, Germany, India, Israel, Japan, the Netherlands, Poland, Slovenia, South Korea, Spain, Switzerland, the UK, and the USA) who attended the workshop this year.

The Program Committee of this year's WG workshop had 27 members from 13 countries (Argentina, Canada, the Czech Republic, France, Germany, Greece, Japan, the Netherlands, Norway, Poland, Slovenia, the UK, and Turkey). The Committee received 89 high-quality submissions and selected 31 of them for presentation at the workshop using a single-blind review process with 3 reviews per submission; this resulted in an acceptance rate of 35%. The extended abstracts of the 31 contributions that were selected for presentation at the workshop form this LNCS volume. In addition to these 31 contributions, the volume also contains a survey on approximation algorithms for tree width, path width, and tree depth prepared by Hans Bodlaender, who delivered the Test of Time Award talk at this year's WG workshop.

The scientific program of the workshop consisted of two invited talks, the Test of Time Award talk, and the talks based on the 31 contributions presented at the workshop. One of the invited talks was delivered by Shiri Chechik from Tel Aviv University, who spoke on approximating shortest paths and distance oracles, and the other by Michał Pilipczuk from the University of Warsaw, who spoke on the interplay of algorithms, graph classes, and logic. The WG 2024 Test of Time Award, which was decided by a committee consisting of Daniel Paulusma, Jan Arne Telle, and Dimitrios Thilikos, was given to the paper *Approximating Treewidth, Pathwidth, and Minimum Elimination Tree Height* by Hans L. Bodlaender, John R. Gilbert, Ton Kloks, and Hjálmtyr Hafsteinsson presented at WG 1991; the Test of Time Award talk was given by Hans Bodlaender. Among the 31 contributions presented at the workshop, the Program Committee selected the paper *Many Views of Planar Point Sets* by Jan Kynčl and Jan Soukup for the Best Paper Award. The Best Student Paper Award was given to two papers: the paper *Face-hitting Dominating Sets in Planar Graphs* by P. Francis, Abraham M. Illickan, Lijo M. Jose, and Deepak Rajendraprasad, and the paper

Degreewidth on Semi-complete Digraphs by Ryan Keeney and Daniel Lokshtanov. We would like to thank Springer-Verlag for sponsoring the Best Paper Award and the Best Student Paper Award. In addition, we would like to thank the European Mathematical Society, the Institute of Mathematics, Physics and Mechanics, and the University of Primorska for additional financial support and the Slovenian Discrete and Applied Mathematics Society for providing logistic support for the workshop. We are indebted to the WG Steering Committee and particularly to Daniel Paulusma, the Steering Committee's chair, for guidance and advice. We would also like to thank all speakers and the authors of the contributions presented at the workshop as well as the members of the Program Committee and the 154 external reviewers. Finally, our very special thanks go to Nina Chiarelli and the members of the Organizing Committee whose excellent work ensured the smooth running of the workshop.

August 2024

Daniel Kráľ
Martin Milanič

Organization

Program Committee

Isolde Adler	Otto-Friedrich-Universität Bamberg, Germany
Édouard Bonnet	LIP, ENS Lyon, France
Flavia Bonomo-Braberman	University of Buenos Aires, Argentina
Sergio Cabello	University of Ljubljana and IMFM, Slovenia
Tınaz Ekim	Boğaziçi University, Turkey
Jessica Enright	University of Glasgow, UK
Henning Fernau	University of Trier, Germany
Jakub Gajarský	University of Warsaw, Poland
Archontia C. Giannopoulou	National and Kapodistrian University of Athens, Greece
Thekla L.S. Hamm	Utrecht University, the Netherlands
Meike Hatzel	National Institute of Informatics, Japan
Philipp Kindermann	University of Trier, Germany
Ekkehard Köhler	Brandenburg University of Technology Cottbus–Senftenberg, Germany
Jan Kratochvíl	Charles University, Czech Republic
Daniel Král'	Masaryk University, Czech Republic
Matjaž Krnc	University of Primorska, Slovenia
O-Joung Kwon	Hanyang University, Republic of Korea
Barnaby Martin	Durham University, UK
Dániel Marx	CISPA, Germany
Martin Milanič	University of Primorska, Slovenia
Bojan Mohar	Simon Fraser University, Canada and IMFM, Slovenia
Torsten Mütze	Universität Kassel, Germany
Christophe Paul	CNRS, Université de Montpellier, France
Pawel Rzążewski	Warsaw University of Technology and University of Warsaw, Poland
Jan Arne Telle	University of Bergen, Norway
Viktor Zamaraev	University Liverpool, UK
Stanislav Živný	University of Oxford, UK

Additional Reviewers

Abu-Khzam, Faisal
Ahn, Jungho
Arrighi, Emmanuel
Baghirova, Narmina
Bahadir, Selim
Bandopadhyay, Susobhan
Behrooznia, Nastaran
Beisegel, Jesse
von Bergen, Hanno
Bessy, Stéphane

Blažej, Václav
Bose, Prosenjit
Bougeret, Marin
Bousquet, Nicolas
Brandes, Ulrik
Cao, Yixin
Castelli Aleardi, Luca
de Castro Mendes Gomes, Guilherme
Cavallaro, Dario
Chudigiewitsch, Florian
Cohen, Johanne
Conte, Alessio
Cook, Linda
Dabrowski, Konrad K.
Dallard, Clément
Davot, Tom
Dettlaff, Magda
Diestel, Reinhard
Disser, Yann
Dragan, Feodor
Ducoffe, Guillaume
Eagling-Vose, Tala
Eiben, Eduard
Erdem, Burak Nur
Feghali, Carl
Felsner, Stefan
Feng, Zhidan
Fiala, Jiri
Focke, Jacob
Foucaud, Florent
Francis, Mathew
Garlet Milani, Marcelo
Gimbel, John
Giroudeau, Rodolphe
Gonçalves, Daniel
Gorsky, Maximilian
Gottlieb, Eric
Groenland, Carla
Gupta, Siddharth
Gurski, Frank
Guzman Pro, Santiago
Haase, Carolina
Hanaka, Tesshu
Harris, David
Hilaire, Claire
Hoang, Hung
Inkulu, R.
Ishii, Toshimasa
Jacob, Hugo
Jaffke, Lars
Kanesh, Lawqueen
Kellerhals, Leon
Kim, Donggyu
Kinnersley, Bill
Kiss, Peter
Klobas, Nina
Klute, Fabian
Knop, Dušan
Kobayashi, Yasuaki
Kolay, Sudeshna
Komusiewicz, Christian
Kunz, Pascal
Kutner, David
Labarre, Anthony
Laekhanukit, Bundit
Lehtilä, Tuomo
de Lima, Paloma
Lisowski, Grzegorz
Livieratos, John
Lužar, Borut
Mann, Kevin
Marković, Petar
Masařík, Tomáš
Mavropoulos, Filippos
McConnell, Ross
Memoli, Facundo
Merino, Arturo
Micka, Ondrei
Miklavič, Štefko
Misra, Pranabendu
Miura, Kazuyuki
Mnich, Matthias
Molter, Hendrik
Morelle, Laure
Muzi, Irene
Myrvold, Wendy
Mömke, Tobias
Müller, Haiko
N. R., Aravind
Namrata, Namrata
Nedela, Roman
Nichterlein, André

Nikabadi, Amir
Nimbhorkar, Prajakta
Oh, Eunjin
Oostveen, Jelle
Ordyniak, Sebastian
Özdeş, Çelik
Peczarski, Marcin
Pekárek, Jakub
Pettersson, William
Pfretzschner, Matthias
Picouleau, Christophe
Piecyk, Marta
Piselli, Tommaso
Pérez-Lantero, Pablo
Ratajczak, Fabienne
Rauch, Johannes
Rollin, Jonathan
Rote, Günter
Rutter, Ignaz
Salehi, Özlem
Sandeep, R.B.
dos Santos, Vinicius F.
Saumell, Maria
Scheffler, Robert
Scheucher, Manfred
Schirneck, Martin
Schlipf, Lena
Schlöter, Jens
Schulz, André
Siggers, Mark
Smith, Siani
Souza, Uéverton
Speckmann, Bettina
Stacho, Ladislav
Stamoulis, Giannos
Steiner, Raphael
Strehler, Martin
Stumpf, Peter
Subramanian, Sathyawageeswar
Šámal, Robert
Špacapan, Simon
Tale, Prafullkumar
Tennenhouse, Craig
Tokar, Nizamettin
Valla, Tomáš
Verbeek, Kevin
Wellnitz, Philip
Wiederrecht, Sebastian
Wiese, Andreas
Yamanaka, Katsuhisa
Yildiz, Mehmet Akif
Zoros, Dimitris

Contents

Test of Time Award Survey

Approximation Algorithms for Treewidth, Pathwidth, and Treedepth—A Short Survey

Hans L. Bodlaender(✉)

Department of Information and Computing Sciences, Utrecht University, Utrecht, The Netherlands
h.l.bodlaender@uu.nl

Abstract. This short survey discusses old and new approximation algorithms for treewidth, and for the related parameters pathwidth and treedepth.

Keywords: Treewidth · Pathwidth · Treedepth · Approximation · Graph Algorithms

1 Introduction

Treewidth is one of the most studied graph parameters. The interest in this parameter comes from several applications, some in a theoretical setting, and some in an applied setting. Related graph parameters are pathwidth and treedepth. Each of these three parameters is NP-complete [3,27,41]. In this short survey, we discuss approximation algorithms for treewidth, pathwidth, and treedepth. We focus on algorithms with a guaranteed approximation ratio. We also discuss lower bound proofs for approximating the treewidth of graphs. For surveys on heuristic algorithms (without guaranteed ratios) for treewidth, see e.g., [14,46].

This paper is organised as follows. In Sect. 2, we briefly introduce some notations. Section 3 introduces the parameters treewidth, pathwidth, and treedepth, and discusses equivalent definitions, and a relation with Gaussian elimination. Section 4 discusses algorithms that approximate the treewidth and give tree decompositions of the input graphs with a guaranteed bound on the width. Most of these algorithms have a similar shape, and are described in the setting of a general framework. The most recent algorithms use a different method. After that, Sect. 5 looks at lower bound results on polynomial time approximation of treewidth. Section 6 discusses approximations of treedepth and pathwidth. Some concluding remarks are made in Sect. 7.

This survey is on the occasion of the Test-of-Time award for the paper *Approximating Treewidth, Pathwidth, and Minimum Elimination Tree Height*. This paper was first presented at the 17th International Workshop on Graph Theoretic Concepts in Computer Science, WG 1991 [13] in Fischbachau, Germany, June 17–19, 1991; the award was given at the 50th International Workshop on Graph Theoretic Concepts in Computer Science, WG 2024 in Gozd Martuljek, Slovenia, June 19–21, 2024.

D. Král' and M. Milanič (Eds.): WG 2024, LNCS 14760, pp. 3–18, 2025.
https://doi.org/10.1007/978-3-031-75409-8_1

2 Some Notation

We sometimes use interval notation to denote intervals of integers, e.g., $[1, 3] = \{1, 2, 3\}$. When not differently specified, the variable n denotes the number of vertices of graph $G = (V, E)$.

For a graph $G = (V, E)$ and a vertex set $W \subseteq V$, the subgraph of G induced by W is $G[W] = (W, \{\{v, w\} \in E \mid v, w \in W\})$. With $G \setminus W$, we denote the graph $G[V \setminus W]$, i.e., the subgraph induced by all vertices not in W.

The operation to *eliminate* a vertex v in a graph G is the following: first, turn the neighbourhood of v into a clique (add an edge between each pair of non-adjacent neighbours of v), and then remove v. An example is given in Fig. 1.

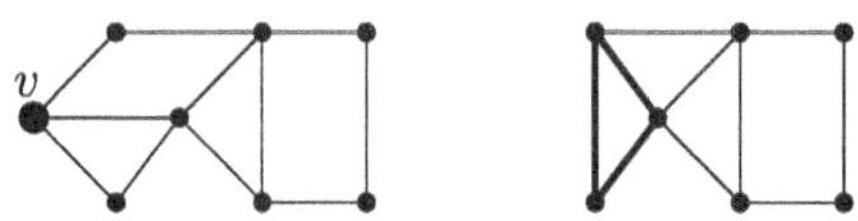

Fig. 1. A graph before and after eliminating vertex v

Most of approximation algorithms that we discuss in this paper can be divided in two categories: polynomial time approximation algorithms (with a running time of the form $O(n^{O(1)})$), and FPT approximation algorithms (with a running time of the form $O(f(k)n^{O(1)})$ where f is a computable function, k the parameter at hand). We say that an algorithm for parameter p has approximation ratio $f(p, n)$, or, equivalently, is a $f(p, n)$-approximation, if there is a constant c, such that for each graph G with n vertices, the algorithm outputs a solution with value at most $f(p(G), n) \cdot p(G) + c$. E.g., if we output tree decompositions of width at most $4\mathtt{tw}(G) + 3$, then we say that the algorithm has a approximation ratio 4; if we output tree decompositions of width at most $\sqrt{\log \mathtt{tw}(G)} \cdot \mathtt{tw}(G)$, then the algorithm has a approximation ratio $\sqrt{\log \mathtt{tw}}$.

3 Graph Parameters

Treewidth. Several groups of authors independently invented graph parameters that are equivalent or similar to treewidth. Already in 1972, Bertelè and Briochi [7] introduced the equivalent concept of dimension. For some time, different terminologies were used, e.g., the notion of partial k-trees [3], until the concepts of treewidth and tree decompositions, introduced by Robertson and Seymour [44] became the standard terminology.

Definition 1. *A tree decomposition of a graph $G = (V, E)$ is a pair $(T, \mathcal{X})$, where $T = (I, F)$ is a tree, and $\mathcal{X}$ is a collection of subsets of V, with for each $i \in I$ a set X_i, such that the following three properties hold:*

1. $\bigcup_{i \in I} X_i = V$.

2. *For all $\{v, w\} \in E$, there is an $i \in I$ with $v, w \in X_i$.*
3. *For all $v \in V$, the nodes $i \in I$ with $v \in X_i$ form a connected subtree of $\mathcal{T}$.*

The width *of a tree decomposition $(\mathcal{T}, \mathcal{X})$ is* $\max_{i \in I} |X_i| - 1$. *The* treewidth *of a graph G, denoted* $\mathtt{tw}(G)$, *is the minimum width of a tree decomposition of G.*

There are many definitions that are equivalent to treewidth; see [9] for an overview of several of these. Some are related to chordal graphs and the different equivalent characterisations of a chordal graph. An elimination scheme of a graph $G = (V, E)$ is a permutation π of the vertices. Suppose we eliminate the vertices of G one by one, in the order of elimination scheme π. The *width* of π is the maximum over all vertices of the degree when it is eliminated.

Theorem 1. *For a graph $G = (V, E)$ and integer k, the following are equivalent.*

1. $\mathtt{tw}(G) \leq k$.
2. *G is a subgraph of a chordal graph with maximum clique size at most $k + 1$.*
3. *G has an elimination scheme of width at most k.*

Proof. (Sketch.) The equivalence of (1) and (2) follows from the fact that a chordal graph has a tree decomposition where each bag is a clique. The equivalence of (2) and (3) follows from the fact that chordal graphs have an elimination scheme where the neighbours of a vertex that is eliminated always form a clique. (For more background on chordal graphs, see [21].) □

Definition 2. *Let $G = (V, E)$ be a graph, $W \subseteq V$ a set of vertices, and $\alpha \in (0, 1)$ a real number between 0 and 1. A set $S \subseteq V$ is an α-balanced separator for W in G, if each connected component of $G \setminus S$ has at most $\alpha \cdot |W|$ vertices in W.*

Central to many of the approximation algorithms discussed in this paper is the following well known result. The proof below is based on Robertson and Seymour [44, Theorem 2.5].

Theorem 2. *Let $G = (V, E)$ be a graph, and $W \subseteq V$ a set of vertices. There is a $\frac{1}{2}$-balanced separator of W in G of size at most* $\mathtt{tw}(G) + 1$.

Proof. (Sketch.) Take a tree decomposition of minimum width, $(\mathcal{T}, \mathcal{X})$. Take some node i_0 in T. If X_{i_0} is not a $\frac{1}{2}$-separator of W, then suppose W_1 is a connected component of $G \setminus X_{i_0}$ with $|W_1| > |W|/2$. Look at the subtrees of $\mathcal{T}$ obtained by removing i_0 from $\mathcal{T}$. All vertices in W_1 are in bags of exactly one of these subtrees – this follows from the properties of tree decompositions. Let i_1 be the neighbour of i_0 in this subtree. When i_1 is not a $\frac{1}{2}$-separator of W, we repeat this step, and continue doing so until we find the desired balanced separator. After moving from i_j to i_{j+1}, we cannot go back to i_j in the next step, as this would give two disjoint subsets of W that both have size more than $|W|/2$. So the process stops, and we find a bag X_i that is a $\frac{1}{2}$-separator of W. □

Pathwidth. A *path decomposition* is a tree decomposition $(\mathcal{T}, \mathcal{X})$ with $\mathcal{T}$ a path. The *pathwidth* of a graph G, denoted $\mathtt{pw}(G)$, is the minimum width of a path decomposition of G.

Clearly, the treewidth of a graph is at most its pathwidth. Several equivalent definitions of pathwidth exist, see [9] for a partial survey.

Treedepth. There are several equivalent definitions of treedepth. We give below a commonly used one.

Definition 3. *An* elimination forest *of a graph $G = (V, E)$ is a forest $F = (V, F)$ with each tree in the forest rooted, such that for each edge $\{v, w\} \in E$, v is an ancestor of w in F or w is an ancestor of v in F. The* treedepth *of a graph G is the minimum depth over all elimination forests of G.*

One equivalent notion is that of the *vertex ranking number* (see e.g., [10]). A vertex k-ranking of a graph $G = (V, E)$ is a function $f : V \to [1, k]$, such that for each $v, w \in V$ with $f(v) = f(w)$, on each path from v to w in G, there is a vertex x with $f(x) < f(v)$. The vertex ranking number of G is the smallest k for which G has a vertex k-ranking. Another is that of *minimum elimination tree height*: for an elimination ordering π of G, eliminate the vertices in order of π, and build a rooted forest by G by setting the parent of a vertex the lowest numbered neighbour it has when it is eliminated. Several other equivalent definitions can be found in a survey by Nešetřil and Ossona de Mendez [40], see also e.g., [39].

Theorem 3. *For each graph G, the treedepth, vertex ranking number of minimum elimination tree height are equal.*

Gaussian Elimination and Graphs. We now look at the relation between sparse symmetric matrices, Gaussian elimination, and graphs.

Suppose we have an n by n matrix A. We say that A is *symmetric* if for all $i, j \in [1, n]$, $a_{ij} = a_{ji}$. We can associate an undirected graph G_A with n vertices to a symmetric matrix A. The vertex set of G_A is $[1, n]$, i.e., we have a vertex for each pair of a row and its corresponding column. The edges are formed by the non-zero's: for all $i, j \in [1, n]$, we have an edge $\{i, j\}$ in G_A, if and only if $i \neq j$ and $a_{ij} \neq 0$.

If a matrix A has only non-zero entries (or only zero's on the main diagonal), then G_A is a complete graph. If we have a sparse matrix (a matrix with many entries that are 0), then G_A can reveal interesting structural insights.

In one step of the Gaussian elimination process, we create a column with only one non-zero by subtracting one of the rows a number of times from other rows. Suppose $a_{11} \neq 0$. Now, in the Gaussian elimination process, we can create a new matrix A' with a first column with all $a'_{i1} = 0$ for $i \neq 1$: for all $i \neq 1$ with $a_{i1} \neq 0$, we subtract $\frac{a_{i1}}{a_{11}}$ times the first row from the ith row. I.e., $a'_{ij} = a_{ij} - \frac{a_{i1}}{a_{11}} \cdot a_{j1}$. After this, the Gaussian elimination process continues, but the row and corresponding column of the first step are ignored; e.g., when we sweep with the first row, we continue on the $n-1$ by $n-1$ submatrix without the first row and first column.

Consider the sparse matrix A at the left side of (1); suppose entries are only 0 when explicitly shown as such. If we perform a Gaussian elimination step with the first row, then we obtain a 0 at the first element of the fourth and sixth row. Only the four entries marked with (*) change their value; all others are unchanged.

$$A = \begin{pmatrix} a_{11} & 0 & 0 & a_{14} & 0 & a_{16} \\ 0 & a_{22} & 0 & a_{24} & a_{25} & 0 \\ 0 & 0 & a_{33} & a_{34} & 0 & a_{36} \\ a_{14} & a_{24} & a_{34} & a_{44} & a_{45} & 0 \\ 0 & a_{25} & 0 & a_{45} & a_{55} & a_{56} \\ a_{16} & 0 & a_{36} & 0 & a_{56} & a_{66} \end{pmatrix} \rightarrow \begin{pmatrix} a_{11} & 0 & 0 & a_{14} & 0 & a_{16} \\ 0 & a_{22} & 0 & a_{24} & a_{25} & 0 \\ 0 & 0 & a_{33} & a_{34} & 0 & a_{36} \\ \mathbf{0} & a_{24} & a_{34} & (*) & a_{45} & (*) \\ 0 & a_{25} & 0 & a_{45} & a_{55} & a_{56} \\ \mathbf{0} & 0 & a_{36} & (*) & a_{56} & (*) \end{pmatrix} \quad (1)$$

It is interesting to see what happens to G_A when we eliminate a row with Gaussian elimination. We can create new non-zero elements: if we sweep with row k, then that a_{ij} only can become a new non-zero element when $a_{1k} \neq 0$ and $a_{kj} \neq 0$. In graph terms, this means that we can edges between neighbours of vertices of k. A In many applications (when matrix entries are real values from some application), it is not likely that a non-zero element becomes 0. Continuing on the submatrix without row and column corresponds to removing vertex k from the graph. Thus, one step of the Gaussian elimination process precisely corresponds to the elimination of a vertex (as described in Sect. 2). Figure 2 gives G_A, and the graph corresponding to the last five columns and rows of the matrix at the right side in (1).

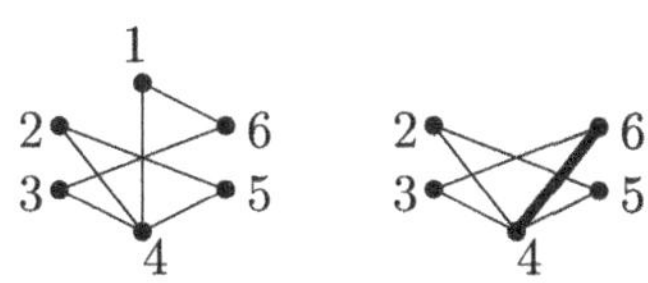

Fig. 2. Left: the graph G_A. Right: graph obtained from G_A by eliminating vertex 1.

The *minimum front-size* of matrix M is the minimum over all orderings in which we eliminate variables of the maximum number of non-zero's in one elimination step [12,19]. For a fast execution of Gaussian elimination, it can be useful to eliminate rows in an order such that the front-size is small. Observe that this number of non-zero's equals the degree of the corresponding vertex when it is eliminated from $G[M]$, and thus the minimum front-size of M equals the treewidth of $G[M]$—see e.g. [12] and Theorem 1.

The treedepth of a graph $G[M]$ reflects the time of a parallel algorithm for the Gaussian elimination: we can eliminate simultaneously a set of rows that corresponds to an independent set in $G[M]$ (and, in later steps, the graph resulting from $G[M]$ from earlier eliminations). Thus, the height of the elimination tree gives the number of such rounds of eliminating sets of rows (see Theorem 3 and [35,36]).

4 Approximation of Treewidth

Many approximation algorithms for TREEWIDTH have a similar structure. We first describe an algorithmic framework, that given an algorithm that finds balanced separators, gives an algorithm that approximates treewidth. After this, we briefly describe a number of approximation algorithms for treewidth by pointing out some specific characteristics and how they fit in the framework. At the end of this section, we discuss recent approximation algorithms for treewidth that are based on a different algorithmic paradigm.

An Algorithmic Framework. The main idea of the framework algorithm is that a tree decomposition is constructed top-down, where at each step, we use a balanced separator of the parent bag. Algorithm 1 gives pseudo-code of the framework. The framework has three steps that need further specification. In particular, different treewidth approximation algorithms implement Step 6 (where a balanced separator is found) in a different way. Possibly, the balanced separator algorithm can tell that the treewidth of the graph is larger than k, and we halt the entire procedure (Step 7). The bound in Step 3 can be set to the maximum width we aim to obtain (e.g., to $4k+3$ if we aim to obtain a 4-approximation algorithm). We first will discuss the correctness of the procedure, and what approximation ratios we will obtain.

Algorithm 1. Treewidth-Approximate(G, W, k)

1: Input: Graph $G = (V, E)$, vertex set $W \subseteq V$.
2: Output: rooted tree decomposition with W subset of root bag, or '$\texttt{tw}(G) > k$'.
3: **if** $|V| <$ bound **then**
4:     Output a tree decomposition with one bag V.
5: **else**
6:     Find a balanced separator S of W in G. ▷ *How? See discussion!*
7:     **if** No 'good' balanced separator exists **then**
8:         Output $\texttt{tw}(G) > k$; stop.
9:     **end if**
10:     Compute the connected components of $G \setminus S$.
11:     Suppose the connected components of $G \setminus S$ have vertex sets $V_1, \ldots, V_r$.
12:     **for** $i = 1, \ldots r$ **do**
13:         $W_i \leftarrow V_i \cap W$.
14:         Recursively call Treewidth-Approximate($G[V_i \cup S]$, $W_i \cup S$, k)
15:     **end for**
16:     Output a tree decomposition with new root bag i_r with $X_{i_r} = W \cup S$, and the disjoint union of all tree decompositions of the graphs $G[V_i \cup S]$, with i_r the parent of all root bags of these tree decompositions. ▷ See Figure 3.
17: **end if**

Lemma 1. *Suppose that at Step 6 of the Treewidth-Approximate algorithm, the algorithm either correctly determines that the treewidth of G is larger than k,*

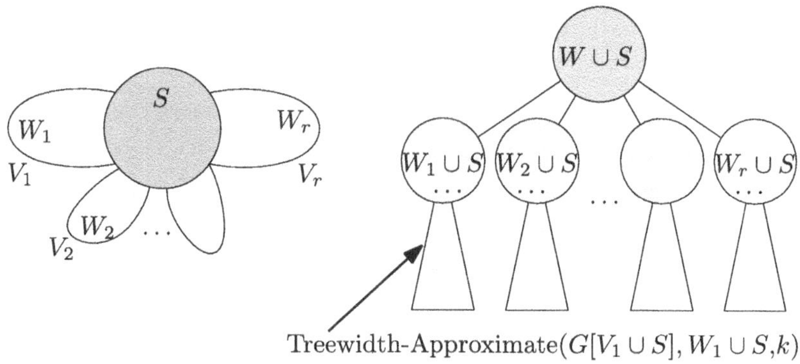

Fig. 3. Illustration of the last step of Treewidth-Approximate

or it finds an α-balanced separator of W in G of size $f(n,k)$. Then if we call Treewidth-Approximate(G, W, k) with $|W| \leq \frac{1}{1-\alpha} f(n,k)$, then it either outputs a tree decomposition of G, with W a subset of the root bag of width at most $\max\{\text{bound}, (1 + \frac{1}{1-\alpha}) \cdot f(n,k)\}$ *or correctly outputs that* $\mathtt{tw}(G) > k$.

Proof. It is not hard to verify that when the algorithm outputs a tree decomposition, this indeed satisfies the requirements of tree decompositions of G.

With induction, we have that for each recursive call done by the algorithm, the set in the second argument has size at most $\frac{1}{1-\alpha} f(n,k)$. This holds for the main call. If we recurse, then we have for each i that $|W_i| \leq \alpha \cdot |W|$, so $|W_i \cup S| \leq \alpha |W| + f(n,k) \leq \alpha \cdot \frac{1}{1-\alpha} f(n,k) + f(n,k) = \frac{1}{1-\alpha} \cdot f(n,k)$. Each bag is the union of a separator of size at most $f(n,k)$ and a set of size $\frac{1}{1-\alpha} f(n,k)$, so the width of the tree decomposition is at most $f(n,k) + \frac{1}{1-\alpha} f(n,k)$. □

If the separator algorithm in Step 6 uses $T(n,k)$ time, then the entire procedure uses $O(n \cdot T(n,k))$ time, as the number of recursive calls in total is $O(n)$.

The First Treewidth Approximation. We now will discuss a number of different approximation algorithms for treewidth that fit (with variations) in the framework discussed above. The first such algorithm is based on the work by Robertson and Seymour [45] from the 1980s.

In the thirteenth paper in their monumental series of papers on Graph Minors, Robertson and Seymour [45] give an algorithm that given a graph G and an integer k, either decides that the branchwidth of G is at most k, or give a branch decomposition for G of width at most $3k$. The algorithm uses $O(3^{3k} n^2)$ time. For each graph G with branchwidth $\mathtt{bw}$ and treewidth $\mathtt{tw}$, it holds that

$$\mathtt{bw} - 1 \leq \mathtt{tw} \leq \left\lfloor \frac{3}{2} \mathtt{bw} \right\rfloor - 1.$$

The combination of these results directly implies a 4.5-approximation algorithm for treewidth that uses $O(3^{3\mathtt{tw}} n^2)$ time. More close observation reveals that we have a 4-approximation algorithm.

Lemma 2. *Let $G = (V, E)$ be a graph with treewidth at most k, and $W \subseteq V$. Then we can partition V in three sets V_1, V_2, V_3, with*

1. *no vertex in V_1 is adjacent to a vertex in V_3;*
2. $|V_1 \cap W| \leq \frac{2}{3}|W|$; $|V_2 \cap W| \leq \frac{2}{3}|W|$;
3. $|V_2| \leq k + 1$.

Proof. Let V_2 be a $\frac{1}{2}$-balanced separator of W in G. We can group the connected components of $G \setminus V_2$ in two sets such that each contains at most two-thirds of the vertices in W. □

The set V_2 functions as a $\frac{2}{3}$-separator of W. Applying the framework, we will use sets W of size at most $\frac{1}{1-\frac{2}{3}} \cdot (k+1) = 3k+3$. Each vertex in W either is in V_1, V_2, V_3: we try all 3^{3k+3} choices for this. For each, find a minimum size vertex separator S in the graph $G \setminus (W \cap V_2)$ that separates the vertices in $V_1 \cap W$ from the vertices in $V_2 \cap W$. If S has size larger than k, then $\mathtt{tw}(G) > k$. S can be found by standard network flow techniques. As the size of the separator is at most k, we can employ at most k rounds of a flow augmentation in the Ford-Fulkerson algorithm, and thus each of the 3^{3k+3} tries can be done in $O(k \cdot (|V| + |E|)) = O(k^2 n)$ time.

Theorem 4 (Folklore, after Robertson, Seymour [45]). *There is an algorithm that, given a graph G, and integer k, either decides that $\mathtt{tw}(G) > k$ or finds a tree decomposition of G of width at most $4k + 3$ in $O(3^{3k} k^2 n)$ time.*

Various Separator Algorithms. We now discuss several other treewidth approximation algorithms, that all can be seen as special cases of Algorithm 1. In [12,13], it was noted that we can use the following result.

Theorem 5 (Leighton, Rao, 1988 [33,34]). *There is a polynomial time algorithm, that given a graph $G = (V, E)$ and a vertex set $W \subseteq V$, finds a $\frac{2}{3}$-balanced separator of W in G of size $O(r \log n)$, where r is the minimum size of a $\frac{1}{2}$-balanced separator of W.*

If we use the algorithm of Theorem 5 in the framework, then $r \leq \mathtt{tw}(G) + 1$, and thus we find separators of size $O(\mathtt{tw}(G) \cdot \log n)$, and the algorithm outputs tree decomposition of width $O(\mathtt{tw}(G) \cdot \log n)$, i.e., we have a polynomial time algorithm with approximation ratio $O(\log n)$.

A number of papers give results that give FPT approximation or polynomial time approximation algorithms, with better approximation ratios or better running times.

- Matoušek and Thomas [37] use a probabilistic algorithm to find separators, resulting in an algorithm that uses $O(\mathtt{tw}^{O(\mathtt{tw})} n \log^2 n + n \log n \log p)$ time and has ratio 6, with a probability p to give an incorrect answer.
- Lagergren [31,32] gives a parallel algorithm whose sequential variant uses $O(\mathtt{tw}^{O(\mathtt{tw})} \cdot n \log^2 n)$ time with a approximation ratio of 8.

- Reed [43] gives an algorithm that uses $O(\mathtt{tw}^{O(\mathtt{tw})} \cdot n \log n)$ time with a approximation ratio of 8.
- Amir [1,2] gives several algorithms, including an FPT algorithm with ratio $3\frac{2}{3}$ and a polynomial time algorithm with ratio $O(\log \mathtt{tw})$.
- Bouchitté et al. [16] use an separator approximation algorithm by Even [20] and obtain a polynomial time algorithm with approximation ratio $O(\log \mathtt{tw})$.
- Feige, Hajiaghay, and Lee [23] give an improved algorithm for finding separators, which gives a polynomial time algorithm with approximation ratio $O(\sqrt{\log \mathtt{tw}})$.
- Fomin et al. [25] have a fast $O(\mathtt{tw}^7 n \log n)$ (polynomial) time algorithm with an approximation ratio $O(\mathtt{tw})$.
- A version of Reed's algorithm with a running time of the form $O(2^{O(\mathtt{tw})} n \log n)$ was given by Belbasi and Fürer [5].

Double Balanced Separators. We now briefly discuss a variant that can help to speed up the algorithm, and has an interesting connection to treedepth. Suppose that we add to the separator found in Step 6 of Algorithm 1 a second balanced separator, but now one that is an α-balanced separator for the entire (current) vertex set V, for some $\alpha \in (0,1)$. This increases the width of the obtained tree decomposition, but also ensures that the recursion depth of the algorithm, and hence the height of the tree $\mathcal{T}$ of the obtained tree decomposition are bounded by $O(\log n)$.

For instance, Reed [43] gives a clever algorithm that finds such a balanced separator of the vertex set of size $O(\mathtt{tw})$ in $O(\mathtt{tw}^{O(\mathtt{tw})} n)$ time. When going in recursion, all but $O(\mathtt{tw})$ vertices only participate in one branch, and with this, it follows that the total running time is bounded by $O(\mathtt{tw}^{O(\mathtt{tw})} n \log n)$.

We thus see that each graph G has a tree decomposition $(\mathcal{T}, \mathcal{X})$ of width $O(\mathtt{tw}(G))$ with $\mathcal{T}$ a tree with height $O(\log n)$. With the following lemma it follows that $\mathtt{td}(G) = O(\mathtt{tw}(G) \cdot \log n)$.

Lemma 3. *Suppose G has a tree decomposition $(\mathcal{T}, \mathcal{X})$ of width k and with T a rooted tree of height t. Then the treedepth of G is at most $(k+1) \cdot (t+1)$.*

Proof. Take an arbitrary node of $\mathcal{T}$ as root. For each vertex v, let i_v be the highest node in $\mathcal{T}$ with $v \in X_{i_v}$. Assume some arbitrary ordering on the vertices V. We can rank the vertices top-down in the tree decomposition: set $f(v) = (k+1) \cdot d + j$, if i_v has distance d to the root of $\mathcal{T}$, and v is the jth vertex in X_{i_v}. One can check that this is a vertex $(k+1) \cdot (t+1)$-ranking of G, □

An $O(c^{\mathtt{tw}} n)$ 5-Approximation Algorithm. The first algorithm with a constant approximation ratio for treewidth with a running time of the form $c^{\mathtt{tw}} n$ for some constant c was given in [11]. The algorithm uses a combination of techniques to speed up the steps in the framework. One of these is the following: with a variation of a technique from [8], we obtain (by running recursively the same algorithm on a smaller graph) a tree decomposition of the input graph G of width $10k + O(1)$. This tree decomposition can be used as a data structure that allows queries that allow to find balanced separators in less time. The paper also gives a 3-approximation algorithm with running time of the form $O(c^{\mathtt{tw}} n \log n)$.

Local Improvement. In 2021, Korhonen [29] gave a novel algorithm that improved upon the result from [11] in several ways. His algorithm has running time of the form $O(c^{\mathtt{tw}}n)$ and gives a 2-approximation for treewidth. The algorithm not only has a better approximation ratio, but also is much simpler and it has a significantly better constant c at the base of the exponent.

The algorithm also uses a new and very interesting technique. Instead of building a tree decomposition top-down as in Algorithm 1, Korhonen's algorithm starts with a tree decomposition of larger width and then improves this tree decomposition in a number of successive steps.

First, by recursively calling the algorithm on a smaller graph using a modification of a technique from [8], we get a tree decomposition of width at most $4k+3$. Then, as long as the current tree decomposition has a bag with $2k+3$ or more vertices, the algorithm uses an improvement step: employing a generalisation of a result by Bellenbaum and Diestel [6], Korhonen shows that this large bag can be split and the vertices can be rearranged such that the resulting tree decomposition has fewer large bags. When the process stops, all bags have size at most $2k+2$. Possibly, the algorithm finds that no improvement of a large bag is possible, and then it concludes that $\mathtt{tw}(G) > k$.

Theorem 6 (Korhonen, 2021 [29]). *There is an algorithm using $2^{O(k)}n$ time, that given a graph G and integer k, either correctly outputs that $\mathtt{tw}(G) > k$, or gives a tree decomposition of G of width at most $2k+1$.*

A later paper by Korhonen and Lokshtanov [30] use this local improvement method to obtain for each $\epsilon > 0$, $(1+\epsilon)$-approximation algorithms that run in $\mathtt{tw}^{\mathtt{tw}/\epsilon}n^{O(1)}$ time.

5 Lower Bounds for Treewidth Approximation

In the previous section, we have seen a number of polynomial time and FPT approximation algorithms for the treewidth of graphs. Notably, we do not know of a polynomial time approximation algorithm for treewidth with a constant approximation ratio. Some complexity theoretic results indicate that the existence of such algorithms is unlikely.

The first published lower bound result was the relatively simple observation that polynomial time approximations with an additive term are unlikely. See [12] for similar results for pathwidth and treedepth.

Fig. 4. Replace each vertex by a clique of size $c+1$. In this example $c = 1$

Theorem 7 (Bodlaender, Gilbert, Hafsteinsson, Kloks [12,13]**).** *If $P \neq NP$, then for each $c \geq 0$, there is no polynomial time algorithm A_c, that given a graph G and integer k, either correctly outputs that the treewidth of G is more than k, or that the treewidth of G is at most $k + c$.*

Proof. Suppose on the contrary that we have such an algorithm A_c. Take an input $(G = (V, E), k)$ for the NP-complete TREEWIDTH problem. Build a graph H by replacing each vertex $v \in V$ by a clique with $c + 1$ vertices, with copies of v and w adjacent in H, if and only if v and w are adjacent in G. See Fig. 4. One can show that $\mathtt{tw}(H) + 1 = (c + 1)(\mathtt{tw}(G) + 1)$. So, if we apply A_c to H with target value $(k + 1)(c + 1) - 1$, then one of the following two cases holds. If A_c tells that the treewidth of H is at most $(k+1)(c+1) - 1 + c$, then the treewidth of H is at most $(k + 1)(c + 1) - 1$, and the treewidth of G is at most k. If A_c tells that the treewidth of H is larger than $(k+1)(c+1) - 1$, then the treewidth of G is larger than k. So, we have a polynomial time algorithm that decides on the treewidth of G, contradicting $P \neq NP$. □

For many years, it was a major open problem whether approximations of treewidth with a constant approximation ratio existed. The first breakthrough was obtained by Wu et al. [47].

Theorem 8 (Wu, Austrin, Pitassi, Liu, 2014 [47]**).** *Suppose the Small Set Expansion hypothesis holds. Then, for any $\alpha > 1$, there is a $c > 0$, such that there is no polynomial time algorithm that for any given graph G with n vertices, either correctly outputs that $\mathtt{pw}(G) \leq c \cdot n$, or correctly outputs that $\mathtt{tw}(G) \geq \alpha \cdot c \cdot n$.*

The theorem implies that when we assume that the Small Set Expansion hypothesis holds, then there is no polynomial time approximation algorithm for TREEWIDTH and PATHWIDTH with a constant approximation ratio. Wu et al. [47] show that a similar inapproximability hold for a number of other graph parameters, but do not discuss treedepth.

The Small Set Expansion (SSE) hypothesis was introduced by Raghavendra and Steurer [42] in 2010. The hypothesis implies the Unique Games Conjecture, and is still topic of debate whether it should be considered to be holding. Very recently, Bonnet [15] gave an inapproximality result for treewidth for the much more common assumption that $P \neq NP$. He also gave a time bound based on the Exponential Time Hypothesis. The results are based on a clever transformation from 3-SATISFIABILITY.

Theorem 9 (Bonnet, 2024 [15]**).** *If $P \neq NP$, then there is no polynomial time approximation algorithm that approximates treewidth with approximation ratio 1.00005. If the Exponential Time Hypothesis holds, then there is a $\delta > 1$, such that an approximation algorithm with ratio δ takes time $2^{n^{1-o(1)}}$.*

6 Approximations of Treedepth and Pathwidth

All existing positive results for treewidth approximation directly transfer to approximations for treedepth and pathwidth with an additional factor $O(\log n)$, as for each graph G, $\mathtt{pw}(G) \leq \mathtt{td}(G)$, and $\mathtt{td}(G) = O(\log n \cdot \mathtt{tw}(G))$. (For the former result, we can build a path decomposition by visiting the leaves of $\mathcal{T}$ 'from left to right', and for each, taking a bag with the leaf and all its ancestors in $\mathcal{T}$. For the latter result, see Lemma 3.)

Thus, in [13], $O(\log^2 n)$-approximations of PATHWIDTH and TREEDEPTH were given. Using the algorithm by Feige et al. [23], we get approximations for these problems with ratio $O(\sqrt{\log \mathtt{tw}} \cdot \log n)$.

Some recent results give better approximations. Many of these are based on a 'win-win strategy': either we find a certain substructure in the graph that shows a lower bound, or we find an approximate solution. We start by a very simple example, that is based on a technique by Fellows and Langston [24, Theorem 9], and the discussion in [38, Section 6.2].

Lemma 4. *There is a linear time algorithm, that, when given a connected graph $G = (V, E)$ and an integer k, either correctly outputs that the treedepth of G is more than k, or gives an elimination forest for G of depth at most 2^k.*

Proof. Make a depth-first search spanning tree T of G. Such a tree has the property that each edge is between a pair with one vertex an ancestor of the other vertex. If F has depth at most 2^k, we output T; otherwise, G has a path of length more than 2^k, so the treedepth of G is more than k ([38, Eq. 6.2].) □

A related stronger result was given by Kawarabayashi and Rossman [28]. This result was further improved by Czerwiński, Nadara, and Marcin Pilipczuk [18], who showed that each graph of treedepth $\Omega(k^3)$ either contains as subgraph a path of length 2^k, or a subdivision of a full binary tree of depth k, or has treewidth at least k. Based on this structural insight, Czerwiński et al.' [18] obtain a polynomial time approximation algorithm for treedepth. Note that the result also implies a ratio $\mathtt{td} \log^{1.5} \mathtt{td}$, as $\mathtt{tw}(G) \leq \mathtt{td}(G)$.

Theorem 10 (Czerwiński, Nadara, and Marcin Pilipczuk [18]). *There is a polynomial time algorithm that approximates the treedepth of an input graph G with a ratio $\mathtt{tw} \log^{1.5} \mathtt{tw}$.*

For pathwidth, an algorithm with bounds similar to those for treedepth given in Lemma 4 was obtained in 1996 by Cattell, Dinneen and Fellows [17]. This algorithm exploits that graphs of small pathwidth do not contain a large complete binary tree as subgraph.

Theorem 11 (Cattell, Dinneen, Fellows, 1996 [17]). *There is a polynomial time algorithm that when given a graph G and an integer k, either correctly outputs that the pathwidth of G is more than k, or gives a path decomposition of width at most 2^k.*

Recently, much stronger results were obtained by Groenland et al. [26], who give a combinatorial result that tells that each graph G has pathwidth at most $(\mathtt{tw}(G)+1)(h+1)$ or has a subdivision of a complete binary tree of height $h+1$ as subgraph. The latter of these gives a lower bound h for the pathwidth. Groenland et al. [26] combine these insights with the treewidth approximation algorithm of Feige et al. [22] and obtain the following result on the approximation of pathwidth. Note that this result also implies a polynomial time approximation of pathwidth with a ratio $O(\texttt{pw}(G)\sqrt{\log(\texttt{pw}(G))})$.

Theorem 12 (Groenland, Joret, Nadara, Walczak, 2023 [26]). *There is a polynomial time algorithm that approximates the pathwidth of an input graph G with a ratio $O(\texttt{tw}(G)\sqrt{\log(\texttt{tw}(G))})$.*

Recently, Bansal et al. [4] gave a polynomial time algorithm to approximate the pathwidth of a graph with an approximation ratio of the form $O(\log^{1+o(1)} n)$. They use a different form of 'recursive partitioning' that is guided by a solution of a linear program.

7 Conclusions

In this short survey, older and newer developments on approximation algorithms for treewidth and the related parameters pathwidth and treedepth were discussed. In addition to theoretical studies with a focus to guarantees on running time and provable quality of approximations, several experimental studies on treewidth heuristics have been executed; for surveys, see [14] or [46]. Given the many theoretical and practical applications of the notions, these topics are of great interest. In this survey, we also did not discuss exact algorithms—of course, the running times of (in particular parameterized) approximation algorithms should be compared with that of exact algorithms. There also are several papers on treewidth approximation for special graph classes, but this short survey does not discuss these. We end with some open problems.

- The algorithms with currently best known (time or ratio) bounds for treewidth approximation are:
  - A polynomial time approximation algorithm with ratio $O(\sqrt{\log \texttt{tw}})$ by Feige et al. [23].
  - A 2-approximation algorithm with running time $2^{O(\texttt{tw})}n$ by Korhonen [29].
  - An $(1+\epsilon)$-approximation algorithm with running time $\texttt{tw}^{O(\texttt{tw}/\epsilon)}n^{O(1)}$ by Korhonen and Lokshtanov [30].
  - An $O(\texttt{tw}^7 n \log n)$ time algorithm with ratio $O(\texttt{tw})$ Fomin et al. [25].

  Are further improvements possible?
- Bonnet [15] showed a lower bound of 1.00005 for the approximation ratio for polynomial time treewidth approximation algorithms, conditional on $P \neq NP$. Can this bound be improved, and to what value? Is it possible to prove a similar result for pathwidth, and/or for treedepth?
- Are there polynomial time approximation algorithms for treedepth with a constant approximation ratio? Is it possible to show a negative result based on the Small Set Expansion hypothesis, similar to the results by Wu et al. [47]?

Acknowledgement. I thank John Gilbert, Hjalli Hafsteinsson, and Ton Kloks for great memories and great cooperation.

References

1. Amir, E.: Efficient approximation for triangulation of minimum treewidth. In: Breese, J.S., Koller, D. (eds.) Proceedings of the 17th Annual Conference on Uncertainty in Artificial Intelligence, UAI 2001, pp. 7–15. Morgan Kaufmann (2001)
2. Amir, E.: Approximation algorithms for treewidth. Algorithmica **56**, 448–479 (2010). https://doi.org/10.1007/S00453-008-9180-4
3. Arnborg, S., Corneil, D.G., Proskurowski, A.: Complexity of finding embeddings in a k-tree. SIAM J. Algebraic Discrete Methods **8**, 277–284 (1987). https://doi.org/10.1137/0608024
4. Bansal, N., Katzelnick, D., Schwartz, R.: Almost logarithmic approximation for cutwidth and pathwidth (2023). https://doi.org/10.48550/ARXIV.2311.15639
5. Belbasi, M., Fürer, M.: An improvement of Reed's treewidth approximation. J. Graph Algorithms Appl. **26**(2), 257–282 (2022). https://doi.org/10.7155/JGAA.00593
6. Bellenbaum, P., Diestel, R.: Two short proofs concerning tree-decompositions. Comb. Prob. Comput. **11**(6), 541–547 (2002). https://doi.org/10.1017/S0963548302005369
7. Bertele, U., Brioschi, F.: Nonserial Dynamic Programming. Academic Press, New York (1972)
8. Bodlaender, H.L.: A linear time algorithm for finding tree-decompositions of small treewidth. SIAM J. Comput. **25**, 1305–1317 (1996). https://doi.org/10.1137/S0097539793251219
9. Bodlaender, H.L.: A partial k-arboretum of graphs with bounded treewidth. Theoret. Comput. Sci. **209**, 1–45 (1998). https://doi.org/10.1016/S0304-3975(97)00228-4
10. Bodlaender, H.L., et al.: Rankings of graphs. SIAM J. Discrete Math. **11**, 168–181 (1998). https://doi.org/10.1137/S0895480195282550
11. Bodlaender, H.L., Drange, P.G., Dregi, M.S., Fomin, F.V., Lokshtanov, D., Pilipczuk, M.: A $c^k n$ n 5-approximation algorithm for treewidth. SIAM J. Comput. **45**(2), 317–378 (2016). https://doi.org/10.1137/130947374
12. Bodlaender, H.L., Gilbert, J.R., Hafsteinsson, H., Kloks, T.: Approximating treewidth, pathwidth, frontsize, and minimum elimination tree height. J. Algorithms **18**, 238–255 (1995). https://doi.org/10.1006/JAGM.1995.1009
13. Bodlaender, H.L., Gilbert, J.R., Hafsteinsson, H., Kloks, T.: Approximating treewidth, pathwidth, and minimum elimination tree height. In: Schmidt, G., Berghammer, R. (eds.) WG 1991. LNCS, vol. 570, pp. 1–12. Springer, Heidelberg (1992). https://doi.org/10.1007/3-540-55121-2_1
14. Bodlaender, H.L., Koster, A.M.C.A.: Treewidth computations I. Upper bounds. Inform. Comput. **208**, 259–275 (2010). https://doi.org/10.1016/J.IC.2009.03.008
15. Édouard Bonnet: Treewidth inapproximability and tight ETH lower bound (2024). https://arxiv.org/abs/2406.11628
16. Bouchitté, V., Kratsch, D., Müller, H., Todinca, I.: On treewidth approximations. Discret. Appl. Math. **136**, 183–196 (2004). https://doi.org/10.1016/S0166-218X(03)00440-2
17. Cattell, K., Dinneen, M.J., Fellows, M.R.: A simple linear-time algorithm for finding path-decompositions of small width. Inform. Process. Lett. **57**, 197–204 (1996). https://doi.org/10.1016/0020-0190(95)00190-5
18. Czerwinski, W., Nadara, W., Pilipczuk, M.: Improved bounds for the excluded-minor approximation of treedepth. SIAM J. Discret. Math. **35**(2), 934–947 (2021). https://doi.org/10.1137/19M128819X

19. Duff, I.S., Reid, J.K.: The multifrontal solution of indefinite sparse symmetric linear equations. ACM Trans. Math. Softw. **9**, 302–325 (1983). https://doi.org/10.1145/356044.356047
20. Even, G., Naor, J., Rao, S., Schieber, B.: Fast approximate graph partitioning algorithms. SIAM J. Comput. **28**, 2187–2214 (1999). https://doi.org/10.1137/S0097539796308217
21. Even, S.: Graph Algorithms. Pitman, London (1979)
22. Feige, U., Hajiaghayi, M., Lee, J.R.: Improved approximation algorithms for minimum-weight vertex separators. In: Proceedings of the 37th Annual Symposium on Theory of Computing, STOC 2005, pp. 563–572. ACM Press (2005). https://doi.org/10.1145/1060590.1060674
23. Feige, U., Hajiaghayi, M., Lee, J.R.: Improved approximation algorithms for minimum weight vertex separators. SIAM J. Comput. **38**, 629–657 (2008). https://doi.org/10.1137/05064299X
24. Fellows, M.R., Langston, M.A.: On search, decision and the efficiency of polynomial-time algorithms. In: Proceedings of the 21st Annual Symposium on Theory of Computing, STOC 1989, pp. 501–512 (1989). https://doi.org/10.1145/73007.73055
25. Fomin, F.V., Lokshtanov, D., Saurabh, S., Pilipczuk, M., Wrochna, M.: Fully polynomial-time parameterized computations for graphs and matrices of low treewidth. ACM Trans. Algorithms **14**(3), 34:1–34:45 (2018). https://doi.org/10.1145/3186898
26. Groenland, C., Joret, G., Nadara, W., Walczak, B.: Approximating pathwidth for graphs of small treewidth. ACM Trans. Algorithms **19**(2), 16:1–16:19 (2023). https://doi.org/10.1145/3576044
27. Kashiwabara, T., Fujisawa, T.: An NP-complete problem on interval graphs. In: Proceedings IEEE Symposium on Circuits and Systems, pp. 82–83 (1979)
28. Kawarabayashi, K., Rossman, B.: A polynomial excluded-minor approximation of treedepth. J. Eur. Math. Soc. **24**(4), 1449–1470 (2022). https://doi.org/10.4171/JEMS/1133
29. Korhonen, T.: A single-exponential time 2-approximation algorithm for treewidth. In: 62nd IEEE Annual Symposium on Foundations of Computer Science, FOCS 2021, pp. 184–192. IEEE (2021). https://doi.org/10.1109/FOCS52979.2021.00026
30. Korhonen, T., Lokshtanov, D.: An improved parameterized algorithm for treewidth. In: Saha, B., Servedio, R.A. (eds.) Proceedings of the 55th Annual ACM Symposium on Theory of Computing, STOC 2023, pp. 528–541. ACM (2023). https://doi.org/10.1145/3564246.3585245
31. Lagergren, J.: Efficient parallel algorithms for tree-decomposition and related problems. In: Proceedings of the 31st Annual Symposium on Foundations of Computer Science, FOCS 1990, pp. 173–182 (1990). https://doi.org/10.1109/FSCS.1990.89536
32. Lagergren, J.: Efficient parallel algorithms for graphs of bounded tree-width. J. Algorithms **20**, 20–44 (1996). https://doi.org/10.1006/JAGM.1996.0002
33. Leighton, F.T., Rao, S.: An approximate max-flow min-cut theorem for uniform multi-commodity flow problems with applications to approximate algorithms. In: Proceedings of the 29th Annual Symposium on Foundations of Computer Science, FOCS 1988, pp. 422–431. IEEE (1988). https://doi.org/10.1109/SFCS.1988.21958
34. Leighton, F.T., Rao, S.: Multicommodity max-flow min-cut theorems and their use in designing approximation algorithms. J. ACM **46**(6), 787–832 (1999). https://doi.org/10.1145/331524.331526

35. Liu, J.W.H.: The role of elimination trees in sparse factorization. SIAM J. Matrix Anal. Appl. **11**, 134–172 (1990). https://doi.org/10.1137/0611010
36. Liu, J.W.H.: The multifrontal method for sparse matrix solution: theory and practice. SIAM Rev. **34**(1), 82–109 (1992). https://doi.org/10.1137/1034004
37. Matoušek, J., Thomas, R.: Algorithms for finding tree-decompositions of graphs. J. Algorithms **12**, 1–22 (1991). https://doi.org/10.1016/0196-6774(91)90020-Y
38. Nešetřil, J., Ossona de Mendez, P.: Sparsity. AC, vol. 28. Springer, Heidelberg (2012). https://doi.org/10.1007/978-3-642-27875-4
39. Nešetřil, J., de Mendez, P.O.: Tree-depth, subgraph coloring and homomorphism bounds. Eur. J. Comb. **27**, 1022–1041 (2006). https://doi.org/10.1016/J.EJC.2005.01.010
40. Nešetřil, J., de Mendez, P.O.: On low tree-depth decompositions. Graphs Comb. **31**(6), 1941–1963 (2015). https://doi.org/10.1007/S00373-015-1569-7
41. Pothen, A.: The complexity of optimal elimination trees. Technical Report CS-88-13, Pennsylvania State University (1988). https://www.cs.purdue.edu/homes/apothen/Papers/shortest-etree1988.pdf
42. Raghavendra, P., Steurer, D.: Graph expansion and the unique games conjecture. In: Schulman, L.J. (ed.) Proceedings of the 42nd ACM Symposium on Theory of Computing, STOC 2010, pp. 755–764. ACM (2010). https://doi.org/10.1145/1806689.1806792
43. Reed, B.: Finding approximate separators and computing tree-width quickly. In: Proceedings of the 24th Annual Symposium on Theory of Computing, STOC 1992. pp. 221–228. ACM Press, New York (1992). https://doi.org/10.1145/129712.129734
44. Robertson, N., Seymour, P.D.: Graph minors. II. Algorithmic aspects of tree-width. J. Algorithms **7**, 309–322 (1986). https://doi.org/10.1016/0196-6774(86)90023-4
45. Robertson, N., Seymour, P.D.: Graph minors. XIII. The disjoint paths problem. J. Comb. Theor. Ser. B **63**, 65–110 (1995). https://doi.org/10.1006/JCTB.1995.1006
46. Tamaki, H.: Experimental analysis of treewidth. In: Fomin, F.V., Kratsch, S., van Leeuwen, E.J. (eds.) Treewidth, Kernels, and Algorithms. LNCS, vol. 12160, pp. 214–221. Springer, Cham (2020). https://doi.org/10.1007/978-3-030-42071-0_15
47. Wu, Y.L., Austrin, P., Pitassi, T., Liu, D.: Inapproximability of treewidth and related problems. J. Artif. Intell. Res. **49**, 569–600 (2014). https://doi.org/10.1613/JAIR.4030

Contributed Talks

Fast Winning Strategies for the Attacker in Eternal Domination

Guillaume Bagan, Nicolas Bousquet, Nacim Oijid, and Théo Pierron[✉]

Universite Claude Bernard Lyon 1, CNRS, INSA Lyon, LIRIS, UMR5205, 69622 Villeurbanne, France
theo.pierron@univ-lyon1.fr

Abstract. Dominating sets in graphs are often used to model monitoring problems, by posting guards on the vertices of the dominating set. If an (unguarded) vertex is attacked, at least one guard can then react by moving there. This yields a new set of guards, which may not be dominating anymore. A dominating set is *eternal* if one can endlessly resist to attacks.

From the attacker's perspective, if we are given a non-eternal dominating set, the question is to determine how fast can we provoke an attack that cannot be handled by a neighboring guard. We investigate this question from a computational complexity point of view, by showing that this question is PSPACE-hard, even for graph classes where finding a minimum eternal dominating set is in P.

We then complement this result by giving polynomial time algorithms for cographs and trees, and showing a connection with treedepth for the latter. We also investigate the problem from a parameterized complexity perspective, mainly considering two parameters: the number of guards and the number of steps.

Keywords: eternal dominating set · treedepth · PSPACE-completeness · parameterized complexity

1 Introduction

Let $G = (V, E)$ be a graph and $D \subseteq V$ be a set of vertices called *guards*. The *mobile domination game* is played by two players: Attacker and Defender. At each turn, Attacker chooses a non-guarded vertex v (we say that Attacker *attacks* v). Then, Defender has to move a guard along an edge uv incident to v (we say that Defender defends against the attack); see Fig. 1 for an illustration. If she cannot move such a guard (that is v had no neighbor u in D), Attacker wins. Otherwise, the game continues with the set $D \cup \{v\} \setminus \{u\}$ of guards. If Defender can defend against any infinite sequence of attacks, the initial set of guards D is called an *eternal dominating set*. The *eternal domination number* $\gamma^\infty(G)$ [10] is the size of the smallest eternal dominating set of G. This notion

This research was supported by the ANR project P-GASE (ANR-21-CE48-0001-01).

D. Král' and M. Milanič (Eds.): WG 2024, LNCS 14760, pp. 21–34, 2025.
https://doi.org/10.1007/978-3-031-75409-8_2

has been widely studied, see e.g. [23] for a recent survey. Many variants have been introduced in the literature, for example when all the guards can move at each attack [16]. This leads to similar notions of m-eternal dominating set and m-eternal domination number $\gamma_m^\infty(G)$ (where m stands for "multiple" and is not a parameter).

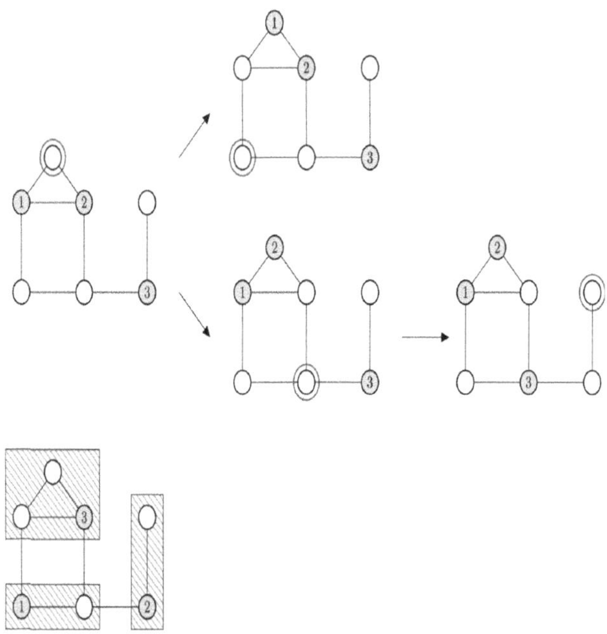

Fig. 1. Above, an example of a winning strategy for Attacker in 3 turns, depending on the answer of Defender. Each number represents a guard, and the attack is circled. Below, an eternal dominating set with 3 guards. Each guard protects its own clique.

For every graph G, we have $\alpha(G) \leq \gamma^\infty(G) \leq \theta(G)$ [10] and $\gamma^\infty(G) \leq \binom{\alpha(G)+1}{2}$ [21], where $\alpha(G)$ is the maximum size of an independent set in G, and $\theta(G)$ is the minimum size of a clique cover of G. The latter inequality has been proved to be tight for an infinite class of graphs [17]. Several existing works also focus on classes of graphs for which $\gamma^\infty(G) = \theta(G)$, see e.g. [1,10,22].

One can naturally wonder if these parameters can be easily computed. This motivates the introduction of the following problem.

Eternal-Dominating-Number
Input: A graph G, an integer $k \geq 1$
Question: $\gamma^\infty(G) \leq k$?

This problem lies in EXP and is coNP-hard [3] but the exact complexity class this problem belongs to remains open. However, on restricted graph classes, there exist some polynomial time algorithms such as on perfect graphs. Indeed, for perfect graphs, we have $\alpha(G) = \gamma^{\infty}(G) = \theta(G)$, which yields a polynomial algorithm for ETERNAL-DOMINATING-NUMBER [10].

Deciding the m-variant of ETERNAL-DOMINATING-NUMBER (where every guard is allowed to move at each turn) is harder since deciding if $\gamma_m^{\infty}(G) \leq k$ is already NP-hard even on Hamiltonian split graphs [4]. However, polynomial time algorithms have been proposed on trees [22], unit interval graphs [7], interval graphs [26] and cactus graphs [6]. Moreover, great attention has been paid to the study of lower and upper bounds for the m-eternal dominating number in grids. See for example [19,24]. The spy game, a generalization of m-eternal domination has been proved PSPACE-hard on directed graphs and NP-hard on (non-directed) graphs [11]. Another generalization of m-eternal domination, the guarding game, has been proved EXP-complete [27].

Surprisingly, the situation is quite different for the seemingly close problem:

ETERNAL-DOMINATING-SET
Input: A graph G, a set of guards $D \subseteq V$
Question: Is D an eternal dominating set of G?

This problem has been recently proved EXP-complete [28]. It is not clear that this problem is easier than the first. Indeed, there is nothing preventing graphs to have an easy-to-find eternal dominating set of size k while it may be hard to tell whether some given vertices form an eternal dominating set. In particular, on perfect graphs, the complexity of ETERNAL-DOMINATING-SET is still open contrary to the complexity of ETERNAL-DOMINATING-NUMBER.

In this paper, we are not only interested in determining whether a configuration of guards is an eternal dominating set but also to find, in the negative case, a winning strategy for Attacker minimizing the number of steps. This question naturally arises for board games, like Chess or Go, but has also attracted a lot of attention in the combinatorial game theory literature: for the domination game [8], maker-breaker games [15,18], the non-planarity game [2] and the spy game [11]. Recently, this problem has been studied for m-eternal domination on trees in [5], where Blažej et al. prove that if Defender does not have enough guards to m-eternally dominate a graph G, then Attacker can win in at most d turns, where d is the diameter of G.

Informally, we say that Attacker wins in t turns on (G, D) if he has a strategy that makes Defender lose before the t-th turn, regardless of her strategy. We denote by $t_G(D)$ the minimum t such that Attacker wins in t turns on (G, D) (We postpone the formal definitions to the next section.) In particular, $t_G(D)$ is $+\infty$ if and only if D is an eternal dominating set of G.

The goal of this paper is to study the complexity of computing $t_G(D)$, that is, of the following problem.

FAST-STRATEGY
Input: A graph $G = (V, E)$, a set of guards $D \subseteq V$, an integer $t \geq 1$
Question: is $t_G(D) \leq t$?

Our Results and Organization of the Paper. In Sect. 3, we show that this problem is coNP-hard on bipartite and split graphs. On the positive side, we show that ETERNAL-DOMINATING-SET can be decided in polynomial time on bipartite graphs. We also show that FAST-STRATEGY is PSPACE-hard even restricted to perfect graphs (recall that ETERNAL-DOMINATING-NUMBER can be decided in polynomial time on perfect graphs). More specifically, we prove that the problem is PSPACE-hard on 2-unimodal graphs reducing from the UNORDERED-CNF problem. Due to space constraints, this proof is deferred to the full version of the article.

Then, we give two positive results: we prove that FAST-STRATEGY can be decided in polynomial time on trees (Sect. 4) and on cographs (Sect. 5). For trees, the main step of the proof consists in introducing arenas, which are special subtrees and proving that Attacker can win in at most k steps if and only if there exists an arena for which some treedepth-related parameter is at most k. We finally prove that the polynomial time algorithm for computing the treedepth on trees can be adapted for this new parameter. For cographs, we use their decomposition as disjoint unions or joins of smaller cographs and analyze the strategies of Attacker and Defender in each case to provide a recursive algorithm for FAST-STRATEGY. Finally, we conclude by stating some results on the parameterized complexity of FAST-STRATEGY, whose details can be found in the full version of the article.

Related Work. The notion of k-secure dominating set was introduced by Burger et al. [9]. In their setting, a dominating set is *k-secure* if it remains dominating after a sequence of k attacks, which looks like the problem we are considering in this paper. However, in their setting, the sequence of attacks is known in advance (oblivious adversary) whereas, in our case, Attacker can adapt his moves according to how Defender defends the attacks (adaptive adversary).

2 Preliminaries

For an integer $n > 0$, $[n]$ represents the set $\{1, 2, \dots, n\}$. All graphs considered in this paper are finite, loopless, and simple. Let $G = (V, E)$ be a graph. Given a vertex $v \in V$, $N(v)$ denotes the *neighborhood* of v, i.e., the set $\{y \in V : vy \in E\}$. The *degree* $d(v)$ of v is $|N(v)|$. A set $S \subseteq V$ is an *independent set* of G if there is no edge uv for every u, v in S. The *independence number* $\alpha(G)$ is the size of a largest independent set of G. A *clique* is a subset of pairwise adjacent vertices. The *clique covering number* $\theta(G)$ is the minimum number of cliques in which V can be partitioned. A *vertex cover* of G is a set $S \subseteq V$ such that every edge e in E has an endpoint in S. A *matching* of G is a set of edges of G that pairwise do not share an endpoint. A graph G is *perfect* if $\alpha(H) = \theta(H)$ for every induced subgraph H of G.

A *rooted tree* T is a tree with a distinguished vertex r called the root. Consider the orientation of T such that every vertex is reachable from r. u is a *child* of v if there is an arc vu in this orientation and u is a *descendant* of v if there is an oriented path from v to u along this orientation. A rooted subtree of T is a subtree of T induced by a node and all its descendants. The *height* of T is the number of vertices in a longest directed path starting at r.

A *td-decomposition* of a connected graph $G = (V, E)$ is a rooted tree T with set of vertices V such that u is a descendant of v or v is a descendant of u for every $uv \in E$. The *treedepth* of G is the minimum height of a td-decomposition of G (see Fig. 2).

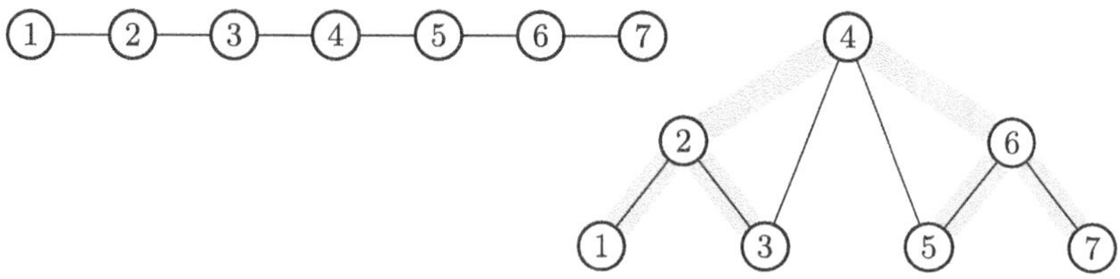

Fig. 2. On the left, the graph P_7 of treedepth 3. On the right, an optimal td-decomposition of P_7

Let $G = (V, E)$ be a graph and $D \subseteq V$ be a set of g vertices called *guards*. In the eternal domination game, a *strategy for Attacker* is a function mapping each set S of g vertices to a vertex of $V \setminus S$. A *strategy for Defender* is a partial function mapping pairs (S, v) where S is a set of g vertices and $v \in V \setminus S$ to some vertex of $N(v) \cap S$ if it exists.

Given two strategies f, g for Attacker and Defender, we say that Defender can *resist t attacks* on (G, D) using (f, g) if either $t = 0$, or $g(D, f(D))$ is defined, and she can resist $t-1$ turns on $(G, D \cup \{g(D, f(D))\} \setminus \{f(D)\})$ using (f, g). We extend this notion: Defender can resist t attacks on (G, D) when Attacker plays according to f if she has a strategy g so that the previous condition is satisfied.

We are interested in the *fastest way* for Attacker to win (if he can), that is the minimum t such that Attacker has a strategy f such that Defender cannot resist t attacks on (G, D) when Attacker plays according to f. We denote this number t by $t_G(D)$. Note that $t_G(D)$ might be $+\infty$ if Defender can eternally answer to Attacker's move, *i.e.* if Attacker has no winning strategy.

3 Bipartite and Split Graphs

The goal of this section is to study the complexity of the problems Eternal-Dominating-Set and Fast-Strategy on bipartite graphs. Recall that bipartite graphs are perfect, hence determining the smallest eternal dominating set can be done in polynomial time [16]. We extend this result to Eternal-Dominating-Set.

Theorem 1. *Eternal-Dominating-Set is in PTime on bipartite graphs.*

On the opposite, we show that determining the smallest number of moves in a winning strategy is significantly harder, as summarized by the following statement.

Theorem 2. *FAST-STRATEGY is in* PSPACE *and* coNP-*hard on bipartite graphs.*

Both these statements are based on the following lemma, which gives some insights about the shape of winning strategies for the eternal domination game on bipartite graphs. Its proof can be found in the full version of the article.

Lemma 1. *Let G be a bipartite graph and D be a set of guards. If Attacker wins in t turns on (G, D), then he can win in at most t turns by always attacking the same side of the bipartition.*

The main consequence of this result is the following preprocessing statement, proved in the full version of the article.

Lemma 2. *Let G be a bipartite graph and D be a set of guards. Then, one can compute in polynomial time a bipartite graph G' on the same set of vertices such that $t_G(D) = t_{G'}(D)$ where D and $V(G') \setminus D$ form a bipartition of $V(G')$.*

We are now ready to prove Theorem 1.

Proof. (Proof of Theorem 1). Let $G = ((A, B), E)$ be a bipartite graph and D be a set of guards. By Lemma 2, we can modify the instance in order to assume that $B = D$. We claim that D is an eternal dominating set of G if and only if G admits a matching saturating A (i.e. such that all the vertices of A are matched). Proving this claim is enough to conclude since this criterion can be tested in polynomial time.

Let us now prove the claim. If G has a matching M saturating A then the edges of M and the unmatched vertices in B form a clique cover of G where each clique contains a guarded vertex. Hence, G can be eternally dominated (Defender always defends an attacked vertex v with a guard in the same clique as v in the clique cover).

For the converse direction, assume there is no matching saturating A. By Hall's marriage theorem (see e.g.[13]), there is a set of vertices $S \subseteq A$ such that $|N(S)| < |S|$. In particular, Attacker wins in at most $|N(S)| + 1$ turns by successively attacking vertices in S: they can only be defended by vertices of $N(S)$.

Regarding Theorem 2, the membership of FAST-STRATEGY in PSPACE is actually an easy consequence of Lemma 2, since it gives a polynomial bound (namely the number of guards plus one) on the minimum number of turns of a winning strategy for Attacker. The hardness proof is left out to the full version of the paper. A simple modification of our reduction actually yields a similar result on split graphs, whose proof is also given in the full version.

Theorem 3. *FAST-STRATEGY is* coNP-*hard on split graphs.*

4 Trees

Theorem 4. *Fast-Strategy can be solved in polynomial time on trees.*

In a first part, given G and D, we will introduce the concept of arenas, and we will show that, when G is a tree, $t_G(D)$ is related to the maximum possible treedepth of an arena of G. We will then prove that this parameter is exactly the parameter of interest to decide Fast-Strategy on trees. We then recall an algorithm from Iyer et al. [20] to compute the treedepth of a tree, and adapt it in a last part to compute $t_G(D)$ in polynomial time. The main concepts introduced in this section are arenas. Roughly speaking, an arena X of a tree T and a set of guards D is a minimal subtree of T where Attacker wins and where guards outside X cannot help the defense of X.

Definition 1. *Let T be a tree and $D \subseteq V(T)$ be a set of guards. An* arena *of (T, D) is a subtree X of T such that:*

1. *$V(X) \cap D$ and $V(X) \setminus D$ are both independent;*
2. *the guarded vertices in X have degree exactly 2 in X (in particular no leaf is guarded);*
3. *there is no vertex in $V(X) \setminus D$ with a neighbor in $D \setminus V(X)$.*

For the ease of notation, we identify a subtree X with its set of vertices, so that $X \cap D$ denotes the set of guarded vertices in X. Equivalently (even if we will not prove it since we will not use it), an arena is a set satisfying the following:

- $|D \cap X| < \alpha(T[X])$;
- there is no unguarded vertex in X with a guarded neighbor outside X (that is, there is no edge between $X \setminus D$ and $D \setminus X$);
- X is minimal for these properties.

Definition 2. *Let T be a tree, $D \subset V(T)$ and X an arena of (T, D). The* contracted tree *of X is obtained from X by replacing each vertex from $V(X) \cap D$ by an edge between its neighbors. The* contracted treedepth *of X, denoted by* $\mathrm{ctd}(X)$*, is the treedepth of this contracted tree (Fig. 3).*

In the rest of this part, we will prove the following:

Theorem 5. *Given a tree T and a set of guards D, $t_T(D)$ is the minimum value of* $\mathrm{ctd}(X)$ *over all arenas X of (T, D).*

In particular, Theorem 5 ensures that $t_T(D) = +\infty$ if and only if (T, D) has no arena. Let T be a tree and D be a set of guards. We split the proof of Theorem 5 into two independent lemmas, whose proofs can be found in the full version of the article.

Lemma 3. *If (T, D) has an arena X of contracted treedepth t, then $t_T(D) \leq t$.*

Lemma 4. *If $t_T(D) = t < +\infty$, then there exists an arena of contracted treedepth at most t.*

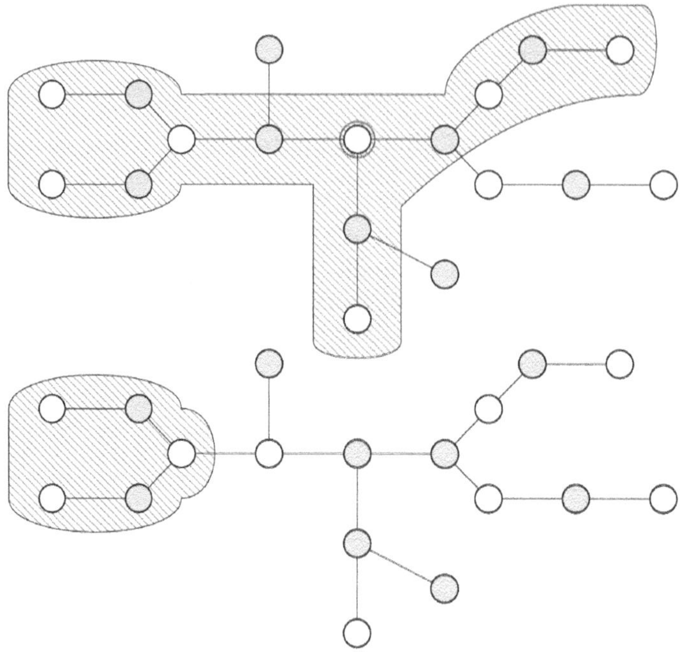

Fig. 3. Above, a tree with an arena of contracted treedepth 3. Each grayed vertex is guarded, and the attack is circled. Below, the same tree after the Defender has moved the guard from left to the attack. A smaller arena of contracted treedepth 2 is created (its contracted tree is a path on three vertices).

In particular, Attacker wins the eternal domination game if and only if there is an arena. This concludes the proof of Theorem 5. Note in particular that, since the treedepth of a tree is logarithmic in its number of vertices [25], we get the following corollary.

Corollary 1. *If T is an n-vertex tree and D a set of guards, then $t_T(D)$ is either ∞ or $O(\log n)$.*

It remains to show that we can compute the minimum contracted treedepth among all arenas in polynomial time. This polynomial time algorithm is an adaptation of a polynomial time algorithm that computes the treedepth of trees. The description of this algorithm and its correctness can be found in the full version of this article.

5 Cographs

Theorem 6. FAST-STRATEGY *is computable in polynomial time on cographs.*

A cograph is either an isolated vertex, or is obtained by taking the disjoint union (denoted by $\cup$) or the complete join (denoted by $\bowtie$) of two smaller cographs. This inductive definition naturally provides a representation of each cograph as a decomposition tree called *cotree*, whose leaves are the vertices of the cograph, and the internal nodes are either *join nodes* or *union nodes*[1]. We can recover the adjacency between the vertices of a cograph G from its cotree as follows: two vertices uv are adjacent in G if and only if their closest ancestor $\text{lca}(u, v)$ in the cotree is a join node.

Our polynomial time algorithm for FAST-STRATEGY on cographs will compute $t_G(D)$ using a recursive procedure applied to the cotree of G. Note that this is not restrictive since one can compute the cotree of a given cograph in linear time [12]. We actually consider a more general game, the *eternal domination game with reservists*, defined as follows. This notion is similar to firefighter reserve [14].

This game is played on a board (G, D, r) where G is a graph, D is a set of guards and $r \geq 1$ is a number of reservists. Attacker plays as usual by attacking a non-guarded vertex v. Now Defender defends by either moving a guard from a neighbor of v to v, or by adding a new guard on v and decrease r by one. Attacker wins when r reaches 0, that is when Defender moves her last reservist on G. We denote by $t_G(D, r)$ the smallest number of turns required for Attacker to win. In particular, observe that $t_G(D, 1) = t_G(D)$. By convention, we assume that $t_G(D, 0) = 0$.

Theorem 6 relies on the two following results, that handle respectively the leaves and the nodes of a cotree. For the base case, observe that Attacker can only win on an isolated vertex if it is not guarded and there is only one reservist, and in that case he wins in one turn. Therefore we get the following.

Observation 7. *Let G be a graph containing only an isolated vertex. Then, $t_G(D, r) = 0$ if $r = 0$, 1 if $D = \emptyset$ and $r = 1$ and $+\infty$ otherwise.*

We now state two recursive formulas satisfied by $t_G(D, r)$ that allow us to compute $t_G(D)$. Note that the computation only uses the values of the form $t_H(D \cap V(H), r)$ for some cograph H whose cotree is a subtree of the cotree of G and for some integer r between 0 and $|V(G)|$. Therefore, the computation can be done in polynomial time, which proves Theorem 6.

Proposition 1. *Let G_1, G_2 be two cographs and D_1, D_2 be sets of guards on G_1 and G_2. We have:*

$$t_{G_1 \cup G_2}(D_1 \cup D_2, r) = \min_{0 \leqslant i \leqslant r} (t_{G_1}(D_1, i) + t_{G_2}(D_2, r - i))$$
$$t_{G_1 \bowtie G_2}(D_1 \cup D_2, r) = \min\{t_{G_1}(D_1, r + |D_2|), t_{G_2}(D_2, r + |D_1|)\}.$$

The rest of this section is devoted to proving this statement. Each of the two equalities is proved by showing two inequalities. We postpone the proof of the easiest ones, stated in the two following lemmas, to the full version of this article.

[1] Note that this decomposition is not necessarily unique but, for algorithmic purposes, we only need that such a decomposition tree exists and can be computed in polynomial time which is indeed the case.

Lemma 5. *Let G_1, G_2 be two cographs, $r \geqslant 0$ and D_1, D_2 be sets of guards on G_1 and G_2. We have:*

$$t_{G_1 \cup G_2}(D_1 \cup D_2, r) \leq \min_{0 \leqslant i \leqslant r} (t_{G_1}(D_1, i) + t_{G_2}(D_2, r - i)).$$

Lemma 6. *Let G_1, G_2 be two cographs, $r \geqslant 0$ and D_1, D_2 be sets of guards on G_1 and G_2. We have:*

$$t_{G_1 \bowtie G_2}(D_1 \cup D_2, r) \leq \min\{t_{G_1}(D_1, r + |D_2|), t_{G_2}(D_2, r + |D_1|)\}.$$

To prove the remaining inequalities, we consider an arbitrary order $\leq_G$ over the vertices of G and introduce $\mathcal{S}^*$ as the following strategy of Defender: if Attacker attacks a vertex v, then Defender answers (if possible) with a guarded vertex $u \in N(v)$ whose common ancestor $\text{lca}(u, v)$ in the cotree of G is the deepest. If several such vertices exist, she picks the smallest with respect to the order $\leq_G$. If no such vertex exists (*i.e.* if v has no guarded neighbor), Defender calls a reservist. We denote by $t^*_G(D, r)$ the smallest number of turns required for Attacker to win on (G, D, r) against this strategy $\mathcal{S}^*$. Notice that this strategy does not depend on the number of available reservists, nor on the connected components of G that do not contain v.

Observe that if Attacker wins in k turns on (G, D, r), then he also wins in at most k turns against $\mathcal{S}^*$. Therefore, we get the following.

Observation 8. *Every cograph G satisfies $t^*_G \leq t_G$.*

In the following, we will show by induction that this inequality is actually an equality, meaning that $\mathcal{S}^*$ is an optimal strategy for Defender.

For the base case, Observation 7 clearly remains true when Attacker plays against the strategy $\mathcal{S}^*$.

Observation 9. *Let G be a graph containing only an isolated vertex. Then, $t^*_G(D, r) = 0$ if $r = 0$, 1 if $D = \emptyset$ and $r = 1$ and $+\infty$ otherwise.*

To prove the inductive step, we will show that the converse inequalities of Lemmas 5 and 6 hold when replacing t_G by t^*_G. More precisely, we show the two following lemmas.

Lemma 7. *Let G_1, G_2 be two cographs, $r \geqslant 0$ and D_1, D_2 be sets of guards on G_1 and G_2. We have:*

$$t^*_{G_1 \cup G_2}(D_1 \cup D_2, r) \geq \min_{0 \leqslant i \leqslant r} (t^*_{G_1}(D_1, i) + t^*_{G_2}(D_2, r - i)).$$

Lemma 8. *Let G_1, G_2 be two cographs, $r \geqslant 0$ and D_1, D_2 be sets of guards on G_1 and G_2. We have:*

$$t^*_{G_1 \bowtie G_2}(D_1 \cup D_2, r) \geq \min\{t^*_{G_1}(D_1, r + |D_2|), t^*_{G_2}(D_2, r + |D_1|)\}.$$

Before sketching the proof of these results, let us show how to use them to conclude the proof of Proposition 1.

Proof (Proof of Proposition 1). We prove by induction on cographs that every cograph G satisfies $t_G(D,r) = t^*_G(D,r)$ for any set of guards D and $r \geqslant 0$. The base case is already provided by Observation 7 and Observation 9.

Now, let G be a cograph defined as the union of two cographs G_1 and G_2. Let $r \geqslant 0$ and D be a set of guards, and $D_i = D \cap V(G_i)$ for $i = 1, 2$. Applying successively Lemma 5, the induction hypothesis on G_1, G_2 and Lemma 7, we get

$$\begin{aligned} t_G(D,r) &\leq \min_{0 \leqslant i \leqslant r} (t_{G_1}(D_1,i) + t_{G_2}(D_2,r-i)) \\ &= \min_{0 \leqslant i \leqslant r} (t^*_{G_1}(D_1,i) + t^*_{G_2}(D_2,r-i)) \leq t^*_G(D,r) \end{aligned}$$

This concludes using Observation 8. The case of joins is similar, we only use Lemmas 6 and 8 instead.

It thus remains to prove Lemmas 7 and 8. We sketch the proof of the latter, the rest being available in the full version of this article. Lemma 8 relies on the following result.

Lemma 9. *Let $G = G_1 \bowtie G_2$ be a cograph, D be a set of guards and $r \geq 1$ be an integer. Then, Attacker can win on (G, D, r) in $t^*_G(D,r)$ steps against the strategy $\mathcal{S}^*$ by only playing on G_1 or by only playing on G_2.*

Lemma 9 states that (up to exchanging G_1 and G_2) Attacker has a winning strategy in $t^*_{G_1 \bowtie G_2}(D,r)$ turns on $(G_1 \bowtie G_2, D, r)$ against the strategy $\mathcal{S}^*$ where he plays only on G_1. Since vertices of G_2 have the highest possible common ancestors with vertices of G_1 in the cotree of $G_1 \bowtie G_2$, this is equivalent to playing on $(G_1, D_1, r + |D_2|)$ against the strategy $\mathcal{S}^*$ where $D_i = D \cap V(G_i)$ for $i = 1, 2$. In particular, $t^*_G(D,r) \geqslant t^*_{G_1}(D_1, r + |D_2|)$, and we get Lemma 8 by symmetry.

The proof of Lemma 9 is based on this trade-off between increasing the number of reservists and playing only on one side of a join. We first summarize this in the following observation.

Observation 10. *Let G_1 and G_2 be two cographs, each one having a set of guards D_1 and D_2. Assume that Attacker wins in k turns against $\mathcal{S}^*$ on $(G_1 \bowtie G_2, D_1 \cup D_2, r)$ by playing only on G_2. Then $k \geqslant t^*_{G_2}(D_2, r + |D_1|)$.*

Another observation is given below. It formalizes the intuition that defending using a reservist instead of a guard already in a graph does not slow down Attacker.

Lemma 10. *Let G be a graph, D a set of guards on G and $r \geq 1$. For every $x \in D$, we have $t^*_G(D,r) \leq t^*_G(D \setminus \{x\}, r+1)$.*

We are now ready to conclude the proof of Lemma 9.

Proof (Proof of Lemma 9).

We proceed by induction on $t^* = t^*_G(D,r)$. If $t^* = 1$, the result follows since Attacker wins against $\mathcal{S}^*$ in a single move.

Assume now that $t^* \geq 2$, and let u be the first vertex attacked by Attacker in a winning strategy in t^* turns against $\mathcal{S}^*$. Without loss of generality, assume that u is in G_1 and let v be the answer from Defender according to $\mathcal{S}^*$. On the obtained instance $(G, D \setminus \{v\} \cup \{u\}, r)$ (or $(G, D \cup \{u\}, r-1)$ if a reservist has been used), Attacker has a winning strategy in $t^* - 1$ turns when Defender uses $\mathcal{S}^*$. So, by induction, Attacker has a winning strategy against $\mathcal{S}^*$ playing only in G_1 or in G_2. We can assume that it is in G_2 (since otherwise we are done), and we may thus apply Observation 10.

Setting $D_i = D \cap V(G_i)$, we consider three cases depending on v:

- If v is a reservist, then $t^* - 1 \geqslant t^*_{G_2}((D \cup \{u\}) \cap V(G_2), r - 1 + |(D \cup \{u\}) \cap V(G_1)|) = t^*_{G_2}(D_2, r + |D_1|)$.
- If v is in G_1, we similarly get $t^* - 1 \geqslant t^*_{G_2}(D_2, r + |D_1|)$.
- Otherwise v is in G_2 and we get $t^* - 1 \geqslant t^*_{G_2}(D_2 \setminus \{v\}, r + |D_1| + 1)$. By Lemma 10, this is at least $t^*_{G_2}(D_2, r + |D_1|)$.

In each case, there is a winning strategy for Attacker on $(G_2, D_2, r + |D_1|)$ against $\mathcal{S}^*$ in at most $t^* - 1$ turns. But this is also a strategy for Attacker on (G, D, r) against $\mathcal{S}^*$ where he plays only on G_2. This is a contradiction by definition of t^*, which concludes the proof.

6 Conclusion

In addition to the complexity results above, we considered the complexity of FAST-STRATEGY parameterized by (a combination of) the number of turns t, the number of guards g or the twin-width tww of the graph. We provide another reduction from INDEPENDENT-SET for general graphs. For bipartite graphs, the reduction from Sect. 3 gives the lower bounds, and we use other results of this section to design an algorithm and a kernel. Finally, we show that the problem can be encoded in first-order logic, allowing us to use well-known meta-theorems. The results are summarized in the table below.

class of graphs	parameter	complexity
General	t	AW[*]
General	g	XP
General	$g + t$	W[1]-hard
Bipartite	t	co-W[1]-hard
Bipartite	g	FPT
Bipartite	$n - g$	linear kernel
Nowhere dense	t	FPT
Bounded twin-width	$t + tww$	FPT

Quite often in the literature, game problems are either in PTIME or PSPACE-hard. Motivated by Theorem 2, we propose the following conjecture.

Conjecture 1. FAST-STRATEGY is PSPACE-complete on bipartite graphs.

Besides Conjecture 1, we enumerate some open questions for future work.

1. Are there other graph classes where FAST-STRATEGY is polynomial? Good candidates are unit interval graphs. We think that the notion of arena (defined for trees) can be adapted to these graphs.
2. In this paper, we have only considered the case where there is at most one guard per vertex. One can allow multiple guards on the same vertex *i.e.* the guards now form a multiset.
   There is a generic reduction to the case considered here, roughly consisting in adding twins. More precisely, given a graph G and a multiset of guards D, consider the graph G' and set of guards D' obtained by replacing k guards on a vertex v into a set of k true twins, all of them being guarded. One can see that $t_G(D) = t_{G'}(D')$. Cographs are preserved by this transformation since they are closed under adding true twins. Thus, Theorem 6 still holds. However, this is not true for trees and adapting the proof of Theorem 4 does not seem straightforward.
3. For perfect graphs, we know that ETERNAL-DOMINATING-NUMBER is in P and MOBILE-DOMINATION-GAME is PSPACE-hard. What is the complexity of ETERNAL-DOMINATING-SET on such graphs?
4. From a parameterized complexity perspective, we showed that MOBILE-DOMINATION-GAME parameterized by the number of guards is FPT on bipartite graphs. It is then natural to look for a polynomial kernel.
5. Can our results be adapted in the "all guards move" variant of eternal domination (where Defender can move more than one guard at each turn)? An upper bound on the number of moves for Attacker to win is given in [5].

References

1. Anderson, M., Barrientos, C., Brigham, R., Carrington, R., Vitray, J., Yellen, J.: Maximum demand graphs for eternal security. J. Combin. Math. Combin. Comput. **61**, 111–128 (2007)
2. Anuradha, V., Jain, C., Snoeyink, J., Szabó, T.: How long can a graph be kept planar? Electron. J. Combin. **15**(1) (2008)
3. Bagan, G., Joffard, A., Kheddouci, H.: Eternal dominating sets on digraphs and orientations of graphs. Discret. Appl. Math. **291**, 99–115 (2021)
4. Bard, S., Duffy, C., Edwards, M., Macgillivray, G., Yang, F.: Eternal domination in split graphs. J. Comb. Math. Comb. Comput. **101**, 121–130 (2017)
5. Blažej, V., Křišťan, J.M., Valla, T.: Efficient attack sequences in m-eternal domination. arXiv:2204.02720 (2022)
6. Blažej, V., Křišťan, J.M., Valla, T.: Computing m-eternal domination number of cactus graphs in linear time. arxiv:2301.05155 (2023)
7. Braga, A., de Souza, C.C., Lee, O.: The eternal dominating set problem for proper interval graphs. Inf. Process. Lett. **115**(6), 582–587 (2015)
8. Brešar, B., Dorbec, P., Klavžar, S., Košmrlj, G., Renault, G.: Complexity of the game domination problem. Theoret. Comput. Sci. **648**, 1–7 (2016)

9. Burger, A., Cockayne, E., Grundlingh, W., Mynhardt, C., Van Vuuren, J., Winterbach, W.: Finite order domination in graphs. J. Comb. Math. Comb. Comput. **49**, 159–176 (2004)
10. Burger, A., Cockayne, E., Gründlingh, W., Mynhardt, C., Vuuren, J.V., Winterbach, W.: Infinite order domination in graphs. J. Comb. Math. Comb. Comput. **50**, 179–194 (2004)
11. Cohen, N., Martins, N.A., Mc Inerney, F., Nisse, N., Pérennes, S., Sampaio, R.: Spy-game on graphs: complexity and simple topologies. Theoret. Comput. Sci. **725**, 1–15 (2018)
12. Corneil, D.G., Perl, Y., Stewart, L.K.: A linear recognition algorithm for cographs. SIAM J. Comput. **14**(4), 926–934 (1985)
13. Diestel, R.: Graph Theory: Springer Graduate Text GTM 173. Springer Graduate Texts in Mathematics (GTM). Springer (2012)
14. Fomin, F.V., Heggernes, P., van Leeuwen, E.J.: The firefighter problem on graph classes. Theoret. Comput. Sci. **613**, 38–50 (2016)
15. Gledel, V., Iršič, V., Klavžar, S.: Fast winning strategies for the maker-breaker domination game. Electron. Notes Theor. Comput. Sci. **346**, 473–484 (2019)
16. Goddard, W., Hedetniemi, S., Hedetniemi, S.: Eternal security in graphs. J. Combin. Math. Combin. Comput. **52** (2005)
17. Goldwasser, J.L., Klostermeyer, W.F.: Tight bounds for eternal dominating sets in graphs. Discret. Math. **308**(12), 2589–2593 (2008)
18. Hefetz, D., Krivelevich, M., Stojaković, M., Szabó, T.: Fast winning strategies in maker-breaker games. J. Combin. Theory Ser. B **99**(1), 39–47 (2009)
19. Inerney, F.M., Nisse, N., Pérennes, S.: Eternal domination: D-dimensional cartesian and strong grids and everything in between. Algorithmica **83**(5), 1459–1492 (2021)
20. Iyer, A.V., Ratliff, H.D., Vijayan, G.: Optimal node ranking of trees. Inf. Process. Lett. **28**(5), 225–229 (1988)
21. Klostermeyer, W., MacGillivray, G.: Eternal security in graphs of fixed independence number. J. Comb. Math. Comb. Comput. **63**, 11 (2007)
22. Klostermeyer, W., MacGillivray, G.: Eternal dominating sets in graphs. J. Combin. Math. Combin. Comput. **68**, 97–111 (2009)
23. Klostermeyer, W.F., Mynhardt, C.: Eternal and secure domination in graphs. In: Topics in Domination in Graphs, pp. 445–478. Springer (2020)
24. Lamprou, I., Martin, R., Schewe, S.: Eternally dominating large grids. Theoret. Comput. Sci. **794**, 27–46 (2019)
25. Nešetřil, J., De Mendez, P.O.: Sparsity: Graphs, Structures, and Algorithms, vol. 28. Springer, Heidelberg (2012)
26. Rinemberg, M., Soulignac, F.: The eternal dominating set problem for interval graphs. Inf. Process. Lett. **146**, 06 (2019)
27. Šámal, R., Valla, T.: The guarding game is E-complete. Theoret. Comput. Sci. **521**, 92–106 (2014)
28. Virgile, V.: Mobile guards' strategies for graph surveillance and protection. Ph.D. thesis, University of Victoria (2024)

Independent Set Reconfiguration in $\boldsymbol{H}$-Free Graphs

Valentin Bartier[1], Nicolas Bousquet[1], and Moritz Mühlenthaler[2(✉)]

[1] Univ. Lyon, Université Lyon 1, CNRS, LIRIS UMR CNRS 5205, 69621 Lyon, France

[2] G-SCOP, Université Grenoble-Alpes, Saint-Martin-d'Hères, France
moritz.muhlenthaler@grenoble-inp.fr

Abstract. Given a graph G and two independent sets of G, the independent set reconfiguration problem asks whether one independent set can be transformed into the other by moving a single vertex at a time, such that at each intermediate step we have an independent set of G. We study the complexity of this problem for H-free graphs under the token sliding and token jumping rule. Our contribution is twofold. First, we prove a reconfiguration analogue of Alekseev's theorem for connected graphs H, showing that the problem is PSPACE-complete unless H is a path or a subdivision of the claw. We then show that under the token sliding rule the problem admits a polynomial-time algorithm if the input graph is fork-free, generalizing known results for P_4-free graphs and claw-free graphs. This implies a complete classification of the complexity of token sliding in H-free graphs, H being connected or not.

1 Introduction

For some nominal combinatorial problem, such as the satisfiability problem or the independent set problem, a corresponding *reconfiguration problem* asks the following question: Given two solutions to some instance of the nominal problem, can one be transformed into the other by a sequence of allowed modifications, such that each intermediate solution is feasible? Such problems appear in many domains, such as combinatorial puzzles (Rubik's cube, Sokoban), motion planning, or in the context of phylogenetic trees, see [23,34] for an overview of recent results. The complexity of reconfiguration problems on graphs has received considerable attention recently, for example the reconfiguration of vertex covers [32,33], dominating sets [9,21,36], colorings [8,10,14,17], and independent sets [4,12]. The natural complexity class of reconfiguration problems is PSPACE and we may ask under which conditions we can obtain polynomial-time algorithms. We consider this question for the reconfiguration of independent sets in graphs that are H-free, that is, graphs that do not contain a fixed graph H as an induced subgraph.

Our study of independent set reconfiguration in H-free graphs is largely motivated by a similar line of research on the nominal problem of finding a

D. Král' and M. Milanič (Eds.): WG 2024, LNCS 14760, pp. 35–49, 2025.
https://doi.org/10.1007/978-3-031-75409-8_3

maximum independent set (MIS) in H-free graphs. The input is an H-free graph G and the task is to find an independent set of G of maximum size. Despite roughly 40 years of effort, a complete classification of the complexity is not known. Alekseev showed that the problem remains NP-hard for a large class of connected graphs H.

Theorem 1 ([1]). *The maximum independent set problem on H-free graphs is* NP*-hard, unless H is a path, the claw, or a subdivision of the claw.*

Let P_k denote the path graph on k vertices and see Fig. 1 for illustrations of P_4, the claw and the fork. Minty [31] and Sbihi [35] showed independently that a maximum independent set of a claw-free graph can be computed in polynomial time. Twenty years later, Alekseev generalized this result to fork-free graphs [2]. Noticing that a maximum independent set can be computed in polynomial time in P_4-free graphs, the smallest interesting case left open by Theorem 1 is $H = P_5$. The complexity of this problem remained open for three decades, but recently, polynomial-time algorithms were obtained for $H = P_5$ [30] and $H = P_6$ [20]. The complexity remains open for $H = P_\ell$ for $\ell \geq 7$.

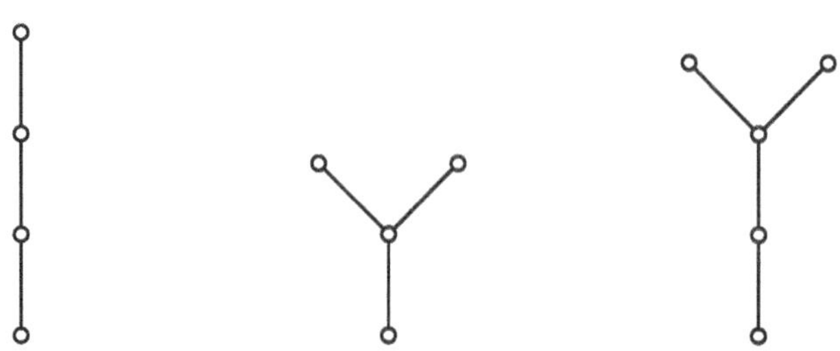

(a) P_4 (b) claw (c) fork

Fig. 1. Small graphs of interest.

We investigate the complexity of INDEPENDENT SET RECONFIGURATION (ISR) in H-free graphs. The ISR problem has been introduced in [24] and it has been studied widely since, see e.g., [4,7,12,16,18,25,26]. An instance of ISR is a graph G and two independent sets I and J of G of the same size, which are considered as *tokens* placed on the vertices of G. The question is whether I can be transformed into J by a sequence of *moves*, such that in each intermediate step the tokens form an independent set. Two main types of moves have been considered: *token jumping*, where a token can move to any other vertex of the graph (provided that it is not adjacent to a token), and *token sliding*, where a token can slide along an edge to a neighboring vertex. We denote by TOKEN JUMPING and TOKEN SLIDING and the corresponding variants of ISR, where the only allowed moves are token jumping and token sliding, respectively. Both problems are equivalent for maximum independent sets and are known to be PSPACE-complete, even for planar graphs of maximum degree three [22,24].

Polynomial-time algorithms are known for example on trees [16], interval graphs [7], bipartite permutation and bipartite distance hereditary graphs [18].

We prove the following reconfiguration analogue of Theorem 1 for ISR in H-free graphs.

Theorem 2. *For any connected graph H that is not a path, the claw, or a subdivision of the claw, Token Sliding and Token Jumping in H-free graphs are* PSPACE*-complete.*

On the positive side, it is known that Token Sliding and Token Jumping admit polynomial-time algorithms on P_4-free graphs [11,28] and claw-free graphs [12]. We generalize these results for Token Sliding, showing that the problem admits a polynomial-time algorithm in fork-free graphs.

Theorem 3. *Token Sliding in fork-free graphs admits a polynomial-time algorithm.*

Let us note that the complexity of Token Jumping on fork-free graphs remains open. Combining Theorems 2 and 3 with a result of Belmonte et al., who showed that Token Sliding is PSPACE-complete on split graphs, we obtain a complete classification of the complexity of Token Sliding in H-free graphs. Notice that here, the graph H is not necessarily connected.

Corollary 1. *Token Sliding in H-free graphs admits a polynomial-time algorithm if and only if H is a subgraph of the union of a fork and isolated vertices.*

Overview of the Proofs. In order to prove Theorem 2, we use similar ideas to those that Alekseev used to prove Theorem 1: Let G_t be the graph obtained from a graph $G = (V, E)$ by subdividing each edge of G exactly t times. The main idea is to relate maximum independent sets of a graph G with those of G_t. In [1] Alekseev showed that $\alpha(G_{2t}) = \alpha(G) + t|E|$, where $\alpha(G)$ denotes the size of a maximum independent set of G. We may therefore assume that the input graph has arbitrarily large girth and that any two vertices of degree at least three are arbitrarily far apart. Theorem 1 then follows from the fact that MIS is NP-hard on graph of maximum degree three. To prove Theorem 2, we associate each maximum independent set of G in a canonical way with a maximum independent set of G_{2t}. Let I and J be two maximum independent sets of G and let I' and J' be the corresponding canonical independent sets of G_{2t}. We show that there exists a transformation from I to J if and only if there exists a transformation from I' to J'. Combining this with the fact that Token Sliding and Token Jumping are PSPACE-complete on graphs of maximum degree three [22,24] yields Theorem 2.

We then show that Token Sliding in H-free admits a polynomial-time algorithm on fork-free graphs and for maximum independent sets (Theorem 3). Let G be a fork-free graph and I and J be independent sets of G of the same size. We notice that a vertex v of G is *irrelevant* for Token Sliding if no independent set reachable from I contains v. Throughout the algorithm, we reduce the graph by deleting irrelevant vertices. For instance, any vertex with at least three tokens in its neighborhood is irrelevant. In general, we distinguish two cases, depending on whether I and J are maximum or not. In the first case, we provide a simple reduction to Token Sliding in claw-free graphs: We show that if I is maximum, then any induced claw of G contains precisely one token on a vertex of the claw that is not the center. This means that the center of any claw is irrelevant and may be removed from the graph. We thus obtain an equivalent instance on a claw-free graph. Notice that by the equivalence of Token Sliding and Token Jumping for maximum independent sets, this also yields a polynomial-time algorithm for Token Jumping in fork-free graphs if the independent sets are maximum.

The case where I and J are not maximum is more involved. We observe that, after removing irrelevant vertices, each connected component of the graph induced by the vertices I and J is either a path or a cycle. Sliding tokens on induced paths is easy, so it remains to deal with the cycles and isolated vertices. Similar obstructions occur for alternating cycles in Matching Reconfiguration [24] and cycles induced by the symmetric difference of independent sets of claw-free graphs [12]). We show that by sliding tokens along an I-augmenting path (I is not maximum), we may assume that G contains an I-free vertex, that is, a vertex f, such that $I + f$ is an independent set. We then show that, for any vertex c of a cycle in the graph induced by the symmetric difference of the two independent sets, either we can transform I into $I - c + f$ or find an irrelevant vertex, which can be removed (Theorem 12). In the former case, this allows us to move the remaining tokens on the cycle to their target positions in J. To prove Theorem 12, we first show that we may assume that every connected component of G is prime with respect to the modular decomposition. Notice that the modular decomposition was used "in disguise" in the form of a *cotree* in the polynomial-time algorithm for ISR in cographs (P_4-free graphs) in [11]. In order to move the token on c to the free vertex f, we imagine sliding it on a shortest path P from c to f. Then c may be blocked by other tokens on P or in the neighborhood of P. Using a structural characterization of claws in fork-free prime graphs due to Brandstädt et al. [15], we show we may move c and every vertex that is blocking c forward on the path, one by one in reverse order, or find an irrelevant vertex.

Finally, to obtain Corollary 1, we observe that Token Sliding is PSPACE-complete on $2K_2$-free graphs, since the problem is PSPACE-complete on split graphs [6]. By this observation and Theorem 3, the choice of H is limited to the union of a fork and and $\ell \geq 1$ independent vertices. Then, either the size of the independent sets is bounded by a function of ℓ, or there is a fork with many tokens in the neighborhood of its center. In the former case there are polynomially many independent sets and we may therefore conclude in polynomial time. In the latter case, the center vertex is irrelevant and may be removed. Iterating this argument yields a fork-free graph and we may conclude by Theorem 3.

Related Work. The complexity of independent set reconfiguration has been widely studied in the literature, for example in particular graph classes [7,16,18] and in the context of parameterized complexity [27,34]. For H-free graphs we are aware of only two results: Bonsma [11] showed that TOKEN SLIDING is solvable in polynomial on in P_4-free graphs. In [12], Bonsma et al. showed that TOKEN SLIDING and TOKEN JUMPING admit a polynomial-time algorithm on claw-free graphs [12]. In the case that more than one graph is forbidden (as induced subgraph), Kamiński et al. [28] showed that TOKEN JUMPING admits a polynomial-time algorithm on even-hole free graphs. Interestingly, the complexity of MIS is open for this graph class. The complexity of TOKEN SLIDING and TOKEN JUMPING can be very different, for example, Lokshtanov and Mouawad [29] showed that TOKEN JUMPING is NP-complete on bipartite graphs (which are odd-hole free), whereas TOKEN SLIDING remains PSPACE-complete. Similarly, TOKEN JUMPING admits a linear-time algorithm on split graphs, while TOKEN SLIDING is PSPACE-complete on this graph class [6]. The reader is referred to [3] for a complete and recent bibliography on the topic. ISR has also been studied from the point of view of parameterized complexity, where the parameter is for example the size of the independent sets or the length of the reconfiguration sequence. The reader is referred to [13] for an overview of results in this direction. Many other reconfiguration problems have been studied when an induced subgraph is forbidden, especially from a structural point of view. For single vertex recoloring for instance, a complete dichotomy theorem was recently obtained by Belavadi, Cameron, Merkel in [5], who characterized exactly for which graphs H, the graph of $(\chi(G)+1)$-colorings of G is connected for every H-free graph G (where $\chi(G)$ is the chromatic number of G).

Organization. Section 2 contains notation and definitions. In Sect. 3, we prove the reconfiguration analogue of Alekseev's theorem (Theorem 2). We discuss irrelevant vertices of fork-free graphs in Sect. 4. We propose a polynomial-time algorithm for TOKEN SLIDING for maximum independent sets in fork-free graphs in Sect. 5 and for non-maximum independent sets in Sect. 6. Section 7 concludes the paper with a discussion of open questions.

Some proofs are omitted due to space limitations.

2 Preliminaries

For two sets U, W we denote by $U \triangle W$ their symmetric difference $(U \setminus W) \cup (W \setminus U)$. Graphs in this paper are undirected and finite. Let $G = (V, E)$ be a graph. For a vertex $v \in V$ we denote by $N(v) := \{w \in V \mid vw \in E\}$ be the *open neighborhood of* v and by $N[v] = \{v\} \cup N(v)$ the *closed neighborhood of* v. Similarly, for a set $X \subseteq V$, we denote by $N(X) \subseteq V \setminus X$ the set of vertices that have a neighbor in X and let $N[X] := N(X) \cup X$. A graph G' is an *induced subgraph* of G if it can be obtained from G be deleting vertices (and their incident edges). For $X \subseteq V$, we denote by $G[X]$ the induced subgraph of G on the vertex set X. A set $I \subseteq V$ is *independent* in G if the vertices in I are pairwise non-adjacent in G. We denote by $\alpha(G)$ the size of a maximum independent set of G. For an independent set I of G and $v \in V$, we say that v is *I-free* if $N[v] \cap I = \emptyset$.

For $t \geq 1$ we denote by P_t the path graph on t vertices. A vertex of degree one of P_t is called *endvertex* and a vertex of degree two of P_t is called *internal vertex*. Given graph G, the graph $\overline{G}$ is the *complement graph* of G and is obtained by replacing every edge of G by a non-edge and vice versa.

A *claw* is a graph on four vertices that has one vertex of degree three (the *center*) and three vertices of degree one (the *leaves*), see Fig. 1b. A *fork* is a graph obtained from a claw by subdividing one edge, that is, by replacing an edge by a path on three vertices, see Fig. 1c. The *center* of a fork F is the vertex of F of degree 3. For $t \geq 1$, the *t-subdivision* of G is the graph obtained from G by replacing each edge uv of G with a copy of P_{t+2} and identifying its endvertices with u and v. We say that a t-subdivision of G is *even* if t is even.

Independent Set Reconfiguration. In the following we define two adjacency relations on the independent sets of G that correspond to three natural reconfiguration steps on independent sets: *token sliding* (TS) and *token jumping* (TJ). Let $I, J \subseteq V$ be two independent sets of G. We say that I and J are TJ-adjacent ($I \leftrightarrow_{\mathrm{TJ}} J$) if there is precisely one vertex in I that is not in J and vice versa. Furthermore, we say that I and J are TS-*adjacent* ($I \leftrightarrow_{\mathrm{TS}} J$) if $I \leftrightarrow_{\mathrm{TJ}} J$ and the vertex u in $I \setminus J$ and the vertex v in $J \setminus I$ are adjacent in G. For each rule $R \in \{\mathrm{TS}, \mathrm{TJ}\}$ we say that I and J admit a *R-reconfiguration sequence* $I_0, I_1, \ldots, I_\ell$ and write $I \leftrightsquigarrow_R J$ if $I = I_0$, $J = I_\ell$ and for $0 \leq i < \ell$ we have $I_i \leftrightarrow_R I_{i+1}$. If it is clear from the context we may omit the rule R. Given a graph G and two independent sets $I, J \subseteq V(G)$, the problem Token Sliding (resp., Token Jumping) asks whether I and J admit a TS-reconfiguration sequence (resp., TJ-reconfiguration sequence).

Note that Token Sliding and Token Jumping are equivalent on maximum independent sets since, if a token can jump on a non-adjacent vertex, the independent set is indeed not maximum.

3 A Reconfiguration Analogue of Alekseev's Theorem

In this section we prove Theorem 2, a reconfiguration analogue of Alekseev's Theorem 1. In the following, let $G = (V, E)$ be a graph and G_t be the t-subdivision of G for some even $t \geq 1$. For any independent set I of G, we can construct in a canonical way an independent set I' of G_t by placing for each edge uv of G on the internal vertices of the copy of P_{t+2} in G_t that corresponds to uv exactly $t/2$ additional tokens. The independent set I' is called the *extension of* I. Alekseev proved the following:

Theorem 4 ([1]). *The graph G has a maximum independent set of size k if and only if G_t has maximum independent set of size $k + t \cdot |E(G)|/2$.*

Theorem 4 implies immediately that for any maximum independent set I of G, the extension I' of I in G is also maximum. Our goal in this section is to adapt Theorem 4 to Token Sliding and Token Jumping as follows.

Theorem 5. *Let I and J be two maximum independent sets of G and let $R \in \{\mathrm{TS}, \mathrm{TJ}\}$. Then $I \leftrightsquigarrow_R J$ with respect to G if and only if $I' \leftrightsquigarrow_R J'$ with respect to G_t.*

Recall that TOKEN SLIDING and TOKEN JUMPING are PSPACE-complete in graphs of maximum degree three [22,24]. By considering subdivisions of G, we may assume that G has arbitrarily high girth. Therefore, just as Alekseev's Theorem 4 implies Theorem 1, it suffices to prove Theorem 5 to obtain Theorem 2. The remainder of this section is devoted to the proof of Theorem 5. Observe that by Theorem 4 the independent sets I' and J' of Theorem 5 are maximum independent sets of G_t. Recall that for maximum independent sets, TOKEN SLIDING and TOKEN JUMPING are equivalent[1], so it suffices to consider the TS version.

For an edge uv of G let $S_{uv} = \{s_{uv}^1, s_{uv}^2, \ldots, s_{uv}^t\}$ be the set of internal vertices of the copy of P_{t+2} that replaces the edge uv in the construction of G_t. We assume that s_1 is adjacent to u and s_t is adjacent to v and that for $1 < i < t$, the vertex s_i is adjacent to s_{i+1} and s_{i-1}.

Claim 1. *Let uv be an edge of G and let I' be a maximum independent set of G_t. If $\{u, v\} \subseteq I'$ then $|I' \cap S_{uv}| = (t-2)/2$ and $|I' \cap S_{uv}| = t/2$ otherwise.*

By abuse of notation, we will denote by V the subset of vertices of G_t corresponding to vertices of G. The following claim, which ensures that a single reconfiguration step in G can be adapted in G_t, will immediately imply the "if"-direction of Theorem 5.

Claim 2. *Let I_1, I_2 be two maximum independent sets of G such that $I_1 \leftrightarrow_{TS} I_2$ and let I_1' and I_2' be the extensions of I_1 and I_2 in G_t. Then $I_1' \leftrightsquigarrow_{TS} I_2'$.*

Let us now prove the converse direction of Theorem 5. We start with the following claim:

Claim 3. *Let I' be a maximum independent set of G_t. Then there are no three vertices $u, v, w \in I' \cap V$ such that $uv, vw \in E(G)$.*

Let I' be a maximum independent set of G_t. By the previous claim, the graph $G[I' \cap V]$ consists of isolated vertices and edges. Let $v(I')$ (resp, $e(I')$) denote the number of isolated vertices (resp., isolated edges) of $G[I' \cap V]$.

Claim 4. $v(I') + e(I') = \alpha(G)$.

To complete the proof of Theorem 5, consider a TS-reconfiguration sequence $I_1' \leftrightarrow_{\mathrm{TS}} I_2' \leftrightarrow_{\mathrm{TS}} \ldots \leftrightarrow_{\mathrm{TS}} I_\ell'$ of maximum independent sets of G_t such that $I' = I_1'$ and $J' = I_\ell$, where I' (resp., J') is the extension of I (resp. J) in G_t. Notice that $G[I' \cap V]$ and $G[J' \cap V]$ contain no edge. And by Claim 4, we can associate to each independent set I' an induced subgraph consisting of $\alpha(G)$ isolated vertices or edges. We obtain from I' a maximum independent set I_i of G by picking all

[1] If a token can jump to a non-neighbor then the independent set is not maximum.

isolated vertices and exactly one endpoint of each edge of $G[I_i' \cap V]$, say the first vertex with respect to an arbitrary fixed ordering of V. To conclude the proof it is enough to show that for each $i \in \{1, 2, \ldots, \ell-1\}$, we have $I_i \leftrightsquigarrow_{\mathrm{TS}} I_{i+1}$.

Claim 5. $|(I_i' \cap V) \Delta (I_{i+1}' \cap V)| \leq 1$.

By Claims 4 and 5, we just have to distinguish the following cases to complete the proof of the converse direction:

1. $e(I_i') = e(I_{i+1}')$. In this case we have $I_i = I_{i+1}$ and we are done.
2. $e(I_i') = e(I_{i+1}') + 1$. Let uv be the edge of G that is not present in $G[I_{i+1}' \cap V]$ and assume without loss of generality that $u \in I_{i+1}$. It follows that I_{i+1}' is obtained from I_i' by sliding a token from v to S_{uv}^p. Then either $u \in I_i$, so we have $I_i = I_{i+1}$ or $v \in I_i$ and we have $I_i \leftrightarrow_{\mathrm{TS}} I_{i+1}$.
3. $e(I_i') = e(I_{i+1}') - 1$. The same argument can be applied by exchanging I_i' and I_{i+1}'.

So, for every $1 \leq i < \ell$, we have $I_i \leftrightsquigarrow_{\mathrm{TS}} I_{i+1}$. And then $I \leftrightsquigarrow_{\mathrm{TS}} J$ which completes the proof.

4 Irrelevant Vertices in Fork Free Graphs

In the following, let $G = (V, E)$ be a fork-free graph and I and J be two independent sets of G. We will identify vertices of G that are irrelevant in any transformation from I to J in the sense that they cannot contain a token. We show that vertices adjacent to at least two tokens are irrelevant if they satisfy a certain neighborhood-condition. As a consequence, all vertices adjacent to at least three tokens are useless and can be safely deleted. We say that a set $X \subseteq V$ of vertices is *locally blocked* (with respect to I) if any vertex in X is adjacent to at least two tokens in I. We are interested in locally blocked vertices that remain locally blocked under any token move. Let $X \subseteq V$ be a locally blocked subset of vertices. We say that the set $B_X := N(X) \cap I$ is X-*blocking*: no token can slide from B_X to X, since there is another token in B_X blocking it.

Let $u \in V$ and $X \subseteq V - u$. A vertex v of $G - (X + u)$ is an X-*twin of* u if uv is an edge and $N(v) \cap X = N(u) \cap X$. A set $X \subseteq V$ of vertices *permanently blocked* if it is locally blocked (with respect to I) and for each independent set I' such that $I \leftrightsquigarrow_{\mathrm{TS}} I'$ and for each vertex u in the blocking set $B_X = N(X) \cap I'$, any I'-free neighbor v of u is an X-twin of u. (It means that if we move a token in the neighborhood of X from u to another vertex v then the u and v have the same neighbords in X). We first introduce the following reduction rule, which states that permanently blocked vertices can be safely removed from the graph.

Reduction Rule Z. Let $X \subseteq V$ be permanently blocked. Return $(G - X, I, J)$ if $X \cap J = \emptyset$ and a trivial No-instance otherwise.

A reduction rule is *safe* if applying it yields an equivalent instance.

Lemma 1. *Reduction Rule Z is safe.*

At this point it is not quite clear how to use Reduction Rule Z algorithmically. We will later prove that certain subsets of vertices are permanently blocked (and can be found in polynomial time). For now, we show that any vertex of G adjacent to at least three tokens of I is permanently blocked and can thus be safely deleted according to Lemma 1.

Lemma 2. *Let $c \in V$ be a vertex that is adjacent to exactly $k \geq 3$ tokens in I and let I' be an independent set such that $I' \leftrightsquigarrow_{TS} I$. Then c is adjacent to exactly k tokens of I'.*

Lemma 2 immediately implies the following sufficient condition for a vertex to be permanently blocked.

Observation 6. *Let $c \in V$ be a vertex that is adjacent to at least three tokens in I. Then c is permanently blocked.*

From the safeness of Reduction Rule Z and the previous observation we conclude that the following Reduction Rule is safe.

Reduction Rule A. Let $c \in V$ such that c is adjacent to at least three tokens in I. Return $(G - c, I, J)$ if $c \notin J$, otherwise return a trivial No-instance.

We say that G is I-reduced if Reduction Rule A is not applicable with respect to G and I. That is, G contains no vertex that is adjacent to at least three I-tokens. Notice that by applying Reduction Rule A exhaustively, we may obtain from G in polynomial time a fork-free graph G' that is I-reduced and J-reduced. The resulting equivalent instance is either (G', I, J) or a trivial No-instance. The following observations are consequences of Lemma 2.

Observation 7. *If G is I-reduced then G is I'-reduced for every independent set I' such that $I \leftrightsquigarrow_{TS} I'$.*

Observation 8. *Let G be I-reduced and let I' be an independent set such that $I \leftrightsquigarrow_{TS} I'$. Then for any vertex v of G we have $|N(v) \cap I'| \leq 2$.*

5 Maximum Independent Set Reconfiguration in Fork-Free Graphs

As a warm-up, let us prove the following theorem.

Theorem 9. *Token Sliding and Token Jumping on fork-free graphs for maximum independent sets admits a polynomial-time algorithm.*

To do so, we provide a simple reduction to the same problem in claw-free graphs, which admits a polynomial-time algorithm [12]. In the following, let $G = (V, E)$ be a fork-free graph and I and J be two maximum independent sets of G. Note that it suffices to consider the token sliding rule only since I and J are maximum. By Lemma 1 and Observation 6 we may assume that G is both I-reduced and J-reduced. Therefore, G contains no induced claw that contains three tokens of I. Since I is maximum, the following stronger property holds.

Lemma 3. *Let $\{c, x, y, z\}$ induce a claw in G with center c. Then $|\{x, y, z\} \cap I| = 1$.*

By the previous lemma, no maximum independent set of G has a token on the center of any induced claw of G. Therefore, the following reduction rule is safe.

Reduction Rule MIS. Let $\{c, x, y, z\}$ induce a claw in G with center c. Return $(G - c, I, J)$.

Proof (of Theorem 9). The induced claws of G can be enumerated in polynomial time (by brute-force), so Reduction Rule MIS can be applied exhaustively in polynomial time. We obtain an equivalent instance (G', I, J) of Token Sliding such that the graph G' is claw-free. It remains to apply the algorithm from [12] on (G', I, J) to decide the initial instance (G, I, J). Notice that any reconfiguration sequence from I to J in G' is a reconfiguration sequence from I to J in G. □

6 Non-maximum Token Sliding in Fork-Free Graphs

We provide a polynomial-time algorithm for Token Sliding on fork-free graphs, generalizing the algorithms for claw-free graphs [12] and P_4-free graphs [11]. In the following, let $G = (V, E)$ be a fork-free graph and let $I, J \subseteq V$ be two independent sets of G of the same size. If I and J are maximum, we may invoke Theorem 9 and are done. Otherwise, we show that we may assume that G is I-reduced and J-reduced and that each connected component of G is prime.

Theorem 10. *Let (G, I, J) be an instance of Token Sliding such that G is fork-free. Then we may compute in polynomial from (G, I, J) an equivalent instance (G', I', J') of Token Sliding on a fork-free graph G' that is I'-reduced and J'-reduced, such that each connected component of G' is prime.*

A *complex* is a complete bipartite graph minus a matching. We recall the following characterization of bipartite fork-free graphs due to Alekseev, which implies that each connected component of $G[I \triangle J]$ is a path or an (even) cycle.

Theorem 11 ([2])**.** *A bipartite and connected fork-free graph G is either a path, a cycle or a complex.*

Lemma 4. *If G is I-reduced and J-reduced then $I \triangle J$ is a collection of disjoint paths, cycles and isolated vertices.*

Proof. Since Reduction Rule A is not applicable, each vertex of I has at most two neighbors in J and vice versa. Noticing that $G[I \triangle J]$ is a bipartite fork-free graph, we conflude that each connected component of $G[I \triangle J]$ is a cycle or a path by Theorem 11. □

The algorithm proceeds as follows. We treat each connected component of $G[I \triangle J]$ independently. Let C be a non-trivial connected component of $G[I \triangle J]$ and let $I' := (I \setminus V(G)) \cup (J \cap V(C))$. We say that C can be *resolved* if $I \leftrightsquigarrow_{\mathrm{TS}} I'$. That is, C can be resolved if there exists a reconfiguration sequence that replaces on C the tokens of I with the tokens of J. By abuse of notation we say that an isolated vertex $u \in I$ of $G[I \triangle J]$ can be resolved if there exists a reconfiguration sequence that slides u to an isolated vertex $v \in J \setminus I$ of $G[J \triangle I]$.

Resolving paths in $G[I \triangle J]$ is easy while cycles and isolated vertices require more work. In order to deal with the latter ones we provide the following theorem. The proof is non-trivial and uses, among other things, a characterization of fork-free prime extensions of the claw due to Brandstädt et al. [15].

Theorem 12. *Let $v \in I$ and let u be an I-free vertex of G. Then $J := I - v + u$ is an independent set of G. If G is prime then there is a polynomial-time algorithm that computes a reconfiguration sequence $I \leftrightsquigarrow_{TS} J$ or finds a permanently blocked vertex of G.*

Let C be a connected component of $G[I \triangle J]$. If C is a path it can be resolved simply by sliding tokens one after the other along the path. Recall that if I and J are not maximum. In order to resolve a cycle or an isolated vertex we need an I-free vertex in order to apply Theorem 12. We show that we can always produce an I-free vertex by finding an I-augmenting path. Let G be a graph and A be an independent set of G. An independent set X of $G - A$ is called *magnifier of* A if the independent set $(A \cup X) \setminus (N(X) \cap A)$ is strictly larger that A and the corresponding subgraph $G[X \cup (N(X) \cap A)]$ is called *A-augmenting graph* [2]. Notice that augmenting graphs are bipartite.

Lemma 5. *If G is I-reduced and I is maximal but not maximum then there exists an I-augmenting path.*

Proof. Since I is not maximum, there exists an I-augmenting graph H and since I is maximal, H has at least two vertices. Since G is fork-free, Theorem 11 ensures that H is either a path or a complex. Since furthermore Reduction Rule A is not applicable, we have that H is an I-augmenting path. □

We use the following special case of the result of Gerber et al., who showed that I-augmenting paths can be computed in polynomial time in graphs with no induced skew stars.

Theorem 13 ([19]). *There is a polynomial-time algorithm that finds, if it exists, an I-augmenting path in a graph that contains no skew-star.*

By moving tokens along an I-augmenting path, we may assume that G contains an I-free vertex. If C is an isolated vertex then we can resolve C by Theorem 12 or remove a permanently blocked vertex. Similarly, if C is a cycle, we then pick an arbitrary token on $u \in C \cap I$ and apply Theorem 12 to obtain a transformation $I' \leftrightsquigarrow_{\mathrm{TS}} I' - u + v$ or a permanently blocked vertex. In the former case we slide the remaining tokens on $C \cap I$ to $C \cap J$ just as for paths and then slide the token on v back to C (again according to Theorem 12).

Lemma 6. *Assume that G is prime, I-reduced and J-reduced, and that I and J are not maximum. Then we can find in polynomial-time either a C-resolving sequence or a permanently blocked vertex of G.*

We now have all the ingredients of the polynomial-time algorithm for Token Sliding in fork-free graphs.

Theorem 14. *Token Sliding in fork-free graphs admits a polynomial-time algorithm.*

Proof. Let (G, I, J) be an instance of Token Sliding such that G is fork-free. We may assume that I and J are not maximum, since otherwise Theorem 9 applies. This can be tested in polynomial time by the algorithm given in [2]. Otherwise, we invoke Theorem 10 and obtain in polynomial time an equivalent instance (G', I', J'), such that G' is fork-free and each connected component of G' is prime (or (G', I', J') is a trivial No-instance).

We proceed by trying to resolve each connected component C of $G'[I' \triangle J']$. By Lemma 4, the graph $G'[I' \triangle J']$ is a collection of disjoint paths, cycles and isolated vertices. If C is a path we return the sequence sliding the tokens from $I' \cap C$ to $J' \cap C$ one by one. If $G'[I' \triangle J']$ contains two isolated vertices $u \in I$ and $v \in J$, Theorem 12 ensures that we can either find a reconfiguration sequence $I \leftrightsquigarrow_{\mathrm{TS}} I - u + v$ or a permanently blocked vertex polynomial time. If C is a cycle, by Lemma 6, we can either find a resolving sequence for C or a permanently blocked vertex. If at any step we find a permanently blocked vertex, we can delete it from G' by Lemma 1 and restart the algorithm. Since each time we restart the algorithm at least one vertex is removed from the graph, the overall running time is polynomial. □

7 Conclusion

For connected graphs H, we showed that Token Sliding and Token Jumping are PSPACE-complete on H-free graphs, unless H is a path, a claw, or a subdivision of the claw. Furthermore, we showed that Token Sliding admits a polynomial-time algorithm on fork-free graphs. Combining the two results with PSPACE-hardness of Token Sliding on split graphs due to Belmonte et al. [6], we obtain a complete classification of the complexity of Token Sliding on H-free graphs. In contrast, a complexity dichotomy for Token Jumping seems out of reach. An interesting first step in this direction is to investigate, whether Token Jumping admits a polynomial-time algorithm on fork-free graphs if the independent sets are not maximum. There are two main reasons why our approach for Token Sliding does not work for Token Jumping in this case. First, we cannot assume that tokens stay in their respective modules, so the modular decomposition seems less useful. Secondly, the irrelevant vertices for Token Sliding may not be irrelevant for Token Jumping, for example, vertices that are adjacent to three tokens. This means in particular that in order to resolve a cycle, it is not sufficient to consider augmenting chains, but it may be necessary to deal with large complexes induced by the symmetric difference of the independent sets. A second interesting question is whether Token Jumping admits a polynomial-time algorithm on P_5-free graphs. As a first step, the case of maximum independent sets could be considered.

Acknowledgements. We would like to thank Martin Milanic for his insightful comments, in particular for pointing out that combining Theorems 2 and 3 with the result of [6] directly yields a complexity dichotomy for Token Sliding on H-free graphs for connected graphs H.

References

1. Alekseev, V.E.: The effect of local constraints on the complexity of determination of the graph independence number. In: Combinatorial-Algebraic Methods in Applied Mathematics, pp. 3–13 (1982)
2. Alekseev, V.E.: Polynomial algorithm for finding the largest independent sets in graphs without forks. Discrete Appl. Math. **135**(1–3), 3–16 (2004). https://doi.org/10.1016/S0166-218X(02)00290-1
3. Bartier, V.: Combinatorial and algorithmic aspects of reconfiguration. Aspects combinatoires et algorithmiques de la Reconfiguration. Ph.D. thesis, Grenoble Alpes University, France (2021)
4. Bartier, V., Bousquet, N., Dallard, C., Lomer, K., Mouawad, A.E.: On girth and the parameterized complexity of token sliding and token jumping. Algorithmica **83**(9), 2914–2951 (2021). https://doi.org/10.1007/S00453-021-00848-1
5. Belavadi, M., Cameron, K., Merkel, O.: Reconfiguration of vertex colouring and forbidden induced subgraphs (2023)
6. Belmonte, R., Kim, E.J., Lampis, M., Mitsou, V., Otachi, Y., Sikora, F.: Token sliding on split graphs. Theory Comput. Syst. **65**(4), 662–686 (2021). https://doi.org/10.1007/S00224-020-09967-8
7. Bonamy, M., Bousquet, N.: Token sliding on chordal graphs. In: Bodlaender, H.L., Woeginger, G.J. (eds.) WG 2017. LNCS, vol. 10520, pp. 127–139. Springer, Cham (2017). https://doi.org/10.1007/978-3-319-68705-6_10
8. Bonamy, M., Defrain, O., Klimosová, T., Lagoutte, A., Narboni, J.: On Vizing's edge colouring question. J. Comb. Theory Ser. B **159**, 126–139 (2023). https://doi.org/10.1016/J.JCTB.2022.10.005
9. Bonamy, M., Dorbec, P., Ouvrard, P.: Dominating sets reconfiguration under token sliding. Discret. Appl. Math. **301**, 6–18 (2021). https://doi.org/10.1016/J.DAM.2021.05.014
10. Bonamy, M., et al.: Diameter of colorings under Kempe changes. Theor. Comput. Sci. **838**, 45–57 (2020). https://doi.org/10.1016/J.TCS.2020.05.033
11. Bonsma, P.S.: Independent set reconfiguration in cographs and their generalizations. J. Graph Theory **83**(2), 164–195 (2016). https://doi.org/10.1002/JGT.21992
12. Bonsma, P., Kamiński, M., Wrochna, M.: Reconfiguring independent sets in claw-free graphs. In: Ravi, R., Gørtz, I.L. (eds.) SWAT 2014. LNCS, vol. 8503, pp. 86–97. Springer, Cham (2014). https://doi.org/10.1007/978-3-319-08404-6_8
13. Bousquet, N., Mouawad, A.E., Nishimura, N., Siebertz, S.: A survey on the parameterized complexity of the independent set and (connected) dominating set reconfiguration problems. CoRR abs/2204.10526 (2022). https://doi.org/10.48550/ARXIV.2204.10526

14. Bousquet, N., Perarnau, G.: Fast recoloring of sparse graphs. Eur. J. Comb. **52**, 1–11 (2016). https://doi.org/10.1016/J.EJC.2015.08.001
15. Brandstädt, A., Hoàng, C.T., Vanherpe, J.: On minimal prime extensions of a four-vertex graph in a prime graph. Discret. Math. **288**(1–3), 9–17 (2004). https://doi.org/10.1016/J.DISC.2004.06.019
16. Demaine, E.D., et al.: Linear-time algorithm for sliding tokens on trees. Theor. Comput. Sci. **600**, 132–142 (2015). https://doi.org/10.1016/J.TCS.2015.07.037
17. Dvorák, Z., Feghali, C.: A Thomassen-type method for planar graph recoloring. Eur. J. Comb. **95**, 103319 (2021). https://doi.org/10.1016/J.EJC.2021.103319
18. Fox-Epstein, E., Hoang, D.A., Otachi, Y., Uehara, R.: Sliding token on bipartite permutation graphs. In: Elbassioni, K., Makino, K. (eds.) ISAAC 2015. LNCS, vol. 9472, pp. 237–247. Springer, Heidelberg (2015). https://doi.org/10.1007/978-3-662-48971-0_21
19. Gerber, M.U., Hertz, A., Lozin, V.V.: Augmenting chains in graphs without a skew star. J. Comb. Theory Ser. B **96**(3), 352–366 (2006). https://doi.org/10.1016/J.JCTB.2005.09.007
20. Grzesik, A., Klimosová, T., Pilipczuk, M., Pilipczuk, M.: Polynomial-time algorithm for maximum weight independent set on p_6-free graphs. ACM Trans. Algorithms **18**(1), 4:1–4:57 (2022). https://doi.org/10.1145/3414473
21. Haddadan, A., Ito, T., Mouawad, A.E., Nishimura, N., Ono, H., Suzuki, A., Tebbal, Y.: The complexity of dominating set reconfiguration. Theor. Comput. Sci. **651**, 37–49 (2016). https://doi.org/10.1016/J.TCS.2016.08.016
22. Hearn, R.A., Demaine, E.D.: PSPACE-completeness of sliding-block puzzles and other problems through the nondeterministic constraint logic model of computation. Theor. Comput. Sci. **343**(1–2), 72–96 (2005). https://doi.org/10.1016/J.TCS.2005.05.008
23. van den Heuvel, J.: The complexity of change. In: Blackburn, S.R., Gerke, S., Wildon, M. (eds.) Surveys in Combinatorics 2013. London Mathematical Society Lecture Note Series, vol. 409, pp. 127–160. Cambridge University Press (2013). https://doi.org/10.1017/CBO9781139506748.005
24. Ito, T., et al.: On the complexity of reconfiguration problems. Theor. Comput. Sci. **412**(12–14), 1054–1065 (2011). https://doi.org/10.1016/J.TCS.2010.12.005
25. Ito, T., Kamiński, M., Ono, H.: Fixed-parameter tractability of token jumping on planar graphs. In: Ahn, H.-K., Shin, C.-S. (eds.) ISAAC 2014. LNCS, vol. 8889, pp. 208–219. Springer, Cham (2014). https://doi.org/10.1007/978-3-319-13075-0_17
26. Ito, T., Kamiński, M., Ono, H., Suzuki, A., Uehara, R., Yamanaka, K.: On the parameterized complexity for token jumping on graphs. In: Gopal, T.V., Agrawal, M., Li, A., Cooper, S.B. (eds.) TAMC 2014. LNCS, vol. 8402, pp. 341–351. Springer, Cham (2014). https://doi.org/10.1007/978-3-319-06089-7_24
27. Ito, T., Kamiński, M., Ono, H., Suzuki, A., Uehara, R., Yamanaka, K.: Parameterized complexity of independent set reconfiguration problems. Discret. Appl. Math. **283**, 336–345 (2020). https://doi.org/10.1016/J.DAM.2020.01.022
28. Kamiński, M., Medvedev, P., Milanič, M.: Complexity of independent set reconfigurability problems. Theor. Comput. Sci. **439**, 9–15 (2012). https://doi.org/10.1016/J.TCS.2012.03.004
29. Lokshtanov, D., Mouawad, A.E.: The complexity of independent set reconfiguration on bipartite graphs. ACM Trans. Algorithms **15**(1), 1–19 (2019). https://doi.org/10.1145/3280825
30. Lokshtanov, D., Vatshelle, M., Villanger, Y.: Independent set in P_5-free graphs in polynomial time. In: Proceedings of the Twenty-Fifth Annual ACM-SIAM Sym-

posium on Discrete Algorithms, SODA 2014, pp. 570–581. SIAM (2014). https://doi.org/10.1137/1.9781611973402.43
31. Minty, G.J.: On maximal independent sets of vertices in claw-free graphs. J. Comb. Theory Ser. B **28**(3), 284–304 (1980). https://doi.org/10.1016/0095-8956(80)90074-X
32. Mouawad, A.E., Nishimura, N., Raman, V., Siebertz, S.: Vertex cover reconfiguration and beyond. Algorithms **11**(2), 20 (2018). https://doi.org/10.3390/A11020020
33. Mouawad, A.E., Nishimura, N., Raman, V., Simjour, N., Suzuki, A.: On the parameterized complexity of reconfiguration problems. Algorithmica **78**(1), 274–297 (2017). https://doi.org/10.1007/S00453-016-0159-2
34. Nishimura, N.: Introduction to reconfiguration. Algorithms **11**(4), 52 (2018). https://doi.org/10.3390/A11040052
35. Sbihi, N.: Algorithme de recherche d'un stable de cardinalite maximum dans un graphe sans etoile. Discret. Math. **29**(1), 53–76 (1980). https://doi.org/10.1016/0012-365X(90)90287-R
36. Suzuki, A., Mouawad, A.E., Nishimura, N.: Reconfiguration of dominating sets. J. Comb. Optim. **32**(4), 1182–1195 (2016). https://doi.org/10.1007/s10878-015-9947-x

Enumerating Minimal Solution Sets for Metric Graph Problems

Benjamin Bergougnoux[1(✉)], Oscar Defrain[2], and Fionn Mc Inerney[3]

[1] Institute of Informatics, University of Warsaw, Warsaw, Poland
benjamin.bergougnoux@mimuw.edu.pl

[2] Aix Marseille Université, CNRS, LIS, 13009 Marseille, France
oscar.defrain@lis-lab.fr

[3] Algorithms and Complexity Group, Technische Universität Wien, Vienna, Austria

Abstract. Problems from metric graph theory like Metric Dimension, Geodetic Set, and Strong Metric Dimension have recently had a strong impact in parameterized complexity by being the first known problems in NP to admit double-exponential lower bounds in the treewidth, and even in the vertex cover number for the latter, assuming the Exponential Time Hypothesis. We initiate the study of enumerating minimal solution sets for these problems and show that they are also of great interest in enumeration. Specifically, we show that enumerating minimal resolving sets in graphs and minimal geodetic sets in split graphs are equivalent to enumerating minimal transversals in hypergraphs (denoted Trans-Enum), whose solvability in total-polynomial time is one of the most important open problems in algorithmic enumeration. This provides two new natural examples to a question that emerged in recent works: for which vertex (or edge) set graph property Π is the enumeration of minimal (or maximal) subsets satisfying Π equivalent to Trans-Enum? As very few properties are known to fit within this context—namely, those related to minimal domination—our results make significant progress in characterizing such properties, and provide new angles to approach Trans-Enum. In contrast, we observe that minimal strong resolving sets can be enumerated with polynomial delay. Additionally, we consider cases where our reductions do not apply, namely graphs with no long induced paths, and show both positive and negative results related to the enumeration and extension of partial solutions.

Keywords: Algorithmic enumeration · Hypergraph dualization · Metric dimension · Geodetic sets · Resolving sets · Matroids

1 Introduction

Metric graph theory is a central topic in mathematics and computer science that is the subject of many books and articles, with far-reaching applications such

---

This work is an extended abstract of [10].

D. Král' and M. Milanič (Eds.): WG 2024, LNCS 14760, pp. 50–64, 2025.
https://doi.org/10.1007/978-3-031-75409-8_4

as in group theory [3], matroid theory [5], learning theory [18], and computational biology [6]. Two very well-studied metric graph problems that arise in the context of network design and monitoring are the metric dimension [42,55] and geodetic set [43] problems. There is a rich literature concerning these problems and their variants, see, e.g., [8,50,54,56], with applications being found in areas like network verification [7], chemistry [45], and genomics [58]. In this paper, we study the algorithmic enumeration of *all* minimal solution sets for the metric dimension, geodetic set, and strong metric dimension problems.

In the METRIC DIMENSION problem, given an undirected and unweighted graph G and a positive integer k, the question is whether there exists a subset $S \subseteq V(G)$ of at most k vertices such that, for any two distinct vertices $u, v \in V(G)$, there exists a vertex $w \in S$ with $\mathsf{dist}(u, w) \neq \mathsf{dist}(v, w)$. A set of vertices $S \subseteq V(G)$ that satisfies the latter property is a *resolving set* of G. This problem was shown to be NP-complete in Garey and Johnson's book [40]. In the last decade, its complexity was greatly refined. It was proven to be NP-hard in, e.g., planar graphs [26], interval graphs of diameter 2 [38], and split graphs [35]. However, it is polynomial-time solvable in, e.g., trees [55], cographs [35], outerplanar graphs [26], and bounded feedback edge number graphs [35].

In the GEODETIC SET problem, given a graph G and a positive integer k, the question is whether there exists a subset $S \subseteq V(G)$ of at most k vertices such that every vertex in G is on a shortest path between two vertices of S. A set of vertices $S \subseteq V(G)$ that satisfies the latter property is known as a *geodetic set* of G. It is NP-complete in, e.g., co-bipartite graphs [31] and interval graphs [16], but polynomial-time solvable in, e.g., well-partitioned chordal graphs [4], outerplanar graphs [52], and distance-hereditary graphs [49].

In the STRONG METRIC DIMENSION problem, given a graph G and a positive integer k, the question is whether there exists a subset $S \subseteq V(G)$ of at most k vertices such that, for any pair of vertices $u, v \in V(G)$, there exists a vertex $w \in S$ with either u belonging to a shortest w–v path or v belonging to a shortest w–u path. A set of vertices $S \subseteq V(G)$ that satisfies the latter property is known as a *strong resolving set* of G. There is a polynomial-time reduction from an instance (G, k) of STRONG METRIC DIMENSION to an instance (G', k) of VERTEX COVER, where $V(G) = V(G')$ and the edges of G' connect pairs of vertices that are so-called "mutually maximally distant" in G [53].

Recently, the three above problems were shown to be important well beyond the field of metric graph theory by being the first known problems in NP to admit conditional double-exponential lower bounds in the treewidth, and even the vertex cover number (vc) for STRONG METRIC DIMENSION [37]. Further, in [36,37] they proved that, unless the Exponential Time Hypothesis (ETH) fails, all three problems do not admit kernelization algorithms that output a kernel with $2^{o(\mathsf{vc})}$ vertices. Such kernelization lower bounds were priorly only known for two other problems [19,23]. Lastly, matching upper bounds were also provided in [36,37], and the lower bound technique developed in [37] was used to obtain similar results for other NP-complete problems [17,18].

Despite these problems being well-studied from an algorithmic complexity point of view, they have yet to be studied from the perspective of enumeration. We remedy this by initiating the study of enumerating minimal solution sets—the gold standard for enumeration—for the following problems.

Minimal Resolving Sets Enumeration (MinResolving)
Input: A graph G.
Output: The set of (inclusion-wise) minimal resolving sets of G.

Minimal Strong Resolving Sets Enumeration (MinStrongResolving)
Input: A graph G.
Output: The set of (inclusion-wise) minimal strong resolving sets of G.

Minimal Geodetic Sets Enumeration (MinGeodetic)
Input: A graph G.
Output: The set of (inclusion-wise) minimal geodetic sets of G.

In the same manner that these problems had a strong impact in parameterized complexity [36,37], we show that they are also of great interest in enumeration. Specifically, they relate to classical problems such as the enumeration of maximal independent sets in graphs, which admits a polynomial-delay[1] algorithm [59], and the enumeration of minimal transversals in hypergraphs, whose solvability in total-polynomial time is one of the most important open problems in algorithmic enumeration [28,30].

In the minimal transversals enumeration problem, denoted Trans-Enum and also known as hypergraph dualization, given a hypergraph $\mathcal{H}$, the goal is to list all (inclusion-wise) minimal subsets of vertices that hit every edge of $\mathcal{H}$. The best-known algorithm for Trans-Enum runs in incremental quasi-polynomial time by generating the ith minimal transversal of $\mathcal{H}$ in time $N^{o(\log N)}$, where $N = |\mathcal{H}| + i$ [39]. Ever since, a lot of effort has been made to solve the problem in total-polynomial time in restricted cases. Notably, there are polynomial-delay algorithms for β-acyclic hypergraphs [28] and hypergraphs of bounded degeneracy [29] or without small holes [46], and incremental-polynomial-time algorithms for bounded conformality hypergraphs [51] and geometric instances [33].

Due to the inherent difficult nature of the problem, and since no substantial progress has been made in the general case since [39], Trans-Enum has gained the status of a "landmark problem" in terms of tractability, in between problems admitting total-polynomial-time algorithms, and those for which the existence of such algorithms is impossible unless $\mathsf{P} = \mathsf{NP}$. This has motivated the study of particular cases of problems that are known to be at least as hard[2] as Trans-

[1] The different notions from enumeration complexity are defined in Sect. 2.

[2] An enumeration problem is *at least as hard as* another enumeration problem if a total-polynomial-time algorithm for the first implies a total-polynomial-time algorithm for the second; the problems are (polynomially) *equivalent* if the reverse direction also holds.

ENUM; see, e.g., [20,28,47]. One of the most successful examples is the case of minimal dominating sets enumeration, with many particular cases shown to admit total-polynomial-time algorithms [13,41,47,48]. On the other hand, for problems that are notably harder than TRANS-ENUM and for which the existence of total-sub-exponential-time algorithms is open, adapting the algorithm of [39] as in [32,34], or using it as a subroutine as in [2,24,25] has also proved fruitful.

In light of these results, a line of research that emerged from [47] (see also, e.g., [21,57]) consists of exploring the following question.

Question 1. For which vertex (or edge) set graph property Π is the enumeration of minimal (or maximal) subsets satisfying Π equivalent to TRANS-ENUM?

We make progress on Question 1 by surprisingly proving that TRANS-ENUM is equivalent to MINRESOLVING (in general graphs) and MINGEODETIC in split graphs. Notably, this adds two natural problems to the short list of those known to have this property. Curiously, this contrasts with MINSTRONGRESOLVING that we show can be solved with polynomial delay using a relationship with VERTEX COVER established in [53].

Interestingly, we additionally show that MINGEODETIC is a particular case of enumerating the minimal flats of the graphic matroid associated to K_n that are transversals of a given hypergraph. To the best of our knowledge, the latter problem is open, and thus, this encloses MINGEODETIC by two generation problems whose complexity statuses are unsettled. Hence, disproving the equivalence between MINGEODETIC and TRANS-ENUM by, e.g., showing that the first problem does not admit a total-polynomial-time algorithm unless $\mathsf{P} = \mathsf{NP}$,[3] would imply that the aforementioned variant of flats enumeration is intractable.

Finally, we observe that our reductions prove that MINGEODETIC and MINRESOLVING are at least as hard as TRANS-ENUM on P_5 and P_6-free graphs, respectively. This motivates the study of these problems on P_k-free graphs for smaller values of k. In the long version of this work [10], we show that both problems admit linear-delay algorithms in P_4-free graphs using a variant of Courcelle's theorem for enumeration and clique-width [22]. Except for MINGEODETIC that we completely characterize with respect to TRANS-ENUM, the case of MINRESOLVING for $k = 5$ is yet to be classified. While it would be interesting to extend our hardness result (with respect to TRANS-ENUM) for MINRESOLVING to P_5-free graphs or even split or co-bipartite graphs, it is challenging as our construction in the proof of Theorem 8 seems impossible to extend to these cases. We note that the case of co-bipartite graphs is also open for MINGEODETIC. For these graph classes, we were not able to devise total-polynomial-time algorithms; we however prove in [10] that the extension problems for MINRESOLVING and MINGEODETIC are NP-hard on split and co-bipartite graphs, respectively, which suggests that the generation is non-trivial in those cases.

Concerning the difficulties and novelties of our techniques, our main results are based on several reductions which—unlike classical NP-hardness reductions—preserve as much as possible the set of solutions of the original instance. For

[3] As TRANS-ENUM is in QP (quasi-polynomial time) and it is believed that $\mathsf{NP} \not\subseteq \mathsf{QP}$.

example, two main difficulties with our reduction from Trans-Enum to Min-Resolving are to ensure that there are (1) at most a polynomial number of minimal resolving sets that do not correspond to minimal transversals, and (2) only a few minimal resolving sets corresponding to the same minimal transversal.

The paper is organized as follows. We introduce notions from metric graph theory and algorithmic enumeration in Sect. 2, the reductions are presented in Sects. 3 and 4, and perspectives for further research are discussed in Sect. 5. Due to space constraints, the missing proofs, as well as the tractability results on P_4-free graphs and the hardness for extensions problems, can be found in [10].

2 Preliminaries

For $n \in \mathbb{Z}^+$, let $[n] := \{1, \dots, n\}$. In this paper, logarithms are binary. We begin with definitions from enumeration complexity and refer the reader to [44,57]. An algorithm runs in *total-polynomial* time if it outputs every solution exactly once and stops in a time which is polynomial in the size of the input plus the output. Moreover, if the algorithm outputs the ith solution in a time which is polynomial in the size of the input plus i, then it runs in *incremental-polynomial* time. An enumeration algorithm runs with *polynomial delay* if before the first output, between consecutive outputs, and after the last output it runs in a time which is polynomial in the size of the input. Clearly, an algorithm running with polynomial delay runs in incremental-polynomial time, and an incremental-polynomial-time algorithm runs in total-polynomial time. Moreover, reductions between enumeration problems (as described in Footnote 5) can sometimes preserve polynomial delay; this will be the case in this paper and we refer to [57] for more details on these aspects.

For graphs and hypergraphs, see [27] and [9] for definitions that are not recalled below. A *hypergraph* $\mathcal{H}$ is a set of vertices $V(\mathcal{H})$ together with a family of edges $E(\mathcal{H}) \subseteq 2^{V(\mathcal{H})}$. It is a *graph* when each of its edges has size precisely two, and *Sperner* if no two distinct edges $E, F \in \mathcal{H}$ are such that $E \subseteq F$. A *transversal* of $\mathcal{H}$ is a subset $T \subseteq V(\mathcal{H})$ such that $E \cap T \neq \emptyset$ for all $E \in E(\mathcal{H})$. It is *minimal* if it is inclusion-wise minimal. The set of minimal transversals of $\mathcal{H}$ is denoted by $Tr(\mathcal{H})$, and the problem of listing $Tr(\mathcal{H})$ (given $\mathcal{H}$) by Trans-Enum.

For a graph G and two vertices x, y, $\mathsf{dist}(x, y)$ is the length of a shortest x-y path in G. Two vertices u, v are *false twins* if their open neighborhoods $N(u)$ and $N(v)$ are equal, and *twins* if they are also adjacent. For $k \in \mathbb{Z}^+$, we call P_k a path on k vertices. We say that a vertex is *complete* (*anti-complete*, resp.) to a subset $S \subseteq V(G)$ if it is adjacent (non-adjacent, resp.) to each vertex in S; a set $A \subseteq V(G)$ is said to be complete to S if every vertex $x \in A$ is. For resolving sets, a vertex x is said to *distinguish* a pair a, b of vertices if $\mathsf{dist}(a, x) \neq \mathsf{dist}(b, x)$. For geodetic sets, a pair x, y of vertices is said to *cover* a vertex v if v lies on an x–y shortest path. We say that a (strong) resolving set (geodetic set, resp.) is minimal if it is inclusion-wise minimal.

Given a hypergraph $\mathcal{H}$ on vertex set $\{v_1, \dots, v_n\}$ and edge set $\{E_1, \dots, E_m\}$, the *incidence bipartite graph* of $\mathcal{H}$ is the bipartite graph with bipartition $V =$

$\{v_1, \ldots, v_n\}$ and $H = \{e_1, \ldots, e_m\}$, with an edge between $v_i \in V$ and $e_j \in H$ if v_i belongs to E_j. The *non-incidence bipartite graph* of $\mathcal{H}$ is the graph with the same vertices, but where there is an edge between $v_i \in V$ and $e_j \in H$ if v_i does *not* belong to E_j. Finally, the *(non-)incidence co-bipartite graph* of $\mathcal{H}$ is the (non-)incidence bipartite graph of $\mathcal{H}$ where V and H are completed into cliques. For illustrations of these constructions, see the reductions in the next sections.

3 Resolving Sets

In this section, we prove that Trans-Enum and MinResolving are equivalent, and that our reductions preserve polynomial delay. As for MinStrongResolving, we show that it is equivalent to the enumeration of maximal independent sets in graphs, and hence, that it admits a polynomial-delay algorithm.

We first deal with the reduction from MinResolving. It is clear from the definition of distinguishing a pair of vertices that the resolving sets of a graph G are exactly the transversals of the hypergraph $\mathcal{H}$ with the same vertex set and with an edge $E_{ab} := \{v \in V(G) : \mathsf{dist}(a, v) \neq \mathsf{dist}(b, v)\}$ for every pair a, b of distinct vertices in G. Since $\mathcal{H}$ has n vertices and $O(n^2)$ edges, and as it can be constructed in polynomial time in n, we derive the following.

Theorem 2. *There is a polynomial-delay algorithm for* MinResolving *whenever there is one for* Trans-Enum.

Let us now deal with the reduction from Trans-Enum. Let $\mathcal{H}$ be a hypergraph on vertex set $\{v_1, \ldots, v_n\}$ and edge set $\{E_1, \ldots, E_m\}$. For convenience, in our proof, we will furthermore assume that n and m are powers of 2 greater than 2, and that no edge of $\mathcal{H}$ contains the full set of vertices. Note that these assumptions can be conducted without loss of generality, in particular since it can be assumed that $\mathcal{H}$ is Sperner and an edge containing the full set of vertices would imply it is the only edge of $\mathcal{H}$. We describe the construction of a graph on $O(n + m)$ vertices and $O(n^2 + m^2)$ edges whose set of minimal resolving sets can be partitioned into two families where the first one has size $O(nm^2)$, and where the second roughly consists of $O(nm)$ copies of the minimal transversals of $\mathcal{H}$. See Fig. 1 for an illustration of the construction.

We start from the non-incidence co-bipartite graph of $\mathcal{H}$ with bipartition $V := \{v_1, \ldots, v_n\}$ and $H := \{e_1, \ldots, e_m\}$, to which we add a clique $H' := \{e'_1, \ldots, e'_m\}$ that we make complete to V. Then V, H, and H' are cliques, v_i is adjacent to every $e' \in H'$, and it is adjacent to e_j if and only if $v_i \notin E_j$. We construct two additional sets $U := \{u_1, u'_1, \ldots, u_{(\log n)+1}, u'_{(\log n)+1}\}$ and $W := \{w_1, w'_1, \ldots, w_{(\log m)+1}, w'_{(\log m)+1}\}$ on $2(\log n)+2$ and $2(\log m)+2$ vertices, respectively. We complete U into a clique minus each of the edges $u_i u'_i$, $i \in [(\log n) + 1]$, and add to W each of the edges $w_j w'_j$, $j \in [(\log m) + 1]$. For an integer $j \in \mathbb{N}$, we shall note $I(j)$ the set of indices (starting from 1) of bits of value 1 in the binary representation of j. Then, we connect each v_i, $i \in [n]$, to the vertices u_k and u'_k for every $k \in I(i)$, and each of e_j and e'_j, $j \in [m]$, to the vertices w_k and w'_k for every $k \in I(j)$. Observe that, by the nature of the binary coding, no element of V is complete or anti-complete to U, and the same can be said for $H \cup H'$ and W. Note that this binary representation gadget is derived from ideas used in, e.g., [37]. Finally, we connect every vertex of U to every vertex of $H \cup H'$, and connect every vertex of W to every vertex of V. This concludes the construction of our graph G. We start with easy observations. The following exploits that the sets U and W are constituted of pairs of twins.

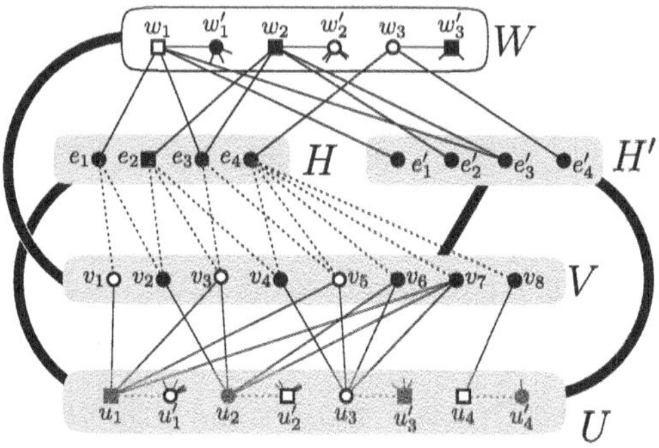

Fig. 1. Illustration of the reduction from TRANS-ENUM to MINRESOLVING with $\mathcal{H}$ consisting of $E_1 = \{v_1, v_2\}$, $E_2 = \{v_2, v_3, v_4\}$, $E_3 = \{v_3, v_5\}$, and $E_4 = \{v_4, v_5, v_6, v_7, v_8\}$. Dashed lines represent non-edges, and a bold line between two sets of vertices A, B means that A is complete to B. For legibility, we do not show the edges of $G[U]$, which is almost a clique, nor the edges of the cliques H, H', and V. We only show the non-edges between V and H. We also do not fully represent some of the edges incident to the vertices u'_i and w'_j. The set of white vertices is one of the $O(nm)$ minimal resolving sets associated to the minimal transversal $\{v_1, v_3, v_5\}$ of $\mathcal{H}$. The set of square vertices is one of the $O(nm^2)$ minimal resolving sets not associated with a minimal transversal.

Lemma 3. *Let S be a minimal resolving set of G. Then, S intersects each of $\{u_i, u'_i\}$ ($i \in [(\log n) + 1]$) and $\{w_j, w'_j\}$ ($j \in [(\log m) + 1]$) on one vertex.*

In the following, let us consider arbitrary
$X \in \left\{\{x_1, \ldots, x_{(\log n)+1}\} : (x_1, \ldots, x_{(\log n)+1}) \in \prod_{i=1}^{(\log n)+1} \{u_i, u'_i\}\right\}$,
$Y \in \left\{\{y_1, \ldots, y_{(\log m)+1}\} : (y_1, \ldots, y_{(\log m)+1}) \in \prod_{j=1}^{(\log m)+1} \{w_j, w'_j\}\right\}$,
and denote by $\mathcal{Z}$ the set of all possible unions $Z := X \cup Y$. There are $4nm$ possible choices for Z and, by Lemma 3, any minimal resolving set contains one such Z as a subset. We characterize the pairs of vertices these sets distinguish.

Lemma 4. *Let $Z \in \mathcal{Z}$ and let P be the set of all pairs $\{e_j, e'_j\}$ with $j \in [m]$. Then, Z distinguishes a pair a, b of distinct vertices if and only if $\{a, b\} \notin P$.*

Since by Lemma 3 every minimal resolving set contains a choice of Z as above, we get that the non-trivial part of minimal resolving sets in G is dedicated to distinguishing pairs in P. We characterize these non-trivial parts in the following.

Lemma 5. *If S is a minimal resolving set of G such that $S \cap (H \cup H') \neq \emptyset$, then $S = Z \cup \{e\}$ for some $Z \in \mathcal{Z}$ and $e \in H \cup H'$.*

Lemma 6. *If S is a minimal resolving set of G such that $S \cap (H \cup H') = \emptyset$, then $S = Z \cup T$ for some $Z \in \mathcal{Z}$ and some minimal transversal T of $\mathcal{H}$.*

Lemma 7. *If T is a minimal transversal of $\mathcal{H}$, then $Z \cup T$ is a minimal resolving set of G for any $Z \in \mathcal{Z}$.*

Note that, to every $T \in Tr(\mathcal{H})$, there are $4\,\mathrm{nm}$ corresponding distinct minimal resolving sets in G obtained by extending T with every possible $Z \in \mathcal{Z}$. We show that our reduction still preserves polynomial delay at the cost of (potentially exponential) space using a folklore trick on regularizing the outputs.

Theorem 8. *There is a polynomial-delay algorithm for TRANS-ENUM whenever there is one for MINRESOLVING.*

Proof. Let A be an algorithm for MINRESOLVING running with polynomial delay $f(n)$ for some function $f : \mathbb{N} \to \mathbb{N}$, where n is the number of vertices in G. We first describe an incremental-polynomial-time algorithm B for TRANS-ENUM generating the ith solution in $O(i \cdot (nm^2 \cdot f(n)))$ time. We start by constructing G as above. Clearly, this can be done in polynomial time in $n + m$. Then, we simulate A on G. Each time A produces a set of the form $S = Z \cup T$ with $|T| \geq 2$, we check whether T has already been obtained before by keeping every such T in memory, and output it as a solution for TRANS-ENUM if not. This concludes the description of B. Its correctness follows from Lemmas 5, 6, and 7.

Let us analyze the complexity of B. By Lemma 5, A generates at most $O(nm^2)$ solutions in total time $O(nm^2 \cdot f(n))$ before generating a first solution of the form $Z \cup T$ with $T \in Tr(\mathcal{H})$. Hence, the first solution of B is obtained in $O(nm^2 \cdot f(n))$ time, as required. Suppose now that B has produced i solutions $T_1, \ldots, T_i \in Tr(\mathcal{H})$ in $O(i \cdot (nm^2 \cdot f(n)))$ time. By Lemmas 5 and 6, when B produces the $(i+1)^{\text{th}}$ solution for TRANS-ENUM, the simulation of A has generated at most $i \cdot 4nm + O(nm^2)$ solutions of the form $Z \cup \{e\}$ with $e \in H \cup H'$ or $Z \cup T$ for $T \in \{T_1, \ldots, T_i\}$. This takes $O\big((i \cdot 4nm + nm^2) \cdot f(n)\big)$ time by assumption, after which B produces the next solution. Thus, in total, B has spent $O(i \cdot 4nm \cdot f(n) + nm^2 \cdot f(n))$ time outputting the $(i+1)^{\text{th}}$ solution of TRANS-ENUM as desired.

In the incremental time of B, the dependence on i is linear. Using a folklore trick on regularizing the delay of such algorithms (see, e.g., [15, Proposition 3]), we can regularize algorithm B to polynomial-delay by keeping each new set T in a queue, and pulling a new set from the queue every $O(nm^2 \cdot f(n))$ steps. □

The space needed for the reduction of Theorem 8 to hold is potentially exponential, as every obtained minimal transversal is stored in a queue. However, using another folklore trick (see, e.g., [14, Section 3.3]) on checking whether solutions have already been obtained by running the same algorithm on the same number of steps minus one, the reduction could preserve incremental-polynomial time and polynomial space at the cost of a worse dependence on the number of solutions. We end this section by dealing with MinStrongResolving.

Theorem 9. *MinStrongResolving can be solved with polynomial delay.*

Proof. It is known that, given a graph G, another graph G' can be constructed in polynomial time such that the vertex covers of G' are exactly the strong resolving sets of G [53, Theorem 2.1]. Moreover, the size of G' is polynomial in the size of G, namely it satisfies $V(G') = V(G)$. As the vertex covers of a graph are the complements of its independent sets, we deduce a polynomial-delay algorithm for MinStrongResolving by the algorithm of Tsukiyama et al. [59]. □

4 Geodetic Sets

In this section, we prove that Trans-Enum and MinGeodetic on split graphs are equivalent, and that our reductions preserve polynomial delay. As for the general case, we show MinGeodetic to be a particular case of enumerating all the minimal flats of the graphic matroid associated to K_n that are transversals of a given hypergraph, whose complexity status is unsettled to date.

We first deal with the reduction from Trans-Enum. Let $\mathcal{H}$ be a hypergraph on vertex set $\{v_1, \ldots, v_n\}$ and edge set $\{E_1, \ldots, E_m\}$. We furthermore assume that $n, m \geq 1$ and that no vertex of $\mathcal{H}$ appears in every edge. Note that these assumptions can be conducted without loss of generality. In particular, if a vertex v appears in every edge, then $Tr(\mathcal{H})$ consists of $\{v\}$ and the minimal transversals of $\mathcal{H}' := \{E \setminus \{v\} : E \in \mathcal{H}\}$, and so, solving Trans-Enum on $\mathcal{H}$ is essentially equivalent to solving it on $\mathcal{H}'$, and thus, we can recursively remove such vertices.

We describe the construction of a split graph G on $O(n+m)$ and $O(n^2m^2)$ edges whose set of minimal geodetic sets is partitioned into two families where the first has size $O(m)$ and the second is in bijection with the set of minimal transversals of $\mathcal{H}$. See Fig. 2 for an illustration of the construction. We start from the non-incidence bipartite graph of $\mathcal{H}$ with bipartition $V := \{v_1, \ldots, v_n\}$ and $H := \{e_1, \ldots, e_m\}$, to which we add a set of vertices $U := \{u_1, \ldots, u_m\}$ with u_j only adjacent to e_j for each $j \in [m]$. We then complete $U \cup V$ into a clique and add a vertex e^* adjacent to every vertex in V. Finally, we add a vertex u^* adjacent to every vertex in G. This completes the construction. We note that G is a split graph with clique $K := U \cup V \cup \{u^*\}$ and independent set $I := H \cup \{e^*\}$.

Since u^* is a universal vertex of G, the diameter of G is at most 2, and we may reformulate x being on a shortest a–b path with $a \neq x \neq b$ as x being the middle vertex of a P_3 in G (as an induced subgraph). We derive easy observations.

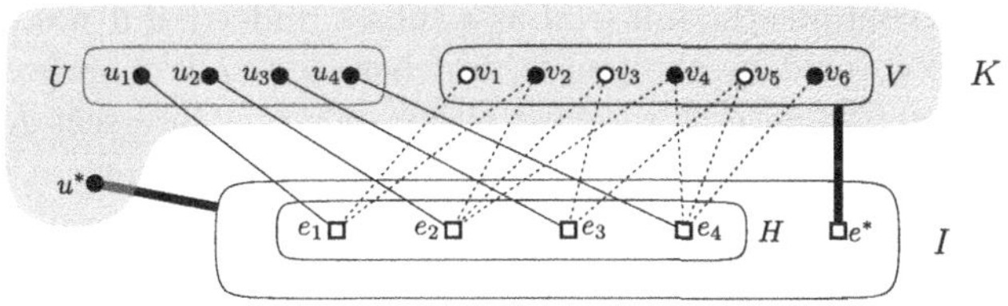

Fig. 2. Illustration of the reduction from TRANS-ENUM to MINGEODETIC with $\mathcal{H}$ consisting of $E_1 = \{v_1, v_2\}$, $E_2 = \{v_2, v_3, v_4\}$, $E_3 = \{v_3, v_5\}$, and $E_4 = \{v_4, v_5, v_6\}$. Dashed lines represent non-edges and the bold lines incident to u^* and e^* mean these two vertices are complete to I and V, respectively. For legibility, the edges of the clique K are omitted. Square vertices belong to any geodetic set. The set of white vertices is a minimal geodetic set obtained from the minimal transversal $\{v_1, v_3, v_5\}$ of $\mathcal{H}$.

Lemma 10. *The set I is contained in every geodetic set of G.*

Proof. This holds since no vertex in I is the middle vertex of a P_3 in G. □

Lemma 11. *Only U is not covered by the pairs of vertices in I.*

Lemma 12. *The set $I \cup \{u\}$ is a minimal geodetic set of G for every $u \in U$.*

We may now characterize minimal geodetic sets that are of interest as far as the transversality of $\mathcal{H}$ is concerned.

Lemma 13. *Let S be a minimal geodetic set of G such that $S \cap U = \emptyset$. Then, $S \cap K$ is a minimal transversal of $\mathcal{H}$.*

Lemma 14. *If T is a minimal transversal of $\mathcal{H}$, then $T \cup I$ is a minimal geodetic set of G.*

Theorem 15. *There is a polynomial-delay algorithm for TRANS-ENUM whenever there is one for MINGEODETIC on split graphs.*

Proof. This is a consequence of the fact that the graph G can be constructed in polynomial time in the size of $\mathcal{H}$, has polynomial size, and that it contains m minimal geodetic sets $I \cup \{u\}$ with $u \in U$. All the other minimal geodetic sets of G are of the form $I \cup T$ where T is A Minimal Transversal of $\mathcal{H}$. Hence, a polynomial-delay algorithm for MINGEODETIC would take at most m times its delay between two consecutive minimal geodetic sets of the form $I \cup T$. □

We now argue that a polynomial-delay algorithm for TRANS-ENUM yields one for MINGEODETIC on split graphs. Let G be a split graph of bipartition (K, I) with K the clique and I the independent set. Among all such partitions we consider one that maximizes the size of I and we further assume that $|I| \geq 2$, as otherwise the instance is trivial. As in Lemma 10, let us first note that $I \subseteq S$ for any geodetic set S of G as the neighborhood of every vertex $x \in I$ is a clique. By the maximality of I, every vertex $v \in K$ that is not covered by a pair of vertices in I has precisely one neighbor $u \in I$ and the set of vertices at distance at most 2 from u contains the full set I as a subset. Indeed, if it was not the case, then v would be covered by the pair u, w, where $w \in I$ is a vertex at distance three from u. Thus, to cover v we must either pick v or intersect $K \setminus N(u)$ the non-neighborhood of u in K. We identify all such vertices $v_1, \ldots, v_k$ and their only neighbors $u_1, \ldots, u_k$ in I to construct a hypergraph $\mathcal{H}$ on vertex set K with an edge $E_i = \{K \setminus N(u_i)\} \cup \{v_i\}$ for every $i \in [k]$. We note that possibly $u_i = u_j$ for distinct $i, j \in [k]$, which is of no concern in the following. Clearly, the construction can be achieved in polynomial time in the size of G, and by the above remarks we obtain a bijection between the minimal transversals of $\mathcal{H}$ and the minimal geodetic sets of G. This yields the next theorem.

Theorem 16. *There is a polynomial-delay algorithm for* MINGEODETIC *on split graphs whenever there is one for* TRANS-ENUM.

We now end the section by showing that the general case of MINGEODETIC reduces to enumerating all the minimal flats of the graphic matroid associated to the clique K_n that are transversals of a given n-vertex hypergraph.

Let us start with the construction. We consider a graph G and construct a hypergraph $\mathcal{H}$ whose vertices are (unordered) pairs of distinct vertices of G, denoted uv instead of $\{u, v\}$ for convenience, and where every vertex v of G gives rise to an edge $E_v := \{xy : v \text{ is on a shortest } x\text{ - }y \text{ path}\}$ in $\mathcal{H}$. To avoid ambiguity, we shall refer to the vertices of $\mathcal{H}$ as *nodes*, and use variables r, s, t for nodes in the following. Then, $\mathcal{H}$ has $O(n^2)$ nodes and $O(n)$ edges. Clearly, every transversal T of $\mathcal{H}$ induces a geodetic set $\bigcup_{t \in T} t$ of G, as every E_v is hit by a pair t in T, and that pair covers v in G. Unfortunately, minimal transversals of $\mathcal{H}$ do not necessarily define minimal geodetic sets of G in that way, and not every minimal geodetic set of G defines a minimal transversal of $\mathcal{H}$ by considering all the pairs of elements contained in it. Consider for example the graph G obtained from a triangle abc by adding a pendent vertex d adjacent to c. Then, $\{a, b, d\}$ is the only minimal geodetic set of G, while $\{ab, ad, bd\}$ is easily verified to be a transversal of $\mathcal{H}$ that is *not* minimal as it contains the transversal $\{ad, bd\}$ as a subset. On the other hand, $\{ac, bd\}$ can be checked to be a minimal transversal of $\mathcal{H}$, while $\{a, b, c, d\}$ is not a minimal geodetic set of G. We nevertheless show that consistent sets that are transversals of $\mathcal{H}$ are in bijection with the geodetic sets of G for an appropriate notion of consistency.

A subset U of nodes of $\mathcal{H}$ is *consistent* if, whenever two distinct nodes $r, s \in U$ are such that $r \cap s \neq \emptyset$, then the unique other node t such that $r \cup s = s \cup t = r \cup t$ is also part of U. In other words, a set of pairs is consistent if and only if it is the family of all pairs of some set. For example, a subset U containing ab and bc but not ac is not consistent, while the set $U = \{ab, ad, bd\}$ or the set of all nodes of $\mathcal{H}$ are consistent. The aforementioned correspondence is the following.

Theorem 17. *There is a bijection between the minimal geodetic sets of G and the minimal consistent subsets of nodes that are transversals of $\mathcal{H}$.*

We now discuss the implications of Theorem 17. Observe that the consistency of a subset of vertices of $\mathcal{H}$ as defined above may also be expressed as satisfying a set of implications $\Sigma := \{r, s \to t : r \cup s = s \cup t = r \cup t\}$ in the sense that any subset containing the premise of an implication in Σ must contain its conclusion. It is well known that the consistent sets in that context are the closed sets of a lattice [11,60]. In the particular case of the rules defined above, the lattice is in fact known to be the lattice of flats (subsets of edges that are maximal with respect to the size of their spanning trees) of the graphic matroid associated to the clique on n vertices, or equivalently, to be the lattice of partitions of a finite n-element set [11]. Consequently, listing minimal consistent transversals in our context may be reformulated as the enumeration of the minimal flats of the matroid associated to the clique K_n that are transversals of $\mathcal{H}$. To the best of our knowledge, no output-quasi-polynomial-time algorithm is known for that problem. It should however be noted that in the more general setting where Σ is allowed to contain any implications with premises of size at most two, the enumeration is intractable as it generalizes the dualization in lattices given by implicational bases of size at most two [24].

5 Perspectives for Further Research

We conclude with some open problems. It would be interesting to know the complexity of MinResolving in the natural graph classes for which neither our reductions nor our positive results apply, this includes split and co-bipartite graphs. Other open directions are to know whether MinGeodetic admits a total-quasi-polynomial-time algorithm, or how it relates to problems known to be harder than Trans-Enum that admit sub-exponential-time algorithms. Problems of interest include the dualization of products of posets [34] or the dualization in distributive lattices [32]. Among candidates for Question 1, the case of connected domination is of interest. While it has been shown equivalent to Trans-Enum in split graphs [1,47], the status of the problem is open in general graphs; this later (general) case was however conjectured not to be equivalent to Trans-Enum by Kanté at the 2015 Lorentz Workshop on enumeration algorithms [12].

Acknowledgements. This work was supported by the ANR project DISTANCIA (ANR-17-CE40-0015) and the Austrian Science Fund (FWF, project Y1329). We would like to thank the anonymous reviewers for their careful reading.

References

1. Abu-Khzam, F.N., Fernau, H., Gras, B., Liedloff, M., Mann, K.: Enumerating minimal connected dominating sets. In: 30th Annual European Symposium on Algorithms (ESA 2022). Schloss Dagstuhl-Leibniz-Zentrum für Informatik (2022)
2. Adaricheva, K.V., Nation, J.B.: Discovery of the D-basis in binary tables based on hypergraph dualization. Theoret. Comput. Sci. **658**, 307–315 (2017)
3. Agol, I.: The virtual Haken conjecture (with an appendix by Ian Agol, Daniel Groves and Jason Manning). Doc. Math. **18**, 1045–1087 (2013)
4. Ahn, J., Jaffke, L., Kwon, O., Lima, P.T.: Well-partitioned chordal graphs. Discret. Math. **345**(10), 112985 (2022)

5. Bandelt, H.-J., Chepoi, V., Knauer, K.: COMs: complexes of oriented matroids. J. Comb. Theory Ser. A **156**, 195–237 (2018)
6. Bandelt, H.-J., Dress, A.W.M.: Split decomposition: a new and useful approach to phylogenetic analysis of distance data. Mol. Phylogenet. Evol. **1**(3), 242–252 (1992)
7. Beerliova, Z., et al.: Network discovery and verification. IEEE J. Sel. Area Commun. **24**(12), 2168–2181 (2006)
8. Bensmail, J., Mazauric, D., Mc Inerney, F., Nisse, N., Pérennes, S.: Sequential metric dimension. Algorithmica **82**(10), 2867–2901 (2020)
9. Berge, C.: Hypergraphs: Combinatorics of Finite Sets, vol. 45. Elsevier, Amsterdam (1984)
10. Bergougnoux, B., Defrain, O., Mc Inerney, F.: Enumerating minimal solution sets for metric graph problems. arXiv preprint arXiv:2309.17419 (2023)
11. Birkhoff, G.: Lattice Theory, vol. 25. American Mathematical Society (1940)
12. Bodlaender, H., Boros, E., Heggernes, P., Kratsch, D.: Open problems of the Lorentz workshop, "Enumeration algorithms using structure". Department of Information and Computing Sciences Utrecht University, Utrecht, The Netherlands (2015)
13. Bonamy, M., Defrain, O., Heinrich, M., Pilipczuk, M., Raymond, J.-F.: Enumerating minimal dominating sets in K_t-free and variants. ACM Trans. Algorithms **16**(3), 1–23 (2020)
14. Bonamy, M., Defrain, O., Micek, P., Nourine, L.: Enumerating minimal dominating sets in the (in)comparability graphs of bounded dimension posets. arXiv preprint arXiv:2004.07214 (2020)
15. Capelli, F., Strozecki, Y.: Geometric amortization of enumeration algorithms. In: 40th International Symposium on Theoretical Aspects of Computer Science (STACS 2023). LIPIcs, vol. 254, pp. 18:1–18:22. Schloss Dagstuhl - Leibniz-Zentrum für Informatik (2023)
16. Chakraborty, D., Das, S., Foucaud, F., Gahlawat, H., Lajou, D., Roy, B.: Algorithms and complexity for geodetic Sets on planar and chordal graphs. In: 31st International Symposium on Algorithms and Computation (ISAAC 2020). LIPIcs. Schloss Dagstuhl–Leibniz-Zentrum für Informatik (2020)
17. Chakraborty, D., Foucaud, F., Majumdar, D., Tale, P.: Tight (double) exponential bounds for identification problems: locating-dominating set and test cover. arXiv preprint arXiv:2402.08346 (2024)
18. Chalopin, J., Chepoi, V., Mc Inerney, F., Ratel, S.: Non-clashing teaching maps for balls in graphs. In: Proceedings of the 37th Annual Conference on Learning Theory (COLT 2024) (2024)
19. Chandran, S., Issac, D., Karrenbauer, A.: On the parameterized complexity of biclique cover and partition. In: 11th International Symposium on Parameterized and Exact Computation (IPEC 2016). LIPIcs, vol. 63, pp. 11:1–11:13. Schloss Dagstuhl - Leibniz-Zentrum für Informatik (2016)
20. Conte, A., Grossi, R., Kanté, M.M., Marino, A., Uno, T., Wasa, K.: Listing induced steiner subgraphs as a compact way to discover steiner trees in graphs. In: 44th International Symposium on Mathematical Foundations of Computer Science (MFCS 2019). Schloss Dagstuhl-Leibniz-Zentrum fuer Informatik (2019)
21. Conte, A., Kanté, M.M., Marino, A., Uno, T.: Maximal irredundant set enumeration in bounded-degeneracy and bounded-degree hypergraphs. In: Colbourn, C.J., Grossi, R., Pisanti, N. (eds.) IWOCA 2019. LNCS, vol. 11638, pp. 148–159. Springer, Cham (2019). https://doi.org/10.1007/978-3-030-25005-8_13

22. Courcelle, B.: Linear delay enumeration and monadic second-order logic. Discret. Appl. Math. **157**(12), 2675–2700 (2009)
23. Cygan, M., Pilipczuk, M., Pilipczuk, M.: Known algorithms for edge clique cover are probably optimal. SIAM J. Comput. **45**(1), 67–83 (2016)
24. Defrain, O., Nourine, L.: Dualization in lattices given by implicational bases. Theoret. Comput. Sci. **814**, 169–176 (2020)
25. Defrain, O., Nourine, L., Vilmin, S.: Translating between the representations of a ranked convex geometry. Discret. Math. **344**(7), 112399 (2021)
26. Díaz, J., Pottonen, O., Serna, M., van Leeuwen, E.J.: Complexity of metric dimension on planar graphs. J. Comput. Syst. Sci. **83**(1), 132–158 (2017)
27. Diestel, R.: Graph Theory. Graduate Texts in Mathematics, vol. 173. Springer (2012)
28. Eiter, T., Gottlob, G.: Identifying the minimal transversals of a hypergraph and related problems. SIAM J. Comput. **24**(6), 1278–1304 (1995)
29. Eiter, T., Gottlob, G., Makino, K.: New results on monotone dualization and generating hypergraph transversals. SIAM J. Comput. **32**(2), 514–537 (2003)
30. Eiter, T., Makino, K., Gottlob, G.: Computational aspects of monotone dualization: a brief survey. Discret. Appl. Math. **156**(11), 2035–2049 (2008)
31. Ekim, T., Erey, A., Heggernes, P., van 't Hof, P., Meister, D.: Computing minimum geodetic sets of proper interval graphs. In: Fernández-Baca, D. (ed.) LATIN 2012. LNCS, vol. 7256, pp. 279–290. Springer, Heidelberg (2012). https://doi.org/10.1007/978-3-642-29344-3_24
32. Elbassioni, K.: On dualization over distributive lattices. Discrete Math. Theor. Comput. Sci. **24**(Discrete Algorithms) (2022)
33. Elbassioni, K., Rauf, I., Ray, S.: A global parallel algorithm for enumerating minimal transversals of geometric hypergraphs. Theoret. Comput. Sci. **767**, 26–33 (2019)
34. Elbassioni, K.M.: Algorithms for dualization over products of partially ordered sets. SIAM J. Discrete Math. **23**(1), 487–510 (2009)
35. Epstein, L., Levin, A., Woeginger, G.J.: The (weighted) metric dimension of graphs: hard and easy cases. Algorithmica **72**(4), 1130–1171 (2015)
36. Foucaud, F., et al.: Metric dimension and geodetic set parameterized by vertex cover. arXiv preprint arXiv:2405.01344 (2024)
37. Foucaud, F., et al.: Problems in NP can admit double-exponential lower bounds when parameterized by treewidth or vertex cover. In: Proceedings of the 51st International Colloquium on Automata, Languages and Programming (ICALP 2024) (2024)
38. Foucaud, F., Mertzios, G.B., Naserasr, R., Parreau, A., Valicov, P.: Identification, location-domination and metric dimension on interval and permutation graphs. II. Algorithms and complexity. Algorithmica **78**, 914–944 (2017)
39. Fredman, M.L., Khachiyan, L.: On the complexity of dualization of monotone disjunctive normal forms. J. Algorithms **21**(3), 618–628 (1996)
40. Garey, M.R., Johnson, D.S.: Computers and Intractability: A Guide to the Theory of NP-Completeness. W. H. Freeman (1979)
41. Golovach, P.A., Heggernes, P., Kanté, M.M., Kratsch, D., Sæther, S.H., Villanger, Y.: Output-polynomial enumeration on graphs of bounded (local) linear MIM-width. Algorithmica **80**(2), 714–741 (2018)
42. Harary, F., Melter, R.A.: On the metric dimension of a graph. Ars Combin. **2**, 191–195 (1976)
43. Harary, F., Loukakis, E., Tsouros, C.: The geodetic number of a graph. Math. Comput. Model. **17**(11), 89–95 (1993)

44. Johnson, D.S., Yannakakis, M., Papadimitriou, C.H.: On generating all maximal independent sets. Inf. Process. Lett. **27**(3), 119–123 (1988)
45. Johnson, M.: Structure-activity maps for visualizing the graph variables arising in drug design. J. Biopharm. Statist. **3**, 203–236 (1993)
46. Kanté, M.M., Khoshkhah, K., Pourmoradnasseri, M.: Enumerating minimal transversals of hypergraphs without small holes. In: 43rd International Symposium on Mathematical Foundations of Computer Science (MFCS 2018). Schloss Dagstuhl-Leibniz-Zentrum fuer Informatik (2018)
47. Kanté, M.M., Limouzy, V., Mary, A., Nourine, L.: On the enumeration of minimal dominating sets and related notions. SIAM J. Discret. Math. **28**(4), 1916–1929 (2014)
48. Kanté, M.M., Limouzy, V., Mary, A., Nourine, L., Uno, T.: A polynomial delay algorithm for enumerating minimal dominating sets in chordal graphs. In: International Workshop on Graph-Theoretic Concepts in Computer Science (WG 2015), pp. 138–153. Springer (2015)
49. Kanté, M.M., Nourine, L.: Polynomial time algorithms for computing a minimum hull set in distance-hereditary and chordal graphs. SIAM J. Discret. Math. **30**(1), 311–326 (2016)
50. Karpovsky, M.G., Chakrabarty, K., Levitin, L.B.: On a new class of codes for identifying vertices in graphs. IEEE Trans. Information Theory **44**(2), 599–611 (1998)
51. Khachiyan, L., Boros, E., Elbassioni, K., Gurvich, V.: On the dualization of hypergraphs with bounded edge-intersections and other related classes of hypergraphs. Theoret. Comput. Sci. **382**(2), 139–150 (2007)
52. Mezzini, M.: Polynomial time algorithm for computing a minimum geodetic set in outerplanar graphs. Theoret. Comput. Sci. **745**, 63–74 (2018)
53. Oellermann, O.R., Peters-Fransen, J.: The strong metric dimension of graphs and digraphs. Discret. Appl. Math. **155**(3), 356–364 (2007)
54. Sebő, A., Tannier, E.: On metric generators of graphs. Math. Oper. Res. **29**(2), 383–393 (2004)
55. Slater, P.J.: Leaves of trees. In: Proceedings of the Sixth Southeastern Conference on Combinatorics, Graph Theory, and Computing, pp. 549–559. Congressus Numerantium, No. XIV. Utilitas Mathematica (1975)
56. Slater, P.J.: Domination and location in acyclic graphs. Networks **17**(1), 55–64 (1987)
57. Strozecki, Y.: Enumeration complexity. Bull. EATCS **1**(129) (2019)
58. Tillquist, R.C., Lladser, M.E.: Low-dimensional representation of genomic sequences. J. Math. Biol. **79**(1), 1–29 (2019). https://doi.org/10.1007/s00285-019-01348-1
59. Tsukiyama, S., Ide, M., Ariyoshi, H., Shirakawa, I.: A new algorithm for generating all the maximal independent sets. SIAM J. Comput. **6**(3), 505–517 (1977)
60. Wild, M.: The joy of implications, aka pure horn formulas: mainly a survey. Theoret. Comput. Sci. **658**, 264–292 (2017)

Feedback Vertex Set for Pseudo-disk Graphs in Subexponential FPT Time

Gaétan Berthe[1(✉)], Marin Bougeret[1], Daniel Gonçalves[1], and Jean-Florent Raymond[2]

[1] LIRMM, Université de Montpellier, CNRS, Montpellier, France
gaetan.berthe@lirmm.fr
[2] CNRS, LIP, Lyon, France

Abstract. In this paper we investigate the existence of parameterized algorithms running in subexponential time for two fundamental cycle-hitting problems: Feedback Vertex Set and Triangle Hitting. We focus on the class of pseudo-disk graphs, which forms a common generalization of several graph classes where such results exist, like disk graphs and square graphs. In these graphs we show that given a geometric representation FVS can be solved in time $2^{\mathcal{O}(k^{9/10}\log k)}n^{\mathcal{O}(1)}$ and TH in time $2^{\mathcal{O}(k^{3/4}\log k)}n^{\mathcal{O}(1)}$.

Keywords: geometric intersection graphs · subexponential FPT algorithms · cycle-hitting problems · pseudo-disk graphs

1 Introduction

Context. The purpose of Parameterized Complexity is to provide an accurate view of the algorithmic complexity of a (typically NP-hard) decision problem and to understand the different contributions to the running time of the parameters of the instance. In this framework, a standard objective is to find FPT algorithms whose time complexities have the form $f(k) \cdot n^{\mathcal{O}(1)}$ where n and k respectively denote the size and some parameter of the instance, and f is some computable function. Hence, the potential super-polynomial part of the running time is confined to the $f(k)$ term.

In this paper we mainly focus on FEEDBACK VERTEX SET (FVS for short) which is the problem of deciding, given a graph G and an integer k, whether G has a set of k vertices whose deletion yields a forest. Some NP-hard problems like FVS cannot be solved in time $2^{o(k)}n^{\mathcal{O}(1)}$ in general graphs [3] (assuming the Exponential Time Hypothesis) but nevertheless admit algorithms with such running times (called *subexponential FPT algorithms*) when the inputs are restricted to certain graph classes. This was initially proved for particular problems in planar graphs and related classes (like map graphs) and later unified by Demaine el al. [4] in a general framework called *Bidimensionality Theory*, which

D. Král' and M. Milanič (Eds.): WG 2024, LNCS 14760, pp. 65–78, 2025.
https://doi.org/10.1007/978-3-031-75409-8_5

in essence states that every *bidimensional*[1] problem (like FVS) can be solved in subexponential FPT time on any graph class excluding a minor.

The next step has then been to move away from minor-closed graph classes and investigate in which other classes the basic NP-hard graph problems admit subexponential FPT algorithms. Natural candidates in this direction are intersection graphs of objects in the plane. Indeed, while such graphs are not planar, the underlying planarity may allow to lift techniques and ideas from the bidimensionality theory. This is not straightforward in general and required new ideas as explained for example in [12] which discusses the extension of bidimensionality to (unit) disk graphs.

These recent developments led to subexponential FPT algorithms in disk graphs [12] running in time $2^{\mathcal{O}(k^{13/14}\log k)}n^{\mathcal{O}(1)}$ for FVS, and in $2^{\mathcal{O}(k^{9/10}\log k)}n^{\mathcal{O}(1)}$ for TH (the TRIANGLE HITTING problem where given a graph G and integer k one has to decide if there is a set S of k vertices whose deletion yields to a triangle-free graph). Recently [1], these running times have been improved to $2^{\mathcal{O}(k^{7/8}\log k)}n^{\mathcal{O}(1)}$ for FVS and $2^{\mathcal{O}(k^{2/3}\log k)}n^{\mathcal{O}(1)}$ for TH when a disk representation is given, and a $2^{\mathcal{O}(k^{9/10}\log k)}n^{\mathcal{O}(1)}$ for FVS and $2^{\mathcal{O}(k^{4/5}\log k)}n^{\mathcal{O}(1)}$ for TH otherwise. On the other hand, we proved in a companion paper [2] that under the Exponential Time Hypothesis, neither TH nor FVS do admit algorithms running in time $2^{o(\sqrt{n})}$ in $K_{2,2}$-free contact-2-DIR graphs of maximum degree 6, a very restricted subclass of pseudo-disk graphs.

Contribution and Techniques. In this paper we consider pseudo-disk graphs, a classical generalization of disk graphs where informally each vertex is now a pseudo-disk (a subset of the plane that is homeomorphic to a disk), and such that for any two intersecting pseudo-disks, their boundaries intersect on at most two points. We prove that, given a pseudo-disk representation of a graph, one can solve FVS in time $2^{\mathcal{O}(k^{9/10}\log k)}n^{\mathcal{O}(1)}$ (Theorem 2), which is our main result, and TH in time $2^{\mathcal{O}(k^{3/4}\log k)}n^{\mathcal{O}(1)}$.

For FVS, we apply the following strategy. Given an input (G,k), the objective is to reduce the treewidth of G to $o(k)$ in order to solve FVS in $2^{\mathcal{O}(\mathsf{tw}(G))}n^{\mathcal{O}(1)}$ using a standard dynamic programming algorithm [3]. To do so, one has to delete from G any structure that could make $\mathsf{tw}(G)$ large (close to k). A classical approach is to start from a $\mathcal{O}(1)$-approximation M_0 (with $|M_0| = \mathcal{O}(k)$) of a minimum FVS. Observe that as FVS is bidimensional, the existence of M_0 implies that $\boxplus(G)$ (the size of the largest square grid contained as a minor in G) is $\mathcal{O}(\sqrt{k})$. However, unlike in planar graphs where $\mathsf{tw}(G) = \mathcal{O}(\boxplus(G))$ (a property known under the name of *subquadratic grid minor property*), having $\boxplus(G) = \mathcal{O}(\sqrt{k})$ is not enough to ensure that $\mathsf{tw}(G) = o(k)$, as for example a clique K_k has $\boxplus(K_k) \leq \sqrt{k}$ but $\mathsf{tw}(K_k) = k-1$. Thus, as first step, we use a folklore preprocessing routine that produces $2^{\mathcal{O}(\frac{k}{p}\log p)}n^{\mathcal{O}(1)}$ instances (G',k') where each graph has clique number at most $p = \mathcal{O}(k^{\epsilon})$.

[1] Informally: yes-instances are minor-closed and a solution on the (r,r)-grid has size $\Omega(r^2)$.

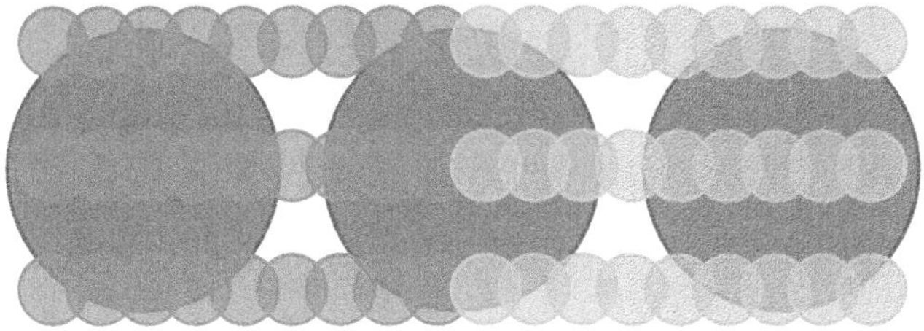

Fig. 1. In this graph G' (from [8]), $\omega(G')$ is constant, $\mathsf{tw}(G') \geq k'$ (where $k' = 3$ here) as it contains $K_{k',k'}$ as a minor, and $\boxplus(G) = \mathcal{O}(\sqrt{k'})$ as the set M_0 of the k' big disks is a FVS.

The problem now is that (even in disk-graphs) this additional property on the clique number is still not enough to force small treewidth, as shown for instance by the graph G' of Fig. 1. Thus, we apply a second preliminary branching step (described in the context of disk graphs in [12]). This produces again $2^{\mathcal{O}(\frac{k}{p}\log p)}n^{\mathcal{O}(1)}$ instances (G'', M, k''), where M is a feedback vertex set of G'' with $|M| = \mathcal{O}(pk') = \mathcal{O}(pk)$, with the additional property that for any $v \in M$, $N(v)\backslash M$ is an independent set. This property may look like a technical assumption, but observe that it is necessary to ensure that the treewidth is $o(k)$. For example with G' and M_0 like in Fig. 1, we have $\mathsf{tw}(G') \geq k'$, but this case is no longer possible after the second branching step as the property that for any $v \in M_0$, $N(v)\backslash M_0$ is an independent set does *not* hold.

Once these two preliminaries branching steps are done, our main technical challenge is to kernelize each instance (G'', M, k'') to obtain a smaller equivalent instance $\left(\tilde{G}, \tilde{k}\right)$ such that $\left|V\left(\tilde{G}\right)\right| = \mathcal{O}\left(p^6|M|\right) = \mathcal{O}\left(p^7 k\right)$ (see Lemma 2). This is the crux of the paper. The techniques used to achieve this kernel are sketched in Sect. 4. Once the number of vertices of $\tilde{G}$ is small, we can show (Lemma 1) that $\mathsf{tw}(\tilde{G}) = \mathcal{O}\left(\sqrt{p\left|V\left(\tilde{G}\right)\right|}\right) = \mathcal{O}\left(\sqrt{k^{1+8\epsilon}}\right)$ and solve $\tilde{G}$ by classic dynamic programming, leading to our subexponential algorithm for FVS in pseudo-disk graphs.[2]

For TH, we revisit the strategy of [1] (originally designed for disk graphs) that produces a subexponential number of instances of order quasi-linear in the parameter. To improve the running time we show (Lemma 1) that the number of edges in a n-vertex pseudo-disk graph of ply[3] p is $\mathcal{O}(pn)$, which implies that the produced instances have sublinear treewidth $\mathcal{O}(\sqrt{pn})$. This improves the treewidth bound $f(p)\sqrt{n}$ (for some function f) given by Dvorák et al. in [6]. The problem is then solved by dynamic programming on approximate tree-decompositions.

[2] We have been told in a private communication that it might be possible to extend the arguments of [12] for FVS in pseudo-disk graphs without a geometrical representation, and that the time complexity would be worse than the one we obtain.

[3] The ply is the maximum number of pseudo-disks sharing a common point.

Organization of the Paper. In Sect. 2 we give the necessary definitions and properties and present the aforementioned preprocessing step. Section 3 reduces our main result to a technical kernelization lemma. Section 4 is devoted to sketching the first part of the proof of this lemma. We conclude with directions for future research in Sect. 5. Due to space constraints, the proofs of the statements marked with the ✂ symbol can be found in the full version of the paper.

2 Preliminaries

2.1 Basics

In this paper logarithms are binary and all graphs are simple, loopless and undirected. Unless otherwise specified we use standard graph theory terminology, as in [5] for instance. Given a graph G, we denote by $\omega(G)$ the maximum order of a clique in G. We denote by $d_G(v)$ the degree of $v \in V(G)$, or simply $d(v)$ when G is clear from the context. We use the notation $\Delta(G)$ for the maximum degree of the vertices of G. We denote by $\mathsf{tw}(G)$ the treewidth of G.

2.2 Graph Classes

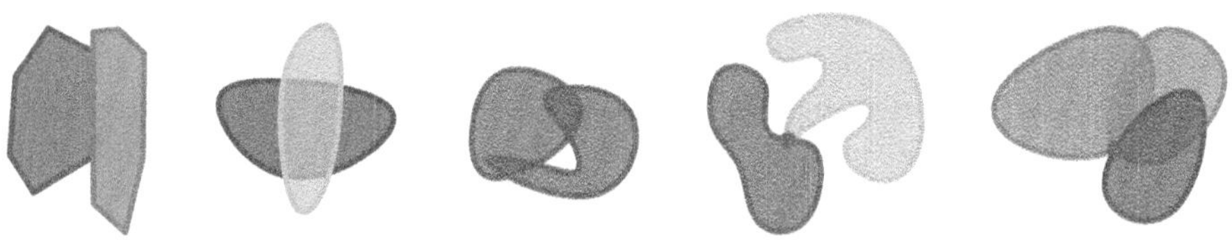

Fig. 2. Three forbidden intersections between pseudo-disks, and two avoidable examples.

In this article, we are mainly concerned with geometric graphs described by the intersection or contact of objects in the Euclidean plane. The most general class we consider are *string graphs*, which are intersection graphs of strings (a.k.a. Jordan arcs).

Pseudo-disk Graphs. The focus on this paper is on *pseudo-disk graphs.* A *pseudo-disk* $\mathcal{D}$ is a subset of the plane that is homeomorphic to a disk. We denote its boundary by $\partial\mathcal{D}$ and call *internal points* any point of $\mathcal{D}$ that does not belong to $\partial\mathcal{D}$. A set of pseudo-disks $\mathcal{S}$ forms a *system of pseudo-disks* if, for any two intersecting elements $\mathcal{D}_1, \mathcal{D}_2 \in \mathcal{S}$,

their borders, $\partial\mathcal{D}_1$ and $\partial\mathcal{D}_2$, intersect on at most two points. Under minor perturbation, any system of pseudo-disk, can be such that for any two intersecting elements $\mathcal{D}_1, \mathcal{D}_2 \in \mathcal{S}$, either their borders do not intersect but in that case one is contained in the other, or their borders intersect on exactly two points while $\mathcal{D}_1 \cap \mathcal{D}_2$ contains internal points. Similarly, we can require that no point belongs to more than two boundaries (see Fig. 2). Given a system of pseudo-disks $\mathcal{S}$, we denote by $G_{\mathcal{S}}$ the corresponding intersection graph, and this defines the class of pseudo-disk graphs.

Note that pseudo-disk graphs are in particular string graphs and they form a common generalization of various classes of "fat" intersection graphs such as disk graphs, intersection graphs of axis-parallel squares, and more generally any intersection graph obtained from homothetic copies of a given convex set, but they also generalize the contact graphs of segments (see full version).

Given $\mathcal{S}$ a system of pseudo-disks and z a point in the plane, the *ply* of z (wrt. $\mathcal{S}$) is the number of pseudo-disks of $\mathcal{S}$ containing z. The *ply* of a maximal connected region $\mathcal{R}$ of $\mathbb{R}^2 \setminus \bigcup_{\mathcal{D}\in\mathcal{S}} \partial\mathcal{D}$, is the ply of any of its points. The *ply* of $\mathcal{S}$ is the maximum ply of a point of the plane wrt. $\mathcal{S}$. A pseudo-disk graph G has *ply* p if it is the intersection graph of a system of pseudo-disks of ply p.

Representation of Pseudo-disk Graphs. A system of pseudo-disks $\mathcal{S}$ is represented by the directed plane multi-graph $\overrightarrow{P}_{\mathcal{S}}$ defined as follows (see also Fig. 3). For any two $\mathcal{D}_1, \mathcal{D}_2 \in \mathcal{S}$, every point in $\partial\mathcal{D}_1 \cap \partial\mathcal{D}_2$ is a vertex of $\overrightarrow{P}_{\mathcal{S}}$. For any $\mathcal{D} \in \mathcal{S}$, the Jordan arcs in $\partial\mathcal{D}$ joining any two such points form the arcs of $\overrightarrow{P}_{\mathcal{S}}$. Those arcs are oriented in such a way that $\partial\mathcal{D}$ corresponds to a clockwise cycle around $\mathcal{D}$. It may remain disks $\mathcal{D} \in \mathcal{S}$ with uncrossed boundaries (i.e. $\partial\mathcal{D} \cap \partial\mathcal{D}' = \emptyset$ for any $\partial\mathcal{D}' \in \mathcal{S}$). For such a disk, we pick an arbitrary point in $\partial\mathcal{D}$ as a vertex for $\overrightarrow{P}_{\mathcal{S}}$, and the rest of $\partial\mathcal{D}$ corresponds to a clockwise loop on this vertex. Note that $\overrightarrow{P}_{\mathcal{S}}$ has at most $2|E(G_{\mathcal{S}})| + |V(G_{\mathcal{S}})|$ vertices, and at most $4|E(G_{\mathcal{S}})| + |V(G_{\mathcal{S}})|$ edges. This is a convenient feature of pseudo-disks system to admit a polynomial (in terms of $G_{\mathcal{S}}$) space data structure representing them. This graph $\overrightarrow{P}_{\mathcal{S}}$ is called the *pseudo-disk representation* of $\mathcal{S}$ and $G_{\mathcal{S}}$. We denote $\overrightarrow{P}^*_{\mathcal{S}}$ the dual graph of $\overrightarrow{P}_{\mathcal{S}}$. Observe that any arc is oriented from a region (of $\mathcal{S}$) with lower ply towards one with higher ply.

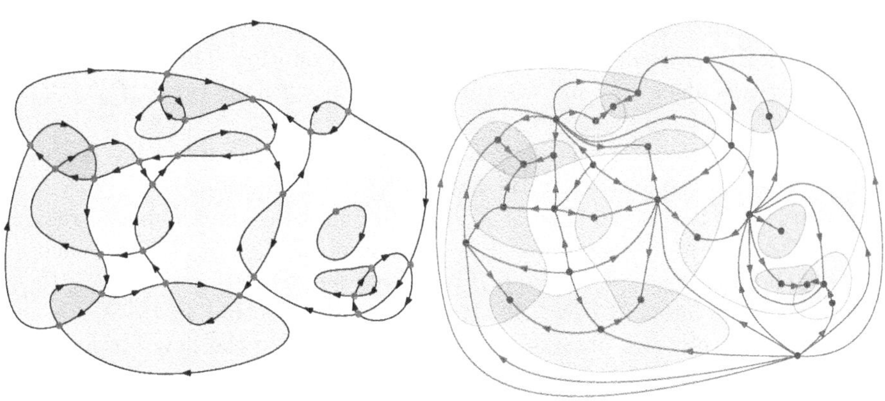

Fig. 3. A pseudo-disk representation $\overrightarrow{P}_{\mathcal{S}}$ and its dual $\overrightarrow{P}^*_{\mathcal{S}}$.

2.3 Branching Steps and Properties of Pseudo-disk Graphs

Dvořák et al. proved in [6] that n-vertex pseudo-disk graphs with ply p have treewidth at most $f(p)\sqrt{n}$ for some function f. In Lemma 1 below we improve this bound to $\mathcal{O}(\sqrt{pn})$. Our proof requires the following consequence of the results on balanced separators of string graphs of Lee [11] and the links between separators and treewidth of Dvořák and Norin [7].

Theorem 1 **([11] and [7]).** *Any m-edge string graph has treewidth $\mathcal{O}(\sqrt{m})$.*

Lemma 1. *If a graph G on n vertices admits a pseudo-disk representation with ply p, then G has at most $3epn$ edges. Furthermore, $\mathsf{tw}(G) = \mathcal{O}(\sqrt{pn})$.*

Proof. To bound the number of edges of G, we follow the same approach as [9] for contact of strings. Consider a pseudo-disk graph G with n vertices and m edges having a representation with ply p. Then, pick a subset $V' \subseteq V(G)$ of pseudo-disks by picking each one, randomly and independently, with probability $1/p$. The expected size of V' is n/p. For each $uv \in E(G)$, the probability that at an arbitrarily chosen point $p_{uv} \in \mathcal{D}_u \cap \mathcal{D}_v$, only $\mathcal{D}_u$ and $\mathcal{D}_v$ remain is $(1-1/p)^{q-2}/p^2 \geq (1-1/p)^{p-2}/p^2$, where q is the ply of p_{uv}. By linearity of expectation, the expected number of edges uv for which only u and v remain at p_{uv} is at least $m(1-1/p)^{p-2}/p^2$. The following claim implies that $m(1-1/p)^{p-2}/p^2 \leq 3n/p$.

Claim 1. If a graph H has an edge subset $E' \subseteq E(H)$ such that H admits a pseudo-disk representation where for any edge $uv \in E'$ there is a point in $\mathcal{D}_u \cap \mathcal{D}_v$ with ply two, then $|E'| \leq 3|V(H)|$.

Proof. We show this by proving that the graph $(V(H), E')$ is planar, and this is obtained by constructing a pseudo-disk representation with ply at most two [10]. We start with the representation of H and we show how to modify it in order to delete at least one edge of $E(H)\backslash E'$. Iterating this process yields the desired representation.

If there is a pseudo-disk $\mathcal{D}_v$, whose points all have ply at least three, one can simply move $\mathcal{D}_v$ far away from the rest of the representation. The obtained graph has at least two edges less, and we still have a point with ply two in $\mathcal{D}_u \cap \mathcal{D}_v$ for any $uv \in E'$.

Similarly if there is an edge ab of $E(H)\backslash E'$ with $\mathcal{D}_a \subset \mathcal{D}_b$ and a has degree 1, we can move its disk away from the rest of the representation.

Otherwise, let ab be an edge of $E(H)\backslash E'$ minimizing the number of regions in $\mathcal{D}_a \cap \mathcal{D}_b$. Note that by the first rule there is no $\mathcal{D}_c \subseteq \mathcal{D}_a \cap \mathcal{D}_b$. For such an edge ab, and for every $c \in V(H)\backslash\{a, b\}$, the border $\partial\mathcal{D}_c$ intersects $\partial\mathcal{D}_a \cap \mathcal{D}_b$ as many times as it intersect $\partial\mathcal{D}_b \cap \mathcal{D}_a$. Hence, replacing $\mathcal{D}_a$ and $\mathcal{D}_b$ by $\mathcal{D}_a\backslash\mathcal{D}_b$ and $\mathcal{D}_b\backslash\mathcal{D}_a$ respectively, one obtains a representation where the new borders $\partial\mathcal{D}_a$ and $\partial\mathcal{D}_b$ intersect the same borders (exactly twice) as the old one except for the edge ab. Finally, since there was no pseudo-disk contained in $\mathcal{D}_a \cap \mathcal{D}_b$, ab is the only deleted edge. ⌟

Since $1/e \leq (1 - 1/p)^{p-2}$ for $p \geq 3$, we get $m \leq 3epn$ as claimed. The second part of the statement follows by Theorem 1. □

As first step of our algorithms we make use of two preprocessing routines: the first one is a folklore branching that allows to reduce cliques larger than a chosen size p (where typically $p = k^{\epsilon}$) and the second, given a (possibly non-optimal) solution M to the problem, allows to consider subcases where for any $v \in M$, $N(v) \backslash M$ is an independent set. These steps are described in [12] for disk graphs. We combine them in the following routine.

Corollary 1 (✂). *Let Π be FVS or TH. There is a $2^{\mathcal{O}(\frac{k}{p}\log k)} n^{\mathcal{O}(1)}$-time algorithm that, given a pseudo-disk graph G with a representation, a parameter k, and an integer $p \in [6, k]$, returns a collection $\mathcal{C}$ of size $2^{\mathcal{O}(\frac{k}{p}\log k)}$ of quadruples $(G', \overrightarrow{P'}, M, k')$ such that $\overrightarrow{P'}$ is a pseudo-disk representation of G' and:*

1. *(G', k') is an instance of Π where G' is an induced subgraph of G, $\omega(G') \leq p$, and $k' \leq k$.*
2. *$|M| = \mathcal{O}(pk)$, $G' - M$ is triangle-free (when Π is TH) or is a forest (when Π is FVS), and for any $v \in M$, $N(v) \backslash M$ is an independent set.*
3. *(G, k) is a yes-instance of Π if and only if there exists $(G', \overrightarrow{P'}, M, k') \in \mathcal{C}$ such that (G', k') is a yes-instance.*

3 Hitting Cycles in Pseudo-disk Graphs

The main result of this paper is the following.

Theorem 2. *There is an algorithm that, given an n-vertex pseudo-disk graph with a representation and a parameter k, solves FVS in time $2^{\mathcal{O}(k^{\frac{9}{10}}\log k)} n^{\mathcal{O}(1)}$.*

To achieve this, we proceed in three steps. We first use the preprocessing of Corollary 1. Then, we kernelize each instance provided by the branching process in Lemma 2. We show that these kernelized instances have small treewidth, and thus we can conclude with a dynamic algorithm to solve each of them. The main technical ingredient of this proof is the following lemma, whose proof is deferred to the full version due to space constraints, and for which we provide a sketch in Sect. 4.

Lemma 2 (✂). *Given a quadruple $(G, \overrightarrow{P}_{\mathcal{S}}, M, k)$ as given by Corollary 1 (for Π = FVS), there is a polynomial time algorithm that returns an equivalent instance (G', k') where $k' \leq k$ and G' is a pseudo-disk graph with ply at most p and $\mathcal{O}(p^7 k)$ vertices.*

As we will apply Corollary 1 for $p = \mathcal{O}(k^{\epsilon})$, the above will directly imply that $|V(G')|$ is almost linear, and, using Lemma 1, that G' has sublinear treewidth. Thus, the whole technical difficulty is to prove Lemma 2. Now that the lemma is stated, we can proceed with the proof of the theorem.

Proof (of Theorem 2). We apply Corollary 1 on an instance (G, k) provided with a representation $\overrightarrow{P}_{\mathcal{S}}$ and on a value $p \in [6, k]$ to be set later. As a result we obtain a collection $\mathcal{C}$ of $2^{\mathcal{O}(\frac{k}{p}\log k)}$ instances that are each provided with a ply p pseudo-disk representation, and with a feedback vertex set M of size $\mathcal{O}(pk)$, such that for any $v \in M$, $N(v)\backslash M$ is an independent set. Furthermore, solving these instances of FVS is enough to get a solution to our initial instance (G, k). This first stage is done in time $2^{\mathcal{O}(\frac{k}{p}\log k)}n^{\mathcal{O}(1)}$.

Then for each of the $2^{\mathcal{O}(\frac{k}{p}\log k)}$ obtained instances, we apply the kernelization described in Lemma 2, and we get a pseudo-disk graph G' with ply at most p and with only $\mathcal{O}(p^7k)$ vertices. By Lemma 1, this kernel has treewidth at most $\mathcal{O}(\sqrt{p^8k})$, and can thus be solved in time $2^{\mathcal{O}(\mathrm{tw}(G'))}n^{\mathcal{O}(1)} = 2^{\mathcal{O}\left(p^4k^{\frac{1}{2}}\right)}n^{\mathcal{O}(1)}$ with classical algorithms [3]. The overall time complexity of this algorithm is

$$2^{\mathcal{O}(\frac{k}{p}\log k)}n^{\mathcal{O}(1)} \cdot 2^{\mathcal{O}(\frac{k}{p}\log k)} \cdot n^{\mathcal{O}(1)} \cdot 2^{\mathcal{O}\left(p^4k^{\frac{1}{2}}\right)}n^{\mathcal{O}(1)}.$$

Thus setting $p = k^{\frac{1}{10}}$ the time complexity is indeed $2^{\mathcal{O}\left(k^{9/10}\log k\right)}n^{\mathcal{O}(1)}$. □

4 Overview of the First Part of Proof of Lemma 2

Recall that in Lemma 2 we are given a quadruple $(G, \overrightarrow{P}_{\mathcal{S}}, M, k)$ as computed by Corollary 1. The equivalent instance computed in Lemma 2 will be obtained using reduction rules. Before applying these rules, we define the set M' by adding to M every pseudo-disk $\mathcal{D}$ that contains a point of $\partial\mathcal{D}_u \cap \partial\mathcal{D}_v$, for $u, v \in M$ (see Fig. 4, left). We denote $\mathcal{R}_{M'}$ the region $\mathcal{R}_{M'} = (\cup_{u \in M'}\mathcal{D}_u)$. Note that this region (and its complement) may not be connected. Let us now partition the vertices of $V(G)\backslash M'$ into three types.

Definition 1. *A vertex $v \in V(G)\backslash M'$ is an* inner-M'-vertex *if $\mathcal{D}_v \subsetneq \mathcal{R}_{M'}$, it is a* border-$M'$-vertex *if $\mathcal{D}_v$ intersects $\mathcal{R}_{M'}$ without being included, and it is an* outer-M'-vertex *if $\mathcal{D}_v$ does not intersect $\mathcal{R}_{M'}$. We denote $I_{M'}$, $B_{M'}$, and $O_{M'}$ the vertex sets with these three types of vertices (see Fig. 4 (right)).*

Let us state a few properties about these vertices.

Claim 2 (✂).

(a) $|M'| = \mathcal{O}(p|M|)) = \mathcal{O}(p^2k)$.
(b) For any edge uv of $G - M'$, the intersection $\mathcal{D}_u \cap \mathcal{D}_v$ does not intersect $\mathcal{R}_{M'}$.
(c) For every $u \in V(G)\backslash M'$, the pseudo-disk $\mathcal{D}_u$ does not contain any point of $\partial\mathcal{D}_v \cap \partial\mathcal{D}_w$, for any two vertices $v, w \in M'$.

Note that Property (b) implies that every inner-M'-vertex u is isolated in $G - M'$. The kernelization is divided into two parts. The first part (see Subsect. 4.1) deals with these inner-M'-vertices, while the second part (not included here due to lack of space) deals with border- and outer-M'-vertices.

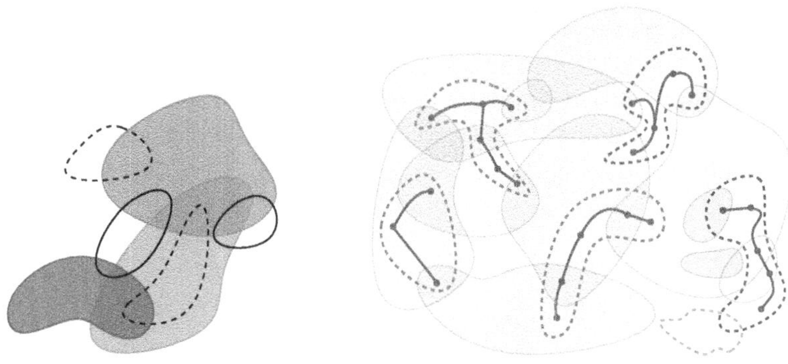

Fig. 4. (left) From M to M'. The pseudo-disks of M are filled. Among the others, those added to M' have solid border. (right) Here, the filled pseudo-disks belong to M'. In dashed green, the inner pseudo-disks of $I_{M'}$. In dashed red, the border pseudo-disks of $B_{M'}$. In dashed orange, the only pseudo-disk $v \in O_{M'}$. For any pseudo-disk u not in M', we depicted its hosted graph H_u, which is defined in Definition 2. (Color figure online)

4.1 Sketch of Proof for Inner-M'-vertices

The purpose of this section is to reduce the number of inner-M'-vertices. Let $\mathcal{S}'$ be the restriction of $\mathcal{S}$ to vertices of M'. Let us denote $\overrightarrow{P}_{\mathcal{S}'}$ the representation of $G[M']$. Since the pseudo-disk system has ply at most p, the graph $G[M']$ has $\mathcal{O}(p|M'|) = \mathcal{O}(p^3k)$ edges by Lemma 1. Each of such edge induces at most two vertices in $\overrightarrow{P}_{\mathcal{S}'}$. Hence, $\overrightarrow{P}_{\mathcal{S}'}$ has $\mathcal{O}(p^3k)$ vertices, and thus by planarity of $\overrightarrow{P}_{\mathcal{S}'}$, the number of edges, and faces in $\overrightarrow{P}_{\mathcal{S}'}$ is also $\mathcal{O}(p^3k)$.

Definition 2. *Given a vertex $u \in V(G)\backslash M'$, we define the* hosted graph H_u *as a plane graph drawn within $\mathcal{D}_u$, with a (single) vertex in a face f of $\overrightarrow{P}_{\mathcal{S}'}$ if and only if $\mathcal{D}_u$ and f intersect, and with edges between vertices lying in adjacent faces of $\overrightarrow{P}_{\mathcal{S}'}$, if their common border intersects $\mathcal{D}_u$ (see Fig. 4, right).*

Claim 3. For any inner-M'-vertex u, the hosted graph H_u is a tree.

Proof. By Properties (b) and (c), $\mathcal{D}_u$ does not contain a crossing, meaning that there is no v_1, v_2 such that $\mathcal{D}_u$ contains a point in $\partial\mathcal{D}_{v_1} \cap \partial\mathcal{D}_{v_2}$. Hence, the family $\{\mathcal{D}_v \cap \mathcal{D}_u, v \in N(u)\}$ is a laminar set family. This implies that H_u is a tree. ⌟

In order to count inner vertices, we distinguish two cases according to the maximum degree of H_u.

Claim 4. The number of vertices $u \in I_{M'}$ such that $\Delta(H_u) \geq 3$ is $\mathcal{O}(p^3k)$.

Proof. To see this, let us construct a bipartite pseudo-disk graph as follows. In one part, we consider the arcs of $\overrightarrow{P}_{\mathcal{S}'}$. These arcs are slightly shortened (so that they form an independent set) and slightly thickened (so that they intersect the adjacent regions we define below), see Fig. 5. In the other part, we define a pseudo-disk $\mathcal{D}'_u$ for each vertex $u \in I_{M'}$ such that $\Delta(H_u) \geq 3$ as follows. Consider a vertex x of H_u such that $d_{H_u}(x) \geq 3$, and let $\mathcal{D}'_u$ be the intersection of $\mathcal{D}_u$ and the face of $\overrightarrow{P}_{\mathcal{S}'}$ containing x. Since triangle-free pseudo-disk graphs are planar, by [3, Lemma 9.24], the second part has size $\mathcal{O}(|E(\overrightarrow{P}_{\mathcal{S}'})|) = \mathcal{O}(p|M'|) = \mathcal{O}(p^3k)$. ⌟

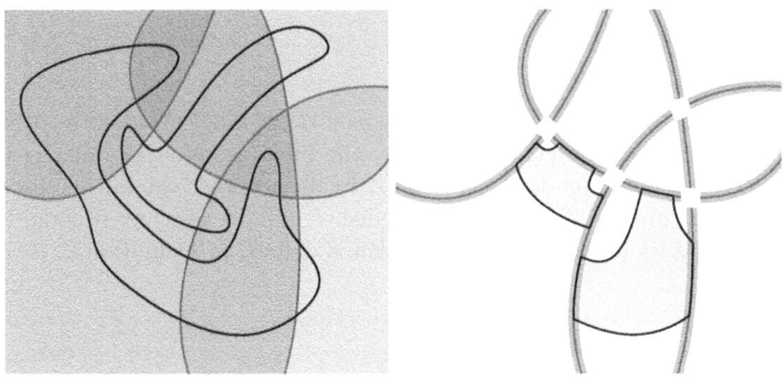

Fig. 5. Two inner-M' pseudo-disks with $\Delta(H_u) \geq 3$, and the bipartite pseudo-disk graph constructed to prove Claim 4.

Now, we focus on vertices $u \in I_{M'}$ such that $\Delta(H_u) \leq 2$. In that case, H_u is a path with possibly only one vertex, and as we show with bounded length.

Claim 5. For any vertex $u \in I_{M'}$ such that H_u is a path, this path has length at most $d(u)$, and $d(u) \leq 2p$. Furthermore, $N(u)$ can be split into two cliques of size at most p each.

Proof. For any neighbor v of u, $\mathcal{D}_v$ contains one of the two faces f_1 and f_2 of $\overrightarrow{P}_{\mathcal{S}'}$ containing the endpoints of the path H_u (see Fig. 6 (left)). Otherwise, $\partial\mathcal{D}_u$ and $\partial\mathcal{D}_v$ would intersect in at least four points. Since the representation has ply at most p, there are at most $2p$ pseudo-disks containing one (or two) of these faces. Among those, the pseudo-disks containing f_1 clearly induce a clique, such as the remaining ones, since all of them contain f_2. ⌟

Even if these paths have bounded length, the number of such paths can still be arbitrarily large as we could draw them one "parallel" to each other, creating large class of twins (see Fig. 6). Thus, we need the following kernelization rule for the feedback vertex set problem.

(twins) Consider an independent set S of $V(G)\setminus M'$ such that all the vertices in S have the same neighborhood N. If N admits a partition into two cliques and if $|S| > 4$, delete $|S| - 4$ vertices of S and keep the same parameter k.

Proof. (Safeness of the rule) Denote $v_1, \ldots, v_s$ the vertices in S, for some $s > 4$, and denote C_1, C_2 the two cliques partitioning N. Let $G' = G - \{v_5, \ldots, v_s\}$. Clearly G' cannot have a minimum feedback vertex set larger than G. Hence, it is sufficient to show that G' has a feedback vertex set X, that is also a feedback vertex set for G.

Consider any minimum feedback vertex set X of G', and let us transform it (if needed) into a minimum feedback vertex set intersecting N on at least $|N| - 1$ vertices. Observe first that for any i, $|C_i \backslash X| \leq 2$. If $|C_1 \backslash X| = 2$, then $\{v_1, v_2, v_3, v_4\} \subseteq X$, as otherwise the two vertices in $C_1 \backslash X$ and any v_i not in X would form a triangle in $G - X$. In such case, defining X' by replacing in X the vertices v_1, v_2, v_3, v_4 with the vertices in $C_1 \backslash X$ and in $C_2 \backslash X$ would result in the desired feedback vertex set. Indeed, the vertices v_1, v_2, v_3, v_4 are isolated in $G' - X'$, and $G' - \{v_1, v_2, v_3, v_4\} - X'$ is a forest, as $X \subseteq X' \cup \{v_1, v_2, v_3, v_4\}$.

Hence, we consider that $|C_1 \backslash X| \leq 1$ and $|C_2 \backslash X| \leq 1$. If $|C_1 \backslash X| = 1$ and $|C_2 \backslash X| = 1$, then $|\{v_1, v_2, v_3, v_4\} \cap X| \geq 3$, as otherwise the two vertices in $\{v_1, v_2, v_3, v_4\} \backslash X$, the vertex in $C_1 \backslash X$ and the vertex in $C_2 \backslash X$ would form a 4-cycle in $G - X$. In such case, defining X' by replacing in X two vertices in $\{v_1, v_2, v_3, v_4\} \cap X$ with the vertices in $C_1 \backslash X$ and in $C_2 \backslash X$ would result in the desired feedback vertex set. Indeed, the vertices v_1, v_2, v_3, v_4 not in X' are isolated in $G' - X'$, and $G' - \{v_1, v_2, v_3, v_4\} - X'$ is a forest, as $X \subseteq X' \cup \{v_1, v_2, v_3, v_4\}$.

We thus have a minimum feedback vertex set of G' intersecting N on at least $|N| - 1$ vertices. Since the vertices $v_5, \ldots, v_s$ are leaves or are isolated in $G - X$, we have that $G - X$ is the forest $G' - X$ with $s - 4$ leaves or isolated vertices added, and it is thus a forest. ⌟

Identifying the configuration required by the previous twins rule and updating the representation of G when we delete vertices can be done in polynomial time, implying that this kernelization can be performed in time polynomial in $|V(G)|$.

From now on, among the vertices $u \in I_{M'}$ such that H_u is a path, there are at most four vertices with the same neighborhood. We can thus focus on bounding the number of such vertices u with distinct neighborhoods. Indeed, if $I^*_{M'}$ is a maximal subset without twins (among vertices of $I_{M'}$ such that H_u is a path), then $|I_{M'}| \leq 4|I^*_{M'}| + \mathcal{O}(p^3 k)$ (by Claim 4). In the following subsection, we sketch how to bound $|I^*_{M'}|$ in order to obtain the following claim.

Claim 6. After the kernelization of Subsect. 4.1, the number of inner-M'-vertices is $\mathcal{O}(p^5 |M'|)$, that is $|I_{M'}| = \mathcal{O}(p^7 k)$.

4.2 Sketch for Bounding the Size of $I^*_{M'}$

Given two hosted graphs H_u and H_v that are paths, we say that H_u and H_v are *identical* if their vertices are drawn in the same faces of $\overrightarrow{P}_{\mathcal{S}'}$, and if these faces are visited in the same order. Let $H^{\neq}_{M'}$ be a maximum set of pairwise non-identical paths H_u, for vertices $u \in I_{M'}$. As vertices v of $I^*_{M'}$ are non twins, their

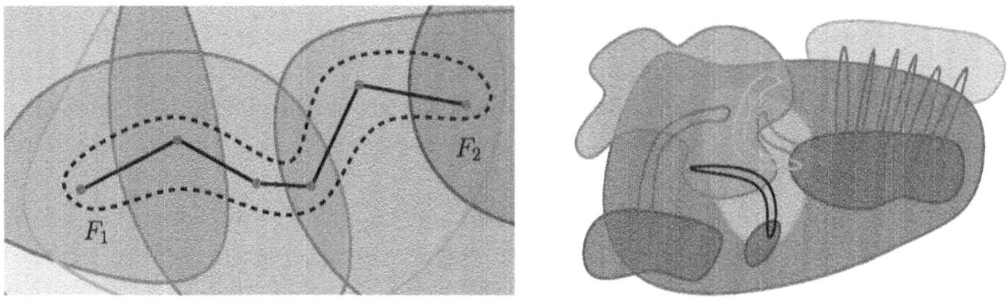

Fig. 6. Left: a hosted graph H_u that is a path. Right: some vertices $u \in I_{M'}$ such that H_u is a path. Some of these vertices have many twins.

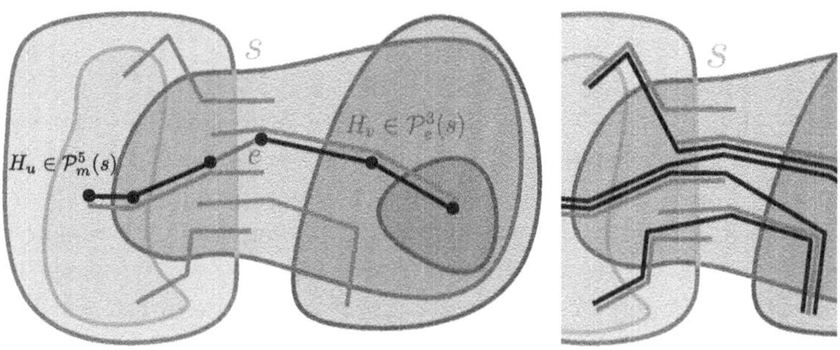

Fig. 7. In green the paths of $P^3_e(s)$. Left: in black and brown a path of $\mathcal{P}^5_m(s)$. Right: in black the paths of $P^5_m(s)$. (Color figure online)

paths H_v are not identical, and thus $|I^*_{M'}| \leq |H^{\neq}_{M'}|$. Bounding the size of this set $H^{\neq}_{M'}$ thus bounds the sizes of $I^*_{M'}$ and $I_{M'}$. As a worst case, we assume that for any subpath P' of some path $P_1 \in H^{\neq}_{M'}$, there is a path $P_2 \in H^{\neq}_{M'}$ that is identical to P'.

We prove that $|H^{\neq}_{M'}| = \mathcal{O}(p^4 l)$, where l denotes the number of faces in $\overrightarrow{P}_{\mathcal{S}'}$, by showing that $|H^{\neq,i}_{M'}| = \mathcal{O}(i^3 l)$, where $H^{\neq,i}_{M'}$ is the set of paths $P \in H^{\neq}_{M'}$ of length i. Observe that $|H^{\neq,1}_{M'}| \leq |E(\overrightarrow{P}_{\mathcal{S}'})| = \mathcal{O}(l)$ as these paths cross a distinct arc of $\overrightarrow{P}_{\mathcal{S}'}$.

Due to lack of space, we now only sketch how to bound $|H^{\neq,i}_{M'}|$ for some odd $i \geq 3$. Let s be an edge of $\overrightarrow{P}_{\mathcal{S}'}$. Let $\mathcal{P}^i_m(s)$ be the set of paths of $H^{\neq,i}_{M'}$ whose middle edge e crosses s (see Fig. 7 left), and $\mathcal{P}^{\lceil i/2 \rceil}_e(s)$ be the set of paths of $H^{\neq,\lceil i/2 \rceil}_{M'}$ whose first or last edge crosses s. The simple observation that any path from $\mathcal{P}^i_m(s)$ can be created by "gluing" two paths from $\mathcal{P}^{\lceil i/2 \rceil}_e(s)$ leads to the trivial bound $|\mathcal{P}^i_m(s)| \leq |\mathcal{P}^{\lceil i/2 \rceil}_e(s)|^2$. However, since these paths do not intersect each other, we can prove the key property that $|\mathcal{P}^i_m(s)| \leq 3|\mathcal{P}^{\lceil i/2 \rceil}_e(s)|$ (see Figure 7 right). By summing over all s, we get $|H^{\neq,i}_{M'}| \leq 6|H^{\neq,\lceil i/2 \rceil}_{M'}|$ (we get an extra factor 2 as each path in $H^{\neq,\lceil i/2 \rceil}_{M'}$ is counted twice, one time for each of its endpoint). This induction resolves in $|H^{\neq,i}_{M'}| \leq 6^{\log(i)}|H^{\neq,1}_{M'}| = \mathcal{O}(i^3 l)$.

5 Discussion

In this paper we gave algorithms for FEEDBACK VERTEX SET and TRIANGLE HITTING in pseudo-disk graphs running in times $2^{\mathcal{O}(k^{9/10} \log k)} n^{\mathcal{O}(1)}$ and $2^{\mathcal{O}(k^{3/4} \log k)} n^{\mathcal{O}(1)}$, respectively. This generalizes the previous results [12] on disk graphs, with an improvement on the running time. On the other hand, as noted in the introduction, there is for both problems an ETH-based lower-bound of $2^{o(\sqrt{n})}$ [2]. So a natural problem is to get matching upper- and lower-bounds. We have no evidence to believe that our upper-bounds could be tight. Besides, our algorithms require a pseudo-disk representation of the input graph. So a second open problem is to provide a robust subexponential FPT algorithm for FVS in pseudo-disk graphs. The bulk of our algorithm for FVS consists in applying reduction rules to obtain an instance of size polynomial in the parameter. However this is not strictly speaking a kernel as we do not reduce the input instance but the subexponential number of instances produced by the preprocessing step. It could however be interesting to investigate if the ideas from our reduction steps could be useful for kernelization in pseudo-disk graphs.

References

1. An, S., Cho, K., Oh, E.: Faster algorithms for cycle hitting problems on disk graphs. In: Morin, P., Suri, S. (eds.) WADS 2023. LNCS, vol. 14079, pp. 29–42. Springer, Cham (2023). https://doi.org/10.1007/978-3-031-38906-1_3
2. Berthe, G., Bougeret, M., Gonçalves, D., Raymond, J.F.: Subexponential algorithms in geometric graphs via the subquadratic grid minor property: the role of local radius. In: Bodlaender, H.L. (ed.) 19th Scandinavian Symposium and Workshops on Algorithm Theory (SWAT 2024). Leibniz International Proceedings in Informatics (LIPIcs), vol. 294, pp. 11:1–11:18. Schloss Dagstuhl – Leibniz-Zentrum für Informatik, Dagstuhl, Germany (2024). https://doi.org/10.4230/LIPIcs.SWAT.2024.11, https://drops.dagstuhl.de/entities/document/10.4230/LIPIcs.SWAT.2024.11
3. Cygan, M., et al.: Parameterized Algorithms, 1st edn. Springer, Cham (2015)
4. Demaine, E.D., Fomin, F.V., Hajiaghayi, M., Thilikos, D.M.: Subexponential parameterized algorithms on bounded-genus graphs and h-minor-free graphs. J. ACM **52**(6), 866–893 (2005). https://doi.org/10.1145/1101821.1101823
5. Diestel, R.: Graph Theory. Graduate Texts in Mathematics. 3rd edn, vol. 173. no. 33, p. 12 (2005)
6. Dvořák, Z., Lokshtanov, D., Panolan, F., Saurabh, S., Xue, J., Zehavi, M.: Efficient approximation for subgraph-hitting problems in sparse graphs and geometric intersection graphs. arXiv e-prints pp. arXiv–2304 (2023)
7. Dvořák, Z., Norin, S.: Treewidth of graphs with balanced separations. J. Comb. Theory, Series B **137**, 137–144 (2019). https://doi.org/10.1016/j.jctb.2018.12.007
8. Fomin, F.V., Lokshtanov, D., Saurabh, S.: Excluded grid minors and efficient polynomial-time approximation schemes. J. ACM (JACM) **65**(2), 1–44 (2018)
9. Fox, J., Pach, J.: Touching stings (2012). private communication
10. Kratochvíl, J.: Intersection graphs of noncrossing arc-connected sets in the plane. In: North, S. (ed.) GD 1996. LNCS, vol. 1190, pp. 257–270. Springer, Heidelberg (1997). https://doi.org/10.1007/3-540-62495-3_53

11. Lee, J.R.: Separators in region intersection graphs. In: Papadimitriou, C.H. (ed.) 8th Innovations in Theoretical Computer Science Conference (ITCS 2017). Leibniz International Proceedings in Informatics (LIPIcs), vol. 67, pp. 1:1–1:8. Schloss Dagstuhl–Leibniz-Zentrum fuer Informatik, Dagstuhl, Germany (2017). https://doi.org/10.4230/LIPIcs.ITCS.2017.1
12. Lokshtanov, D., Panolan, F., Saurabh, S., Xue, J., Zehavi, M.: Subexponential parameterized algorithms on disk graphs (extended abstract). In: Proceedings of the 2022 Annual ACM-SIAM Symposium on Discrete Algorithms (SODA), pp. 2005–2031. SIAM (2022)

Oriented Trees in $O(k\sqrt{k})$-Chromatic Digraphs, a Subquadratic Bound for Burr's Conjecture

Stéphane Bessy, Daniel Gonçalves, and Amadeus Reinald(✉)

LIRMM, Université de Montpellier, CNRS, Montpellier, France
{stephane.bessy,daniel.goncalves,amadeus.reinald}@lirmm.fr

Abstract. In 1980, Burr conjectured that every directed graph with chromatic number $2k-2$ contains any oriented tree of order k as a subgraph. Burr showed that chromatic number $(k-1)^2$ suffices, which was improved in 2013 to $\frac{k^2}{2} - \frac{k}{2} + 1$ by Addario-Berry et al.

We give the first subquadratic bound for Burr's conjecture, by showing that every directed graph with chromatic number $8\sqrt{\frac{2}{15}}k\sqrt{k} + O(k)$ contains any oriented tree of order k. Moreover, we provide improved bounds of $\sqrt{\frac{4}{3}}k\sqrt{k} + O(k)$ for arborescences, and $(b-1)(k-3)+3$ for paths on b blocks, with $b \geq 2$.

Keywords: Directed Graphs · Graph Colouring

1 Introduction

In essence, many of the questions in structural graph theory ask for necessary conditions for a (di)graph to contain some other given (di)graph. A natural way to specify this question is to ask, when imposing that a (di)graph G has chromatic number c, which (di)graphs are necessarily contained in G. A well-known result by Erdős [9] states that there exist graphs with arbitrarily large girth and chromatic number. Therefore, all (di)graphs one can hope to be necessarily contained in (di)graphs of chromatic number c are (directed) trees. This question is easily settled when the graph is undirected, as all k-chromatic graphs contain all trees of size k, which is tight by considering the clique K_k. When asking for *induced* subtrees instead, and additionally forbidding a fixed clique, this is the notorious Gyárfás–Sumner conjecture [11,19], which is still wide open.

In this paper, we investigate the directed variant of the question, asking which directed trees are contained as (non induced) subgraphs in digraphs with sufficiently large chromatic number. We refer the reader to [14] for lecture notes on the topic. A digraph is *c-universal* for some value $c > 0$ if every digraph with chromatic number at least c contains it as a subdigraph. Since the chromatic number of a digraph does not decrease when replacing digons with arcs, if H is c-universal in the class of oriented graphs, it is also c-universal. Moreover,

D. Král' and M. Milanič (Eds.): WG 2024, LNCS 14760, pp. 79–91, 2025.
https://doi.org/10.1007/978-3-031-75409-8_6

considering oriented graphs with arbitrarily large girth and chromatic number yields that H must be an oriented tree. We are thus interested in upper bounds on the universality of oriented trees. Burr posed the following conjecture in 1980.

Conjecture 1 (Burr [4]*).* Every oriented tree of order k is $(2k-2)$-universal.

Observe that this is the best one can ask for, as the out-star on k vertices S_k^+ is not contained in the regular tournament of order $2k-3$. In the same paper, Burr showed all oriented trees are $(k-1)^2$-universal. This stood as the only bound for the problem for decades, until Addario-Berry, Havet, Linhares-Sales, Reed and Thomassé gave the following improvement in 2013.

Theorem 2 (Addario-Berry et al. [1]). *Every oriented tree on k vertices is $(\frac{k^2}{2}-\frac{k}{2}+1)$-universal.*

Until now, this has remained the best bound for the general case. Improvements have only been obtained when asking for particular oriented trees, or by further restricting the directed graphs on which one imposes the chromatic number requirement. We provide the first subquadratic bound for Burr's conjecture, by showing the following (with $8\sqrt{2/15} \simeq 2.92$).

Theorem 3. *Every k-vertex oriented tree is $(8\sqrt{\frac{2}{15}}k\sqrt{k}+\frac{11}{3}k+\sqrt{\frac{5}{6}}\sqrt{k}+1)$-universal.*

Moreover, we give improved functions for the case of arborescences, showing they are $(\sqrt{4/3}\cdot k\sqrt{k}+k/2)$-universal (where $\sqrt{4/3}\simeq 1.15$).

Regarding results on specific oriented trees, a folklore observation yields the $(2k-2)$-universality of out-stars and in-stars. Indeed, a digraph forbidding S_k^+ (or S_k^-) is necessarily $(2k-4)$-degenerate, yielding chromatic number at most $2k-3$. One of the standard results on the matter is the Gallai-Roy-Hasse-Vitaver Theorem [10,12,18,20], dating from 1962, and stating that the directed path of order k are k-universal. A natural extension is the case of *b-blocks* paths, which can be arc-partitioned into b maximal directed paths. In 2007, Addario-Berry et al. proved the following, improving on a result by El Sahili [7].

Theorem 4 (Addario-Berry et al. [2]). *Paths on k vertices with 2 blocks are k-universal.*

A linear bound has also been achieved for 3-block paths by El Joubbeh [5], and for 4-block paths by El Joubbeh and Ghazal [6]. In the same paper, they improve the general bound for b-block paths when b is small compared to k. We provide the first general linear bounds, with b fixed, for paths with b blocks.

Theorem 5. *For $b\geq 2$, every b-block path on k vertices is $((b-1)(k-3)+3)$-universal.*

Along the way, this settles the conjecture for 3-block paths, giving us a bound of $2k-3$. Nevertheless, reflecting the idea that the case of oriented paths appears easier, Havet [13] conjectured that every oriented tree of order k with ℓ-leaves is $k + \ell + 1$-universal. Another notable example where the bound is known to be linear is when looking for antidirected trees [1].

When considering the question over restricted classes of graphs, only a few cases are known. For instance, acyclic digraphs [1], or more generally bikernel-perfect digraphs, satisfy the conjecture. This is the building block for the $k^2/2 + O(k)$ bound of [1], and will also be key in our proof of Theorem 3. Perhaps the most fruitful restriction to date is Sumner's conjecture [19], predating Burr's conjecture, and stating that every tournament of order $2k - 2$ contains every oriented tree of order k. A linear bound of $3k - 3$ was achieved by El Sahili [8], and the conjecture has been proved for sufficiently large k by Kühn and Osthus [15]. Another restriction of interest are digraphs with sufficiently large chromatic number compared to their order, as shown by Naia [16].

Roadmap of Our Strategy. A natural technique when trying to establish bounds for the universality of trees is to proceed by induction, usually starting from a star or a path. A step of the induction goes as follows: given an oriented tree T' which we know to be c'-universal, we consider a tree T obtained by "gluing" simple subtrees to T'. Here, gluing or appending a tree T'' to the tree T' means considering the disjoint union of T' and T'' and identifying one vertex of T'' to one vertex of T'. We then derive a bound for the universality of T in terms of $|T'|$, c', and the structure being glued. It was shown in [1] that augmenting T' with ℓ out-neighbours appended to arbitrary vertices yields a tree T that is $(c' + 2|T| - 4)$-universal. This already gives a quadratic bound for the conjecture, where the worst cases are given by orientations of paths, and more generally trees with few leaves.

The cornerstone of our contributions is the establishment of such a gluing lemma appending oriented paths, and an improved version for directed paths. We leverage this in the following win-win strategy when deriving universality bounds for some T by induction. If T has many leaves (more than $O(\sqrt{k})$), we apply the gluing lemma of [1] to the tree obtained by removing the leaves of T. Otherwise, we observe T is decomposable into few oriented paths, which we can glue one by one a limited number of times through Lemma 14.

Further Directions. Our ability to control the growth of our universality bound for trees as $O(k\sqrt{k})$ stems from our construction of trees and corresponding bounds by gluing both paths and leaves. In the extreme cases where a tree can be built by adding many leaves at each step, or by gluing few paths, better bounds can be obtained (see Proposition 8 in [1] for leaves). Nevertheless, each gluing operation to some tree T increases the bound by at least $|T|$. Thus, if a tree falls between these two categories, such as a star of degree $\sqrt{k}$ subdivided $\sqrt{k}$ times, our current tools do not allow us to improve on the $O(k\sqrt{k})$ bound. Adapting one of the gluing lemmas to get rid of this $O(|T|)$ cost may skew our win-win strategy towards that type of gluing, and yield a better bound.

Structure of the Paper. In Sect. 2, we introduce various technical tools, and the gluing lemma for leaves. In Sect. 3, we prove our gluing lemma for directed paths, deriving a $(b-1)(k-3)+3$ bound for b-block paths in Theorem 5, and a $\sqrt{4/3} \cdot k\sqrt{k}+O(k)$ bound for arborescences in Theorem 12. In Sect. 4, we prove our gluing lemma for oriented paths and derive Theorem 3, showing that oriented trees of order k are $8\sqrt{2/15} \cdot k\sqrt{k}+O(k)$-universal. The proofs for arborescences and general oriented trees follow the same line, with only the technical lemmas differing in nature.

2 Preliminaries

Throughout the paper, all graphs and digraphs we consider are simple. In general, we follow standard terminology, which can be found in Bang-Jensen and Gutin [3]. The *order* $|G|$ of a (di)graph G is the size of its vertex set. A *digon* in a digraph is an oriented cycle of length 2. Digraphs may contain digons, whereas *oriented graphs* do not contain digons. A path with b blocks is an oriented path whose arcs can be partitioned into b maximal directed subpaths, or blocks. When clear from the context, we may say tree or path to refer to an oriented tree or oriented path. A (proper) *colouring* of a (di)graph G is an assignment of colours to V such that any pair of adjacent vertices is coloured differently. Then, the *chromatic number* of G, denoted $\chi(G)$, is the minimal number of colours required in such a colouring. A *directed acyclic graph*, *DAG* for short, is a directed graph containing no directed cycles. Given a digraph D, an *in-kernel* (respectively *out-kernel*) of D is an independent K set that in-dominates (respectively out-dominates) D. That is, every vertex in $V(D)-K$ admits an in-neighbour (respectively out-neighbour) in K. It is well-know that any DAG contains an in-kernel (and an out-kernel) [17].

2.1 Universality in Directed Acyclic Graphs

The following lemma is an easy extension of a result from [1] stating that oriented trees on k vertices are k-universal in the class of acyclic digraphs. For the sake of completeness, we provide its proof below.

Lemma 6. *Let T be an oriented tree on k vertices where one vertex r is specified as the root of T. Let also D be an acyclic digraph with chromatic number at least k. Then D admits a partition (X, K) of its vertex set with $\chi(D[K]) \leq k-1$ and such that for every vertex x of X, there exists a copy of T in D where r is identified to x and all other vertices of T lie in K.*

Proof. Let us prove the result by induction on k. If $k=1$, then we choose $X=V(D)$ and $K=\emptyset$. Now, consider $k>1$ and a tree T on k vertices with root r. As $k>1$, there exists a leaf l in T different from r. Denote $T \setminus \{l\}$ by T' and let l' be the (only) neighbour of l in T. Without loss of generality, we can assume that l' is an in-neighbour of l, the case of an out-neighbour being symmetrical. Let K_0 be an in-kernel of D, which exists as D is acyclic. Since K_0 is independent,

$\chi(D \setminus K_0) \geq k-1$, for otherwise colouring K_0 with a single colour would yield $\chi(D) \leq k-1$. Then, the induction hypothesis provides a partition (X', K') of $V(D \setminus K_0)$ for the tree T'. Now, let us show that $(X = X', K = K' \cup K_0)$ is the desired partition of D for T.

First, note that $\chi(D[K]) \leq \chi(D[K']) + \chi(D[K_0]) \leq (k-2) + 1 = k-1$. Moreover, for any vertex x in X, there exists a copy of T' in $D \setminus K_0$ where r is identified to x and the other vertices of T' lie in K'. In particular l' is in K' and as K_0 is an in-kernel of D, l' has an out-neighbour in K_0 to which we can identify l. □

2.2 Gluing Leaves

We introduce the gluing lemma of [1], the first example of deriving a universality bound for a tree constructed by some other tree for which we have a bound. Let T be an oriented tree. Let u be a leaf of T and u' its unique neighbour in T. If the arc between u and u' is oriented toward u, we say that u is an *out-leaf* of T, otherwise, we say that u is an *in-leaf* of T. Denote by $Out(T)$ (resp. $In(T)$) the set of all the out-leaves (resp. in-leaves) of T.

Lemma 7 (Addario-Berry et al. [1]). *For $k \geq 3$, let T be an oriented tree of order k other than S_k^+ or S_k^-. If $T - Out(T)$, respectively $T - In(T)$ is c-universal, then T is $(c + 2k - 4)$ universal.*

The next corollary follows by performing two gluing operations in a row, one for $Out(T)$ and one for $In(T)$, and since S_k^+ and S_k^- are $2k-2$-universal.

Corollary 8. *Let T be an oriented tree of order k other than S_k^+ or S_k^-. Let $L(T)$ denote the set of all the leaves of T. If $T - L(T)$ is c-universal, then T is $(c + 4k - 8)$-universal.*

2.3 Decomposing Trees Into Paths

The following decomposition is rather classical, we state it for sake of completeness. Let T be an undirected tree, rooted in some vertex r. A path $P = v_1, \ldots, v_\ell$ of T is *descending* if v_i is the parent of v_{i+1} in T for $i = 1, \ldots, \ell - 1$. Then, we consider v_1 as the *beginning* of P.

Lemma 9. *Let T be an unoriented rooted tree with p leaves. Then, there exist p descending paths $P_1, \ldots, P_p$ of T such that for every $i = 1, \ldots, p-1$ the path P_{i+1} and $T_i = \bigcup_{j=1}^{i} P_j$ intersect in the beginning of P_{i+1}, and $T = T_p = \bigcup_{j=1}^{p} P_j$.*

Proof. To start, consider a path P_1 from the root of T to any of its leaf l_1. It forms a descending path of T Then, $P_1, \ldots, P_i$ being defined, with $i < p$, there exists a leaf l_{i+1} of T not contained in $T_i = \bigcup_{j=1}^{i} P_j$. As T_i contains the root of T, there exists a descending path P_{i+1} from T_i to l_{i+1}. Notice that T_i and P_{i+1} only intersect on the beginning of P_{i+1}. When all $P_1, \ldots, P_p$ have been defined, it is clear that $T_p = T$. □

3 Arborescences and b-block Paths

In this section, we tackle the restriction of T to an arborescence or a path on b blocks. Consider a tree T' for which we have a universality bound c'. We wish to obtain a universality bound for the tree T obtained by appending a directed path P_ℓ to T'. It turns out that gluing directed paths as such is considerably simpler and "cheaper" than arbitrary orientations of paths. This improves on the general bound for the cases above, and more generally for any oriented tree that can be decomposed into few directed paths.

3.1 Gluing a Directed Path

We first show a technical result, which provides us with all the structure required to prove the gluing lemma for directed paths.

Lemma 10. *For every integer $\ell \geq 0$, and every digraph D, there exists a partition of $V(D)$ into sets X, Y, Z such that:*

(X) For every vertex $x \in X$ there exists $y^+ \in N^+(x) \cap Y$, and $y^- \in N^-(x) \cap Y$ such that there exists a y^+y^--directed path whose vertices are all in Y.
(Y) $D[Y]$ is a DAG, and for every sink y of $D[Y]$, there exists a directed path of length ℓ starting at y and whose remaining vertices are all in Z.
(Z) $\chi(D[Z]) \leq \ell$.

Symmetrically, there also exists a partition (X, Y, Z) for D satisfying (X), (Z) and the following in place of (Y):

$(\overleftarrow{Y})$ $D[Y]$ is a DAG, and for every source y of $D[Y]$, there exists a directed path of length ℓ ending at y and whose remaining vertices are all in Z.

Proof. Throughout this proof, we replace D with an oriented subgraph obtained by replacing each digon of D with an arc arbitrarily. The resulting partition will also satisfy the conditions in the initial digraph. Note that to obtain the second partition, it is enough to reverse all the arcs of D and consider the first partition on this digraph. Therefore, we focus on obtaining the first partition and proceed by induction on ℓ to do so. For $\ell = 0$, let $Z = \emptyset$, let Y be the vertex set of a maximal induced acyclic subdigraph of D, and let $X = V(D) \setminus Y$. Properties (Y) and (Z) hold trivially. For property (X), note that by maximality of Y, for any vertex $x \in X$ the graph $D[Y \cup \{x\}]$ is not acyclic. It thus contains a cycle passing through x. Removing x from this cycle, and taking y^-, y^+ its neighbours on it provides the directed path of property (X).

Assume by induction that the lemma holds for a fixed $\ell \geq 0$, and let us denote X', Y', Z' the corresponding subsets of $V(D)$, satisfying properties (X'), (Y') and (Z'). We will build X, Y, Z satisfying the lemma for $\ell + 1$. Let $S' \subseteq Y'$ be the set of vertices that are sinks in $D[Y']$, note that S' is a stable set in $D[Y']$, and thus also in D. Let $Z = Z' \cup S'$, and note that $\chi(D[Z]) \leq \chi(D[Z']) + 1 \leq \ell + 1$, ensuring (Z) holds.

Note that $D[Y' \setminus S']$ remains acyclic, and that all its sinks have an out-neighbour in S' which was necessarily a sink in $D[Y']$. Now, all such out-neighbours are the beginning of a directed path of length ℓ whose vertices are all contained in Z', by induction. Therefore, every sink of $D[Y' \setminus S']$ is the beginning of some directed path of length $\ell+1$, whose remaining vertices are all in Z. Now, observe that $Y' \setminus S'$ is an in-dominating set of X'. Indeed, for every $x \in X'$ the vertex $y^+ \in Y'$ given by property (X') is actually a vertex of $Y' \setminus S'$. Indeed, we assumed D is oriented, so $y^+ \neq y^-$ and the y^+y^--directed path ensures that y^+ is not a sink in $D[Y']$.

We define Y as a vertex set containing $Y' \setminus S'$, and which induces a maximal directed acyclic graph in $D[V \setminus Z]$. We have already noted that $D[Y' \setminus S']$ is acyclic, thus Y can be obtained by iterative addition of vertices in X' while maintaining that $D[Y]$ is acyclic. After having considered all the vertices of X', we are guaranteed that for every vertex x of $X = X' \setminus Y$, $D[Y \cup \{x\}]$ contains a cycle going through x. Removing x from it yields the directed path between an out-neighbour y^+ and an in-neighbour y^- of x, ensuring (X).

For property (Y), we have to note that no vertex in $Y \cap X'$ is a sink in $D[Y]$. Indeed, as $Y' \setminus S'$ in-dominates X', every vertex of $Y \cap X'$ has an out-neighbour in $Y' \setminus S' \subseteq Y$. Now, since the sinks of $D[Y]$ were already sinks in $D[Y' \setminus S']$, and since, as noted above, these vertices begin a directed path of length $\ell + 1$ with all remaining vertices in Z, property (Y) holds. □

We are now ready to prove the gluing lemma for directed paths.

Lemma 11. *Let T' be an oriented tree of order $k' \geq 1$, and let T be the oriented tree obtained by appending a directed path of length ℓ from any of its endpoints to some vertex of T'. If T' is c'-universal, then T is $(c' + k' + 2\ell - 3)$-universal.*

Proof. Let T' be any oriented tree of order k', and consider a digraph D with chromatic number at least $c' + k' + 2\ell - 3$. We will first show how to append a directed path of length ℓ to T' by identifying the source of the path to some vertex of T'. We start by applying Lemma 10 to D with $\ell - 2$ in place of ℓ, and let X, Y, Z be the corresponding partition of $V(D)$. If $D[Y]$ has chromatic number at least $k' + \ell = |T|$, then by Lemma 6, it contains the tree T.

Otherwise, $D[Y]$ has chromatic number at most $k' + \ell - 1$. Since $D[Z]$ has chromatic number at most $\ell - 2$, we have that $D[X]$ has chromatic number at least $c' + k' + 2\ell - 3 - (k' + \ell - 1) - (\ell - 2) = c'$. By hypothesis $D[X]$ must contain a copy of T', which we identify with T' implicitly, and we denote by x the vertex on this copy corresponding to the vertex of T' to which we wish to identify the source of the directed path. Lemma 10 ensures us that x has an out-neighbour $y^+ \in Y$, and a y^+y^--directed path in $D[Y]$, giving us the path $(x, y^+, ..., y^-)$. If y^- is not a sink, we may follow a directed path from y^- towards a sink s of $D[Y]$. Then, property (Y) yields a path P of length $\ell - 2$ starting in s and wholly contained in Z. Now, $(x, y^+, ..., y^-, ..., s)$ along with the path P yield a directed path of length at least ℓ starting in x. Considering T' in X along with the path obtained from x yields the desired copy of T in D. The case is symmetrical when appending a directed path from its sink, using the second kind of partition and property $(\overleftarrow{Y})$ in Lemma 10. This achieves to show the lemma. □

3.2 Growing Arborescences and b-block Paths

With the gluing lemma for directed paths in hand, the proof for b-block paths is now straightforward. We start with the bound for 2-block paths via Theorem 4, and glue $b-2$ paths one by one using Lemma 11.

Theorem 5. *For $b \geq 2$, every b-block path on k vertices is $((b-1)(k-3)+3)$-universal.*

Proof. We proceed by induction on b. The case $b = 2$ holds by Theorem 4. We thus assume that P has $b \geq 3$ blocks, and let P' be a subpath of P obtained by deleting one of the end-blocks of P, that is, a block containing an end-vertex of P. We let ℓ be the length of this end-block, and note that P' has $k-\ell$ vertices.

By induction hypothesis, every graph with chromatic number at least $(b-2)(k-\ell-3)+3$ contains a copy of P'. As P is obtained from P' by appending a directed path of length ℓ, Lemma 11 yields that every graph with chromatic number at least $(b-2)(k-\ell-3)+3+(k-\ell)+2\ell-3$ contains a copy of P. Then, we have $(b-2)(k-\ell-3)+3+(k-\ell)+2\ell-3 \leq (b-1)(k-\ell-3)+2\ell+3$. Since $2\ell+3 \leq (b-1)\ell+3$ as $b \geq 3$, we finally obtain $(b-1)(k-\ell-3)+2\ell+3 \leq (b-1)(k-3)+3$. This achieves to show that b-block paths on k vertices are $((b-1)(k-3)+3)$ universal. □

We now move on to the case of arborescences. The proof is by induction, and follows the same win-win argument as the general case. At a given step, if there are more than $O(\sqrt{k})$ leaves, we use the gluing lemma for leaves, otherwise we leverage our gluing lemma for paths.

Theorem 12. *Every k-vertex arborescence is $(\sqrt{\frac{4}{3}}k\sqrt{k}+\frac{k}{2})$-universal.*

Proof. We show the result for out-arborescences, the case of in-arborescences being symmetrical. Let us proceed by induction on k, the number of vertices in T, distinguishing between two cases with respect to p the number of leaves of T.

For any k, denote $\sqrt{4/3}\cdot k\sqrt{k}+k/2$ by $g(k)$. Notice we can trivially assume that $k \geq 2$. And if T is the out-star S_k^+, recall it is $2k-2$-universal, and thus $g(k)$-universal (since $2k-2 \leq g(k)$). Now, consider a digraph D with $\chi(D) \geq g(k)$ and T an out-arborescence on $k \geq 3$ vertices and p leaves, different from S_k^+.

Assume first that $p < \sqrt{4k/3}$. Using Lemma 9, T decomposes into p descending paths $P_1, \ldots, P_p$. As T is an out-arborescence here, every P_i is a directed path. Again, we denote $\bigcup_{j=1}^{i} P_j$ by T_i and know that T_i and P_{i+1} only intersect at the beginning of P_{i+1}. Now, we show by induction on i that T_i is $(i \cdot |T_i|)$-universal. For $i=1$, as T_1 is a directed path, it is $|T_1|$-universal by the Gallai-Roy-Hasse-Vitaver Theorem. Assuming induction step for some $i < p$, Lemma 11 implies that T_{i+1} is $(i \cdot |T_i| + |T_i| + 2l_{i+1} - 3)$-universal, where l_{i+1} is the length of the path P_{i+1}. As $|T_{i+1}| = |T_i| + l_{i+1}$, we obtain that $i \cdot |T_i| + |T_i| + 2l_{i+1} - 3 \leq (i-1) \cdot |T_i| + 2(|T_i| + l_{i+1}) \leq (i-1) \cdot |T_{i+1}| + 2(|T_{i+1}|) = (i+1) \cdot |T_{i+1}|$. Thus we conclude that T_{i+1} is $(i+1) \cdot |T_{i+1}|$-universal, proving step $i+1$ of the induction. Finally we obtain that T_p is $p \cdot k$-universal. As the chromatic number of D is at least $g(k)$ and $p \cdot k < (\sqrt{4k/3})k < g(k)$ we find a copy of T in D in this case.

Now, assume that $p \geq \sqrt{4k/3}$. Let T' be the tree obtained from T by removing its leaves, that is $T' = T - Out(T)$. By induction, every digraph with chromatic number greater than $g(k - \sqrt{4k/3}) \geq g(k')$ contains a copy of T', where k' is the number of vertices of T'. Hence, by Lemma 7, as $T \neq S_k^+$, every digraph with chromatic number at least $g(k - \sqrt{4k/3}) + 2k - 4$ contains a copy of T. To conclude this case, we have to check that $g(k - \sqrt{4k/3}) + 2k - 4 \leq g(k)$. This is numerically verified for all $k \geq 2$, concluding the proof. □

4 Oriented Trees

In this section, we adapt the previous techniques to append any oriented path to a tree, generalizing the case of directed paths above.

4.1 Gluing an Oriented Path

Given an oriented path Q, we obtain a *rooted oriented path* from Q by selecting one of the extremities of Q, called the *root* of Q. Now, given a rooted oriented path Q, with root r, a digraph D and a vertex x of D, we say that D contains a copy of Q *starting at* x if D contains a copy of Q where r is identified to x.

The next two lemmas are the respective counterparts of Lemmas 10 and 11. In both of those, the obtained bounds now depend quadratically on the length of the path being appended, where this dependency was linear in the last section.

Lemma 13. *Let $\ell \geq 0$ be an integer, and Q a rooted oriented path of length ℓ. Then, for every digraph D, there exists a partition of $V(D)$ into sets X, Y, Z such that:*

(X) Every vertex $x \in X$ admits both an in-neighbour and an out-neighbour in Y,
(Y) Y induces a directed acyclic graph, and for every vertex $y \in Y$, there exists a copy of Q starting at y and contained in $D[Y \cup Z]$,
(Z) $\chi(D[Z]) \leq \frac{\ell(\ell+1)}{2}$.

Proof. We prove the result by induction on ℓ. For $\ell = 0$, let $Z = \emptyset$, Y be the vertex set of a maximal acyclic subdigraph of D, and $X = V(D) \setminus Y$. Properties (Y) and (Z) hold trivially, while property (X) follows from the maximality of $D[Y]$ as an acyclic subdigraph of D.

Assume the result holds for any rooted oriented path P of length $\ell \geq 0$. To show it holds for $\ell + 1$, let us consider a rooted oriented path Q of length $\ell + 1$ rooted in r. Let r' be the unique neighbour of r in Q and Q' be the oriented path

$Q - r$ rooted in r'. We will ensure property (Y) for $\ell + 1$ by appending either an in or out neighbour to some copy of Q' obtained by induction. Consider subsets $X', Y', Z' \subseteq V(D)$ satisfying the properties of the lemma for Q' by induction. Let us sketch the strategy of an inductive step. To exhibit the sets X, Y, Z satisfying the properties for Q, we first define Z by starting from Z', and adding a subset of vertices from Y' such that every vertex remaining in Y' is the endpoint of a subpath Q (in $D[Z \cup Y]$). Then, we will build Y by starting from the vertices remaining in Y' (later Y''), then adding a subset of vertices of X' ensuring that the remaining vertices in X', now forming X, all admit an in- and an out-neighbour towards Y. Finally, we must also ensure that these vertices newly added to Y are also the beginning of some Q (in $Z \cup Y$).

Let us start by defining Z. If $\chi(D[Y']) \leq \ell + 1$, we let $Z = Z' \cup Y'$, and define $Y'' = \emptyset$ the remaining vertices of Y'. Otherwise, Y' induces an acyclic digraph with $\chi(D[Y']) \geq \ell + 2$. We may now apply Lemma 6 to $D[Y']$, with Q as the oriented tree T rooted in r (note that $|Q| = \ell + 2$). This gives us a subset $K \subseteq Y'$ with $\chi(D[K]) \leq \ell + 1$ and such that every $y \in Y' \setminus K$ is the beginning of some Q whose remaining vertices belong to K. We let $Z = Z' \cup K$, and define $Y'' = Y' \setminus K$ to be the vertices left in Y'. Observe that in both cases above, Z already satisfies property (Z). Indeed, by induction we have $\chi(D[Z']) \leq \frac{\ell(\ell+1)}{2}$, and we have either $\chi(D[Y']) \leq \ell + 1$ or $\chi(D[K]) \leq \ell + 1$, ensuring $\chi(D[Z]) \leq \frac{\ell(\ell+1)}{2} + \ell + 1 = \frac{(\ell+1)(\ell+2)}{2}$.

We now turn to defining Y and X. Consider a subset $W \subseteq X'$ chosen to be maximal such that $Y'' \cup W$ still induces an acyclic digraph, and let $Y = Y'' \cup W$. Then, we let $X = X' \setminus W$. At this point, observe that X already satisfies property (X). Indeed, any $x \in X$ having only out-neighbours or in-neighbours in Y could be added to W, then Y, without creating any directed cycles in Y, contradicting the maximality of W.

What is left to show is that property (Y) holds. To do so, consider any $y \in Y$, recall $Y = Y'' \cup W$ and let us deal with two cases. If $y \in Y''$, we are in the case where $\chi(D[Y']) \geq \ell + 1$. Then, Lemma 6 yields that y must be the beginning of a subpath Q whose remaining vertices lie in $K \subseteq Z$, so Q is contained in $D[Y \cup Z]$, ensuring (Y) here. Otherwise, $y \in W$, and we know that $y \in X'$. Therefore y admits both an in-neighbour y^- and an out-neighbour y^+ in Y' by induction. Now, we leverage the induction to yield a copy of Q', starting in either y^- or y^+, whose vertices all belong to $Y' \cup Z'$. Note in particular that this subpath cannot contain y since $W \cap (Y'' \cup Z') = \emptyset$. Therefore, according to the orientation of the arc between r and r' in Q, either (y, y^+) or (y, y^-) along with the copy of Q' yields the desired copy of Q, ensuring (Y). □

Lemma 14. *Let T' be an oriented tree of order $k' \geq 1$, and let T be the oriented tree obtained by appending a rooted oriented path of length ℓ to any vertex of T'. If T' is c'-universal, then T is $(c' + k' + \frac{\ell(\ell+1)}{2} - 1)$-universal.*

Proof. Let us consider a digraph D with chromatic number at least $c' + k' + \ell(\ell + 1)/2 - 1$, and Q the rooted oriented path of length ℓ we want to append to T. We denote by r the root of Q and by r' its unique neighbour in Q. Let Q'

be the oriented path of length $\ell - 1$ obtained by deleting r from Q and rooted in r'. We start by applying Lemma 13 to D with the rooted path Q', letting X, Y, Z be the corresponding partition of $V(D)$. If $D[Y]$ has chromatic number at least $k' + \ell = |T|$, since it is a directed acyclic graph, it contains the tree T by Lemma 6.

Otherwise, $\chi(D[Y]) \leq k' + \ell - 1$, and we also know $\chi(D[Z]) \leq (\ell - 1)\ell/2$ by Lemma 13. Therefore, $D[X]$ has chromatic number at least $(c' + k' + \ell(\ell + 1)/2 - 1) - (k' + \ell - 1) - (\ell - 1)\ell/2 = c'$. By hypothesis $D[X]$ must contain a copy of T', and we denote x the vertex on this copy where Q has to start to form T. Lemma 10 ensures us that x has both an in-neighbour y^- and an out-neighbour y^+ in Y. Then, both y^- and y^+ are the beginning of a rooted copy of Q' contained in $Y \cup Z$. Therefore, according to the orientation of the arc between r and r' in Q, we obtain a rooted copy of Q starting in x with either (y^-, x) or (y, x^+), with all remaining vertices in $Y \cup Z$. Except from x, this copy of Q is disjoint from the copy of T' obtained in x, yielding a copy of T in D. □

4.2 Growing Oriented Trees

We are now ready to prove the main result by leveraging the gluing lemma for oriented paths.

Theorem 3. *Every k-vertex oriented tree is $(8\sqrt{\frac{2}{15}}k\sqrt{k} + \frac{11}{3}k + \sqrt{\frac{5}{6}}\sqrt{k} + 1)$-universal.*

Proof. We set $f(k) = 8\sqrt{2/15} \cdot k\sqrt{k} + 11k/3 + \sqrt{5/6}\sqrt{k} + 1$.
First notice that the statement is trivially true for $k = 1$ and $k = 2$. Indeed, we have $f(1) \geq 1$ and $f(2) \geq 2$. On the other hand, if T is an out-star S_k^+ or an in-star S_k^-, it is $(2k - 2)$-universal by degeneracy. One can verify numerically that $f(k) \geq 2k - 2$ which yields $f(k)$-universality.

Let us now proceed by induction on $k = |T|$, the number of vertices in T. We can thus assume $k \geq 3$, and take any T of size k different from S_k^+ and S_k^-. As in Theorem 12, we use two strategies according to the number p of leaves in T.

Assume first that the number p of leaves in T is at least $\sqrt{5k/6}$. Consider the subtree T' of T obtained by removing all of its leaves, and let $k' = |T'| \leq k - \sqrt{5k/6}$. By induction we know that T' is $f(k - \sqrt{5k/6})$-universal, yielding via Corollary 8 that T is $(f(k - 5k/6) + 4k - 8)$-universal. To show that T is $f(k)$-universal, it thus suffices to satisfy $f(k) \geq f(k - \sqrt{5k/6}) + 4k - 8$, which we numerically check for all $k \geq 1$.

From now on, we may thus assume that T contains less than $\sqrt{5k/6}$ leaves, that is $p < \sqrt{5k/6}$. Our goal is to split T into paths, which we use to build T by using Lemma 14. Let us first apply Lemma 9, taking any vertex of T as the root, to obtain a decomposition of the graph underlying T into descending paths. This yields descending paths $Q_1, \ldots, Q_p$, which we consider with their initial orientation in T, such that $T = \bigcup_{j=1}^{p} Q_j$, and which we root in their endpoint closest to r in T. Note then that T can be obtained recursively by starting from $T_1 = Q_1$, and building T_{i+1} by identifying the root of Q_{i+1} to some vertex of T_i.

At this point, the number of paths is bounded by p, but we also need to bound their length before applying Lemma 14. We therefore cut paths of length exceeding $\ell = \lceil\sqrt{6k/5}\rceil \leq \sqrt{6k/5}+1$ into smaller ones as follows. For $j \in [1,p]$, we partition the arcs of Q_j into a minimal number of consecutive subpaths $(Q_{j,h})_h$, such that all have length exactly ℓ except for a possible last path having length strictly less than ℓ. Let us consider $(P_i)_{i\in[1,m]}$ to be the m resulting paths.

To bound the number m of newly created paths $(P_i)_i$, we count those of length strictly less than ℓ separately from those with length exactly ℓ. By construction, at most one path of length strictly less than ℓ is created per Q_j, so there are at most $p < \sqrt{5k/6}$ of them. To bound the number of paths of length exactly $\ell = \lceil\sqrt{6k/5}\rceil$, since $|A(T)| = k-1$ and since the paths $(P_i)_i$ are arc-disjoint, there must be at most $\lfloor(k-1)/\lceil\sqrt{6k/5}\rceil\rfloor \leq \sqrt{5k/6}$ of them. Finally, $(P_i)_i$ consists of $m \leq 2\sqrt{5k/6}$ paths at most, each of length at most $\ell \leq \sqrt{6k/5}+1$. Observe that T can still be constructed recursively using $(P_i)_i$. Now, we have a way to build T starting with $T_1 = P_1$, and constructing T_{i+1} from T_i by identifying an endpoint of P_{i+1} closest to r with a vertex of T_i.

To show T is $f(k)$-universal, we will recursively compute bounds $(c_i)_i$ such that T_i is c_i-universal. In turn, showing $f(k) \geq c_m$ will yield the result since $T_m = T$. Assume T_i, of order k_i, is c_i-universal. Recall T_{i+1} is obtained from T_i by appending a path of length at most ℓ to one of its vertices. According to Lemma 14, T_{i+1} is then c_{i+1}-universal for $c_{i+1} = c_i + k_i + \ell(\ell+1)/2 - 1$. Recall we ensured that $k \geq 3$ and T is different from S_k^+ and S_k^-, allowing us to apply the lemma. Now, since T_i is obtained from P_1, of order at most $\ell+1$, by i identifications with paths of length at most ℓ, we have $k_i \leq i\ell+1$. This yields:

$$c_{i+1} \leq c_i + i\ell + 1 + \frac{\ell(\ell+1)}{2} - 1 = c_i + i\ell + \frac{\ell(\ell+1)}{2}$$

$$c_{i+1} \leq c_1 + \ell(\sum_{j=1}^{i} j) + \frac{i\ell(\ell+1)}{2} \quad \text{by an immediate induction}$$

$$c_{i+1} \leq c_1 + \frac{1}{2}\ell i(\ell+i+2)$$

Now, since $|T_1| \leq \ell+1$ we have $c_1 \leq \ell(\ell+1)/2+1$ by Theorem 2. We set $i = m-1$, and with the inequalities $\ell \leq \sqrt{6k/5}+1$ and $m \leq 2\sqrt{5k/6}$ we obtain:

$$c_m \leq \frac{\ell(\ell+1)}{2} + 1 + \frac{1}{2}\ell(m-1)(\ell+m+1) \leq \frac{\ell m}{2}(\ell+m)+1$$

$$c_m \leq 8\sqrt{\frac{2}{15}}k^{3/2} + \frac{11}{3}k + \sqrt{\frac{5}{6}}\sqrt{k} + 1$$

By definition, $f(k) = c_m$, which achieves to prove that T is $f(k)$-universal and concludes the proof. □

Acknowledgments. This paper was partially supported by the ANR DIGRAPHS (ANR-19-CE48-0013). We would like to thank the anonymous referees for the constructive feedback and suggestions on an earlier version of the paper.

References

1. Addario-Berry, L., Havet, F., Linhares Sales, C., Reed, B., Thomassé, S.: Oriented trees in digraphs. Discret. Math. **313**(8), 967–974 (2013). https://doi.org/10.1016/j.disc.2013.01.011
2. Addario-Berry, L., Havet, F., Thomassé, S.: Paths with two blocks in n-chromatic digraphs. J. Comb. Theo. Ser. B **97**(4), 620–626 (2007). https://doi.org/10.1016/j.jctb.2006.10.001
3. Bang-Jensen, J., Gutin, G.: Digraphs. Springer Monographs in Mathematics. Springer, London (2009). https://doi.org/10.1007/978-1-84800-998-1
4. Burr, S.A.: Subtrees of directed graphs and hypergraphs. In: Proceedings of the Eleventh Southeastern Conference on Combinatorics, Graph Theory and Computing, Congressus Numerantium, vol. 28, pp. 227–239 (1980)
5. El Joubbeh, M.: On three blocks paths p(k, l, r). Discret. Appl. Math. **322**, 237–239 (2022). https://doi.org/10.1016/j.dam.2022.08.021, https://www.sciencedirect.com/science/article/pii/S0166218X22003250
6. El Joubbeh, M., Ghazal, S.: Existence of paths with t blocks in k (t)-chromatic digraph. Discret. Appl. Math. **342**, 381–384 (2024)
7. El Sahili, A.: Paths with two blocks in k-chromatic digraphs. Discret. Math. **287**(1), 151–153 (2004). https://doi.org/10.1016/j.disc.2004.03.013
8. El Sahili, A.: Trees in tournaments. J. Comb. Theo. Ser. B **92**(1), 183–187 (2004). https://doi.org/10.1016/j.jctb.2004.04.002
9. Erdős, P.: Graph theory and probability. Can. J. Math. **11**, 34–38 (1959). https://doi.org/10.4153/CJM-1959-003-9
10. Gallai, T.: On directed paths and circuits. In: Proceedings of the Colloquim held at Tihany, pp. 115–118 (1966)
11. Gyárfás, A.: On Ramsey covering-numbers. Infinite Finite Sets **2**, 801–816 (1975). https://users.renyi.hu/~gyarfas/Cikkek/05_Gyarfas_OnRamseyCoveringNumbers.pdf
12. Hasse, M.: Zur algebraischen begründung der graphentheorie. i. Mathematische Nachrichten **28**(5–6), 275–290 (1965)
13. Havet, F.: Trees in tournaments. Discrete Math. **243**(1–3), 121–134 (2002)
14. Havet, F.: Orientations and colouring of graphs (2013). https://oc.g-scop.grenoble-inp.fr/conf/sgt2013/oleron.pdf
15. Kühn, D., Mycroft, R., Osthus, D.: A proof of sumner's universal tournament conjecture for large tournaments. Proc. Lond. Math. Soc. **102**(4), 731–766 (2011). https://doi.org/10.1112/plms/pdq035, https://londmathsoc.onlinelibrary.wiley.com/doi/abs/10.1112/plms/pdq035
16. Naia, T.: Trees contained in every orientation of a graph. Electron. J. Combinatorics **29**(2), P2.26 (2022). https://doi.org/10.37236/10487
17. von Neumann, J., Morgenstern, O.: Theory of Games and Economic Behavior. Priceton University Press (1944)
18. Roy, B.: Nombre chromatique et plus longs chemins d'un graphe. Revue française d'informatique et de recherche opérationnelle **1**(5), 129–132 (1967)
19. Sumner, D.: Subtrees of a graph and the chromatic number. In: The Theory and Applications of Graphs: Fourth International Conference, May 6–9, 1980, Western Michigan University, Kalamazoo, Michigan. Wiley, New York (1981)
20. Vitaver, L.: Determination of minimal coloring of vertices of a graph by means of boolean powers of the incidence matrix. Doklady Akademii Nauk **147, 4**(4), 758–759 (1962)

Improved Outerplanarity Bounds for Planar Graphs

Therese Biedl[1(✉)] and Debajyoti Mondal[2]

[1] David R. Cheriton School of Computer Science, University of Waterloo, Waterloo, Canada
biedl@uwaterloo.ca

[2] Department of Computer Science, University of Saskatchewan, Saskatoon, Canada
d.mondal@usask.ca

Abstract. In this paper, we study the outerplanarity of planar graphs, i.e., the number of times that we must (in a planar embedding that we can initially freely choose) remove the outerface vertices until the graph is empty. It is well-known that there are n-vertex graphs with outerplanarity $\frac{n}{6} + \Theta(1)$, and not difficult to show that the outerplanarity can never be bigger. We give here improved bounds of the form $\frac{n}{2g} + 2g + O(1)$, where g is the *fence-girth*, i.e., the length of the shortest cycle with vertices on both sides. This parameter g is at least the connectivity of the graph, and often bigger; for example, our results imply that planar bipartite graphs have outerplanarity $\frac{n}{8} + O(1)$. We also show that the outerplanarity of a planar graph G is at most $\frac{1}{2}\text{diam}(G) + O(\sqrt{n})$, where $\text{diam}(G)$ is the diameter of the graph. All our bounds are tight up to smaller-order terms, and a planar embedding that achieves the outerplanarity bound can be found in linear time.

Keywords: Planar graphs · Outerplanarity · Fence girth · Diameter

1 Introduction

The *outerplanarity* of a planar graph is a well-known tool, both for deriving efficient algorithms and for proving lower bounds for graph drawings. It measures how often we have to remove the vertices on the outerface (a *peel*) until the graph is empty. (Detailed definitions are in Sect. 2.) In this paper, we obtain better upper bounds on the outerplanarity of a planar graph, which is important from the perspective of the following two application areas.

The first application of outerplanarity is to design faster algorithms for various problems in planar graphs. Baker [3] showed that for a planar graph with constant outerplanarity, numerous graph problems, such as independent set, vertex cover, dominating set can all be solved in linear time. (There are numerous generalizations, see e.g. [12,15,16,19,20].) The running times of many such algorithms have an exponential dependency on the outerplanarity or related parameters. Hence an upper bound on the outerplanarity with respect to the size of the graph can provide an estimate of how large of a graph these algorithms may be able to process in practice.

D. Kráľ and M. Milanič (Eds.): WG 2024, LNCS 14760, pp. 92–106, 2025.
https://doi.org/10.1007/978-3-031-75409-8_7

Another major application of outerplanarity is to derive lower bounds for various optimization criteria in graph drawing. For example, there exists a planar graph with a fixed planar embedding (known as nested triangles graph) that requires at least a $\frac{2}{3}n \times \frac{2}{3}n$-grid in any of its straight-line grid drawings that respect the given embedding (attributed to Leiserson [23] by Dolev, Trickey and Leighton [14]). Here a grid drawing maps each vertex to a grid point and each edge to a straight line segment between its end vertices. The crucial ingredient to their proof is that the nested triangles graph has $\frac{n}{3}$ peels (in this embedding), and any embedding-preserving planar straight-line grid-drawing of a planar graph with k peels requires at least a $2k \times 2k$-grid. (In fact, this lower bound holds for many other planar graph drawing styles [1,18,30].) Nested triangles graphs have outerplanarity $\frac{n}{6}$, and thus gives a lower bound of an $\frac{n}{3} \times \frac{n}{3}$-grid for the planar straight-line grid drawing even when one can freely choose an embedding to draw the graph. This raises a natural question of whether $\frac{n}{6}$ is the largest outerplanarity (perhaps up to lower-order terms) that a planar graph can have. This turns out to be true, via a detour into the radius, which we discuss next.

The *eccentricity* of a vertex v in G is the smallest integer k such that the shortest-path distance from v to any other vertex in G is at most k. The *radius* of G (denoted $\text{rad}(G)$) is the smallest eccentricity over all the vertices of G, while the *diameter* of G (denoted $\text{diam}(G)$) is the largest eccentricity. For 3-connected planar graphs, Harant [21] proved an upper bound of $\text{rad}(G) \leq \frac{n}{6} + \Delta^* + \frac{3}{2}$, where Δ^* is the maximum degree of the dual graph, i.e., the maximum length of a face. Ali et al. [2] improved the upper bound to $\frac{n}{6} + \frac{5\Delta^*}{6} + \frac{5}{6}$ and more generally $\frac{n}{2\kappa} + O(\Delta^*)$, where κ is the connectivity of the graph; these bounds are tight within an additive constant. This easily implies upper bounds on the outerplanarity. (Proofs of results marked with ($\star$) are in the full version [5].)

Observation 1 ($\star$). *Every planar graph G has outerplanarity at most* $\min\{1+\text{rad}(G), \frac{n+26}{6}\}$, *and this bound holds even if the spherical embedding of G is fixed.*

A *triangulated graph* G is a maximal planar graph; in any planar embedding, faces then have length 3. It is folklore that for a triangulated graph, the difference between radius and outerplanarity is at most 1. But for graphs with greater face-lengths, the two parameters become very different (consider a cycle). The radius-bound on 3-connected graphs by Ali et al. *increases* as the faces get bigger, while one would expect the outerplanarity to *decrease* as faces get bigger. So our goal in this paper is to find bounds on the outerplanarity that do not depend on the face-lengths and improve on $\frac{n}{6}$ for some graphs. We use a parameter that we call the *fence-girth*: In a planar graph G with a fixed embedding, a *fence* is a cycle C with other vertices both strictly inside and strictly outside C, and the *fence-girth* is the shortest length of a fence. (For a graph without cycles, the fence-girth is ∞.) Our main result is the following:

C1. Every planar graph has outerplanarity at most $\lfloor\frac{n-2}{2g}\rfloor + O(g)$ for any integer $g \geq 3$ that is at most the fence-girth. We can find a planar embedding with this number of peels in linear time. Some graphs with fence-girth g have outerplanarity at least $\lfloor\frac{n-2}{2g}\rfloor$. (Sect. 4).

We are not aware of prior results for outerplanarity-bounds, but since the radius is closely related to it for triangulated graphs, we contrast our result to the best radius-bound of $\frac{n}{2\kappa}+O(\Delta^*)$ by Ali et al. [2]. The fence-girth is never less than the connectivity κ, so our theorem implies outerplanarity $\frac{n}{2\kappa}+O(1)$, where $\kappa \leq 5$. Hence up to small constant terms our bound is never worse than Ali et al.'s, and often it will be better. For example, for bipartite planar graphs the fence-girth is at least 4, so with $g = 4$ we obtain a bound of $\frac{n}{8} + O(1)$, whereas Ali et al.'s bound is only $\frac{n}{6} + O(1)$. Secondly, the prior bound held only for 3-connected planar graphs, while we make no such restrictions. Finally, we can find a suitable embedding in linear time while all existing algorithms for outerplanarity [6,22] take quadratic time or more.

For a triangulated graph G, the fence-girth is the same as the connectivity $\kappa \leq 5$. Result **C1.** hence implies that $\text{rad}(G) \leq \frac{n}{2\kappa} + O(1)$. This bound was previously known [2], but our result comes with a linear-time algorithm to find a vertex with this eccentricity, which is new:

C2. For a κ-connected triangulated graph G, we can find a vertex s with $d_G(s, z) \leq \lfloor\frac{n-2}{2\kappa}\rfloor + O(1)$ for all $z \in V(G)$ in linear time (Sect. 4).

Ali et al. did not study the run-time to find a vertex of small eccentricity; while their proof could be turned into an algorithm, its run-time would be $O(n \cdot \text{rad}(G))$, hence quadratic. The known subquadratic algorithms for computing the radius of a planar graph are far from being linear [8,17,29], and algorithms that provide $(1+\epsilon)$-approximation have running time of the form $O(f(1/\epsilon)n \log^2 n)$ [9, 28], where f is a polynomial function on $(1/\epsilon)$. Linear-time algorithms for the radius are only known for special subclasses of planar graph classes [11,15].

Since the outerplanarity (for triangulated graphs) is closely related to the radius, and the radius is closely related to the diameter, it is natural to ask to bound the outerplanarity in terms of the diameter. We can show the following:

C3. Every planar graph G has outerplanarity at most $\frac{1}{2}\text{diam}(G)+O(\sqrt{n})$, and a corresponding embedding can be found in linear time. Every triangulated graph G has radius $\frac{1}{2}\text{diam}(G) + O(\sqrt{n})$, and a vertex of this eccentricity can be found in linear time (Sect. 5).

Similar results with a 'correction term' of $O(\sqrt{n})$ have been studied before, for example, Boitmanis et al. [7] gave an algorithm that computes the diameter and radius within such an error term in $O(|E(G)|\sqrt{n})$ time. So our contribution is that we can find a vertex of eccentricity $\frac{1}{2}\text{diam}(G) + O(\sqrt{n})$ in linear time, hence faster than Boitmanis et al. [7].

We also show that this bound is tight and that the correction-term $O(\sqrt{n})$ cannot be avoided, not even for triangulated graphs. In particular, this answers (negatively) a question on `MathOverflow` [27] whether $\text{rad}(G) \leq \frac{1}{2}\text{diam}(G) + O(1)$ for all triangulated graphs; such a relationship does hold for interval graphs [25], chordal graphs [26], and various grid graphs and generalizations [11].

C4. There exists a triangulated graph G with radius $\frac{1}{2}\text{diam}(G) + \Omega(\sqrt{n})$ (Sect. 5).

2 Definitions

We assume familiarity with graph theory and planar graphs (see for example [13]) and fix throughout a planar graph G with n vertices. For a path π in G, the *length* $|\pi|$ is its number of edges. For two vertices y, z, write $d_G(y, z)$ for the length of the shortest path between them; we only need undirected graph distance, i.e., if G has directed edges then $d_G(y, z)$ measures the distance in the underlying undirected graph. For a set of vertices L, write $G \backslash L$ for the graph obtained by deleting the vertices in L and $G[L] := G \backslash (V \backslash L)$ for the graph induced by L. We need the following *separator theorem for trees*:

Theorem 2. *[24] Let T be a tree with non-negative node-weights $w(\cdot)$. Then in linear time we can find a node S such that for every subtree T' of $T \backslash S$ we have $w(T') \leq \frac{1}{2} w(T)$, where $w(T')$ denotes the sum of weights of nodes in T'.*

One easily derived consequence of the separator theorem is the following:

Observation 3 ($\star$). *Any connected graph G has a vertex with eccentricity at most $\lfloor \frac{n}{2} \rfloor$ that we can find in linear time.*

A *spherical embedding* of G describes a drawing Γ of G on a sphere Σ by listing for each *face* (maximal region of $\Sigma \backslash \Gamma$) the closed walk(s) of G that bound the face. Graph G is called *triangulated* if all faces are triangles; the spherical embedding is then unique. A *planar embedding* of G is a drawing of G in the plane described by giving a spherical embedding Γ and fixing one face F (the *outerface*) which becomes the infinite face in the planar drawing.

For the following definition, assume that G is *plane* (comes with a fixed planar embedding (Γ, F)). Define the *peels* [22] of G as follows: L_1 consists of all vertices on the outerface F. For $i > 1$, L_i consists of all vertices on the outerface of $G \backslash (L_1 \cup \ldots \cup L_{i-1})$, where this graph uses as planar embedding the one inherited from G. The *number of peels* (which depends on Γ and F) is the minimum number k such that $G \backslash (L_1 \cup \ldots \cup L_k)$ is the empty graph. We use the term *fixed-spherical-embedding (fse) outerplanarity* of G for the minimum number of peels over all choices of outerface F (but keeping the same spherical embedding Γ). The (unrestricted) *outerplanarity* of G is the minimum number of peels over all choices of spherical embedding Γ and outerface F of Γ.

3 Toolbox

In this section, we give some definitions and methods that will be used by multiple proofs later. Throughout, we assume that the input graph G comes with a fixed spherical embedding which we will never change. We also assume that G is connected, for if it is not then we can add edges between components that share a face until G is connected. This does not add cycles (so does not change the fence-girth) and it can only decrease the diameter (hence improve the outerplanarity bound), therefore adding such edges does not affect our results.

The Tree of Peels: We will compute a tree $\mathcal{T}$ that stores, roughly speaking, the hierarchy of peels for some outerface, see also Fig. 1. Formally, pick a *root-vertex* r arbitrarily, except that it should not be a cutvertex. Choose as outerface of G a face incident to r. Define $L_0 := \{r\}$ and compute the peels $L_1, \ldots, L_k$ of $G \backslash L_0$. These *layers* $L_0, L_1, \ldots, L_k$ are not quite the peels of G (because we start with one vertex rather than a face), and not quite the layers of a breadth-first search (BFS) tree (because we include in the next layer all vertices that share a face with vertices of the previous layer, whether they are adjacent or not). We direct each edge (y, z) from the higher-indexed to the lower-indexed layer; edges connecting vertices within a layer remain undirected.

We organize the layers $L_0, \ldots, L_k$ into a tree $\mathcal{T}$ (the *tree of peels*) as follows: The root of $\mathcal{T}$ is a node R that corresponds to the entire graph G; define $V(R) := \{r\}$. For $i = 1, 2, 3, \ldots$, add a node N_K to $\mathcal{T}$ for each connected component K of $G \backslash (L_0 \cup \ldots \cup L_{i-1})$. Component K is part of one connected component P of $G \backslash (L_0 \cup \ldots \cup L_{i-2})$; make N_K a child of the node N_P corresponding to P in $\mathcal{T}$. Define $V(N_K)$ to be the outerface vertices of K.

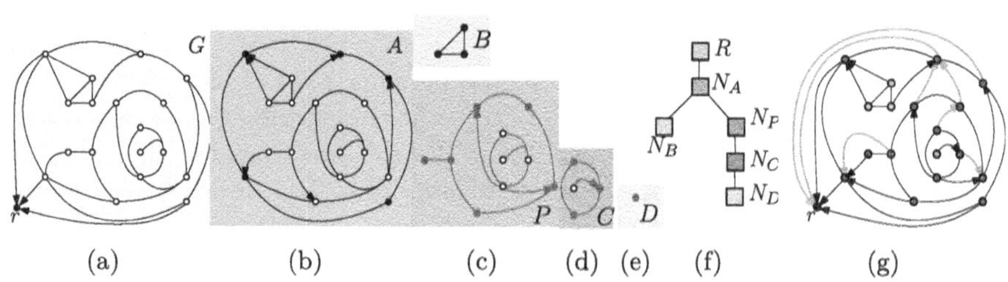

Fig. 1. (a) A plane graph G. (b)–(e) The graphs obtained by deleting $L_0, L_1, \ldots, L_3$, respectively. Solid vertices are the set $V(N)$ of the corresponding node N. (f) The tree of peels $\mathcal{T}$. (g) The augmentation H.

Throughout this paper, we will use 'node' (and upper-case letters) for the elements of $\mathcal{T}$ while we reserve 'vertex' (and lower-case letters) for G. An *interior node* of $\mathcal{T}$ is a node that is neither the root nor a leaf. We think of each node N as 'storing' the vertices in $V(N)$ and observe that these vertices induce a connected subgraph. Also, every vertex of G is stored at exactly one node of $\mathcal{T}$. We need a few easy observations:

Observation 4 ($\star$). *The following holds for the tree of peels $\mathcal{T}$:*

1. *The root R of $\mathcal{T}$ has a single child.*
2. *Let Y, Z be the nodes that store the ends y, z of an edge e. Then either e is undirected and $Y{=}Z$, or e is directed (say $y \to z$) and Z is the parent of Y.*
3. *For any interior node N of $\mathcal{T}$, the size $|V(N)|$ is at least the fence-girth.*

Augmenting G: It will be helpful if every vertex except root-vertex r has an outgoing edge. In general, this need not hold for our input graph G. We therefore augment G with further edges. The following result was shown in [4]; the result there was for the peels while our definition of layers $L_0, \ldots, L_k$ is slightly different, but one easily verifies that the proof carries over.

Claim 5. *(based on Observation 2 in [4]) We can add edges to G (while maintaining planarity) such that for all $i \geq 1$ every vertex in L_i has a neighbour in L_{i-1}.*

Let H be the graph obtained by adding a set of directed edges such every vertex except r has an outgoing edge in H (Fig. 1g). Because we only add edges between adjacent layers, the following is easily shown.

Observation 6 (⋆). *The augmented graph H has the same tree of peels as G (assuming we start with the inherited planar embedding and outerface and use the same root-vertex).*

Adding edges to G may decrease the fence-girth, but not the node-sizes of $\mathcal{T}$, which is all that will be used below. We will in the following *only* consider graph H, so every vertex $z \neq r$ has an outgoing edge.

The Detour-Method: We need the following method to connect two given vertices of H that are stored in the same node of $\mathcal{T}$ (Fig. 2a).

Definition 1. *Fix a node A_0 of $\mathcal{T}$, and two vertices $s_0, z_0 \in V(A_0)$, as well as* exit conditions *$\xi_0, \xi_1, \xi_2, \ldots$ which are (possibly negative) integers that will be specified by each application. The* detour method *at node A_0 finds a path connecting s_0 and z_0 as follows:*

- *Initialize $i = 0$; we have $s_i, z_i \in V(A_i)$. We will also maintain paths τ_s (from s_0 to s_i) and τ_z (from z_0 to z_i); initially these are simply $\langle s_0 \rangle$ and $\langle z_0 \rangle$.*
- *If $d_H(s_i, z_i) \leq \xi_i$, then set σ to be a path from s_i to z_i that has distance at most ξ_i. Exit with 'success' and return i and τ_s, τ_z, σ.*
- *If $d_H(s_i, z_i) > \xi_i$, and A_i was the root, then return 'fail'.*
- *Otherwise let A_{i+1} be the parent of A_i. Find directed edges $s_i \to s_{i+1}$ and $z_i \to z_{i+1}$, append them to paths τ_s and τ_z, update $i \leftarrow i+1$ and repeat.*

Since each iteration gets us closer to the root, the algorithm must terminate. If it exits successfully, say at index i^*, then we get a walk $s_0 \overset{\tau_s}{\rightsquigarrow} s_{i^*} \overset{\sigma}{\rightsquigarrow} z_{i^*} \overset{\tau_z}{\leftsquigarrow} z_0$. To bound the length of this walk, the following observation will be useful:

Observation 7 (⋆). *If the detour-method does not succeed at index $i > 0$, and A_i is not the root, then $\lfloor \frac{1}{2}|V(A_i)| \rfloor \geq \xi_i + 1$.*

Proof. (Sketch) s_i and z_i lie on a cycle within $V(A_i)$, hence $d_H(s_i, t_i) \leq \lceil \frac{1}{2}|V(A_i)| \rceil$.

We demonstrate how to use the detour-method with the following result that will be needed later. For any node N, write $\mathbf{a}(N)$ for the number of vertices stored at strict ancestors of N.

Lemma 1 (⋆). *Assume that $|V(N)| \geq 3$ for all interior nodes N. For any node A_0 and any $s_0, t_0 \in V(A_0)$ we have $d_H(s_0, t_0) \leq \max\{2\lceil\sqrt{\mathbf{a}(A_0)}\rceil - 2, 4\}$.*

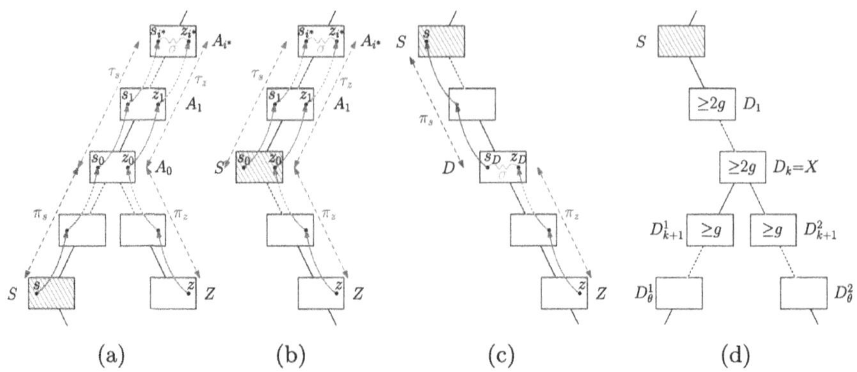

Fig. 2. (a) Path-finding for nodes in $V(\mathcal{T}_R)$ when Z is not a descendent of S. (b) Path-finding by using a detour at S when Z is a descendent of S. (c) Path-finding for deep descendants of S. (d) Finding a small common ancestor of deep nodes.

Proof. (Sketch) We are done if $A_0 = R$, for then $s_0 = z_0 = r$. So assume that $A_0 \neq R$ and apply the detour-method at A_0, using exit condition $\xi_i = \max\{2\lceil\sqrt{\mathbf{a}(A_0)}\rceil - 2, 4\} - 2i$ for $i \geq 0$. If the method exits successfully at index i^*, then combining the three returned paths τ_s, τ_z, σ gives a path of the desired length from s_0 to z_0 since τ_s, τ_z have length i^* each. If the method were to fail, then the size of the ancestors of A_0 would grow linearly with the distance from R by Observation 7. Summing over them one could then show that more than $\mathbf{a}(A_0)$ vertices stored at strict ancestors of A_0, an impossibility. □

4 Outerplanarity and Fence-Girth

In this section, we first prove that any planar graph G has outerplanarity at most $\lfloor \frac{n-2}{2g} \rfloor + O(g)$ where $g \geq 3$ is a (user-given) integer that is supposed to be at most the fence-girth.[1] Then we discuss implications and lower bounds.

Separator-Node S: To prove the upper bound, define the layers $L_0, L_1, \ldots$, augmentation H and tree of peels $\mathcal{T}$ as in Sect. 3. If any interior node of $\mathcal{T}$ stores fewer than g vertices, then repoert that g was too big (cf. Observation 4(3)) and abort. Otherwise, apply the separator theorem (Theorem 2) to the tree $\mathcal{T}$, using node-weight $w(N) := |V(N)|$, i.e., the number of stored vertices. Let S be a node such that any subtree $\mathcal{T}'$ of $\mathcal{T}\setminus S$ stores at most $\frac{n}{2}$ vertices; we write $V(\mathcal{T}')$ for these stored vertices. We know that $S \neq R$, because the root has only one child, and $w(\mathcal{T}\setminus R) = n - 1 > \frac{n}{2}$.

Crucially, S is close to all other nodes of $\mathcal{T}$. Since we will frequently need the following upper bound, we introduce a convenient notation δ for it.

[1] We use a parameter g that is separate from the fence-girth since the latter can be $\Theta(n)$; to minimize the upper bound one should set g to be $\min\{\text{fence-girth}, \frac{1}{2}\sqrt{n-2}\}$.

Observation 8. *For any node Z of $\mathcal{T}$ we have $d_{\mathcal{T}}(S,Z) \leq \delta := \lfloor \frac{n-2}{2g} \rfloor + 1$.*

Proof. Let Π be the path from S to Z in $\mathcal{T}$. Every interior node N of Π belongs to the same subtree $\mathcal{T}'$ of $\mathcal{T}\backslash S$, and satisfies $|V(N)| \geq g$ since it is neither leaf nor root. Node $Z \neq S$ also belongs to $\mathcal{T}'$ and $|V(Z)| \geq 1$. Altogether therefore $\frac{n}{2} \geq |V(\mathcal{T}')| \geq g(|\Pi|-1)+1$ or $|\Pi| \leq \frac{n-2}{2g}+1$, which implies $d_{\mathcal{T}}(S,Z) = |\Pi| \leq \delta$ by integrality. □

The overall idea of our proof is now to pick a vertex $s \in V(S)$, and to argue that $d_H(s,z) \leq \delta + 2g - 2$ for all vertices z. Actually, for most cases below, this will hold for *any* choice of $s \in S$.

Vertices Stored in $\mathcal{T}_R$: Let $\mathcal{T}_R$ be the subtree of $\mathcal{T}\backslash S$ that contains root R. For vertices in this subtree, the detour method proves the distance-bound.

Claim 9. *Assume that $g \geq 3$. For any $s \in V(S)$ and any $z \in V(\mathcal{T}_R)$ we have $d_H(s,z) \leq \delta + g$.*

Proof. Figure 2a illustrate this proof. Let Z be the node that stores z, and let A_0 be the least common ancestor of S and Z; this is a strict ancestor of S since $Z \in \mathcal{T}_R$. Follow directed edges from s to some vertex $s_0 \in V(A_0)$; the resulting path π_s has length $d_{\mathcal{T}}(S,A_0)$ since every directed edge gets us closer to the root. Likewise we can get a path π_z of length $d_{\mathcal{T}}(Z,A_0)$ from z to some node $z_0 \in V(A_0)$ (possibly $z_0 = z$).

Now apply the detour-method with A_0, s_0, z_0, using $\xi_i = g-1$. This will always exit with 'success' at a node A_{i^*} that is not the root, because any two vertices stored at the child of the root have outgoing edges towards the root-vertex r and hence distance $2 \leq g-1$. Let τ_s, τ_z, σ be the paths, then $\pi := s \overset{\pi_s}{\rightsquigarrow} s_0 \overset{\tau_s}{\rightsquigarrow} s_{i^*} \overset{\sigma}{\rightsquigarrow} z_{i^*} \overset{\tau_z}{\leftsquigarrow} z_0 \overset{\pi_z}{\leftsquigarrow} z$ has length $d_{\mathcal{T}}(S,Z) + 2i^* + |\sigma|$.

To bound $|\pi|$, consider the path Π from S to Z in $\mathcal{T}$, which goes through A_0, and let $A_1, \ldots, A_{i^*}$ be the ancestors of A_0 that were visited by the detour-method. If $i^* > 0$, then by Observation 7 we have $|V(A_j)| \geq 2(\xi_j+1) = 2g$ for $j = 1, \ldots, i^*-1$, node A_{i^*} and the interior nodes of Π store at least g vertices each which nodes Z and R store at least one vertex. Therefore

$$\tfrac{n}{2} \geq |V(\mathcal{T}_R)| \geq 2g(i^*-1) + g + (|\Pi|-1)g + 1 + 1 \geq g(|\Pi| + 2i^* - 2) + 2,$$

hence $|\Pi| + 2i^* \leq \frac{n-4}{2g} + 2$ which is at most $\delta+1$ by integrality. If $i^* = 0$ then $|\Pi| + 2i^* \leq \delta$ by Observation 8. Either way $|\pi| = |\Pi| + 2i^* + |\sigma| \leq \delta + 1 + (g-1)$. □

Vertices at Descendants of S: In light of Claim 9, we only need to worry about vertices that are stored at descendants of S. We first introduce two methods to find short paths for these in special situations. Recall that $\mathbf{a}(S)$ denotes the number of vertices stored at strict ancestors of node S.

Claim 10. *Assume that $\mathbf{a}(S) \leq g^2$ and $g \geq 3$. Then for any $s \in V(S)$ and any vertex z stored at a descendant Z of S we have $d_H(s,z) \leq \delta + 2g - 2$.*

Proof. See also Fig. 2b. Use directed edges to find a path π_z of length $d_{\mathcal{T}}(Z,S) \le \delta$ from z to some vertex $z_0 \in V(S)$. We also know that $d_H(s,z_0) \le \max\{2\lfloor\sqrt{\mathbf{a}(S)}\rfloor-2, 4\} \le 2g-2$ by Lemma 1. Combining the paths gives the desired length: $d_H(z,s) \le d_H(z,z_0) + d_H(z_0,s) \le \delta + 2g - 2$. □

Claim 11. *Let D be a descendant of S. Let $s_D \in V(D)$ be a vertex of eccentricity at most $\lfloor\frac{|V(D)|}{2}\rfloor$ in the connected graph $H[V(D)]$ (Observation 3). Follow directed edges from s_D to reach a vertex $s \in V(S)$. Then for any vertex z stored at a descendant Z of D we have $d_H(s,z) \le d_{\mathcal{T}}(Z,S) + \lfloor\frac{|V(D)|}{2}\rfloor$.*

Proof. See Fig. 2c. Use directed edges to go from z to a vertex $z_D \in V(D)$ along a path π_z of length $d_{\mathcal{T}}(Z,D)$. Find a path σ of length at most $\lfloor\frac{|V(D)|}{2}\rfloor$ to connect z_D to s_D, and then follow $d_{\mathcal{T}}(D,S)$ directed edges to get to s. Combining the paths gives the desired length since $d_{\mathcal{T}}(Z,D) + d_{\mathcal{T}}(D,S) = d_{\mathcal{T}}(Z,S)$. □

Corollary 1. *Assume that $|V(S)| \le 4g-3$. Then there exists an $s \in V(S)$ such that for any vertex z stored at a descendant of S we have $d_H(s,z) \le \delta + 2g - 2$.*

So we are done if $\mathbf{a}(S) \le g^2$ or $|V(S)| \le 4g-3$. Otherwise we distinguish the descendants of S by their depth, using Claim 11 (for a carefully chosen D) for the 'deep' ones and the method of Claim 10 for the others. Define the *threshold-value* $\theta = \lceil\frac{n-\mathbf{a}(S)}{2g}\rceil - 1$, call a node Z *deep* if it is a descendant of S with $d_{\mathcal{T}}(S,Z) \ge \theta$, and call a vertex *deep* if it is stored at a deep node. A straight-forward math manipulation gives the following upper bound.

Observation 12 (⋆). *We have $\theta \le \delta - 2\lceil\sqrt{\mathbf{a}(S)}\rceil + 2g + 1$.*

Claim 13. *Assume that $g \ge 3$ and $|V(S)| \ge 4g-2$ and $\mathbf{a}(S) > g^2$. Then for any $s \in V(S)$ and any z stored at a descendant Z of S that is not deep, we have $d_H(s,z) \le \delta + 2g - 2$.*

Proof. The method is exactly the same as in the proof of Claim 10 (walk from z to a vertex $z_0 \in V(S)$ and apply Lemma 1 to connect z_0 to s, see also Fig. 2b), but the analysis is different. Since Z is not deep, we have $d_H(z,z_0) = d_{\mathcal{T}}(Z,S) \le \theta-1 \le \delta-2\lceil\sqrt{\mathbf{a}(S)}\rceil+2g$. By $\mathbf{a}(S) > g^2 \ge 9$ we have $d_H(z_0,s) \le 2\lceil\sqrt{\mathbf{a}(S)}\rceil-2$ and so $d_H(z,s) \le d_H(z,z_0) + d(z_0,s) \le \delta + 2g - 2$. □

For deep nodes we show that there exists a suitable node with which to apply Claim 11. This is proved via a counting-argument: due to the (carefully chosen) threshold θ otherwise more than n vertices would be stored in tree $\mathcal{T}$.

Claim 14. *Assume that at least $4g - |V(S)| + 1$ vertices are deep. Then there exists a descendant D of S (possibly S itself) such that D is an ancestor of all deep nodes, and $|V(D)| \le 2g-1$.*

Proof. Let X be the least common ancestor of all deep nodes. This is a descendant of S as well (possibly $X = S$), so enumerate the path from S to X as $S{=}D_0, D_1, \dots, D_k{=}X$ for some $k \ge 0$. We are done if $|V(D_i)| \le 2g-1$ for some

$0 \leq i \leq k$, so assume (for contradiction) that the k nodes $D_1, \ldots, D_k$ store at least $2g$ vertices each. If X were deep (so $k \geq \theta$) then $D_1, \ldots, D_k$ would store at least $\theta \cdot 2g \geq n - \mathbf{a}(S) - 2g$ vertices. We also store $\mathbf{a}(S)$ vertices at strict ancestors at S, $|V(S)|$ vertices at S and at least $4g - |V(S)| + 1$ deep vertices, in total hence more than n, impossible.

So X is not deep, see also Fig. 2d. Since X is the least common ancestor of deep descendants, therefore it must have at least two children X^1, X^2 that are ancestors of deep descendants. For $j = 1, 2$, enumerate the path from X^j to a deep descendant of X^j as $D^j_{k+1}, \ldots, D^j_\theta$. Then for $i = k{+}1, \ldots, \theta{-}1$ node D^j_i is not a leaf and stores at least g vertices, so $|V(D^1_i)| + |V(D^2_i)| \geq 2g$. So for $i = 1, \ldots, \theta - 1$ we can find $2g$ vertices stored at not-deep descendants of distance i from S. In total these not-deep strict descendants of S hence store at least $(\theta{-}1)2g \geq n - \mathbf{a}(S) - 4g$ vertices. As above therefore more than n vertices are stored in the tree of peels, impossible. □

Now we put everything together.

Theorem 15. *Let G be a planar graph with $n \geq 3$ vertices and let $g \geq 3$ be an integer that is at most the fence-girth of G. Then G has a planar supergraph H with* $\mathrm{rad}(H) \leq \lfloor \frac{n-2}{2g} \rfloor + 2g - 1$. *Furthermore, a vertex of H with this eccentricity can be found in linear time.*

Proof. Compute layers $L_0, L_1, \ldots$, augmentation H, and tree of peels $\mathcal{T}$ as in Sect. 3. Find separator-node S and compute $|V(S)|$, $\mathbf{a}(S), \theta$, and the number of deep vertices. If $|V(S)| \geq 4g{-}2$ and there are at least $4g{-}|V(S)|{+}1$ deep vertices, then arbitrarily fix a deep node Z and find node D of Claim 14 by walking from S towards Z until we encounter a node that stores at most $2g - 1$ vertices. In all other cases set $D := S$. Pick s as in Claim 11 applied to D. Each of these steps takes linear time (we elaborate on this in the full version [5]).

Applying various cases we show that $d_H(s, z) \leq \delta + 2g - 2$ for all z (which implies the result by $\delta = \lfloor \frac{n-2}{2g} \rfloor + 1$). This holds for all $z \in V(\mathcal{T}_R)$ by Claim 9, so consider a vertex z stored at a descendant Z of S. The bound holds by Claim 10 if $\mathbf{a}(S) \leq g^2$ and by Corollary 1 if $|V(S)| \leq 4g - 3$, so assume neither. If Z is not deep, then apply Claim 13. If Z is deep and there are at least $4g - |V(S)| + 1$ deep vertices, then combine Claim 14 with Claim 11 to get the bound. The only remaining case is that $\mathbf{a}(S) > g^2$, $|V(S)| \geq 4g - 2$, Z is deep, and at most $4g - |V(S)|$ vertices are deep (so $|V(S)| \leq 4g - 1$ since there is a deep vertex at Z). In this case, $d_{\mathcal{T}}(Z, S) = \theta$, for otherwise Z's parent would also be deep and store $g \geq 3 \geq 4g{-}|V(S)|{+}1$ deep vertices. Also $\theta \leq \delta{-}2\lceil\sqrt{\mathbf{a}(S)}\rceil{+}2g{+}1 \leq \delta{-}1$ by $\lceil\sqrt{\mathbf{a}(S)}\rceil \geq g + 1$. We used $D = S$ and picked $s \in V(S)$ as in Claim 11, so this gives $d_H(z, s) \leq d_{\mathcal{T}}(Z, S) + \lceil\frac{V(S)}{2}\rceil \leq \delta - 1 + 2g - 1$ as desired. □

Theorem 15 implies **C2** from the introduction, for if G is triangulated then necessarily $H = G$ and so the radius-bound holds for the input-graph G as well. It also implies **C1**:

Corollary 2. *Let G be a spherically-embedded graph with $n \geq 3$ vertices and let $g \geq 3$ be an integer that is at most the fence-girth of G. Then G has fse-outerplanarity at most $\lfloor \frac{n-2}{2g} \rfloor + 2g$. Furthermore, an outerface of G that achieves this outerplanarity can be found in linear time.*

Proof. Without changing the spherical embedding of G, compute the supergraph H with $\text{rad}(H) \leq \lfloor \frac{n-2}{2g} \rfloor + 2g - 1$; the outerplanarity bound holds by Observation 1 and the outerface can be found by picking any face incident to the vertex s that achieves this eccentricity. □

Discussion: Theorem 15 requires $g \geq 3$. We can always choose such a g if G is simple, but if G has parallel edges or loops (which we did not exclude) then in the fixed spherical embedding the fence-girth may only be 2 or 1. We hence briefly discuss the case $g \in \{1, 2\}$. Going through all proofs where $g \geq 3$ or $|V(N)| \geq 3$ is actually used (Lemma 1, Claim 9 and 10), one sees that the results hold for $g = 2$ if we increase the permitted distance-bound by 2. (Likewise we need to raise the permitted distance-bound in Claim 13 since it uses Lemma 1.) Therefore $\text{rad}(H) \leq \delta + 2g = \lfloor \frac{n-2}{2g} \rfloor + 1$ for $g = 2$. For $g = 1$, $\text{rad}(H) \leq \lfloor \frac{n}{2} \rfloor = \lfloor \frac{n-2}{2g} \rfloor + 1$ by Observation 3.

Theorem 15 also assumes that we are *given* g. However, the computation of $\mathcal{T}$ does not depend on g, and the only thing we require is that all its interior nodes store at least g vertices. So if we are not given g, then we can compute $\mathcal{T}$, define $g^* = \min\{|V(N)| : N \text{ is an interior node}\}$ and use $g := \min\{g^*, \sqrt{n-2}/2\}$ as parameter for Theorem 15.

Tightness: We now design graphs with large outerplanarity (relative to the fence-girth g). (These graphs actually have *girth* g, i.e., *any* cycle (not just those that are fences) has length at least g.) We do this first for the fixed-spherical-embedding outerplanarity, where the lower bounds hold even for $g \in \{1, 2\}$. Then, for $g \geq 3$ and at a slight decrease of the lower bound, we give bounds for the (unrestricted) outerplanarity. Roughly speaking, the graphs consist of nested cycles of length g, with a single vertex or path inside the innermost/outside the outermost cycle, and (if desired) with edges added to ensure that all spherical embeddings have the same fse-outerplanarity. See Fig. 3.

Lemma 2 ($\star$). *For $g \geq 1$ there exists an infinite class $\mathcal{G}_g = \{G_g^k : k \geq 1 \text{ odd}\}$ of spherically embedded graphs of girth and fence-girth g for which the fse-outerplanarity is at least $\frac{k+3}{2} = \frac{n-2}{2g} + \frac{3}{2}$.*

Theorem 16 ($\star$). *For $g \geq 3$ there exists an infinite class $\mathcal{H}_g = \{H_g^k : k \geq 3 \text{ odd}\}$ of planar graphs of girth and fence-girth g that have outerplanarity at least $\frac{n-2}{2g} + 1 + \frac{3+\chi(g \text{ even})}{2g}$.*

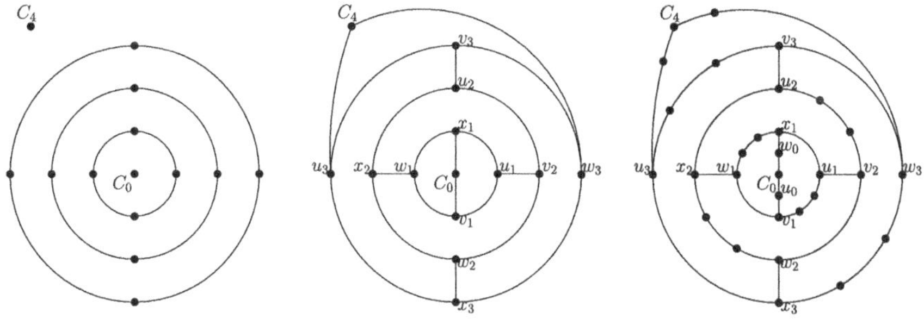

Fig. 3. Graphs G_3^4, H_3^4 and H_3^8.

5 Outerplanarity and Diameter

With much the same techniques as for Theorem 15, we can also bound the outerplanarity in terms of the diameter, as long as we permit a 'correction-term' of $O(\sqrt{n})$.

Theorem 17. *Any simple plane graph G with $n \geq 14$ vertices has a plane supergraph H with* $\text{rad}(H) \leq \lceil \frac{1}{2}\text{diam}(G) \rceil + 2\sqrt{n-4} - 2$.

Proof. Compute layers $L_0, L_1, \ldots$, augmentation H and the tree of peels $\mathcal{T}$ as in Sect. 3. If $\text{diam}(\mathcal{T}) = 1$ then $\mathcal{T}$ consists only of root R and its unique child N that stores all vertices of $V(G)\backslash r$; in consequence G has outerplanarity at most $2 \leq 1 + 2\sqrt{n-4} - 1$ by $n \geq 5$. So assume that $\text{diam}(\mathcal{T}) \geq 2$ and let U, V be two nodes of $\mathcal{T}$ with $d_{\mathcal{T}}(U, V) = \text{diam}(\mathcal{T})$. Let S be the node 'halfway between them', i.e., of distance $\lceil \text{diam}(\mathcal{T})/2 \rceil$ from U along the unique path from U to V in $\mathcal{T}$. One easily verifies that $d_{\mathcal{T}}(S, Z) \leq \lceil \text{diam}(\mathcal{T})/2 \rceil$ for all nodes Z of $\mathcal{T}$, otherwise Z would be too far away from either U or V since $\mathcal{T}$ is a tree. Also note that $\text{diam}(\mathcal{T}) \leq \text{diam}(H)$ by Observation 4(2) and $\text{diam}(H) \leq \text{diam}(G)$ since H is a supergraph. Finally observe that $\mathbf{a}(S) \leq n-4$, for S is an interior node by $\text{diam}(\mathcal{T}) \geq 2$ and stores at least three vertices by simplicity while at least one of U, V is not an ancestor of S and stores at least one vertex.

Pick $s \in S$ arbitrarily. For any $z \in V(G)$ (say z is stored at node Z), we find a path from s to z as follows (see also Fig. 2b and 2c): Let A_0 be the least common ancestor of S and Z (quite possibly $A_0 = S$ or $A_0 = Z$). Follow directed edges from s and z to reach vertices $s_0, z_0 \in A_0$; the total number of these edges is $d_{\mathcal{T}}(S, Z) \leq \lceil \frac{1}{2}\text{diam}(\mathcal{T}) \rceil \leq \lceil \frac{1}{2}\text{diam}(G) \rceil$. Observe that $4 \leq \lceil \sqrt{n-4} \rceil$ and also $\sqrt{\mathbf{a}(A_0)} \leq \sqrt{\mathbf{a}(S)} \leq \sqrt{n-4}$. Using Lemma 1 therefore $d_H(s_0, z_0) \leq 2\lceil \sqrt{n-4} \rceil - 2$ and so $d_H(s, z) \leq \lceil \frac{1}{2}\text{diam}(G) \rceil + 2\sqrt{n-4} - 2$. □

Theorem 17 implies **C3** from the introduction: The fse-outerplanarity of G is at most the fse-outerplanarity of H, which is at most $\text{rad}(H) + 1$. If G is triangulated then necessarily $H = G$ and so $\text{rad}(G) \leq \lfloor \frac{1}{2}\text{diam}(G) \rfloor + O(\sqrt{n})$. Theorem 18 implies that the 'correction-term' of $O(\sqrt{n})$ cannot be avoided for the graph in Fig. 4.

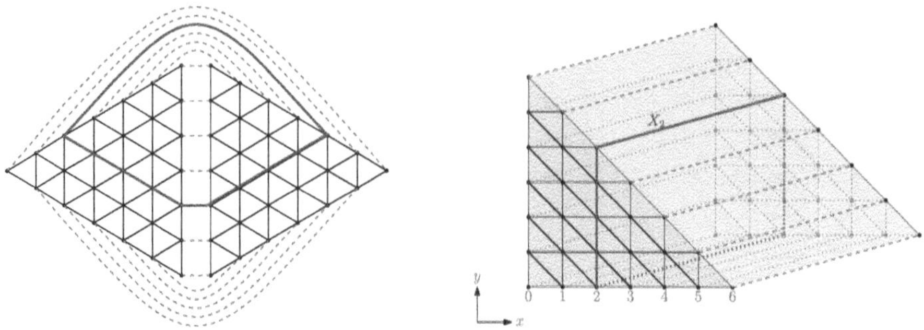

Fig. 4. Graph H (for $k = 2$) for the proof of Theorem 18, both in a planar drawing and embedded on the triangular prism. Connector-edges are blue/dashed, X_2 is green/bold. (Color figure online)

Theorem 18 ($\star$). *For every positive integer k, there exists a triangulated graph G with $n = (3k+1)(3k+2)$ vertices that has diameter at most $3k + 1$ and radius at least $2k = \frac{1}{2}\text{diam}(G) + \Omega(\sqrt{n})$.*

6 Remarks

While the '$\frac{n}{2g}$'-part of our bound in Theorem 15 is tight, the '$+2g$' part could use improvement. We can easily prove (with the same techniques as in Theorem 17) a bound of $\frac{n-2}{2g} + O(\sqrt{n})$, but does every planar graph with fence-girth g have outerplanarity $\frac{n}{2g} + O(1)$?

Also, our linear-time algorithm carefully side-steps the question of how to compute the fence-girth (it instead uses a parameter g for which the node-sizes of $\mathcal{T}$ are big enough). Testing whether the fence-girth is at most k is easily done if the spherical embedding is fixed and k is a constant, using the subgraph isomorphism algorithm by Eppstein [15]. But the fence-girth need not be constant and Eppstein's algorithm does not work if the embedding can be changed. Algorithms to compute the girth [10] do not seem transferrable to the fence-girth. How easy is it to compute the fence-girth, both when the spherical embedding is fixed and when it can be chosen freely?

Acknowledgments. Research by TB supported by NSERC; FRN RGPIN-2020-03958. Research by DM supported by NSERC; FRN RGPIN-2018-05023.

References

1. Alam, M.J., Bläsius, T., Rutter, I., Ueckerdt, T., Wolff, A.: Pixel and voxel representations of graphs. In: Di Giacomo, E., Lubiw, A. (eds.) GD 2015. LNCS, vol. 9411, pp. 472–486. Springer, Cham (2015). https://doi.org/10.1007/978-3-319-27261-0_39

2. Ali, P., Dankelmann, P., Mukwembi, S.: The radius of k-connected planar graphs with bounded faces. Discret. Math. **312**(24), 3636–3642 (2012). https://doi.org/10.1016/j.disc.2012.08.019
3. Baker, B.S.: Approximation algorithms for np-complete problems on planar graphs (preliminary version). In: 24th Annual Symposium on Foundations of Computer Science, Tucson, Arizona, USA, 7-9 November 1983, pp. 265–273. IEEE Computer Society (1983). https://doi.org/10.1109/SFCS.1983.7
4. Biedl, T.: On triangulating k-outerplanar graphs. Discret. Appl. Math. **181**, 275–279 (2015)
5. Biedl, T., Mondal, D.: Improved outerplanarity bounds for planar graphs. arXiv preprint arXiv:2407.04282 (2024). https://arxiv.org/abs/2407.04282
6. Bienstock, D., Monma, C.L.: On the complexity of embedding planar graphs to minimize certain distance measures. Algorithmica **5**(1), 93–109 (1990). https://doi.org/10.1007/BF01840379
7. Boitmanis, K., Freivalds, K., Ledins, P., Opmanis, R.: Fast and simple approximation of the diameter and radius of a graph. In: Àlvarez, C., Serna, M. (eds.) WEA 2006. LNCS, vol. 4007, pp. 98–108. Springer, Heidelberg (2006). https://doi.org/10.1007/11764298_9
8. Cabello, S.: Subquadratic algorithms for the diameter and the sum of pairwise distances in planar graphs. ACM Trans. Algorithms (TALG) **15**(2), 1–38 (2018)
9. Chan, T.M., Skrepetos, D.: Faster approximate diameter and distance oracles in planar graphs. Algorithmica **81**(8), 3075–3098 (2019). https://doi.org/10.1007/s00453-019-00570-z
10. Chang, H., Lu, H.: Computing the girth of a planar graph in linear time. SIAM J. Comput. **42**(3), 1077–1094 (2013). https://doi.org/10.1137/110832033
11. Chepoi, V., Dragan, F.F., Vaxès, Y.: Center and diameter problems in plane triangulations and quadrangulations. In: Eppstein, D. (ed.) Proceedings of the Thirteenth Annual ACM-SIAM Symposium on Discrete Algorithms, January 6-8, 2002, San Francisco, CA, USA, pp. 346–355. ACM/SIAM (2002). http://dl.acm.org/citation.cfm?id=545381.545427
12. Demaine, E.D., Hajiaghayi, M.T., Nishimura, N., Ragde, P., Thilikos, D.M.: Approximation algorithms for classes of graphs excluding single-crossing graphs as minors. J. Comput. Syst. Sci. **69**(2), 166–195 (2004). https://doi.org/10.1016/j.jcss.2003.12.001
13. Diestel, R.: Graph Theory, Graduate Texts in Mathematics, vol. 173. 4th Edition, Springer (2012)
14. Dolev, D., Leighton, F.T., Trickey, H.: Planar embedding of planar graphs. Technical report Massachusetts Inst of Tech Cambridge lab for Computer Science (1983)
15. Eppstein, D.: Subgraph isomorphism in planar graphs and related problems. J. Graph Algorithms Appl. **3**(3), 1–27 (1999). https://doi.org/10.7155/jgaa.00014
16. Eppstein, D.: Diameter and treewidth in minor-closed graph families. Algorithmica **27**(3), 275–291 (2000). https://doi.org/10.1007/s004530010020
17. Gawrychowski, P., Kaplan, H., Mozes, S., Sharir, M., Weimann, O.: Voronoi diagrams on planar graphs, and computing the diameter in deterministic $\tilde{\mathrm{O}}(\mathrm{n}^{5/3})$ time. SIAM J. Comput. **50**(2), 509–554 (2021). https://doi.org/10.1137/18M1193402
18. Giacomo, E.D., Didimo, W., Liotta, G., Meijer, H.: Computing radial drawings on the minimum number of circles. J. Graph Algorithms Appl. **9**(3), 365–389 (2005). https://doi.org/10.7155/jgaa.00114
19. Grohe, M.: Local tree-width, excluded minors, and approximation algorithms. Combinatorica **23**(4), 613–632 (2003). https://doi.org/10.1007/s00493-003-0037-9

20. Hajiaghayi, M.T., Nishimura, N., Ragde, P., Thilikos, D.M.: Fast approximation schemes for $k_{3,3}$-minor-free or k_5-minor-free graphs. Electron. Notes Discret. Math. **10**, 137–142 (2001). https://doi.org/10.1016/S1571-0653(04)00379-8
21. Harant, J.: An upper bound for the radius of a 3-connected planar graph with bounded faces. Contemporary methods in graph theory (Bibliographisches Inst., Mannheim, 1990) **353**, 358 (1990)
22. Kammer, F.: Determining the smallest k such that G is k-outerplanar. In: Arge, L., Hoffmann, M., Welzl, E. (eds.) ESA 2007. LNCS, vol. 4698, pp. 359–370. Springer, Heidelberg (2007). https://doi.org/10.1007/978-3-540-75520-3_33
23. Leiserson, C.E.: Area-efficient graph layouts (for VLSI). In: 21st Annual Symposium on Foundations of Computer Science, pp. 270–281. IEEE Computer Society (1980)
24. Lipton, R., Tarjan, R.: A separator theorem for planar graphs. SIAM J. Appl. Math. **36**(2), 177–189 (1979)
25. Pramanik, T., Mondal, S., Pal, M.: The diameter of an interval graph is twice of its radius. Int. J. Math. Comput. Sci. **5**(8), 1412–1417 (2011)
26. Shook, J.M., Wei, B.: A characterization of the centers of chordal graphs. arXiv preprint arXiv:2210.00039 (2022)
27. username 'verifying.' (2013). (https://mathoverflow.net/users/82650/verifying) : Diameter vs radius in maximal planar graphs. MathOverflow https://mathoverflow.net/questions/227888/diameter-vs-radius-in-maximal-planar-graphs. Accessed 19 June 2023
28. Weimann, O., Yuster, R.: Approximating the diameter of planar graphs in near linear time. ACM Trans. Algorithms **12**(1), 12:1–12:13 (2016). https://doi.org/10.1145/2764910
29. Wulff-Nilsen, C.: Wiener Index, Diameter, and Stretch Factor of a Weighted Planar Graph in Subquadratic Time. Ph.D. thesis, Department of Computer Science, University of Copenhagen (2008)
30. Zhang, H., He, X.: Visibility representation of plane graphs via canonical ordering tree. Inf. Process. Lett. **96**(2), 41–48 (2005)

XNLP-Hardness of Parameterized Problems on Planar Graphs

Hans L. Bodlaender and Krisztina Szilágyi(✉)

Utrecht University, Utrecht, The Netherlands
{h.l.bodlaender,k.szilagyi}@uu.nl

Abstract. The class XNLP consists of (parameterized) problems that can be solved nondeterministically in $f(k)n^{O(1)}$ time and $g(k)\log n$ space, where n is the size of the input instance and k the parameter. The class XALP consists of problems that can be solved in the above time and space with access to an additional stack. These two classes are a "natural home" for many standard graph problems and their generalizations.

In this paper, we show the hardness of several problems on planar graphs, parameterized by outerplanarity, treewidth and pathwidth, thus strengthening several existing results. In particular, we show XNLP-hardness of the following problems parameterized by outerplanarity: All-or-Nothing Flow, Target Outdegree Orientation, Capacitated (Red-Blue) Dominating Set, Target Set Selection etc. We also show the XNLP-completeness of Scattered Set parameterized by pathwidth and XALP-completeness parameterized by treewidth and outerplanarity.

Keywords: Parameterized Complexity · XNLP · XALP · Planar Graphs · Outerplanarity

1 Introduction

In classical complexity theory, we can classify problems depending on their space complexity or their time complexity. These two classifications are intertwined as follows: $L \subseteq NL \subseteq P \subseteq NP \subseteq PSPACE = NPSPACE \subseteq EXP$. In particular, the class NL (Nondeterministic Logarithmic-space) is contained in the class P (Polynomial-time). It is natural to ask whether there is a similar chain of inclusions for the parameterized analogues of these classes. In particular, what can we say about the relationship between FPT (analogue of P) and XNL (analogue of NL)?

This question was first posed by Elberfeld et al. [12]. They defined the class $N[fpoly, flog]$ (here called XNLP) as the class of (parameterized) problems that can be solved with a nondeterministic Turing machine in $f(k)n^{O(1)}$ time and

K. Szilágyi—Supported by the project CRACKNP that has received funding from the European Research Council (ERC) under the European Union's Horizon 2020 research and innovation programme (grant agreement No 853234).

D. Král' and M. Milanič (Eds.): WG 2024, LNCS 14760, pp. 107–120, 2025.
https://doi.org/10.1007/978-3-031-75409-8_8

$g(k) \log n$ space and gave the first problems complete for this class. They also present several problems that are complete for this class, stating: *The real challenge lies in finding problems together with* natural *parameterizations that are complete.*

In the last decade, this question has been resolved: many standard graph problems and their generalizations were shown to be XNLP-complete (see Table 1 for examples of such problems). This class turned out to be a natural home for several problems that are $W[t]$-hard for all t but unlikely to belong to $W[P]$, such as Bandwidth and List Coloring (parameterized by pathwidth) [8].

Recently, in [7] a "tree variant" of the class XNLP was introduced. This class is called XALP, and many problems that are XNLP-complete parameterized by pathwidth are shown to be XALP-complete parameterized by treewidth (see Table 1). The class XALP contains problems that can be solved by a nondeterministic Turing machine with access to an auxiliary stack in $f(k)n^{O(1)}$ time and $g(k) \log n$ space.

A natural question to ask is whether we can say something about hardness for other graph parameters. Several XNLP-hard problems for treewidth are known to be easy (i.e. in FPT) parameterized by the so-called stable gonality parameter [5]. In this paper, we study the hardness of problems on planar graphs, parameterized by outerplanarity, treewidth and pathwidth.

Outerplanarity is a natural parameter to study for problems on planar graphs. As graphs of outerplanarity k have treewidth at most $3k-1$ (see e.g. [4]), algorithmic results for graphs of bounded treewidth carry over to planar graphs of bounded outerplanarity. This has also been exploited in the well-known and often applied layering technique of Baker [3], resulting in approximation schemes for many problems on planar graphs.

There are several celebrated results in the field of parameterized complexity where the complexity of a problem significantly decreases when we move from general graphs to planar graphs, for instance Dominating Set parameterized by solution size then drops from being W[2]-complete [11] to a fixed parameter tractable problem with a linear kernel [1]. The results of this paper show that for several problems that are in XP for treewidth as parameter, we have no complexity drop when we move to the realm of planar graphs.

Our Results. An overview of the results of this paper can be seen in Table 1. In addition to these results, we also show the XNLP-completeness of Binary CSP for $k \times n$-grids (parameterized by k).

Organization. In Sect. 2 we introduce some definitions and notations. In Sect. 3 we give an overview of our results on Binary CSP and two related problems. In Sect. 4, we consider the complexity of the All-or-Nothing Flow problem, and prove it to be XNLP-hard with outerplanarity as parameter. In Sect. 5 we state several problems whose hardness can be shown by a reduction from All-of-Nothing Flow. Section 6 discusses the Scattered Set problem. We finish with some concluding remarks in Sect. 7. The missing proofs can be found in the arXiv version of this paper [10].

Table 1. An overview of results of this paper and results from the literature. Results of this paper are stated with the corresponding section numbers.

	outerplanarity	treewidth	pathwidth
BINARY CSP	XALP-c. (3)	XALP-c. [7]	XNLP-c. [8]
LIST COLORING	XALP-c. (3)	XALP-c. [7]	XNLP-c. [8]
PRECOLORING EXTENSION	XALP-c. (3)	XALP-c. [7]	XNLP-c. [8]
SCATTERED SET	XALP-c. (6)	XALP-c. (6)	XNLP-c. (6)
ALL-OR-NOTHING FLOW	XNLP-h. (4)	XALP-c. [7]	XNLP-c. [5]
ALL-OR-NOTHING FLOW with cap. ≤ 2	XNLP-h. (5)	XALP-c. (5)	XNLP-c. (5)
TARGET OUTDEGREE ORIENTATION	XNLP-h. (5)	XALP-c. [7]	XNLP-c. [5]
CAPACITATED (RED-BLUE) DOMINATING SET	XNLP-h. (5)	XALP-c. [7]	XNLP-c. [6]
CAPACITATED VERTEX COVER	XNLP-h. (5)	XALP-c. [7]	XNLP-c. [6]
f-DOMINATING SET	XNLP-h. (5)	XALP-c. (5)	XNLP-c. (5)
k-DOMINATING SET	XNLP-h. (5)	XALP-c. (5)	XNLP-c. (5)
TARGET SET SELECTION	XNLP-h. (5)	XALP-c. (5)	XNLP-c. (5)

2 Definitions and Notation

2.1 Graph Notions

Given an integer n, we define $[n] = \{1, 2, \ldots, n\}$. Throughout this paper, graphs are undirected unless specified otherwise. Given a graph $G = (V, E)$, we denote by $N(v)$ the open neighbourhood of a vertex v, i.e. $N(v) = \{u \in V : \{u, v\} \in E\}$. A planar graph is outerplanar (or 1-outerplanar) if it has na embedding such that all its vertices are on the outer face. Informally speaking, a planar graph G is k-outerplanar if its vertices are on k "layers". Formally, k-outerplanar graphs are defined as follows.

Definition 1. *An embedding of a planar graph G is k-*outerplanar *if for every vertex v there is a sequence $F_1, v_1, F_2, v_2, \ldots, F_\ell, v_\ell$ such that:*

- *F_1 is the unbounded face and $v_\ell = v$;*
- *For each $i \in [\ell]$, F_i contains v_i;*
- *For each $i \in [\ell - 1]$, v_i belongs to F_{i+1};*
- *$\ell \leq k$.*

*A planar graph is k-*outerplanar *if it has a k-outerplanar embedding.*

In other words, a graph is k-outerplanar if after k operations consisting of removing all the vertices on the outer face, we obtain an empty graph. The outerplanarity of a graph can be computed in polynomial time [14].

The definition of treewidth and pathwidth can be found in standard textbooks or in the arXiv version [10]. Throughout the paper, we assume that we are given the tree decomposition and path decomposition of the input graph.

Kloks [15] introduced the notion of nice tree and path decompositions. We will use a variant of nice path decompositions, which we will describe with a sequence of operations on terminal graphs. Here, a terminal graph is a triple (V, E, X), together with a binary relation $\prec \subset X \times X$ which is astrict total ordering on X. We call elements of X terminals. We consider the following operations on terminal graphs:

- **Introduce**. Given a terminal graph $G = (V, E, X)$, the introduce operation adds a new isolated vertex v to V and to X, with v the smallest vertex regarding the ordering in X: $G' = (V \cup \{v\}, E, \{v\} \cup X)$, with $v \prec x$ for all $x \in X$.
- **Forget**. Given a terminal graph $G = (V, E, X)$ with $X \neq \emptyset$, the largest element of X is no longer a terminal: $G' = (V, E, X \setminus \{x\})$, with $y \prec x$ for all $y \in X \setminus \{x\}$.
- **Add-Edge**(i). Given a terminal graph $G = (V, E, X)$ with $|X| > i$, we add an edge between the ith and the $(i + 1)$st terminal: V and X are unchanged, and one edge is added to E.
- **Swap**(i). Given a terminal graph $G = (V, E, X)$ with $|X| > i$, swap in the ordering in X the ith and the $(i + 1)$st vertex.

To build graphs of small treewidth, we need one more operation, that now has two terminal graphs as input.

- **Join**. Given two terminal graphs $G_1 = (V_1, E_1, X)$ and $G_2 = (V_2, E_2, X)$ that only intersect in their terminal nodes, i.e., $V_1 \cap V_2 = X$ and $E_1 \cap E_2 = \emptyset$, we build the graph $G = (V_1 \cup V_2, E_1 \cup E_2, X)$. The operation 'fuses' the two terminal graphs, identifying for each i, the ith terminal node of the first graph with the ith terminal node of the second graph.

Note that with a sequence of Swap operations, we can order the terminals in any manner; one can thus easily observe that a nice path decomposition of a graph $G = (V, E)$ of pathwidth at most k (see [15]) can be transformed into a sequence of Introduce, Forget, Add-Edge and Swap operations that creates the terminal graph $G = (V, E, \emptyset)$ with $O(kn)$ operations, and with each intermediate terminal graph having at most $k + 1$ terminals. Similarly, a nice tree decomposition can be transformed to a similar sequence of Introduce, Forget, Add-Edge, Swap and Join operations. From this sequence of operations, we can go back to a *generalized tree decomposition*, i.e., we have a rooted tree, with each bag consisting of the set of terminals, and of one of the following types: Introduce, Forget, Add-Edge, Swap and Join. We omit the simple details; one can also carry out these transformations in logarithmic space.

2.2 The Classes XNLP and XALP

A *parameterized problem* is a language $L \subseteq \Sigma^* \times \mathbb{N}$ for some finite alphabet Σ. The class XNLP consists of parameterized problems that can be solved by a non-deterministic algorithm which uses $f(k)n^{O(1)}$ time and $g(k) \log n$ space for some computable function f (where k is the parameter and n is the size of the input instance).

As in the classical complexity setting, in order to define XNLP-hardness and XNLP-completeness, we need the notion of *reduction.*

Definition 2. *A* parameterized reduction *from a problem* $L_1 \subseteq \Sigma_1^* \times \mathbb{N}$ *to a problem* $L_2 \subseteq \Sigma_2^* \times \mathbb{N}$ *is a function* $f : \Sigma_1^* \times \mathbb{N} \to \Sigma_2^* \times \mathbb{N}$ *such that the following two conditions are satisfied:*

- *For all* $(x, k) \in \Sigma_1^* \times \mathbb{N}$, $(x, k) \in L_1$ *if and only if* $f((x, k)) \in L_2$,
- *There is a computable function* $g : \mathbb{N} \to \mathbb{N}$ *such that for all* $(x, k) \in \Sigma_1^* \times \mathbb{N}$ *we have* $k' \leq g(k)$, *where* $f((x, k)) = (y, k')$.

We call f *a* parameterized logspace reduction *or* pl-reduction *if there is an algorithm that computes* $f((x, k))$ *in space* $O(g'(k) + \log |x|)$, *where* $g' : \mathbb{N} \to \mathbb{N}$ *is a computable function and* $|x|$ *the size of* x *(i.e. number of bits needed to store* x*).*

We call a problem XNLP-hard if any other problem in XNLP can be pl-reduced to it, and XNLP-complete if it is XNLP-hard and in XNLP.

We define XALP as the class of problems that can be solved by a non-deterministic Turing machine with a stack in $f(k)n^{O(1)}$ time and $g(k) \log n$ space. For other equivalent definitions of XALP we refer the reader to [7].

Membership in XNLP (respectively XALP) can usually be shown by standard dynamic programming techniques on path decompositions (respectively tree decompositions). Intuitively, the log space complexity can be achieved by guessing the entry (rather than storing all the entries). Examples of membership proofs can be found in [6–8].

In our hardness proofs, we skip the details that show the logarithmic space bounds for the pl-reductions. The standard technique to realize such space is to recompute needed values instead of storing them. For example, if the reduction uses a Sidon set, we do not store the Sidon set, but each time we need the ith element of the set, we recompute it. This increases the time of the reduction by a polynomial factor, but keeps the used space small.

3 Binary CSP

The Constraint Satisfaction Problems (CSP) form a well-known class of problems with a wide set of applications, ranging from artificial intelligence to operations research. In this paper, we study Binary CSP, a CSP where all constraints are binary. Formally, the problem is defined as follows.

Binary CSP
Input: A graph $G = (V, E)$, a set $\mathcal{C}$, set $C(v) \subseteq \mathcal{C}$ for each $v \in V$, set $C(u, v) \subseteq \mathcal{C} \times \mathcal{C}$ for each ordered pair $(u, v) \in V \times V$ such that $\{u, v\} \in E$
Task: Is there a function $f : V \to \mathcal{C}$ such that for every $v \in V$, $f(v) \in C(v)$ and for every $uv \in E$, we have $(f(u), f(v)) \in C(u, v)$?

We call the elements of $\mathcal{C}$ colors, the sets $C(v)$ domains (or vertex constraints) and the sets $C(u, v)$ edge constraints. We emphasize that $C(u, v)$ contains ordered pairs, i.e. even though the graph is undirected, the ordering of vertices matters for the edge constraints.

List Coloring and Precoloring Extension are special cases of Binary CSP (for the definitions, we refer to [10]). In this paper, we strengthen two results from the literature (i.e. we show completeness for a smaller graph class): XALP-completeness parameterized by treewidth [7] and XNLP-completeness parameterized by pathwidth [8].

We show that Binary CSP is XNLP-complete for $k \times n$-grids parameterized by k. The proof is by reduction from Binary CSP parameterized by pathwidth, and the main idea is to embed a modified interval representation (where each interval has a vertical line at its left endpoint) of the given graph into a grid, and use their crossings to check whether the edge constraints are satisfied.

We also show that Binary CSP is XALP-complete parameterized by outerplanarity by reducing from Binary CSP parameterized by treewidth. The main idea is to draw the nice tree decomposition on the plane and use crossover gadgets for edge crossings. For the full proofs, we refer to [10].

4 All-or-Nothing Flow

The All-or-Nothing Flow problem asks if there is an all-or-nothing flow (i.e. a flow where every edge has either zero flow or flow equal to its capacity) with a given value. For precise definitions, see [10] and standard textbooks.

In this section, we show that All-or-Nothing Flow is XNLP-hard with the outerplanarity plus the pathwidth of the graph as parameter. The problem is known to be NP-complete [2], XNLP-complete with pathwidth as parameter [5] and XALP-complete with treewidth as parameter [7]. The problem is a good starting point for further hardness proofs, and a natural generalisation of network flow.

All-or-Nothing Flow
Input: A flow network (G, c, s, t), and an integer r.
Task: Is there an all-or-nothing flow from s to t with value exactly r in G?

We remark that the variant where we ask if there is a flow whose value is at least r is equally hard, since we can add a new source s' and an arc from s' to the old source s with capacity r.

Theorem 1. All-or-Nothing Flow *is XNLP-hard for planar graphs with outerplanarity as parameter.*

Proof. We reduce from Binary CSP with pathwidth as parameter, and employ the technique of modelling the choice of a colour for a vertex by choosing a number from a Sidon set (or Golomb ruler), see e.g. [13].

A *Sidon set* is a set of positive integers $\{a_1, \ldots, a_n\}$ with the property that each different pair of integers from the set has a different sum: $a_{i_1} + a_{i_2} = a_{j_1} + a_{j_2}$ implies $\{i_1, i_2\} = \{j_1, j_2\}$. Sidon sets are also known as Golomb rulers. Erdős and Turán [13] have shown that for each n, there is a Sidon set of size n with all numbers in the set at most $4n^2$; one can also observe from their proof that this set can be constructed with logarithmic space.

Now, suppose we are given an instance of BINARY CSP with pathwidth as parameter. That is, we have a graph $G = (V, E)$, a set D of colours, for each vertex $v \in V$, a subset $C(v) \subseteq D$, and for each edge $\{v, w\} \in E$, for the ordered pair vw, a set of allowed colour pairs $C(v, w) \subseteq C(v) \times C(w)$. Also, we are given a path decomposition of G of width at most k.

Our aim is to construct an instance of ALL-OR-NOTHING FLOW with underlying graph H such that it has a solution if and only if the given BINARY CSP instance has a solution.

First, we modify the instance as follows. By a simple modification of the colour sets, we can assume that the sets of allowed colours for vertices are disjoint, i.e., for $v \neq w$, $C(v) \cap C(w) = \emptyset$. Second, we change the path decomposition to a series of Introduce, Forget, Swap, and Add-Edge operations that give $G = (V, E, \emptyset)$ when starting with the empty graph.

Without loss of generality, we may assume that $D = \bigcup_{v \in V} C(v)$. Take a Sidon set with $|D|$ positive integers in $O(|D|^2)$. Suppose these are $a_1, \ldots, a_{|D|}$. Let $L = \max\{a_1, \ldots, a_{|D|}\} + 1$. Note that the set $\{a_1 + L, \ldots, a_{|D|} + L\}$ is also a Sidon set. Now, we assign to each colour c in D a unique element $s(c)$ from $\{a_1 + L, \ldots, a_{|D|} + L\}$.

For a vertex $v \in V(G)$, we write $S(v) = \{s(c) : c \in C(v)\}$: the set of Sidon numbers of the colours we can give to v. For a pair of vertices $v, w \in V(G)$, we write $S(v) + S(w) = \{s(c) + s(c') : c \in C(v) \wedge c' \in C(w)\}$. For an ordered pair of endpoints of an edge $\{v, w\} \in E$, we write $S(v, w) = \{s(c) + s(c') : c \in C(v) \wedge c' \in C(w) \wedge (c, c') \in C(v, w)\}$.

Gadgets. Before describing the construction of H, we introduce three auxiliary gadgets, *XOR*, *Edge* and *Swap*. In order to simplify the drawing, instead of parallel arcs, we draw one arc whose capacity is denoted by the (multi)set of capacities of the parallel arcs. We denote the multiset consisting of t ones by $\mathbf{1}_t$.

Informally, given a vertex $v \in V(G)$, the $XOR(v)$ gadget will "select" exactly one value $s(c) \in S(v)$, which corresponds to coloring v with color c. Formally, given a vertex $v \in V(G)$, $XOR(v)$ is a graph that consists of three nodes, v_1, v_2, v_3 and has $2L$ edges of capacity 1 from v_1 to v_2 and for each element of $S(v)$ an arc from v_2 to v_3 with that capacity (see Fig. 1). Note that, since all elements of $S(v)$ are between L and $2L - 1$ and the incoming flow to v_2 is at most $2L$, we cannot have more than one arc from v_2 to v_3 with nonzero flow.

The $Edge(u, v)$ gadget checks whether the constraints of an edge $\{u, v\} \in E(G)$ are satisfied and creates a "copy" of u and v. It consists of vertices $u', v', e_1, e_2, u'', v''$ which are connected as follows (see Fig. 2). The arcs from u' (respectively v') to e_1 have capacities $S(v)$ (respectively $S(u)$) and the arcs from e_1 to e_2 have capacities $S(u, v)$. We add an $XOR(u)$ (respectively $XOR(v)$) gadget

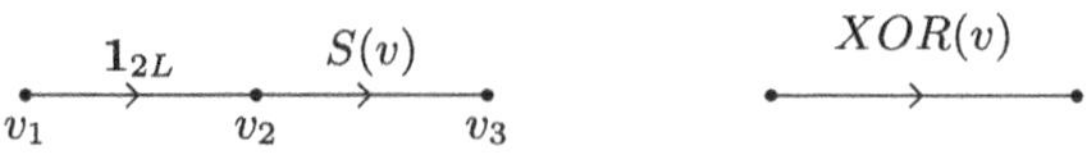

Fig. 1. $XOR(v)$ gadget and its schematic representation.

from e_2 to u'' (respectively v''). In our construction, the vertices u', v' will have no other outgoing arcs, and the vertices u'', v'' will have no other incoming arcs. The following claim follows from the Sidon set property (for the proof, refer to [10]).

Claim 1. If there is nonzero incoming flow to u' or v', then the outgoing flow of u'' (v'' respectively) is the same as the incoming flow to u' (v' respectively).

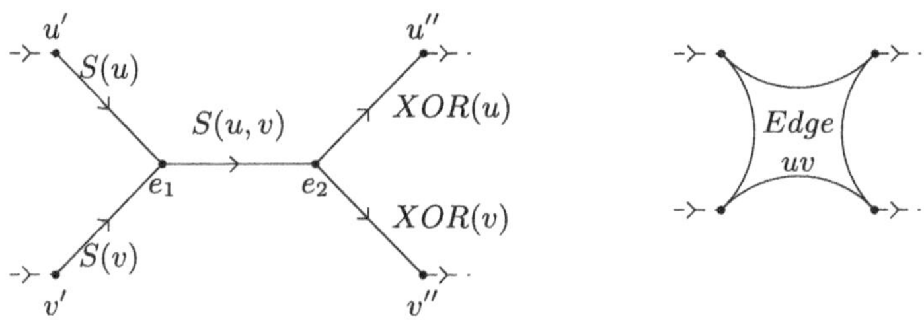

Fig. 2. The *Edge* gadget and its schematic representation.

The *Swap* gadget is identical to the *Edge* gadget with no constraints and u' and v' swapped (see Fig. 3). Analogously, the outgoing flow from u'' (respectively v'') is the same as the incoming flow to u (respectively v).

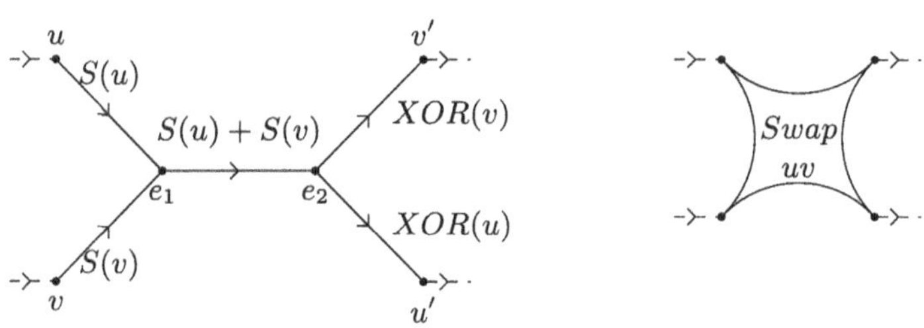

Fig. 3. The *Swap* gadget and its schematic representation.

Construction of H. We will now construct the graph H step by step, according to the series $\mathcal{S}$ of Introduce, Forget, Swap and Add-Edge operations that give $G = (V, E, \emptyset)$. We will denote the ith operation in this sequence by σ_i. We will denote by $X_i \subseteq V(G)$ the set of terminals after step i and by $\prec_i$ the ordering of X_i.

Informally, the graph H will consist of several parts. For each operation σ_i, we add a column C_i containing $|X_i|$ vertices to H. Each vertex $v \in C_i$ will have a *label* $\ell(v) \in X_i$, and the vertices in C_i will be ordered from bottom to top according to the ordering $\prec_i$ of their labels.

If σ_i is an Add-Edge (respectively Swap) operation, we will add the Edge (respectively Swap) gadget between the corresponding vertices in C_{i-1} and C_i. We will also add arcs between the remaining vertices in C_{i-1} and C_i with the same label, which will ensure that all the vertices with the same label have the same flow. The graph H will also contain a path on the bottom which will start at the source and will introduce vertices according to the order in which they are introduced in $\mathcal{S}$. Lastly, we will add arcs to the sink to enforce flow conservation (see Fig. 4).

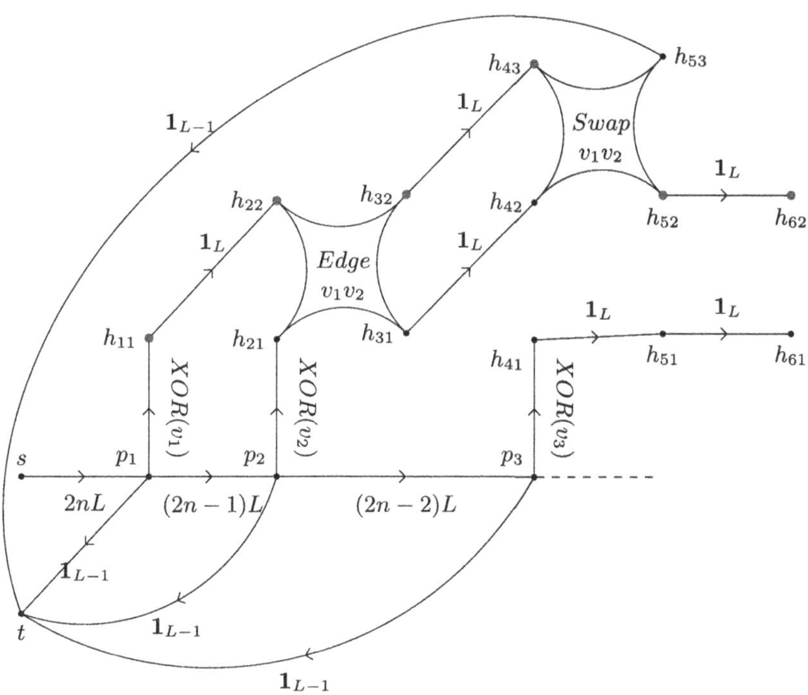

Fig. 4. Graph H corresponding to the following sequence of operations: $Introduce(v_1)$, $Introduce(v_2)$, $Add-Edge(v_1, v_2)$, $Introduce(v_3)$, $Swap(v_1, v_2)$, $Forget(v_2)$. The blue vertices are labelled with v_1. (Color figure online)

Let us now formally describe H and its embedding in the plane. To facilitate the description, we will specify the coordinates of some vertices. Firstly, we draw the source vertex s at $(0, 0)$ and the sink t at $(0, -1)$. For each $i \in \{1, ..., |\mathcal{S}|\}$

and $j \in \{1, ..., |X_i|\}$, we draw a vertex h_{ij} at (i, j). We denote the ith column by $C_i = \{h_{i1}, \ldots, h_{i|X_i|}\}$. We set $\ell(h_{ij})$ to be the jth smallest element of X_i according to $\prec_i$.

We introduce an auxiliary variable *last* and set $last = 1$. Set $p_0 = s$. For $i = 1, ..., |S|$, we add the following vertices and arcs depending on σ_i:

- **Insert** a vertex $v \in V(G)$: construct a vertex p_{last} at $(i, 0)$ and add an arc of capacity $(2n - last + 1)L$ from p_{last-1} to p_{last}. Add a gadget $XOR(v)$ starting from p_{last} and ending at h_{i1} (recall that $\ell(h_{i1}) = v$). Add $L - 1$ arcs of capacity 1 from p_j to t. For $j \in \{1, ..., |X_{i-1}|\}$, add L arcs of capacity 1 from h_{ij} to $h_{i+1,j+1}$. Increase *last* by 1.
- **Forget** a vertex $v \in V(G)$: let j be such that $\ell(h_{ij}) = v$. Add L arcs from h_{ij} to t of capacity 1. Since v is the maximum element of X_i, there are no vertices above h_{ij}, which implies that the arcs from h_{ij} to t can be drawn without intersecting any other arcs.
- **Add-Edge** $\{u, v\} \in E(G)$: let j be such that $\ell(h_{i-1,j}) = u$, $\ell(h_{i-1,j+1}) = v$. Add the gadget Edge between the vertices $h_{i-1,j}, h_{i-1,j+1}, h_{i,j}, h_{i,j+1}$ as in Fig. 4.
- **Swap** u and v: analogous to the Add-Edge case, except that we add the Swap gadget instead.

Finally, we add nL arcs of capacity 1 from p_n to t. We also add L arcs of capacity 1 from each vertex in the last column (i.e. $h_{|S|,j}$) to t.

Claim 2. For all $j \in \{1, \ldots, n\}$, the arc p_{j-1} to p_j has nonzero flow, and there is nonzero flow going out from p_j via the corresponding XOR gadget.

Proof of Claim. We will prove the claim by induction. Clearly, the arc from $p_0 = s$ to p_1 is used, since that is the only outgoing arc from the source. Suppose now that for all $j' < j$, the arc from $p_{j'-1}$ to $p_{j'}$ is used. Let the label of p_j be $\ell(p_j) = v$ and let p_j be introduced in the operation σ_i. By induction, the incoming flow of p_j is equal to $(2n - j + 1)L$. There are three ways for the flow to go out from p_j: through the $XOR(v)$ gadget, through the arc to p_{j+1} and through the arcs to t. The total capacity of the arcs in the gadget and the arcs to t is at most $2L$, so the arc from p_j to p_{j+1} has to be used. Suppose that the outgoing flow through the $XOR(v)$ is zero. In that case, the maximum possible outgoing flow is $(2n-j)L + L - 1 = (2n - j + 1)L - 1$, which leads to a contradiction. By the properties of the XOR gadget, the outgoing flow from p_j through $XOR(v)$ is equal to $s(c)$ for some color c, which proves the claim.

Note that this also implies that the incoming flow to $h_{i,1}$ is equal to $s(c)$. It is easy to see that the same amount of flow will go through all the other vertices with the same label. Thus we can assign the color c to the vertex $v \in V(G)$. For each edge in G, the corresponding *Edge* gadget checks whether the edge constraint is satisfied, so we obtain a valid coloring of G.

It remains to show that the pathwidth and outerplanarity of H are bounded by $O(k)$. The ith bag of the path decomposition will contain t and all the vertices in the strip $\{(x,y) \in \mathbb{R}^2 : i-1 \leq x \leq i\}$. It is easy to see that this gives a valid path decomposition of width at most $2k+10$.

The vertex t is on the outer face of H. After deleting t, the vertices s and $p_1, \ldots, p_n$ are on the outer face, as well as the highest vertex in each column C_i. In each of the following steps, the highest remaining vertex in each column will be on the outer face (together with the gadgets adjacent to it). Since each column has at most k elements, the outerplanarity of H is at most $k+1$. □

5 Reductions from All-or-Nothing Flow

The All-or-Nothing Flow problem appears to be a good starting point for many other hardness proofs. In all cases, the reductions replace edges or arcs by a small gadget with outerplanarity at most 2, and thus the graph stays planar and the outerplanarity increases by at most 2. This way, XNLP-hardness is obtained for each of the following problems on planar graphs with outerplanarity as parameter. Values are always given in unary. Details and background can be found in [10].

- All-or-Nothing Flow with capacities 1 or 2
- Target Outdegree Orientation (Maximum Outdegree Orientation, Minimum Outdegree Orientation): given an undirected graph, where each edge has a positive integer weight, and each vertex a positive integer target, can we orient the edges such that the outdegree of each vertex equals (respectively is at most, is at least) its target? We consider a special case where edge weights are 1 or 2.
- Capacitated Dominating Set: given a graph G, an integer k and a function $c: V \to \mathbb{N}$, is there a set $S \subseteq V$ of size at most k such that each vertex $v \in S$ is "responsible" for dominating at most $c(v)$ vertices in $V \setminus S$?
- Capacitated Red-Blue Dominating Set: given a bipartite graph $(R \cup B, E)$ (consisting of red and blue vertices), an integer k and a function $c: b \to \mathbb{N}$, is there a set $S \subseteq B$ of size at most k such that each vertex in $v \in S$ is "responsible" for dominating at most $c(v)$ red vertices?
- Capacitated Vertex Cover: given a graph G, integer k and a function $c: V \to \mathbb{N}$, is there a set $S \subseteq V$ of size k such that each vertex in $v \in S$ is "responsible" for covering at most $c(v)$ edges?
- Target Set Selection: given a graph G where each vertex v has a threshold $t(v) \in \mathbf{N}$, and integer ℓ, can we activate ℓ vertices, such that G becomes activated in the cascading process where each vertex becomes and stays activated when at least $t(v)$ neighbors are activated?
- f-Dominating Set: given a graph G, for each vertex v a value $f(v) \in \mathbb{N}$, and an integer $\ell \in \mathbb{N}$, is there a set S of k vertices, such each vertex $v \notin S$ has at least $f(v)$ neighbors in S?
- k-Dominating Set: given a graph G, integers $k, \ell \in \mathbb{N}$, is there a set S of ℓ vertices, such each vertex $v \notin S$ has at least k neighbors in S?

6 Scattered Set

We consider the Scattered Set problem, which is a generalization of the Independent Set problem and is defined as follows.

Scattered set
Input: $G = (V, E)$, $k \in \mathbb{N}$, $d \in \mathbb{N}$
Task: Is there a set $I \subseteq V$ of size k such that for any pair of distinct vertices $u, v \in I$ the distance between u and v is at least d?

We prove that Scattered Set is XALP-complete parameterized by treewidth and by outerplanarity, and XNLP-complete parameterized by treewidth. We reduce from Binary CSP. The main idea of the proof is to "expand" vertices and edges, i.e. replace vertices by cycles and edges by several paths connecting the corresponding cycles. The complete proof can be found in [10].

7 Conclusion

In this paper, we showed XALP-completeness or XNLP-hardness for several problems on planar graphs parameterized by outerplanarity. In a number of cases, these were problems not yet established to be XALP-complete for treewidth, and thus for these, we also obtained new results for the general class of graphs. For several "flow-like" problems, we could only show XNLP-hardness; we conjecture that these are also XALP-complete parameterized by outerplanarity. It would be sufficient to show that All-or-Nothing Flow is XALP-hard for outerplanarity; the follow up transformations can be used to show several other problems including Capacitated Dominating Set, Target Set Selection are XALP-complete.

The results are also interesting in the light of a conjecture by Pilipczuk and Wrochna [16], which claims that an XNLP-hard problem has no XP algorithm that uses $O(f(k)n^{O(1)})$ space. Our results are negative in the sense that it appears that for the problems studied in this setting, going from graphs of bounded treewidth to graphs of bounded outerplanarity, the complexity does not drop.

We expect that more problems that are in XP with outerplanarity as parameter are complete for XALP, and leave finding more examples as an interesting open problem. One candidate is Multicommodity Flow: in [9], it was shown that it is XALP-complete for treewidth and XNLP-complete for pathwidth; capacities are given in unary. The result already holds for 2 commodities. The gadgets used in that proof are non-planar, so establishing the complexity of Multicommodity Flow with outerplanarity seems to need new techniques.

Another interesting direction for further research is the following. Problems with an algorithm with running time of the form $O(2^{O(\ell)}n^{O(1)})$ with ℓ the treewidth or outerplanarity are in XP when we take 'logarithmic treewidth' or 'logarithmic outerplanarity' as parameter, i.e., the parameter is $\frac{\ell}{\log n}$. In [8], it was shown that Dominating Set and Independent Set are XNLP-complete with logarithmic pathwidth as parameter; this translates to XALP-completeness for these problems with logarithmic treewidth as parameter [7]. As these proofs produce non-planar graphs, it is open whether Dominating Set and Independent Set are XALP-complete when we take logarithmic outerplanarity as parameter.

References

1. Alber, J., Fellows, M.R., Niedermeier, R.: Polynomial-time data reduction for dominating set. J. ACM **51**(3), 363–384 (2004). https://doi.org/10.1145/990308.990309
2. Alexandersson, P.: NP-complete variants of some classical graph problems. arXiv, abs/2001.04120 (2020). http://arxiv.org/abs/2001.04120
3. Baker, B.S.: Approximation algorithms for NP-complete problems on planar graphs. J. ACM **41**(1), 153–180 (1994). https://doi.org/10.1145/174644.174650
4. Bodlaender, H.L.: A partial k-arboretum of graphs with bounded treewidth. Theoret. Comput. Sci. **209**(1–2), 1–45 (1998). https://doi.org/10.1016/S0304-3975(97)00228-4
5. Bodlaender, H.L., Cornelissen, G., van der Wegen, M.: Problems hard for treewidth but easy for stable gonality. In: Bekos, M.A., Kaufmann, M. (eds.) Proceedings 48th International Workshop on Graph-Theoretic Concepts in Computer Science, WG 2022, volume 13453 of Lecture Notes in Computer Science, pp. 84–97. Springer (2022). https://doi.org/10.1007/978-3-031-15914-5_7
6. Bodlaender, H.L., Groenland, C., Jacob, H., Jaffke, L., Lima, P.T.: XNLP-completeness for parameterized problems on graphs with a linear structure. In: Dell, H., Nederlof, J. (eds.)Proceedings 17th International Symposium on Parameterized and Exact Computation, IPEC 2022, volume 249 of LIPIcs, pp. 8:1–8:18. Schloss Dagstuhl - Leibniz-Zentrum für Informatik (2022). https://doi.org/10.4230/LIPIcs.IPEC.2022.8
7. Bodlaender, H.L., Groenland, C., Jacob, H., Pilipczuk, M., Pilipczuk, M.: On the complexity of problems on tree-structured graphs. In: Dell, H., Nederlof, J., (eds.) Proceedings 17th International Symposium on Parameterized and Exact Computation, IPEC 2022, volume 249 of LIPIcs, pp. 6:1–6:17. Schloss Dagstuhl - Leibniz-Zentrum für Informatik (2022). https://doi.org/10.4230/LIPIcs.IPEC.2022.6
8. Bodlaender, H.L., Groenland, C., Nederlof, J., Swennenhuis, C.M.F.: Parameterized problems complete for nondeterministic FPT time and logarithmic space. In: Proceedings 62nd IEEE Annual Symposium on Foundations of Computer Science, FOCS 2021, pp. 193–204. IEEE (2022). https://doi.org/10.1109/FOCS52979.2021.00027
9. Bodlaender, H.L., Mannens, I., Oostveen, J.J., Pandey, S., van Leeuwen, E.J.: The parameterised complexity of integer multicommodity flow. In: Misra, N., Wahlström, M. (eds.)Proceedings 18th International Symposium on Parameterized and Exact Computation, IPEC 2023, volume 285 of LIPIcs, pp. 6:1–6:19. Schloss Dagstuhl - Leibniz-Zentrum für Informatik (2023). https://doi.org/10.4230/LIPICS.IPEC.2023.6
10. Bodlaender, H.L., Szilágyi, K.: XNLP-hardness of parameterized problems on planar graphs. arXiv, abs/1004.2642 (2024). https://doi.org/10.48550/arXiv.2402.03087

11. Downey, R.G., Fellows, M.R.: Fixed-parameter tractability and completeness I: basic results. SIAM J. Comput. **24**(4), 873–921 (1995). https://doi.org/10.1137/S0097539792228228
12. Elberfeld, M., Stockhusen, C., Tantau, T.: On the space and circuit complexity of parameterized problems: classes and completeness. Algorithmica **71**(3), 661–701 (2014). https://doi.org/10.1007/s00453-014-9944-y
13. Erdös, P., Turán, P.: On a problem of Sidon in additive number theory, and on some related problems. J. London Math. Soc. **s1-16**(4), 212–215 (1941). https://doi.org/10.1112/jlms/s1-16.4.212
14. Kammer, F.: Determining the smallest k Such That G Is k-Outerplanar. In: Arge, L., Hoffmann, M., Welzl, E. (eds.) ESA 2007. LNCS, vol. 4698, pp. 359–370. Springer, Heidelberg (2007). https://doi.org/10.1007/978-3-540-75520-3_33
15. Kloks, T.: Treewidth, Computations and Approximations, volume 842 of Lecture Notes in Computer Science. Springer (1994). https://doi.org/10.1007/BFB0045375
16. Pilipczuk, M., Wrochna, M.: On space efficiency of algorithms working on structural decompositions of graphs. ACM Trans. Comput. Theo. **9**(4), 18:1–18:36 (2018). https://doi.org/10.1145/3154856

Augmenting Plane Straight-Line Graphs to Meet Parity Constraints

Aleksander Bjørn Grodt Christiansen[1], Linda Kleist[2], Irene Parada[3(✉)], and Eva Rotenberg[1]

[1] Technical University of Denmark, Lyngby, Denmark
{abgch,erot}@dtu.dk
[2] Universität Potsdam, Potsdam, Germany
kleist@cs.uni-Potsdam.de
[3] Universitat Politècnica de Catalunya, Barcelona, Spain
irene.parada@upc.edu

Abstract. Given a plane geometric graph G on n vertices, we want to augment it so that given parity constraints of the vertex degrees are met. In other words, given a subset R of the vertices, we are interested in a plane geometric supergraph G' such that exactly the vertices of R have odd degree in $G' \setminus G$. We show that the question of whether such a supergraph exists can be decided in polynomial time for two interesting cases. First, when the vertices are in convex position, we present a linear-time algorithm. Building on this insight, we solve the case when G is a plane geometric path in $O(n \log n)$ time. This solves an open problem posed by Catana, Olaverri, Tejel, and Urrutia (Appl. Math. Comput. 2020).

Keywords: Plane geometric graphs · Augmentation problems · Parity constraints · Geometric paths · Convex geometric graphs

1 Introduction

A fundamental class of problems in graph drawing concerns augmenting existing embedded graphs, such that the resulting graph has some desired properties. In this paper, we approach the natural question of augmenting graphs to meet

A.B.G.C. supported by the VILLUM Foundation grant 37507 "Efficient Recomputations for Changeful Problems". I.P. is a Serra Húnter fellow and acknowledges the support of Independent Research Fund Denmark grant 2020-2023 (9131-00044B) "Dynamic Network Analysis", the Margarita Salas Fellowship funded by the Ministry of Universities of Spain and the European Union (NextGenerationEU), and grant PID2019-104129GB-I00 funded by MICIU/AEI/10.13039/501100011033. E.R. partially supported by the VILLUM Foundation grant 37507 "Efficient Recomputations for Changeful Problems" and the Independent Research Fund Denmark grant 2020-2023 (9131-00044B) "Dynamic Network Analysis".

D. Král' and M. Milanič (Eds.): WG 2024, LNCS 14760, pp. 121–135, 2025.
https://doi.org/10.1007/978-3-031-75409-8_9

degree constraints. Given a geometric graph, that is, a graph together with a planar straight-line embedding, the problem is to augment the graph with straight-line edges such that constraints are met concerning the degrees of vertices in the resulting plane geometric graph. Even in the simplest version of this problem, where the degree constraints are modulo two, even or odd degree, the problem is NP-hard for general graphs [9], and conjectured to be NP-hard even for trees [9]. In this paper, we resolve the question about the tractability of parity constraint augmentation for geometric paths, showing it to be near-linear time solvable.

Our solution uses a connection between the parity constraint satisfaction problem and the problem of finding a compatible geometric spanning tree of a geometric graph. We use this connection to characterize a class of geometric paths which always admit an augmentation (except for obvious negative cases). The remaining paths share enough structural properties with paths on convex point sets, that we can use the same approach for these as for graphs on points in convex position. For such graphs, we give a surprisingly simple solution that computes the answer via a linear bottom-up traversal of the (arbitrarily rooted) weak dual graph.

The Planar Parity Constraint Satisfaction Problem. A *geometric graph* $G = (V, E)$ is a graph drawn in the plane such that its vertex set V is a point set in general position (no three points are collinear) and its edge set E is a set of straight-line segments between those points. A geometric graph is *plane* if no two of its edges cross and it is *convex* if its vertices are in convex position (regardless of its edges). The *visibility graph* of a plane geometric graph G, denoted by $\text{Vis}(G)$, is a geometric graph that has V as its vertex set and two vertices u and v share an edge in $\text{Vis}(G)$ if and only if $uv \notin E$ and uv does not cross any edge in E (so G can be augmented by uv). Note that $\text{Vis}(G)$ might have many edges and can be far from being plane; see Fig. 1a for an example.

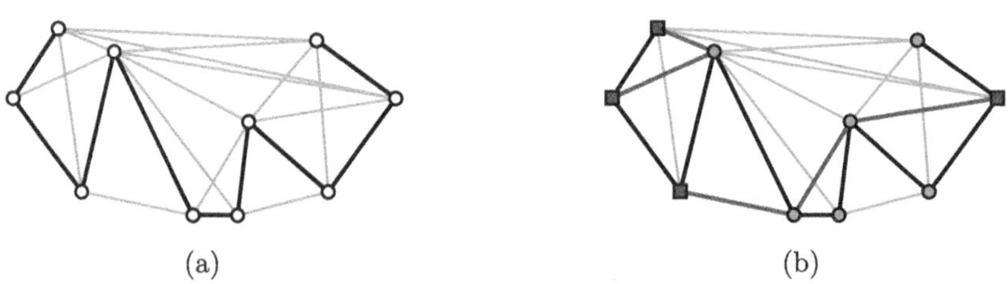

Fig. 1. (a) A plane geometric graph G (in black) and its visibility graph $\text{Vis}(G)$ (in gray). (b) A solution set for G and the four unhappy vertices (red squares).

We are interested in plane subgraphs G^+ of $\text{Vis}(G)$ such that $G \cup G^+$ satisfies given parity constraints for the vertex degrees. These constraints can be interpreted as a set of *unhappy* vertices R that would like to change the parity of their degree in G. We refer to vertices that are not unhappy as *happy*. A set of edges H in Vis(G) is called a *solution set* for (G, R) if H is crossing-free and the vertices that have odd degree in H are exactly the vertices in R; see Fig. 1b. Note that by the handshaking lemma, a solution set can only exist if $|R|$ is even.

Our Contribution. In Sect. 2, we present a linear-time algorithm that, given a convex plane geometric graph $G = (V, E)$ and any set R of unhappy vertices, outputs whether there exists a solution set. Building on these insights, in Sect. 3, we present a conceptually elegant $O(|V| \log |V|)$-time algorithm that decides the existence of solution sets for plane geometric paths, solving an open problem by Catana et al. [9]. We also characterize those paths that admit a positive answer for any even subset R of unhappy vertices.

Related Work. In graph augmentation problems, one is typically interested in adding (few) new edges to a given graph in order to achieve some desired property; for instance, 2-connectivity, a reduced diameter, to obtain an Eulerian graph, or meet other desired vertex degrees. These problems arise in many applications like transportation and telecommunication networks and constitute an active research topic in graph theory and algorithms. Augmenting graphs to meet connectivity constraints is well-studied in the general case [8,10,15,16,28] as well as in the planar case [22,29], and in the plane geometric case [1,4,5,19,23,29,30]. Reducing the diameter has also received considerable attention [2,6,12,17].

The study of augmentation to meet parity constraints and, in particular, structures involving the set of odd-degree vertices has a long history. *T-joins* are subgraphs in which the set of vertices of odd degree is exactly T. The minimum-weight T-join problem can be efficiently solved and generalizes several problems including the Chinese postman problem [14,24]. The concept of odd-vertex pairing also appears in classic results [21,26].

Augmentation to meet parity constraints has also been considered as a relaxed version of the more general problem of augmenting graphs to meet a certain degree sequence. For abstract graphs, Dabrowski et al. [13] provide a polynomial-time algorithm to achieve parity constraints. For both the plane topological setting, where edges are not required to be straight lines, and the plane geometric setting, Catana et al. [9] showed that deciding whether a general plane (geometric) graph can be augmented to meet a set of parity constraints is NP-complete; even in more restricted cases where the set of unhappy vertices R is V or all odd-degree vertices (i.e., the goal is to make the graph Eulerian). For maximal outerplane graphs in the topological setting, they show how to decide in polynomial time if the parity constrains can be met in all of the cases above. They also show that the same holds if one can only augment by adding a matching. In contrast, deciding whether a plane geometric cycle can be augmented with a (perfect) geometric matching so that all vertices change parities is known to be NP-complete [27].

The geometric setting has also been studied in a slightly different light [3,7]. Given a set of vertices along with a set of parity constraints for these vertices, Aichholzer et al. [3] showed that one can always construct a plane spanning tree and a 2-connected outerplane graph satisfying all of the constraints. García et al. [18] study the following question: Given a plane geometric tree T on a point set S, find a T-compatible plane spanning tree on S sharing the minimum number of edges with T. They show that this corresponds to finding a spanning forest of the visibility graph of T with the minimum number of components. For certain plane geometric paths, they prove that one can always find a spanning tree of the visibility graph.

2 Geometric Graphs with a Hugging Cycle

For a plane geometric graph $G = (V, E)$, a *hugging cycle* C is a plane geometric (simple) cycle with vertex set V such that the edges of G lie in the (closed) interior of C. Note that a plane geometric graph (or path) might not admit a hugging cycle or might admit more than one. We denote by $\text{Vis}^C(G)$ the restriction of $\text{Vis}(G)$ to the edges that lie in the (closed) interior of C.

An important concept towards our result is a *convexly hugging cycle*; a hugging cycle C of a geometric plane graph G such that all bounded faces of $G \cup C$ are convex. The main result in this section is the following theorem.

Theorem 1. *Let $G = (V, E)$ be a plane geometric graph and let C be a convexly hugging cycle of G, and let $R \subset V$ be a set of vertices. Then, there is an $O(|V|)$-time algorithm for deciding whether $\text{Vis}^C(G)$ contains a subgraph where exactly the vertices of R have odd degree.*

If G is a convex plane geometric graph, then the edges on the convex hull constitute a (in fact, unique) hugging cycle C and $\text{Vis}^C(G) = \text{Vis}(G)$, which, furthermore, is convexly hugging. We thus directly obtain the following corollary:

Corollary 1. *Let $G = (V, E)$ be a convex plane geometric graph, and let $R \subseteq V$ be the set of unhappy vertices. There exists an $O(|V|)$-time algorithm to decide whether (G, R) admits a solution, assuming the convex hull is part of the input.*

We remark that if G is a connected convex plane geometric graph, the convex hull can be computed in linear time [25]. In contrast, for disconnected convex plane geometric graphs, computing the hugging cycle is lower bounded by $\Omega(|V| \log(|V|))$ in the decision tree model of computing.

Now we present the proof of Theorem 1. Let C denote the convexly hugging cycle for G. In a first step, we consider a geometric subgraph G_f of G induced by a bounded (convex) face f of $G \cup C$, i.e., all edges are on the convex hull. To characterize the existence of solution sets for these cases, we distinguish three scenarios: (i) all edges are present, i.e., G_f is a cycle (Lemma 1); (ii) exactly one edge is missing, i.e., G_f is a path (Lemma 2); and (iii) at least two edges are missing, i.e., G_f is a collection of paths and isolated vertices (Lemma 3). The proofs of Lemmas 1 to 3 can be found in the full version.

Lemma 1 ($\star$). *Let $C = (V, E)$ be a convex plane geometric cycle and $R \subseteq V$. Then (C, R) admits a solution set if and only if $|R| = 0$ or $|R|$ is even and $V \setminus R$ contains two non-adjacent vertices.*

Lemma 2 **(⋆).** *Let $G = (V, E)$ be a convex geometric path with $|V| \geq 3$ and all edges on the boundary of the convex hull. Given $R \subseteq V$, there exists a solution set for (G, R) if and only if $|R|$ is even and $V \setminus R$ contains an internal vertex of G.*

Lemma 3 **(⋆).** *Let $G = (V, E)$ be a collection of at least two geometric paths (could be singleton sets) on a convex point set with $|V| \geq 3$ and all edges on the boundary of the convex hull. Given $R \subseteq V$, there exists a solution set for (G, R) if and only if $|R|$ is even.*

Using Lemmas 1 and 3, we present a linear-time algorithm for any plane geometric graph with a convexly hugging cycle. To this end, we use the concept of the weak dual. The *weak dual* of a plane graph is the plane graph that has a vertex for each bounded face and an edge between two vertices if and only if the corresponding faces share an edge. For a plane geometric graph $G = (V, E)$ with a convexly hugging cycle C, the weak dual graph of $G \cup C$ is a tree since it is the weak dual of a biconnected outerplanar graph. For clarity, we sometimes refer to the vertices of the weak dual as the faces they correspond to.

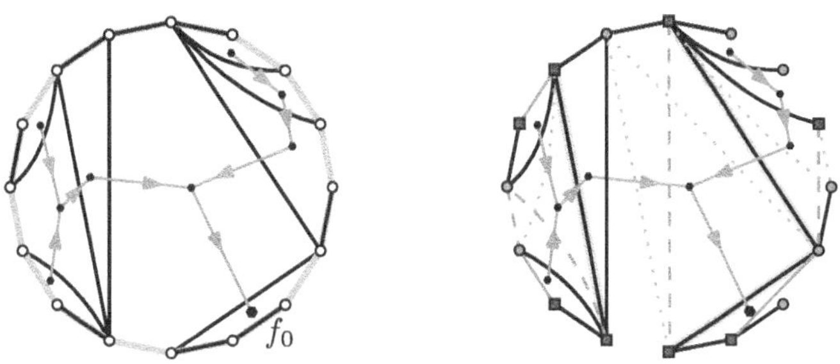

(a) (b)

Fig. 2. (a) A plane geometric graph G, a convexly hugging cycle (gray), and their dual tree $D(G)$ rooted at f_0. (b) A solution set (dashed and solid blue). The highlighted edges (pink) are marked flexible in the execution of the algorithm. (Color figure online)

Proof (of Theorem 1). Consider the convex weak dual graph $D(G)$ of $G \cup C$. As remarked above, $G \cup C$ is a biconnected outerplanar geometric graph and thus, $D(G)$ is a tree. We root $D(G)$ at a leaf f_0 and orient all edges towards f_0, making $D(G)$ an in-arborescence. See Fig. 2a for an illustration on a convex point set, and note that our proof holds in the more general case of plane geometric graphs with a convexly hugging cycle.

For every face f that is not the root, we define the *connector edge* uv to be the edge of G that f shares with its parent in $D(G)$. At the root f_0, let the connector edge be an edge of C incident to this face. Note that since f_0 is a leaf of $D(G)$ such an edge exists and it is not shared with the child face.

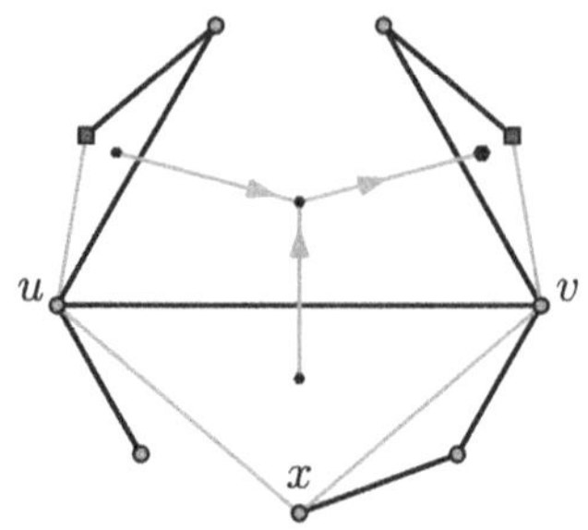

Fig. 3. A convex plane geometric tree T (black), a set R (red), and the unique solution set (in gray). In it, u and v are not made happy within the bottom face, while it is trivial to satisfy that face locally. This shows that just answering whether a face can be satisfied locally is not enough information. (Color figure online)

The idea of the algorithm is to consider $D(G)$ bottom up, taking care at each step of a face f of $G \cup C$ whose children have already been processed. For a face f (different from the root) with connector edge uv, we define $\Pi(f, p_u, p_v)$ to be the problem asking whether the geometric graph corresponding to the subtree of $D(G)$ rooted at f has a solution that satisfies the given parity constraints, where the parity constraints at u and v are replaced by p_u and p_v.

To solve $\Pi(f, p_u, p_v)$, we use bottom-up dynamic programming in the tree $D(G)$. Note that the graph G_f restricted to face f has all edges on its convex hull. Hence, Lemmas 1 to 3 characterize when there exists a solution set. Note that these lemmas are constructive and provide a solution set in linear time when it exists. Consider a child f' of f with connector edge wx; we assume that problem $\Pi(f', p_w, p_x)$ has already been solved for all possible values of p_w, p_x. We call f' and its connector edge *inflexible* if there is only one set of parities p_w, p_x such that $\Pi(f', p_w, p_x)$ has a solution, otherwise we call them *flexible*. The example in Fig. 3 shows that it is necessary to do the book-keeping of flexible faces; otherwise we may falsely conclude that there exists no solution set.

Case (i): If f has no flexible child, then the vertices on face f have no choice. Thus, we can immediately determine whether $\Pi(f, p_u, p_v)$ has a solution by applying Lemmas 1 and 3 on the instance in f to fulfill the parity constraints from $\Pi(f, p_u, p_v)$.

Case (ii): If f has at most two flexible children, say with connector edges wx and yz, we consider all at most sixteen choices for $p_w, p_x, p_y, p_z \in \{0, 1\}$. For each of them, we first check whether there is a solution in the children with these parity constraints. (Note that we only need to consider four choices, because the number of unhappy vertices in any problem needs to be even for there to exist a solution.) Then, for each of these choices that is valid for the children, we produce an instance for f where we adjust the happiness status of w, x, y, z so that the combination with p_w, p_x, p_y, p_z gives the desired parities. This instance can be checked using Lemmas 1 and 3.

Case (iii): If f has at least three flexible children, we show that $\Pi(f, p_u, p_v)$ always has a solution set if p_u, p_v are such that the total number of unhappy vertices is even; otherwise there cannot be a solution. If f has multiple flexible children, trying all parity combinations for the vertices could lead to exponentially many cases. We are saved by the following observation: when a face contains at least three flexible edges, then there is a choice for them that produces two non-consecutive vertices on f that do not need to change their parities in f to satisfy $\Pi(f, p_u, p_v)$. Lemma 1 then ensures that any of the two options of parity constraints for u and v such that the total number of unhappy vertices in the problem is even, are satisfiable in f.

For a face f, if $\Pi(f, p_u, p_v)$ admits a solution for more than one value of p_u, p_v (there are at most two such problems that can be solvable), f and the connector edge uv are marked flexible and, if $f \neq f_0$, we proceed to a new face. For an example consider Fig. 2b. If at any face none of the problems admits a solution we conclude that no solution set for (G, R) exists. If we reached the root f_0 and the problem $\Pi(f_0, p_u, p_v)$ where p_u, p_v match the original parity constraints has a solution, we conclude that (G, R) admits a solution set. Note that using this algorithm it is easy to efficiently construct a solution set for (G, R) if one exists.

The correctness of the algorithm is based on the following fact. Let $uv \in E$ be the connector edge for f. Then, uv separates G into two subgraphs, one of which contains the vertices in f and all the descendant faces. Thus, uv defines two subinstances of the problem that only overlap in u and v. There exists a solution set for (G, R) if and only if those subinstances admit solutions that together satisfy the parity constraints of u and v. This implies that the algorithm checks all the possibilities, as it keeps all possible parities of u and v in valid solutions of the subinstance considered so far.

It remains to show that the constructed solutions are valid. Processing face f requires to satisfy all the parity constraints from R on all the vertices in f except for u and v. The last check in the root face f_0 enforces the parity constraints from R on all its vertices. Thus, for each vertex, the last face containing it enforces the correct parity.

The time spent by the algorithm in each face is linear in its size. Thus, given the convexly hugging cycle C, $D(G)$ can be computed in linear time and the algorithm runs in $O(|V|)$ time. □

3 Plane Geometric Paths

In this section, we efficiently solve the decision problem for plane geometric paths.

Theorem 2. *Let $G = (V, E)$ be a plane geometric path and let $R \subseteq V$. There exists an algorithm to decide whether (G, R) admits a solution set in $O(|V| \log |V|)$ time.*

To prove this result, we consider two cases. Either the path is *universally happy*, i.e., there exist solution sets for any unhappy subset R of even cardinality, or the path has a particular structure (called *pseudo-convexity*) which allows us to use our algorithm for Theorem 1.

3.1 Universal Happiness

Universal happiness requires Vis(G) to be connected: otherwise there exists no solution set when R consist of an odd number of vertices from two different components of Vis(G); see Fig. 4.

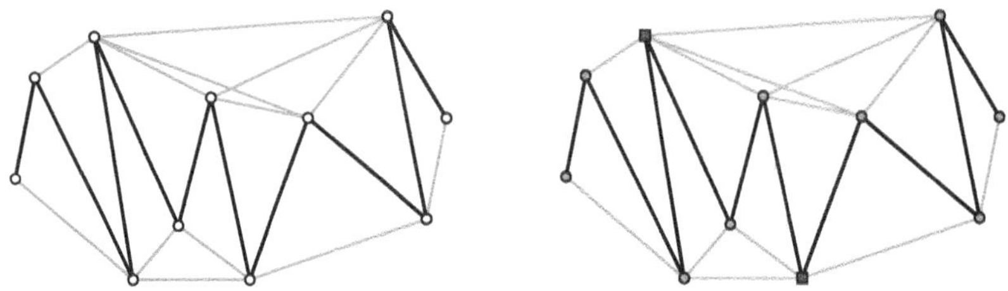

(a) (b)

Fig. 4. An example in which Vis(G) is not connected and an even set of two unhappy vertices (square) that does not admit a solution.

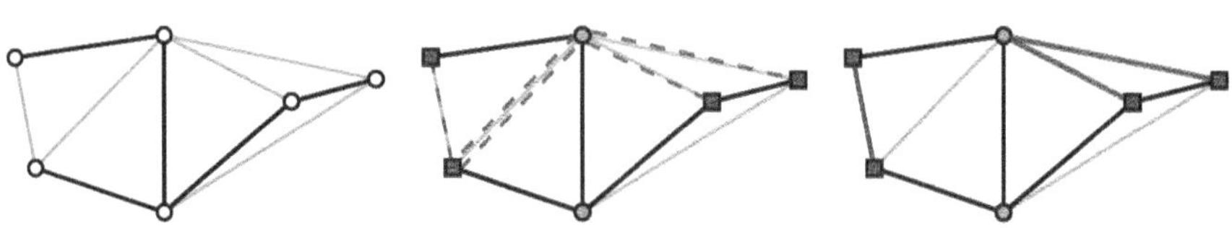

(a) (b) (c)

Fig. 5. Illustration for the proof of Lemma 4. (a) A plane geometric graph G (in black) and a plane spanning tree T of Vis(G) (in gray). (b) A set R and paths (in dashed red) between (an arbitrary partition of R into) pairs of R in T. (c) The symmetric difference of the paths yields a solution set (in red). (Color figure online)

In the next lemma we show that the existence of a plane spanning tree in Vis(G) is sufficient as long as $|R|$ is even. The proof uses standard arguments (for example, in the context of T-joins, and also similar to Theorem 3 in [21]).

Lemma 4. *Let* $G = (V, E)$ *be a plane geometric graph such that* Vis(G) *has a plane spanning tree* T. *Then for all* $R \subseteq V$ *with* $|R|$ *even* (G, R) *admits a solution set.*

Proof. Consider a fixed but arbitrary $R \subseteq V$ with $|R|$ even. We partition R into pairs. For each such pair (u, v), we consider the unique uv-path P_{uv} in T. Augmenting G by P_{uv} satisfies the parity constraints for u and v without altering the happiness status of the other vertices. To construct a solution set we can iterate over the paths P_{uv} swapping their edges. More formally, let H^m denote the edge multi-set of the union of all these paths P_{uv}. Deleting two copies of an edge maintains the parities of the degrees of all vertices. Consequently, we can define a solution set H for (G, R) as the subset of H^m consisting on the edges of odd multiplicity; see Fig. 5 for an illustration. □

García et al. [18] give a characterization of the existence of a plane spanning tree in the visibility graph for the case of a path. To use it, we introduce some notation. Let $G = (V, E)$ be a plane geometric path, and let $q_0, \ldots, q_{k-1}$ be the vertices on the convex hull of V ordered clockwise. Adding an edge $e = q_i q_{i+1} \in \text{Vis}(G)$ to G creates a unique face (indices are considered modulo k). Its boundary is a geometric graph that we call *pocket*: it consists of the edge e together with the subpath of G connecting the endpoints of e. We say that a pocket $P = v_0, v_1, \ldots, v_s$ with $v_0 = q_i, v_s = q_{i+1}$ for some i is *pseudo-convex* if it does not contain an endpoint of G in its interior and it is either convex or every reflex vertex $v_j \in P$ satisfies the condition that the rays obtained by extending the edges $v_{j-1}v_j$ and $v_{j+1}v_j$ (indices considered modulo s) intersect P for the first time at $e = q_i q_{i+1}$. Note that if G is pseudo-convex, its two endpoints lie on the convex hull. Figure 6 illustrates a path and its pockets. If every pocket of G is pseudo-convex, we say that G is pseudo-convex.

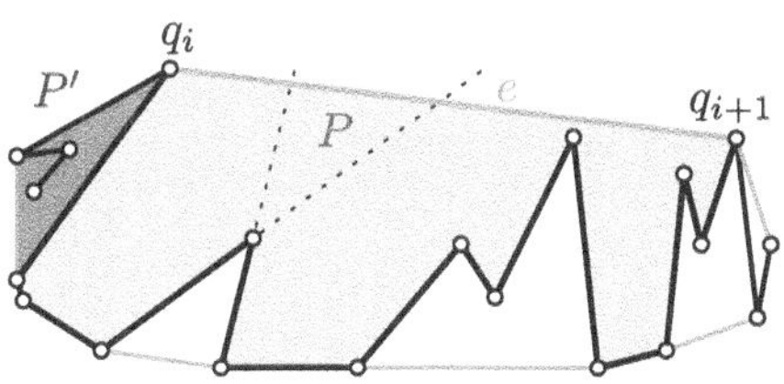

Fig. 6. Example of a path and its pockets. While pocket P (with blue interior) is pseudo-convex, pocket P' (with red interior) is not. (Color figure online)

We exploit the following fact which has been showed by García et al. [18] using a slightly different language.

Lemma 5 (Theorem 3 and Remark 2 in [18]). *Let $G = (V, E)$ be a plane geometric path.* Vis(G) *has a plane spanning tree if and only if G is not pseudo-convex.*

Using Theorem 5 and our tools of the proof for Theorem 2, we show in Sect. 3.3 that the sufficient condition in Lemma 4 is also necessary.

Theorem 3. *Let $G = (V, E)$ be a plane geometric path. This geometric graph G is universally happy if and only if* Vis(G) *has a plane spanning tree, or equivalently, G is not pseudo-convex.*

3.2 Tight Hulls and Proof of Theorem 2

Our main tools to prove Theorems 2 are 1 and Lemmas 5 and 4. The last ingredient is to bridge the gap between pseudo-convexity and convexity.

Let $G = (V, E)$ be a plane geometric pseudo-convex path and consider a (pseudo-convex) pocket P induced by some edge $e = q_i q_{i+1} \in \text{Vis}(G)$. Let $r_1, \ldots, r_s$ be the reflex vertices of P ordered counterclockwise. The *tight hull* of P consists of the edges $q_i r_1, r_1 r_2, \ldots, r_{s-1} r_s, r_s q_{i+1}$; see Fig. 7 for an illustration. Each edge on the tight hull of P defines a unique face, creating a *subpocket* of P. García et al. [18] show that this tight hull always exists and each subpocket is convex (Lemma 4.i in [18]). Replacing the convex hull edge of every pocket of G by the tight hull of the pocket, one gets the tight hull of G.

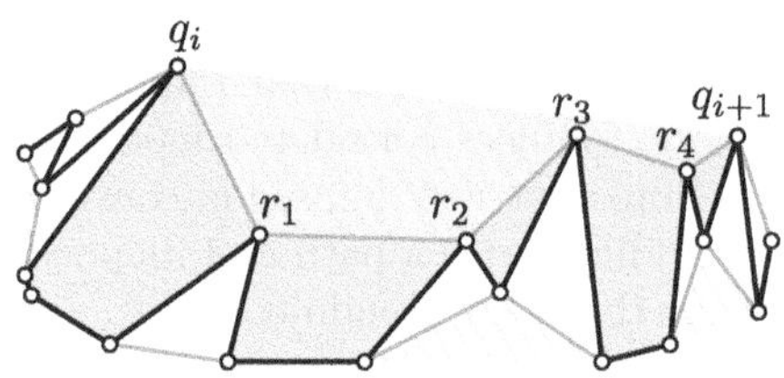

Fig. 7. Example of a pseudo-convex path and the tight hull of a pocket (with blue interior), its subpockets (with darker blue interior), and the tight hull of G forming a hugging cycle (in gray). (Color figure online)

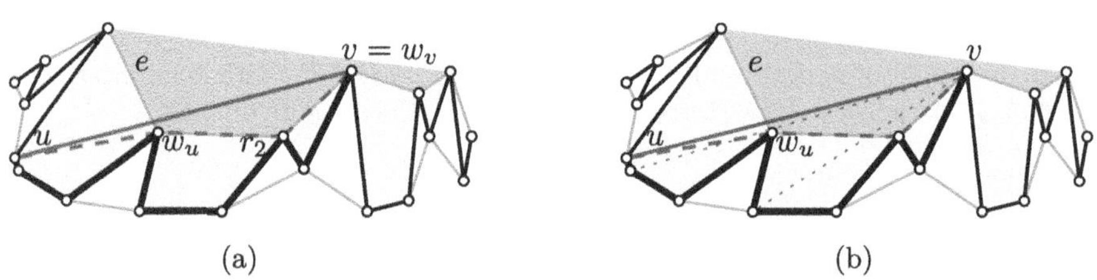

Fig. 8. Illustration for the proof of Lemma 7. (a) An edge uv of Type (i) crossing the tight hull (in solid red) and the edges replacing it (in dashed red). (b) Edges from H crossing replacement edges would be of Type (i) and define a shorter subpath in G, contradicting the choice of uv. (Color figure online)

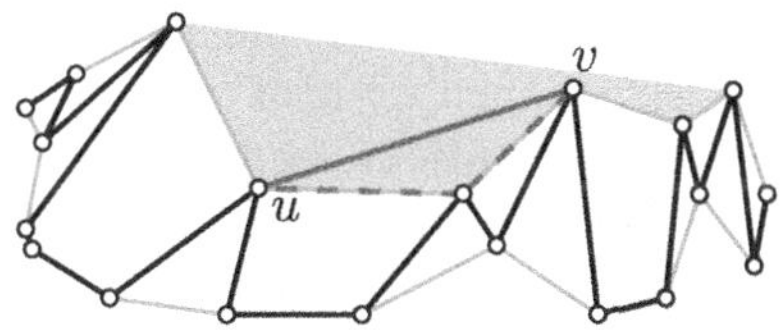

Fig. 9. Illustration for the proof of Lemma 7. An edge uv of Type (ii) in the interior of the tight hull.

Lemma 6. *Let $G = (V, E)$ be a plane geometric pseudo-convex path. Then the tight hull T of G is a convexly hugging cycle.*

Proof. By definition, the edges of G lie in the (closed) interior of T. We now show that T contains all vertices. Suppose there exists a vertex v in the strict interior of T. Then v is contained in the interior of some pocket of G, guaranteeing that the pocket and thus G are not pseudo-convex. We note that T is a cycle: by construction it is connected and each vertex has degree 2. Finally, the convexity of all bounded faces of $G \cup T$ follows from the convexity of the subpockets. $\square$

Looking for a solution set of a pseudo-convex path, we now show that we may restrict our attention to the tight hull T, more precisely, to $\text{Vis}^T(G)$, the restriction of $\text{Vis}(G)$ to the tight hull and its interior.

Lemma 7. *Let $G = (V, E)$ be a plane geometric pseudo-convex path, T its tight hull, and let $R \subseteq V$. If (G, R) allows a solution set H then there also exists a solution set H' within* $\text{Vis}^T(G)$.

Proof. We distinguish two types of *unwanted* edges in H: (i) edges crossing the tight hull and (ii) edges in the exterior of the tight hull (not on its boundary).

Let uv be an unwanted edge of Type (i) in a pocket P of G. Recall that subpockets are convex. Since uv is part of a solution set, it can cross at most two tight hull edges. Let e be the edge of the tight hull crossed by uv such that the subpocket P° of e includes u. By convexity of P°, vertex u is convex in P° (and in P); the other vertex v lies outside P°. For an illustration see Fig. 8a.

Among all such edges of Type (i), we consider the edge uv that minimizes the number of vertices on the subpath π of G between u and v. Let w_u be the first reflex vertex (of P) encountered when walking along π from u towards v, and let w_v be the first reflex vertex (of P) encountered when walking along π from v towards u. If v itself is a reflex vertex, then let $w_v = v$. Note that, by convexity of the subpockets, we have $uw_u \in \text{Vis}(G)$ and $vw_v \in \text{Vis}(G)$.

The idea is as follows. Let $w_u = r_1, r_2, \ldots, r_\ell = w_v$ be the reflex vertices of P belonging to π, see also Fig. 8a. We aim to replace the solution set H by $H \triangle \{uw_u, r_1r_2, \ldots, r_{\ell-1}r_\ell, w_vv, vu\}$, where $\triangle$ denotes the symmetric difference. We show that none of the replacement edges intersect an edge of H.

To this end, we show that by the choice of uv, the edges uw_u and vw_v do not intersect any edge of H. Any edge of H crossing uw_u but not uv has to have one endpoint $x \in \pi[u, w_u]$ distinct from both u and w_u, and one endpoint y not in $\pi[u, w_u]$. If y belongs to $\pi[w_u, v]$, then xy is of of Type (i) and thus contradicts the minimality of uv. For an illustration, see Fig. 8b. If y is outside of π, then xy crosses uv, which is not possible since both edges were part of the solution set H. Analogously, we have $vw_v \in \text{Vis}(G)$.

The situation is similar when considering a replacement edge $r_i r_{i+1}$. Suppose that edge $h \in H$ intersects $r_i r_{i+1}$. Then h is of Type (i). If both vertices of h belong to π then h contradicts the minimality of uv. Otherwise, h has exactly one endpoint not belonging to π, but then h must cross the closed curve $uv \cup \pi$, contradicting the properties of H.

Thus, we replace the solution set H by $H \triangle \{uw_u, r_1 r_2, \ldots, r_{\ell-1} r_\ell, w_v v, vu\}$ and obtain a non-crossing edge set. Because $\{uw_u, r_1 r_2, \ldots, r_{\ell-1} r_\ell, w_v v, vu\}$ forms a cycle, all degree parities are maintained. Moreover, H contains fewer unwanted edges of Type (i). Repeating this procedure, we obtain a solution set H without unwanted edges of Type (i).

Suppose now that H contains an unwanted edge of Type (ii) between u and v in a pocket P; see Fig. 9.

Let π denote the path on the tight hull between u and v that lies inside P. Since uv is not on the tight hull, π must contain other vertices apart from u and v. The absence of Type (i) edges guarantees that π does not cross any edge from H. Thus, since $\pi \cup \{uv\}$ forms a cycle, replacing H by $H \triangle (\pi \cup \{uv\})$ results in a solution set with fewer edges of Type (ii). Repeating this procedure, yields the lemma. □

Proof (of Theorem 2). First we check whether G is pseudo-convex: Using existing ray shooting techniques in polygons [11,20], the first intersection point on a ray can be computed in $O(\log |V|)$ time. For $O(|V|)$ reflex vertices, we obtain $O(|V| \log |V|)$ time, which also bounds the preprocessing time required. If G is not pseudo-convex, then Theorem 5 guarantees the existence of a plane spanning tree in $\text{Vis}(G)$ and G is universally happy by Lemma 4. If G is pseudo-convex, the tight hull T can be computed in linear time by considering the vertices in the order given by the path. By Lemma 6, T is a convexly hugging cycle and Lemma 7 guarantees a solution set in $\text{Vis}^T(G)$ if there exists a solution set in $\text{Vis}(G)$. Hence, we may use the linear-time algorithm given in the proof of Theorem 1 to decide whether there exists a solution set. □

3.3 Proof of Theorem 3

We are now ready to show that a plane geometric path is universally happy if and only if its visibility graph has a plane spanning tree.

Proof (of Theorem 3). By Lemma 4, we know that the existence of a plane spanning tree in $\text{Vis}(G)$ implies universal happiness of G. It remains to show that it is a necessary condition.

Suppose $\text{Vis}(G)$ does not contain a plane spanning tree. Then, by Theorem 5, G is pseudo-convex. By Lemma 6, the tight hull T of G is a hugging cycle

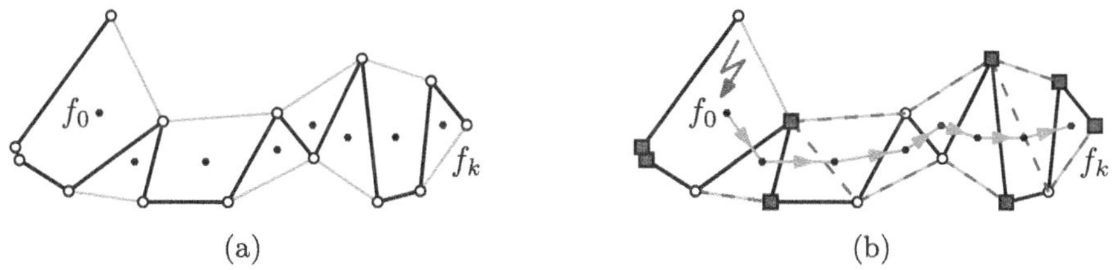

Fig. 10. Illustration for the proof of Theorem 3. (a) A pseudo-convex path G and its weak dual tree $D(G)$ of $G \cup T$. (b) A set of unhappy vertices R that has no solution set.

and by Lemma 7, we may restrict our attention to edges in $\text{Vis}^T(G)$. We now construct a set R without a solution set by considering the weak dual tree $D(G)$ of the outerplane graph $G \cup T$. Root $D(G)$ at an arbitrary leaf face f_0. Note that $D(G)$ is a path on vertices $f_0, f_1, \dots, f_k$. For an illustration, consider Fig. 10. In particular, we consider the faces in reverse order, beginning from the other leaf face f_k (as the algorithm of Theorem 1 would do in order to compute a solution set or decide that none exist).

As before, we denote the dual edge of $f_i f_{i-1}$ as the connector edge of face f_i. For a face $f \neq f_0$, let V_f be the set of vertices on its boundary and uv its connector edge. By renaming, we may assume that u is incident to the unique edge shared by f and the tight hull. For $f = f_0$, let u denote the endpoint of G in f_0 and v its neighbor in G. For $f = f_k$, we define $R := V_f$ if $|V_f|$ is even and $R := V_f \setminus \{u\}$ if $|V_f|$ is odd. Clearly, the vertices $V_f \setminus \{u, v\}$ can only be satisfied by edges inside f. By Lemma 2, there exists a solution set inside f satisfying $V_f \setminus \{u, v\}$ only if v serves as the happy inner vertex. More precisely, the unique way to satisfy $V_f \setminus \{u, v\}$ inside f is by adding the star with center u and leaves $V_f \setminus \{u, v\}$. To see that, note that any plane triangulation of f has at least two vertices of degree 2. The only choices for these vertices are v and the neighbor of u on the tight hull of f. It follows that the connector vertices u and v are thus guaranteed to be unhappy after inserting edges in f_k to satisfy the vertices of $V_{f_k} \setminus \{u, v\}$.

Processing the faces in the order $f_k, f_{k-1}, f_{k-2}, \dots, f_0$, we now add vertices of G to R. To this end, let $U_f := \bigcup_{f' \text{ after } f \text{ in } D(G)} V_{f'}$ denote the set of all previously considered vertices. For a face f, we add $V_f \setminus U_f$ to R if $|V_f|$ is even and $V_f \setminus \{u\} \cup U_f$ otherwise. This maintains the invariant that u and v of a face f are unhappy in any solution set satisfying $U_f \setminus \{u, v\}$ in the later faces. It follows that all vertices of V_f are unhappy and thus for any face $f \neq f_0$ in any solution set, all vertices in $V_f \setminus \{v\}$ must be connected to u. Finally, in f_0 every vertex in $V_f \setminus \{u\}$ is unhappy. Consequently, no inner vertex is happy, and by Lemma 2, there exists no solution set. $\square$

4 Conclusion

We showed how to decide in polynomial time whether a plane geometric path can be augmented (with straight-line edges and without creating crossings) to meet some given parity constraints of the vertices. The main question that remains is whether an analogous result exists for plane geometric trees. More restricted versions of this problem are also open, in particular when set of unhappy vertices consists of all the vertices or all the odd-degree vertices.

References

1. Abellanas, M., Olaverri, A.G., Hurtado, F., Tejel, J., Urrutia, J.: Augmenting the connectivity of geometric graphs. Comput. Geom. **40**(3), 220–230 (2008). https://doi.org/10.1016/j.comgeo.2007.09.001
2. Adriaens, F., Gionis, A.: Diameter minimization by shortcutting with degree constraints. In: Zhu, X., Ranka, S., Thai, M.T., Washio, T., Wu, X. (eds.) Proceedings of the IEEE International Conference on Data Mining (ICDM 2022), pp. 843–848. IEEE (2022). https://doi.org/10.1109/ICDM54844.2022.00095
3. Aichholzer, O., et al.: Plane graphs with parity constraints. Graphs and Combinatorics **30**(1), 47–69 (2012). https://doi.org/10.1007/s00373-012-1247-y
4. Akitaya, H.A., Inkulu, R., Nichols, T.L., Souvaine, D.L., Tóth, C.D., Winston, C.R.: Minimum weight connectivity augmentation for planar straight-line graphs. Theor. Comput. Sci. **789**, 50–63 (2019). https://doi.org/10.1016/j.tcs.2018.05.031
5. Al-Jubeh, M., Ishaque, M., Rédei, K., Souvaine, D.L., Tóth, C.D., Valtr, P.: Augmenting the edge connectivity of planar straight line graphs to three. Algorithmica **61**(4), 971–999 (2011). https://doi.org/10.1007/s00453-011-9551-0
6. Alon, N., Gyárfás, A., Ruszinkó, M.: Decreasing the diameter of bounded degree graphs. J. Graph Theory **35**(3), 161–172 (2000)
7. Alvarez, V.: Parity-constrained triangulations with steiner points. Graphs and Combinatorics **31**(1), 35–57 (2013). https://doi.org/10.1007/s00373-013-1389-6
8. Benczúr, A.A., Karger, D.R.: Augmenting undirected edge connectivity in $\tilde{O}(n^2)$ time. J. Algorithms **37**(1), 2–36 (2000). https://doi.org/10.1006/jagm.2000.1093
9. Catana, J.C., Olaverri, A.G., Tejel, J., Urrutia, J.: Plane augmentation of plane graphs to meet parity constraints. Appl. Math. Comput. **386**, 125513 (2020). https://doi.org/10.1016/j.amc.2020.125513
10. Cen, R., Li, J., Panigrahi, D.: Augmenting edge connectivity via isolating cuts. In: Proceedings of the 2022 ACM-SIAM Symposium on Discrete Algorithms (SODA 2022), pp. 3237–3252. SIAM (2022). https://doi.org/10.1137/1.9781611977073.127
11. Chazelle, B., et al.: Ray shooting in polygons using geodesic triangulations. Algorithmica **12**(1), 54–68 (1994). https://doi.org/10.1007/BF01377183
12. Chung, F.R.K., Garey, M.R.: Diameter bounds for altered graphs. J. Graph Theory **8**(4), 511–534 (1984). https://doi.org/10.1002/jgt.3190080408
13. Dabrowski, K.K., Golovach, P.A., van 't Hof, P., Paulusma, D.: Editing to Eulerian graphs. J. Comput. Syst. Sci. **82**(2), 213–228 (2016). https://doi.org/10.1016/j.jcss.2015.10.003
14. Edmonds, J., Johnson, E.L.: Matching, euler tours and the Chinese postman. Math. Program. **5**(1), 88–124 (1973). https://doi.org/10.1007/BF01580113
15. Eswaran, K.P., Tarjan, R.E.: Augmentation problems. SIAM J. Comput. **5**(4), 653–665 (1976). https://doi.org/10.1137/0205044

16. Frank, A.: Augmenting graphs to meet edge-connectivity requirements. SIAM J. Discret. Math. **5**(1), 25–53 (1992). https://doi.org/10.1137/0405003
17. Frati, F., Gaspers, S., Gudmundsson, J., Mathieson, L.: Augmenting graphs to minimize the diameter. Algorithmica **72**(4), 995–1010 (2014). https://doi.org/10.1007/s00453-014-9886-4
18. García, A., Huemer, C., Hurtado, F., Tejel, J.: Compatible spanning trees. Comput. Geom. **47**(5), 563–584 (2014). https://doi.org/10.1016/j.comgeo.2013.12.009
19. García, A., et al.: Geometric biplane graphs II: graph augmentation. Graphs and Combinatorics **31**(2), 427–452 (2015). https://doi.org/10.1007/s00373-015-1547-0
20. Guibas, L.J., Hershberger, J., Snoeyink, J.: Compact interval trees: a data structure for convex hulls. Int. J. Comput. Geom. Appl. **1**(1), 1–22 (1991). https://doi.org/10.1142/S0218195991000025
21. Hadlock, F.: Finding a maximum cut of a planar graph in polynomial time. SIAM J. Comput. **4**(3), 221–225 (1975). https://doi.org/10.1137/0204019
22. Kant, G., Bodlaender, H.L.: Planar graph augmentation problems. In: Dehne, F., Sack, J.-R., Santoro, N. (eds.) WADS 1991. LNCS, vol. 519, pp. 286–298. Springer, Heidelberg (1991). https://doi.org/10.1007/BFb0028270
23. Kranakis, E., Krizanc, D., Morales-Ponce, O., Stacho, L.: Bounded length, 2-edge augmentation of geometric planar graphs. Discret. Math. Algorithms Appl. **4**(3) (2012). https://doi.org/10.1142/S179383091250036X
24. Kwan, M.K.: Graphic programming using odd or even points. Chinese Math **1**, 273–277 (1960)
25. Melkman, A.A.: On-line construction of the convex hull of a simple polyline. Inf. Process. Lett. **25**(1), 11–12 (1987). https://doi.org/10.1016/0020-0190(87)90086-X
26. Nash-Williams, C.S.J.A.: On orientations, connectivity and odd-vertex-pairings in finite graphs. Can. J. Math. **12**, 555–567 (1960). https://doi.org/10.4153/CJM-1960-049-6
27. Pilz, A., Rollin, J., Schlipf, L., Schulz, A.: Augmenting geometric graphs with matchings. In: Proceedings of the 28th International Symposium on Graph Drawing and Network Visualization (GD 2020). LNCS, vol. 12590, pp. 490–504. Springer (2020). https://doi.org/10.1007/978-3-030-68766-3_38
28. Plesník, J.: Minimum block containing a given graph. Arch. Math. **27**(6), 668–672 (1976)
29. Rutter, I., Wolff, A.: Augmenting the connectivity of planar and geometric graphs. J. Graph Algorithms Appl. **16**(2), 599–628 (2012). https://doi.org/10.7155/jgaa.00275
30. Tóth, C.D.: Connectivity augmentation in planar straight line graphs. Eur. J. Comb. **33**(3), 408–425 (2012). https://doi.org/10.1016/j.ejc.2011.09.002

Covering a Graph with Minimal Local Sets

Nathan Claudet(✉) and Simon Perdrix

Inria Mocqua, LORIA, CNRS, Université de Lorraine, 54000 Nancy, France
nathan.claudet@inria.fr

Abstract. Local sets, a graph structure invariant under local complementation, have been originally introduced in the context of quantum computing for the study of quantum entanglement within the so-called graph state formalism. A local set in a graph is made of a non-empty set of vertices together with its odd neighborhood. We show that any graph can be covered by minimal local sets, i.e. that every vertex is contained in at least one local set that is minimal by inclusion. More precisely, we introduce an algorithm for finding a minimal local set cover in polynomial time. This result is proved by exploring the link between local sets and cut-rank. We prove some additional results on minimal local sets: we give tight bounds on their size, and we show that there can be exponentially many of them in a graph. Finally, we provide an extension of our definitions and our main result to q-multigraphs, the graphical counterpart of quantum qudit graph states.

Keywords: Local complementation · Cut-rank · Graph states

1 Introduction

Context and Contributions. Local complementation, introduced by Kotzig [40], is a transformation that consists in complementing the neighborhood of a given vertex of a graph. Local complementation is strongly related to the notion of vertex minor [13,38,47], has various applications in the study particular families of graphs including circle graphs [3,15], and plays a central role in the study of graph parameters like rank-width [48] and cut-rank [5,49]. Local complementation finds significant applications in quantum computing, ranging from addressing fundamental questions on quantum entanglement [26,46] to its utilization in various protocols [9,10,41,42], and computational models like ZX-calculus [12,16–18,21] or measurement-based quantum computation [14,43,44].

We consider in this paper a structure invariant under local complementation called *local set* [30]. Given a simple, undirected graph G, a non-empty set of vertices is said *local* if it is of the form $D \cup Odd_G(D)$, where $Odd_G(D)$ is the set of vertices which share an odd number of edges with vertices of D. Minimal local sets are local sets that are minimal by inclusion.

D. Král' and M. Milanič (Eds.): WG 2024, LNCS 14760, pp. 136–150, 2025.
https://doi.org/10.1007/978-3-031-75409-8_10

Local sets are related to the local minimal degree $\delta_{loc}(G)$ of a graph G which is the minimum degree of its vertices, up to local complementation[1]. It has been proved in [30] that the local minimal degree of a graph is the size of its smallest (minimal) local set minus one. The local minimal degree can also be characterized using the cut-rank function, which associates with any cut of a graph the rank of the matrix describing the edges of the cut.

Foliage and foliage partition [8,25,54] have been recently introduced: the foliage of a graph can be defined as the union of the size-2 minimal local sets; and two vertices are in the same component of the foliage partition if they belong to a same minimal local set of size two. Notice that minimal local sets of size two are composed of two non-isolated vertices, which can be either a leaf and its neighbor or twins. In [8], the question of the extension of foliation is raised as size-2 local sets do not cover all the vertices of a graph in general. One can naturally relax the size restriction, and hence consider minimal-local-set covers – MLS covers for short. One of our main result is to show that any vertex can be covered by a minimal local set, and thus that any graph has an MLS cover.

Theorem 1. *Any graph has an MLS cover.*

This result relies on our cut-rank-based characterisation of minimal local sets together with the peculiar properties of the cut-rank which, as a set function, is symmetric, linearly bounded and submodular. We also introduce an efficient algorithm which, given a graph, produces a family of a minimal local sets that cover any vertex of the graph. Furthermore, we investigate properties related to the size and quantity of minimal local sets in a graph, showing in particular that some graphs have an exponential number of minimal local sets.

Related Work on Quantum Graph States. Graph states are resources for quantum computing, which are in one-to-one correspondence with mathematical graphs. The survey [26] by Hein et al. provides an excellent introduction to graph states. Local sets were introduced in [30] as a combinatorial tool in the context of the preparation of graph states.

Within the graph state formalism, two locally equivalent graphs[2] represent the same entanglement[3], and two graphs representing the same entanglement have the same cut-rank functions. Notice that the converses of these two assertions are false[4], and little is known on non locally-equivalent graphs that represent the same entanglement except that their minimal local sets must satisfy some strong constraints [45,53].

---

[1] i.e. $\delta_{loc}(G) = \min_{H \equiv G} \delta(H)$ where $H \equiv G$ if H can be transformed in G by a sequence of local complementations.

[2] i.e. equal up to local complementation.

[3] More precisely the corresponding graph states are equal up to local unitaries.

[4] Two graphs representing the same entanglement does not generally imply that they are locally equivalent: a counterexample of order 27 has been discovered using computer assisted methods [35]. Moreover, two graphs having the same cut-rank function does not generally imply that they represent the same entanglement: one counterexample involves two isomorphic Petersen graphs [20,26].

The local minimum degree is strongly related to the notion of k-uniformity [23,24,39,50–52], which is defined as follows: a quantum state is k-uniform if for any subset of k qubits, the corresponding reduced state is maximally mixed, i.e. roughly speaking contains no information. Notice that a graph state is k-uniform if and only if the minimal local sets of the corresponding graph contain at least $k+1$ vertices[5], i.e. its local minimum degree is k or more. The $\lfloor n/2 \rfloor$-uniform n-qubit states are called absolutely maximally entangled [1,19,22,27–29,31]. A graph state is absolutely maximally entangled if and only if the (minimal) local sets of the corresponding graph are of size at least $\lfloor n/2 \rfloor + 1$. Graphs that satisfy this property have been classified, and exist only for $n = 2, 3, 5, 6$ [32,52], there exists however an infinite family of graphs whose minimal local sets are of size at least linear in their order, with constant 0.189 [34].

Structure of the Paper. First, we define local sets and minimal local sets on a graph, and provide some examples, in Sect. 2. Then we show how they can be defined alternatively using the cut-rank function, and prove additional results on the size and number of minimal local sets. Namely, we provide a tight bound on the size of minimal local sets, and show that a lower bound on the size of the minimal local sets implies a lower bound on their number. In Sect. 3, we prove the main result of this paper. For any graph, every vertex is contained in at least one minimal local set. We then give a polynomial-time algorithm that emerges from the proof of our main result, that finds a family of minimal local sets that cover all vertices of the graph. Finally, in Sect. 4, we extend the notion of local sets to q-multigraphs, the graphical counterpart of quantum qudit graph states, and show that our main result extends to any prime dimension.

2 Minimal Local Sets

This work focuses on the notion on minimal local sets. We give their definition and some basic properties in Sect. 2.1, then give an alternative definition using the cut-rank function in Sect. 2.2. We study the size and number of minimal local sets in Sect. 2.3.

2.1 Preliminaries

Let us first give some notations and basic definitions. A graph G is a pair (V, E), where V is the set of vertices, and $E \subseteq V^2$ is the set of edges. Here we only consider graphs that are undirected (if $(u, v) \in E$, $(v, u) \in E$) and simple ($\forall u \in V, (u, u) \notin E$). A cut is a bipartition of the vertices of the graph; with a slight abuse of notation we use A to denote the cut $\{A, V \setminus A\}$. The set $N_G(u) = \{v \mid (u, v) \in E\}$ is the neighborhood of u. For any $D \subseteq V$, $Odd_G(D) = \Delta_{u\in D} N_G(u) = \{v \in V \mid |N_G(v) \cap D| = 1 \bmod 2\}$ is the odd neighborhood of D,

[5] Local sets are nothing but the support of the stabilizers of the state, the reduced state on k qubits [26] is a maximally mixed state if it does not contain any local set.

where Δ denotes the symmetric difference on vertices. Informally, $Odd_G(D)$ is the set of vertices that are the neighbors of an odd number of vertices in D. A local complementation according to a given vertex u consists in complementing the subgraph induced by the neighborhood of u, leading to the graph $G \star u = G \Delta K_{N_G(u)}$ where Δ denotes the symmetric difference on edges and K_A is the complete graph on the vertices of A. Two graphs are said locally equivalent if there are related by a sequence of local complementations.

Definition 1 (Local set). *Given $G = (V, E)$, a* local set *L is a non-empty subset of V of the form $L = D \cup Odd_G(D)$ for some $D \subseteq V$ called a* generator.

Local sets are invariant under local complementation - hence their name: if L is a local set in a graph, so is in any locally equivalent graph, but possibly with a distinct generator [30].

Definition 2 (Minimal local set). *A minimal local set is a local set that is minimal by inclusion.*

Minimal local sets satisfy the following property: under local complementation, any vertex of a minimal local set can be made its generator.

Proposition 1 ([30]). *For any minimal local set L defined on a graph G, for any $x \in L$, there exists G' locally equivalent to G such that $L = \{x\} \cup N_{G'}(x)$.*

We are interested in families of minimal local sets such that each vertex is contained in at least one minimal local set. We call such a family an *MLS cover.*

Definition 3 (MLS cover). *Given a graph $G = (V, E)$, $\mathcal{L} \subseteq 2^V$ is an MLS cover if*

- *$\forall L \in \mathcal{L}$, L is a minimal local set of G,*
- *$\forall u \in V$, $\exists L \in \mathcal{L}$ such that $u \in L$.*

Local sets, minimal local sets, and MLS covers are illustrated in Fig. 1.

The main goal of this paper is to prove that any graph has an MLS Cover, i.e. any vertex is contained in at least one minimal local set, and to provide an efficient algorithm to compute an MLS cover, given any graph. This is done in Sect. 3. First, for this purpose, we exhibit the links between (minimal) local sets and the cut-rank function.

2.2 Links with the Cut-Rank Function

Given a graph G and a cut A, one can define the map $\lambda_A : 2^A \to 2^{V \setminus A} = D \mapsto Odd_G(D) \setminus A$, which is linear with respect to the symmetric difference[6]. Notice that D is in the kernel of λ_A, i.e. $\lambda_A(D) = \emptyset$, if and only if $Odd_G(D) \subseteq A$, that is D is a generator of a local set included in A. The rank of λ_A is nothing but the so-called cut-rank of A. The cut-rank function, introduced by Bouchet under the name "connectivity function" [4,6], and coined "cut-rank" by Oum and Seymour [49], is usually defined as follows:

[6] $\forall D, D' \subseteq A, \lambda_G(D \Delta D') = \lambda_A(D) \Delta \lambda_A(D')$.

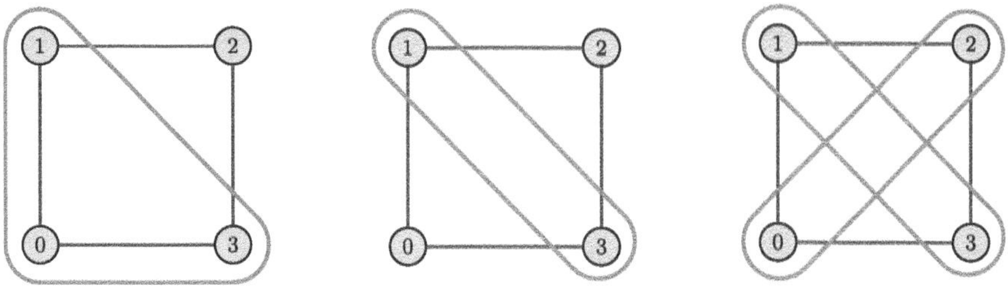

Fig. 1. (Left) A local set generated by $D = \{0\}$: $Odd_G(D) = \{1,3\}$. This is not a minimal local set. (Middle) A local set generated by $D = \{1,3\}$: $Odd_G(D) = \emptyset$. In particular, this is a minimal local set, as neither $\{1\}$, $\{3\}$ nor $\emptyset$ is a local set. (Right) an MLS cover of the graph, i.e. a set of minimal local sets such that each vertex is contained in at least one of them.

Definition 4 (Cut-rank function). *For $A \subseteq V$, let the cut-matrix $\Gamma_A = ((\Gamma_A)_{ab} : a \in A,\ b \in V \setminus A)$ be the matrix with coefficients in $\mathbb{F}_2$ (the finite field of order 2) such that $\Gamma_{ab} = 1$ if and only if $(a,b) \in E$. The cut-rank function of G is defined as*

$$\begin{aligned} \text{cutrk}\colon 2^V &\longrightarrow \mathbb{N} \\ A &\longmapsto \textit{rank}(\Gamma_A) \end{aligned}$$

A set A is said *full cut-rank* when $\text{cutrk}(A) = |A|$. The proposition below lists essential properties of the cut-rank function.

Proposition 2 ([49]). *The cut-rank function satisfies the following properties:*

- ***symmetry:*** $\forall A \subseteq V,\ \text{cutrk}(V \setminus A) = \text{cutrk}(A)$,
- ***linear boundedness:*** $\forall A \subseteq V,\ \text{cutrk}(A) \leqslant |A|$,
- ***submodularity:*** $\forall A, B \subseteq V,\ \text{cutrk}(A \cup B) + \text{cutrk}(A \cap B) \leqslant \text{cutrk}(A) + \text{cutrk}(B)$.

The conjunction of symmetry, linear boundedness and submodularity, together with the fact that cutrk has values in $\mathbb{N}$, implies the following useful properties:

Proposition 3. *For any graph $G = (V, E)$ of order n,*

(i) $\text{cutrk}(\emptyset) = \text{cutrk}(V) = 0$,
(ii) $\forall A \subseteq V, \text{cutrk}(A) \leqslant \lfloor n/2 \rfloor$, *so a full-cut-rank set is of size at most* $\lfloor n/2 \rfloor$,
(iii) If $\text{cutrk}(K) = |K|$ *then for any* $A \subseteq K$, $\text{cutrk}(A) = |A|$, *i.e. any subset of a full-cut-rank set is full cut-rank.*

Proof. *(i)* Using linear boundedness, $\text{cutrk}(\emptyset) \leqslant |\emptyset| = 0$ so $\text{cutrk}(\emptyset) = 0$ as cutrk has values in $\mathbb{N}$. Then, using symmetry, $\text{cutrk}(V) = \text{cutrk}(\emptyset) = 0$. *(ii)* Let $A \subseteq V$. Using linear boundedness, $\text{cutrk}(A) \leqslant |A|$. Also, using symmetry, $\text{cutrk}(A) = \text{cutrk}(V \setminus A) \leqslant |V \setminus A| = n - |A|$. Thus, $\text{cutrk}(A) \leqslant \min(|A|, n - |A|) \leqslant \lfloor n/2 \rfloor$.

(iii) Let $A \subseteq K$. Using submodularity, $\text{cutrk}(A) + \text{cutrk}(K \setminus A) \geqslant \text{cutrk}(K) + \text{cutrk}(\emptyset) = |K| + 0$. Moreover, using linear boundedness, $\text{cutrk}(A) \leqslant |A|$ and $\text{cutrk}(K \setminus A) \leqslant |K| - |A|$, so $\text{cutrk}(A) = |A|$. □

Local sets can alternatively be defined using uniquely the cut-rank function.

Proposition 4. *Given a graph $G = (V, E)$ and $A \subseteq V$,*

- *A is a local set if and only if $\forall a \in A, \text{cutrk}(A) \leqslant \text{cutrk}(A \setminus \{a\})$,*
- *A is a minimal local set if and only if A is not full cut-rank, but each of its proper subset is, i.e. $\forall a \in A, \text{cutrk}(A) \leqslant \text{cutrk}(A \setminus \{a\}) = |A| - 1$.*

Proof. Notice that $|A| - \text{cutrk}(A)$, the dimension of the kernel of λ_A, counts the number of sets that generate a local set in A: $2^{|A|-\text{cutrk}(A)} = |\{D \subseteq A \mid D \cup Odd_G(D) \subseteq A\}|$. As a consequence, $A \subseteq V$ being a local set means that for any $a \in A$, there is more local sets (counted with their generators) in A than in $A \setminus \{a\}$. Then $|A \setminus \{a\}| - \text{cutrk}(A \setminus \{a\}) < |A| - \text{cutrk}(A)$. This translates to $\text{cutrk}(A) \leqslant \text{cutrk}(A \setminus \{a\})$. □

As a direct consequence, two graphs that have the same cut-rank function have the same local sets[7]. This is an equivalence with the additional condition that the local sets have the same number of generators:

Proposition 5. *Two graphs have the same cut-rank function if and only if they have the same local sets with the same number of generators.*

Proof. The cut-rank function can be computed from the number of generators of each local set: for every $A \subseteq V$,

$$\text{cutrk}(A) = |A| - log_2\left(1 + \sum_{L \text{ local set in } A} |\{D \subseteq L \mid D \cup Odd_G(D) = L\}|\right)$$

Conversely, the number of generators of each local set can be recursively computed from the values of the cut-rank function:

$$|\{D \subseteq A | D \cup Odd_G(D) = A\}| = 2^{|A|-\text{cutrk}(A)} - \sum_{B \subsetneq A} |\{D \subseteq B | D \cup Odd_G(D) = B\}|$$

□

2.3 Size and Number of Minimal Local Sets

According to Propositions 3 and 4, minimal local sets are of size at most half the order of the graph. We slightly refine this bound and show it is tight:

[7] As local complementation does not change the rank of cut-matrices, this provides an alternative proof of the invariance of local sets under local complementation.

Proposition 6. *For any minimal local set A in a graph of order n,*

$$|A| \leq \begin{cases} n/2 & \text{if } n = 0 \bmod 4 \\ \lfloor n/2 \rfloor + 1 & \text{otherwise} \end{cases}$$

This bound is tight in the sense that for every $n > 0$ there exists a graph of order n that contains a minimal local set of this particular size.

Proof. If $n \neq 0 \bmod 4$, the bound derives from Proposition 4 along with Proposition 3. Otherwise, the bound is slightly stronger due to constraints of the cut-rank function. See details, along with explicit constructions of graphs that attain the bound, in the extended version of this paper [11]. □

Small and large minimal local sets can coexist in a graph, this is for instance the case for a path graph P_n of order $n > 2$ which has minimal local sets of any size ranging from 2 to $\lceil n/2 \rceil$ (see the extended version of this paper [11]). Complete graphs, and bipartite complete graphs only have small minimal local sets (which are respectively any pair of vertices and any pair of vertices in the same partition). There exist also graphs with only 'large' minimal local sets, namely at least $0.189n$ where n is the order of the graph. Indeed, the size of the smallest local set is nothing but the local minimal degree of the graph plus one [30], and it has been proved that there exist graphs with such a large local minimal degree [34].

The size of the local sets is also, to some extent, related to the number of minimal local sets in a graph. For instance, complete graphs and bipartite complete graphs have a number of minimal local sets quadratic in their order. More generally, if the size of the minimal local sets is upper bounded by k then there are obviously $O(n^k)$ minimal local sets in a graph of order n. Maybe more surprisingly, a lower bound on the size of the minimal local sets implies a lower bound on their number:

Proposition 7. *Given a graph G of order n, if all minimal local sets of G are of size at least m, then the number of minimal local sets in G is at least*

$$\frac{1-2r}{3\sqrt{n}} 2^{n\left(1-(1-r)H_2\left(\frac{1}{2(1-r)}\right)\right)}$$

where $r = \frac{m}{n}$ and $H_2(x) = -x\log_2(x) - (1-x)\log_2(1-x)$ is the binary entropy.

Proof. Proposition 3 along with Proposition 4 imply that any set with more than $n/2$ vertices contains a minimal local set. One can upper bound the number of sets of size $n/2$ in which a given fixed minimal local set is contained. Roughly speaking, a large minimal local set is contained in fewer sets of size $n/2$ than a small minimal local set. Thus, a counting argument implies the lower bound on the number of minimal local sets. See details in the extended version of this paper [11]. □

Proposition 7 implies that a graph of order n, in which the size of the minimal local sets is lower bounded by cn for some constant $c > 0$, has an exponential number of minimal local sets. In particular, graphs of order n and local minimal degree of size at least $0.189n$ have at least 1.165^n minimal local sets. Notice that there also exist explicit graphs with exponentially many minimal local sets, this is for instance the case of the graph $K_{k,k}\Delta M_k$ defined as the symmetric difference of a complete bipartite graph and a matching, which is of order $n = 2k$ and has at least $2^{k-1} = \frac{1}{2}\sqrt{2}^n$ minimal local sets (see the extended version of this paper [11]).

As the number of minimal local sets can be exponential in the order of a graph, we focus, in the next section, on MLS covers, which are representative families of minimal local sets covering any vertex of a graph. Notice that if a graph has an MLS cover, then it admits an MLS cover made of at most a linear number of minimal local sets.

3 MLS Cover

To cover a vertex u of a graph G with a minimal local set, one can consider the local set generated by u, i.e. its closed neighborhood $N_G[u] = \{u\} \cup N_G(u)$. However, such a local set is not always minimal, worse yet, it does not necessarily contain a minimal local set that includes u. Indeed, in the cycle C_4 of order 4, for every vertex u, $N_G[u]$ contains a single minimal local set that does not include u (see Fig. 1).

In this particular C_4 example however, one can notice that, roughly speaking, minimizing local sets of the form $N_G[u]$ is enough to get an MLS cover: considering the minimal local sets included in closed neighborhoods is enough to produce an MLS cover.

This strategy does not work in general. There exist graphs (see the extended version of this paper [11]) for which an MLS cover requires minimal local sets that are not subsets of any closed neighborhood. Thus, one can wonder what is the best strategy to produce an MLS cover and even whether it exists for every graph. We show in this section that every graph admits an MLS cover (Theorem 1, proved in Sect. 3.1) and then introduce an efficient algorithm for finding an MLS cover (Sect. 3.2).

Theorem 2. *Any graph has an MLS cover.*

3.1 Proof of Theorem 1

In this subsection, we prove Theorem 1. First, we introduce a cut-rank-based characterisation of the existence of a minimal local set covering a given vertex:

Lemma 1. *Given a graph $G = (V, E)$, $a \in V$ is contained in a minimal local set if and only if there exists $A \subseteq V$ such that A is full cut-rank but $A \cup \{a\}$ is not.*

Proof. Let $a \in V$. Suppose a is contained in a minimal local set $A \subseteq V$. Then $A \setminus \{a\}$ is full cut-rank but A is not. Conversely, suppose that there exists $A \subseteq V$ such that A is full cut-rank but $A \cup \{a\}$ is not (obviously, $a \notin A$). According to Proposition 3, any set $B \subseteq A$ is full cut-rank. Among all sets $B \subseteq A$ such that $B \cup \{a\}$ is not full cut-rank (there is at least one: A), take one of them which is minimal by inclusion. Call it C. Let us show that $C \cup \{a\}$ is a minimal local set. By hypothesis, $C \cup \{a\}$ is not full cut-rank. Every subset of C is full cut-rank, because such a set is also a subset of A. Also, every subset of $C \cup \{a\}$ containing a is full cut-rank, by minimality of C. So, according to Proposition 4, $C \cup \{a\}$ is a minimal local set. So a is contained in some minimal local set. □

Notice that a set is full cut-rank when the cut-rank function is locally strictly increasing in the following sense:

Lemma 2. *Given a graph* $G = (V, E)$, $C \subseteq V$ *is full cut-rank if and only if* $\forall a \in C$, $\mathrm{cutrk}(C \setminus \{a\}) < \mathrm{cutrk}(C)$.

Proof. ($\Rightarrow$) follows from Proposition 3. The proof of ($\Leftarrow$) is by induction on the size of C. The property is obviously true when $|C| \leqslant 1$. Assume $|C| \geqslant 2$, for any $b \in C$ and any $a \in C \setminus \{b\}$, using submodularity,

$$\begin{aligned} &\mathrm{cutrk}((C \setminus \{a\}) \cup (C \setminus \{b\})) \\ &+ \mathrm{cutrk}((C \setminus \{a\}) \cap (C \setminus \{b\})) \leqslant \mathrm{cutrk}(C \setminus \{a\}) + \mathrm{cutrk}(C \setminus \{b\}) \\ &\mathrm{cutrk}(C) + \mathrm{cutrk}(C \setminus \{a, b\}) < \mathrm{cutrk}(C) + \mathrm{cutrk}(C \setminus \{b\}) \\ &\mathrm{cutrk}((C \setminus \{b\}) \setminus \{a\}) < \mathrm{cutrk}(C \setminus \{b\}) \end{aligned}$$

By induction hypothesis, $C \setminus \{b\}$ is full cut-rank, so C is also full cut-rank as $\mathrm{cutrk}(C) > \mathrm{cutrk}(C \setminus \{b\}) = |C \setminus \{b\}| = |C| - 1$. □

When a set is not full cut-rank, one can find a sequence of nested subsets with a larger cut-rank, leading to a full cut-rank subset:

Lemma 3. *Given a graph* $G = (V, E)$, *for any* $A \subseteq V$, *there exist* $B_0 \subsetneq \ldots \subsetneq B_{|A|-\mathrm{cutrk}(A)-1} \subsetneq B_{|A|-\mathrm{cutrk}(A)} = A$ *such that* $\forall i$, $|B_i| = \mathrm{cutrk}(A) + i$ *and* $\mathrm{cutrk}(B_i) \geqslant \mathrm{cutrk}(A)$. *In particular,* B_0 *is full cut-rank.*

Proof. Note this makes sense because of the linear boundedness of cutrk, which ensures that $|A| - \mathrm{cutrk}(A) \geqslant 0$. It is enough to show that for any $B \subseteq A$, if $\mathrm{cutrk}(B) \geqslant \mathrm{cutrk}(A)$ and $|B| > \mathrm{cutrk}(A)$ then $\exists a \in B$ such that $\mathrm{cutrk}(B \setminus \{a\}) \geqslant \mathrm{cutrk}(A)$. There are two cases: (i) if B is full cut-rank then according to Proposition 3 for any $a \in B$, $B \setminus \{a\}$ is also full cut-rank so $\mathrm{cutrk}(B \setminus \{a\}) = |B| - 1 \geqslant \mathrm{cutrk}(A)$; (ii) if B is not full cut-rank then according to Proposition 2, $\exists a \in B$ such that $\mathrm{cutrk}(B \setminus \{a\}) \geqslant \mathrm{cutrk}(B)$, so $\mathrm{cutrk}(B \setminus \{a\}) \geqslant \mathrm{cutrk}(A)$. □

An interesting consequence of Proposition 3 is that for any full-cut-rank set, there exists a disjoint full-cut-rank set of same cardinality.

Corollary 1. *Given a graph $G = (V, E)$, for any full-cut-rank set $A \subseteq V$, $\exists B \in \binom{V\setminus A}{|A|}$ which is full cut-rank*[8].

Proof. Let $A \subseteq V$ be a full-cut-rank set, i.e. $\text{cutrk}(A) = |A|$. Using symmetry, $\text{cutrk}(V \setminus A) = |A|$. Then applying Proposition 3 on $V \setminus A$ yields the result. □

We are now ready to conclude our reasoning and prove that any vertex a of an arbitrary graph $G = (V, E)$ is contained in some minimal local set. Let r be the maximal cardinality of a full-cut-rank set in V, i.e. $r = \max\{|A| \mid \text{cutrk}(A) = |A|\}$, and let $A \subseteq V$ be a full-cut-rank set of size r. For any vertex $a \in V$,

- If $a \notin A$ then by maximality of r, $A \cup \{a\}$ is not full cut-rank, thus by Proposition 1, a is contained in a minimal local set.
- If $a \in A$, then according to Corollary 1, there exists a full-cut-rank set $B \in \binom{V\setminus A}{r}$. By maximality of r, $B \cup \{a\}$ is not full cut-rank, so, according to Proposition 1, a is contained in a minimal local set.

This concludes the proof.

3.2 A Polynomial-Time Algorithm for Finding an MLS Cover

In this subsection, we turn Theorem 1 into a polynomial-time algorithm that generates an MLS cover for any given input graph.

Theorem 3. *There exists an algorithm that computes an MLS cover in time polynomial in the order of the graph.*

We provide in the following a textual description of the algorithm. The corresponding pseudo-code can be found in the extended version of this paper [11]. In the following, n refers to the order of the graph.

Description of the Algorithm. The core of the algorithm consists, given a graph G and a vertex a, in producing a minimal local set that contains a. One can then iterate on uncovered vertices to produce an MLS cover of the graph.

Given a vertex a, a minimal local set that contains a is produced as follows, using essentially two stages:

(i) First, the algorithm produces a full-cut-rank set A such that $A \cup \{a\}$ is not full cut-rank. The procedure consists in starting with an empty set A – which is full cut-rank – and then increase the size of A, in a full-cut-rank preserving manner, until $A \cup \{a\}$ is not full cut-rank. To increase the size of A, notice that if $A \cup \{a\}$ is full cut-rank then, according to Proposition 3, there exists a disjoint set A' – so in particular $a \notin A'$ – such that $|A'| = |A| + 1$ and A' is full cut-rank. The proof of Proposition 3 is constructive: starting from $C = V \setminus (A \cup \{a\})$ some vertices are removed one-by-one from C to get the set A'. To produce such a set A' from C, there are $O(n)$ vertices to remove, at each step there are

[8] Given a set K and an integer k, $\binom{K}{k}$ refers to $\{B \subseteq K \mid |B| = k\}$.

$O(n)$ candidates, and deciding whether a vertex can be removed costs a constant number of evaluations of the cut-rank function.

(ii) Given a full-cut-rank set A such that $A \cup \{a\}$ is not full cut-rank, Proposition 1 guarantees the existence of a minimal local set that contains a. Notice that the proof of Proposition 1 consists in finding, among all subsets B of A such that B is full cut-rank but $B \cup \{a\}$ is not, one which is minimal by inclusion. To do so, one can start from A and remove one-by-one vertices from A until reaching a set which is minimal by inclusion. It takes $O(n)$ steps[9], each step involving a constant number of evaluations of the cut-rank function.

Complexity. Stage (i) uses $O(n^3)$ evaluations of the cut-rank function, and stage (ii) only $O(n)$, so the overall algorithm uses $O(n^4)$ evaluations of the cut-rank function. Using Gaussian elimination, the cut-rank can be computed in $O(n^\omega)$ field operations [7,33] where $\omega < 2.38$.

3.3 Generalization to Any Symmetric, Linearly Bounded and Submodular Function with Values in $\mathbb{N}$

It has to be noted that the existence of an MLS cover and the fact that one can be computed efficiently remains true for any notion equivalent to minimal local sets defined with a function that shares some properties of the cut-rank function. Precisely, given a set V, and a function $\mu : 2^V \to \mathbb{N}$ that satisfies symmetry, linear boundedness, and submodularity, define a μ-minimal local set $A \subseteq V$ as a set which satisfies $\mu(A) \leqslant |A| - 1$ and $\forall B \subsetneq A$, $\mu(B) = |B|$. Then V is covered by its μ-minimal local sets, and a "μ-MLS cover" can be computed efficiently (given that the values of μ can be computed efficiently). Indeed, every proof we give in Sect. 3 (as the proof of Proposition 3) uses only the fact that $\text{cutrk} : 2^V \to \mathbb{N}$ satisfies symmetry, linear boundedness, and submodularity.

An example of a function satisfying such properties is the connectivity function of a matroid. Given a matroid (E, r) where E is the ground set and r is the rank function, the connectivity function is the function λ that maps any set $X \subseteq E$ to $\lambda(X) = r(X) + r(E \setminus X) - r(E)$.

4 Extension to q-Multigraphs

While (simple, undirected) graphs correspond to quantum qubit graphs states, their natural higher dimension extension are multigraphs, which correspond to quantum qudit graph states [2,37]. A quick introduction to multigraphs and qudit graph states can be found in [41]. Here we explain how our main result extends to q-multigraphs.

Definition 5 (q-multigraphs). *Given a prime number q, a q-multigraph G is a pair (V, Γ) where V is the set of vertices and $\Gamma : V \times V \longrightarrow \mathbb{F}_q$ is the adjacency matrix of G: for any $u, v \in G$, $\Gamma(u, v)$ is the multiplicity of the edge (u, v) in G.*

[9] We only need to consider each vertex at most once. Indeed, at each step, if $A \setminus \{b\}$ is full cut-rank, for any $B \subseteq A$, $B \setminus \{b\}$ is also full cut-rank by Proposition 3.

Here we consider undirected simple q-multigraphs, i.e. $\forall u, v \in V$, $\Gamma(u,v) = \Gamma(v,u)$ and $\Gamma(u,u) = 0$. The cut-rank function on q-multigraphs is defined similar to graphs (which correspond to 2-multigraphs):

Definition 6 (Cut-rank function on q-multigraphs). *Let $G = (V, \Gamma)$ be a q-multigraph. For $A \subseteq V$, let the cut-matrix $\Gamma_A = ((\Gamma_A)_{ab} : a \in A,\ b \in V \setminus A)$ be the matrix with coefficients in $\mathbb{F}_q$ such that $\Gamma_{ab} = \Gamma(a,b)$. The cut-rank function of G is defined as*

$$\begin{aligned} \text{cutrk}\colon 2^V &\longrightarrow \mathbb{N} \\ A &\longmapsto \boldsymbol{rank}(\Gamma_A) \end{aligned}$$

It is not hard to see that the cut-rank function on q-multigraphs satisfies the properties of Proposition 2, namely symmetry, linear boundedness, and submodularity. As explained in Sect. 3.3, Theorem 1 extends to any function with values in $\mathbb{N}$ that satisfies these properties. To extend local sets to q-multigraphs, the easier is to use their characterisation in terms of the cut-rank function (see Proposition 4).

Definition 7 (Local sets, Minimal local sets, and MLS covers on q-multigraphs). *Given $G = (V, \Gamma)$ a q-multigraph,*

- *$A \subseteq V$ is a* local set *if $\forall a \in A, \text{cutrk}(A) \leqslant \text{cutrk}(A \setminus a)$,*
- *$A \subseteq V$ is a* minimal local *set if A is a non-full-cut-rank set whose proper subsets are all full cut-rank, i.e. $\forall a \in A, \text{cutrk}(A) \leqslant \text{cutrk}(A \setminus a) = |A| - 1$,*
- *$\mathcal{L} \subseteq 2^V$ is an* MLS cover *if $\forall L \in \mathcal{L}$, L is a minimal local set of G, and $\forall u \in V$, $\exists L \in \mathcal{L}$ such that $u \in L$.*

Local complementation naturally extends to q-multigraphs (see [41] for a definition), and two locally equivalent q-multigraphs have the same cut-rank function [36]. This implies that again, in the case of q-multigraphs, local sets are invariant under local complementation. Our main result also extends to q-multigraphs.

Theorem 4. *Any q-multigraph has an MLS cover.*

5 Conclusion

In this work, we have shown that any graph admits an MLS cover, thus every vertex is contained in a minimal local set. We have also introduced a polynomial time algorithm for producing an MLS cover of a given graph. Our results rely on a cut-rank-based characterisation of minimal local sets, and the peculiar properties of the cut-rank function. As a consequence, our results extend to any set function which is symmetric, linearly bounded, and submodular; this is how we generalized our results to q-multigraphs for instance. We have also proved a few results on the size and the number of minimal local sets in a graph, which open a few directions for future work:

- One can define the minimal local set number of a graph as the smallest number of minimal local sets needed to cover all vertices of a graph; the complexity of the associated decision problem is unknown, up to our knowledge.
- Given an MLS cover $\{L_i\}$ of a graph, one can define its intersection graph which vertices are the minimal locals sets L_i and two vertices are connected if the corresponding minimal local sets have a non-empty intersection. What is the structure of such intersection graph? Notice for instance that the cycle of order 4 has a single MLS cover which intersection graph is empty. Is there always an MLS cover which intersection graph is a forest? Can it be exploited to speed up some algorithms? Could it be related to the rank-width parameter?
- As the number of minimal local sets of a graph can be exponentially large, one can wonder whether there exists a polynomial delay algorithm for enumerating all minimal local sets of a graph.

Acknowledgments. The authors thank Mathilde Bouvel and Noé Delorme for fruitful discussions. This work is supported by the PEPR integrated project EPiQ ANR-22-PETQ-0007 part of Plan France 2030, by the STIC-AmSud project Qapla' 21-STIC-10, and by the European projects NEASQC and HPCQS.

References

1. Arnaud, L., Cerf, N.J.: Exploring pure quantum states with maximally mixed reductions. Phys. Rev. A **87**(1) (2013)
2. Bahramgiri, M., Beigi, S.: Graph states under the action of local clifford group in non-binary case. arXiv:quant-ph/0610267 (2006)
3. Bouchet, A.: Circle graph obstructions. J. Comb. Theory Ser. B **60**(1), 107–144 (1994)
4. Bouchet, A.: Digraph decompositions and eulerian systems. SIAM J. Algebraic Discrete Methods **8**(3), 323–337 (1987)
5. Bouchet, A.: Connectivity of isotropic systems. Ann. N. Y. Acad. Sci. **555**(1), 81–93 (1989)
6. Bouchet, A.: Recognizing locally equivalent graphs. Discret. Math. **114**(1), 75–86 (1993)
7. Bunch, J.R., Hopcroft, J.E.: Triangular factorization and inversion by fast matrix multiplication. Math. Comput. **28**(125), 231–236 (1974)
8. Burchardt, A., Hahn, F.: The foliage partition: an easy-to-compute LC-invariant for graph states. arXiv:2305.07645 (2023)
9. Cautrès, M., Claudet, N., Mhalla, M., Perdrix, S., Savin, V., Thomassé, S.: Vertex-minor universal graphs for generating entangled quantum subsystems. In: 51st International Colloquium on Automata, Languages, and Programming (ICALP 2024). Leibniz International Proceedings in Informatics (LIPIcs), vol. 297, pp. 36:1–36:18 (2024)
10. Claudet, N., Mhalla, M., Perdrix, S.: Small k-pairable states. arXiv:2309.09956 (2023)
11. Claudet, N., Perdrix, S.: Covering a graph with minimal local sets. arXiv:2402.10678 (2024)

12. Coecke, B., Duncan, R.: Interacting quantum observables: categorical algebra and diagrammatics. New J. Phys. **13**(4), 043016 (2011)
13. Dahlberg, A., Helsen, J., Wehner, S.: How to transform graph states using single-qubit operations: computational complexity and algorithms. Quant. Sci. Technol. **5**(4), 045016 (2020)
14. Danos, V., Kashefi, E., Panangaden, P., Perdrix, S.: Extended measurement calculus. Semant. Tech. Quant. Comput. 235–310 (2009)
15. de Fraysseix, H.: Local complementation and interlacement graphs. Discret. Math. **33**(1), 29–35 (1981)
16. Duncan, R., Kissinger, A., Perdrix, S., Van De Wetering, J.: Graph-theoretic simplification of quantum circuits with the ZX-calculus. Quantum **4**, 279 (2020)
17. Duncan, R., Perdrix, S.: Graph states and the necessity of euler decomposition. In: Ambos-Spies, K., Löwe, B., Merkle, W. (eds.) CiE 2009. LNCS, vol. 5635, pp. 167–177. Springer, Heidelberg (2009). https://doi.org/10.1007/978-3-642-03073-4_18
18. Duncan, R., Perdrix, S.: Pivoting makes the ZX-calculus complete for real stabilizers. In: Proceedings of the 10th International Workshop on Quantum Physics and Logic. Electronic Proceedings in Theoretical Computer Science, vol. 171, pp. 50–62. Open Publishing Association (2014)
19. Facchi, P., Florio, G., Parisi, G., Pascazio, S.: Maximally multipartite entangled states. Phys. Rev. A **77**, 060304 (2008)
20. Fon-Der-Flaass, D.G.: Local Complementations of Simple and Directed Graphs, pp. 15–34. Springer Netherlands, Dordrecht (1996)
21. Gong, X., Wang, Q.: Equivalence of local complementation and euler decomposition in the qutrit ZX-calculus. arXiv:1704.05955 (2017)
22. Goyeneche, D., Alsina, D., Latorre, J.I., Riera, A., Życzkowski, K.: Absolutely maximally entangled states, combinatorial designs, and multiunitary matrices. Phys. Rev. A **92**, 032316 (2015)
23. Goyeneche, D., Raissi, Z., Di Martino, S., Życzkowski, K.: Entanglement and quantum combinatorial designs. Phys. Rev. A **97**, 062326 (2018)
24. Goyeneche, D., Życzkowski, K.: Genuinely multipartite entangled states and orthogonal arrays. Phys. Rev. A **90**, 022316 (2014)
25. Hahn, F., Dahlberg, A., Eisert, J., Pappa, A.: Limitations of nearest-neighbor quantum networks. Phys. Rev. A **106**, L010401 (2022)
26. Hein, M., Dür, W., Eisert, J., Raussendorf, R., Van den Nest, M., Briegel, H.: Entanglement in graph states and its applications. arXiv:quant-ph/0602096 (2006)
27. Helwig, W.: Absolutely maximally entangled qudit graph states. arXiv:1306.2879 (2013)
28. Helwig, W., Cui, W.: Absolutely maximally entangled states: existence and applications. arXiv:1306.2536 (2013)
29. Helwig, W., Cui, W., Latorre, J.I., Riera, A., Lo, H.K.: Absolute maximal entanglement and quantum secret sharing. Phys. Rev. A **86**, 052335 (2012)
30. Høyer, P., Mhalla, M., Perdrix, S.: Resources required for preparing graph states. In: Algorithms and Computation: 17th International Symposium, ISAAC 2006, Kolkata, India, December 18–20, 2006. Proceedings 17, pp. 638–649. Springer, Berlin Heidelberg (2006)
31. Huber, F., Eltschka, C., Siewert, J., Gühne, O.: Bounds on absolutely maximally entangled states from shadow inequalities, and the quantum macwilliams identity. J. Phys. A: Math. Theor. **51**(17), 175301 (2018)
32. Huber, F., Gühne, O., Siewert, J.: Absolutely maximally entangled states of seven qubits do not exist. Phys. Rev. Lett. **118**, 200502 (2017)

33. Ibarra, O.H., Moran, S., Hui, R.: A generalization of the fast LUP matrix decomposition algorithm and applications. J. Algorithms **3**(1), 45–56 (1982)
34. Javelle, J., Mhalla, M., Perdrix, S.: On the minimum degree up to local complementation: bounds and complexity. In: Graph-Theoretic Concepts in Computer Science, pp. 138–147. Springer Berlin Heidelberg (2012)
35. Ji, Z., Chen, J., Wei, Z., Ying, M.: The LU-LC conjecture is false. Quantum Inf. Comput. **10**(1), 97–108 (2010)
36. Kante, M., Rao, M.: The rank-width of edge-coloured graphs. Theory Comput. Syst. **52** (2007)
37. Ketkar, A., Klappenecker, A., Kumar, S., Sarvepalli, P.: Nonbinary stabilizer codes over finite fields. IEEE Trans. Inf. Theory **52**, 4892–4914 (2006)
38. Kim, D., Oum, S.I.: Vertex-minors of graphs: a survey. Discrete Appl. Math. **351**, 54–73 (2024)
39. Kłobus, W., et al.: k-uniform mixed states. Phys. Rev. A **100**, 032112 (2019)
40. Kotzig, A.: Eulerian lines in finite 4-valent graphs and their transformations. Colloqium Graph Theory Tihany **1966**, 219–230 (1968)
41. Marin, A., Markham, D., Perdrix, S.: Access structure in graphs in high dimension and application to secret sharing. arXiv:1304.7105 (2013)
42. Markham, D., Sanders, B.C.: Graph states for quantum secret sharing. Phys. Rev. A **78**, 042309 (2008)
43. Mhalla, M., Murao, M., Perdrix, S., Someya, M., Turner, P.S.: Which graph states are useful for quantum information processing? In: Theory of Quantum Computation, Communication, and Cryptography: 6th Conference, TQC 2011, Madrid, Spain, May 24–26, 2011, Revised Selected Papers 6, pp. 174–187. Springer (2014)
44. Mhalla, M., Perdrix, S.: Graph states, pivot minor, and universality of (X, Z)-measurements. Int. J. Unconv. Comput. **9**(1–2), 153–171 (2013)
45. Van den Nest, M., Dehaene, J., De Moor, B.: Local unitary versus local clifford equivalence of stabilizer states. Phys. Rev. A **71**(6) (2005)
46. Van den Nest, M., Dehaene, J., Moor, B.D.: Graphical description of the action of local clifford transformations on graph states. Phys. Rev. A **69**(2) (2004)
47. Oum, S.I.: Rank-width and vertex-minors. J. Comb. Theory Ser. B **95**(1), 79–100 (2005)
48. Oum, S.i.: Rank-width: algorithmic and structural results. Discrete Appl. Math. **231**, 15–24 (2017), algorithmic Graph Theory on the Adriatic Coast
49. Oum, S.I., Seymour, P.: Approximating clique-width and branch-width. J. Comb. Theory Ser. B **96**(4), 514–528 (2006)
50. Raissi, Z., Burchardt, A., Barnes, E.: General stabilizer approach for constructing highly entangled graph states. Phys. Rev. A **106**, 062424 (2022)
51. Raissi, Z., Teixidó, A., Gogolin, C., Acín, A.: Constructions of k-uniform and absolutely maximally entangled states beyond maximum distance codes. Phys. Rev. Res. **2**, 033411 (2020)
52. Scott, A.J.: Multipartite entanglement, quantum-error-correcting codes, and entangling power of quantum evolutions. Phys. Rev. A **69**(5) (2004)
53. Zeng, B., Chung, H., Cross, A.W., Chuang, I.L.: Local unitary versus local clifford equivalence of stabilizer and graph states. Phys. Rev. A **75**(3) (2007)
54. Zhang, D.: Bell pair extraction using graph foliage techniques. arXiv:2311.16188 (2023)

A New Approach for Approximating Directed Rooted Networks

Sarel Cohen[1], Lior Kamma[1(✉)], and Aikaterini Niklanovits[2]

[1] The Academic College of Tel Aviv-Yaffo, Tel Aviv, Israel
{sarelco,liorkm}@mta.ac.il

[2] Hasso Plattner Institute, University of Potsdam, Potsdam, Germany
Aikaterini.Niklanovits@hpi.de

Abstract. We consider the *k-outconnected directed Steiner tree* problem (k-DST). Given a directed edge-weighted graph $G = (V, E, w)$, where $V = \{r\} \cup S \cup T$, and an integer k, the goal is to find a minimum cost subgraph of G in which there are k edge-disjoint rt-paths for every terminal $t \in T$. The problem is known to be NP-Hard. Furthermore, the question on whether a polynomial time, subpolynomial approximation algorithm exists for k-DST was answered negatively by Grandoni *et al.* (2018), by proving an approximation hardness of $\Omega(|T|/\log|T|)$ under NP≠ZPP.

Inspired by modern day applications, we focus on developing efficient algorithms for k-DST in graphs where terminals have out-degree 0, and furthermore constitute the vast majority in the graph. We provide the first approximation algorithm for k-DST on such graphs, in which the approximation ratio depends (primarily) on the size of S. We present a randomized algorithm that finds a solution of weight at most $O(k|S|\log|T|)$ times the optimal weight, and with high probability runs in polynomial time.

1 Introduction

Network design problems deal with finding, inside a large network G, a cheap subnetwork satisfying certain criteria. This class of problems captures a wide variety of theoretical problems, among which are the minimum spanning tree problem, minimum Steiner tree (or forest) problem or minimum k-connected subgraph in both directed and undirected graphs (see, e.g. [15,19,22]). In addition to its theoretical importance, network design research is of great interest for developers of practical networks such as telecommunication networks.

Our focus in this paper is the *minimum k-outconnected directed Steiner tree* problem (k-DST). An instance of k-DST consists of a directed edge-weighted graph $G = (V, E, w)$, a designated node $r \in V$ called the *root*, a subset $T \subseteq V \setminus \{r\}$ of vertices called *terminals* and an integer k. Nodes in $V \setminus (T \cup \{r\})$ are referred to as *Steiner nodes*, and the set of Steiner nodes will be denoted by S. The goal is to find a minimum-weight subgraph of G, in which for every $t \in T$ there are at least k edge-disjoint rt-paths. From a theoretical perspective, k-DST is one

D. Král' and M. Milanič (Eds.): WG 2024, LNCS 14760, pp. 151–165, 2025.
https://doi.org/10.1007/978-3-031-75409-8_11

of the most fundamental and intensively studied problems in computer science. The most extensively studied variation of the problem is the *minimum directed Steiner tree* (DST), in which $k = 1$. This is, in part, due to the fact that many combinatorial problems are known to reduce to DST. Notable such examples among network design problems are the minimum node-weighted Steiner tree [21] in both directed and undirected graphs and the group Steiner problem [12].

Moreover, k-DST has many applications in network design, circuit layouts, and phylogenetic tree reconstruction [16] and specifically in designing *multicast* schemes [28,30]. When designing a multicast routing scheme, the goal is to distribute information from a single source to multiple destinations, while minimizing the usage of network resources (e.g. cost of links). To ensure that the network is robust against link failures, we require that the network contains several source-destinations routes that are, in a sense, mutually independent. The connection to k-DST is straightforward. Modeling the network as a directed graph, where edges model point-to-point links, a multicast routing scheme which can survive $k - 1$ failures is a k-outconnected Steiner tree.

Networks Consisting of Mostly Terminals. Upon approaching k-DST, one might be inclined to think of T as a "small" designated subset of V. Indeed, in the majority of previous work, the approximation factors of approximation algorithms for k-DST mainly depend on the size of the terminal set T. In general, however, this need not be the case. In fact, modern day applications present networks in which the lion's share of the network consists of terminals. Consider, for example the problem of efficiently and reliably distributing critical healthcare information, including patient data and medical resources, ensuring minimal transmission costs. In particular, the root in this case is a central health database, the Steiner nodes are regional hospitals and clinics and the terminal set consists of remote clinics in rural areas, which are the majority of health sites and do not transmit any information.

From a theoretical point of view, k-DST is still an interesting problem when restricted to instances in which most of the nodes of G are terminals and have no outgoing edges. Specifically, known hardness results for the problem hold even when $|V \setminus T| = O(|V|^{1/d})$ for some constant $d \in \mathbb{N}$. Moreover, there is no evidence that the problem is easier (even to approximate) when restricted to instances where $|V \setminus T|$ is polylogarithmic in $|V|$.

Related Work. k-DST is known to be NP-hard, as it models, for example, the Steiner tree in undirected graphs as a special case [17]. However, many special cases are widely studied with positive results. For example, the case where $T = V \setminus \{r\}$, known as *minimum k-outconnected spanning subgraph problem* is proven to be polynomially time solvable [7], as was also generalized by Frank and Tardos [10] (see also [9]). Their algorithm was used as a subroutine to obtain a 2-approximation algorithm for the undirected variant and its generalization [20], which is proven to be NP-hard.

When the existence of non-terminal vertices in the graph is also considered, the complexity of k-DST is increased, and the focus hence swifts to approximation algorithms. For the case where $k = 1$ for example, Charikar *et al.* [3] presented a family of greedy approximation algorithms $\mathcal{A}_d$, indexed by an integer $d > 1$. By employing a technique introduced by Zelikovsky [31], for every $d > 1$ they showed a reduction of the instance G to a new instance G' consisting of a d-layered graph. The algorithm $\mathcal{A}_d$ returns a feasible solution (in G) of weight at most $O(d^3|T|^{1/d})$ times the optimal weight, and runs in time $|V|^{O(d)}$. A similar result has been given by Rothvoß [29], who applied the Lasserre hierarchy in order to obtain a family of algorithms giving similar approximation guarantees. This result was recently extended to the case $k = 2$ by Grandoni and Laekhanukit [14]. For every $d > 1$, they develop a polynomial time approximation algorithm of ratio $\tilde{O}(d^3|T|^{1/d})$ and running time $O(|V|^{O(d)})$. For the undirected case of the problem when restricting the value of k to be 1, Byrka *et al.* provided a 1.39-approximation algorithm [1].

Moreover, the *directed Steiner tree* problem (i.e. $k = 1$), is solved by a trivial algorithm in time $2^{|S|} \cdot |V|^{O(1)}$, thus if S is "small" (i.e. $|S| \leq O(\log |V|)$) the problem is solvable in polynomial time. On the other hand, the Dreyfus-Wagner algorithm [6] solves the problem in time $O(3^{|T|} \cdot |V|^2)$, thus showing that it can also be solved in polynomial time if the number of terminals is "small". It is hence natural to consider developing algorithms for k-DST whose running time or approximation ratio is based on the amount of terminals. Such approximation algorithms for k-DST are developed only for the special cases where G is an L-layered graph, by Laekhanukit who gave a $\mathcal{O}(k^L Llog|T|)$-approximation algorithm [24], and for the case where G is a quasi-bipartite graph for which Chan *et al.* gave a polynomial time $\mathcal{O}(\log|T|\log k)$-approximation algorithm [2].

When it comes to hardness of k-DST much more results have been obtained. Cheriyan *et al.* [4] proved it to be at least as hard to approximate as the label cover problem. Therefore there is no $2^{\log^{1-\varepsilon}|V|}$-approximation for k-DST for any fixed $\varepsilon > 0$ unless $NP \subseteq DTIME(n^{polylog(n)})$. Laekhanukit [23] showed that unless $NP = ZPP$, the k-DST problem admits no $k^{1/4-\varepsilon}$-approximation for any $\varepsilon > 0$. We note that this hardness result holds even when restricted to instances in which $|V \setminus T| = O(|V|^{1/d})$ for some constant $d \in \mathbb{N}$ via a simple polynomial reduction. Another polynomial reduction of k-DST that provides us with a lower bound for approximating it is the one to the *set cover* problem. Nelson [26] showed that for every $c < 0.5$, set cover cannot be approximated in polynomial time within $2^{\log^{1-\delta_c(m)}(m)}$, unless SAT on m variables can be solved in time $2^{O(2^{\log^{1-\delta_c(m)}(m)})}$, where $\delta_c(m) = 1 - \frac{1}{(\log\log m)^c}$. Through a simple polynomial time reduction we are able to obtain a lower bound for approximating k-DST (in fact, even DST) in terms of $|S|$. In particular we conclude that k-DST cannot be approximated in polynomial time to within $2^{\log^{1-\delta_c(|S|)}(|S|)}$, unless SAT on m variables can be solved in time $2^{O(2^{\log^{1-\delta_c(m)}(m)})}$. One of the most recent results is the one from Liao *et al.* [25] where the improved the prreviously known lower approximation hardness to $\Omega(|T|/\log|T|))$ under NP$\neq$ ZPP, $\Omega(2^{k/2}/k)$ under NP$\neq$ ZPP and $\Omega(k/L)^{L/4}$ on L-layered graphs for $L \leq O(\log n)$.

1.1 Our Results

Let $G = (\{r\} \cup S \cup T, E, w)$, $k \in \mathbb{N}$ be an instance of k-DST. Assume in addition that every edge $uv \in E$ satisfies that $u \notin T$. The main contribution of this paper is providing the first approximation algorithm for k-DST on graphs where the terminals do not transmit any information. Our algorithm is a randomized Las Vegas algorithm, which runs, with high probability, in polynomial time.

Theorem 1. *There is a randomized $O(k|S|\log|T|)$-approximation algorithm for k-DST when the outdegree of the vertices in T is 0, that runs in polynomial time with high probability.*

This approximation guarantee depends primarily on the size of S, and up to a logarithmic factor is independent of the size of T. If $|S| = O(|T|^{1/d})$ for some constant $d > 1$, the approximation ratio of our algorithm is comparable with that of Charikar *et al.* [3]. Moreover, if $|S|$ is polylogarithmic in $|T|$, we achieve a polylogarithmic approximation ratio (in terms of $|V|$) in polynomial time.

Overview and Techniques. Throughout this paper let $G = (\{r\} \cup S \cup T, E, w), k$ be an instance of k-DST, where $V = \{r\} \cup S \cup T$, and $\mathcal{U} := \{U \subseteq V \setminus \{r\} : U \cap T \neq \emptyset\}$. An $e = xy \in E$ *covers* $U \in \mathcal{U}$ if e enters U, i.e. $x \notin U$ and $y \in U$. Given a set $F \subseteq E$, denote by $\varrho_F(U) := \{xy \in F : x \notin U,\ y \in U\}$ the set of edges in F covering U. By the minimum-cut maximum-flow theorem, a set $F \subseteq E$ is a feasible solution for k-DST if and only if every $U \in \mathcal{U}$ is covered by at least k edges, i.e. $|\varrho_F(U)| \geq k$.

Connectivity Augmentation. An intermediate problem between DST and k-DST is the one of increasing the rooted outconnectivity of a graph by one (connectivity augmentation problem). Formally, let $\ell \in \mathbb{N}$. Given a graph $G = (V, E, w)$, a root $r \in V$, a set of terminals $T \subseteq V \setminus \{r\}$, and a set $E_\ell \subseteq E$ of edges such that in the subgraph (V, E_ℓ) of G there are ℓ edge-disjoint rt-paths for every $t \in T$, we seek a minimum cost set $F \subseteq E \setminus E_\ell$ such that in $(V, E_\ell \cup F)$ there are $\ell + 1$ edge disjoint rt-paths for every $t \in T$. An additional result presented in this paper is the following.

Theorem 2. *There is a randomized approximation algorithm for the connectivity augmentation problem when the outdegree of the vertices in T is 0, that constructs an $O(|S|\log|T|)$-approximate solution. Also, with probability at least $1 - \frac{\log|T|}{2^{|E|}}$ the algorithm runs in polynomial time.*

The connection to k-DST is quite clear. Starting with an empty set of edges, our approximation algorithm iteratively finds small-weight solutions to the connectivity augmentation problem k times, and produces edge sets $E_1, \ldots, E_k \subseteq E$ whose union is a small-weight feasible solution to k-DST, incurring an additional factor of k in the approximation ratio. We note that very often when solving connectivity problems by repeatedly invoking a connectivity augmentation mechanism, it is possible to reduce the additional factor to the approximation ratio from k to $\log k$, using a standard linear-programming scaling technique introduced by Goemans *et al.* [13]. The algorithm presented in this paper for the augmentation problem, however, is invoked on an auxiliary graph, rather than on G. This auxiliary graph, and thus the induced linear program, change between invocations, and hence it is not clear whether the scaling technique is applicable.

Given an instance to the connectivity augmentation problem, and applying the minimum-cut maximum-flow theorem once more, we deduce that a set $F \subseteq E \setminus E_\ell$ is a feasible solution to the problem if and only if $|\varrho_{E_\ell \cup F}(U)| \geq \ell + 1$ for every $U \in \mathcal{U}$. From our assumptions on E_ℓ, we have $|\varrho_{E_\ell}(U)| \geq \ell$ for every $U \in \mathcal{U}$. In addition $F \cap E_\ell = \emptyset$, and therefore F is a feasible solution if and only if $\varrho_F(U) \neq \emptyset$ for all $U \in \mathcal{U}_\ell$, where $\mathcal{U}_\ell := \{U \in \mathcal{U} : |\varrho_{E_\ell}(U)| = \ell\}$. Hence, the augmentation problem is formulated as the following integer program.

$$\begin{aligned} &min \textstyle\sum_{e \in E} w_e x_e \qquad s.t. \textstyle\sum_{e \in \varrho_{E \setminus E_\ell}(U)} x_e \geq 1 \qquad \forall U \in \mathcal{U}_\ell, \\ &\qquad\qquad\qquad\qquad\quad x_e \in \{0,1\} \qquad \forall e \in E \end{aligned} \tag{IP}$$

Implicit Hitting Set. In the *hitting set* problem, we are given a ground set E, with weights $\{w_e\}_{e \in E}$, and a collection $\mathcal{S} \subseteq 2^E$ of subsets of E. The goal is to find a minimum weight subset $F \subseteq E$ such that $F \cap U \neq \emptyset$ for all $U \in \mathcal{S}$. The problem is known to be NP-hard [17], and can be approximated by an $O(\log |\mathcal{S}|)$-factor via a greedy algorithm. Karp *et al.* [18] observed that many combinatorial optimization problems possess the following property. Given an instance Π, we construct in polynomial time an instance $E, w, \mathcal{S}$ to the hitting set problem, where E, w are given explicitly, and $\mathcal{S}$, whose size might be exponential in the size of Π, is given implicitly by a membership oracle. Karp *et al.* named this setting the *implicit hitting set* problem, and observed that the well-known greedy algorithm for the hitting-set problem cannot be used to approximate this type of problems, as it examines all sets in $\mathcal{S}$.

We show that under reasonable assumptions we still achieve a randomized polynomial time $O(\log |\mathcal{S}|)$-approximation algorithm. Specifically, we slightly modify the well-known randomized rounding algorithm for hitting set, and devise a $O(\log |\mathcal{S}|)$-approximation algorithm that, with very high probability, say $\geq 1 - 2^{-|E|}$, runs in time polynomial to $|E|$. Details can be found in the full version of the paper [5]. The connectivity augmentation problem can be viewed as an implicit hitting set instance, where $\mathcal{S} := \{\varrho_{E \setminus E_\ell}(U) : U \in \mathcal{U}_\ell\}$. However, even for $\ell = 0$ we get $\log |\mathcal{S}| = \Omega(|V|)$. Achieving a $|T|$-approximation for the problem is immediate (increase the rt-connectivity by one for each $t \in T$ separately), hence we seek to exploit its structure, and specifically the structure of $\mathcal{U}_\ell$ to improve our guarantees.

A set-family $\mathcal{G} \subseteq 2^V$ is called *intersecting*, if when $X, Y \in \mathcal{G}$ and $X \cap Y \neq \emptyset$, then $X \cap Y, X \cup Y \in \mathcal{G}$. Frank [11] observed that if $T = V \setminus \{r\}$, the family $\mathcal{U}_\ell$ is intersecting, and also showed that the problem of covering intersecting set-families by directed edges can be solved optimally efficiently. In general, however, $\mathcal{U}_\ell$ is not intersecting. In a following paper [8] Frank coined the refined notion of T-Intersecting Families.

Definition 1. *A set family $\mathcal{G} \subseteq 2^V$ is called* T-intersecting *if for every $X \in \mathcal{G}$, $X \cap T \neq \emptyset$; and for every $X, Y \in \mathcal{G}$, if $X \cap Y \cap T \neq \emptyset$ then $X \cap Y, X \cup Y \in \mathcal{G}$.*

The proof of Proposition 1 can be found in the full version of the paper [5].

Proposition 1. *$\mathcal{U}_\ell$ is T-intersecting.*

Frank considered the problem of covering T-intersecting families, and showed that if every edge of G enters T, then a T-intersecting family can be covered efficiently using the primal dual approach. For arbitrary graphs, however, the problem of optimally covering a T-intersecting family of sets is NP-hard as it models the directed Steiner tree problem as a special case. By taking advantage of several structural properties of T-intersecting families, we present a mechanism for constructing a small-weight cover of $\mathcal{U}_\ell$. The mechanism is presented in detail in Sect. 3. Loosely speaking, we iteratively prune $\mathcal{U}_\ell$, covering small sub-families of $\mathcal{U}_\ell$ one at a time. Since the approximation ratio of the hitting set randomized algorithm depends on the size of the family to be covered, the problem of covering a small sub-family of $\mathcal{U}_\ell$ can be better approximated. We show that by pruning $\mathcal{U}_\ell$ carefully, the process ends after a small number of iterations, thus constructing a feasible solution with better approximation guarantees.

Strict Cores of T-Intersecting Families. In [27] Nutov studied rooted connectivity problems and defined the notion of a *core* of a set family. Given a set family $\mathcal{G} \subseteq 2^V$, a set $X \in \mathcal{G}$ is a core of $\mathcal{G}$ if it contains exactly one inclusion-minimal set of $\mathcal{G}$. Nutov gave an algorithm for covering so-called bi-uncrossable set families by iteratively covering the sub-families of cores. Formally, the iterative process is as follows. Given a set family $\mathcal{G}$, find a small-weight edge set $F_1 \subseteq E$ that covers all cores of $\mathcal{G}$. Denote by $\mathcal{G}^{F_1}$ the family of all sets in $\mathcal{G}$ not covered by F_1. Continue iteratively by finding $F_2, \ldots, F_t$ such that for every $j \in [t]$, F_j covers all cores of $\mathcal{G}^{F_1 \cup \ldots \cup F_{j-1}}$. Nutov additionally bounded the number of iterations needed to cover bi-uncrossable set families, when iteratively covering the sub-families of cores. In Sect. 3 we show that by iteratively covering the sub-family of cores of the T-intersecting family $\mathcal{U}_\ell$, we can guarantee that no more than $\log |T|$ iterations are needed. However we also show, the sub-family of cores can be as large as the entire $\mathcal{U}_\ell$, and hence our approximation guarantees are still too large.

Therefore, in this paper we present the new notion of *strict cores*. A strict core of $\mathcal{U}_\ell$ is a set $X \in \mathcal{U}_\ell$ that contains exactly one inclusion-minimal set $C \in \mathcal{U}_\ell$, and satisfies $X \cap T = C \cap T$. We show that the sub-family of strict cores of $\mathcal{U}_\ell$ is significantly smaller than the sub-family of cores, and give a polynomial time approximation algorithm for covering the sub-family of strict cores. As we show in Sect. 3.1 we can make no guarantee, however, on the number of iterations needed to iteratively cover the sub-family of strict cores. The main technical crux of this paper is therefore the construction of a new auxiliary graph in which every edge cover to the family of strict cores also covers the family of cores.

2 Approximating k-Outconnected Steiner Tree

We begin the proof of Theorem 1 by showing that it is implied by Theorem 2. To this end, let $G = (\{r\} \cup S \cup T, E, w)$ and k be an instance of k-DST. Denote by $\text{opt}(G)$ the value of an optimal solution, and let $\mathcal{U} = \{U \subseteq V \setminus \{r\} : U \cap T \neq \emptyset\}$. We present an iterative algorithm that finds a feasible solution for k-DST, by iteratively increasing the connectivity between the root and the terminals. The algorithm performs k iterations numbered $0, 1, \ldots, k-1$. For every $0 \leq \ell \leq k-1$, in the beginning of the ℓth iteration, the algorithm holds a set $E_\ell \subseteq E$ of edges, starting with $E_0 = \emptyset$, such that in (V, E_ℓ), there are at least ℓ edge-disjoint rt-paths for every $t \in T$. During the ℓth iteration, the algorithm invokes the $O(|S| \log |T|)$-approximation algorithm whose existence is implied by Theorem 2, to find an approximate solution $F \subseteq E \setminus E_\ell$ to the connectivity augmentation instance given by G, E_ℓ. The algorithm is described in detail as Algorithm 1.

1: $E_0 = \emptyset$.
2: **for** $\ell = 0, 1, \ldots, k-1$ **do**
3: let $F \subseteq E \setminus E_\ell$ be an $O(|S| \log |T|)$-approximate solution to the connectivity augmentation problem on G, E_ℓ (apply the algorithm in Theorem 2).
4: let $E_{\ell+1} = E_\ell \cup F$.
5: **return** E_k.

Algorithm 1: Approximation Algorithm for k-DST

Through a simple induction we are able to see that for every $t \in T$, there are k edge-disjoint rt-paths in (V, E_k). This also implies that the edge set constructed by Algorithm 1 is a feasible solution to k-DST. The following lemma shows that the weight of this set is at most $O(k|S| \log |T|)$ times the optimum.

Lemma 1. $w(E_k) \leq O(k|S| \log |T|) \cdot \text{opt}(G)$.

Proof. Let $0 \leq \ell \leq k-1$. Consider an optimal solution $F^* \subseteq E$ to the k-DST problem on G. Then for every $U \in \mathcal{U}$, $|\varrho_{F^*}(U)| \geq k$. Let $U \in \mathcal{U}$ be such that $|\varrho_{E_\ell}(U)| = \ell < k$. Then $|\varrho_{F^* \setminus E_\ell}(U)| \geq k - \ell \geq 1$. Therefore $F^* \setminus E_\ell$ is a feasible solution for the augmentation problem defined by G, E_ℓ. We conclude that the set F constructed in line 3 satisfies $w(F) \leq O(|S| \log |T|) \,\text{opt}(G)$. The lemma follows.

To conclude the proof of Theorem 1 it remains to show that the algorithm runs in polynomial time with high probability. Theorem 2 guarantees that line 3 always returns an approximate solution, and it runs in polynomial time with probability at least $1 - \frac{\log |T|}{2^{|E|}}$. Since Algorithm 1 performs k iterations, we get by a union bound that Algorithm 1 runs in polynomial time with probability at least $1 - \frac{k \log |T|}{2^{|E|}}$, and since $k \log |T| \leq |E|^2$, this probability is very high. The remainder of the paper is devoted to the proof of Theorem 2.

3 Increasing Rooted Connectivity By One

In this section, we prove Theorem 2 by presenting an $O(|S|\log|T|)$-approximation algorithm for the connectivity augmentation problem. Let $G = (\{r\} \cup S \cup T, E, w)$ be the instance graph, and $E_\ell \subseteq E$ be a set of edges such that in (V, E_ℓ) there are ℓ edge-disjoint rt-paths for every $t \in T$. Let $\mathcal{F} := \{U \subseteq V \setminus \{r\} : U \cap T \neq \emptyset \; and \; |\varrho_{E_\ell}(U)| = \ell\}$, then Proposition 1 implies $\mathcal{F}$ is T-intersecting. By the min-cut max-flow theorem, a set $F \subseteq E \setminus E_\ell$ is a feasible solution for the augmentation problem if and only if $\varrho_F(U) \neq \emptyset$ for every $U \in \mathcal{F}$. The linear program below can be viewed as a fractional relaxation of the augmentation problem.

$$min \sum_{e \in E} w_e x_e \qquad s.t. \begin{array}{ll} \sum_{e \in \varrho_{E\setminus E_\ell}(U)} x_e \geq 1 & \forall U \in \mathcal{F}, \\ x_e \geq 0 & \forall e \in E \end{array} . \tag{LP}$$

Fix some $x \in \mathbb{R}_+^E$. For every $e \in E$, we think of x_e as the capacity of e in G. Define $\hat{x} \in \mathbb{R}_+^E$ by $\hat{x}_e = 1$ for all $e \in E_\ell$ and $\hat{x}_e = x_e$ otherwise. Then x is a feasible solution for (LP) if and only if the capacity of a minimum rt-cut in $(G, \hat{x})$ is at least $\ell + 1$ for every $t \in T$. Therefore, the feasibility of x can be verified in time polynomial in the size of G. Moreover, if x is not feasible, there is some $t \in T$ such that in $(G, \hat{x})$ there is an rt-cut of capacity less than $\ell + 1$. Such a cut $U \subseteq V$ is a violated constraint and is found in polynomial time. The program (LP) can therefore be solved efficiently using a separation oracle. With high probability we thus efficiently obtain an $O(\log|\mathcal{F}|)$-approximate solution for the connectivity augmentation problem. However, in general $\log|\mathcal{F}| = \Omega(|V|)$, hence the approximation factor might be too large, since obtaining approximation of factor $|T|$ times the optimal value is trivial (increase the rt-connectivity for every $t \in T$ separately). Note that we did not use any properties of $\mathcal{F}$, other than finding an efficient separation oracle for (LP). Specifically, we did not use the structural properties of T-intersecting families.

In what follows, we present a mechanism for covering the T-intersecting family $\mathcal{F}$ by edges. Loosely speaking, we iteratively prune $\mathcal{F}$ by covering small sub-families. For each such sub-family we efficiently find a small weight cover. By choosing the sub-families carefully, the process ends after a small number of iterations. More formally, we look for sub-families $\mathcal{F}_1, \ldots, \mathcal{F}_t \subseteq \mathcal{F}$ such that (i) if $F \subseteq E \setminus E_\ell$ covers $\mathcal{F}_1, \ldots, \mathcal{F}_t$ then F covers $\mathcal{F}$; (ii) for every $j \in [t-1]$, given a cover $F \subseteq E \setminus E_\ell$ for $\mathcal{F}_1, \ldots, \mathcal{F}_j$, we can efficiently find a small-weight cover $F' \subseteq E \setminus E_\ell$ for $\mathcal{F}_{j+1}$; and (iii) t is small.

We start by exhibiting several fundamental properties of $\mathcal{F}$. Given a set $F \subseteq E \setminus E_\ell$, a set $X \in \mathcal{F}$ is called *F-tight* if F does not cover X, i.e. $\varrho_F(X) = \emptyset$. Denote by $\mathcal{F}^F$ the family of F-tight elements of $\mathcal{F}$. A crucial property of $\mathcal{F}$ is that the structural property of being T-intersecting is, in a sense, hereditary. In particular it is easy to prove that $\mathcal{F}^F$ is T-intersecting as follows. Let $X, Y \in \mathcal{F}^F$ be such that $X \cap Y \cap T \neq \emptyset$. Then $X \cap Y, X \cup Y \in \mathcal{F}$. Furthermore, by submodularity of the cut function, $0 = |\varrho_F(X)| + |\varrho_F(Y)| \geq |\varrho_F(X \cup Y)| + |\varrho_F(X \cap Y)| \geq 0$. Therefore $|\varrho_F(X \cup Y)| = |\varrho_F(X \cap Y)| = 0$, and thus $X \cap Y$ and $X \cup Y$ are F-tight elements of $\mathcal{F}$, that is $X \cup Y, X \cap Y \in \mathcal{F}^F$.

The following definition, and the two lemmas that follow it are inspired by a similar observation by Nutov [27] for problems regarding undirected graphs. The proofs of Lemmas 2 and 3 can be found in the Appendix of the full version of the paper [5].

Definition 2. *Let $\mathcal{M}(\mathcal{F}^F)$ denote the family of inclusion-minimal elements of $\mathcal{F}^F$. A set $X \in \mathcal{F}^F$ is called an $\mathcal{F}^F$-*core *if it contains exactly one inclusion-minimal element of $\mathcal{F}^F$.*

Lemma 2. *Let $X \in \mathcal{F}^F$. Then for every $C \in \mathcal{M}(\mathcal{F}^F)$, either $C \cap X \cap T = \emptyset$ or $C \subseteq X$. In particular, for every distinct $C, D \in \mathcal{M}(\mathcal{F}^F)$, $C \cap D \cap T = \emptyset$.*

Lemma 3. *If $F' \subseteq E \backslash E_\ell$ covers all $\mathcal{F}^F$-cores. Then $|\mathcal{M}(\mathcal{F}^{F \cup F'})| \leq \frac{1}{2}|\mathcal{M}(\mathcal{F}^F)|$.*

Lemma 3 implies that by iteratively covering the sub-family of tight cores, we cover $\mathcal{F}$ after at most $\log |\mathcal{M}(\mathcal{F})|$ iterations. Since $|\mathcal{M}(\mathcal{F})| \leq |T|$, we conclude that the process terminates after at most $\log |T|$ iterations. The next section constitutes the technical crux in the proof of Theorem 2.

3.1 Covering Cores Cheaply and Efficiently

In this section we conclude proof of Theorem 2 by giving an $O(|S|)$-approximation algorithm for the problem of covering the family of $\mathcal{F}^F$-cores. A naïve approach suggested by the preceding discussion is to simply apply the implicit hitting set approximation algorithm to cover the family of $\mathcal{F}^F$-cores. However, the sub-family of $\mathcal{F}^F$-cores can be almost as large as the entire $\mathcal{F}^F$. Hence the approximation factor might be as large as $\Omega(|V|)$. We therefore refine the definition of cores. This is where we diverge from [27], and from all previous work.

Definition 3. *A set $X \in \mathcal{F}^F$ is called a* strict $\mathcal{F}^F$-core *if there exists $C \in \mathcal{M}(\mathcal{F}^F)$ such that $X \cap T = C \cap T$, that is, X and C have the same set of terminals.*

One can easily verify, using Lemma 2 that every strict $\mathcal{F}^F$-core contains exactly one inclusion-minimal set of $\mathcal{F}^F$. Thus every strict $\mathcal{F}^F$-core is also an $\mathcal{F}^F$-core. As opposed to the family of tight cores, the family of strict cores is significantly small.

Lemma 4. *The number of strict $\mathcal{F}^F$-cores is at most $2^{|S|} \cdot |T|$.*

Proof. Let X be a strict $\mathcal{F}^F$-core. Then there is a unique $C \in \mathcal{M}(\mathcal{F}^F)$ such that $C \subseteq X$ and $C \cap T = X \cap T$. Therefore $X \setminus C \subseteq S$, and hence the family of strict $\mathcal{F}^F$-cores is contained in $\{C \cup X' : C \in \mathcal{M}(\mathcal{F}^F),\ X' \in 2^S\}$.

By Lemma 4, if we can model the problem of covering all strict $\mathcal{F}^F$-cores as an implicit hitting set problem, the approximation factor of the randomized rounding algorithm is reduced to $O(|S| + \log |T|) = O(|S|)$. However, if $F' \subseteq$

$E \setminus E_\ell$ covers all strict cores, it does not necessarily cover all cores. Nevertheless, not all is lost.

In a sense, the essence of the inadequacy of covering only strict cores lies in the following scenario. Consider an edge e leaving a covered core X and entering a non-covered strict core C covers the strict core, but does not reduce the number of non-covered minimal cores. Covering all strict cores with such edges attains no advancement to the algorithm. We thus turn to construct an auxiliary graph, in which this scenario cannot occur.

More formally, given G, E_ℓ and F, let $\mathrm{opt}_{\mathrm{aug}}(G, E_\ell)$ denote the weight of an optimal solution for the augmentation problem. We show that we can construct a new graph G^F on the same vertex set V, in which (i) we can find in G^F an edge set $A \subseteq E(G^F)$ covering all $\mathcal{F}^F$-cores such that $w(A) \leq O(|S|)\,\mathrm{opt}_{\mathrm{aug}}(G, E_\ell)$; and (ii) given a cover $A \subseteq E(G^F)$ for the family of $\mathcal{F}^F$-cores, we can construct a cover $F' \subseteq E \setminus E_\ell$ of the $\mathcal{F}^F$-cores in G such that $w(F') \leq w(A)$. Moreover, the construction of G^F, as well as (i) and (ii), can be done in polynomial time with high probability. Note that unlike simple reductions commonly used in dealing with the directed Steiner tree problem (e.g. reducing the instance to an acyclic graph or a layered graph), the construction we present preserves the vertex set intact, and specifically does not change $|S|$.

Following the discussion above, we want the weights to satisfy the triangle inequality, and therefore we first define a special form of metric completion of G. In classical literature, for every $u, v \in V$, the weight of the edge uv in the metric completion of a weighted graph G is defined to be the length of a shortest uv-path in G. By the minimum-cut maximum-flow theorem, for every $u, v \in V$, a shortest uv-path in G is a minimum-weight edge-cover of the set $\{U \subseteq V \setminus \{u\} : u \in U\}$. Therefore an analogous way of viewing the definition of a metric completion is the following. For every $u, v \in V$, the weight of the edge uv in the metric completion of G is defined to be the minimum weight of an edge cover for the set $\{U \subseteq V \setminus \{u\} : u \in U\}$. In our setting not all such subsets $U \subseteq V \setminus \{u\}$ need to be covered. We therefore wish to refine the classical notion of a metric completion of G. Specifically, if every $X \in \mathcal{F}^F$ such that $v \in X$ satisfies $u \in X$, then in a sense we do not need to cover any uv-cut, since all *relevant* cuts (i.e. those belonging to $\mathcal{F}^F$) have been covered. We can therefore set the weight of the edge uv to zero.

Formally we define a graph $G^0 = (V, (E \setminus E_\ell) \cup E^0, w^0)$ as $E^0 = \{uv :$ for every $X \in \mathcal{F}^F$, if $v \in X$ then $u \in X\}$ and constructed through the following procedure. For every $e \in (E \setminus E_\ell) \cup E^0$, we define $w^0_e = 0$ if $e \in F \cup E^0$, and $w^0_e = w_e$ otherwise. Next, let G^1 be the standard metric completion of G^0. That is, G^1 is the complete directed graph on V, where w^1_{uv} is the length of a shortest uv-path in G^0. The weight assignment w^1 satisfies the triangle inequality, and is therefore an asymmetric metric.

Next, following the discussion above, we construct G^F by removing from G^1 edges, which we can assert will not belong to any inclusion-minimal solution. Formally, G^F is the subgraph of G^1 constructed as follows. For every $s \in \bigcup_{C \in \mathcal{M}(\mathcal{F}^F)} (C \cap T)$ and for every $u \in V \setminus \left(\bigcup_{C \in \mathcal{M}(\mathcal{F}^F)} (C \cap T)\right)$, if every

$X \in \mathcal{F}^F$ that contains s also contains u, remove from G^1 all edges outgoing from u. For the proof of the claim that G^F can be constructed in polynomial time given G, E_ℓ, F we refer the reader to the full version of the paper [5].

Lemma 5. *For every $A \subseteq E(G^F)$, if A covers all $\mathcal{F}^F$-cores, we can construct in polynomial time an edge set $F' \subseteq E \setminus E_\ell$ that covers all $\mathcal{F}^F$-cores such that $w(F') \leq w^1(A)$.*

Proof. Let $A \subseteq E(G^F)$ be an $\mathcal{F}^F$-cores cover in G^F, $e = uv \in A$, and $P_e \subseteq (E \setminus E_\ell) \cup E^0$ be a shortest uv-path in G^0. Then $w^1_e = w^0(P_e) = \sum_{e' \in P_e} w^0_{e'} = \sum_{e' \in P_e \cap (E \setminus (E_\ell \cup F))} w_{e'}$. Let $X \in \mathcal{F}^F$ be an $\mathcal{F}^F$-core covered by e. Then $u \notin X$ and $v \in X$. Hence there is an edge $e' = u'v' \in P_e$ that covers X. Since $X \in \mathcal{F}^F$ and $v' \in X$ and $u' \notin X$, it follows that $e' = u'v' \notin F \cup E^0$, and hence $e' \in E \setminus (E_\ell \cup F)$. We conclude that $F' := \bigcup_{e \in A} (P_e \cap (E \setminus (E_\ell \cup F))$ covers all $\mathcal{F}^F$-cores, and $w(F') = \sum_{e \in F'} w_e \leq \sum_{e \in A} \sum_{e' \in P_e \cap (E \setminus (E_\ell \cup F))} w_{e'} = \sum_{e \in A} w^1_e = w^1(A)$.

Due to Lemma 5, in order to find a low-cost cover $F' \subseteq E \setminus E_\ell$ for all $\mathcal{F}^F$-cores it suffices to find a low-cost cover $A \subseteq E(G^F)$. The following two lemmas show that such a cover exists, and that it is enough for A to cover the strict $\mathcal{F}^F$-cores.

Lemma 6. *There exists an edge set $A \subseteq E(G^F)$ such that A covers all $\mathcal{F}^F$-cores and $w^1(A) \leq \mathrm{opt}_{\mathrm{aug}}(G, E_\ell)$.*

Proof. For every set $E' \subseteq E \setminus E_\ell$, from our construction we get that $w^1(E') \leq w(E')$, and therefore $\mathrm{opt}_{\mathrm{aug}}(G^1, E_\ell) \leq \mathrm{opt}_{\mathrm{aug}}(G, E_\ell)$. Hence, there is an edge set $A' \subseteq E(G^1)$ such that in the subgraph $(V, E_\ell \cup A')$ there are $\ell + 1$ edge-disjoint rt-paths for all $t \in T$, and such that $w^1(A') \leq \mathrm{opt}_{\mathrm{aug}}(G, E_\ell)$. We show that there is a set $A \subseteq E(G^F)$ such that in the subgraph $(V, E_\ell \cup A)$ there are $\ell + 1$ edge-disjoint rt-paths for all $t \in T$ and $w^1(A) \leq w^1(A')$. Clearly, such A covers all strict $\mathcal{F}^F$-cores in G^F and $w^1(A) \leq \mathrm{opt}_{\mathrm{aug}}(G, E_\ell)$. We continue by induction on $|A' \setminus E(G^F)|$. If $|A' \setminus E(G^F)| = 0$, take $A = A'$, and the result follows. Otherwise, let $uv \in A' \setminus E(G^F)$. Then $u \in V \setminus \left(\bigcup_{C \in \mathcal{M}(\mathcal{F}^F)} (C \cap T)\right)$ and there is $s \in \bigcup_{C \in \mathcal{M}(\mathcal{F}^F)} (C \cap T)$ such that every $Z \in \mathcal{F}^F$ containing u also contains s. By the definition of A', every set $Z \in \mathcal{F}^F$ containing s is covered by A'. Let $A'' = (A' \setminus \{uv\}) \bigcup \{sv\}$. By triangle inequality, $w^1_{sv} \leq w^1_{su} + w^1_{uv}$. Since $su \in E^0$, $w^1_{su} = 0$ we get that $w^1_{sv} \leq w^1_{uv}$. Hence, $w^1(A'') \leq w^1(A')$. Since $s \in \bigcup_{C \in \mathcal{M}(\mathcal{F}^F)} (C \cap T)$, then $sv \in E(G^F)$, we get $|A'' \setminus E(G^F)| = |A' \setminus E(G^F)| - 1$.

Lemma 7. *For every $A \subseteq E(G^F)$, A covers all $\mathcal{F}^F$-cores if and only if A covers all strict $\mathcal{F}^F$-cores.*

Proof. Let $A \subseteq E(G^F)$. Clearly, if A covers all $\mathcal{F}^F$-cores, then it covers all strict $\mathcal{F}^F$-cores. Assume, therefore, that A covers all strict $\mathcal{F}^F$-cores. Let $X \in \mathcal{F}^F$ be an $\mathcal{F}^F$-core that is not a strict core. There is a unique $C \in \mathcal{M}(\mathcal{F}^F)$ such that $C \subseteq X$. Let $X' = X \setminus (T \setminus C)$. First note that since for every $u \in T \setminus C$, no edge leaves u, then by removing from X all terminals in $T \setminus C$, no new edges enter X'.

Therefore $X' \in \mathcal{F}$. To see that $X' \in \mathcal{F}^F$, let $xy \in F$ be some edge, and assume $y \in X' \subseteq X$. Assume $y \in C$. Since $C \in \mathcal{F}^F$, then $x \in C \subseteq X'$. Otherwise, since $x \notin T$, it follows that $x \notin X \setminus X'$. Since $X \in \mathcal{F}^F$, it follows that $x \in X$, and thus $x \in X'$. Therefore xy does not cover X', and thus $X' \in \mathcal{F}^F$. Next we note that $X' \cap T = C \cap T$, and thus X' is a strict core.

By assumption on A, there is an edge $uv \in A$ such that $v \in X' \subseteq X$ and $u \notin X'$. If $u \in X$, then $u \in T$. Therefore $u \notin X$, and thus X is covered by A.

The following lemma shows that in G^F we can find with high probability an approximate cover for the set of $\mathcal{F}^F$-cores.

Lemma 8. *There is a randomized algorithm that finds a set $A \subseteq E(G^F)$ such that A covers all $\mathcal{F}^F$-cores and $w^1(A) \leq O(|S|)\,\mathrm{opt}_{\mathrm{aug}}(G, E_\ell)$. Moreover, with probability at least $1 - 2^{-|E|}$ the algorithm runs in polynomial time.*

The proof of Lemma 8 is technically involved, however the main idea is quite straightforward. First, note that by Lemma 7, it suffices to show an algorithm that finds an approximate cover for the set of strict $\mathcal{F}^F$-cores. We show that a linear program relaxation of the latter has a polynomial time separation oracle, hence the randomized rounding algorithm for the hitting set finds such a cover in polynomial time with high probability. The proof is given in detail in the full version of the paper [5].

We are now ready to prove Theorem 2, by describing an $O(|S| \log |T|)$-approximation algorithm for the augmentation problem. The algorithm iteratively prunes $\mathcal{F}$ by covering the set of tight cores in every iteration. Starting with $F = \emptyset$, and during every iteration until F covers $\mathcal{F}$, we add to F a set $F' \subseteq E \setminus E_\ell$ of weight at most $O(|S|)\,\mathrm{opt}_{\mathrm{aug}}(G, E_\ell)$ that covers the family of $\mathcal{F}^F$-cores. The mechanism is presented in detail as Algorithm 2.

1: $j \leftarrow 1$
2: $F_j \leftarrow \emptyset$
3: **while** $\mathcal{F}^F \neq \emptyset$ **do**
4:     *(implicitly)* let $\mathcal{F}_j$ be the set of $\mathcal{F}^F$-cores.
5:     let $F' \subseteq E \setminus E_\ell$ be an edge set covering $\mathcal{F}_j$ such that $w(F') \leq O(|S|)\,\mathrm{opt}_{\mathrm{aug}}(G, E_\ell)$.
6:     $F \leftarrow F \cup F'$.
7:     $j \leftarrow j + 1$.
8: **return** F.

Algorithm 2: $O(|S| \log |T|)$-Approximation Algorithm for the Connectivity Augmentation Problem

Proof (of Theorem 2). We first show that each iteration is executed in polynomial time with probability at least $1 - 2^{-|E|}$. Given a set $F \subseteq E \setminus E_\ell$, we efficiently (deterministically) check whether $\mathcal{F}^F = \emptyset$, by verifying that for every $t \in T$, there are $\ell + 1$ edge-disjoint rt-paths in $(V, E_\ell \cup F)$. To see that line 5 runs in polynomial time with high probability, first note that given G, E_ℓ, F, we can construct G^F in polynomial time. By Lemma 8, there is a randomized algorithm that finds a set $A \subseteq E(G^F)$ such that A covers all $\mathcal{F}^F$-cores and $w^1(A) \leq O(|S|)\,\mathrm{opt}_{\mathrm{aug}}(G, E_\ell)$. Moreover, with probability at least $1 - 2^{-|E|}$ the algorithm runs in polynomial time. Lemma 5 then guarantees that we can construct from A an edge set $F' \subseteq E \setminus E_\ell$ that covers all $\mathcal{F}^F$ cores, and such that $w(F') \leq w^1(A) \leq O(|S|)\,\mathrm{opt}_{\mathrm{aug}}(G, E_\ell)$.

Therefore each iteration can be done in polynomial time with probability at least $1-2^{-|E|}$. Lemma 3 ensures that Algorithm 2 terminates after at most $\log |T|$ iterations. Applying a union bound, we get that Algorithm 2 is an $O(|S| \log |T|)$-approximation algorithm for the connectivity augmentation problem, that runs in polynomial time with probability at least $1 - (2^{-|E|} \log |T|)$.

References

1. Byrka, J., Grandoni, F., Rothvoß, T., Sanità, L.: Steiner tree approximation via iterative randomized rounding. J. ACM (JACM) **60**(1), 1–33 (2013)
2. Chan, C., Laekhanukit, B., Wei, H., Zhang, Y.: Polylogarithmic approximation algorithm for k-connected directed steiner tree on quasi-bipartite graphs. In: Byrka, J., Meka, R. (eds.) Approximation, Randomization, and Combinatorial Optimization. Algorithms and Techniques, APPROX/RANDOM 2020, August 17–19, 2020, Virtual Conference. LIPIcs, vol. 176, pp. 63:1–63:20. Schloss Dagstuhl - Leibniz-Zentrum für Informatik (2020). https://doi.org/10.4230/LIPICS.APPROX/RANDOM.2020.63
3. Charikar, M., et al.: Approximation algorithms for directed steiner problems. J. Algorithms **33**(1), 73–91 (1999)
4. Cheriyan, J., Laekhanukit, B., Naves, G., Vetta, A.: Approximating rooted steiner networks. ACM Trans. Algorithms **11**(2), 8:1–8:22 (2014).https://doi.org/10.1145/2650183
5. Cohen, S., Kamma, L., Niklanovits, A.: A new approach for approximating directed rooted networks. arXiv:2407.07543 (2024)
6. Dreyfus, S.E., Wagner, R.A.: The steiner problem in graphs. Networks **1**(3), 195–207 (1971). https://doi.org/10.1002/NET.3230010302
7. Edmonds, J.: Optimum branchings. J. Res. Natl. Bur. Stand. **71B**, 233–240 (1967)
8. Frank, A.: Increasing the rooted-connectivity of a digraph by one. Math. Program. **84**(3), 565–576 (1999). https://doi.org/10.1007/s101070050040
9. Frank, A.: Rooted k-connections in digraphs. Discret. Appl. Math. **157**(6), 1242–1254 (2009). https://doi.org/10.1016/J.DAM.2008.03.040
10. Frank, A., Tardos, É.: Generalized polymatroids and submodular flows. Math. Program. **42**(1-3), 489–563 (1988). https://doi.org/10.1007/BF01589418
11. Frank, A.: Kernel systems of directed graphs. Acta Sci. Math **41**(1), 63–76 (1979)
12. Garg, N., Konjevod, G., Ravi, R.: A polylogarithmic approximation algorithm for the group steiner tree problem. J. Algorithms **37**(1), 66–84 (2000). https://doi.org/10.1006/JAGM.2000.1096
13. Goemans, M.X., Goldberg, A.V., Plotkin, S.A., Shmoys, D.B., Tardos, É., Williamson, D.P.: Improved approximation algorithms for network design problems. In: Sleator, D.D. (ed.) Proceedings of the Fifth Annual ACM-SIAM Symposium on Discrete Algorithms. 23–25 January 1994, Arlington, Virginia, USA, pp. 223–232. ACM/SIAM (1994). http://dl.acm.org/citation.cfm?id=314464.314497

14. Grandoni, F., Laekhanukit, B.: Surviving in directed graphs: a quasi-polynomial-time polylogarithmic approximation for two-connected directed steiner tree. In: Hatami, H., McKenzie, P., King, V. (eds.) Proceedings of the 49th Annual ACM SIGACT Symposium on Theory of Computing, STOC 2017, Montreal, QC, Canada, June 19–23, 2017, pp. 420–428. ACM (2017). https://doi.org/10.1145/3055399.3055445
15. Gupta, A., Könemann, J.: Approximation algorithms for network design: a survey. Surv. Oper. Res. Manage. Sci. **16**(1), 3–20 (2011)
16. Hwang, F.K., Richards, D.S.: Steiner tree problems. Networks **22**(1), 55–89 (1992). https://doi.org/10.1002/NET.3230220105
17. Karp, R.M.: Reducibility among combinatorial problems. In: Miller, R.E., Thatcher, J.W. (eds.) Proceedings of a symposium on the Complexity of Computer Computations, held March 20–22, 1972, at the IBM Thomas J. Watson Research Center, Yorktown Heights, New York, USA, pp. 85–103. The IBM Research Symposia Series, Plenum Press, New York (1972). https://doi.org/10.1007/978-1-4684-2001-2_9
18. Karp, R.M., Moreno-Centeno, E.: The implicit hitting set approach to solve combinatorial optimization problems with an application to multigenome alignment. Oper. Res. **61**(2), 453–468 (2013)
19. Kerivin, H., Mahjoub, A.R.: Design of survivable networks: a survey. Networks **46**(1), 1–21 (2005). https://doi.org/10.1002/NET.20072
20. Khuller, S., Vishkin, U.: Biconnectivity approximations and graph carvings. J. ACM **41**(2), 214–235 (1994). https://doi.org/10.1145/174652.174654
21. Klein, P., Ravi, R.: A nearly best-possible approximation algorithm for node-weighted Steiner trees. J. Algorithms **19**(1), 104–115 (1995)
22. Kortsarz, G., Nutov, Z.: Approximating minimum-cost connectivity problems. In: Gonzalez, T.F. (ed.) Handbook of Approximation Algorithms and Metaheuristics. Chapman and Hall/CRC (2007). https://doi.org/10.1201/9781420010749.CH58
23. Laekhanukit, B.: Parameters of two-prover-one-round game and the hardness of connectivity problems. In: Chekuri, C. (ed.) Proceedings of the Twenty-Fifth Annual ACM-SIAM Symposium on Discrete Algorithms, SODA 2014, Portland, Oregon, USA, January 5–7, 2014, pp. 1626–1643. SIAM (2014). https://doi.org/10.1137/1.9781611973402.118
24. Laekhanukit, B.: Approximating directed steiner problems via tree embedding. In: Chatzigiannakis, I., Mitzenmacher, M., Rabani, Y., Sangiorgi, D. (eds.) 43rd International Colloquium on Automata, Languages, and Programming, ICALP 2016, July 11–15, 2016, Rome, Italy. LIPIcs, vol. 55, pp. 74:1–74:13. Schloss Dagstuhl - Leibniz-Zentrum für Informatik (2016). https://doi.org/10.4230/LIPICS.ICALP.2016.74
25. Liao, C., Chen, Q., Laekhanukit, B., Zhang, Y.: Almost tight approximation hardness for single-source directed k-edge-connectivity. In: Bojanczyk, M., Merelli, E., Woodruff, D.P. (eds.) 49th International Colloquium on Automata, Languages, and Programming, ICALP 2022, July 4–8, 2022, Paris, France. LIPIcs, vol. 229, pp. 89:1–89:17. Schloss Dagstuhl - Leibniz-Zentrum für Informatik (2022). https://doi.org/10.4230/LIPICS.ICALP.2022.89
26. Nelson, J.: A note on set cover inapproximability independent of universe size. Electron. Colloquium Comput. Complex. (ECCC) **14**(105) (2007)
27. Nutov, Z.: A note on rooted survivable networks. Inf. Process. Lett. **109**(19), 1114–1119 (2009)

28. Oliveira, C.A.S., Pardalos, P.M.: A survey of combinatorial optimization problems in multicast routing. Comput. Oper. Res. **32**, 1953–1981 (2005). https://doi.org/10.1016/J.COR.2003.12.007
29. Rothvoß, T.: Directed steiner tree and the lasserre hierarchy. CoRR arXiv:1111.5473 (2011)
30. Voß, S.: Steiner tree problems in telecommunications. In: Resende, M.G.C., Pardalos, P.M. (eds.) Handbook of Optimization in Telecommunications, pp. 459–492. Springer (2006). https://doi.org/10.1007/978-0-387-30165-5_18
31. Zelikovsky, A.: A series of approximation algorithms for the acyclic directed steiner tree problem. Algorithmica **18**(1), 99–110 (1997). https://doi.org/10.1007/BF02523690

Approximations and Hardness of Covering and Packing Partially Ordered Items

Ilan Doron-Arad[1], Guy Kortsarz[2], Joseph Naor[1], Baruch Schieber[3], and Hadas Shachnai[1](✉)

[1] Computer Science Department, Technion, Haifa, Israel
{idoron-arad,naor,hadas}@cs.technion.ac.il
[2] Computer Science Department, Rutgers University-Camden, Camden, NJ, USA
guyk@camden.rutgers.edu
[3] Computer Science Department, New Jersey Institute of Technology, Newark, NJ, USA
sbar@njit.edu

Abstract. Motivated by applications in production planning and storage allocation in hierarchical databases, we initiate the study of *covering partially ordered items* (CPO). Given a value $k \in \mathbb{N}$, and a directed graph $G = (V, E)$ where each vertex has a size in $\{0, 1, \ldots, k\}$, we seek a collection of subsets of vertices $C_1, \ldots, C_t$ that *cover* all the vertices, such that for any $1 \leq j \leq t$, the total size of vertices in C_j is bounded by k, and there are no edges from $V \setminus C_j$ to C_j. The objective is to minimize the number of subsets t. CPO is closely related to the *rule caching problem* (RCP) that has been widely studied in the networking area. The input for RCP is a directed graph $G = (V, E)$, a profit function $p : V \rightarrow \mathbb{Z}_0^+$, and $k \in \mathbb{N}$. The output is a subset $S \subseteq V$ of maximum profit such that $|S| \leq k$ and there are no edges from $V \setminus S$ to S.

Our main result is a 2-approximation algorithm for CPO on out-trees, complemented by an asymptotic 1.5-hardness of approximation result. We also give a two-way reduction between RCP and the *densest k-subhypergraph* problem, surprisingly showing that the problems are equivalent with respect to polynomial-time approximation within any factor $\rho \geq 1$. This implies that RCP cannot be approximated within factor $|V|^{1-\varepsilon}$ for any fixed $\varepsilon > 0$, under standard complexity assumptions. Prior to this work, RCP was only known to be strongly NP-hard. We further show that there is no EPTAS for the special case of RCP where the profits are uniform, assuming Gap-ETH. Since this variant admits a PTAS, we essentially resolve the complexity status of this problem.

H. Shachnai—This work was partly carried out during a visit to NJIT in association with the DIMACS Special Focus on Bridging Continuous and Discrete Optimization.
J. Naor—Supported in part by Israel Science Foundation grant 2233/19 and United States - Israel Binational Science Foundation (BSF) grant 2033185, 2022418.

D. Král′ and M. Milanič (Eds.): WG 2024, LNCS 14760, pp. 166–180, 2025.
https://doi.org/10.1007/978-3-031-75409-8_12

1 Introduction

Partially ordered entities are ubiquitous in the mathematical modeling of scheduling problems, distributed storage allocation, production planning, and unified language models. Often, the partial order represents either precedence constraints or dependencies among entities (or items). Motivated by applications in production planning [1,25] and distributed storage allocation in hierarchical databases [29], we introduce the *covering partially ordered items* (CPO) problem. An instance of CPO consists of a directed graph $G = (V, E)$, a value $k \in \mathbb{N}$, and a size function $s : V \to [0 : k]$.[1] A *configuration* is a subset of vertices $U \subseteq V$ such that $s(U) \leq k$,[2] and U is closed under precedence constraints; that is, for any $u \in U$ and $(z, u) \in E$ it holds that $z \in U$. A *feasible solution* is a set of configurations $C_1, \ldots, C_t$ that *covers* V, namely $\bigcup_{j \in [1:t]} C_j = V$. The *cardinality* of the solution is t, the number of configurations. The goal is to find a feasible solution of minimum cardinality.

CPO can be applied to optimize the distributed storage of large hierarchical data in *unified medical language systems* (UMLS) [29]. UMLS data is often distributed over several databases of bounded size. Due to the hierarchical nature of the medical taxonomy, each database needs to be closed under this hierarchy relation. The problem of minimizing the number of distributed databases of the UMLS data translates to a CPO problem instance.

Another application of CPO arises in production planning for steel mills that employ continuous casting [1,25]. The steel-making process has high energy consumption. One way to save energy is by employing continuous casting and direct charging. In this routine, the molten steel is solidified into slabs and rolled into finished products of various sizes continuously, with no need to reheat the steel in the process. Each finished product requires specific casting, rolling, and thermal treatments in a given order, which can be modeled by a *directed acyclic graph* (DAG). A main challenge is to assign the finished products to batches whose size is dictated by the size of the ladle furnace while minimizing the amount of repeated operations. This gives rise to an instance of CPO.

A natural greedy approach for solving CPO is to repeatedly find, among all subsets of vertices that can be feasibly assigned to a single configuration, a subset that maximizes the size of yet unassigned vertices. This single configuration problem is a variant of the well-known *rule caching problem* (RCP) that has been studied extensively [4,7,9,10,14–17,20–24,26–28]. An instance of RCP consists of a directed graph $G = (V, E)$, a profit function $p : V \to \mathbb{Z}_0^+$, and a value $k \in \mathbb{N}$. We seek a subset of vertices $U \subseteq V$ which is closed under precedence constraints, such that $|U| \leq k$, and $p(U) = \sum_{u \in U} p(u)$ is maximized. In [8] we describe central applications of RCP in networking and the blockchain technology.

Prior to this work RCP was just known to be strongly NP-hard [3,19]. Our initial attempt towards solving CPO was to find a good approximation for RCP. Surprisingly, we were able to show an equivalence between RCP and the *densest*

---

[1] For $i, j \in \mathbb{Z}_0^+$, we denote by $[i : j]$ the set of integers $\{i, i + 1, \ldots, j\}$.

[2] For a set A, a function $f : A \to \mathbb{X}$, and $B \subseteq A$, define $f(B) = \sum_{b \in B} f(b)$.

k-subhypergraph (DkSH) problem with respect to approximability. The input for DkSH consists of a hypergraph $G = (V, E)$ and a value $k \in \mathbb{N}$. The goal is to find a subset of vertices $S \subseteq V$ of cardinality k that maximizes the number (or weight) of induced hyperedges (a more formal definition is given in [8]).[3]

Unfortunately, DkSH is known to be hard to approximate within a factor of $|V|^{1-\varepsilon}$, for $\varepsilon \in (0, 1)$, assuming the Small Set Expansion Conjecture (by combining the results of [18] and [12]). This implies the same hardness of approximation for RCP (see Sect. 1.1). Given this hardness result, we expect CPO to be hard to approximate on general graphs. Thus, we consider the special case of CPO where G is an out-tree (a directed tree where all edges are directed away from the root). We call this problem *covering partially ordered items on out-trees (*CT*)*. To the best of our knowledge, CT is studied here for the first time. We note that when G is an in-tree CPO is trivial since the problem has a feasible (and unique) solution *iff* the total size of the vertices is at most k.

1.1 Our Results

Our first result is an approximation algorithm for CT. Recall that, for $\alpha \geq 1$, $\mathcal{A}$ is an α-approximation algorithm for a minimization (maximization) problem Π if, for any instance of Π, the output of $\mathcal{A}$ is at most α (at least $1/\alpha$) times the optimum.

Theorem 1.1. *There is a polynomial time* 2*-approximation algorithm for* CT.

While out-trees have a simple structure, allowing for a greedy-based bottom-up approach in solving CT, the analysis of our approximation algorithm is non-trivial and requires extra care to make sure that the approximation bound has no additive terms (see below).

CT generalizes the classical *bin packing* (BP) problem. The input for BP is a set of items and a value $k \in \mathbb{N}$. Each item has a size in $[0 : k]$, and the goal is to assign the items into a minimum number of bins of capacity k.[4] An instance of BP is reduced to an instance of CT on a star graph by generating a leaf for each item of the BP instance and adding a root vertex of size zero. This trivial reduction implies that CT is strongly NP-hard [11].

Interestingly, we show that in contrast to BP, CT does not admit an *asymptotic polynomial-time approximation scheme* (APTAS), or even an asymptotic approximation strictly better than $\frac{3}{2}$. This separates CT from BP which admits also an additive logarithmic approximation [13].

Theorem 1.2. *For any* $\alpha < \frac{3}{2}$, *there is no asymptotic* α*-approximation for* CT *unless* P = NP.

Next, we study the hardness of RCP.

[3] DkSH has also been widely studied (see, e.g., [5,6,12] and the references therein).

[4] We use the definition of BP as given in [11]. In an alternative definition found in the literature, bin capacities are normalized to one, and item sizes are in $[0, 1]$.

Theorem 1.3. *For any $\rho \geq 1$, there is a ρ-approximation for* RCP *if and only if there is a ρ-approximation for* DkSH.

Corollary 1.4. *Assuming the Small Set Expansion Hypothesis (SSEH) and* NP $\neq$ BPP, *for any $\varepsilon > 0$ there is no $|V|^{1-\varepsilon}$-approximation for* RCP.

We give a tight lower bound also for the previously studied special case of *uniform* RCP (U-RCP) [2,3].[5] In U-RCP the vertices have uniform (unit) profits (i.e., $p(v) = 1 \; \forall v \in V$). While U-RCP is known to admit a PTAS [2,3], the question of whether the problem admits an EPTAS or even an FPTAS remained open (we give formal definitions relating to approximation schemes in [8]). Our next result gives a negative answer to both of these questions, posed in [2,3].

Theorem 1.5. *Assuming* Gap-ETH, *there is no* EPTAS *for* U-RCP.

Finally, we show that RCP remains essentially just as hard when the in-degrees and out-degrees are bounded.

Theorem 1.6. *A ρ-approximation algorithm for* RCP *instances with in-degrees and out-degrees bounded by* 2, *for any $\rho \geq 1$, implies a ρ-approximation for* RCP.

Due to space constraints, we include in this version only the proof of Theorem 1.1 and defer the proofs of the other theorems to the full version of the paper [8].

Techniques: Our algorithm for CT covers the vertices in a given out-tree T in a bottom-up fashion, starting from the leaves. The key players in this process are vertices called *anchors* which define the candidate subtrees for covering in each iteration. Interestingly, we show that the subtree associated with a specific anchor a (including also all of a's ancestors) can be covered efficiently by using the naive NextFit algorithm.

To eliminate *additive* terms in the approximation guarantee (i.e., obtain an *absolute* ratio of 2), a crucial step in the algorithm is to distinguish in each call to NextFit between the case where NextFit outputs an *even* vs. *odd* number of configurations. In the latter case, we discard the last configuration and cover the corresponding *leftover* vertices in a later iteration of the algorithm.

The crux of the analysis is to charge the number of subsets (i.e., configurations) used by the algorithm separately to each anchor. Consider a subtree of an anchor a, of total size $\text{SA}(a)$, covered at some iteration. Observing that each subset including vertices in this subtree must include also all the ancestors of a, we are able to show that the total number of subsets used is at most twice $\left\lfloor \frac{\text{SA}(a)}{k-h(a)+1} \right\rfloor$, where $h(a)$ is the total size of the ancestors of a in T (including a). To complete the analysis, we lower bound the number of subsets used in *any* feasible solution. This is done via an intricate calculation bounding the *number of occurrences* of each vertex v in the subtree of an anchor a in any feasible cover, which is the heart of the analysis. Our Greedy approach may be useful for other

[5] In [3], U-RCP is called uniform directed all-neighbor knapsack problem.

CPO classes of instances in which the input graph G has a *tree-like* structure (e.g., graphs of bounded treewidth).

Our proofs of hardness for CT and RCP use sophisticated constructions, most notably, to show a two-way reduction between RCP and DkSH and the hardness of RCP with bounded degrees (see [8]).

Organization: Section 2 presents our approximation algorithm for CT and the proof of Theorem 1.1, and Sect. 3 includes some open problems.

2 Approximation Algorithm for CT

In this section, we present our approximation algorithm for CT. We start with some definitions and notations. Let $T = (V, E)$ be an out-tree rooted at a vertex $r \in V$. Recall that in an out-tree all edges are oriented outwards from r. Thus, for an edge $(u, v) \in E$, vertex u precedes v on the (unique) path from r to v. We say that u is the *parent* of v and v is a *child* of u. More generally, if u is on the (unique) path from r to v then u is an *ancestor* of v and v is a *descendant* of u. A vertex v is considered an ancestor of itself but *not* a descendant of itself. Define $h(v)$ to be the total size of the vertices on the path from r to v, which equals the total size of the ancestors of v.

For $U \subseteq V$, let $T[U]$ be the subgraph of T induced by U. If $T[U]$ is connected, then we say that $T[U]$ is a subtree of T. Note that in this case $T[U]$ is also an out-tree. From now on, we consider only induced subgraphs that are connected, namely subtrees of T. If $r \in U$, then $T[U]$ is a subtree of T rooted at r.

For an out-tree $T = (V, E)$ and a subset of vertices $U \subseteq V$, let $\textsc{Ancs}_T(U)$ be the set of the ancestors in T of the vertices in U, and let $\textsc{Desc}_T(U)$ be the set of the descendants in T of the vertices in U. Note that if $T[U]$ is a subtree of T rooted at r, then $\textsc{Ancs}_T(U) = U$. In case U is a singleton set, we omit the set notation; that is, for $v \in V$, let $\textsc{Ancs}_T(v)$ be the set of the ancestors of v in T, and let $\textsc{Desc}_T(v)$ be the set of the descendants of v in T.

We note that if there is a vertex $v \in V$ for which $h(v) > k$ then there is no feasible solution. Also, if there is a leaf ℓ of T for which $h(\ell) = k$ then any solution must include the set $\textsc{Ancs}_T(\ell)$ (of size k), and after adding this set to the solution, we can remove ℓ and all of its ancestors which are not ancestors of any other leaf. Thus, without loss of generality, we assume that for any vertex $v \in V$ it holds that $h(v) < k$. Also, we note that if there is a leaf ℓ of T of size $s(\ell) = 0$ then we can remove ℓ, solve for the resulting tree and then add ℓ to a subset in the cover that includes the parent of ℓ in T. Thus, without the loss of generality, we assume that for any leaf ℓ of T, $s(\ell) > 0$.

The algorithm for computing a cover is iterative. In each iteration, we compute a partial cover as described below. We then continue to the next iteration with the subtree rooted at r induced by the uncovered vertices and their ancestors. The algorithm terminates when either the set of uncovered vertices is empty or the total size of the vertices of the remaining subtree (rooted at r) is at most k, in which case these vertices form the last set in the cover.

In each iteration t of the algorithm, we compute a subset of vertices $A_t \subset V$ that we call *anchors*. We then compute a cover of *some* (potentially all) descendants of the anchors in A_t, and proceed to the next iteration.

Algorithm 1 is the pseudo code of the iterative algorithm. Initially, $V_1 = V$. Consider the t-th iteration, for $t \geq 1$. If $s(V_t) \leq k$ then the algorithm terminates. Otherwise, define A_t as the set of all the vertices $v \in V_t$ such that

(i) The total size of the descendants of v in $T[V_t]$ is more than $k - h(v)$, and
(ii) The total size of the descendants of every child u of v in $T[V_t]$ is at most $k - h(u) = k - h(v) - s(u)$.

Procedure NextFit given in Algorithm 2 is called for every $a \in A_t$. The input to Procedure NextFit is the tree T_a defined as the rooted subtree that consists of the path from r to a and the descendants of a in the subtree $T[V_{t(a)}]$ (see Fig. 1a). When called for an anchor a, Procedure NextFit (Algorithm 2) computes a cover of some (potentially all) descendants of a. The number of sets returned in this procedure call is even, and the total size of the descendants of a that are not covered by the sets returned by Procedure NextFit is at most $k - h(a)$. Let $U_t \subseteq V_t$ be the set of all descendants of anchors in A_t that were covered in iteration t, together with all their ancestors. If $V_t = U_t$ then the algorithm terminates. Otherwise, we let V_{t+1} be the set of ancestors of the vertices $V_t \setminus U_t$ in $T[V_t]$ and continue to iteration $t + 1$. Let $A = \bigcup_t A_t = \{a_1, a_2, \ldots\}$ be the set of anchors computed in all the iterations. For an anchor $a \in A$, let $t(a)$ be the iteration in which a was added to the set of anchors. Note that any leaf ℓ of T

appears in exactly one subset in $\mathcal{C}$. Thus, the iteration in which ℓ is *covered* is uniquely defined.

Algorithm 1 Feasible cover computation

Input: An out-tree $T = (V, E)$ rooted at r and an integer $k > 0$
Output: A feasible cover $\mathcal{C} = C_1, \dots, C_c$

$V_1 \leftarrow V$
$\mathcal{C} \leftarrow \emptyset$
$t \leftarrow 1$
while $s(V_t) > k$ **do**
  $X_t \leftarrow \{u \in V_t \mid s(\textsc{Desc}_{T[V_t]}(u)) \le k - h(u)\}$
  $A_t \leftarrow \{v \in V_t \setminus X_t \mid$ all the children of v in $T[V_t]$ are in $X_t\}$
  $U_t \leftarrow \emptyset$ ▷ *U_t stores the vertices covered in iteration t*
  for $a \in A_t$ **do**
    $T_a \leftarrow T[\textsc{Ancs}_{T[V_t]}(a) \cup \textsc{Desc}_{T[V_t]}(a)]$
    $Q_1, \dots, Q_m \leftarrow \textsc{NextFit}(a, T_a, k)$
    Add $Q_1, \dots, Q_m$ to $\mathcal{C}$ ▷ *Add the partial cover computed by* $\textsc{NextFit}$
    $U_t \leftarrow U_t \cup Q_1 \cup \dots \cup Q_m$
  if $V_t \setminus U_t \ne \emptyset$ **then**
    $V_{t+1} \leftarrow \textsc{Ancs}_{T[V_t]}(V_t \setminus U_t)$
  else
    $V_{t+1} \leftarrow \emptyset$
  $t \leftarrow t + 1$
if $V_t \ne \emptyset$ **then** ▷ *The last set in the cover*
  Add V_t to $\mathcal{C}$
return $\mathcal{C}$

Algorithm 2 Next-Fit packing

procedure $\textsc{NextFit}(a, T_a, k)$
  Input: An anchor $a \in A$, the subtree T_a and an integer $k > 0$
  Output: A feasible cover $Q_1, \dots, Q_m$ of some (potentially all) vertices in $\textsc{Desc}_{T_a}(a)$
  Let $u_1, \dots, u_d$ be the children of a in T_a
  $m \leftarrow 1$
  $Q_m \leftarrow \textsc{Ancs}_{T_a}(a)$
  for $s = 1$ to d **do**
    if $s(Q_m) + s(\textsc{Desc}_{T_a}(u_s)) \le k$ **then**
      $Q_m \leftarrow Q_m \cup \textsc{Desc}_{T_a}(u_s)$
    else
      $m \leftarrow m + 1$
      $Q_m \leftarrow \textsc{Ancs}_{T_a}(a) \cup \textsc{Desc}_{T_a}(u_s)$
  if m is odd **then** ▷ *Remove the subset Q_m if m is odd*
    $m \leftarrow m - 1$ ▷ *Note that $m > 1$*
  return $Q_1, \dots, Q_m$

Definition 2.1. *Let $a \in A$ be an anchor.*

- *If $v \in \textsc{Desc}_{T_a}(a)$ is an ancestor of a leaf ℓ of T that is covered in iteration $t(a)$ then we say that v is* anchored *at a.*
- *If $v \in \textsc{Desc}_{T_a}(a)$ is not anchored at a then we say that v is a* leftover vertex *of a.*

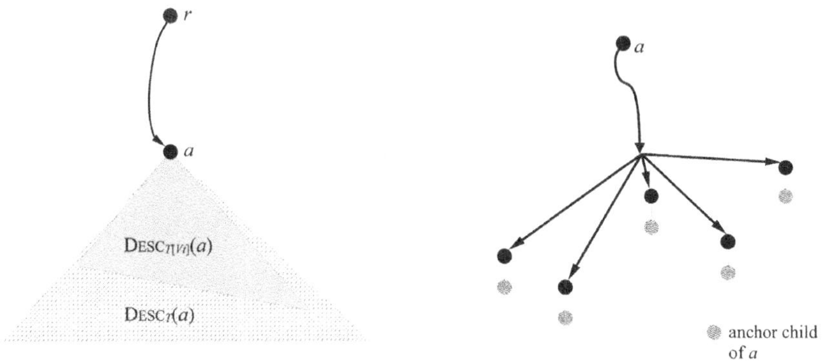

(a) The subtree T_a (b) The subtree S_a

Fig. 1. The subtrees T_a defined in Algorithm 1, and S_a defined in the proof of Lemma 2.5.

- *Let* $\text{SA}(a)$ *denote the total size of the vertices that are anchored at* a*, and* $\text{LO}(a)$ *denote the total size of the leftover vertices of* a*.*

Clearly, $\text{SA}(a) + \text{LO}(a) = s(\text{DESC}_{T_a}(a))$. Our assumption that for every leaf ℓ of T, $s(\ell) > 0$, implies that (i) $\text{SA}(a) > 0$ and (ii) if there are leftover vertices then $\text{LO}(a) > 0$.

The proofs of the next lemmas are given in [8].

Lemma 2.2. *Let* $v \in \text{DESC}_{T_a}(a)$*, and let* u_s *be the (unique) child of* a *that is also an ancestor of* v*. If* v *is a leftover vertex of* a *then all the vertices in the subtree of* T_a *rooted at* v*, as well as the vertices along the path from* u_s *to* v*, are also leftover vertices of* a*. If* v *is anchored at* a *then all the vertices in the subtree of* T_a *rooted at* v*, as well as the vertices along the path from* u_s *to* v*, are also anchored at* a*.*

Lemma 2.3. *For any two anchors* $a, a' \in A$*, the sets of vertices anchored at* a *and* a' *are disjoint.*

Define a "parent-child" relation among anchors as follows. For two anchors a and b, we say that a is the *anchor-parent* of b and b is the *anchor-child* of a if (i) a is an ancestor of b in T, and (ii) the path from a to b (in T) does not contain any anchors other than a and b. Note that if anchor a is an anchor-parent of b then $t(a) > t(b)$; that is, the iteration $t(a)$ in which a is added to the set of anchors is after iteration $t(b)$. This follows from the definition of A_t. For anchor $a \in A$, let $\text{AC}(a) \subset \text{DESC}_T(a) \cap A$ be the set of anchor-children of a. We extend this definition for all $v \in V$, and let $\text{AC}(v) \subset \text{DESC}_T(v) \cap A$ be all the anchors $b \in \text{DESC}_T(v) \cap A$ such that the path from v to b (in T) does not contain any anchors other than b and (possibly) v. For $v \in V$, let $\text{AD}(v) = \text{DESC}_T(v) \cap A$ be the set of anchors that are also descendants of v. A *top* anchor is an anchor that is not an anchor-child of any other anchor. Let $\text{TOP}A \subseteq A$ denote the set of top anchors. Note that if the root r is an anchor then $\text{TOP}A = \{r\}$.

Lemma 2.4. *The number of subsets in the solution computed by Algorithm 1 is upper bounded by*

$$\alpha + \sum_{a \in A} 2 \left\lfloor \frac{\text{SA}(a)}{k - h(a) + 1} \right\rfloor,$$

where

$$\alpha = \begin{cases} 1 & \exists a \in \text{TOP}A \ s.t. \ \text{LO}(a) > 0 \\ 1 & \exists \ leaf \ \ell \in V \ s.t. \ \text{ANCS}_T(\ell) \cap \text{TOP}A = \emptyset \\ 0 & otherwise \end{cases}$$

Proof. Let $\mathcal{Q} = Q_1, \ldots, Q_d$ be the solution computed by the algorithm. Fix $a \in A$, and let $\mathcal{Q}_a$ be the subsets in $\mathcal{Q}$ that were returned by Procedure NextFit (Algorithm 2) when it computed a feasible cover of the vertices in $\text{DESC}_{T_a}(a)$. Note that the union of all the subsets in $\mathcal{Q}_a$ is the set of vertices anchored at a (whose total size is $\text{SA}(a)$) together with all the ancestors of a. Also, $\mathcal{Q}_a$ consists of at least two subsets, and a vertex anchored at a cannot appear in more than one subset in $\mathcal{Q}_a$. Consider the subsets in $\mathcal{Q}_a$ in the order in which they were computed by Algorithm 2. It follows from Procedure NextFit that the total size of the vertices in any pair of consecutive subsets in this ordered list is at least $k - h(a) + 1$. Since the total size of vertices anchored at a is $\text{SA}(a)$, the number of such disjoint pairs is upper bounded by $\left\lfloor \frac{\text{SA}(a)}{k-h(a)+1} \right\rfloor$. By Line 11 of Procedure NextFit, the number of sets in Q_a is even, and thus the total number of subsets in $\mathcal{Q}_a$ is upper bounded by $2\left\lfloor \frac{\text{SA}(a)}{k-h(a)+1} \right\rfloor$. The total upper bound is given by summing this bound over all anchors $a \in A$. We may have one additional subset if the algorithm is terminated when $|V_t| > 0$. By our construction and Lemma 2.2, this may happen *iff* there exists a leaf ℓ that is not anchored at any anchor. If such a leaf ℓ exists then one of the following two conditions must be satisfied: (i) ℓ has no ancestor that is an anchor, or (ii) ℓ is a leftover vertex of the (unique) top anchor a that is an ancestor of ℓ, in which case $\text{LO}(a) > 0$. The lemma follows. □

We now prove a lower bound on the number of subsets in any feasible solution and in particular in the optimal solution. Let $\mathcal{P} = P_1, \ldots, P_p$ be a feasible solution. Since every subset in $\mathcal{P}$ is closed under ancestor relation, some vertices may appear in multiple subsets. We refer to each such appearance of a vertex v as an *occurrence* of v, and associate the size $s(v)$ to each of its occurrences. For an anchor $a \in A$, let $\mathcal{P}(a)$ be the set of all the subsets in $\mathcal{P}$ that contain vertices anchored at either a or a descendant of a. For an anchor $a \in A$, let $\mathcal{U}(a) \subseteq \text{DESC}_T(a)$ be the set of all vertices such that each vertex is both a descendant of a and an ancestor of a vertex anchored at either a or a descendant of a.

Lemma 2.5. *For every $a \in A$, the number of subsets in $\mathcal{P}(a)$ is at least*

$$\mathrm{LB}(a) = \sum_{b \in \mathrm{AD}(a) \cup \{a\}} \left\lfloor \frac{\mathrm{SA}(b)}{k - h(b)} \right\rfloor .$$

If the lower bound is tight then all the leaves that are in the subsets in $\mathcal{P}(a)$ must be anchored at either a or a descendant of a.

Proof. To prove the lower bound of $\mathrm{LB}(a)$ on the *number* of subsets in $\mathcal{P}(a)$ for every $a \in A$, we prove a slightly stronger lower bound of $(k - h(a))\,\mathrm{LB}(a)$ on the *total size* of the occurrences of vertices in $\mathcal{U}(a)$ in subsets in $\mathcal{P}(a)$. Since any subset that contains a descendant of a must contain also the ancestors of a (including a), whose total size is $h(a)$, the total size of the vertices in $\mathcal{U}(a)$ that can be in a single subset in $\mathcal{P}(a)$ is no more than $k - h(a)$. Thus, a lower bound of $(k - h(a))\,\mathrm{LB}(a)$ on the total size of the occurrences of vertices in $\mathcal{U}(a) \subseteq \mathrm{Desc}_T(a)$ in subsets in $\mathcal{P}(a)$ implies a lower bound of $\mathrm{LB}(a)$ on the number of subsets in $\mathcal{P}(a)$ (and on the number of occurrences of anchor a).

The lower bound on the total size of the occurrences of vertices in $\mathcal{U}(a)$ in subsets in $\mathcal{P}(a)$ also implies that if the lower bound is tight then all the leaves that are in the subsets in $\mathcal{P}(a)$ must be anchored at either a or a descendant of a. To see this, note that if any subset in $\mathcal{P}(a)$ contains a leaf ℓ that is not anchored at an anchor in $\mathrm{AD}(a) \cup \{a\}$ then $\ell \notin \mathcal{U}(a)$; also, $s(\ell) > 0$ by our assumption. It follows that the total size of the subsets in $\mathcal{P}(a)$ is strictly more than $(k - h(a))\,\mathrm{LB}(a)$. Clearly, this implies that the number of subsets in $\mathcal{P}(a)$ is strictly more than $\mathrm{LB}(a)$.

The proof is by induction starting from the *bottom* anchors in T, which are the anchors with no anchor-children. For the induction base, consider a bottom anchor a. Note that in this case $\mathcal{U}(a)$ is the set of all vertices anchored at a. The subsets in $\mathcal{P}(a)$ cover all the vertices anchored at a; thus, the total size of the occurrences of these vertices in the subsets in $\mathcal{P}(a)$ is at least $\mathrm{SA}(a)$. Clearly, $\mathrm{SA}(a) \geq (k - h(a)) \left\lfloor \frac{\mathrm{SA}(a)}{k-h(a)} \right\rfloor = (k - h(a))\,\mathrm{LB}(a)$.

For the inductive step, consider an anchor a, and assume the lemma holds for every anchor $b \in \mathrm{AC}(a)$. Specifically, for every anchor $b \in \mathrm{AC}(a)$, the total size of occurrences of vertices in $\mathcal{U}(b)$ in subsets in $\mathcal{P}(b)$ is at least $(k - h(b))\,\mathrm{LB}(b)$. Note that $\mathcal{P}(b) \subseteq \mathcal{P}(a)$ and $\mathcal{U}(b) \subseteq \mathcal{U}(a)$. Let S_a the subtree of T rooted at a, given by the union of the paths from a to each of its anchor-children, *excluding* the anchor-children (see Fig. 1b). Note that the vertices of S_a as well as the vertices in $\mathrm{AC}(v)$ are in $\mathcal{U}(a)$.

Claim 2.6. *For every vertex v of S_a, the total size of the occurrences of vertices in $\mathrm{Desc}_T(v) \cap \mathcal{U}(a)$ in subsets in $\bigcup_{b \in \mathrm{AC}(v)} \mathcal{P}(b)$ is at least $(k - h(v)) \sum_{b \in \mathrm{AC}(v)} \mathrm{LB}(b)$.*

Proof. We prove the claim vertex by vertex, scanning the vertices of S_a bottom-up. Consider a leaf v of S_a. By the definition of S_a, its children are anchors in $\mathrm{AC}(v)$. By the induction hypothesis of Lemma 2.5, for every anchor $b \in \mathrm{AC}(v)$

the total size of the occurrences of vertices in $\mathcal{U}(b)$ in the subsets in $\mathcal{P}(b)$ is at least $(k - h(b))\,\mathrm{LB}(b)$. The total size of such occurrences that are contained in any single subset of $\mathcal{P}(b)$ is at most $k - h(b)$, since any such subset must also contain the ancestors of b (including b) whose size is $h(b)$. It follows that the *number* of occurrences of b in these subsets in $\mathcal{P}(b)$ is at least $\mathrm{LB}(b)$, and the total size of these occurrences is at least $s(b) \cdot \mathrm{LB}(b)$. Note that for any pair of anchors $b, b' \in \mathrm{AC}(v)$, the sets $\mathcal{U}(b)$ and $\mathcal{U}(b')$ are disjoint. Summing over all the anchor-children of v, we have that the total size of the occurrences of vertices in $\textsc{Desc}_T(v) \cap \mathcal{U}(a)$ in the subsets in $\bigcup_{b \in \mathrm{AC}(v)} \mathcal{P}(b)$ is at least $\sum_{b \in \mathrm{AC}(v)} \mathrm{LB}(b)\,((k - h(b)) + s(b)) = (k - h(v)) \sum_{b \in \mathrm{AC}(v)} \mathrm{LB}(b)$. The last equality holds since for every $b \in \mathrm{AC}(v)$, $h(v) + s(b) = h(b)$.

The lower bound for an internal vertex of S_a is obtained similarly. Note that a child u of v is either an anchor or a vertex of S_a. If u is an anchor, that is $u \in \mathrm{AC}(v)$, then as shown above, the total size of the occurrences of u and its descendants in $\mathcal{U}(u)$ in the subsets in $\mathcal{P}(u)$ is $(k - h(u) + s(u))\,\mathrm{LB}(u) = (k - h(v))\,\mathrm{LB}(u)$. Suppose that u is a vertex of S_a. Since u is a child of v and the vertices are scanned bottom-up the lower bound holds for u, and the total size of the occurrences of vertices in $\textsc{Desc}_T(u) \cap \mathcal{U}(a)$ in the subsets in $\bigcup_{b \in \mathrm{AC}(u)} \mathcal{P}(b)$ is $(k - h(u)) \sum_{b \in \mathrm{AC}(u)} \mathrm{LB}(b)$. The total size of such occurrences that is contained in any single subset in $\bigcup_{b \in \mathrm{AC}(u)} \mathcal{P}(b)$ is at most $k - h(u)$. Thus, the *number* of occurrences of u in these subsets is at least $\sum_{b \in \mathrm{AC}(u)} \mathrm{LB}(b)$, and the total size of these occurrences is at least $s(u) \cdot \sum_{b \in \mathrm{AC}(u)} \mathrm{LB}(b)$. We get that the total size of occurrences of u and the vertices in $\textsc{Desc}_T(u) \cap \mathcal{U}(a)$ in the subsets in $\bigcup_{b \in \mathrm{AC}(u)} \mathcal{P}(b)$ is

$$(k - h(u) + s(u)) \sum_{b \in \mathrm{AC}(u)} \mathrm{LB}(b) = (k - h(v)) \sum_{b \in \mathrm{AC}(u)} \mathrm{LB}(b).$$

For any pair u, u' of children of v, the sets $\textsc{Desc}_T(u)$ and $\textsc{Desc}_T(u')$ are disjoint. Summing over all the children of v, we get that the total size of the occurrences of vertices in $\textsc{Desc}_T(v) \cap \mathcal{U}(a)$ in the subsets in $\bigcup_{b \in \mathrm{AC}(v)} \mathcal{P}(b)$ is at least $(k - h(v)) \sum_{b \in \mathrm{AC}(v)} \mathrm{LB}(b)$. □

Next, we consider vertices that are anchored at a. By the definition of $\mathcal{P}(a)$, each such vertex v must occur at least once in subsets in $\mathcal{P}(a)$; also, $v \in \mathcal{U}(a)$. Note that v may be an ancestor of an anchor $b \in \mathrm{AD}(a)$. This may happen in case v is a vertex of S_a, and also in case a leftover vertex of an anchor $b \in \mathrm{AD}(a)$ is anchored at a, and v is on the path from a to b. In case v is an ancestor of an anchor $b \in \mathrm{AD}(a)$, our induction hypothesis and Claim 2.6 already imply a lower bound on the number of its occurrences in subsets in $\mathcal{P}(a)$. Specifically, in case $v \in \mathrm{AD}(a)$, our induction hypothesis implies a lower bound of $\mathrm{LB}(v)$ on the number of its occurrences, and in case $v \notin \mathrm{AD}(a)$ and $\mathrm{AC}(v) \neq \emptyset$, Claim 2.6 implies a lower bound of $\sum_{b \in \mathrm{AC}(v)} \mathrm{LB}(b)$ on the number of its occurrences. We prove that v must occur at least *once more* in subsets in $\mathcal{P}(a)$, *in addition* to this implied lower bound. This results in addition of $\textsc{sa}(a)$ to the total size of the occurrences of vertices anchored at a in the subsets in $\mathcal{P}(a)$.

Claim 2.7. *For every vertex v anchored at a, the number of occurrences of v in the subsets in $\mathcal{P}(a)$ is at least*

$$\begin{cases} 1 & v \notin \textsc{Ancs}_T(\mathrm{AD}(a)) \\ 1 + \mathrm{LB}(v) & v \in \mathrm{AD}(a) \\ 1 + \sum_{b \in \mathrm{AC}(v)} \mathrm{LB}(b) & \textit{otherwise} \end{cases}$$

Proof. If v is anchored at a then it must be an ancestor of a leaf ℓ of T that is anchored at a. Certainly, v must occur in the subset in $\mathcal{P}(a)$ that covers ℓ. If v is not an ancestor of an anchor $b \in \mathrm{AD}(a)$, we are done.

If v is an anchor and thus $v \in \mathrm{AD}(a)$, and the number of occurrences of v in the subsets in $\mathcal{P}(v)$ is strictly more than $\mathrm{LB}(v)$, then we are done. Otherwise, the lower bound $\mathrm{LB}(v)$ is tight, and by the induction hypothesis of Lemma 2.5, all the leaves that are in the subsets in $\mathcal{P}(v)$ must be anchored at v or a descendant of v. Thus, none of these subsets can cover ℓ. It follows that v must occur in at least one more subset in $\mathcal{P}(a)$ that covers ℓ.

A similar argument applies also if $v \notin \mathrm{AD}(a)$ and $\mathrm{AC}(v) \neq \emptyset$. Let $a' \in \mathrm{AD}(a) \cup \{a\}$ be the nearest ancestor of v that is an anchor. By Claim 2.6, the total size of the occurrences of vertices in $\textsc{Desc}_T(v) \cap \mathcal{U}(a')$ in subsets in $\bigcup_{b \in \mathrm{AC}(v)} \mathcal{P}(b)$ is at least $(k - h(v)) \sum_{b \in \mathrm{AC}(v)} \mathrm{LB}(b)$. It follows that the number of occurrences of v in the subsets in $\bigcup_{b \in \mathrm{AC}(v)} \mathcal{P}(b)$ is at least $\sum_{b \in \mathrm{AC}(v)} \mathrm{LB}(b)$. If ℓ is not in any of the subsets in $\bigcup_{b \in \mathrm{AC}(v)} \mathcal{P}(b)$, then v must occur in at least one more subset in $\mathcal{P}(a)$ that covers ℓ, and we are done. Suppose that this is not the case, and ℓ is in a subset in $\bigcup_{b \in \mathrm{AC}(v)} \mathcal{P}(b)$. It is not difficult to verify that the proof of Claim 2.6 implies the lower bound on the total size of the vertices in a *subset* of $\textsc{Desc}_T(v) \cap \mathcal{U}(a')$. This subset is the union of three sets: $\textsc{Desc}_T(v) \cap \left(\bigcup_{b \in \mathrm{AC}(v)} \mathcal{U}(b)\right)$, $\mathrm{AC}(v)$, and $\textsc{Desc}_{S_{a'}}(v)$. Clearly, ℓ is not in any of these three sets. Thus, the total size of the occurrences of vertices in $\textsc{Desc}_T(v) \cap \mathcal{U}(a)$ in the subsets in $\bigcup_{b \in \mathrm{AC}(v)} \mathcal{P}(b) \subset \mathcal{P}(a)$ is strictly more than $(k - h(v)) \sum_{b \in \mathrm{AC}(v)} \mathrm{LB}(b)$. Hence, the number of occurrences of v in the subsets in $\bigcup_{b \in \mathrm{AC}(v)} \mathcal{P}(b)$ is strictly more than $\sum_{b \in \mathrm{AC}(v)} \mathrm{LB}(b)$. □

By Claims 2.6 and 2.7 and our induction hypothesis, we get that the total size of the occurrences of vertices in $\textsc{Desc}_T(a) \cap \mathcal{U}(a)$ in the subsets in $\mathcal{P}(a)$ is at least

$$\begin{aligned} \mathrm{SA}(a) + (k - h(a)) \sum_{b \in \mathrm{AC}(a)} \mathrm{LB}(b) &\geq (k - h(a)) \left(\left\lfloor \frac{\mathrm{SA}(a)}{k - h(a)} \right\rfloor + \sum_{b \in \mathrm{AC}(a)} \mathrm{LB}(b) \right) \\ &= (k - h(a)) \sum_{c \in \mathrm{AD}(a) \cup \{a\}} \left\lfloor \frac{\mathrm{SA}(c)}{k - h(c)} \right\rfloor = (k - h(a))\, \mathrm{LB}(a) \end{aligned}$$

The first equality holds since $\mathrm{LB}(b) = \sum_{c \in \mathrm{AD}(b) \cup \{b\}} \left\lfloor \frac{\mathrm{SA}(c)}{k - h(c)} \right\rfloor$. □

Corollary 2.8. *The number of subsets in any feasible solution is at least*

$$\alpha + \sum_{a \in A} \left\lfloor \frac{\textsc{sa}(a)}{k - h(a)} \right\rfloor,$$

where α is defined in Lemma 2.4.

Proof. If $\textsc{top}A = \{r\}$ then by Lemma 2.5 the number of occurrences of r is at least $\text{LB}(r) = \sum_{a \in A} \left\lfloor \frac{\textsc{sa}(a)}{k-h(a)} \right\rfloor$. If this lower bound is tight then all the leaves in subsets in $\mathcal{P}(r)$ are anchored at some vertex. If $\textsc{lo}(r) > 0$, then there is a leaf of T that is a leftover vertex of r and thus not anchored at any vertex. In this case, at least one additional subset is needed to cover this leaf. If $\textsc{top}A \neq \{r\}$ then $\text{AC}(r) = \textsc{top}A$. In this case, following the proof of Claim 2.6, we get that the total size of the occurrences of the descendants of r in subsets in $\mathcal{P}$ is at least $(k - s(r)) \sum_{a \in \textsc{top}A} \text{LB}(a)$. This implies that r occurs in at least $\sum_{a \in \textsc{top}A} \text{LB}(a) = \sum_{a \in A} \left\lfloor \frac{\textsc{sa}(a)}{k-h(a)} \right\rfloor$ subsets in $\mathcal{P}$. If the bound is tight then all the leaves in these subsets are anchored vertices. Thus, if there exists a vertex that is not anchored at any vertex, at least one additional subset is needed. This occurs when either $\exists a \in \textsc{top}A$ s.t. $\textsc{lo}(a) > 0$, or $\exists$ leaf $\ell \in V$ s.t. none of the ancestors of ℓ is an anchor. □

Corollary 2.8 and Lemma 2.4 imply a factor 2 approximation.

3 Open Problems

An intriguing open problem is to bridge the gap between our 2-approximation and 1.5-inapproximability result for CT. Recall that CT is the special case of CPO on out-trees. While we expect CPO to be hard to approximate on general graphs (as mentioned above), exploring further the hardness of CPO on various graph classes remains open.

Another appealing line of research is to investigate the connections between CPO and a natural covering variant of the DkSH problem defined as follows. Given a hypergraph $G = (V, E)$ and an integer k, find the minimum number of vertex sets, each of cardinality at most k, such that every hyperedge is fully contained in one of the sets. We are not aware of earlier studies of this problem, even in the special case where G is a graph. One interesting direction is to derive nontrivial hardness results for this problem and show possible implications for CPO.

References

1. Biondi, M., Saliba, S., Harjunkoski, I.: Production optimization and scheduling in a steel plant: hot rolling mill. In: 18th World Congress of the International Federation of Automatic Control, pp. 11750–11754 (2011)
2. Bonsma, P.: Most balanced minimum cuts. Discret. Appl. Math. **158**(4), 261–276 (2010)

3. Borradaile, G., Heeringa, B., Wilfong, G.: The knapsack problem with neighbour constraints. J. Discret. Algorithms **16**, 224–235 (2012)
4. Cheng, T., Wang, K., Wang, L.C., Lee, C.W.: An in-switch rule caching and replacement algorithm in software defined networks. In: 2018 IEEE International Conference on Communications (ICC), pp. 1–6. IEEE (2018)
5. Chlamtác, E., Dinitz, M., Konrad, C., Kortsarz, G., Rabanca, G.: The densest k-subhypergraph problem. SIAM J. Discret. Math. **32**(2), 1458–1477 (2018)
6. Chlamtáč, E., Dinitz, M., Makarychev, Y.: Minimizing the union: tight approximations for small set bipartite vertex expansion. In: Proceedings of the 28th Annual ACM-SIAM Symposium on Discrete Algorithms, pp. 881–899 (2017)
7. Dong, M., Li, H., Ota, K., Xiao, J.: Rule caching in SDN-enabled mobile access networks. IEEE Netw. **29**(4), 40–45 (2015)
8. Doron-Arad, I., Kortsarz, G., Naor, J., Schieber, B., Shachnai, H.: Approximations and hardness of packing partially ordered items. arXiv preprint arXiv:2403.01568 (2024)
9. Gamage, S., Pasqual, A.: High-performance parallel packet classification architecture with popular rule caching. In: 2012 18th IEEE International Conference on Networks (ICON), pp. 52–57. IEEE (2012)
10. Gao, P., Xu, Y., Chao, H.J.: OVS-cab: efficient rule-caching for open vswitch hardware offloading. Comput. Netw. **188**, 107844 (2021)
11. Garey, M.R., Johnson, D.S.: Computers and Intractability: A Guide to the Theory of NP-Completeness. W. H. Freeman (1979)
12. Hajiaghayi, M., et al.: The minimum k-colored subgraph problem in haplotyping and DNA primer selection. In: Proceedings of the International Workshop on Bioinformatics Research and Applications (IWBRA), pp. 1–12 (2006)
13. Hoberg, R., Rothvoss, T.: A logarithmic additive integrality gap for bin packing. In: Proceedings of the 28th Annual ACM-SIAM Symposium on Discrete Algorithms, pp. 2616–2625 (2017)
14. Huang, H., Guo, S., Li, P., Liang, W., Zomaya, A.Y.: Cost minimization for rule caching in software defined networking. IEEE Trans. Parallel Distrib. Syst. **27**(4), 1007–1016 (2015)
15. Li, H., Guo, S., Wu, C., Li, J.: FDRC: flow-driven rule caching optimization in software defined networking. In: 2015 IEEE International Conference on Communications (ICC), pp. 5777–5782. IEEE (2015)
16. Li, R., Pang, Y., Zhao, J., Wang, X.: A tale of two (flow) tables: demystifying rule caching in openflow switches. In: Proceedings of the 48th International Conference on Parallel Processing, pp. 1–10 (2019)
17. Li, R., Zhao, B., Chen, R., Zhao, J.: Taming the wildcards: towards dependency-free rule caching with freecache. In: 2020 IEEE/ACM 28th International Symposium on Quality of Service (IWQoS), pp. 1–10. IEEE (2020)
18. Manurangsi, P.: Inapproximability of maximum edge biclique, maximum balanced biclique and minimum k-cut from the small set expansion hypothesis. In: 44th International Colloquium on Automata, Languages, and Programming (ICALP 2017) (2017)
19. Pferschy, U., Scatamacchia, R.: Improved dynamic programming and approximation results for the knapsack problem with setups. Int. Trans. Oper. Res. **25**(2), 667–682 (2018)
20. Rastegar, S.H., Abbasfar, A., Shah-Mansouri, V.: Rule caching in SDN-enabled base stations supporting massive IoT devices with bursty traffic. IEEE Internet Things J. **7**(9), 8917–8931 (2020)

21. Rottenstreich, O., Kulik, A., Joshi, A., Rexford, J., Rétvári, G., Menasché, D.S.: Cooperative rule caching for SDN switches. In: 9th International Conference on Cloud Networking (CloudNet), pp. 1–7 (2020)
22. Rottenstreich, O., Tapolcai, J.: Optimal rule caching and lossy compression for longest prefix matching. IEEE/ACM Trans. Netw. **25**(2), 864–878 (2016)
23. Sheu, J.P., Chuo, Y.C.: Wildcard rules caching and cache replacement algorithms in software-defined networking. IEEE Trans. Netw. Serv. Manag. **13**(1), 19–29 (2016)
24. Stonebraker, M., Jhingran, A., Goh, J., Potamianos, S.: On rules, procedure, caching and views in data base systems. ACM SIGMOD Rec. **19**(2), 281–290 (1990)
25. Wang, Y.Z., Zheng, Z., Zhu, M.M., Zhang, K.T., Gao, X.Q.: An integrated production batch planning approach for steelmaking-continuous casting with cast batching plan as the core. Comput. Ind. Eng. **173**, 108636 (2022)
26. Yan, B., Xu, Y., Chao, H.J.: Adaptive wildcard rule cache management for software-defined networks. IEEE/ACM Trans. Netw. **26**(2), 962–975 (2018)
27. Yan, B., Xu, Y., Xing, H., Xi, K., Chao, H.J.: Cab: a reactive wildcard rule caching system for software-defined networks. In: Proceedings of the 3rd Workshop on Hot Topics in Software Defined Networking, pp. 163–168 (2014)
28. Yang, J., Li, T., Yan, J., Li, J., Li, C., Wang, B.: Pipecache: high hit rate rule-caching scheme based on multi-stage cache tables. Electronics **9**(6), 999 (2020)
29. Ye, Y., Jiang, Z., Diao, X., Yang, D., Du, G.: An ontology-based hierarchical semantic modeling approach to clinical pathway workflows. Comput. Biologyand Med. **39**, 722–732 (2009)

Beyond Recognizing Well-Covered Graphs

Carl Feghali, Malory Marin, and Rémi Watrigant(✉)

Univ Lyon, CNRS, ENS de Lyon, Université Claude Bernard Lyon 1, LIP UMR5668, Lyon, France
{carl.feghali,malory.marin,remi.watrigant}@ens-lyon.fr

Abstract. We prove a number of results related to the computational complexity of recognizing well-covered graphs. For instance, let k and s be positive integers and G be a graph. Then G is said

- $\mathbf{W_k}$ if for any k pairwise disjoint independent vertex sets $A_1, \ldots, A_k$ in G, there exist k pairwise disjoint maximum independent sets $S_1, \ldots, S_k$ in G such that $A_i \subseteq S_i$ for $i \in [k]$.
- $\mathbf{E_s}$ if every independent set in G of size at most s is contained in a maximum independent set in G.

Chvátal and Slater (1993) and Sankaranarayana and Stewart (1992) famously showed that recognizing $\mathbf{W_1}$ graphs or, equivalently, well-covered graphs is coNP-complete. We extend this result by showing that recognizing $\mathbf{W_{k+1}}$ graphs in either $\mathbf{W_k}$ or $\mathbf{E_s}$ graphs is coNP-complete. This answers a question of Levit and Tankus (2023) and strengthens a theorem of Feghali and Marin (2024). We also show that recognizing $\mathbf{E_{s+1}}$ graphs is Θ_2^p-complete even in $\mathbf{E_s}$ graphs, where $\Theta_2^p = \mathrm{P}^{\mathrm{NP[log]}}$ is the class of problems solvable in polynomial time using a logarithmic number of calls to a SAT oracle. This strengthens a theorem of Bergé, Busson, Feghali and Watrigant (2023). We also obtain the complete picture of the complexity of recognizing chordal $\mathbf{W_k}$ and $\mathbf{E_s}$ graphs which, in particular, simplifies and generalizes a result of Dettlaff, Henning and Topp (2023).

1 Introduction

In a graph, an *independent set* is a set of pairwise non-adjacent vertices. An independent set is said to be *maximum* if it is of maximum size, and *maximal* if it is not a subset of any other independent set. Merging these two notions, Plummer [19] introduced the class of *well-covered* graphs, defined as those graphs in which every maximal independent set is also maximum. Well-covered graphs have since been extensively studied; for instance, the problem of recognizing well-covered graphs is coNP-complete on general graphs [5,24], but polynomial-time solvable on claw-free graphs [27], bipartite graphs [22], graphs with girth

Part of this work received funding from CNRS via project GALINETTE.

D. Král' and M. Milanič (Eds.): WG 2024, LNCS 14760, pp. 181–195, 2025.
https://doi.org/10.1007/978-3-031-75409-8_13

at least 5 [10], graphs with no cycle of length 4 and 5 [11], graphs of bounded degree [4], chordal graphs [21], cographs [15], and perfect graphs with bounded clique number [6]. See the seminal survey by Plummer for more details [20].

In this paper, we continue the study of well-covered graphs from a complexity viewpoint, though in a broader context. In 1975, Staples defined, for every integer $k \geq 1$, the class $\mathbf{W_k}$, as a generalization of well-covered graphs. For a positive integer k, a graph G is $\mathbf{W_k}$ if for any k pairwise disjoint independent vertex subsets $A_1, \ldots, A_k$ in G, there exist k pairwise disjoint maximum independent sets $S_1, \ldots, S_k$ in G such that $A_i \subseteq S_i$ for $i \in [k]$. Observe, by definition, that a graph is $\mathbf{W_1}$ if and only if it is well-covered, and $\mathbf{W_{k+1}} \subseteq \mathbf{W_k}$ for each $k \geqslant 1$. The class of $\mathbf{W_2}$ graphs was studied more specifically due to its relation to *shedding vertices*: a vertex v in a graph G is *shedding* if there does not exist an independent set in $V(G) \setminus N[v]$ which dominates $N(v)$. In particular, a graph without isolated vertices is $\mathbf{W_2}$ if, and only if, all of its vertices are shedding [16]. Shedding vertices were also used in order to characterize well-covered graphs of high girth [10] (they are called *extendable vertices*). Deciding whether a vertex of a graph is shedding is coNP-complete [17], even in triangle-free graphs. Similarly, recognizing $\mathbf{W_k}$ graphs was recently shown to be coNP-complete (for each fixed $k \geqslant 2$) [9], thus answering a recent question from [17]. However, the more challenging complexity problems (also raised in [17]) of recognizing $\mathbf{W_2}$ graphs and shedding vertices when the input graph is well-covered remained open.

Problem 1 ([17]). Determine the complexity of recognizing $\mathbf{W_2}$-graphs when the input graph is well-covered.

Problem 2 ([17]). Determine the complexity of recognizing shedding vertices when the input graph is well-covered.

In this paper, we resolve Problems 1 and 2 in, respectively, Theorem 1 (with $k = 2$) and Theorem 2 below.

Theorem 1. *Let $k \geqslant 2$ and G be $\mathbf{W_{k-1}}$. Then deciding if G is $\mathbf{W_k}$ is coNP-complete.*

Theorem 2. *Deciding whether a vertex is shedding in a well-covered graph is coNP-complete.*

The proofs are given in Sect. 2.

Another tool to study well-covered graphs is the property of *k-extendability* introduced by Dean and Zito [6]. To see this, let k be a positive integer. A graph is *k-extendable* if every of its independent sets of size k are contained in an MIS. Thus, a graph G is well-covered if and only if G is k-extendable for every $k \in \{1, 2, \ldots, |V(G)|\}$. It is worth noting that 1-extendable graphs were first introduced by Berge [2] as B-graphs, and that have recently been studied in the context of the design of wireless networks [3]. Dean and Zito [6] showed that if G is a bipartite graph or a perfect graph with bounded clique number, then it is well-covered if it is k-extendable for a "range of small values" of k. From a complexity viewpoint, it thus appears natural to ask whether testing well-coveredness can be done in polynomial-time if the input graph is k-extendable for some range of values of k. Precisely, let $\mathbf{E_s}$ be the class of k-extendable graphs for every $k \in [s]$. Then $\mathbf{E_{s+1}} \subseteq \mathbf{E_s}$ and $\mathbf{W_1} \subseteq \mathbf{E_s}$.

We demonstrate that this intuition is unlikely to be true, as stated in our next theorem, which we prove in Sect. 3.

Theorem 3. *Let $k \geqslant 2$ and $s \geqslant 1$ and G be $\mathbf{E_s}$. Then deciding whether G is $\mathbf{W_k}$ is coNP-complete.*

We should note that, unlike recognizing $\mathbf{W_1}$ graphs (which are in coNP [5,24], where a polynomial certificate for *not* being well-covered consists of two maximal independent sets of different sizes), there is no known polynomial certificate for being 1-extendable, though recognizing 1-extendable graphs is NP-hard [3], even on subcubic planar graphs and on unit disk graphs. Moreover, a close examination of the same reduction also establishes that recognizing 1-extendable graphs is coNP-hard (see Sect. 3). So unless NP = coNP, this problem is neither in NP nor in coNP. In our next result, whose proof is given in Sect. 3, we determine the precise complexity of this problem.

Theorem 4. *Let $s \geqslant 1$ and G be $\mathbf{E_{s-1}}$. Then deciding whether G is $\mathbf{E_s}$ is Θ_2^p-complete.*

Note that $\Theta_2^p = \mathrm{P}^{\mathrm{NP[log]}}$ is the class of problems that can be solved by a polynomial time algorithm using a logarithmic number of calls to a SAT oracle.

For tractable graph classes, we note that a characterization of chordal well-covered graphs was established in [21], stating that *a chordal graph G is well-covered if and only if the simplices of G form a partition of $V(G)$*, which, in turn, leads to a linear-time algorithm for deciding if a chordal graph is well-covered. In [3], the authors asked whether an analogous characterization for 1-extendable graphs can be established. This was recently settled in [7] using a new object called *successive clique covers*. Unfortunately, this characterization does not lead to a linear-time algorithm. In our next theorem, whose proof is in Sect. 4.2, we simplify the characterization, and use it to describe a linear-time algorithm.

Theorem 5. *A chordal graph G is 1-extendable if, and only if, there exists a partition of $V(G)$ into $\alpha(G)$ parts such that each part is a maximal clique of G. This can be tested in linear time.*

Finally, in Sect. 5, we conclude the paper with some old and new open problems.

2 Recognizing $\mathbf{W_k}$ Graphs

We start by proving simultaneously Theorem 2 and the case $k = 2$ of Theorem 1, thereby resolving both Problems 1 and 2.

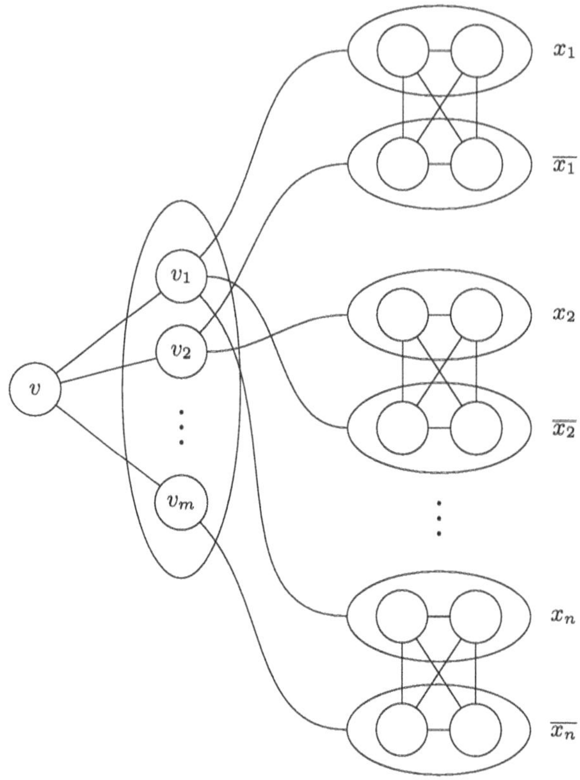

Fig. 1. The graph G'. The clause vertices in an ellipse are pairwise connected. If a clause vertex is connected to an ellipse for some literal, the clause veretx is adjacent to both vertices in the ellipse.

Theorem 6. *Let G be a well-covered graph and $v \in V(G)$. Then the problems of deciding if G is $\mathbf{W_2}$ and if v is shedding are both coNP-complete.*

Proof. Recognizing $\mathbf{W_2}$ graphs is in coNP [9], and deciding if a vertex is shedding is in coNP, a certificate being an independent set S of $V(G) \setminus N[v]$ that dominates $N(v)$. To show the hardness for both problems, we describe a generic reduction from the complement of 3-SAT. Let Φ be a 3-CNF formula with a set C of m clauses $c_1, ..., c_m$ and a set X of n variables $x_1, ..., x_n$. Consider the graph G (with $m + 1 + 4n$ vertices) that is obtained from Φ as follows:

- create a clique K of size m with vertex labels $v_1, \ldots, v_m$;
- add a new vertex v adjacent to all vertices of K;
- for each variable x_i, add the vertices x_i^1, x_i^2, $\overline{x_i}^1$, $\overline{x_i}^2$, and all (six) possible edges between each pair of them;
- if x_i appears in clause c_j, add the edges $x_i^1 v_j$ and $x_i^2 v_j$;
- if $\overline{x_i}$ appears in clause c_j, add the edges $\overline{x_i}^1 v_j$ and $\overline{x_i}^2 v_j$.

That completes the construction of G; see also Fig. 1 for an illustration.

To prove both theorems, we must first show that G is well-covered. For $1 \leq i \leq n$, write $K^i = \{\ell^b\}_{\ell \in \{x_i, \overline{x_i}\}, b \in \{1,2\}}$ and note that K^i is, by construction, isomorphic to the complete graph K_4 on four vertices.

Claim. G is well-covered.

Proof. By the construction, any independent set of G can contain at most vertex from $\{v\} \cup K$ and at most one vertex from each K^i. Thus, $\alpha(G) \leqslant n+1$. However, since we can construct an independent set of size $n+1$ by taking $\{v\} \cup \{x_i^1\}_{1 \leqslant i \leqslant n}$, we have that $\alpha(G) = n+1$. Let S be a maximal independent set of G. To prove the claim, we must show that $|S| = n+1$.

If S contains v, then $S - \{v\}$ is a maximal independent set of $G - (K \cup \{v\})$, which is exactly a disjoint union of n copies of K_4. Thus, S has size $n+1$. So we may assume that S does not contain v. In this case, it must contain, by maximality and construction, a vertex $v_j \in V(K)$ for some $j \in [m]$. Consider the literals ℓ_1, ℓ_2, ℓ_3 appearing in c_j. Then $S - \{v_j\}$ is maximal in $G - (K \cup \{v\} \cup \{\ell_i^b\}_{i \in \{1,2,3\}, b \in \{1,2\}})$, and, as it consists of a disjoint union of $n-3$ copies of K_4 and three copies of K_2, S has again size $n+1$, as required. □

To prove Theorem 1 with $k = 2$, it remains to show that Φ is unsatisfiable if, and only if, G is $\mathbf{W_2}$.

First, suppose that G is $\mathbf{W_2}$. For any truth assignment to X, for $1 \leqslant i \leqslant n$, define a new variable ℓ_i by $\ell_i = x_i$ if x_i is set to true, and by $\ell_i = \overline{x_i}$ otherwise. Let $S_1 = \{\ell_i^1\}_{1 \leqslant i \leqslant n}$ and $S_2 = \{v\}$ and note that these form independent sets in G. Now, since $G \in \mathbf{W_2}$, S_1 is contained in an MIS I of G. By our choice of S_2, $I = S_1 \cup \{v_j\}$ for some $j \in [m]$. Thus, by the definition of S_1, v_j is not adjacent to any vertex that corresponds to a literal that is set to true, and hence the clause c_j is not satisfied by the assignment. Thus, for every truth assignment there is an unsatisfied clause, which means that Φ is unsatisfiable, as needed.

Conversely, suppose that Φ is unsatisfiable. Let S_1 and S_2 be disjoint independent sets of G. We must show that both S_1 and S_2 can be extended to disjoint independent sets of size $n+1$.

We first argue that we may add a distinct vertex from $V(K) \cup \{v\}$ to each of S_1 and S_2. To see this, for each $r \in \{1,2\}$, if $S_r \cap V(K) = \emptyset$, note that there exists a vertex v_{i_r} of K such that v_{i_r} is not adjacent to any vertex of S_r. For otherwise, by setting a literal to true whenever its corresponding vertex is a member of S_r, we can readily infer, by construction, that Φ is satisfied, which is a contradiction. Hence, if $S_1 \cap V(K) = \emptyset$, we can add v_{i_1} to S_1 unless $v_{i_1} \in S_2$, in which case we add v to S_1. An entirely analogous argument can be applied to S_2. Let S_1' and S_2' be the resulting independent sets each containing a distinct vertex from $V(K) \cup \{v\}$. By a similar but simpler argument, we may add a distinct vertex from each K^i to both S_1' and S_2'. Therefore, the resulting two sets are disjoint, independent and, by construction, of size at least $n+1$, as needed. This completes the hardness of deciding if G is $\mathbf{W_2}$. This completes the proof of Theorem 1 with $k = 2$.

Finally, to prove Theorem 2, it remains to show that Φ is unsatisfiable if, and only if, v is shedding.

Suppose Φ is satisfiable by some truth assignment. As before, we define a new variable ℓ by $\ell_i = x_i$ if x_i is set to true and by $\ell_i = \overline{x_i}$ otherwise. Consider the independent set $S = \{\ell_i^1\}_{1 \leqslant i \leqslant n}$. Since the truth assignment satisfies Φ, S dominates $V(K) = N(v)$, and thus v is not shedding.

Conversely, suppose v is not shedding. By definition, there exists an independent set S of $V(G) \setminus N[v]$ that dominates $N(v) = V(K)$. Now, up to the addition of vertices, we may assume that S is maximal in $V(G) \setminus N[v]$; thus, S contains exactly one vertex per K^i for each $1 \leqslant i \leqslant n$. Consider the truth assignment μ on X defined by $\mu(\ell) = 1$ if, and only if, $\ell^1 \in S$ or $\ell^2 \in S$ for some literal ℓ. Since any clause vertex is dominated by S, it follows that μ satisfies Φ. This completes the proof of Theorem 2. □

The complete proof of Theorem 1 is available in the long version of the paper. Since $\mathbf{W_k} \subseteq \mathbf{E_s}$ for any $s, k \geqslant 1$, we have the following immediate consequence.

Corollary 1. *Let $k \geqslant 2$ and $s \geqslant 1$ and G be $\mathbf{E_s}$. Then deciding if G is $\mathbf{W_k}$ is coNP-complete.*

We can also address the missing case in the statement of the previous Corollary (with k replaced by 1), which requires a different proof.

Theorem 7. *Let $s \geqslant 1$ and G be $\mathbf{E_s}$. Then deciding if G is well-covered is coNP-complete.*

Proof. From [5], recognizing well-covered graphs is in coNP. We reduce from the complement of 3-SAT. Let Φ be a 3-CNF formula with a set C of clauses $c_1, ..., c_m$ and a set X of variables $x_1, ..., x_n$.

We first test all $O(n^s \cdot 2^s)$ partial assignments of s variables. If, for one of them, all clauses are satisfied, then it means that Φ is a satisfiable formula, and we output in this case a dummy negative instance of our problem. In the following, we can thus assume that for any partial assignment of s variables, there is at least one clause which is not (yet) satisfied.

Let G be the graph obtained from Φ as follows:

- create a clique K of size m with vertex labels $v_1, \ldots, v_m$;
- for each variable x_i, add three new vertices $u_i, \overline{u_i}, w_i$ that form a triangle;
- if x_i (resp. $\overline{x_i}$) appears in clause c_j, add the edge $u_i v_j$ (resp. $\overline{u_i} v_j$).

Claim. G is $\mathbf{E_s}$.

Proof. A similar argument to the one in the proof of Theorem 6 can be used to show that $\alpha(G) = n + 1$. Thus, it remains to show that any independent set $S \subseteq V(G)$ be of size s is contained in an independent set of size $n + 1$.

Let $I = \{i \mid S \cap \{u_i, \overline{u_i}, w_i\} = \emptyset\}$. Suppose first that S contains a vertex from K, then $S \cup \{w_i\}_{i \in I}$ is an independent set of size $n+1$ that contains S. Suppose finally that $S \cap K = \emptyset$. If there is a vertex c in K that has no neighbor in S, we add c to S and proceed as previously. Otherwise, assigning *true* (resp. *false*) to every variable x_i such that $u_i \in S$ (resp. $\overline{u_i} \in S$) leads to a partial assignment of at most s variables such that all clauses are satisfied, which is impossible. □

To complete the proof, we must show that G is well-covered if and only if Φ is unsatisfiable.

Suppose G is well-covered. For any truth assignment to the variables in X, we form an independent set S in G by adding to S vertex u_i if variable x_i is set to true, and vertex $\overline{u_i}$ if variable x_i is set to false. Note that $|S| = n$; moreover, as G is well-covered and $\alpha(G) = n + 1$, there is a vertex not in S that has no neighbor in S. By construction, this vertex is v_i for some $i \in [m]$, and so clause c_i does not contain a true literal. Therefore, Φ is unsatisfiable.

Conversely, suppose that Φ is unsatisfiable. Let S be any maximal independent set in G. By maximality, $|S \cap \{u_i, \overline{u_i}, w_i\}| = 1$ for each $i \in [n]$. Moreover, observe that $|S \cap K| = 1$ since if $S \cap K = \emptyset$ then, by setting to true the variables in $S \setminus \{w_i\}_{1 \leqslant i \leqslant n}$, one obtains a (possibly partial) truth assignment that satisfies Φ, which is a contradiction. Thus $|S| = n + 1$ as needed. □

3 Recognizing $\mathbf{E_s}$ Graphs

In this section, we prove Theorem 4. Our main motivation behind establishing this result is as follows.

As mentioned, recognizing 1-extendable graphs is NP-hard [3], and the proof is via a reduction from Exact MIS, where, given a graph G and an integer k, one has to determine if $\alpha(G) = k$. That Exact MIS is NP-hard follows in turn via a reduction from the classical Maximum Independent Set problem. On the other hand, Exact MIS belongs to the complexity class DP, introduced by Papadimitriou [18], implying in particular that it is both NP-hard and coNP-hard. Importantly, being DP-complete implies being neither in NP nor in coNP, unless NP = coNP. One might thus wonder if recognizing 1-extendable graphs is in DP. Our aim in Theorem 4 is to show that this is not the case, by proving that recognizing 1-extendable graphs, and more generally $\mathbf{E_s}$ graphs, is Θ_2^p-complete. Here the complexity class $\Theta_2^p = \mathrm{P}^{\mathrm{NP}[O(\log(n)]}$ refers to problems solvable by a polynomial-time deterministic algorithm using a logarithmic number of calls to a Sat oracle. This class has received attention, mainly because it contains various optimization problems across fields such as AI and voting schemes [23,25,28,29]. Interestingly, it also contains an approximation version of well-coveredness [13], where one is interested in the graph class where the greedy algorithm for Maximum Independent Set provides a constant approximation ratio. The Θ_2^p class contains many natural optimization problems implying the comparison of different optimization parameters. In particular, the problem MIS Equality which asks, given two graphs G and H, if $\alpha(G) = \alpha(H)$, is a Θ_2^p-complete problem. The analogous problems for graph coloring, vertex cover and clique are also Θ_2-complete [13,28]. We first show that recognizing $\mathbf{E_s}$ graphs is in Θ_2^p.

Lemma 1. *For any $s \geqslant 1$, recognizing $\mathbf{E_s}$ graphs is in Θ_2^p.*

Proof. We must describe an algorithm that solves the problem with only a logarithmic number of calls to a Sat oracle. To do so, for a fixed $s \geqslant 1$, we introduce the problem Partial s-Extendable, where, given a graph G and an integer r, the task is to determine if each independent set of size at most s can be extended to an independent set of size at least r. This problem is in NP, a certificate being the set of independent sets of G of size r that extend all independent sets of size at most s, and this certificate has a size of at most $O(r \cdot n^s)$. Therefore, PARTIAL s-EXTENDABLE can be solved with a SAT oracle.

The algorithm recognizing $\mathbf{E_s}$ graphs operates as follows on input G:

1. Compute $r = \alpha(G)$ using binary search, requiring $O(\log(n))$ SAT oracle calls to solve MAXIMUM INDEPENDENT SET.
2. Solve PARTIAL s-EXTENDABLE on input (G, r) using only one SAT oracle.

This algorithm straightforwardly recognizes $\mathbf{E_s}$ graphs in polynomial time, utilizing $O(\log(n))$ SAT oracle calls, thus the problem is in Θ_2^p as required. □

Thus, it remains to establish hardness. To do so, we shall proceed by induction on s. We shall need the following construction (extracted from [3]).

For two disjoint graphs G and H, let $\pi(G, H)$ be the graph obtained from G and H by adding a complete bipartite graph $(\Pi_G := \{\pi_u \mid u \in V(G)\}, \Pi_H := \{\pi_u \mid u \in V(H)\})$, and adding the edge $u\pi_u$ for each $u \in V(G) \cup V(H)$, Notice that $\pi(G, H)$ has $2(|V(G)| + |V(H)|)$ vertices and $\alpha(\pi(G, H)) = \max(|V(G)| + \alpha(H), |V(H)| + \alpha(G))$. Let us first address the base case $s = 1$.

Lemma 2. *Recognizing* 1*-extendable graphs is* Θ_2^p*-hard.*

Proof. To prove the lemma, we reduce from MIS EQUALITY.

Let (G, H) be an instance of MIS EQUALITY, where up to the addition of some universal vertices, both G and H have the same number n of vertices. Let $G' := \pi(G, H)$, so that G' has $4n$ vertices and $\alpha(G') = n + \max(\alpha(G), \alpha(H))$. We show that G' is 1-extendable if and only if $\alpha(G) = \alpha(H)$.

Suppose that $\alpha(G) = \alpha(H) = \alpha$. For each vertex $v \in V(G)$, set $S_v = \{v\} \cup (\Pi_G \setminus \{\pi_v\}) \cup S_H$, where S_H is any MIS of H. Then S_v has cardinality $n + \alpha$ and thus is an MIS of G'. Notice that $\Pi_v \cup S_H$ is also an MIS of G', and thus π_v belongs to an MIS too. By symmetry, all vertices from $V(H) \cup \Pi_H$ belong to an MIS of G', thus G' is 1-extendable.

Now, suppose that G' is 1-extendable. Then either $\alpha(G') = \alpha(G) + n$ or $\alpha(G') = \alpha(H) + n$. Assume that $\alpha(G') = n + \alpha(G)$. Let $v \in V(G)$. By 1-extendability of G', π_v is in an MIS S of G'. Let $S_G = (V(G) \cup \Pi_G) \cap S$ and $S_H = (V(H) \cup \Pi_H) \cap S$. Since $\pi_v \in S$ and S is independent, S contains no vertex from Π_H. It follows that S_H is an MIS of H of cardinality $\alpha(H)$. In addition, S_G is an MIS of $G'[V(G) \cup \Pi_v]$ and thus has cardinality n. Altogether, $|S| = n + \alpha(H) = \alpha(G')$, which gives $\alpha(G) = \alpha(H)$. This completes the proof. □

Corollary 2 ([3]). *Recognizing* 1*-extendable graphs is both NP-hard and coNP-hard, and thus neither in NP nor in coNP unless NP = coNP.*

To address the general case $s \geq 1$, we require one more lemma. Given a graph G, let $G^+ = \pi(G_1, G_2)$ where G_1 and G_2 are two copies of G. Notice that $\alpha(G^+) = |V(G)| + \alpha(G)$. For brevity, we write $\Pi_1 = \Pi_{G_1}$ and $\Pi_2 = \Pi_{G_2}$ in G^+.

Lemma 3. *For any graph G and any $s \geqslant 1$, $G^+ \in \mathbf{E_s}$ if and only if $G \in \mathbf{E_{s-1}}$.*

Proof. Suppose first that $G \in \mathbf{E_{s-1}}$. Let $A \subseteq V(G^+)$ be an independent set of size at most s. We must show that A is contained in an MIS of G^+.

Suppose $A \cap (\Pi_1 \cup \Pi_2) = \emptyset$. If $A \subseteq V(G_1)$, then A can be extended into an independent set of size $|V(G)|$ by adding only vertices of Π_1, and then, in turn, into an independent set of size $|V(G)| + \alpha(G)$ by adding any MIS of G_2. Similarly, if $A \subseteq V(G_2)$ then we are done. So we can assume that A intersects both $V(G_1)$ and $V(G_2)$ (notice that this case does not happen if $s = 1$). As $|A| \leq s$, we have that $|A \cap V(G_1)| \leqslant s - 1$. Since $G_1 \in \mathbf{E_{s-1}}$, $A \cap V(G_1)$ can be extended into an MIS $\tilde{A}_1$ of G_1. At the same time, $A \cap V(G_2)$ can be extended to an independent set $\tilde{A}_2$ of size $|V(G)|$ using only new vertices from Π_2. Notice that $\tilde{A}_1 \cup \tilde{A}_2$ is an independent set of size $|V(G)| + \alpha(G)$ that contains A.

So we can assume that $A \cap (\Pi_1 \cup \Pi_2) \neq \emptyset$. Since there are all possible edges between Π_1 and Π_2 in G^+, trivially A intersects at most one of Π_1 and Π_2, say Π_1. Let $A_1 = A \cap (V(G_1) \cup \Pi_1)$, and $A_2 = A \cap V(G_2)$. As in the preceding paragraph, A_1 can be extended to an independent set $\tilde{A}_1$ of size $|V(G)|$ using only new vertices from Π_1, and since $|A_2| \leqslant s - 1$ and $G_2 \in \mathbf{E_{s-1}}$, A_2 can be extended to an independent set $\tilde{A}_2$ of size $\alpha(G)$ using only vertices from $V(G_2)$. Then $\tilde{A}_1 \cup \tilde{A}_2$ is an MIS of G^+ that contains A. Therefore, G^+ is $\mathbf{E_s}$.

Conversely, suppose $G^+ \in \mathbf{E_s}$. Let $A \subseteq V(G)$ be an independent set of G of size at most $s - 1$. We must show that A extends into an MIS in G. Let $v \in \Pi_2$, and consider the independent set $A_1 \cup \{v\}$ where A_1 is exactly A in the copy G_1 of G in G^+. Since $A_1 \cup \{v\}$ is an independent set of G^+ of size at most s, it can be extended to an MIS S of G^+, which has size $|V(G)| + \alpha(G)$. Since $v \in S$, $S \cap \Pi_1 = \emptyset$. Thus, S can be partitioned into two sets S_1 and S_2, where S_1 induce a maximum independent set of G_1 and S_2 a maximum independent set of $G^+[V(G_2) \cup \Pi_2]$. We notice that $|S_2| = |V(G)|$, and thus $|S_1| = \alpha(G)$. In addition, $A_1 \subseteq S_1$ and so A can be extended to an MIS of G. □

We are now ready to prove Theorem 4; in fact, we prove the following more general result.

Theorem 8. *Let $s \geqslant 2$ and G be $\mathbf{E_{s-1}}$. Then deciding whether G is $\mathbf{E_s}$ is Θ_2^p-complete.*

Proof. By Lemma 1, the problem is in Θ_2^p. To show hardness, we proceed by induction on s. For $s = 1$, by Lemma 2, $\mathbf{E_1}$ graphs is Θ_2^p-complete on arbitrary (that is, $\mathbf{E_0}$) graphs. So we can assume that $s \geq 2$ and that, given a graph G in $\mathbf{E_{s-2}}$, deciding whether G is $\mathbf{E_{s-1}}$ is Θ_2^p-hard.

Now, since $G \in \mathbf{E_{s-2}}$, it follows from Lemma 3 that $G^+ \in \mathbf{E_{s-1}}$ and that $G \in \mathbf{E_{s-1}}$ if and only if $G^+ \in \mathbf{E_s}$. This implies the result. □

As said in the introduction, the class of s-extendable graphs and the class $\mathbf{E_s}$ are slightly different. Nevertheless, the same proof, word by word, implies

Corollary 3. *Let $s \geq 1$. Recognizing s-extendable graphs is Θ_2^p-complete.*

4 Chordal Graphs

In this section, we show that, given a chordal graph $G = (V, E)$ and integer $k \geq 2$, deciding if G is $\mathbf{W_k}$ and if G is $\mathbf{E_1}$ can both be done in time $O(|V| + |E|)$, while deciding if G is $\mathbf{E_k}$ is coW[2]-hard when parameterized by k. Before supplying the proofs, we gather the commonly required definitions.

A graph is *chordal* if every chordless cycle of the graph is a triangle. A *perfect elimination ordering* of a graph G is an ordering $v_1, \ldots, v_n$ of $V(G)$ such that, for each $i \in [n]$, the graph induced by $\{v_j : j < i, v_iv_j \in E(G)\}$ is a clique. Dirac [8] famously showed that a graph is chordal if, and only if, it admits a perfect elimination ordering. A vertex v of a graph G is called *simplicial* if $G[N(v)]$ is a clique. Given a simplicial vertex v, we call $N[v]$ a *simplex*. Thus, a graph is chordal if, and only if, every induced subgraph of G has a simplicial vertex.

Let $G = (V, E)$ be a graph, let K be the set of maximal cliques of G, and for $v \in V$, let K_v be the set of maximal cliques of G containing v. A *clique decomposition* T of a (connected) graph G is a tree whose vertex set is K and whose edges are such that $T[K_v]$ is connected (i.e. forms a subtree) for all $v \in V$. The maximal cliques of G are also called bags of T.

The next well-known lemma will be useful.

Lemma 4 ([12]). *A connected graph G is chordal if and only if it has a clique decomposition T; moreover, T can be constructed in time $O(|V| + |E|)$.*

4.1 Recognizing $\mathbf{W_k}$ Chordal Graphs

In this subsection, we prove the following result.

Theorem 9. *Let $k \geq 1$ and G be a chordal graph. Then deciding whether G is $\mathbf{W_k}$ can be done in $O(|V(G)| + |E(G)|)$ time.*

To do so, we require some auxiliary lemmas. The first deals with the (known) case $k = 1$.

Lemma 5 ([21]). *A chordal graph G is well-covered iff there exists a partition of $V(G)$ into simplices, and this can be tested in time $(|V(G)| + |E(G)|)$.*

Lemma 6 ([26]). *A graph G is $\mathbf{W_k}$ iff for every $S \subseteq V(G)$ of size $k - 1$, the graph $G - S$ is well-covered and $\alpha(G) = \alpha(G - S)$.*

The previous lemma immediately gives a naive algorithm for deciding, for an integer $k \geq 1$, whether a chordal G is $\mathbf{W_k}$, by testing, for every subset $V' \subseteq V(G)$ of size $k-1$, whether the graph $G - V'$ is well-covered. This algorithm has running time $O(n^k(n+m))$. We now prove Theorem 9 which improves this running time. We shall need the following generalization of Lemma 5.

Lemma 7. *Let $k \geqslant 1$. A chordal graph G is $\mathbf{W_k}$ iff there exists a partition of $V(G)$ into simplices such that each simplex contains at least k simplicial vertices.*

Proof. Suppose first that G is $\mathbf{W_k}$. Then G is well-covered and so, by Lemma 5, there is a (unique) partition P of $V(G)$ into a number of simplices; moreover, this number must be exactly $\alpha(G)$ (an independent set may be formed only by choosing any one simplicial vertex per simplex). Now, consider any one of those simplices that we denote A, let S be the set of simplicial vertices of G in A and suppose for a contradiction that $|S| \leq k-1$. If $A = S$, then $\alpha(G-A) = \alpha(G)-1$, which contradicts Lemma 6. Otherwise, consider a vertex $v \in A - S$. By Lemma 5, the graph $H = G - S$ is well-covered. Applying Lemma 5 again, there is a (unique) partition of H into simplices. But v can neither be simplicial in H nor a neighbor of a simplicial vertex in H, for P would not exist otherwise. This contradiction implies each simplex in P contains at least k simplicial vertices.

Conversely, suppose there is a partition of $V(G)$ into simplices such that each simplex contains at least k simplicial vertices. Thus, for any set S of $k-1$ vertices in G, the graph $G-S$ has a partition into simplices. By Lemma 5, $G-S$ is well-covered. Together with the fact that $\alpha(G) = \alpha(G - S)$, this implies by Lemma 6 that G is $\mathbf{W_k}$. This completes the proof. □

We are now ready to prove Theorem 9.

Proof of Theorem 9. By Lemma 7, one must check in time $O(|V|+|E|)$ whether a chordal graph G admits a partition into simplices such that each simplex contains at least k simplicial vertices. As this partition, whenever it exists, is unique (by the definition of a simplex), the result is immediate. □

4.2 Recognizing $\mathbf{E_s}$ Chordal Graphs

In this subsection, we prove Theorem 5. Let us first establish the first part, which consists in the characterization.

Theorem 10. *Let G be a chordal graph. G is 1-extendable iff there is a partition of $V(G)$ into $\alpha(G)$ parts such that each of them is a maximal clique in G.*

Proof. Let G be a chordal graph. Assume first that G is 1-extendable. We prove by induction on $\alpha(G)$ that $V(G)$ can be partitioned into $\alpha(G)$ maximal cliques, the statement being obvious if $\alpha(G) = 1$. In the following, we assume $\alpha(G) > 1$. By Lemma 4, we can Let v be a simplicial vertex of G which belongs to a leaf C_v of the tree decomposition of G (it is thus a maximal clique). Let $G' = G[V(G) \setminus C_v]$. Observe that $\alpha(G') = \alpha(G) - 1$, hence we can apply the induction hypothesis on G' and obtain a partition of $V(G')$ into $C_1, \ldots, C_q$ with $q = \alpha(G) - 1$. We claim that for every $i \in [q]$, C_i is also a maximal clique in G. Indeed, if there is $u \in C_v$ such that $C_i \cup \{u\}$ is also a clique, then u would not belong to any maximum independent set in G which violates its 1-extendability, since any maximum independent set of G must intersect C_i.

We also prove the other direction by induction on $\alpha(G)$. Assume that $V(G)$ can be partitioned into $C_1, \ldots, C_{\alpha(G)}$, where each C_i is a maximal clique. In the tree decomposition of G, we choose an arbitrary node of degree at least 2 to be the root, and w.l.o.g. we may assume that C_1 is a leaf of the tree decomposition. Let

P be the maximal clique of G which is the parent of C_1 in the tree decomposition. We denote by v any simplicial vertex of C_1. Let $G' = G[V(G) \setminus C_1]$. We have $\alpha(G') = \alpha(G) - 1$, and by induction G' is 1-extendable. It remains to show that every vertex of C_1 belongs to a maximum independent set of G. This is clearly the case for v since it is only adjacent to vertices in C_1. This is also the case for every $u \in C_1 \setminus \{v\}$ since otherwise it would mean that u is adjacent to every vertex of P, but then P would not be a maximal clique in G. □

The problem of deciding whether a chordal graph G is k-extendable can naively be done in time $O(n^k(n+m))$. Simply enumerate all independent sets of G of size k and then, for each such set S, check the independence number of $G-S$. Then G is k-extendable iff this number is the same for all the independent sets under consideration. However, Theorem 10 may be applied to obtain a linear time algorithm, the proof being available in the long version of the paper.

Theorem 11. *Let $G = (V, E)$ be a chordal graph. Then the problem of deciding if G is 1-extendable can be done in time $O(|V| + |E|)$.*

Contrarily to the analogous problem on the class $\mathbf{W_k}$, this algorithm cannot be extended to a linear-time algorithm for recognizing $\mathbf{E_s}$ graphs for any $s \geqslant 2$, as shows the following rather surprising result.

Theorem 12. *Given a chordal graph G and an integer $s \geqslant 2$, the problem of deciding whether G is in $\mathbf{E_s}$ is coW[2]-hard when parameterized by s.*

Proof. We reduce from the complement of Minimum Dominating Set. Let (G, s) be the input, with $|V(G)| = n$. We may assume that G does not contain a universal vertex, otherwise the problem is trivial. Construct G' as follows: $V(G')$ is composed of a set $C = \{c_v : v \in V(G)\}$ inducing a clique together with two sets $I = \{i_v : v \in V(G)\}$ and $I' = \{i'_v : v \in V(G)\}$ inducing two independent sets, respectively. For every $uv \in E(G)$, we connect i_v to c_v, c_u and i'_v. It can be observed that G' is indeed a chordal graph ($G'[C \cup I]$ is a split graph, and I' is a set of vertices of degree one). Moreover, G' is 1-extendable, as $\alpha(G') = n+1$, and each vertex belongs to an independent set of size $n+1$ since G does not contain a universal vertex. In particular, observe that any maximum independent set of G' contains a vertex of C. We prove that $G' \notin \mathbf{E_s}$ if, and only if G contains a dominating set of size at most s.

If $G' \notin \mathbf{E_s}$, then there exists a set $S \subseteq V(G')$ of size at most s that is not contained in any independent set of size $n+1$. If $S \cap C \neq \emptyset$, then let $S' = S \cup \{i'_v \mid i_v \notin S\}$, and notice that $|S'| = n+1$ and thus S' is an MIS that contains S, which is not possible. So we may assume that $S \cap C = \emptyset$. Let $S' = S \cup \{i'_v \mid i_v \notin S\}$ which an independent set that contains S and for any $v \in V(G)$, S' contains either i_v or i'_v. Since S' cannot be extended to an MIS, it implies that S' dominates C, which implies that $S \cap I$ is in one-to-one correspondence with a dominating set in G of at most s.

Conversely, assume that $G' \in \mathbf{E_s}$, and let D be any subset of $V(G)$ of size at most s. Then the set $S = \{i_v : v \in D\}$ is an independent set of size at most s

in G' which is thus contained in an MIS S' of G', which necessarily intersects C on, say, c_{v^*}. It implies that i_{v^*} does not belong to S and that c_{v^*} is not adjacent to any vertex of S, which means that v^* does not belong to D nor is dominated by D in G, and thus D is not a dominating set in G. □

5 Conclusion and Open Questions

We addressed open questions in the literature by demonstrating that recognizing $\mathbf{W_k}$ graphs and shedding vertices are coNP-complete on well-covered graphs. We closed the complexity gap of testing 1-extendability, proving that the problem is Θ_2^p-complete. Additionally, we provided a simple characterization of 1-extendable chordal graphs, leading to a linear-time algorithm to decide if a graph is 1-extendable, addressing questions left open in [3]. Up to our knowledge, several questions remain open, with two of particular interest.

Problem 3. What is the computational complexity of recognizing well-covered triangle-free graphs?

This problem was indirectly raised by Plummer in his survey [20], where he asked if there exists a characterization of well-covered or $\mathbf{W_2}$ graphs of girth at least 4. In fact, the complexity of recognizing well-covered K_t-free graphs is still open for any constant $t \geqslant 3$, and even the same problem on graphs such that $\chi(G) = t$ for any fixed $t \geqslant 3$ [1]. Notice that recognizing $\mathbf{E_s}$ triangle-free graphs is also Θ_2^p-complete. In the instance (G, H) of MIS Equality, we can assume that both graphs are triangle-free, up to an even subdivision of the edges. Then, one can check that every reduction does not create any triangle. A similar problem, raised by Hujdurović, Milanič, Martin and Ries, concerns co-well-covered graphs, *i.e.*, graphs for which the complement is well-covered.

Problem 4 ([14]). What is the computational complexity of recognizing graphs which are both well-covered and co-well-covered?

Both problems are linked, as any triangle-free graph with no isolated vertex is co-well-covered. Thus, if recognizing triangle-free well-covered graphs is coNP-complete, the same holds for the second problem. Conversely, if there exists a polynomial-time algorithm to check if a graph is well-covered and co-well-covered, one can recognize triangle-free well-covered graphs in polynomial time.

We conclude with three problems related to extendability. The first concerns 1-extendable graphs with high girth. While there is a characterization of well-covered graphs of girth 5, recognizing 1-extendable graphs is NP-hard even for arbitrarily large girth [3]. Can we find some structural property for 1-extendable graphs of large girth, such as the existence of a linear-size independent set (which is the case for well-covered graphs of girth 5)? The second concerns the computational complexity of recognizing $\mathbf{E_s}$ graphs when s is part of the input. The problem is Θ_2^p-hard but not complete. One can show that it belongs to Π_2^p in the polynomial hierarchy; it would be interesting to close this complexity gap.

Finally, note that the hereditary class of graphs for which any induced subgraph is 1-extendable is exactly the class of P_3-free graphs, which can be recognized in polynomial time. We generalize this question for any $k > 1$.

Problem 5. What is the computational complexity of recognizing graphs for which any induced subgraph is k-extendable, k being fixed?

Acknowledgments. We would like to thank Martin Milanič for useful discussions on the subject and for suggesting Problem 5.

Disclosure of Interests. The authors have no competing interests to declare that are relevant to the content of this article.

References

1. Alves, S.R., Dabrowski, K.K., Faria, L., Klein, S., Sau, I., Souza, U.S.: On the (parameterized) complexity of recognizing well-covered (r, ℓ)-graph. Theor. Comput. Sci. **746**, 36–48 (2018)
2. Berge, C.: Some common properties for regularizable graphs, edge-critical graphs and b-graphs. In: Hammer, P.L., Rosa, A., Sabidussi, G., Turgeon, J. (eds.) Theory and Practice of Combinatorics. North-Holland Mathematics Studies, vol. 60, pp. 31–44. North-Holland (1982)
3. Bergé, P., Busson, A., Feghali, C., Watrigant, R.: 1-extendability of independent sets. Algorithmica 1–25 (2023)
4. Caro, Y., Ellingham, M.N., Ramey, J.E.: Local structure when all maximal independent sets have equal weight. SIAM J. Discret. Math. **11**(4), 644–654 (1998)
5. Chvátal, V., Slater, P.J.: A note on well-covered graphs. In: Gimbel, J., Kennedy, J.W., Quintas, L.V. (eds.) Quo Vadis, Graph Theory?. Annals of Discrete Mathematics, vol. 55, pp. 179–181. Elsevier (1993)
6. Dean, N., Zito, J.: Well-covered graphs and extendability. Discret. Math. **126**(1), 67–80 (1994)
7. Dettlaff, M., Henning, M.A., Topp, J.: On α-excellent graphs. Bull. Malaysian Math. Sci. Soc. **46**(2), 65, (2023). ID: Dettlaff2023
8. Dirac, G.A.: On rigid circuit graphs. In: Abhandlungen aus dem Mathematischen Seminar der Universität Hamburg, vol. 25, pp. 71–76. Springer (1961)
9. Feghali, C., Marin, M.: Three remarks on $\mathbf{W_2}$ graphs. Theor. Comput. Sci. **990**, 114403 (2024)
10. Finbow, A., Hartnell, B., Nowakowski, R.J.: A characterization of well covered graphs of girth 5 or greater. J. Comb. Theory Ser. B **57**(1), 44–68 (1993)
11. Finbow, A.S., Hartnell, B.L., Nowakowski, R.J.: A characterization of well-covered graphs that contain neither 4- nor 5-cycles. J. Graph Theory **18**(7), 713–721 (1994)
12. Gavril, F.: The intersection graphs of subtrees in trees are exactly the chordal graphs. J. Comb. Theory Ser. B **16**(1), 47–56 (1974)
13. Hemaspaandra, E., Rothe, J.: Recognizing when greed can approximate maximum independent sets is complete for parallel access to NP. Inf. Process. Lett. **65**(3), 151–156 (1998)
14. Hujdurović, A., Milanič, M., Ries, B.: Graphs vertex-partitionable into strong cliques. Discret. Math. **341**(5), 1392–1405 (2018)

15. Klein, S., de Mello, C.P., Morgana, A.: Recognizing well covered graphs of families with special p_4-components. Graphs Comb. **29**(3), 553–567 (2013)
16. Levit, V.E., Mandrescu, E.: W2-graphs and shedding vertices. Electron. Notes Discret. Math. **61**, 797–803 (2017)
17. Levit, V.E., Tankus, D.: Recognizing $\mathbf{W}_2$ graphs. arXiv preprint arXiv:2306.17272 (2023)
18. Papadimitriou, C.H., Yannakakis, M.: The complexity of facets (and some facets of complexity). In: Proceedings of the Fourteenth Annual ACM Symposium on Theory of Computing, pp. 255–260 (1982)
19. Plummer, M.D.: Some covering concepts in graphs. J. Comb. Theory **8**(1), 91–98 (1970)
20. Plummer, M.D.: Well-covered graphs: a survey. Quaest. Math. **16**(3), 253–287 (1993)
21. Prisner, E., Topp, J., Vestergaard, P.D.: Well covered simplicial, chordal, and circular arc graphs. J. Graph Theory **21**(2), 113–119 (1996)
22. Ravindra, G.: Well-covered graphs. J. Combin. Inform. System Sci. **2**(1), 20–21 (1977)
23. Riege, T., Rothe, J.: Completeness in the Boolean hierarchy: exact-four-colorability, minimal graph uncolorability, and exact domatic number problems-a survey. J. Univers. Comput. Sci. **12**(5), 551–578 (2006)
24. Sankaranarayana, R.S., Stewart, L.K.: Complexity results for well-covered graphs. Networks **22**(3), 247–262 (1992)
25. Spakowski, H., Vogel, J.: θ_2^p-completeness: a classical approach for new results. In: FST TCS 2000: Foundations of Software Technology and Theoretical Computer Science: 20th Conference New Delhi, India, 13–15 December 2000, pp. 348–360. Springer (2000)
26. Staples, J.A.W.: On some subclasses of well-covered graphs. Vanderbilt University (1975)
27. Tankus, D., Tarsi, M.: Well-covered claw-free graphs. J. Comb. Theory Ser. B **66**(2), 293–302 (1996)
28. Wagner, K.W.: More complicated questions about maxima and minima, and some closures of NP. Theor. Comput. Sci. **51**(1–2), 53–80 (1987)
29. Wagner, K.W.: Bounded query classes. SIAM J. Comput. **19**(5), 833–846 (1990)

Untangling Gaussian Mixtures

Eva Fluck[1(✉)], Sandra Kiefer[2], and Christoph Standke[1]

[1] Department of Computer Science, RWTH Aachen University, Ahornstraße 55, 52056 Aachen, Germany
{fluck,standke}@cs.rwth-aachen.de

[2] Department of Computer Science, University of Oxford, Wolfson Building, Parks Road, Oxford OX1 3QD, UK
sandra.kiefer@cs.ox.ac.uk

Abstract. Tangles were originally introduced as a concept to formalize regions of high connectivity in graphs. In recent years, they have also been discovered as a link between structural graph theory and data science: when interpreting similarity in data sets as connectivity between points, finding clusters in the data essentially amounts to finding tangles in the underlying graphs.

This paper explores the potential of tangles in data sets as a means for a formal quantitative study of clusters. Real-world data often follow a normal distribution. Accounting for this, we develop a quantitative theory of tangles in data sets drawn from Gaussian mixtures. To this end, we equip the data with a graph structure that models similarity between the points and allows us to apply tangle theory. We provide explicit conditions under which tangles associated with the marginal Gaussian distributions exist asymptotically almost surely. This can be considered a sufficient formal criterion for the separability of clusters in the data.

Keywords: Graph tangles · Graph connectivity · Gaussian mixture · Soft clustering

1 Introduction

Cluster analysis [6–8,27] can be summarized as the task of grouping data points based on some similarity measurement. There are two central paradigms for clustering algorithms: some, like k-means and k-median, focus on *similarity* and group together data points that are sufficiently close to each other, whereas for example the spectral clustering approach rather uses *dissimilarity* to split the data into sufficiently different groups. Either way, hard clustering, i.e. clustering that partitions the data into disjoint subsets, has its limits when it comes to real-world data. These data sets typically have some inherent fuzziness, and clusters that are most adequate to represent them may be wide-spread and might even overlap. All in all, in real-world data, we cannot always expect a clear answer to the question whether two points should be grouped together or treated as different, and soft-clustering approaches are necessary. The most

D. Král' and M. Milanič (Eds.): WG 2024, LNCS 14760, pp. 196–210, 2025.
https://doi.org/10.1007/978-3-031-75409-8_14

common approach is to assign values to the data points that denote the degree of membership to each cluster. For an overview of soft clustering, see e.g. [9].

Connectivity is a central concept for studying the structure of complex graphs and more general relational structures [13,14,23,35]. It finds its way into cluster analysis when we consider "similar" data points as adjacent or connected (see e.g. [32]). Here, the similarity is usually measured via the distance of the data points in $\mathbb{R}^d$. When we equip a set of data points with a suitable graph structure that reflects their proximity, analyzing similarity of the points becomes a problem of studying the graph connectivity. We can hence borrow also graph-theoretic concepts to analyze the data set. As it turns out, the notion of a *graph tangle* captures exactly the kind of ambiguity described above: it is a precise formal definition of potentially fuzzy but sufficiently distinct clusters in graphs.

Tangles in graphs were first introduced in [35]. They offer a perspective on connectivity that differs from the classical view: instead of focusing on the highly connected regions, they deal with thin parts, i.e., the bottlenecks in the graph. A highly connected region is never located exactly at a bottleneck; it must lie (mostly) on one of the sides defined by a cut along the bottleneck. A tangle is a set of consistent orientations of all the bottlenecks, where the orientation is towards the side on which the major part of the associated highly connected region lies. So a tangle assigns a "signpost" to each low-order cut, which points towards the highly connected region that the tangle represents. Thus, we consider the regions different whenever there is at least one signpost that points into opposite directions in the two tangles. This notion of being sufficiently different is somewhat along the lines of the dissimilarity paradigm, but in a less strict sense than clusters. In fact, the definition via consistent orientations of bottlenecks is robust precisely because it avoids references to an explicit point set.

The publication [35] links tangles closely to branch decompositions of graphs. Furthermore, the authors give a tree decomposition of a graph into its maximal tangles, which was turned into a canonical one in [4]. The original tangle concept was extended to directed graphs [29,34] and tangle-tree decompositions were also found in that setting [24]. In related work, variations of the tangle concept that consider different ways to measure connectivity in graphs [1,3] and in matroids [23] have been explored, see also [26]. The insight that one can define tangles completely without an underlying graph structure was carried further and led to the study of *abstract separation systems* [10,12,15,18], where separations play the role of the previously mentioned cuts.

With Data Science and Machine Learning emerging as central research fields in modern computer science, tangles have caught attention also in the context of clustering. The first transfer of the concept to real-world data happened in digital-image analysis [16], where tangles were used to describe clusters in image data in order to find interesting image parts. As Diestel puts it in [11], "tangles offer a new paradigm for clustering in large data sets". Informally speaking, unlike clusters, tangles come equipped with an *order*, a narrowness threshold that defines how small a cut through the data must be for the corresponding region to be considered a bottleneck. This means that a parameter of the tangle is the "width of the bottlenecks" at the signposts that constitute the tangle. This way, the order is an indicator for the unambiguity of the highly-connected region or the cluster associated with the tangle. Vice versa, branch decompositions serve as witnesses for the absence of tangles above a certain order and may thereby be considered indicators that clusters above a certain quality threshold do not exist within the data. Additionally, the order of the tangles captures hierarchical structure of the data: by decreasing the threshold, some of the previously considered cuts are not classified as small anymore, so the new bottlenecks are a subset of the previous ones and thus, several of the previously distinct dense regions may be united into a single one. It is worth mentioning here that also the tangle-tree decomposition has its correspondence in the realm of clustering, as it yields a canonical decomposition of the data set into clusters, see [13].

So far, the connection between graph tangles and clustering is largely based on intuition, as the numerous figurative expressions used in this introduction indicate. Towards grasping the link formally, [20] establishes an explicit connection between tangles and hierarchical clustering, which can be generalized to a one-to-one correspondence between a certain type of connectivity function together with its tangles and a dendogram obtained via hierarchical clustering [21]. Furthermore, the authors of [17] introduce an algorithmic framework to compute a clustering based on tangles.

1.1 Our Contribution

This paper embarks on a project to make the connection between tangles and clusters more robust, formal, and quantitative, without depending on particular clustering algorithms or paradigms.

Real-world data frequently follow a normal distribution, and clustering algorithms are often tested against data drawn from Gaussian mixtures [5,32]. We develop a quantitative theory of tangles in data sets obtained from sampling from Gaussian mixtures, i.e., we assume that the data points are drawn from various Gaussian distributions. Our goal is to identify incomparable graph tangles, that is, pairs of tangles that orient at least one separation (cut) of the underlying graph differently or rather where none is just a restriction of the other. Our theory allows us to draw conclusions of the kind "With high probability, there exist incomparable tangles in the data" as well as concrete lower bounds on these probabilities. Each tangle will correspond to a marginal distribution of the Gaussian mixture, i.e., to one Gaussian distribution contributing to the mixture, so we can interpret a high probability for the existence of incomparable tangles as a formalization of the distinguishability or separability of the marginals.

In a set of data points drawn from a Gaussian mixture, a clustering algorithm would typically determine clusters that closely correspond to the hidden labels, i.e., it will assign most points correctly to the marginal distribution they have been drawn from [25]. The goal of this work is not to split the data up into partition classes or to actually determine hidden labels. We rather study ways to quantify the existence of clusters (or tangles) in the data set, which could in a next step be interpreted as a quality measure for the reliability of a prediction of the hidden labels. To this end, we take up two standard ways to represent the data as a weighted graph, the δ-neighborhood graph (Fig. 1a) and the fully connected graph (Fig. 1b). In both graphs, similarity of data points is represented in edges and weights on those. We thus use these edge weights as our connectivity measure for candidate bottlenecks to define the tangles.

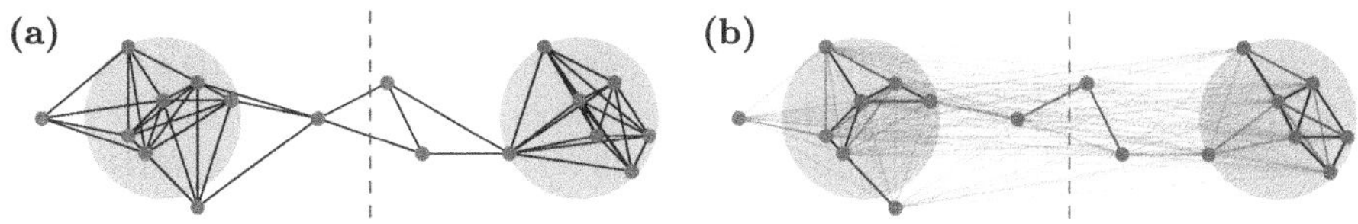

Fig. 1. Data drawn from two Gaussians with hidden labels in red and blue. The dashed line represents a possible low-order cut and the purple circles are candidates for highly connected regions. **(a)** A δ-neighborhood graph on the data. **(b)** The fully connected graph with edge weights represented in the opacity. (Color figure online)

Given a mixture of Gaussian distributions, what is the probability that incomparable tangles associated with the marginals exist? By carefully employing classical tools from probability theory, our analysis provides explicit conditions such that, asymptotically almost surely, the tangles exist. In the δ-neighborhood graph, we further improve our lower bound on the probability: with (explicit) stronger assumptions, we deduce that the probability for the existence of incomparable tangles tends to 1 exponentially fast in the number of data points.

Our theorems are applicable to any Gaussian mixture. But in order to be so versatile, their preconditions are quite technical and it is not trivial to see what choices of parameters and sets fulfill them. We analyze the preconditions of our theorems for specific sets of parameters to showcase what a "good choice" may be as well as to compute the explicit probability bounds in different settings.

Our work differs from [17] in that we consider all possible separations of the data set instead of restricting them. By studying specific well-behaved separations and imposing restrictions on the data sets, the authors of [17] manage to provide theoretical guarantees that the marginal distributions of the mixture are in one-to-one correspondence with the tangles. While their results treat the special case of two Gaussians, our theorems apply to the general setting of any mixture of Gaussian distributions with arbitrary parameters.

Due to space restrictions, details and most of the proofs are omitted. A full version of this extended abstract can be found on arXiv [22].

2 Preliminaries

For $n \in \mathbb{N}_{>0}$, we let $[n] := \{1, \dots, n\}$. For sets $S \subseteq U$, we write $\binom{U}{n}$ to denote all n-element subsets of U, and 2^U to denote the power set of U, and let $S^c := U \setminus S$.

2.1 Graphs and Tangles

This subsection gives a light introduction to the realm of connectivity functions and tangles in graphs. For more background on the topic, we refer to [26].

A *weighted graph* is a triple $G = (V, E, w)$ where the pair (V, E) is an undirected graph with vertex set V and edge set $E = \binom{V}{2}$, and $w\colon E \to \mathbb{R}_{\geq 0}$ is a function that assigns a *weight* to every edge. For $u, v \in V$ with $u \neq v$, we let $w(u, v) := w(\{u, v\})$. If $w(u, v) = 0$, we also say that there is no edge between u and v. In the remainder of this paper, all graphs are assumed to be weighted. If $\min\{w(u, v) \mid u, v \in V, u \neq v\} > 0$, the graph G is a *clique*. For $S, T \subseteq V$, we denote by $G[S]$ the (weighted) subgraph of G induced by the vertices in S, and we let $E(S, T) := \{\{u, v\} \in E \mid u \in S, v \in T\}$. Furthermore, we set $E(S) := E(S, S)$.

For a finite set U, we call $\kappa\colon 2^U \to \mathbb{R}$ a *connectivity function* if and only if:

- κ is *symmetric*, i.e., for all $S \subseteq U$, it holds that $\kappa(S) = \kappa(S^c)$,
- κ is *normalized*, i.e., $\kappa(\emptyset) = \kappa(U) = 0$,
- κ is *submodular*, i.e., $\kappa(S) + \kappa(T) \geq \kappa(S \cap T) + \kappa(S \cup T)$ holds for all $S, T \subseteq U$.

A prominent example for a connectivity function is the edge connectivity, which measures how strongly a vertex set is connected to its complement.

Example 1 (see [26]*).* Let $G = (V, E, w)$ be a weighted graph. For $S \subseteq V$, we define the *edge connectivity* $\kappa_G\colon 2^V \to \mathbb{R}$ via $\kappa_G(S) := \sum_{\{u,v\} \in E(S, S^c)} w(u, v)$.

Using connectivity functions, we can describe highly connected regions in structures formally. In the following, to do so, we introduce tangles based on connectivity functions. Every tangle corresponds to one highly connected region and vice versa. To have a robust notion of high connectedness, instead of describing the region pointwise, the tangle just provides directions towards the region in an unambiguous way. More precisely, for every set with a low value of the connectivity function (i.e., a "low connectivity" to its complement), the tangle contains either the set or its complement, which can be interpreted as determining on which side of the induced separation between the set and its complement the main part of the considered highly connected region lies.

Definition 1. *Let U be a finite set and κ a connectivity function on U. A κ-*tangle *of order $k \geq 0$ is a set $\mathcal{T} \subseteq 2^U$ such that all of the following hold.*

T.0 $\kappa(S) < k$ *for all* $S \in \mathcal{T}$.
T.1 *For all $S \subseteq U$ with $\kappa(S) < k$, either $S \in \mathcal{T}$ or $S^c \in \mathcal{T}$ holds.*
T.2 *For all $S_1, S_2, S_3 \in \mathcal{T}$, it holds that $S_1 \cap S_2 \cap S_3 \neq \emptyset$.*
T.3 *For all $x \in U$, it holds that $\{x\} \notin \mathcal{T}$.*

If κ is clear from the context, we drop it and call $\mathcal{T}$ a tangle.

Conditions (T.0) and (T.1) ensure that a tangle orients all and only the separations of low order. Condition (T.2) ensures that the orientations induced by the tangle all point towards a unique highly connected region. The number 3 might seem arbitrary here, but is necessary and sufficient to give a rich theory (see also [14,26]). Finally, Condition (T.3) means that singletons are not considered to be highly connected regions.

We call tangles $\mathcal{T}_1$, $\mathcal{T}_2$ *incomparable* if there is an $S \subseteq U$ with $S \in \mathcal{T}_1$ and $S^c \in \mathcal{T}_2$, that is, if there is some separation that is oriented differently by them. If $\mathcal{T}_1$ and $\mathcal{T}_2$ are not incomparable, we have $\mathcal{T}_1 \subseteq \mathcal{T}_2$ or $\mathcal{T}_1 \supseteq \mathcal{T}_2$.

As we will see, every clique in a graph induces a tangle. We will use this to prove the existence of tangles in sets of data drawn from Gaussian mixtures.

Notation 1. *Let $G = (V, E, w)$ be a weighted graph. Suppose $W \subseteq V$. If $E(W) = \emptyset$, set $w_W := 1$. Otherwise, let $w_W := \min\{w(u,v) \mid \{u,v\} \in E(W)\}$. We set*

$$\mathcal{T}_G(W) := \left\{ S \subseteq V \;\middle|\; \kappa_G(S) < \frac{2 \cdot |W|^2}{9} \cdot w_W \text{ and } |S \cap W| > |W \setminus S| \right\}.$$

The following lemma provides sufficient conditions for $\mathcal{T}_G(W)$ to be a tangle. The factor $\frac{2}{9}$ is necessary to ensure a non-empty intersection of all triples of sets contained in the tangle.

Lemma 1. *Let $G = (V, E, w)$ be a weighted graph. Let $W \subseteq V$ such that $|W| \geq 2$ and $G[W]$ is a clique. For $w_W := \min\{w(u,v) \mid \{u,v\} \in E(W)\}$, it holds that $\mathcal{T}_G(W)$ is a κ_G-tangle of order $\frac{2}{9} \cdot |W|^2 \cdot w_W$.*

2.2 Probability Theory

We give a brief introduction to the main concepts and notation we use from probability theory. For more background, we refer the reader to [30]. We assume our data to be a subset of some real vector space $\mathbb{R}^d$, which we equip with the Euclidean norm $\|\cdot\|$ as well as the corresponding (standard) topology and Borel σ-algebra, whose elements we call *Borel sets*. For $\boldsymbol{x} \in \mathbb{R}^d$ and $\delta \geq 0$, we define $B_\delta[\boldsymbol{x}] := \{\boldsymbol{x'} \mid \boldsymbol{x'} \in \mathbb{R}^d, \|\boldsymbol{x'} - \boldsymbol{x}\| \leq \delta\}$. For a Borel set $S \subseteq \mathbb{R}^d$, we denote its complement $\mathbb{R}^d \setminus S$ by S^c and its indicator function by $\mathbb{1}_S$. We also set $B_\delta[S] := \bigcup_{\boldsymbol{x} \in S} B_\delta[\boldsymbol{x}]$.[1] The probability spaces we consider consist of some $\mathbb{R}^d$, its Borel σ-algebra, and a probability distribution $\mathbb{D}$. Instead of $\mathbb{D}(\{D \in \mathbb{R}^d \colon D \text{ has property } E\})$, we write $\Pr_{D\sim\mathbb{D}}(E)$ or simply $\Pr(E)$.

All random variables that we consider are real-valued. For a random variable X, we denote its expected value by $\mathbb{E}(X)$ and its variance by $\mathrm{Var}(X)$. The covariance of variables X and Y is denoted by $\mathrm{Cov}(X, Y)$. Random variables $X_1, \ldots, X_n$ are called *independent* if, for all $A_1, \ldots, A_n \in \mathcal{B}(\mathbb{R}^d)$, it holds that $\Pr(\bigwedge_{i=1}^n (X_i \in A_i)) = \prod_{i=1}^n \Pr(X_i \in A_i)$.

Recall that the *Gaussian distribution* $\mathcal{N}(\mu, \sigma^2)$ with mean $\mu \in \mathbb{R}$ and standard deviation $\sigma > 0$ as parameters is a distribution on $\mathbb{R}$ with density

[1] The set $B_\delta[S]$ is not necessarily a Borel set. However, it is Lebesgue-measurable as a continuous image of the Borel set $S \times B_\delta(\mathbf{0})$, which suffices for our purposes.

$\varphi(x \mid \mu, \sigma^2) = \frac{1}{\sigma\sqrt{2\pi}} \exp\left(-\frac{(x-\mu)^2}{2\sigma^2}\right)$. Its distribution function is denoted by Φ_{μ,σ^2}. For $\boldsymbol{\mu} = (\mu_1, \ldots, \mu_d) \in \mathbb{R}^d$ and $\sigma > 0$, the d-dimensional *spherical Gaussian distribution* $\mathcal{N}_{\text{sp}}^d(\boldsymbol{\mu}, \sigma^2)$ with mean $\boldsymbol{\mu}$ and standard deviation σ is the distribution of the random vector $(X_1, \ldots, X_d)$ whose coordinates are independent and the i-th coordinate is $\mathcal{N}(\mu_i, \sigma^2)$-distributed.

To study sums of independent random variables, we use Hoeffding's Inequality [28] and Berry-Esseen's Inequality [2,19,36].

3 The Data Model

We assume that the data points are drawn from two or more spherical Gaussian distributions (which we call the *marginals* or *marginal distributions*) with given parameters.[2] In most of the considered cases, we will have two distributions with equal standard deviations. Let d be the dimension of the marginal distributions. Then the data points will be represented by d-dimensional vectors. To make explicit reference to them, it will be convenient to consider these vectors as labeled with indices $1, \ldots, n$. Therefore, we assume our data encoded as a $(d \times n)$-dimensional matrix D whose i-th column is the data point $\boldsymbol{x}_i$. Accordingly, we also consider D as a transposed tuple of vectors and write $D = (\boldsymbol{x}_1, \ldots, \boldsymbol{x}_n)$.

In the general situation, we cannot assume to draw the same number of data points from each of the participating Gaussians. Instead, we assume that, for each marginal, the number of data points drawn from it is given as a parameter.[3] To this end, we introduce a *hidden labeling*, which assigns to each matrix column the distribution that the corresponding data point is drawn from.

Definition 2. *Let $m \in \mathbb{N}$ and $r_1, \ldots, r_m \in \mathbb{Q}_{>0}$ and suppose $r_1 + \ldots + r_m = 1$. A natural number n is called* compatible *with $(r_1, \ldots, r_m)$ if all $r_k \cdot n$ are in $\mathbb{N}$. In that case, we set $n_k := r_k \cdot n$. A* hidden labeling *for $(r_1, \ldots, r_m, n)$ is a function $\ell\colon [n] \to [m]$ such that, for each $k \in [m]$, it holds that $|\ell^{-1}(k)| = n_k$.*

If we write $r_k = \frac{p_k}{q_k}$ with coprime p_k and q_k, then n is compatible with $(r_1, \ldots, r_m)$ if and only if n is divisible by each q_k. If $r_1, \ldots, r_m$ are clear from the context, we drop them and simply speak about *compatible* n and ℓ.

Definition 3. *Suppose $d, m \in \mathbb{N}$ and $r_1, \ldots, r_m \in \mathbb{Q}_{>0}$ with $r_1 + \ldots + r_m = 1$ and let n be compatible with $(r_1, \ldots, r_m)$. Let $\ell\colon [n] \to [m]$ be a hidden labeling for $(r_1, \ldots, r_m, n)$. Furthermore, let $(\boldsymbol{\mu}_1, \sigma_1), \ldots, (\boldsymbol{\mu}_m, \sigma_m)$ be such that for each $k \in [m]$, it holds that $\boldsymbol{\mu}_k \in \mathbb{R}^d$ and $\sigma_k > 0$. Let $(X_i)_{i \in [n]}$ be a collection of independent d-dimensional random variables such that $X_i \sim \mathcal{N}_{sp}^d(\boldsymbol{\mu}_{\ell(i)}, \sigma_{\ell(i)}^2)$.*

[2] The task of learning these parameters from a set of data is very well studied, see, e.g. [31] or [33].

[3] An alternative model is drawing the data from a Gaussian mixture with given weights (see [27, Chapter 8]). This is mathematically easier to handle due to the independence of the data points. Our results thus translate to this setting. However, the data will have a slightly higher variance, which leads to stronger preconditions in our theorems.

The data distribution defined by $((r_k, \boldsymbol{\mu}_k, \sigma_k)_{k\in[m]}, n, \ell)$ *is the joint distribution of the* X_i *and denoted by* $\mathbb{D}[(r_k, \boldsymbol{\mu}_k, \sigma_k)_{k\in[m]}, n, \ell]$.

Note that the induced data distribution is over the real vector space of $(d\times n)$-dimensional matrices whose colums correspond to d-dimensional data points.

Notation 2. *In the setting of Definition 3, let* f_k *be the density of the* k*-th Gaussian distribution, i.e.,* $f_k(\boldsymbol{x}) = \varphi(\boldsymbol{x} \mid \boldsymbol{\mu}_k, \sigma_k^2)$. *The induced measure is denoted by* ν_k, *so for every Borel set* $A \subseteq \mathbb{R}^d$, *we have* $\nu_k(A) = \int_A f_k(\boldsymbol{x})d\boldsymbol{x}$. *We set* $\bar{f} := \sum_{k\in[m]} r_k f_k$ *and* $\bar{\nu} := \sum_{k\in[m]} r_k \nu_k$.

Note that, for all compatible n and ℓ, it holds that $\bar{f} = \frac{1}{n}\sum_{i=1}^{n} f_{\ell(i)}$.

To analyze the data points, we define an underlying graph, which captures similarity. We interpret the data points as the vertices and consider two ways to define (weighted) adjacency, which are standard in the clustering context [32].

Definition 4. *Let* $n, d \in \mathbb{N}_{>0}$ *and suppose* $w\colon \mathbb{R}^d \times \mathbb{R}^d \to \mathbb{R}_{\geq 0}$ *is symmetric and* $D = (\boldsymbol{x}_i)_{i\in[n]} \in \mathbb{R}^{d\times n}$. *Then* $G(D, w)$ *denotes the weighted graph with vertex set* $[n]$, *edge set* $\big\{\{i,j\} \mid w(\boldsymbol{x}_i, \boldsymbol{x}_j) > 0\big\}$ *and weights* $w'(i,j) = w(\boldsymbol{x}_i, \boldsymbol{x}_j)$.

Example 2. For $c > 0$ and $\hat{w}_c(\boldsymbol{x}, \boldsymbol{y}) := \exp\left(\frac{-\|\boldsymbol{x}-\boldsymbol{y}\|^2}{2c^2}\right)$, the graph $G(D, \hat{w}_c)$ is a *fully connected graph.* In this work, we only have $c = \sigma$, where σ is the standard deviation of the Gaussian distribution at hand (see also [32]). Accordingly, we drop c and speak about *the* fully connected graph. See also Fig. 1b.

Instead of maintaining all distances between vertices, we can also define a threshold for adjacency.

Example 3. For $\delta > 0$ and $w_\delta(\boldsymbol{x}, \boldsymbol{y}) := \begin{cases} 1, & \text{if } \|\boldsymbol{x}-\boldsymbol{y}\| \leq \delta \\ 0, & \text{otherwise} \end{cases}$, the graph $G(D, w_\delta)$ is a *δ-neighborhood graph.* See also Fig. 1a.

Since every graph vertex represents a point in $\mathbb{R}^d$, we can interpret separations of $\mathbb{R}^d$ as separations of the graph vertex set. This motivates the following notation and will ultimately allow us to bound probabilities for events on the graphs by bounding probabilities in $\mathbb{R}^d$ associated with the data distributions.

Notation 3. *For* $d, n \in \mathbb{N}_{>0}$, $A \subseteq \mathbb{R}^d$ *and a data matrix* $D = (\boldsymbol{x}_i)_{i\in[n]} \in \mathbb{R}^{d\times n}$, *we set* $V_A(D) := \{i \in [n] \mid \boldsymbol{x}_i \in A\}$. *If* D *is clear, we only write* V_A.

If $A \subseteq \mathbb{R}^d$ is a Borel set and $D \sim \mathbb{D}$ for some probability distribution $\mathbb{D}$ on $\mathbb{R}^{d\times n}$, then the size of V_A (denoted by $|V_A|$) is a random variable.

4 Results for the δ-Neighborhood Graph

In this section, we study the tangles that occur in the δ-neighborhood graph. We find that the unweighted structure makes it possible to derive strong lower bounds on the probability that incomparable tangles corresponding to the different marginal distributions exist. By Lemma 1, every non-trivial clique $G[W]$ induces a κ_G-tangle. Every hyperball of radius at most $\frac{\delta}{2}$ induces a clique in the δ-neighborhood graph. As our goal is to identify tangles associated with the dense regions around the means, we consider the $\frac{\delta}{2}$-neighborhood $B_{\frac{\delta}{2}}[\boldsymbol{\mu}]$ of μ in the following. We bound the probability that the induced cliques are non-trivial and that the tangles corresponding to different means are indeed incomparable. To do so, we find small separations that are oriented differently by the tangles.

4.1 Probability Bounds

To allow for different geometric objects for the separation, we define it via a Borel set S. In Sect. 4.2, we consider particular choices for S. Often the local minima of the underlying joint distribution will induce a good choice, but, to actually compute the probabilities, other choices can be useful. The preconditions of the theorem ensure that the desired tangle exists in expectation, that is the order of the tangle must be larger in expectation than the edge connectivity of V_S and in expectation at least half of the points of V_S must be part of the clique.

Theorem 1. *Suppose $\delta > 0$, $d, m \in \mathbb{N}_{>0}$ and let $(r_1, \boldsymbol{\mu}_1, \sigma_1), \ldots, (r_m, \boldsymbol{\mu}_m, \sigma_m)$ be as in Definition 3. Choose $k \in [m]$ and define $B := B_{\frac{\delta}{2}}[\boldsymbol{\mu}_k]$. Let $S \subseteq \mathbb{R}^d$ be a Borel set. If*

$$\bar{\nu}(B \cap S) > \frac{1}{2}\bar{\nu}(B) \quad \textit{and} \quad \frac{2}{9}\bar{\nu}(B)^2 > \int_S \int_{S^c} w_\delta(\boldsymbol{x}, \boldsymbol{y}) \bar{f}(\boldsymbol{x}) \bar{f}(\boldsymbol{y}) \, d\boldsymbol{y} \, d\boldsymbol{x},$$

then for compatible n and ℓ and $\mathbb{D} = \mathbb{D}[(r_k, \boldsymbol{\mu}_k, \sigma_k)_{k \in [m]}, n, \ell]$, it holds that

$$\Pr_{D \sim \mathbb{D}} \left(\mathcal{T}_{G(D, w_\delta)}(V_B) \text{ is a tangle that contains } V_S \right) = 1 - O\left(\tfrac{1}{n}\right) .$$

In fact, the proof arguments given in the full-version [22] yield that, for every $\varepsilon > 0$, the order of the tangle is with probability $1 - O\left(\frac{1}{n}\right)$ at least $(1 - \varepsilon) \cdot n^2 \cdot \frac{2}{9}\bar{\nu}(B)^2$.

Theorem 1 can be used to show the existence of incomparable tangles asymptotically almost surely. However, since the hidden constants in the asymptotic expression are large, for small data sets, it only yields a trivial bound. We now make reasonably stricter requirements for the data distribution that allow us to find a probability bound with significantly faster convergence to 1. As a first step, we derive an upper bound on the size of a cut induced by a set $S \subseteq \mathbb{R}^d$.

Lemma 2. *Suppose $\delta > 0$, $d, n \in \mathbb{N}_{>0}$, $D \in \mathbb{R}^{d \times n}$ and let $S \subseteq \mathbb{R}^d$ be a Borel set. We set $A := B_\delta[S] \cap B_\delta[S^c]$. Then $\kappa_{G(D,w_\delta)}(V_S) \leq \frac{1}{4}|V_A|^2$.*

Using the lemma, we can answer the question whether a cut is oriented by the considered tangle by checking if a sum of independent random variables is negative. Again the preconditions ensure that the tangle exists in expectation.

Theorem 2. *Suppose $\delta > 0$, $d, m \in \mathbb{N}_{>0}$ and let $(r_1, \boldsymbol{\mu}_1, \sigma_1), \ldots, (r_m, \boldsymbol{\mu}_m, \sigma_m)$, n, ℓ, and $(X_i)_{i \in [n]}$ be as in Definition 3. Let $k \in [m]$ and define $B := B_{\frac{\delta}{2}}[\boldsymbol{\mu}_k]$. Let S be a Borel set with $B \subseteq S$. We set $A := B_\delta[S] \cap B_\delta[S^c]$ and suppose $2\sqrt{2}\bar{\nu}(B) > 3\bar{\nu}(A)$.*

Let $\mathbb{D} = \mathbb{D}[(r_k, \boldsymbol{\mu}_k, \sigma_k)_{k \in [m]}, n, \ell]$. For $i \in [n]$, define $Y_i := (3 \cdot \mathbb{1}_A - 2\sqrt{2} \cdot \mathbb{1}_B)(X_i)$ and set $\rho_i := \mathbb{E}(|Y_i - \mathbb{E}(Y_i)|^3)$. Then

$$\begin{aligned}
&\Pr_{D \sim \mathbb{D}} \left(\mathcal{T}_{G(D,w_\delta)}(V_B) \text{ is a tangle that contains } V_S \right) \\
&\geq \max \left\{ 1 - \exp\left(-n \cdot \frac{2}{(2\sqrt{2}+3)^2} \cdot \left(3\bar{\nu}(A) - 2\sqrt{2}\bar{\nu}(B)\right)^2 \right), \right. \\
&\qquad \left. \Phi_{0,1}\left(\frac{-\sum_{i=1}^n \mathbb{E}(Y_i)}{\sqrt{\sum_{i=1}^n \mathrm{Var}(Y_i)}} \right) - \frac{C \cdot \sum_{i=1}^n \rho_i}{\left(\sum_{i=1}^n \mathrm{Var}(Y_i)\right)^{\frac{3}{2}}} \right\} \\
&\quad - \left(1 + \sum_{k=1}^m \frac{n_k \cdot \nu_k(B)}{1 - \nu_k(B)} \right) \cdot \prod_{k=1}^m \left(1 - \nu_k(B)\right)^{n_k}
\end{aligned}$$

where C is the constant from Berry-Esseen's Inequality. As a consequence,

$$\Pr_{D \sim \mathbb{D}} \left(\mathcal{T}_{G(D,w_\delta)}(V_B) \text{ is a tangle that contains } V_S \right) = 1 - e^{-\Omega(n)}.$$

The first term of the bound results from Hoeffding's Inequality, the second from Berry-Esseen's Inequality, and the last one is the probability that the clique contains at least two points. Note that the bound derived from Berry-Esseen's Inequality only yields an asymptotic bound of $1 - O(\frac{1}{\sqrt{n}})$. However, if either n or the distance of the means is small, it often gives better results than the bound derived from Hoeffding's Inequality. Using the union bound with the event A_k being "$\mathcal{T}_{G(D,w_\delta)}(V_{B_{\frac{\delta}{2}}[\boldsymbol{\mu}_k]})$ is a tangle that contains V_S", we can apply Theorems 1 and 2 to bound the probability that the tangles we consider are incomparable.

Corollary 1. *Let $\delta > 0$, $d, m \in \mathbb{N}_{>0}$ and let $(r_1, \boldsymbol{\mu}_1, \sigma_1), \ldots, (r_m, \boldsymbol{\mu}_m, \sigma_m)$ be as in Definition 3. Suppose that there are Borel sets S_{k_1,k_2} for all $\{k_1, k_2\} \in \binom{[m]}{2}$ with $S_{k_1,k_2} = S^c_{k_2,k_1}$. If for each pair (k_1, k_2), the conditions of Theorem 1 with $k = k_1$ and $S = S_{k_1,k_2}$ are fulfilled, then*

$$\Pr_{D \sim \mathbb{D}}\left((\mathcal{T}_{G(D,w_\delta)}(V_{B_{\frac{\delta}{2}}[\boldsymbol{\mu}_k]}))_{k\in[m]} \textit{ are pairwise incomparable tangles}\right) = 1 - O\left(\tfrac{1}{n}\right)$$

for compatible n and ℓ and $\mathbb{D} = \mathbb{D}[(r_k, \boldsymbol{\mu}_k, \sigma_k)_{k\in[m]}, n, \ell]$. If all k_1 and S_{k_1,k_2} meet the conditions of Theorem 2, then

$$\Pr_{D \sim \mathbb{D}}\left((\mathcal{T}_{G(D,w_\delta)}(V_{B_{\frac{\delta}{2}}[\boldsymbol{\mu}_k]}))_{k\in[m]} \textit{ are pairwise incomparable tangles}\right) = 1 - e^{-\Omega(n)}\ .$$

4.2 Computational Results

We have seen theorems of the form "If some condition is met, there will be incomparable tangles with at least probability P". However, it is not obvious to see when their preconditions are fulfilled and which bounds they yield. Here, we analyze the implications of our theorems for some explicit parameter sets. For space reasons, the details of our analysis had to be deferred to the full version [22] and we can only give an overview here.

We begin with the case of two Gaussians, starting with an in-depth study of the one-dimensional setting followed by the generalization to higher dimensions. Then, we analyse a case of more than two marginals by studying a mixture of three Gaussians whose means form an equilateral triangle. We explore reasonable choices for the Borel set S that defines the candidate low-order separation such that the conditions for our probability bounds are met also when the means of the marginals are close to each other. Since the structural properties of the data only depend on the ratio of the distances of the means and the standard deviation[4], we can restrict the following analysis to mixtures where all marginals have standard deviation 1 or, in the case of a mixture with different standard deviations, we assume that at least one marginal has standard deviation 1.

We start with the base case of two one-dimensional Gaussians with means 0 and λ, standard deviation $\sigma = 1$ and equally many points drawn from each. Here we need $\lambda > 2\sigma$ to have more than one dense region, because for $\lambda \leq 2\sigma$, the mean density $\bar{f}$ only has one maximum at $\frac{\lambda}{2}$. As the low-order cut used to distinguish the tangles, we choose the cut at $\frac{\lambda}{2}$, so $S = \left(-\infty, \frac{\lambda}{2}\right]$. As Fig. 2a shows, Theorem 2 yields good lower bounds already for small data sets.

This setting can be generalized in three ways. First, we vary the fraction r of data points drawn from one marginal distribution. Here, the considered cut is at the local minimum of the average density function. The obtained probability bounds for this case are shown in Fig. 2b. Next, we vary the ratio α of the standard deviations. Again, we separate the regions at the local minimum of the joint density function. In both of these first two generalizations, we can compute

[4] The ratio $s^2 = \frac{\lambda^2}{\sigma^2}$ is often called signal-to-noise ratio (SNR).

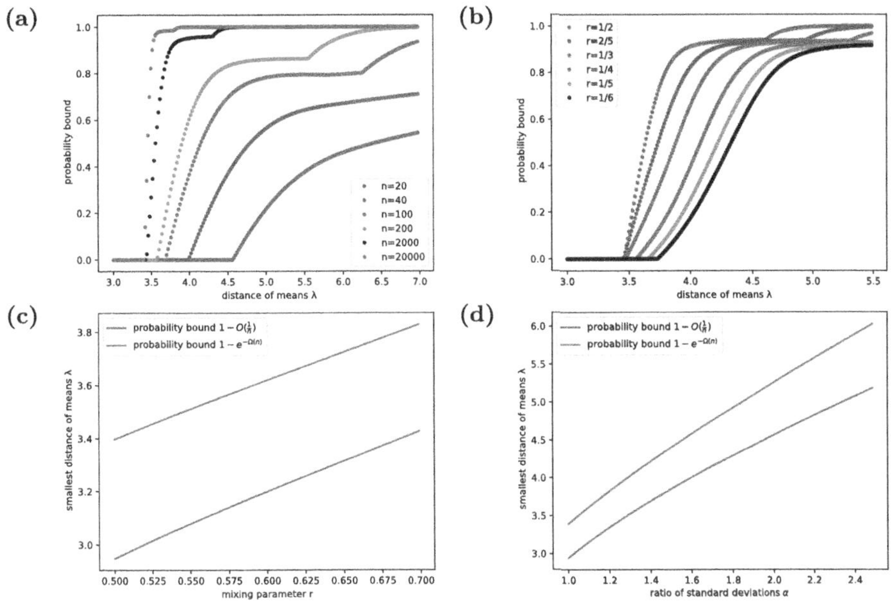

Fig. 2. **(a)**–**(d)** Our computational results in the one-dimensional case. In all plots except for **(d)**, the marginal distributions have $\sigma = 1$. **(a)** shows the probability bound depending on λ for different n. **(b)** shows the bound depending on λ for $n = 900$ and different mixing parameters r. **(c)**, **(d)** show plots of the smallest mean distance λ such that the preconditions of Theorem 3 are met. In **(c)**, we vary the mixing parameter r, and in **(d)**, we vary the ratio α between the standard deviations (i.e., the standard deviation of the second marginal is changed to α). Note that at some point the bound corresponding to Hoeffding's Inequality becomes larger than the bounds relating to Berry-Esseen's Inequality.

the smallest distance of the means where the conditions of Corollary 1 can be met via the following lemma. For $r = \frac{1}{2}$ and $\alpha = 1$, we obtain the bounds $\lambda > 2.948$ and $\lambda > 3.397$, respectively, see Figs. 2c and 2d.

Lemma 3. *Let* $\lambda, \sigma, \alpha > 0$ *and* $r_1 \in (0, 1)$. *Let* $m := 2$, $\mu_1 := 0$, $\mu_2 := \lambda$, $\sigma_1 := \sigma$, $\sigma_2 := \alpha \cdot \sigma$, $r_2 := 1 - r_1$. *Then with* $c := \operatorname{argmin}_{0 \leq x \leq \lambda} \bar{f}(x)$ *and Notation 2, the following hold for compatible* n *and* ℓ *and* $\mathbb{D} = \mathbb{D}[((r, 0, \sigma), (1 - r, \lambda, \alpha\sigma)), n, \ell]$.

1. If $\frac{2}{3} \cdot \min\{\bar{f}(0), \bar{f}(\lambda)\} > \bar{f}(c)$, *there is a* $\delta > 0$ *such that*

$$\Pr_{D \sim \mathbb{D}} \begin{pmatrix} \mathcal{T}_{G(D,w_\delta)}(V_{B_{\delta/2}[0]}) \textit{ and } \mathcal{T}_{G(D,w_\delta)}(V_{B_{\delta/2}[\lambda]}) \\ \textit{are incomparable tangles} \end{pmatrix} = 1 - O\left(\tfrac{1}{n}\right) .$$

2. If $\frac{\sqrt{2}}{3} \cdot \min\{\bar{f}(0), \bar{f}(\lambda)\} > \bar{f}(c)$, *there is a* $\delta > 0$ *such that*

$$\Pr_{D \sim \mathbb{D}} \begin{pmatrix} \mathcal{T}_{G(D,w_\delta)}(V_{B_{\delta/2}[0]}) \textit{ and } \mathcal{T}_{G(D,w_\delta)}(V_{B_{\delta/2}[\lambda]}) \\ \textit{are incomparable tangles} \end{pmatrix} = 1 - e^{-\Omega(n)} .$$

We discuss a generalization to higher dimensions in the full version [22]. There, we also present our results for the fully connected graph. Here, we can prove a similar statement as in Theorem 1 with the same asymptotic probability. However, our bounds on the edge connectivity do not translate directly from the δ-neighborhood graph to the fully connected graph. We prove such a bound in the one-dimensional case and use it to obtain exponentially fast convergence to 1 as in Theorem 5 via Berry-Esseen's Inequality.

5 Conclusion

In [20], a precise connection between tangles, a subject from structural graph theory, and clustering, a central technique in data science, was established. Our work aims at setting the foundation for an in-depth quantitative analysis of tangles in data sets as a means to capture soft cluster assignments. To this end, we have applied tangle theory to graphs for sets of data points drawn from Gaussian mixtures. For the two standard graph models in this context, we have found explicit conditions under which, asymptotically almost surely, incomparable tangles that can be associated with the participating distributions exist. First, we have provided a probability bound with convergence to 1, which becomes meaningful for large data sets. Second, we have given a probability bound with exponential convergence to 1 (but stronger preconditions), which also yields a lower bound on the tangle order.

In a future project, it would be interesting to study the fully connected graph in more detail and possibly with other tools. More refined bounds on the tangle order would lead to stronger probability bounds. As another follow-up study, we leave as an open problem to push our probability bounds (or the conditions on the data distributions) in higher dimensions further. To exploit the tangle potential in the Gaussian setting in more depth, we suggest the theoretical analysis of multiple Gaussian distributions whose means are in arbitrary position or whose standard deviations differ. In a next step, the results could be applied to actually recover (with some bounded error probability) the hidden labels. One can also go beyond Gaussian mixtures and study other "well-clustered" data.

It can also be worthwhile to approach the problem via branch decompositions. The duality between tangle orders and orders of branch decompositions ([35], also [26, Theorem 6.1]) will yield upper bounds on the probability that tangles of high order exist, which may be useful to detect incomparable tangles.

Finally, regarding the benefit of the formalized connection between clustering and tangles, there is not only a potential of tangles to yield theoretical insights about the existence and quality of clusters in data sets. Indeed, exploring the connection further, the field of structural graph theory might benefit from the broad research on fast approximation algorithms for clustering, possibly yielding efficient approximations for computing graph tangles or decompositions. However, note that there may be tangles not stemming from clusters, so clustering algorithms will not necessarily detect *all* tangles in the data.

Disclosure of Interests. The authors have no competing interests to declare that are relevant to the content of this article.

References

1. Adler, I., Gottlob, G., Grohe, M.: Hypertree width and related hypergraph invariants. Eur. J. Comb. **28**(8), 2167–2181 (2007). https://doi.org/10.1016/j.ejc.2007.04.013
2. Berry, A.C.: The accuracy of the Gaussian approximation to the sum of independent variates. Trans. Am. Math. Soc. **49**(1), 122–136 (1941)
3. Bozyk, L., Defrain, O., Okrasa, K., Pilipczuk, M.: On objects dual to tree-cut decompositions. CoRR abs/2103.14667 (2021). https://arxiv.org/abs/2103.14667
4. Carmesin, J., Diestel, R., Hamann, M., Hundertmark, F.: Canonical tree-decompositions of finite graphs I. Existence and algorithms. J. Comb. Theory, Ser. B **116**, 1–24 (2016). https://doi.org/10.1016/j.jctb.2014.04.001
5. Chaudhuri, K., Dasgupta, S., Vattani, A.: Learning mixtures of gaussians using the k-means algorithm. CoRR abs/0912.0086 (2009). http://arxiv.org/abs/0912.0086
6. Cohen-Addad, V., Gupta, A., Kumar, A., Lee, E., Li, J.: Tight FPT approximations for k-median and k-means. In: 46th International Colloquium on Automata, Languages, and Programming, ICALP 2019. LIPIcs, vol. 132, pp. 42:1–42:14. Schloss Dagstuhl - Leibniz-Zentrum für Informatik (2019). https://doi.org/10.4230/LIPIcs.ICALP.2019.42
7. Cohen-Addad, V., Kanade, V., Mallmann-Trenn, F., Mathieu, C.: Hierarchical clustering: objective functions and algorithms. In: Proceedings of the Twenty-Ninth Annual ACM-SIAM Symposium on Discrete Algorithms, SODA 2018, pp. 378–397 (2018). https://doi.org/10.1137/1.9781611975031.26
8. Dasgupta, S.: A cost function for similarity-based hierarchical clustering. In: Proceedings of the 48th Annual ACM SIGACT Symposium on Theory of Computing, STOC 2016, pp. 118–127 (2016). https://doi.org/10.1145/2897518.2897527
9. De Oliveira, J.V., Pedrycz, W.: Advances in Fuzzy Clustering and its Applications. Wiley, Hoboken (2007)
10. Diestel, R.: Abstract separation systems. Order **35**(1), 157–170 (2018). https://doi.org/10.1007/s11083-017-9424-5
11. Diestel, R.: Tangles: a new paradigm for clusters and types. CoRR abs/2006.01830 (2020). https://arxiv.org/abs/2006.01830
12. Diestel, R., Erde, J., Weißauer, D.: Structural submodularity and tangles in abstract separation systems. J. Combin. Theory Ser. A **167**, 155–180 (2019). https://doi.org/10.1016/j.jcta.2019.05.001
13. Diestel, R., Hundertmark, F., Lemanczyk, S.: Profiles of separations: in graphs, matroids, and beyond. Combinatorica **39**(1), 37–75 (2018). https://doi.org/10.1007/s00493-017-3595-y
14. Diestel, R., Oum, S.: Unifying duality theorems for width parameters in graphs and matroids (extended abstract). In: Kratsch, D., Todinca, I. (eds.) WG 2014. LNCS, vol. 8747, pp. 1–14. Springer, Cham (2014). https://doi.org/10.1007/978-3-319-12340-0_1
15. Diestel, R., Oum, S.: Tangle-tree duality in abstract separation systems. Adv. Math. **377**, 107470, 24 (2021). https://doi.org/10.1016/j.aim.2020.107470
16. Diestel, R., Whittle, G.: Tangles and the Mona Lisa. arXiv (2016). arXiv:1603.06652 [math.CO]
17. Elbracht, C., et al.: Clustering with tangles: algorithmic framework and theoretical guarantees. CoRR abs/2006.14444 (2021). https://arxiv.org/abs/2006.14444

18. Klepper, S., et al.: Trees of tangles in abstract separation systems. J. Combin. Theory Ser. A **180**, 105425, 27 (2021). https://doi.org/10.1016/j.jcta.2021.105425
19. Esseen, C.G.: Fourier analysis of distribution functions. A mathematical study of the Laplace-Gaussian law. Acta Mathematica **77**(1), 1–125 (1945)
20. Fluck, E.: Tangles and single linkage hierarchical clustering. In: 44th International Symposium on Mathematical Foundations of Computer Science, MFCS 2019. LIPIcs, vol. 138, pp. 38:1–38:12. Schloss Dagstuhl - Leibniz-Zentrum für Informatik (2019). https://doi.org/10.4230/LIPIcs.MFCS.2019.38
21. Fluck, E.: Tangles and hierarchical clustering. SIAM J. Discret. Math. **38**(1), 75–92 (2024). https://doi.org/10.1137/22M1484936
22. Fluck, E., Kiefer, S., Standke, C.: Untangling Gaussian mixtures. CoRR abs/2403.06671 (2024). https://doi.org/10.48550/ARXIV.2403.06671
23. Geelen, J., Gerards, B., Whittle, G.: Tangles, tree-decompositions and grids in matroids. J. Comb. Theory Ser. B **99**(4), 657–667 (2009). https://doi.org/10.1016/j.jctb.2007.10.008
24. Giannopoulou, A.C., Kawarabayashi, K., Kreutzer, S., Kwon, O.: The canonical directed tree decomposition and its applications to the directed disjoint paths problem. CoRR abs/2009.13184 (2020). https://arxiv.org/abs/2009.13184
25. Giraud, C., Verzelen, N.: Partial recovery bounds for clustering with the relaxed k-means. CoRR abs/1807.07547 (2018). http://arxiv.org/abs/1807.07547
26. Grohe, M.: Tangled up in blue (a survey on connectivity, decompositions, and tangles). arXiv (2016). arXiv:1605.06704 [cs.DM]
27. Hennig, C., Meila, M., Murtagh, F., Rocci, R.: Handbook of Cluster Analysis. CRC Press (2015)
28. Hoeffding, W.: Probability inequalities for sums of bounded random variables. J. Am. Stat. Assoc. **58**, 13–30 (1963), http://links.jstor.org/sici?sici=0162-1459(196303)58:301<13:PIFSOB>2.0.CO;2-D&origin=MSN
29. Johnson, T., Robertson, N., Seymour, P.D., Thomas, R.: Directed tree-width. J. Comb. Theory, Ser. B **82**(1), 138–154 (2001). https://doi.org/10.1006/jctb.2000.2031
30. Klenke, A.: Probability Theory: A Comprehensive Course. Universitext, 2nd edn. Springer, London (2014). https://doi.org/10.1007/978-1-4471-5361-0, translation from the German language edition
31. Liu, A., Moitra, A.: Settling the robust learnability of mixtures of Gaussians. In: Khuller, S., Williams, V.V. (eds.) STOC 2021: 53rd Annual ACM SIGACT Symposium on Theory of Computing, Virtual Event, Italy, 21–25 June 2021, pp. 518–531. ACM (2021). https://doi.org/10.1145/3406325.3451084
32. von Luxburg, U.: A tutorial on spectral clustering. Stat. Comput. **17**(4), 395–416 (2007). https://doi.org/10.1007/s11222-007-9033-z
33. Moitra, A., Valiant, G.: Settling the polynomial learnability of mixtures of gaussians. In: 51th Annual IEEE Symposium on Foundations of Computer Science, FOCS 2010, Las Vegas, Nevada, USA, 23–26 October 2010, pp. 93–102. IEEE Computer Society (2010). https://doi.org/10.1109/FOCS.2010.15
34. Reed, B.A.: Introducing directed tree width. Electron. Notes Discret. Math. **3**, 222–229 (1999). https://doi.org/10.1016/S1571-0653(05)80061-7
35. Robertson, N., Seymour, P.D.: Graph minors. X. Obstructions to tree-decomposition. J. Comb. Theory Ser. B **52**(2), 153–190 (1991). https://doi.org/10.1016/0095-8956(91)90061-N
36. Tyurin, I.S.: A refinement of the remainder in the Lyapunov theorem. Theory Probab. Appl. **56**(4), 693–696 (2012). https://doi.org/10.1137/S0040585X9798572X

Face-Hitting Dominating Sets in Planar Graphs

P. Francis[1], Abraham M. Illickan[2], Lijo M. Jose[3(✉)], and Deepak Rajendraprasad[3]

[1] Department of Mathematics SAS, VIT-AP University, Amaravati, Andhra Pradesh, India
francis@vitap.ac.in

[2] Department of Computer Science, University of California, Irvine, USA
aillicka@uci.edu

[3] Department of Computer Science, Indian Institute of Technology Palakkad, Kanjikode, India
112004005@smail.iitpkd.ac.in, deepak@iitpkd.ac.in

Abstract. A *dominating set* of a graph G is a subset S of its vertices such that each vertex of G not in S has a neighbor in S. A *face-hitting set* of a plane graph G is a set T of vertices in G such that every face of G contains at least one vertex of T. We show that the vertex-set of every plane (multi-)graph without isolated vertices, self-loops or 2-faces can be partitioned into two disjoint sets so that both the sets are dominating and face-hitting. We also show that all the three assumptions above are necessary for the conclusion.

As a corollary, we show that every n-vertex simple plane triangulation has a dominating set of size at most $(1-\alpha)n/2$, where αn is the maximum size of an independent set in the triangulation. Matheson and Tarjan [European J. Combin., 1996] conjectured that every plane triangulation with a sufficiently large number of vertices n has a dominating set of size at most $n/4$. Currently, the best known general bound for this is by Christiansen, Rotenberg and Rutschmann [6] [SODA, 2024] who showed that every plane triangulation on $n > 10$ vertices has a dominating set of size at most $2n/7$. Our corollary improves their bound for n-vertex plane triangulations which contain a maximal independent set of size either less than $2n/7$ or more than $3n/7$.

Keywords: Dominating sets · Face-hitting sets · Domination number · Planar triangulations · Planar graphs

1 Introduction

All graphs considered in this paper are finite and undirected. A *dominating set* of a graph G is a subset S of its vertices such that each vertex of G not in S

Lijo M. Jose—Supported by Half-Time Research Assistantship of Ministry of Education, Government of India.

D. Kráľ and M. Milanič (Eds.): WG 2024, LNCS 14760, pp. 211–219, 2025.
https://doi.org/10.1007/978-3-031-75409-8_15

has a neighbor in S. The *domination number* $\gamma(G)$ of G is the cardinality of a smallest dominating set in G. A k-coloring $f : V(G) \rightarrow [k]$ of G is called *domatic* if each color class of f is a dominating set of G. A *face-hitting set* of a plane graph G is a set T of vertices in G such that every face of G contains at least one vertex of T. Let $\beta(G)$ denote the size of a smallest face-hitting set in G. A k-coloring $f : V(G) \rightarrow [k]$ is called *polychromatic* if each color class of f is a face-hitting set of G. The polychromatic number $p(G)$ of G is the largest number k such that there is a polychromatic k-coloring of G.

Let f be a face in a plane graph G. If G is not connected, the boundary of f could be a disjoint union of closed walks. The *length* of f is the total length of all these walks, where the length of a walk is the number of edges in the walk (counting repetitions). A k-face (resp. k^+-face, k^--face) is a face of length k (resp. at least k, at most k). The *degree* of a vertex is the number of distinct vertices adjacent to it. The main result of this paper is that every plane graph without isolated vertices, self-loops or 2-faces has a 2-coloring which is simultaneously domatic and polychromatic.

Theorem 1. *Every plane graph G without isolated vertices, self-loops or 2-faces, has two disjoint subsets $V_1, V_2 \subseteq V(G)$, such that both V_1 and V_2 are dominating and face-hitting.*

The assumptions made on G in the above theorem are indeed necessary. It follows from Theorem 1 that every n-vertex plane graph G satisfying the premise of the theorem has a subset S of vertices of cardinality at most $n/2$ which is both dominating and face-hitting. It is easy to see that, if we allow isolated vertices or length-1 faces, then $\gamma(G)$ can be as large as n. If we allow 2-faces, then $\beta(G)$ can be as large as $3n/4$. For the second observation, consider G to be a disjoint union of $n/4$ components, each component being a K_4 in which every edge is replaced with a 2-face. We cannot allow self-loops even if all faces have length at least 3 because, in this case, we can still have a 3^+-face which has only one vertex in its boundary. For example consider a vertex v with three self loops l_1, l_2, l_3 such that l_1 and l_2 have disjoint interiors, but both are inside of l_3. We then add a neighbor of v inside the self-loops l_1 and l_2 so that all the faces have length at least 3. Then the face bounded by l_1, l_2, l_3 has only one vertex in its boundary. It is clear that we cannot have a polychromatic 2-coloring in this case.

We can also see that the bound of $n/2$ is tight under these assumptions. A disjoint collection of edges (or 3-length paths, or 4-length cycles) have $\gamma(G) \geq |V(G)|/2$. Bose et al. [4] showed that there exists an infinite family of simple connected plane graphs with $\beta(G) \geq \lfloor |V(G)|/2 \rfloor$.

Theorem 1 has an interesting application. A *near-triangulation* is a simple planar graph embedded in the plane such that all its faces except possibly the outer one are bounded by three edges. A *triangulation* is a near-triangulation in which the outer face is also bounded by three edges. If we remove an independent set from a triangulation G and then apply Theorem 1 on the resultant graph G', we can show that the dominating and face-hitting sets V_1, V_2 of G' will both be dominating sets of G. This observation helps us make some progress towards a conjecture by Matheson and Tarjan. Matheson and Tarjan [18] in 1996 proved that every n-vertex plane triangulation G has a domatic 3-coloring (hence $\gamma(G) \leq n/3$) and conjectured that if n is sufficiently large, then $\gamma(G) \leq n/4$. Špacapan [21] in 2020 showed that every plane triangulation on $n > 6$ vertices has a dominating set of size at most $17n/53$. Very recently in 2024 Christiansen, Rotenberg and Rutschmann [6] showed that every plane triangulation on $n > 10$ vertices has a dominating set of size at most $2n/7$. As a consequence of Theorem 1, we improve this bound for n-vertex plane triangulations which contain a maximal independent set of size either less than $2n/7$ or more than $3n/7$ and verify Matheson-Tarjan conjecture for n-vertex plane triangulations with an independent set of size $n/2$.

The novelty in our result is that we find a single 2-coloring which is both polychromatic and domatic. In fact, it is easy to show that every plane graph G in this class has a domatic 2-coloring and a polychromatic 2-coloring. Any proper 2-coloring of a spanning forest of G will be a domatic 2-coloring. Existence of a polychromatic 2-coloring for G is a solved exercise in Lovász's famous book - *Combinatorial Problems and Exercises* since the first edition [16]. Domatic colorings and polychromatic colorings on their own have been studied extensively in planar graphs.

The monograph by Haynes et al. [11] is a comprehensive reference on Domination. Reed [19], Sohn and Yuan [20], and Xing et al. [22] established upper bounds $3n/8$, $4n/11$ and $5n/14$ respectively on domination number for n-vertex graphs with minimum degree at least three, four and five. For triangulated discs, the upper bound of $n/3$ by Matheson and Tarjan is tighter than the above general bounds. The domination number of planar graphs have received special attention. MacGillivray and Seyffarth [17] established an upper bound of three and ten respectively on the domination number of planar graphs with diameter two and three. Goddard and Henning [10] improved upon this and showed that there is only one planar graph of diameter two with domination number three and also showed that every sufficiently large planar graph of diameter three has domination number at most seven. The dominating set problem on planar graphs is NP-complete. Fomin and Thilikos [8] showed that the k-dominating set problem on planar graphs can be solved in time $O(2^{15.13\sqrt{k}} + n^3)$. Variants of dominating sets like independent dominating sets and total dominating sets in planar triangulations have also been studied. Botler, Fernandes and Gutiérrez [5] proved that every planar triangulation G on n vertices have an independent dominating set of size less than $3n/8$ and this could be improved to $n/3$ if the minimum degree of G is at least five. Claverol et al. [7] showed that any near-triangulation of order n has a total dominating set of size at most $2n/5$, with two exceptions. Francis et al. [9] showed that every near-triangulation with minimum degree at least three, has two disjoint total dominating sets.

Face hitting sets have an interesting connection with terrain guarding problems. Bose et al. [4] used the idea of polychromatic colorings of triangulations in the context of guarding polyhedral terrains. They showed that there is an infinite family of plane triangulations where the smallest face hitting set has size $\lfloor n/2 \rfloor$. For a plane graph G, let $g(G)$ denote the size of the smallest face in G. Alon et al. [1] proved that for any plane graph G with $g(G)$ at least 3, $p(G) \geq \lfloor \frac{3g-5}{4} \rfloor$. They also showed that this bound is nearly tight, as there are plane graphs for which $p(G) \leq \lfloor \frac{3g+1}{4} \rfloor$. It is easy to observe that for every plane triangulation G, $p(G)$ is either 2 or 3. As a consequence of Heawood's theorem [12], $p(G) = 3$ for a plane triangulation G if and only if G is Eulerian (degree of each vertex is even) [1]. Hoffmann and Kriegel [13] showed that every 2-connected bipartite plane graph admits a polychromatic 3-coloring by showing that every such graph can be transformed into an Eulerian triangulation by adding edges only. The decision problem whether a plane graph is polychromatic k-colorable is NP-complete when $k \in \{3, 4\}$ [1]. Horev and Krakovski [15] proved that every plane graph of degree at most 3, other than K_4 and a subdivision of K_4 on five vertices, admits a polychromatic 3-coloring. Horev et al. [14] proved that every 2-connected cubic bipartite plane graph admits a polychromatic 4-coloring. This result is tight, since any such graph must contain a face of size 4.

1.1 Terminology and Notation

Let G be a graph. The vertex-set and the edge-set of G are denoted respectively by $V(G)$ and $E(G)$. The open neighborhood (resp. closed neighborhood) of a vertex v in graph G is denoted by $N_G(v)$ (resp. $N_G[v]$). The *degree* $d_G(v)$ of a vertex v in G is $|N_G(v)|$. A set of vertices I of a graph G is called an *independent set* if no two vertices in I are adjacent. A graph is *planar* if it can be embedded on the plane in such a way that no two edges cross. A plane graph G is a planar graph together with such an embedding. We denote the set of vertices lying on the boundary of a face f as $V(f)$. A cycle of length exactly k, at least k and at most k in G are respectively termed *k-cycle*, *k^+-cycle* and *k^--cycle*.

2 Proof of Theorem 1

In this section, we prove the following theorem, which is a restatement of Theorem 1 in the language of vertex coloring.

Theorem 2. *If G is a plane graph without isolated vertices, self-loops or 2-faces, then G has a 2-coloring which is simultaneously domatic and polychromatic.*

The main part of the paper is devoted to proving the next lemma from which Theorem 2 will follow easily. We call a 2-coloring of a plane graph *3^+-polychromatic* if every 3^+-face contains vertices of both colors.

Lemma 1. *If G is a connected plane graph without self-loops and with minimum degree at least two, then G has a 2-coloring which is simultaneously domatic and 3^+-polychromatic.*

To prove Lemma 1 we construct a supergraph G' from G by adding more edges to G without violating planarity. An edge uw is only added to G if it forms a facial triangle uvw where $\{uv, vw\} \subseteq E(G)$. We call the edges in $E(G)$ as *true edges* and those in $E(G') \setminus E(G)$ as *dummy edges*. Vertices u and v are called *true neighbors* in G' if uv is a true edge. We call a vertex v of G *happy* in G' if there exists a triangle uvw in G' where uv and vw are true edges. Otherwise we call v *unhappy* in G'. A face f in G is *happy* in G' if at least one of the faces of G' contained inside f is a 3-face. Otherwise we call it *unhappy* in G'. A *true angle at* v in G' is an angle between two cyclically consecutive true edges incident on v which connects v to two distinct vertices (ignoring any dummy edges between them). Let $\mathcal{G}$ be the family of all plane multigraphs that can be obtained by adding dummy edges to G which maximizes the sum of the number of happy vertices and the number of happy faces. Let G' be a graph in $\mathcal{G}$ with the smallest number of dummy edges. We make the following observations about G'.

Observation 1. *Every* 3^+*-face of* G *is happy in* G'.

Proof. Suppose f is a 3^+-face of G which is unhappy in G'. Since G is connected, the boundary B of f is a single closed walk. Further since G does not have self loops and f has length at least three, $V(f)$ has size at least 3. Hence we can make f happy by adding a dummy edge inside f between two distinct vertices in $V(f)$ which are at distance two along the walk B. This supergraph G'' of G' is also a planar supergraph of G and the number of faces of G which are happy in G'' is one more than the number of faces of G which are happy in G'. This violates the membership of G' in $\mathcal{G}$.

Remark 1. Notice that both the assumptions on G are necessary for Observation 1. Recall that if we allow self-loops, we can have 3^+-faces bounded by a single vertex, and such faces cannot be made happy. If we allow disconnected graphs, then a 4-face bounded between two 2-cycles cannot be made happy by adding a dummy edge.

Unlike 3^+-faces, we cannot guarantee that all the vertices are happy in G'. The next observation, even though technical, illustrates precisely what happens in the neighborhood of an unhappy vertex.

Observation 2. *If* v *is an unhappy vertex in* G'*, then*

1. *there is exactly one dummy edge incident on* v *through every true angle at* v*, and*
2. *each of these dummy edges makes exactly one true neighbor of* v *happy.*

Proof. Let D_v denote the set of dummy edges incident to v in G' and H_v denote the set of happy true neighbors of v. Since the minimum degree of G is at least two there are at least two true angles at v. If there are no dummy edges incident on v through a true angle uvw, we can add a dummy edge uw without violating planarity, making v happy. This will increase the number of happy vertices contradicting the membership of G' in $\mathcal{G}$. Since the number of true angles at v is at least $|N_G(v)| = d_G(v)$, we have $|D_v| \geq d_G(v)$. Further, since

$|H_v| \leq d_G(v)$, we have $|D_v| \geq |H_v|$. Every vertex in H_v needs at most one edge from D_v to become happy. No vertex outside H_v can be made happy by an edge in D_v. Hence, either if $|D_v| > |H_v|$ or if $|D_v| = |H_v|$ and one of the edges in D_v is making two true neighbors of v happy, then at least one edge in D_v is redundant. That is, we can remove this edge and still leave all the vertices in H_v happy. If this deletion does not leave a true angle at v without a dummy edge, then the happiness of the corresponding face is also intact. This contradicts the choice of G' as a smallest member in $\mathcal{G}$. On the other hand, if this deletion leaves a true angle at v without a dummy edge, we can make v happy as we did in the first case and restore the happiness of the corresponding face. This contradicts the membership of G' in $\mathcal{G}$. Hence $|D_v| = |H_v|$ and each edge in D_v makes exactly one vertex in H_v happy. Since $d_G(v)$ is sandwiched between $|H_v|$ and $|D_v|$, we also have $|H_v| = |D_v| = d_G(v)$.

The next observation is an easy restatement of the equality $|H_v| = d_G(v)$ that we established in Observation 2.

Observation 3. *Two unhappy vertices cannot be true neighbors in G'.*

Observation 4 (Key Observation). *G' has at most one unhappy vertex.*

Proof. Let v be an unhappy vertex and let u be any one of its true neighbors. From Observation 2, it is clear there is a dummy edge e incident to v which makes u happy (and helps no other vertex become happy). If we remove e, the only effect on happiness is that u becomes unhappy and the face f in G containing e may become unhappy. Moreover, one true angle at v, say uvw, becomes free. We can now add the dummy edge uw inside f to make v and f happy. So, the unhappiness of a vertex v can be shifted to any one of its true neighbors without creating any other unhappy vertices or faces. Note that this shifting does not increase the total number of dummy edges.

Suppose there were two unhappy vertices, say v, v' in G'. Since G is connected, there is a path $P = (v = x_0, \ldots, x_k = v')$ in G. Then we can do the above shifting of unhappiness repeatedly from x_i to x_{i+1} for $0 \leq i \leq k-2$. This shifting will end when x_{k-1} and v' are unhappy. This new graph also qualifies to be G' and it contradicts Observation 3. Hence, there can be at most one unhappy vertex in G'.

By the Four Color Theorem [2,3], there exists a proper 4-coloring $\phi : V[G'] \rightarrow \{1,2,3,4\}$ of G'. By Observation 4, we can assume without loss of generality that the unhappy vertex in G' (if one exists) gets color 1 and has a true neighbor of color 2. Obtain a 2-coloring ψ of G' by merging the color classes 1 and 3 to a single color class and 2 and 4 to a single color class different from the previous. This ensures that, in ψ, the unhappy vertex sees both colors of ψ in its closed neighborhood $N_G[v]$. Every happy vertex v is part of a triangle uvw in G' with uv, vw being true edges. Since u, v, w get three different colors in ϕ, $N_G[v] \supseteq \{u, v, w\}$ will contain vertices of both colors in ψ. By Observation 1, every 3^+-face in G contains a 3-face of G'. Hence every 3^+-face of G also sees both the colors in ψ. Hence ψ is a domatic as well as a 3^+-polychromatic 2-coloring of G. This completes the proof of Lemma 1.

To prove Theorem 2, we need to extend Lemma 1 to handle graphs with multiple components and pendant vertices. Let G be a plane graph without isolated vertices, self-loops or 2-faces. Trim the graph by recursively removing vertices of degree one to obtain a subgraph G' with no degree one vertex in it. Since we are only degree one vertices at every stage, we do not create any new components. G' can have only two types of components - components with minimum degree at least two or isolated vertices (trivial components).

Consider a nontrivial component H of G'. Given a (partial) 2-coloring of the vertices of G, a vertex v (resp. face f) is said to be *satisfied*, if $N_G[v]$ (resp. $V(f)$) contains at least one vertex of each color. By Lemma 1 we know that there exists a 2-coloring of H satisfying every 3^+-face and vertex of H. Notice that H may have 2-faces even though G does not.

We start with such a coloring for every nontrivial component in G' and color the isolated vertices in G' with any of the two colors. We then add back the deleted vertices one at a time in the reverse order as we deleted them. Every time we add back a vertex, we color it with the color different from that of its parent. This makes sure that every vertex added back and its parent are satisfied. Since there are no isolated vertices in G, every vertex of G is satisfied in the resulting coloring of G.

Now consider the faces. Since we are adding back degree one vertices to G' recursively to obtain G, we are not creating any new faces even though some of the faces may have a larger boundary due to the new edges added inside them. Note that the vertices which were on the boundary of any face in G' still remain in its boundary in G. So every face which was satisfied in G' is satisfied in this coloring of G. Every face in G' whose boundary contains a closed walk of length at least three is satisfied in G' since that closed walk is the boundary of a 3^+-face in one of the nontrivial components of G'. Let f be an unsatisfied face in G'. By the previous observation, f is bounded by a collection consisting of only 2-cycles and isolated vertices. Since G does not have any 2^--faces, f will either contain a pendant vertex of G inside it or $V(f)$ contains vertices from more than one component of G. In the first case, we colored this pendant vertex differently from its parent and hence f will be satisfied. In the second case, we can flip the colors of one of the interior components and all the components nested inside it, if needed, to ensure that f is satisfied. This completes the proof of Theorem 2.

3 Application to Matheson-Tarjan Conjecture

As a corollary of Theorem 1 we can give a conditional upper bound on the domination number of plane triangulations.

Corollary 1. *Every n-vertex (simple) plane triangulation G with an independent set of size at least αn has $\gamma(G) \leq (1-\alpha)n/2$.*

Proof. Consider a plane triangulation G on n vertices with an independent set I of size at least αn. We delete I from G to get a plane graph G'. Note that since G is a triangulation and every vertex is part of a triangle, removing I from G will not create any isolated vertices in G'. Moreover, G has no 2-faces and since G is a simple triangulation, every vertex removed from G has degree more than two. Hence G' has no 2-faces. Corresponding to every deleted vertex v, a unique face f_v is formed in G' where $V(f_v) = N_G(v)$. Note that $|V(G')| \leq (1 - \alpha)n$. By Theorem 1 we know that there is a face-hitting dominating set S in G' with $|S| \leq (1 - \alpha)n/2$. Since S is a dominating set in G' and G' is a subgraph of G, S dominates all the vertices of $V(G') \setminus S = V(G) \setminus (I \cup S)$ in G as well. Since S is a face-hitting set in G', it dominates every vertex of I in G. Hence S is a dominating set of G.

Corollary 1 proves Matheson-Tarjan Conjecture for plane triangulations which have an independent set of size at least $n/2$. Recall that Christiansen et al. [6] showed that every plane triangulation on $n > 10$ vertices has a dominating set of size at most $2n/7$. Corollary 1 improves this bound when $\alpha > 3/7$. This includes triangulations obtained by adding a new vertex inside every face of a planar graph whose average face length is slightly below $14n/3$ (and connecting it to the vertices on this face). Since a maximal independent set in a graph is a dominating set, our bound improves Christiansen et al.'s bound when G contains a maximal independent set of size either less than $2n/7$ or more than $3n/7$.

References

1. Alon, N., et al.: Polychromatic colorings of plane graphs. Discret. Comput. Geom. **42**(3), 421–442 (2009)
2. Appel, K., Haken, W.: Every planar map is four colorable. I. Discharging. Illinois J. Math. **21**, 429–490 (1977)
3. Appel, K., Haken, W., Koch, J.: Every planar map is four colorable. II. Reducibility. Illinois J. Math. **21**, 491–567 (1977)
4. Bose, P., Shermer, T., Toussaint, G., Zhu, B.: Guarding polyhedral terrains. Comput. Geom. **7**(3), 173–185 (1997)
5. Botler, F., Fernandes, C.G., Gutiérrez, J.: Independent dominating sets in planar triangulations. Electron. J. Comb. P2–12 (2024)
6. Christiansen, A.B., Rotenberg, E., Rutschmann, D.: Triangulations admit dominating sets of size 2 n/7. In: Proceedings of the 2024 Annual ACM-SIAM Symposium on Discrete Algorithms (SODA), pp. 1194–1240. SIAM (2024)
7. Claverol, M., et al.: Total domination in plane triangulations. Discret. Math. **344**(1), 112179 (2021)
8. Fomin, F.V., Thilikos, D.M.: Dominating sets in planar graphs: branch-width and exponential speed-up. SIAM J. Comput. **36**(2), 281–309 (2006)
9. Francis, P., Illickan, A.M., Jose, L.M., Rajendraprasad, D.: Disjoint total dominating sets in near-triangulations. J. Graph Theory **105**(1), 68–77 (2024)
10. Goddard, W., Henning, M.A.: Domination in planar graphs with small diameter. J. Graph Theory **40**(1), 1–25 (2002)

11. Haynes, T.W., Hedetniemi, S., Slater, P.: Fundamentals of Domination in Graphs. CRC Press (2013)
12. Heawood, P.J.: On the four-colour map theorem. Quart. J. Pure Appl. Math. **29**, 270–285 (1898)
13. Hoffmann, F., Kriegel, K.: A graph-coloring result and its consequences for polygon-guarding problems. SIAM J. Discret. Math. **9**(2), 210–224 (1996)
14. Horev, E., Katz, M.J., Krakovski, R., Nakamoto, A.: Polychromatic 4-coloring of cubic bipartite plane graphs. Discret. Math. **312**(4), 715–719 (2012)
15. Horev, E., Krakovski, R.: Polychromatic colorings of bounded degree plane graphs. J. Graph Theory **60**(4), 269–283 (2009)
16. Lovász, L.: Combinatorial Problems and Exercises. North Holland & Akadémiai Kiadó (1979)
17. MacGillivray, G., Seyffarth, K.: Domination numbers of planar graphs. J. Graph Theory **22**(3), 213–229 (1996)
18. Matheson, L.R., Tarjan, R.E.: Dominating sets in planar graphs. Eur. J. Combin. **17**, 565–568 (1996)
19. Reed, B.: Paths, stars and the number three. Comb. Probab. Comput. **5**(3), 277–295 (1996)
20. Sohn, M.Y., Xudong, Y.: Domination in graphs of minimum degree four. J. Korean Math. Soc. **46**(4), 759–773 (2009)
21. Špacapan, S.: The domination number of plane triangulations. J. Combin. Theory Ser. B **143**, 42–64 (2020)
22. Xing, H.M., Sun, L., Chen, X.G.: Domination in graphs of minimum degree five. Graphs Comb. **22**, 127–143 (2006)

The Parameterized Complexity Landscape of the Unsplittable Flow Problem

Robert Ganian, Mathis Rocton(✉), and Daniel Unterberger

Algorithms and Complexity Group, TU Wien, Vienna, Austria
mrocton@ac.tuwien.ac.at

Abstract. We study the well-established problem of finding an optimal routing of unsplittable flows in a graph. While by now there is an extensive body of work targeting the problem on graph classes such as paths and trees, we aim at using the parameterized paradigm to identify its boundaries of tractability on general graphs. We develop novel algorithms and lower bounds which result in a full classification of the parameterized complexity of the problem with respect to natural structural parameterizations for the problem—notably maximum capacity, treewidth, maximum degree, and maximum flow length. In particular, we obtain a fixed-parameter algorithm for the problem when parameterized by all four of these parameters, establish XP-tractability as well as W[1]-hardness with respect to the former three and latter three parameters, and all remaining cases remain paraNP-hard.

1 Introduction

In the Unsplittable Flow problem, we are given an undirected edge-weighted graph G and a set T of *tasks*, where each task consists of a pair of vertex endpoints and two non-negative integers: a *demand* and a *profit*. The goal is to select a subset T' of the tasks and construct a set P of paths such that:

- each task in T' is assigned a path in P which connects its two endpoints;
- for each edge e in G, the total demand of tasks routed via P through e does not exceed the weight of e (i.e., its *capacity*); and
- among all subsets of tasks and sets of paths satisfying the above two conditions, the sum of the profits of tasks in T' is maximized.

Unsplittable Flow has been extensively studied in the literature, and has applications in areas such as resource allocation and scheduling [11,43], multi-commodity routing [14] and caching [15]. It is also known to be notoriously

The authors acknowledge support from the Austrian Science Foundation (FWF, Project 10.55776/Y1329) and the Vienna Science and Technology Fund (WWTF, Project 10.47379/ICT22029). M. Rocton also acknowledges support from the European Union's Horizon 2020 COFUND programme (LogiCS@TUWien, No. 101034440).

D. Kráľ and M. Milanič (Eds.): WG 2024, LNCS 14760, pp. 220–235, 2025.
https://doi.org/10.1007/978-3-031-75409-8_16

intractable. First of all, UNSPLITTABLE FLOW is easily seen to be weakly NP-hard even when G is restricted to be a K_2 via a direct reduction from KNAPSACK—in other words, if we allow the edge capacities to be encoded in binary, then we cannot hope to even obtain an efficient exact algorithm for a single edge. Naturally, in many cases of interest one need not deal with exceedingly large edge capacities, and so in this article we focus solely on the setting where the capacities are encoded in **unary**. While the case of $G = K_2$ is trivial in the unary setting, the problem still remains far from tractable: it is known to be strongly NP-hard on paths even if all tasks have a demand of at most 3 and all edges have the same capacity [9].

Many of the more recent theoretical works targeting UNSPLITTABLE FLOW restrict their attention to specific graph classes such as paths [2, 4, 10, 27–30, 45] or trees [1, 13, 40], and typically deal with the inherent intractability of the problem by aiming for approximation algorithms, parameterized algorithms, or a combination thereof on these graph classes. In this article, we ask a different question: under which conditions can we circumvent the classical intractability of finding an optimal unsplittable flow on **general graphs**? In particular, under which structural parameterizations of the input can we obtain exact fixed-parameter or at least XP algorithms for UNSPLITTABLE FLOW?

Contribution. Since UNSPLITTABLE FLOW generalizes the classical EDGE DISJOINT PATHS problem[1], a natural starting point for our investigation would be to consider the combined parameter of *treewidth* tw [44] plus *maximum degree* Δ—indeed, EDGE DISJOINT PATHS is known to be fixed-parameter tractable under the combined parameterization of $\mathrm{tw} + \Delta$ [24, 25]. Unfortunately, the aforementioned NP-hardness of the problem on paths [9] immediately rules out even XP algorithms under this parameterization. In combination with the known NP-hardness of EDGE DISJOINT PATHS on the class $K_{3,n}$ of complete bipartite graphs [20], one can rule out tractability under all known structural parameterizations of the input graph alone, including *treedepth* [42], the *vertex cover number* [6, 8, 37], *treecut-width* [22, 46, 47], and the *feedback edge number* [3, 12, 34].

However, even in the considered setting of unary-encoded capacities, the NP-hardness reduction on paths inherently requires the edge capacities to be sufficiently large (in contrast to the demands, which are small constants). As our first result, we show that if one also parameterizes by the maximum edge capacity in the graph, UNSPLITTABLE FLOW becomes XP-tractable:

Theorem 1. *UNSPLITTABLE FLOW is in* XP *parameterized by the maximum capacity c, the treewidth* tw *and the maximum degree Δ of the input graph.*

The proof of the above theorem relies on dynamic programming, where we utilize records that store information about how the selected tasks are routed through the given separator (i.e., bag in the decomposition), whereas each task

[1] EDGE DISJOINT PATHS is equivalent to seeking a routing of all tasks in UNSPLITTABLE FLOW under the restriction that all demands, capacities and profits are 1.

may of course also visit the bag multiple times. Interestingly, we use trimmed-down records which suppress seemingly crucial information about such routings—in particular the order in which each selected task visits the bag—to achieve a running time with only single-exponential dependency on each of the parameters.

Next, we show that Theorem 1 is essentially tight. On one hand, we already noted that one cannot drop c from the parameterization, and similarly one can notice that dropping tw or Δ is ruled out by the NP-hardness of EDGE DISJOINT PATHS on grids [41] and $K_{3,n}$ [20]. Naturally, the above still leaves open the existence of a fixed-parameter algorithm under the same parameterization. As our second result, we provide a non-trivial reduction from MULTICOLORED CLIQUE which excludes such an algorithm under standard complexity assumptions. Surprisingly, our reduction even rules out fixed-parameter tractability when restricted to the well-studied class of paths.

Theorem 2. *UNSPLITTABLE FLOW is* W[1]-hard *parameterized by the maximum capacity c, even when restricted to paths.*

While the above provides a seemingly complete complexity-theoretic picture of UNSPLITTABLE FLOW under the considered parameterizations (as well as other structural graph parameters; see also the discussion at the beginning of Sect. 4), the situation is still somewhat unsatisfactory: is there no hope for obtaining a fixed-parameter algorithm by exploiting the structural properties of inputs? In the second part of our article, we identify a restriction on the problem—notably, the maximum length ℓ of the paths that can be used to route the tasks—which turns out to yield fixed-parameter tractability of the problem. We remark that such a restriction is far from unnatural: unsplittable flows with a prescribed bound on their length have already been proposed and studied in the literature [36,39], and avoiding long routes is critical in several application settings (consider, e.g., the routing of perishable goods or the allocation of bandwidth in a network with communication time/delay constraints).

As our third result, we establish fixed-parameter tractability for UNSPLITTABLE FLOW when additionally parameterized by ℓ:

Theorem 3. *UNSPLITTABLE FLOW is fixed-parameter tractable parameterized by the maximum capacity c, the treewidth* tw*, the maximum degree Δ of the input graph and the maximum length ℓ of any admissible flow route.*

While the proof of Theorem 3 also relies on leaf-to-root dynamic programming (as the vast majority of tree-width based algorithms), it is non-standard in the sense that at each point where we compute new information for our dynamic programming table, we need to invoke a separate dynamic-programming subprocedure as a second layer of the computation. Moreover, the algorithm developed for the proof of Theorem 3 also establishes XP-tractability for the problem when one drops c in the parameterization:

Theorem 4. *UNSPLITTABLE FLOW is in* XP *parameterized by the treewidth* tw *and the maximum degree Δ of the input graph along with the maximum length ℓ of any admissible flow route.*

The above shows that ℓ can essentially "replace" c in the parameterization of Theorem 1 in order to circumvent the NP-hardness of UNSPLITTABLE FLOW. On the other hand, the problem remains NP-hard even for fixed values of $c + \text{tw} + \ell$ (due to the aforementioned NP-hardness of EDGE DISJOINT PATHS on $K_{3,n}$ [20]) and also for fixed values of $c + \Delta + \ell$ (due to a different, very recent NP-hardness proof for EDGE DISJOINT PATHS on grids with bounded-length paths [19]). Finally, we complement Theorem 4 with a straightforward reduction from UNARY BIN PACKING [32] that rules out fixed-parameter tractability under the same parameterization as Theorem 4.

Theorem 5. *UNSPLITTABLE FLOW is* W[1]-hard *parameterized by the treewidth* tw *and the maximum degree* Δ *of the input graph along with the maximum length* ℓ *of any admissible flow route.*

Our results are summarized in Fig. 1.

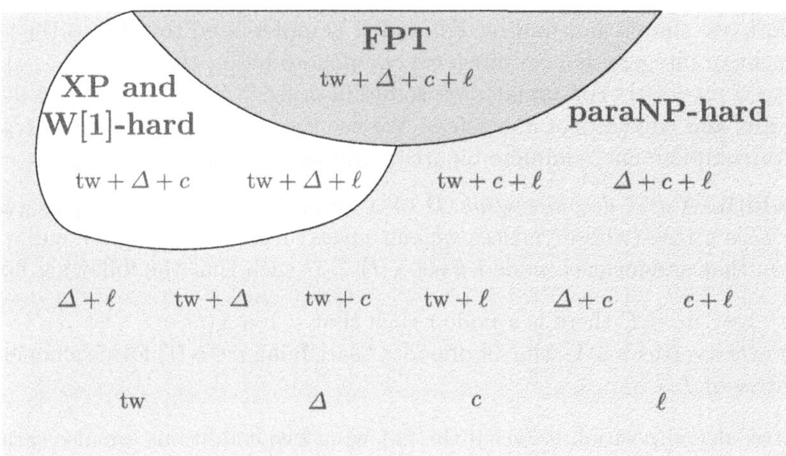

Fig. 1. The complexity landscape of UNSPLITTABLE FLOW under structural parameterizations. Here tw, Δ, c and ℓ denote the treewidth, maximum degree, maximum capacity and maximum length of an admissible route, respectively. A discussion of the (non-)applicability of other major structural parameters is provided in Sect. 4.

Related Work. As mentioned earlier, much of the previous work on UNSPLITTABLE FLOW targeted approximate solutions on paths and trees. Chrobake, Woeginger, Makino and Xu showed that the problem is strongly NP-hard even on paths with uniform profits and capacities [15]. Anagnostopoulos, Grandoni, Leonardi and Wiese obtained a PTAS for UNSPLITTABLE FLOW on paths [2], and this was later improved to a $(1 + \varepsilon)$ PTAS by Grandoni, Momke and Wiese [28].

UNSPLITTABLE FLOW on trees is a generalization of the problem of finding integral multi-commodity flows on trees, which is already APX-hard [26].

Martínez-Muñoz and Wiese obtained exact and approximate fixed-parameter algorithms for the problem on trees when the solution size is included in the parameterization [40]. Chekuri, Ene and Korula showed that UNSPLITTABLE FLOW on trees admits a $\mathcal{O}(\log n)$ approximation when all weights are identical, and $O(\log^2 n)$ otherwise [13].

On general graphs, Guruswami, Khanna, Rajaraman, Shepherd, and Yannakakis established the NP-hardness of approximating UNSPLITTABLE FLOW within a factor of $m^{\frac{1}{2}-\varepsilon}$ for any $\varepsilon > 0$ [31]. Approximation algorithms for UNSPLITTABLE FLOW on general graphs have also been studied, e.g., in the earlier works of Kolman and Scheideler [35,36] or Baveja and Srinivasan [5].

2 Preliminaries

We use standard graph terminology [17] and assume familiarity with the foundations of parameterized complexity theory, including the complexity classes FPT, XP and W[1], and parameterized reductions [16,18]. All graphs considered in this work are simple and undirected, and $\mathbb{N}$ is understood to refer to the set of non-negative integers. For an integer n, we denote by $[n]$ the set $\{1, \ldots, n\}$ and set $[n]_0 = [n] \cup \{0\}$. For brevity, we sometimes use $V(G)$ and $E(G)$ to denote the vertex and edge sets of a graph G. We use the standard notation of $K_\bullet$ and $K_{\bullet,\circ}$ for complete and complete bipartite graphs, respectively.

Treewidth. A *tree-decomposition* $\mathcal{D}$ of a graph $G = (V, E)$ is a pair (D, χ), where D is a tree (whose vertices we call *nodes*) rooted at a node r and χ is a function that assigns each node t a set $\chi(t) \subseteq V$ such that the following holds:

- For every $uv \in E$ there is a node t such that $u, v \in \chi(t)$.
- For every vertex $v \in V$, the set of nodes t satisfying $v \in \chi(t)$ forms a nonempty subtree of T.

A tree-decomposition is *nice* if the following two conditions are also satisfied:

- $|\chi(\ell)| = 1$ for every leaf ℓ of T and $|\chi(r)| = 0$.
- There are only three kinds of non-leaf nodes in T:
  - **Introduce node:** a node t with exactly one child t' such that $\chi(t) = \chi(t') \cup \{v\}$ for some vertex $v \notin \chi(t')$.
  - **Forget node:** a node t with exactly one child t' such that $\chi(t) = \chi(t') \setminus \{v\}$ for some vertex $v \in \chi(t')$.
  - **Join node:** a node t with two children t_1, t_2 where $\chi(t) = \chi(t_1) = \chi(t_2)$.

The set $\chi(t)$ for a node t of D is called a *bag*. The *width* of a tree-decomposition (D, χ) is the size of its largest bag D minus 1, and the *treewidth* of the graph G, denoted $\mathrm{tw}(G)$, is the minimum width of a tree-decomposition of G. It is known that a tree-decomposition can be transformed into a nice tree-decomposition of the same width in linear time. Efficient fixed-parameter algorithms are known for computing a nice tree-decomposition of near-optimal width [7,33,38]:

Proposition 6 ([38]). *There exists an algorithm which, given an n-vertex graph G and an integer k, in time* $2^{\mathcal{O}(k)} \cdot n$ *either outputs a tree-decomposition of G of width at most* $2k+1$ *and* $\mathcal{O}(n)$ *nodes, or determines that* $\mathrm{tw}(G) > k$.

We let D_t denote the subtree of D rooted at a node t, and we use $\chi(D_t)$ to denote the set $\bigcup_{t' \in V(D_t)} \chi(t')$. In the context of leaf-to-root dynamic programming, it will be useful to also define *past*$(t) = \chi(D_t) \setminus \chi(t)$ and *future*$(t) = V \setminus$ *past*(t); to avoid any confusion, we remark that the latter set also contains $\chi(t)$. We also define the edge set *present*(t) containing all edge with precisely one endpoint in *past*(t). These will perform the role of the "boundary" in all of our dynamic programming algorithms, and we observe that $|$*present*$(t)|$ is upper-bounded by a product of $|\chi(t)|$ and the maximum degree Δ of the input graph.

Problem Formulation. For complexity-theoretic reasons and without loss of generality, in this article we formalize UNSPLITTABLE FLOW as a decision problem as opposed to the optimization variant typically considered in works focusing on approximation. To avoid any confusion, we remark that each of the developed algorithms is exact, deterministic and constructive in the sense that it can also output a suitable witness for every yes-instance—and hence fully applicable also to the optimization variant defined in the introduction.

An instance $\mathbb{I}$ of the UNSPLITTABLE FLOW problem studied here consists of:

1. a simple undirected graph $G = (V, E)$,
2. an edge labeling function $\gamma : E \to \mathbb{N}$,
3. a set $\mathcal{T}$ of *tasks* $T \subseteq V \times V \times \mathbb{N} \times \mathbb{N}$,
4. a *profit target* τ, and
5. an integer length bound ℓ.

We call $\gamma(e)$ the *capacity* of the edge e, and for a task $p = (s, t, d, w) \in T$ we call $d = d(p)$ its demand and $w = w(p)$ its profit. Given a subset $T' \subseteq T$ of tasks, we say that a *flow routing* is a mapping which assigns to each task (s, t, d, w) in T' an s-t path of length at most ℓ such that for each edge $e \in E$, the total demand of all paths in the flow routing passing through e is at most $\gamma(e)$. The *profit* of T' is simply the sum of the profits of the tasks in T', i.e., $\sum_{(s,t,d,w) \in T'} w$. The UNSPLITTABLE FLOW problem then asks us to decide whether there exists a subset T' which admits a flow routing and achieves a profit of at least τ.

We note that ℓ is not present in most definitions of UNSPLITTABLE FLOW in the literature, and indeed our first set of results—in Sect. 3—provide lower bounds for the case where ℓ is ignored (i.e., set to a sufficiently large bound such as $\ell = |V|$) along with algorithms that can deal with any choice of ℓ. The size $|\mathbb{I}|$ of an instance $\mathbb{I}$ is simply the sum of the sizes of its components, whereas we assume throughout this work that the capacity function γ is encoded in unary. This condition may be reformulated as requiring that $|\mathbb{I}|$ upper-bounds the maximum capacity of an edge, and is necessary in the sense that UNSPLITTABLE FLOW with binary-encoded capacities is trivially NP-hard even on a K_2 (see Sect. 1).

3 Solving Unsplittable Flow Regardless of Flow Length

This section is dedicated to the complexity of UNSPLITTABLE FLOW when the bound ℓ on the length of the flow routes is not part of the parameterization. As our first result, we prove Theorem 1, which is the only case that is not paraNP-hard in the part of our landscape that excludes ℓ (cf. Fig. 1).

Theorem 1. *UNSPLITTABLE FLOW is in* XP *parameterized by the maximum capacity c, the treewidth* tw *and the maximum degree Δ of the input graph.*

Proof Sketch. We first invoke Proposition 6 to obtain a nice tree-decomposition $\mathcal{D} = (D, \chi)$ of $G = (V, E)$ of width k, where $k \leq 2\text{tw} + 1$. Our proof utilizes a dynamic programming routine that proceeds in a leaf-to-root fashion along $\mathcal{D}$.

We then begin by defining the syntax of the records $R(t)$ we will use for each node t. $R(t)$ contains a set of tuples of the form $(\Lambda, \Theta, \Omega)$, where:

- $\Lambda : \mathit{present}(t) \to \{S, \sum_{t \in S} d(t) \leq c\}$,
- $\Theta : \left(\bigcup_{e \in \mathit{present}(t)} \Lambda(e)\right) \to [\ell]$, and
- $\Omega \in \mathbb{N}$.

Intuitively, these records capture the following information about a hypothetical solution: Λ specifies which tasks are routed through *present*(t), Θ captures their total flow length in *past*(t), and Ω contains information about their profit. We explicitly remark that, (1) unlike one would expect in a similar dynamic program, Λ does not store the order in which the selected tasks cross the boundary *present*(t), and (2) Ω will not be formalized as a simple sum over the profits of all routed tasks but rather a "weighted sum". To simplify our later operations involving Θ, we use $\Lambda(t)$ as shorthand for $\left(\bigcup_{e \in \mathit{present}(t)} \Lambda(e)\right)$, i.e., the set of all tasks mapped by Λ to at least one edge in *present*(t).

In order to formalize the above intuition by defining the semantics of the records, we first need to introduce the notion of a partial solution. A *partial solution* at a node t is a mapping ζ from a subset $Z \subseteq T$ of tasks to a multiset $\mathcal{P}$ of paths—composed entirely of edges with at least one endpoint in *past*(t)—with the following properties:

- for each $z \in Z$ such that both endpoints lie in the future (i.e., in *future*(t)), z is mapped to a (possibly empty) set of vertex disjoint paths whose extremities are the vertices of $\chi(t)$.
- for each $z \in Z$ such that exactly one endpoint u lies in the past (i.e., in *past*(t)), z is mapped to a set of vertex disjoint paths, where exactly one of these connects u to $\chi(t)$ and all of the (possibly 0) others have their extremities in $\chi(t)$.
- for each $z \in Z$ such that both of its endpoints u and v lie in the past, z is either mapped to a single path between u and v, or to a set of vertex disjoint paths such that one of these connects u to $\chi(t)$, another one connects v to $\chi(t)$, and all of the (possibly 0) others have their extremities in $\chi(t)$.
- for each edge $e \in E(G)$, its capacity $\gamma(e)$ is sufficient to accomodate the demand of all tasks of Z containing a path using e.

For a given partial solution $\zeta : Z \mapsto \mathcal{P}$, we define its *partial profit* ω as the sum of the profits of all tasks in Z with both endpoints in *past*(t) plus half of the profits of all tasks in Z with exactly one endpoint in *past*(t). A task is *active* in ζ if it is routed to at least one edge in *present*(t), and we note that there can be at most $|present(t)| \cdot c \leq \text{tw} \cdot \Delta \cdot c$ active tasks. We denote the set of all active tasks in ζ as Z_{act}. For each $z \in Z$, we let θ_z be the total number of edges in $\zeta(z)$.

We are now ready to formalize the semantics of our records. A tuple $(\Lambda, \Theta, \Omega)$ is *admissible* for $R(t)$ if and only if there exists a corresponding partial solution $\zeta : Z \rightarrow \mathcal{P}$ which

1. achieves a partial profit of Ω,
2. satisfies $\theta_z = \Theta(z)$ for each task $z \in Z_{act}$, and
3. for each edge $e \in present(t)$ and task z in Z, $z \in \Lambda(e)$ if and only if e occurs on a path in $\zeta(z)$.

While being admissible is the main prerequisite for being part of our records, we also need to impose two additional conditions which will be important when dealing with join nodes. For a vertex $v \in \chi(t)$, a mapping Λ as above and a task z, we say that v is a *jump-vertex* if v has two incident present edges which are both mapped by Λ to z. We say that a tuple $(\Lambda', \Theta', \Omega)$ *supersedes* a tuple $(\Lambda, \Theta, \Omega)$ if there is a task z such that (a) Λ' can be obtained from Λ by removing z from the edges incident to some set of jump-vertices, and (b) Θ' can be obtained from Θ by reducing the value of $\Theta(z)$ by a positive integer.

We are now ready to define our records: a tuple $(\Lambda, \Theta, \Omega)$ is in $R(t)$ if and only if it is admissible and moreover satisfies the following two Optimality Conditions:

(i) there is no admissible tuple of the form $(\Lambda, \Theta, \Omega')$ such that $\Omega' > \Omega$, and
(ii) there is no admissible tuple that supersedes $(\Lambda, \Theta, \Omega)$.

In other words, the optimality conditions ensure that we do not keep solutions which (i) are clearly sub-optimal or (ii) can achieve the same connections outside of *past*(t) without requiring longer paths.

Having defined our records, we observe that the total size of $R(t)$ is upper-bounded by $|\mathbb{I}|^{\mathcal{O}(k \cdot \Delta \cdot c)}$ simply due to the syntax: for each of the at most $k \cdot \Delta$ edges in *present*(t), there are at most $|\mathbb{I}|^c$ choices of tasks for Λ and at most $|\mathbb{I}|$ choices for Θ. Moreover, since the bag of the root r of $\mathcal{D}$ is empty, we have that $present(r) = \emptyset$; hence $R(r)$ contains a single entry of the form $(\emptyset, \emptyset, \Omega_r)$ where Ω_r is the maximum profit that can be achieved in $\mathbb{I}$. This means that to solve our problem, it now suffices to show how to compute $R(t)$ for each node t in $\mathcal{D}$.

If t is a leaf node, we recall that $present(t) = \emptyset$ and hence $R(t) = \{(\emptyset, 0, \emptyset\}$. For introduce nodes, we observe that the partial solutions and records remain exactly the same as in the child. While the computation for forget nodes is non-trivial, the most interesting case to handle are join nodes, since there it turns out that combining two partial solutions may create closed cycles and thus not result in a partial solution (see Fig. 2). While one could resolve this issue by additionally keeping track of the order in which the present edges are visited in our records (resulting in an additional logarithmic factor in the exponent), careful use of the Optimality Conditions allows us to circumvent it while maintaining a single-exponential running time dependency on our parameters.

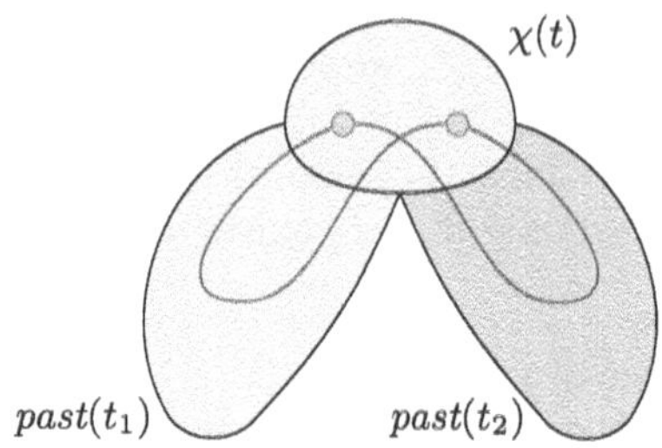

Fig. 2. A join node t with two children t_1, t_2, where two paths between the red vertices—one in each of the partial solutions for t_1 and t_2—create a cycle. By combining a pair of tuples $(\Lambda_1, \Theta_1, \Omega_1) \in R(t_1)$ and $(\Lambda_2, \Theta_2, \Omega_2) \in R(t_2)$ which represent partial solutions containing such paths, the algorithm could produce a tuple which does not represent any partial solution at t. (Color figure online)

The size of $R(t)$ is upper-bounded by $|\mathbb{I}|^{\mathcal{O}(k \cdot \Delta \cdot c)}$, the time required to compute the set of records at each node t of $\mathcal{D}$ is upper-bounded by $|R(t)|^2$, and this computation needs to be performed at most $|\mathbb{I}|$ many times. Hence we can upper-bound the running time of the algorithm by $|\mathbb{I}|^{\mathcal{O}(k \cdot \Delta \cdot c)}$. □

Recalling the discussion in Sect. 1, we note that dropping any of the three parameters listed in Theorem 1 results in paraNP-hardness: for c, tw and Δ this follows by the paraNP-hardness of UNSPLITTABLE FLOW on paths [9], grids with edge-capacity 1 [41] and $K_{3,n}$ with edge-capacity 1 [20], respectively. Hence, the main complexity-theoretic question that remains for the aforementioned three parameters is to exclude the existence of a fixed-parameter algorithm (under standard complexity-theoretic assumptions). We do so by establishing a stronger claim: UNSPLITTABLE FLOW is W[1]-hard parameterized by c even in the well-studied setting [2,15,28] where the input graph is a path. We remark that all of the aforementioned lower-bound results (including Theorem 2) hold even if $\ell = |V|$, i.e., when the additional constraint on the length of the paths is ignored.

Theorem 2. *UNSPLITTABLE FLOW is* W[1]-hard *parameterized by the maximum capacity c, even when restricted to paths.*

Proof Sketch. We reduce from the classical W[1]-hard MULTI-COLORED CLIQUE problem (MCC) parameterized by the number of colors [16,18].

Let $G = (V = (V_1, \ldots, V_k), E)$ be an instance of MCC, and assume without loss of generality that k is even; indeed, if k were odd then we could produce an equivalent instance of MCC by adding a new part V_{k+1} containing a single vertex that is adjacent to all other vertices. Let $(v_1, \ldots, v_n)$ be an ordering on V which respects the partitioning of vertices into the color classes of G—specifically, we require there to exist $i_0 = 0, i_1, \ldots, i_k$ with $V_j = \{v_{i_{j-1}+1}, \ldots v_{i_j}\}$ for all $j \in [k]$.

We construct an instance $\mathbb{I}$ of UNSPLITTABLE FLOW as follows.

First, we create a path $P = (s_1, \ldots, s_k, v_{1s}, v_{1t}, \ldots, v_{ns}, v_{nt}, t_1, \ldots, t_k)$, consisting of $2k$ control vertices at the start and end (forming the subpaths $s_1, \ldots, s_k$ and $t_1, \ldots, t_k$) and 2 consecutive vertices of the form v_{js}, v_{jt} for each $v_j \in V$ which appear in the order constructed above. We denote the corresponding partition of $\{v_{jr} | 1 \leq j \leq n, r \in \{s,t\}\}$ into color classes as $(V'_1, \ldots, V'_k)$.

We set the capacity of each of the edges between control vertices as follows: $c(s_i, s_{i+1}) = c(t_{k-i}, t_{k-i+1}) := \sum_{j=1}^{i} \max(0, k - 2j + 1)$. For all other edges, we set the capacity to p^2 where $p = 0.5k$.

We set τ to equal the sum of the capacities of all edges in P, and set $\ell := |V(P)|$ (or to any other sufficiently large number). Finally, we create the following tasks:

- $T_{sj} = (s_i, v_{js}, \max(0, k - 2i + 1), \bullet)$ for all $v_j \in V_i$,
- $T_i = (v_{is}, v_{it}, \max(k - i, i - 1), \bullet)$ for all $v_i \in V$,
- $T_{tj} = (v_{jt}, t_i, \max(0, 2i - k - 1), \bullet)$ for all $v_j \in V_i$,
- $T_{ij} = (v_{it}, v_{js}, 1, 1)$ for all $v_i v_j \in E$ with $i < j$ and $c(v_i) \neq c(v_j)$,

where the profit of each task is set to its demand times the number of edges between its two endpoints. Observe that since τ is the sum of all edge capacities in P, the instance $\mathbb{I}$ constructed above is a yes-instance if and only if there exists a set of tasks which together use up all of the capacity of all edges in P; indeed, each task provides precisely as much profit as the amount of capacity it takes from $\mathbb{I}$. We call tasks of the form T_{ij} *edge tasks.*

Intuitively, the reduction works by encoding a solution to MCC via a collection of tasks which have endpoints in precisely one consecutive pair of vertices in each V'_i, $i \in [k]$, as illustrated in Fig. 3. To complete the proof, it suffices to formally establish that G contains a multicolored clique if and only if there is a flow on P with total profit τ. □

4 The Missing Ingredient for Fixed-Parameter Tractability

Our results in Sect. 3 show that not even parameterizing by a combination of the treewidth tw, maximum degree Δ and maximum capacity c suffice to achieve fixed-parameter tractability for UNSPLITTABLE FLOW. At this point, one may wonder whether it would be possible to achieve fixed-parameter tractability by strengthening the structural restrictions on the input graph—in particular by replacing treewidth with a different natural graph parameter such as treedepth [42], the vertex cover number [6,8,37], treecut-width [22,46,47], and the feedback edge number [3,12,34]. Recalling that in Sect. 1 we already ruled out any sort of tractability for UNSPLITTABLE FLOW under such parameterizations *alone*, below we also rule out fixed-parameter tractability when they are combined with c:

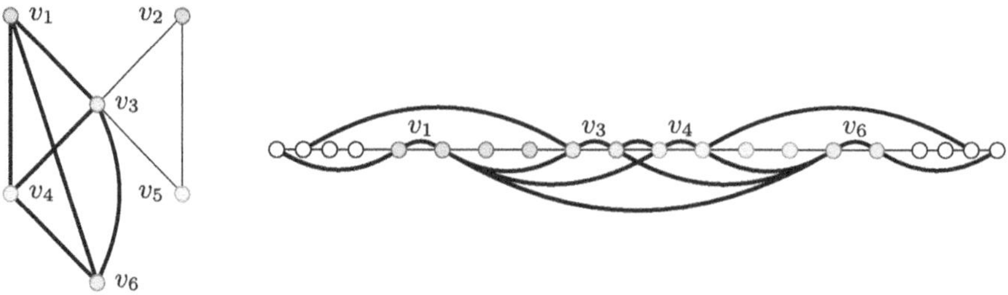

Fig. 3. An illustration of the construction used in the proof of Theorem 2. On the left, we depict an instance of MULTI-COLORED CLIQUE with a solution highlighted in bold. On the right, we show the corresponding instance of UNSPLITTABLE FLOW on the path (the horizontal edges), where the tasks corresponding to the highlighted clique are depicted via bold curves. For ease of presentation, we omit the tasks not chosen in this particular solution from the figure.

- For treedepth and the vertex cover number, it is futile to combine these parameters with Δ as this also bounds the size of the graph. On the other hand, UNSPLITTABLE FLOW parameterized by c plus either of these parameters remains NP-hard due to the previously-mentioned NP-hardness on $K_{3,n}$ [20].
- When parameterizing by treecut-width or the feedback edge number in combination with c and Δ, we can rule out fixed-parameter tractability directly via the established W[1]-hardness on paths (Theorem 2). The same also holds for the recently introduced "slim" variants of tree-cut width [21,23].

But while it seems essentially impossible to achieve fixed-parameter tractability via placing stronger restrictions on the input graph, as the main contribution of this section we show that adding the bound ℓ on the length of admissible flow routes into the parameterization allows us to circumvent the W[1]-hardness of the problem. Indeed, our aim is to prove:

Lemma 3. *UNSPLITTABLE FLOW can be solved in time* $|\mathbb{I}| \cdot (c+1)^{\mathcal{O}(\mathrm{tw}^2 \cdot \Delta^{2\ell})}$.

Before proceeding to the proof of Lemma 3, we observe that it immediately implies the two remaining algorithmic upper bounds depicted in Fig. 1:

Theorem 4. *UNSPLITTABLE FLOW is in* XP *parameterized by the treewidth* tw *and the maximum degree Δ of the input graph along with the maximum length ℓ of any admissible flow route.*

Theorem 5. *UNSPLITTABLE FLOW is* W[1]-hard *parameterized by the treewidth* tw *and the maximum degree Δ of the input graph along with the maximum length ℓ of any admissible flow route.*

Proof Sketch for Lemma 3. We begin by invoking Proposition 6 to compute a nice tree-decomposition (D, χ) of G of width k, where $k \leq 2\mathrm{tw} + 1$. Our proof utilizes a leaf-to-root dynamic programming algorithm working on D, which for each node computes partial solutions for the problem—however the records and the procedures of this dynamic program are entirely different from those used in the proof of Theorem 1.

For a node $t \in V(D)$, we define the set $vis(t)$ of *visible vertices* as the set containing all vertices of G that have distance at most ℓ from $\chi(t)$. The set $E_{vis}(t)$ of *visible edges* then contains all edges of G whose both endpoints are visible vertices. We note that for each node t, $|vis(t)| \leq |\chi(t)| \cdot \Delta^{\ell} \leq \mathcal{O}(\text{tw} \cdot \Delta^{\ell})$, and for visible edges we analogously obtain $|E_{vis}(t)| \leq \mathcal{O}(\text{tw}^2 \cdot \Delta^{2\ell})$.

To better keep track of the different tasks at a node t, we will partition them into two categories: a task is *active* if both of its endpoints lie in $past(t)$, and *future* if it has at least one endpoint in $future(t)$. A *partial solution* at a node t is a mapping ζ from a subset Z of active tasks to paths of length at most ℓ in G such that, for any edge e, its capacity is high enough to accomodate the flow routed through it in the partial solution, i.e., $\mu_{\zeta}(e) := \sum_{z \in Z, e \in \zeta(z)} d(z) \leq \gamma(e)$. The *profit* of ζ is then simply the sum of the profits of tasks in Z, i.e., $\sum_{z \in Z} w(z)$.

The syntax of our records is defined as follows: for each node t, $R(t)$ is a set of tuples of the form (Λ, Ω) where $\Lambda : E_{vis}(t) \to [c]_0$ and $\Omega \in \mathbb{N}$. Intuitively, Λ captures the capacity that has already been consumed by routed active tasks on each visible edge, while Ω is the highest profit that can be achieved for a given Λ. Formally, we define the semantics of $R(t)$ by stipulating that $(\Lambda, \Omega) \in R(t)$ if and only if there exists a partial solution ζ at t with the following properties:

1. $\forall e \in E_{vis}(t), \mu_{\zeta}(e) = \Lambda(e)$, and
2. the profit of ζ is exactly Ω, and
3. every other partial solution ζ' such that $\forall e \in E_{vis}(t), \mu'_{\zeta}(e) = \Lambda(e)$ achieves a profit that is upper-bounded by Ω.

For the root node r of D, we observe that all tasks are active and partial solutions are thus routings of an arbitrary subset of all tasks, and that $E_{vis}(r) = \emptyset$. By the definition of our semantics, we have only a single entry $(\emptyset, \Omega) \in R(r)$ where Ω is the highest profit that can be achieved by any flow routing in $\mathbb{I}$. Hence if we can compute $R(r)$, it suffices to compare this value with the target τ to solve $\mathbb{I}$. Moreover, computing $R(t)$ for a leaf node t is trivial: all tasks are future and hence $R(t) = \{(\emptyset, 0)\}$. Thus, it remains to show how the algorithm can compute the records in a leaf-to-root fashion when t is an introduce, forget or join node.

The main difficulty our algorithm needs to overcome is that for forget and join nodes, the number of "new" active tasks may be very large and hence we cannot exhaustively branch to determine how all of these are routed simultaneously. To deal with this somewhat unusual obstacle, we apply a second layer of dynamic programming in order to compute $R(t)$, where we update the records gradually while processing the new active tasks in an iterative fashion.

Finally, we note that the running time bound can be obtained by verifying that $|R(t)|$ is always upper-bounded by $(c+1)^{|E_{vis}(t)|} \leq (c+1)^{\mathcal{O}(\text{tw}^2 \cdot \Delta^{2\ell})}$. □

As our final result, we complement Theorem 4 by ruling out fixed-parameter tractability under the same parameterization. The proof uses a straightforward reduction from UNARY BIN PACKING parameterized by the number of bins [32].

Theorem 6. *UNSPLITTABLE FLOW is* W[1]-hard *parameterized by the treewidth* tw *and the maximum degree* Δ *of the input graph along with the maximum length* ℓ *of any admissible flow route.*

5 Concluding Remarks

While our results highlight that UNSPLITTABLE FLOW remains highly intractable even in severely restricted settings, they also provide some of the first algorithmic results for the problem on general graphs. Combined, the results also paint a surprisingly rigorous and comprehensive picture of the complexity landscape of UNSPLITTABLE FLOW. For instance, in addition to the discussion at the beginning of Sect. 3, one may note that the XP-tractability arising from Theorem 1 cannot be achieved without a bound on the maximum degree of the input graph even when restricted to the class of trees, since the problem is known to remain NP-hard even on trees of edge capacity at most 2 [26].

A natural question left for future work would be to study the complexity of UNSPLITTABLE FLOW under other parameterizations that are not "graph-structural" in nature, such as when parameterizing by the number of routed tasks in the solution or the target profit τ; previous work has considered this question primarily on specific graph classes such as trees [40].

References

1. Adamaszek, A., Chalermsook, P., Ene, A., Wiese, A.: Submodular unsplittable flow on trees. Math. Program. **172**(1–2), 565–589 (2018)
2. Anagnostopoulos, A., Grandoni, F., Leonardi, S., Wiese, A.: A mazing $2+\epsilon$ approximation for unsplittable flow on a path. ACM Trans. Algorithms **14**(4), 55:1–55:23 (2018)
3. Balabán, J., Ganian, R., Rocton, M.: Computing twin-width parameterized by the feedback edge number. In: 41st International Symposium on Theoretical Aspects of Computer Science, STACS 2024. LIPIcs. Schloss Dagstuhl - Leibniz-Zentrum für Informatik (2024)
4. Batra, J., Garg, N., Kumar, A., Mömke, T., Wiese, A.: New approximation schemes for unsplittable flow on a path. In: Proceedings of the Twenty-Sixth Annual ACM-SIAM Symposium on Discrete Algorithms, SODA, pp. 47–58. SIAM (2015)
5. Baveja, A., Srinivasan, A.: Approximation algorithms for disjoint paths and related routing and packing problems. Math. Oper. Res. **25**(2), 255–280 (2000)
6. Bhore, S., Ganian, R., Montecchiani, F., Nöllenburg, M.: Parameterized algorithms for book embedding problems. J. Graph Algorithms Appl. **24**(4), 603–620 (2020)
7. Bodlaender, H.L., Drange, P.G., Dregi, M.S., Fomin, F.V., Lokshtanov, D., Pilipczuk, M.: A c^k n 5-approximation algorithm for treewidth. SIAM J. Comput. **45**(2), 317–378 (2016)

8. Bodlaender, H.L., Groenland, C., Pilipczuk, M.: Parameterized complexity of binary CSP: vertex cover, treedepth, and related parameters. In: 50th International Colloquium on Automata, Languages, and Programming, ICALP 2023, LIPIcs, vol. 261, pp. 27:1–27:20. Schloss Dagstuhl - Leibniz-Zentrum für Informatik (2023)
9. Bonsma, P.S., Schulz, J., Wiese, A.: A constant factor approximation algorithm for unsplittable flow on paths. In: IEEE 52nd Annual Symposium on Foundations of Computer Science, FOCS 2011, pp. 47–56. IEEE Computer Society (2011)
10. Bonsma, P.S., Schulz, J., Wiese, A.: A constant-factor approximation algorithm for unsplittable flow on paths. SIAM J. Comput. **43**(2), 767–799 (2014)
11. Călinescu, G., Chakrabarti, A., Karloff, H.J., Rabani, Y.: An improved approximation algorithm for resource allocation. ACM Trans. Algorithms **7**(4):48:1–48:7 (2011)
12. Chaplick, S., Di Giacomo, E., Frati, F., Ganian, R., Raftopoulou, C.N., Simonov, K.: Parameterized algorithms for upward planarity. In: 38th International Symposium on Computational Geometry, SoCG 2022. LIPIcs, vol. 224, pp. 26:1–26:16. Schloss Dagstuhl - Leibniz-Zentrum für Informatik (2022)
13. Chekuri, C., Ene, A., Korula, N.: Unsplittable flow in paths and trees and column-restricted packing integer programs. In: Dinur, I., Jansen, K., Naor, J., Rolim, J. (eds.) APPROX/RANDOM -2009. LNCS, vol. 5687, pp. 42–55. Springer, Heidelberg (2009). https://doi.org/10.1007/978-3-642-03685-9_4
14. Chekuri, C., Mydlarz, M., Bruce Shepherd, F.: Multicommodity demand flow in a tree and packing integer programs. ACM Trans. Algorithms **3**(3), 27 (2007)
15. Chrobak, M., Woeginger, G.J., Makino, K., Haifeng, X.: Caching is hard - even in the fault model. Algorithmica **63**(4), 781–794 (2012)
16. Cygan, M., et al.: Parameterized Algorithms. Springer, Cham (2015)
17. Diestel, R.: Graph Theory. Graduate Texts in Mathematics, vol. 173, 4th edn. Springer, Cham (2012)
18. Downey, R.G., Fellows, M.R.: Fundamentals of Parameterized Complexity. Texts in Computer Science, Springer, Cham (2013)
19. Eiben, E., Ganian, R., Kanj, I.: The parameterized complexity of coordinated motion planning. In: 39th International Symposium on Computational Geometry, SoCG 2023. LIPIcs, vol. 258, pp. 28:1–28:16. Schloss Dagstuhl - Leibniz-Zentrum für Informatik (2023)
20. Fleszar, K., Mnich, M., Spoerhase, J.: New algorithms for maximum disjoint paths based on tree-likeness. Math. Program. **171**(1–2), 433–461 (2018)
21. Ganian, R., Hamm, T., Korchemna, V., Okrasa, K., Simonov, K.: The complexity of k-means clustering when little is known. In: International Conference on Machine Learning, ICML 2022. Proceedings of Machine Learning Research, vol. 162, pp. 6960–6987. PMLR (2022)
22. Ganian, R., Kim, E.J., Szeider, S.: Algorithmic applications of tree-cut width. SIAM J. Discret. Math. **36**(4), 2635–2666 (2022)
23. Ganian, R., Korchemna, V.: Slim tree-cut width. In: 17th International Symposium on Parameterized and Exact Computation, IPEC 2022. LIPIcs, vol. 249, pp. 15:1–15:18. Schloss Dagstuhl - Leibniz-Zentrum für Informatik (2022)
24. Ganian, R., Ordyniak, S.: The power of cut-based parameters for computing edge-disjoint paths. Algorithmica **83**(2), 726–752 (2021)
25. Ganian, R., Ordyniak, S., Ramanujan, M.S.: On structural parameterizations of the edge disjoint paths problem. Algorithmica **83**(6), 1605–1637 (2021)
26. Garg, N., Vazirani, V.V., Yannakakis, M.: Primal-dual approximation algorithms for integral flow and multicut in trees. Algorithmica **18**(1), 3–20 (1997)

27. Grandoni, F., Mömke, T., Wiese, A.: Faster (1+ϵ)-approximation for unsplittable flow on a path via resource augmentation and back. In: 29th Annual European Symposium on Algorithms, ESA 2021. LIPIcs, vol. 204, pp. 49:1–49:15. Schloss Dagstuhl - Leibniz-Zentrum für Informatik (2021)
28. Grandoni, F., Mömke, T., Wiese, A.: A PTAS for unsplittable flow on a path. In: STOC 2022: 54th Annual ACM SIGACT Symposium on Theory of Computing, pp. 289–302. ACM (2022)
29. Grandoni, F., Mömke, T., Wiese, A.: Unsplittable flow on a path: the game! In: Proceedings of the 2022 ACM-SIAM Symposium on Discrete Algorithms, SODA, pp. 906–926. SIAM (2022)
30. Grandoni, F., Mömke, T., Wiese, A., Zhou, H.: To augment or not to augment: solving unsplittable flow on a path by creating slack. In: Proceedings of the Twenty-Eighth Annual ACM-SIAM Symposium on Discrete Algorithms, SODA, pp. 2411–2422. SIAM (2017)
31. Guruswami, V., Khanna, S., Rajaraman, R., Bruce Shepherd, F., Yannakakis, M.: Near-optimal hardness results and approximation algorithms for edge-disjoint paths and related problems. J. Comput. Syst. Sci. **67**(3), 473–496 (2003)
32. Jansen, K., Kratsch, S., Marx, D., Schlotter, I.: Bin packing with fixed number of bins revisited. In: Kaplan, H. (ed.) SWAT 2010. LNCS, vol. 6139, pp. 260–272. Springer, Heidelberg (2010). https://doi.org/10.1007/978-3-642-13731-0_25
33. Kloks, T.: Treewidth. Lecture Notes in Computer Science, vol. 842. Computations and Approximations, Springer, Cham (1994)
34. Koana, T., Froese, V., Niedermeier, R.: The complexity of binary matrix completion under diameter constraints. J. Comput. Syst. Sci. **132**, 45–67 (2023)
35. Kolman, P., Scheideler, C.: Simple on-line algorithms for the maximum disjoint paths problem. In: Proceedings of the Thirteenth Annual ACM Symposium on Parallel Algorithms and Architectures, SPAA 2001, pp. 38–47. Association for Computing Machinery, New York (2001). https://doi.org/10.1145/378580.378586
36. Kolman, P., Scheideler, C.: Improved bounds for the unsplittable flow problem. J. Algorithms **61**(1), 20–44 (2006)
37. Korhonen, J.H., Parviainen, P.: Tractable Bayesian network structure learning with bounded vertex cover number. In: Advances in Neural Information Processing Systems 28: Annual Conference on Neural Information Processing Systems 2015, pp. 622–630 (2015)
38. Korhonen, T.: A single-exponential time 2-approximation algorithm for treewidth. In: 62nd IEEE Annual Symposium on Foundations of Computer Science, FOCS 2021, pp. 184–192. IEEE (2021)
39. Martens, M., Skutella, M.: Length-bounded and dynamic k-splittable flows. In: Operations Research Proceedings 2005, Selected Papers of the Annual International Conference of the German Operations Research Society (GOR), pp. 297–302 (2005)
40. Martínez-Muñoz, T., Wiese, A.: FPT and FPT-approximation algorithms for unsplittable flow on trees. In: 29th Annual European Symposium on Algorithms, ESA 2021. LIPIcs, vol. 204, pp. 67:1–67:15. Schloss Dagstuhl - Leibniz-Zentrum für Informatik (2021)
41. Marx, D.: Eulerian disjoint paths problem in grid graphs is NP-complete. Discret. Appl. Math. **143**(1–3), 336–341 (2004)
42. Nesetril, J., de Mendez, P.O.: Sparsity - Graphs, Structures, and Algorithms. Algorithms and Combinatorics, vol. 28. Springer, Cham (2012)

43. Phillips, C.A., Uma, R.N., Wein, J.: Off-line admission control for general scheduling problems. In: Proceedings of the Eleventh Annual ACM-SIAM Symposium on Discrete Algorithms, pp. 879–888. ACM/SIAM (2000)
44. Robertson, N., Seymour, P.D.: Graph minors. II. Algorithmic aspects of tree-width. J. Algorithms **7**(3), 309–322 (1986)
45. Wiese, A.: A (1+epsilon)-approximation for unsplittable flow on a path in fixed-parameter running time. In: 44th International Colloquium on Automata, Languages, and Programming, ICALP 2017. LIPIcs, vol. 80, pp. 67:1–67:13. Schloss Dagstuhl - Leibniz-Zentrum für Informatik (2017)
46. Wollan, P.: The structure of graphs not admitting a fixed immersion. J. Comb. Theory Ser. B **110**, 47–66 (2015)
47. Wollan, P., Marx, D.: Immersions in highly edge connected graphs. SIAM J. Discret. Math. **28**(1), 503–520 (2014)

Lightweight Near-Additive Spanners

Yuval Gitlitz(✉), Ofer Neiman, and Richard Spence

Ben-Gurion University of the Negev, Be'er Sheva, Israel
gitlitz@post.bgu.ac.il, neimano@cs.bgu.ac.il

Abstract. An (α, β)-spanner of a weighted graph $G = (V, E)$, is a subgraph H such that for every $u, v \in V$, $d_G(u, v) \leq d_H(u, v) \leq \alpha \cdot d_G(u, v) + \beta$. The main parameters of interest for spanners are their size (number of edges) and their lightness (the ratio between the total weight of H to the weight of a minimum spanning tree).

In this paper we focus on near-additive spanners, where $\alpha = 1 + \varepsilon$ for arbitrarily small $\varepsilon > 0$. We show the first construction of *light* spanners in this setting. Specifically, for any integer parameter $k \geq 1$, we obtain an $(1+\varepsilon, O(k/\varepsilon)^k \cdot W(\cdot, \cdot))$-spanner with lightness $\widetilde{O}(n^{1/k})$ (where $W(\cdot, \cdot)$ indicates for every pair $u, v \in V$ the heaviest edge in some shortest path between u, v). In addition, we can also bound the number of edges in our spanner by $O(kn^{1+3/k})$.

Keywords: spanners · lightness · shortest path · weighted graph

1 Introduction

Given an undirected, edge-weighted graph $G = (V, E, w)$ with $|V| = n$, and parameters $\alpha \geq 1$ and $\beta \geq 0$, a subgraph $H = (V, E', w)$ is an (α, β)-*spanner* if

$$d_G(x, y) \leq d_H(x, y) \leq \alpha \cdot d_G(x, y) + \beta \tag{1}$$

for all vertex pairs $x, y \in V$, where $d_G(x, y)$ denotes the distance between x and y in G. Spanners and related graph constructs such as emulators (which satisfy (1) but need not be subgraphs of the given graph) have gained attention for their role in distributed computing, approximation of shortest paths and graph compression; see the recent survey [4]. One of the primary goals is to determine the tradeoffs between the multiplicative stretch α, the additive error β, and the size of the spanner H as measured by $|E'|$. An additional (sometimes even more desirable) goal is to construct *lightweight* spanners whose total weight is bounded. Of course, one must pay at least the weight of a minimum spanning tree (MST) to simply ensure H is connected; thus, the lightness of a spanner H is measured as the ratio of the edge weight of H to the edge weight of the MST of G. For algorithmic applications, it is also desired to construct these spanners efficiently.

Spanners were first studied in the 1980s [29] with purely multiplicative stretch, i.e., $\beta = 0$. In [6] it was shown that for all integers $k \geq 1$, all weighted

D. Kráľ and M. Milanič (Eds.): WG 2024, LNCS 14760, pp. 236–250, 2025.
https://doi.org/10.1007/978-3-031-75409-8_17

graphs have a $(2k-1,0)$-spanner with size $O(n^{1+1/k})$ and lightness $O(\frac{n}{k})$ via a simple greedy algorithm. The size bound is optimal assuming Erdős' girth conjecture [21]. Soon thereafter, [13] showed for any $0<\varepsilon<1$, the same greedy algorithm yields a $((2k-1)(1+\varepsilon),0)$-spanner with lightness $O_\varepsilon(kn^{1/k})$ and size $O_\varepsilon(n^{1+1/k})$.[1] Researchers also sought to understand the tradeoffs if one desires a tiny amount $1+\varepsilon$ of multiplicative error, also known as *near-additive spanners*. For unweighted graphs, in a sequence of works, [2,10,18,20,30], $(1+\varepsilon,\beta)$-spanners of size $\approx O(n^{1+1/k})$ where $\beta = O\left(\left(\frac{\log k}{\varepsilon}\right)^{\log k}\right)$ were devised. In [2] it was shown that the additive stretch of near-additive spanners of such size must be $\Omega_k\left(\frac{1}{\varepsilon}\right)^{\log k}$[2]. Near-additive spanners were also studied for weighted graphs. Here, the additive error must depend linearly on the maximum edge weight W_{max}, or the "local" maximum edge weight $W(x,y)$ along a shortest x-y path $P_{x,y}$. The latter is typically more preferred as it is possible for $W_{max} \gg W(x,y)$; in this case, we use $W(\cdot,\cdot)$ to indicate that the additive error for each vertex pair x,y is in terms of $W(x,y)$. In both cases, the best one can hope for is to match the corresponding bounds for unweighted graphs. Indeed, [16] showed that all weighted graphs admit $(1+\varepsilon, \left(\frac{\log k}{\varepsilon}\right)^{O(\log k)} W(\cdot,\cdot))$ spanners of size $O(kn + n^{1+1/k})$.

In the past decade, there has been an surge of interest in light spanners, and algorithms for their construction were devised in various settings. For general graphs, light spanners were studied in several works [9,15,19,27], culminating in lightness $O(n^{1/k}/\varepsilon)$ (with stretch $(2k-1)(1+\varepsilon)$ and size $O(n^{1+1/k}/\varepsilon)$). For Euclidean pointsets of dimension q, [31,35] showed a $(1+\varepsilon)$-spanner of size $n\cdot\varepsilon^{-O(q)}$ and lightness $\varepsilon^{-O(q)}$. For high dimensional pointsets, [23] devised $O(k)$-spanners with lightness $\tilde{O}(n^{1/k^2})$.[3] The low dimensional results were extended for metric spaces of doubling dimension *ddim*: [12,24,25] constructed a $(1+\varepsilon)$-spanner with $n\cdot\varepsilon^{-O(ddim)}$ edges and lightness $\varepsilon^{-O(ddim)}$. For any graph excluding a fixed minor, [11] devised $(1+\varepsilon)$-spanners with lightness $\tilde{O}(\varepsilon^{-3})$. Light prioritized spanners were shown by [7], which were used to get an almost MST with constant average distortion.[4] Recently, [22] devised light reliable spanners.[5] However, despite all these efforts, no previous work gave a meaningful bound on the lightness of near-additive spanners.

1.1 Our Results

In this paper we give the first construction of *light* near-additive spanners for general graphs. The specific bounds are described in the following theorem.

[1] The notation $O_\varepsilon(\cdot)$ hides $(\frac{1}{\varepsilon})^{O(1)}$ factors.

[2] The notation $\Omega_k(\cdot)$ hides $k^{O(1)}$ factors.

[3] The notation $\widetilde{O}(f(n))$ hides polylogarithmic factors in n.

[4] A prioritized spanner receives an arbitrary ranking of the points, and should obtain better stretch for high ranking points.

[5] A reliable spanner is susceptible to massive vertex failures $B \subseteq V$, and still provide meaningful guarantees for all vertex pairs in $V \setminus B^+$, where B^+ is only slightly larger than B.

Theorem 1. *Let G be a weighted graph, let $0 < \varepsilon < 1$, and let k be a positive integer. Then G has a $\left(1+\varepsilon, O(\frac{k}{\varepsilon})^k \cdot W(\cdot,\cdot)\right)$-spanner of size $O(kn^{1+3/k})$ and lightness $\widetilde{O}(\frac{n^{1/k}}{\varepsilon})$.*

We remark that the dependence of β, the additive stretch, on k is somewhat inferior compared to the result of [16]. However, their construction has unbounded lightness. There is an efficient randomized algorithm to construct the spanners of Theorem 1. Note that the lower bound of [2] also holds for near-additive spanners of bounded lightness.

Additionally, if $W_{max} \geq \sqrt{w(MST(G))}$, we can construct a $(1+\varepsilon, 2(1+\varepsilon)W_{max})$-spanner of size $O(n^{3/2})$ and lightness $O(\frac{n^{1/2}}{\varepsilon})$.

1.2 Technical Overview

One of the most popular techniques to design sparse spanners is the method devised by Thorup-Zwick [33]. The basic idea is to sample a sequence of sets $V = A_0 \supseteq A_1 \supseteq \cdots \supseteq A_k$, and define the bunch of $v \in A_i \setminus A_{i+1}$ as all the vertices closer to v than the nearest vertex in A_{i+1} (which is called the pivot). The spanner is constructed by taking all the shortest paths from v to its bunch, and the path to its pivot. There are certain variations of this basic construction that yield the best known (α, β)-spanners [28], in essentially all the regimes of multiplicative stretch α.

Our main technical innovation is an adaptation of this framework that ensures the lightness of the resulting spanner. Specifically, we construct a hierarchy of *nets* in all distance scales of the form 2^i, and define for each vertex a net-point *representative* in each such scale. Now, we replace the connections from each v to each vertex u in its bunch, by connecting v to u's representative in the net of scale $\approx \varepsilon d(u, v)$. In fact, we shall do this only for vertices u which are closer than $\frac{1-\varepsilon}{2}$ times the distance of v to its pivot. Using this variant, we are able to show that any representative participates in $\approx n^{1/k}$ such paths (the basic idea is that the set of all vertices v who connect to a certain representative x, must all lie in a single bunch, whose size is bounded). Then we apply a known relation between nets and the weight of the MST (see Lemma 1) to prove the lightness of these paths.

In addition, to complete the $v - u$ spanner path, we need to guarantee that u is connected to each of its representatives. Directly adding a path from u to the nearest net-point in all scales turns out to be too costly in terms of lightness. Instead, we connect u only to $\log \frac{1}{\varepsilon}$ representatives, and show that these connections are light enough, yet they imply there exist "sufficiently good" representatives in all scales. Finally, we replace the connections to the pivots by taking Shallow Light Trees (SLT) rooted at the pivots. The (quite technical) analysis of the stretch has to take into consideration all these various changes, and we indeed show how they only incur a small factor in the additive stretch.

The selection of the sets A_i in previous works, and also here, is done randomly. Each vertex in A_i advances to A_{i+1} either with uniform $n^{-1/k}$ probability, or with exponentially decreasing $\approx n^{-2^i/k}$ probability (in the latter case there are only $\log k$ sets). The improved dependence of the additive stretch $\beta \approx \left(\frac{\log k}{\varepsilon}\right)^{\log k} W(\cdot,\cdot)$ in [16], is achieved using the latter probability choice. However, a key ingredient in our lightness analysis, that each representative has roughly $n^{1/k}$ paths added to it, crucially relies on the uniform probability choice. For this reason we can only achieve $\beta \approx \left(\frac{k}{\varepsilon}\right)^{k} W(\cdot,\cdot)$, and we leave open the question of obtaining an improved additive stretch.

1.3 Related Work

Pure additive spanners (with $\alpha = 1$) have been widely studied in recent years, mostly in unweighted graphs. It is known that all unweighted graphs have $(1, 2)$-spanners of size $O(n^{3/2})$ [5], $(1, 4)$-spanners of size $\widetilde{O}(n^{7/5})$ [14], and $(1, 6)$-spanners of size $O(n^{4/3})$ [8,36]. Moreover, the upper bound of $O(n^{4/3})$ cannot be improved for pure additive spanners, even if one allows $n^{o(1)}$ additive error [1]. In [32], linear-size spanners were shown with polynomial additive stretch $O(n^{0.403})$.

For weighted graphs [3,17] show constructions of pure additive $(1, O(1) \cdot W(\cdot,\cdot))$-spanners with size guarantees that nearly match their unweighted counterparts. The only previous construction of light spanner with additive stretch was given by [3], who devised a $(1,(4+\varepsilon)W(\cdot,\cdot))$-spanner with lightness $O_\varepsilon(n^{2/3})$. This result has pure additive stretch, and thus is incomparable to our results.

2 Preliminaries

For a graph $G = (V, E)$ and subset $S \subseteq V$, we ordinarily denote $d_G(v, S) = \min_{x \in S} d_G(v, x)$. We use $d(u, v)$ and $d(u, S)$ as shorthand for $d_G(u, v)$ and $d_G(u, S)$ respectively, where G is the input graph. For all constructions, we assume, w.l.o.g., $w(MST(G)) = n$; in other words, the average weight of an MST edge is ≈ 1. This assumption does not lose generality as one can simply scale the edge weights, as well as W_{max} and $W(\cdot,\cdot)$. Also, let $[k] = \{1, 2, 3, \ldots, k\}$.

Definition 1 (Shallow-light tree (SLT)). *Given a weighted graph $G = (V, E)$, parameters $\alpha, \gamma \geq 1$, and a root vertex $s \in V$, a spanning tree $T = (V, E')$ is an (α, γ)-*shallow light tree *(or (α, γ)-SLT) rooted at s if the following are true:*

- *$d_T(s, x) \leq \alpha \cdot d(s, x)$ for all $x \in V$, and*
- *$w(T) \leq \gamma \cdot w(\mathrm{MST}(G))$.*

We use the following result by Khuller et al. [26]:

Theorem 2 ([26]). *Given a weighted graph G, a root vertex $s \in V$, and $\varepsilon > 0$, there exists a polynomial-time constructible $(1 + \varepsilon, O(\frac{1}{\varepsilon}))$-SLT rooted at s.*

Definition 2 (Δ-net). *Given a weighted graph $G = (V, E)$ and a real number $\Delta > 0$, a subset of vertices $N \subseteq V$ is a* Δ-net *if the following hold:*

- *$d(x, N) \leq \Delta$ for all $x \in V$, and*

– *$d(u, v) > \Delta$ for all distinct $u, v \in N$.*

A Δ-net N can be computed via the following simple greedy algorithm: while there exists a vertex whose distance to N is greater than Δ, choose such a vertex and add it to N. The following lemma relates nets to the weight of the MST.

Lemma 1. *For any Δ-net N of a graph G with $|N| \geq 2$ vertices, we necessarily have $|N|\Delta \leq 2w(\text{MST}(G))$.*

Proof. Let T be the MST of G. Note that balls of radius $\frac{\Delta}{2}$ centered at each vertex $v \in N$ are pairwise disjoint, as every pair of vertices in N is of distance greater than Δ. For each $v \in N$, consider the weight of edges in T that lie in $B = B(v, \frac{\Delta}{2})$ (if an edge $\{x, y\} \in T$ leaves the ball, suppose $x \in B$ and $y \notin B$, we consider the relative weight inside, that is $\frac{\Delta}{2} - d(v, x)$). The observation is that every such ball contains at least $\frac{\Delta}{2}$ of the weight of T, which is unique to it. Therefore, $w(T) \geq |N|\frac{\Delta}{2}$. □

3 A Lightweight $(1 + \varepsilon, O(\frac{k}{\varepsilon})^k)$ Near-Additive Spanner

In this section we prove our main result, Theorem 1. The construction of the spanner has three phases; the first phase constructs a hierarchical sequence of $O(\log n)$ Δ-nets (Definition 2), and adds a small set of lightweight paths connecting vertices to their nearby net points, which will be used to define a set of *representative* vertices. The second phase stratify the vertices into k levels, then add paths from each vertex u to the representatives of its bunch. The third and final phase uses SLTs for the (approximate) pivot connections.

First Phase: Δ-Nets. In the first phase, we will construct a nested sequence of Δ-nets (Definition 2), then add shortest paths from each vertex to the closest vertices in some of the Δ-nets. Specifically, let $V = N_{-1} \supseteq N_0 \supseteq \ldots \supseteq N_{\log_2 n}$, where for each $i \geq 0$, N_i is a 2^i-net. In particular, $N_{\log_2 n}$ is an n-net so $|N_{\log_2 n}| = 1$ (as we assumed $w(MST(G)) = n$, so $diam(G) \leq n$), and N_0 is a 1-net. The nets can be constructed using the greedy algorithm following Definition 2: let $N_{\log n}$ consist of a single vertex, and for $i = \log_2 n - 1, \ldots, 0$, use the greedy algorithm to add vertices to N_{i+1} until a 2^i-net N_i is constructed.

Constructing H_0. For each $i \geq -1$, each $v \in N_i \setminus N_{i+1}$, and for all $j \in \{i + 1, \ldots, i + \lceil \log_2 \frac{1}{\varepsilon} \rceil\}$, add a shortest path to H_0 from v to the closest vertex in N_j (if $j > \log_2 n$, simply do nothing). Lastly add H_0 to H.

Lemma 2. *For all $v \in V$ and for all $i \in \{0, \ldots, \log_2 n\}$, there exists $x \in N_i$ such that $d_{H_0}(v, x) \leq (1 + 2\varepsilon)2^i$.*

Proof. The idea is to follow a sequence of paths in H_0 starting from v, where each path leads to a vertex $t := \lceil \log_2 \frac{1}{\varepsilon} \rceil$ levels higher, until we reach a vertex at level i. Write $i = a + bt$ for some integers $0 \leq a < t$ and $b \geq 0$. First we take the path in H_0 from v to the closest vertex $v_0 \in N_a$. Then, we take additional b

paths, the jth one, for $1 \leq j \leq b$, will be from $v_{j-1} \in N_{a+(j-1)t}$ to the closest $v_j \in N_{a+jt}$. All these paths exist in H_0, since every vertex in a net is connected to a vertex in all the t nets above it.

Since N_{a+jt} is a 2^{a+jt}-net, the total length of the concatenation of these paths, that lead from v to a vertex $v_b \in N_i$, is at most

$$\sum_{j=0}^{b} 2^{a+jt} = 2^i \cdot \sum_{j=0}^{b} 2^{-jt} \leq 2^i \cdot \sum_{j=0}^{b} \varepsilon^j \leq 2^i \cdot (1 + 2\varepsilon),$$

where we use that $2^{-t} \leq \varepsilon$, and the last inequality uses that $\varepsilon < 1/10$.

□

Following Lemma 2, we define for each $v \in V$ and $i \in \{0, \ldots, \log_2 n\}$ the *level-i representative* $r_i(v)$ to be a vertex $x \in N_i$ such that $d_{H_0}(v, x) \leq (1+2\varepsilon)2^i$. These representatives will be used in the second phase of the construction. Note that if $v \in N_i$, then $r_i(v) = v$. We also define for $i < 0$, $r_i(v) = v$.

Second Phase: Bunches and Representatives. We describe the second phase of the construction. This phase is similar to those given in [16,34], in which the main idea is to construct a hierarchy of vertex sets $V = A_0 \supseteq A_1 \supseteq A_2 \supseteq \ldots \supseteq A_k \supseteq A_{k+1} = \emptyset$, where each vertex in A_{i-1} is included in A_i with a certain probability. The sets A_i are unrelated to the vertices N_i in the 2^i-net. The main difference is that instead of connecting each vertex u to its bunch, we connect it to a carefully chosen set of representatives of a bunch whose radius is slightly less than half the distance to u's pivot (unlike [16], which used *half-bunches*).

Let $V = A_0 \supseteq A_1 \supseteq A_2 \supseteq \ldots \supseteq A_k \supseteq A_{k+1} = \emptyset$ be sets of vertices. For $i \in [k]$, A_i is obtained by sampling each vertex in A_{i-1} independently with probability $n^{-1/k}$. Given $u \in V$ and $i \in \{0, \ldots, k\}$, the level-i *pivot* of u, denoted $p_i(u)$, is a vertex $x \in A_i$ such that $d(u, x) = d(u, A_i)$; note that if $u \in A_i$, then $p_i(u) = u$. Given $u \in A_i \backslash A_{i+1}$, and a real number $\delta > 0$, let the *δ-bunch* centered at u be defined as follows:

$$B_r(\delta) = \begin{cases} \{v \in A_i : d(u, v) < \delta \cdot d(u, p_{i+1}(u))\} & 0 \leq i \leq k-1 \\ A_k & i = k \end{cases}$$

Adding Paths to Representatives. We now add paths to the spanner H as follows: for all vertices $u \in V$, and for all $v \in B_{\frac{1-\varepsilon}{2}}(u)$, let $j \leq \log n$ be the unique integer such that $\frac{\varepsilon}{8} \cdot d(u, v) \leq 2^j < \frac{\varepsilon}{4} \cdot d(u, v)$. We connect u to the level-j representative $r_j(v)$ of v via a shortest path. Note that if $j < 0$, we add a direct $u - v$ shortest path to H, since we defined $r_j(v) = v$ for such j. On the other hand, we cannot have $j > \log n$, because by our assumption on the MST weight, $d(u, v) \leq diam(G) \leq n$. Denote by $r(u, v)$ the level-j representative of v that u connected to.

Third Phase: SLTs. Lastly, for each $i \in [k]$, consider a "virtual" root vertex s_i which is connected to the vertices in A_i with (virtual) zero-weight edges, and add the edges of a $\left(1+\varepsilon, O\left(\frac{1}{\varepsilon}\right)\right)$-SLT rooted at s_i to the spanner (excluding the zero-weight edges). Let H be the resulting subgraph. This final step effectively creates k forests of SLTs rooted at $A_1, \ldots, A_k$, in which for every vertex u and $i \in \{0, \ldots, k\}$, we have $d_H(u, A_i) \leq (1+\varepsilon)d(u, A_i) = (1+\varepsilon)d(u, p_i(u))$. Note that, unlike the constructions in [16,34], our construction does not necessarily connect centers u to their pivots $p_i(u)$ directly, but instead connects to "approximate" pivots in A_i which are $1+\varepsilon$ times the distance $d(u, A_i)$ to u's actual pivot. We denote this approximate pivot by $p_i'(u)$, such that $p_i'(u) \in A_i$, and $d_H(u, p_i'(u)) \leq (1+\varepsilon)d(u, p_i(u))$ (see Fig. 1).

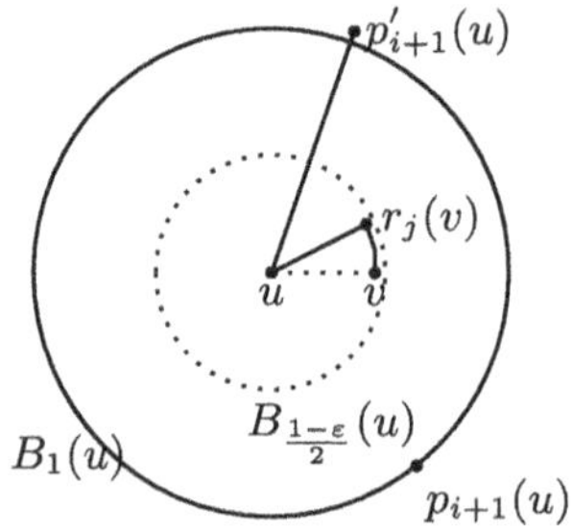

Fig. 1. Illustration of the paths added during the three phases. The dotted and solid circles represent the $\frac{1-\varepsilon}{2}$-bunch, and the 1-bunch of $u \in A_i \setminus A_{i+1}$, respectively. The first phase connects v to its representative $r_j(v)$, the second phase connects u to $r_j(v)$, which is one of the representatives of its $\frac{1-\varepsilon}{2}$-bunch, and the third phase connects u to $p'_{i+1}(u)$, the approximate pivot, via the SLT rooted at s_{i+1}.

3.1 Stretch of the Spanner

In this section, we prove that H is an (α, β)-spanner with α and β as given in Theorem 1. The proofs in this section follow along the lines of the proofs in [16] regarding sparse, near-additive emulators and spanners; the main difference is that by using paths from vertices to "approximate" pivots using SLTs, and by using paths from vertices to the representatives of their $\frac{1-\varepsilon}{2}$-bunches (instead of direct paths to their $\frac{1}{2}$-bunch), we incur slightly weaker upper bounds on the additive error. We first show the following lemma:

Lemma 3. *Let $u \in V$ and $v \in B_{\frac{1-\varepsilon}{2}}(u)$, then $d_H(u,v) \leq (1+\varepsilon)d(u,v)$.*

Proof. Let $r = r(u,v)$ (recall that $r(u,v)$ is the level-j representative of v that u connected to). If $r = v$ then we added a shortest $u-v$ path to H, therefore the statement holds. Otherwise, r is chosen as $r_j(v)$ for $0 \leq j \leq \log n$ satisfying $\frac{\varepsilon}{8}d(u,v) \leq 2^j < \frac{\varepsilon}{4}d(u,v)$. Note that in phase 2 we added a shortest $u-r$ path to H. By Lemma 2 we have that $d_H(v,r) \leq (1+2\varepsilon)2^j < \frac{\varepsilon}{2}d(u,v)$, thus

$$\begin{aligned} d_H(u,v) &\le d_H(u,r) + d_H(v,r) \\ &= d(u,r) + d_H(v,r) \le d(u,v) + d(v,r) + d_H(v,r) \\ &\le d(u,v) + 2d_H(v,r) < (1+\varepsilon)\, d(u,v). \end{aligned} \tag{2}$$

□

The following lemma will be useful for pairs that are sufficiently far apart.

Lemma 4. *Fix* $0 < \varepsilon < \frac{1}{10}$*, let* $\Delta > 7$*, let* $0 \le i \le k$ *and* $x, y \in V$ *such that* $d(x,y) \ge (3\Delta)^i W(x,y)$*. Let* $\eta_i = 1 + \varepsilon + \frac{14}{\Delta-7} i$ *and* $\zeta = \frac{3\Delta}{\Delta-7}$*. Then at least one of the following statements is true:*

- $d_H(x,y) \le \eta_i d(x,y)$
- $d_H(x, A_{i+1}) \le \zeta d(x,y)$.

The proof of Lemma 4 is available in the full version of the paper.

The next lemma is useful for vertex pairs which are relatively nearby.

Lemma 5. *Let* $0 < \varepsilon < \frac{1}{10}$*, let* $0 \le i < k$*, let* $x, y \in V$*, and let* $m = \max\{d(x, A_i), d(y, A_i), d(x,y)\}$*. Then at least one of the following holds:*

- $d_H(x,y) \le 6m$
- $d_H(x, A_{i+1}) \le 10m$ *and* $d_H(y, A_{i+1}) \le 10m$.

Proof. Using the SLT rooted in s_i, we have $d_H(x, p_i'(x)) \le (1+\varepsilon) d(x, A_i)$ and similarly $d_H(y, p_i'(y)) \le (1+\varepsilon) d(y, A_i)$, so that

$$d(p_i'(x), p_i'(y)) \le d(p_i'(x), x) + d(x,y) + d(y, p_i'(y)) \le m + 2(1+\varepsilon)m \le 3.2m. \tag{3}$$

The first case is that $p_i'(y) \in B_{\frac{1-\varepsilon}{2}}(p_i'(x))$. Then by Lemma 3, $d_H(p_i'(x), p_i'(y)) \le (1+\varepsilon) d(p_i'(x), p_i'(y))$.

$$\begin{aligned} d_H(x,y) &\le d_H(x, p_i'(x)) + d_H(p_i'(x), p_i'(y)) + d_H(p_i'(y), y) \\ &\overset{(3)}{\le} (1+\varepsilon)m + (1+\varepsilon)(3.2m) + (1+\varepsilon)m \\ &\le 6m. \end{aligned}$$

If $p_i'(y) \notin B_{\frac{1-\varepsilon}{2}}(p_i'(x))$ then

$$d(p_i'(x), p_{i+1}(p_i'(x))) \le \frac{2}{1-\varepsilon} d(p_i'(x), p_i'(y)) \overset{(3)}{\le} \frac{6.4m}{1-\varepsilon}.$$

Now the second item holds, as

$$\begin{aligned} d_H(x, A_{i+1}) &\le (1+\varepsilon) d(x, A_{i+1}) \\ &\le (1+\varepsilon)[d(x, p_i'(x)) + d(p_i'(x), p_{i+1}(p_i'(x)))] \\ &\le (1+\varepsilon)\left[(1+\varepsilon)m + \frac{6.4m}{1-\varepsilon}\right] \\ &\le 10m. \end{aligned}$$

□

Now we can show the following:

Lemma 6. *H is an (α, β)-spanner with $\alpha = 1 + O(\varepsilon)$ and $\beta = O(\frac{k}{\varepsilon})^k$.*

Proof. We set $\Delta = 7 + \frac{14k}{\varepsilon}$. Let $x, y \in V$ and $W = W(x, y)$. If $d(x, y) \geq (3\Delta)^k W$, then the first item of Lemma 4 must hold, as A_{k+1} is empty by definition.

$$d_H(x,y) \leq \eta_k d(x,y) \leq \left(1 + \varepsilon + \frac{14k}{14k/\varepsilon}\right) = (1+2\varepsilon)d(x,y)$$

Otherwise $d(x, y) < (3\Delta)^k W$. Take the integer $i \in \{0, 1, \ldots, k-1\}$ such that $(3\Delta)^i W \leq d(x, y) < (3\Delta)^{i+1} W$. Apply Lemma 4 on x, y, and i. If the first statement holds, we obtain $d_H(x, y) \leq (1+2\varepsilon)d(x, y)$ as before. Otherwise the second statement holds, in which $d_H(x, A_{i+1}) \leq \frac{3\Delta}{\Delta-7} d(x, y) \leq 4d(x, y)$, and similarly $d_H(y, A_{i+1}) \leq \frac{3\Delta}{\Delta-7} d(x, y) \leq 4d(x, y)$. These together imply $m \leq 4d(x, y)$, with m as in Lemma 5. Set $j = i + 1$ and apply Lemma 5 with x, y, and j.

If the first statement of Lemma 5 holds with x, y, and j, then $d_H(x, y) \leq 6m \leq 24d(x, y)$. Otherwise, the second statement of Lemma 5 holds, and it follows that $\max\{d(x, A_{i+1}), d(y, A_{i+1}), d(x, y)\} \leq 10m$. Repeatedly increase j by 1 until the first statement of Lemma 5 holds. The first statement must hold when $j = k$, in which we have $d_H(x, y) \leq 24 \cdot 10^{k-i-1} d(x, y) \leq 24 \cdot 10^{k-i-1} (3\Delta)^{i+1} W$. This is maximized at $i = k - 1$, thus $d_H(x, y) \leq 24 \cdot (3\Delta)^k W$. We obtain additive stretch at most $24 \cdot (21 + \frac{42k}{\varepsilon})^k W = O\left(\frac{k}{\varepsilon}\right)^k W$ as required. □

3.2 Lightness of the Spanner

To prove the desired lightness, we show that with high probability, the total lightness of the edges added in each phase does not exceed $\widetilde{O}(\frac{n^{1/k}}{\varepsilon})$. The following lemma bounds the lightness of H_0, created in the first phase.

Lemma 7. lightness$(H_0) = O(\frac{\log n}{\varepsilon})$.

Proof. Consider a fixed $i \in \{-1, 0, \ldots, \log_2 n\}$ and recall that $N_{-1} = V$. Each vertex in $N_i \setminus N_{i+1}$ adds shortest paths of total length at most $2^{i+1} + 2^{i+2} + \ldots + 2^{i+\lceil \log(1/\varepsilon) \rceil} = O(\frac{1}{\varepsilon}) \cdot 2^i$. Recall that by Lemma 1, $|N|\Delta \leq 2w(MST(G))$ for any Δ-net N. It follows that the total length of all shortest paths added corresponding to vertices in $N_i \setminus N_{i+1}$ is at most $O(\frac{1}{\varepsilon})w(MST(G))$. We have $O(\log n)$ nets, so the lemma follows. □

In the second phase, we add a shortest path from each vertex to the representatives of its $\frac{1-\varepsilon}{2}$-bunch. We start by bounding the total edge weight from all the paths to the representatives in the nets. Define for any $0 \leq j \leq \log n$, $0 \leq i \leq k-1$, and $x \in N_j$, the set $C_{ij}(x) = \{u \in A_i \setminus A_{i+1} \ : \ \exists v \in B_{\frac{1-\varepsilon}{2}}(u), x = r_j(v)\}$. This is the set of vertices $u \in A_i \setminus A_{i+1}$ that added a path to x because this x is a level-j representative of some $v \in B_{\frac{1-\varepsilon}{2}}(u)$.

Lemma 8. *Let $j \in \{0, \ldots, \log_2 n\}$, $x \in N_j$ and $0 \leq i \leq k-1$. Then $\exists u \in C_{ij}(x)$ such that $C_{ij}(x) \subseteq B_1(u)$ (Fig. 2).*

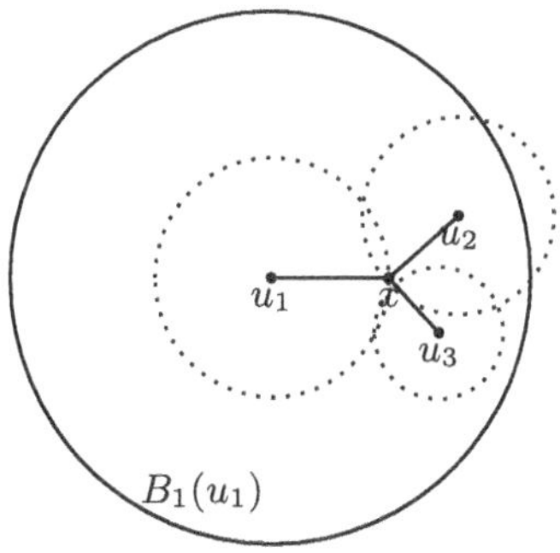

Fig. 2. Illustration of Lemma 8. The dotted circles represent the $\frac{1-\varepsilon}{2}$-bunches centered at u_1, u_2, and u_3 respectively. Lemma 8 states that all the u_i's are contained in a 1-bunch of some vertex in A_i.

Proof. Let $u \in C_{ij}(x)$ be the vertex farthest from x.

As x is a level j representative of a vertex $v \in B_{\frac{1-\varepsilon}{2}}(u)$, then by Lemma 2 we have $d(v,x) \leq d_H(v,x) \leq (1+2\varepsilon)2^j \leq 2 \cdot 2^j$. Recall that j was chosen such that $2^j \leq \frac{\varepsilon}{4}d(u,v)$, hence

$$d(u,x) \leq d(u,v) + d(v,x) \leq (1+\frac{\varepsilon}{2})d(u,v). \tag{4}$$

It follows that for any $u' \in C_{ij}(x)$,

$$\begin{aligned} d(u,u') &\leq d(u,x) + d(x,u') \leq 2d(u,x) \\ &\overset{(4)}{\leq} 2(1+\frac{\varepsilon}{2})d(u,v) \leq 2(1+\frac{\varepsilon}{2})(\frac{1-\varepsilon}{2})d(u,A_{i+1}) < d(u,A_{i+1}). \end{aligned}$$

Therefore by definition $u' \in B_1(u)$. □

Fix $j \in \{0,\ldots,\log_2 n\}$, $x \in N_j$, and $i \in \{0,\ldots,k-1\}$. We claim that each $u \in C_{ij}(x)$ adds a path to x of weight $O(\frac{2^j}{\varepsilon})$. To see this, let $v \in B_{\frac{1-\varepsilon}{2}}(u)$ such that $x = r_j(v)$, then $2^j \geq \frac{\varepsilon}{8}d(u,v)$, and by (4) we have that $d(u,x) \leq (1+\frac{\varepsilon}{2})d(u,v) = O\left(\frac{2^j}{\varepsilon}\right)$.

The following lemma was shown in [33], we give a proof for completeness.

Lemma 9. *With high probability, for every* $u \in V$, $|B_1(u)| \leq O(n^{1/k}\ln n)$.

Proof. Let $0 \leq i \leq k$ be such that $u \in A_i \setminus A_{i+1}$. Suppose first that $i \leq k-1$. Then if we arrange all vertices in A_i by non-decreasing order of distance from u in G, each vertex in A_i is promoted to A_{i+1} independently with probability $n^{-1/k}$. The event that $|B_1(u)| \geq 2n^{1/k}\ln n$ is contained in the event that none of the first $2n^{1/k}\ln n$ vertices in this ordering is promoted. By independence, the probability of the latter event is

$$(1-n^{-1/k})^{2n^{1/k}\ln n} \leq e^{-2\ln n} = \frac{1}{n^2},$$

so by taking a union bound over the n vertices, with probability at least $1 - \frac{1}{n}$ all bunches for $i < k$ are of size at most $O(n^{1/k} \ln n)$. Finally, note that for $u \in A_k$ we have $B_1(u) = A_k$. As each vertex is included in A_k with probability $\frac{1}{n}$ independently, its expected size is 1, and by a simple Chernoff bound with high probability $|A_k| \le O(\ln n)$. □

Assume, as we may, that this high probability event did occur. Thus for any representative $x \in N_j$, by Lemma 9, there are only $O(n^{1/k} \log n)$ paths to x from different $u \in C_{ij}(x)$ for each $0 \le i \le k-1$. Furthermore, each of these paths has length $O(\frac{2^j}{\varepsilon})$. We conclude that the total weight of paths connecting vertices in $V \setminus A_K$ to representatives in N_j is $O\left(k \cdot |N_j| \cdot n^{1/k} \log n \cdot \frac{2^j}{\varepsilon}\right)$. By Lemma 1 we have that $|N_j| \cdot 2^j \le w(MST)$, so the lightness of these paths is $O\left(\frac{kn^{1/k} \log n}{\varepsilon}\right)$. As we have only $O(\log n)$ different choices for $0 \le j \le \log n$ we get lightness $O\left(\frac{kn^{1/k}}{\varepsilon} \cdot \log^2 n\right) = \widetilde{O}(\frac{n^{1/k}}{\varepsilon})$.

It remains to bound the lightness from the paths between vertices in A_k. By our assumption $|A_k| \le O(\log n)$, so there can be at most $O(\log^2 n)$ vertex pairs in A_k. Each such pair may add one path to the spanner of total weight at most $w(MST(G))$, so the lightness of these paths is bounded by $O(\log^2 n)$.

In the second phase we also add direct shortest paths between $u - v$ for any $v \in B_{\frac{1-\varepsilon}{2}}(u)$ such that the unique integer j satisfying $\frac{\varepsilon}{8} \cdot d(u,v) \le 2^j < \frac{\varepsilon}{4} \cdot d(u,v)$ is negative (that is, v is its own representative). In this case, we have $d(u,v) = O(\frac{1}{\varepsilon})$, and by Lemma 9, the total number of such $u - v$ pairs is $O(n^{1+1/k} \log n)$ (once again, assuming the high probability event of the lemma). We conclude that the total lightness of the direct paths is $\widetilde{O}\left(\frac{n^{1/k}}{\varepsilon}\right)$.

Finally, in the third phase we add the edges of k $(1+\varepsilon, O\left(\frac{1}{\varepsilon}\right))$-SLTs to H. Thus, the SLTs have $O(\frac{k}{\varepsilon}) = O(\frac{\log n}{\varepsilon})$ lightness.

Altogether, H has lightness $O(\frac{\log n}{\varepsilon})$ from the first and third phases of the construction (Lemma 7), and with high probability $(\frac{kn^{1/k}}{\varepsilon} \cdot \log^2 n) = \widetilde{O}(\frac{n^{1/k}}{\varepsilon})$ lightness from the second phase. Therefore the lightness of H is $\widetilde{O}(\frac{n^{1/k}}{\varepsilon})$.

3.3 Size of the Spanner

To show the size bound, we must first show that the expected number of edges added at each phase does not exceed $O(kn^{1+3/k})$. In the first phase, we add $O(n \log n)$ edges from the shortest paths to the net points; this follows as, for each $j \in \{0, \dots, \log_2 n\}$, the set of edges added when connecting vertices to the net N_j are a subset of the edges in a shortest path forest rooted at N_j. Since there are $O(\log n)$ nets, we obtain $O(n \log n)$ edges. Note that $n \log n \le kn^{1+3/k}$ for any value of $k \ge 1$.

We now bound the number of edges added in the second phase. The next lemma is an extension of a lemma in [16]:

Lemma 10. *Let $0 \le i \le k$. Let $u, x \in A_i$, and let v (resp. y) be a representative of some vertex in $B_{\frac{1-\varepsilon}{2}}(u)$ (resp. $B_{\frac{1-\varepsilon}{2}}(x)$) for which a shortest u-v path $P_{u,v}$ (resp. x-y path $P_{x,y}$) is added to H. If $P_{u,v} \cap P_{x,y} \neq \emptyset$, then all four points are in $B_1(u)$, or all four points are in $B_1(x)$.*

Proof. Let $z \in V$ be a vertex in the intersection of $P_{u,v}$ and $P_{x,y}$. Assume first that $P_{u,v}$ is not shorter than $P_{x,y}$, and we show all points are in $B_1(u)$ (the other case is symmetric, in which all points are in $B_1(x)$). We have

$$\begin{aligned} d(u,x) &\le d(u,z) + d(z,x) \\ &\le 2d(u,v) \overset{(4)}{\le} 2(1+\varepsilon)\left(\frac{1-\varepsilon}{2}\right) d(u, A_{i+1}) \\ &< d(u, A_{i+1}). \end{aligned}$$

Since $d(z,y) \le d(u,v)$ as well, the same bound holds for $d(u,y)$. □

For fixed $0 \le i \le k-1$, consider the graph G_i containing all shortest paths $P_{u,v}$ where $u \in A_i$ and v is the representative of some vertex in $B_{\frac{1-\varepsilon}{2}}$ that u connects to. The number of edges in G_i is at most $O(n + C_i)$, where C_i is the number of pairwise intersections between these paths.

Lemma 11. $\mathbb{E}(|C_i|) = O(n^{(k+3-i)/k})$.

Proof. By Lemma 10, for every intersecting pair of paths $P_{u,v}$ and $P_{x,y}$ in G_i, we have that all four points u, v, x, and y belong to the same 1-bunch. Each $u \in A_i$ introduces at most $|B_1(u)|^3$ pairwise intersecting paths (since u is fixed and $v, x, y \in B_1(u)$). Note that $|B_1(u)|$ is a random variable distributed geometrically with parameter $n^{-1/k}$, so

$$\begin{aligned} \mathbb{E}(|B_1(u)|^3) &= \sum_{j=1}^{\infty} j^3 n^{-1/k}(1-n^{-1/k})^{j-1} = n^{-1/k}\sum_{j=1}^{\infty} j^3 (1-n^{-1/k})^{j-1} \\ &\le n^{-1/k}\sum_{j=1}^{\infty}(1-n^{-1/k})^{j-1} j(j+1)(j+2) \le \frac{6}{n^{-3/k}} = O(n^{3/k}). \end{aligned}$$

We note that the expected size of A_i is $n^{1-i/k}$, since every vertex joins A_i with probability $n^{-i/k}$. The expected number of intersections at level i, and therefore the expected size of G_i, is

$$C_i = \mathbb{E}\left[\sum_{u \in A_i} |B_1(u)|^3\right] = O(n^{3/k}) \cdot n^{1-i/k} = O(n^{(k+3-i)/k}).$$

□

For $i = k$, the expected size of $\binom{|A_k|}{2}$ is constant, so the expected number of edges added between vertices in A_k is $O(n)$. Lastly, the k SLTs add $O(kn)$ edges. The expected size of H is $O(n \log n + n^{1+3/k} + kn) = O(kn^{1+3/k})$.

3.4 Proof of Theorem 1

By the analysis of Sect. 3.1, H is a $(1+O(\varepsilon), O(\frac{k}{\varepsilon})^k W(\cdot,\cdot))$-spanner. Sections 3.2 and 3.3 showed that with high probability the lightness is $\widetilde{O}\left(\frac{n^{1/k}}{\varepsilon}\right)$, and expected size of H is $O(kn^{1+3/k})$. Reducing the multiplicative stretch to $1+\varepsilon$ can be done by scaling ε by an appropriate constant. So by Markov's inequality, with constant probability we get a spanner H satisfying Theorem 1.

Acknowledgments. This study was funded by Israel Science Foundation (grant No. 970/21).

References

1. Abboud, A., Bodwin, G.: The 4/3 additive spanner exponent is tight. J. ACM **64**(4), 28:1–28:20 (2017). https://doi.org/10.1145/3088511
2. Abboud, A., Bodwin, G., Pettie, S.: A hierarchy of lower bounds for sublinear additive spanners. In: Klein, P.N. (ed.) Proceedings of the Twenty-Eighth Annual ACM-SIAM Symposium on Discrete Algorithms, SODA 2017, Barcelona, Spain, Hotel Porta Fira, 16–19 January, pp. 568–576. SIAM (2017). https://doi.org/10.1137/1.9781611974782.36
3. Ahmed, R., Bodwin, G., Hamm, K., Kobourov, S., Spence, R.: On additive spanners in weighted graphs with local error. In: 47th International Workshop on Graph-Theoretic Concepts in Computer Science (2021)
4. Ahmed, R., et al.: Graph spanners: a tutorial review. Comput. Sci. Rev. **37**, 100–253 (2020). https://doi.org/10.1016/j.cosrev.2020.100253. http://www.sciencedirect.com/science/article/pii/S1574013719302539
5. Aingworth, D., Chekuri, C., Indyk, P., Motwani, R.: Fast estimation of diameter and shortest paths (without matrix multiplication). SIAM J. Comput. **28**, 1167–1181 (1999). https://doi.org/10.1137/S0097539796303421
6. Althöfer, I., Das, G., Dobkin, D., Joseph, D., Soares, J.: On sparse spanners of weighted graphs. Discret. Comput. Geom. **9**(1), 81–100 (1993). https://doi.org/10.1007/BF02189308
7. Bartal, Y., Filtser, A., Neiman, O.: On notions of distortion and an almost minimum spanning tree with constant average distortion. J. Comput. Syst. Sci. **105**, 116–129 (2019). https://doi.org/10.1016/J.JCSS.2019.04.006
8. Baswana, S., Kavitha, T., Mehlhorn, K., Pettie, S.: Additive spanners and (α, β)-spanners. ACM Trans. Algorithms (TALG) **7**(1), 5 (2010)
9. Bodwin, G.: An alternate proof of near-optimal light spanners. CoRR abs/2305.18647 (2023). https://doi.org/10.48550/arXiv.2305.18647
10. Bodwin, G., Williams, V.V.: Very sparse additive spanners and emulators. In: Proceedings of the 2015 Conference on Innovations in Theoretical Computer Science, ITCS 2015, pp. 377–382. Association for Computing Machinery, New York, NY, USA (2015). https://doi.org/10.1145/2688073.2688103
11. Borradaile, G., Le, H., Wulff-Nilsen, C.: Minor-free graphs have light spanners. In: 2017 IEEE 58th Annual Symposium on Foundations of Computer Science, FOCS 2017, pp. 767–778 (2017). https://doi.org/10.1109/FOCS.2017.76

12. Borradaile, G., Le, H., Wulff-Nilsen, C.: Greedy spanners are optimal in doubling metrics. In: Proceedings of the 30th Annual ACM-SIAM Symposium on Discrete Algorithms, SODA 2019, pp. 2371–2379 (2019). https://doi.org/10.1137/1.9781611975482.145
13. Chandra, B., Das, G., Narasimhan, G., Soares, J.: New sparseness results on graph spanners. In: Proceedings of the Eighth Annual Symposium on Computational Geometry, pp. 192–201. ACM (1992)
14. Chechik, S.: New additive spanners. In: Proceedings of the Twenty-Fourth Annual ACM-SIAM Symposium on Discrete Algorithms (SODA), pp. 498–512. Society for Industrial and Applied Mathematics (2013)
15. Chechik, S., Wulff-Nilsen, C.: Near-optimal light spanners. In: Proceedings of the Twenty-Seventh Annual ACM-SIAM Symposium on Discrete Algorithms (SODA), pp. 883–892. Society for Industrial and Applied Mathematics (2016)
16. Elkin, M., Gitlitz, Y., Neiman, O.: Almost shortest paths with near-additive error in weighted graphs. In: Czumaj, A., Xin, Q. (eds.) 18th Scandinavian Symposium and Workshops on Algorithm Theory, SWAT 2022, Tórshavn, Faroe Islands, 27–29 June 2022. LIPIcs, vol. 227, pp. 23:1–23:22. Schloss Dagstuhl - Leibniz-Zentrum für Informatik (2022). https://doi.org/10.4230/LIPICS.SWAT.2022.23
17. Elkin, M., Gitlitz, Y., Neiman, O.: Improved weighted additive spanners. Distrib. Comput. **36**(3), 385–394 (2023). https://doi.org/10.1007/S00446-022-00433-X
18. Elkin, M., Neiman, O.: Efficient algorithms for constructing very sparse spanners and emulators. ACM Trans. Algorithms **15**(1), 4:1–4:29 (2019). https://doi.org/10.1145/3274651
19. Elkin, M., Neiman, O., Solomon, S.: Light spanners. In: International Colloquium on Automata, Languages, and Programming, pp. 442–452. Springer (2014)
20. Elkin, M., Peleg, D.: $(1+\varepsilon, \beta)$-spanner constructions for general graphs. SIAM J. Comput. **33**(3), 608–631 (2004). https://doi.org/10.1137/S0097539701393384
21. Erdős, P.: Extremal problems in graph theory. In: Proceedings of the Symposium on Theory of Graphs and its Applications, pp. 29–36 (1963)
22. Filtser, A., Gitlitz, Y., Neiman, O.: Light, reliable spanners. In: SOCG 2024 (2024, to apear)
23. Filtser, A., Neiman, O.: Light spanners for high dimensional norms via stochastic decompositions. Algorithmica **84**(10), 2987–3007 (2022). https://doi.org/10.1007/S00453-022-00994-0
24. Filtser, A., Solomon, S.: The greedy spanner is existentially optimal. SIAM J. Comput. **49**(2), 429–447 (2020). https://doi.org/10.1137/18M1210678, preliminary version published in PODC 2016
25. Gottlieb, L.: A light metric spanner. In: IEEE 56th Annual Symposium on Foundations of Computer Science, FOCS 2015, Berkeley, CA, USA, 17–20 October 2015, pp. 759–772 (2015). https://doi.org/10.1109/FOCS.2015.52
26. Khuller, S., Raghavachari, B., Young, N.: Balancing minimum spanning trees and shortest-path trees. Algorithmica **14**(4), 305–321 (1995)
27. Le, H., Solomon, S.: A unified framework for light spanners. In: Saha, B., Servedio, R.A. (eds.) Proceedings of the 55th Annual ACM Symposium on Theory of Computing, STOC 2023, Orlando, FL, USA, 20–23 June 2023, pp. 295–308. ACM (2023). https://doi.org/10.1145/3564246.3585185
28. Neiman, O., Shabat, I.: A unified framework for hopsets. In: Chechik, S., Navarro, G., Rotenberg, E., Herman, G. (eds.) 30th Annual European Symposium on Algorithms, ESA 2022, Berlin/Potsdam, Germany, 5–9 September 2022. LIPIcs, vol. 244, pp. 81:1–81:13. Schloss Dagstuhl - Leibniz-Zentrum für Informatik (2022). https://doi.org/10.4230/LIPICS.ESA.2022.81

29. Peleg, D., Schäffer, A.A.: Graph spanners. J. Graph Theory **13**(1), 99–116 (1989)
30. Pettie, S.: Low distortion spanners. ACM Trans. Algorithms **6**(1), 7:1–7:22 (2009). https://doi.org/10.1145/1644015.1644022
31. Salowe, J.S.: Construction of multidimensional spanner graphs, with applications to minimum spanning trees. In: Proceedings of the Seventh Annual Symposium on Computational Geometry, SCG 1991, pp. 256–261. Association for Computing Machinery, New York, NY, USA (1991). https://doi.org/10.1145/109648.109677
32. Tan, Z., Zhang, T.: Almost-optimal sublinear additive spanners. In: Saha, B., Servedio, R.A. (eds.) Proceedings of the 55th Annual ACM Symposium on Theory of Computing, STOC 2023, Orlando, FL, USA, 20–23 June 2023, pp. 281–294. ACM (2023). https://doi.org/10.1145/3564246.3585125
33. Thorup, M., Zwick, U.: Approximate distance oracles. J. ACM **52**(1), 1–24 (2005). https://doi.org/10.1145/1044731.1044732
34. Thorup, M., Zwick, U.: Spanners and emulators with sublinear distance errors. In: SODA 2006, pp. 802–809 (2006). https://doi.org/10.1145/1109557.1109645
35. Vaidya, P.M.: A sparse graph almost as good as the complete graph on points in K dimensions. Discret. Comput. Geom. **6**(3), 369–381 (1991). https://doi.org/10.1007/BF02574695
36. Woodruff, D.: Additive spanners in nearly quadratic time. In: International Colloquium on Automata, Languages, and Programming, pp. 463–474. Springer (2010)

Finite Vertex-Colored Ultrahomogeneous Oriented Graphs

Irene Heinrich(✉), Eda Kaja(✉), and Pascal Schweitzer(✉)

TU Darmstadt, Darmstadt, Germany
{heinrich,kaja}@mathematik.tu-darmstadt.de

Abstract. A relational structure is ultrahomogeneous if every isomorphism of finite induced substructures extends to an automorphism of the structure. We classify the ultrahomogeneous finite binary relational structures with one asymmetric binary relation and arbitrarily many unary relations. In other words, we classify the finite vertex-colored oriented ultrahomogeneous graphs. The classification comprises several general methods with which directed graphs can be combined or extended to create new ultrahomogeneous graphs. Together with explicitly given exceptions, we obtain exactly all vertex-colored oriented ultrahomogeneous graphs this way. Our main technique is a technical tool that characterizes precisely under which conditions two binary relational structures with disjoint unary relations can be combined to form a larger ultrahomogeneous structure.

Keywords: Homogeneous graphs · Automorphism groups · Colored graphs

1 Introduction

A relational structure $\mathcal{R}$ is *ultrahomogeneous* if every isomorphism between finite induced substructures of $\mathcal{R}$ extends to an automorphism of $\mathcal{R}$. Ultrahomogeneity is a natural generalization of transitivity: a graph is vertex-transitive (edge-transitive) if for each two vertices (edges) of the graph there exists an automorphism of the graph which maps the one vertex (edge) to the other. In some sense, ultrahomogeneous structures form the most symmetric structures possible. Indeed, we can think of ultrahomogeneity as stating the following. If two parts of the structure locally look the same then there is a global symmetry of the object demonstrating that the parts are indeed structurally the same, even if the entire structure is taken into account. The emergence and separate handling of highly symmetric structures is unavoidable in various algorithms for symmetry detection and exploitation. For example, Babai's celebrated quasipolynomial time algorithm [2] for the graph isomorphism problem relies on the techniques of local certificates, which treats the case of complete symmetry separately.

Being highly symmetric objects, ultrahomogeneous structures have been extensively studied over the years. In fact, there are numerous books, surveys,

D. Kráľ and M. Milanič (Eds.): WG 2024, LNCS 14760, pp. 251–265, 2025.
https://doi.org/10.1007/978-3-031-75409-8_18

and major results on the matter (see for example [4,6,8,14,16,20–22]). Beyond the intrinsic combinatorial interest in ultrahomogeneous structures, part of their appeal is their applicability in model theory in the form of stability theory, ω-categoricity, and Fraïssé limits [1,8]. They also have natural applications in the study of permutation groups and Ramsey theory [3].

Homogeneity is usually considered for countable structures. Having algorithmic applications in mind, however, in this paper we will focus exclusively on finite structures. In fact, even when only considering finite ultrahomogeneous structures, there exists an extensive body of research. Finite simple graphs have been independently classified by Gardiner [9] and by Gol'fand and Klin [11]. These graphs are, up to taking complements, disjoint unions of complete graphs all of the same order, the 5-cycle, and the line graph of the $K_{3,3}$ (also known as 3×3 rook's graph). Subsequently Lachlan classified finite ultrahomogeneous digraphs [15]. These include further infinite families and some exceptional graphs. The oriented graphs among them are described in Theorem 1 and Fig. 2 below. Apart from graphs, finite ultrahomogeneous groups [5,7,18] and finite ternary relational structures (sometimes called 3-graphs) [17] have been classified.

A major research program initiated by Cherlin that aimed at classifying ultrahomogeneous edge colored graphs (or equivalently structures with only binary relations) recently led to the classification of their automorphism groups (primitive binary permutation groups) [10].

Crucially, none of the structures discussed so far have unary relations (in the terminology of graphs they are without loops or vertex colors). This implies in particular, that the structures are transitive. However, this limits the applicability of the results because no structures with different types of atoms (vertices) can be captured. Such types of vertices can be modeled with vertex colors. For ultrahomogeneous vertex-colored graphs, research is limited. There is a classification when the color classes form independent sets [13,19]. More generally, the ultrahomogeneous vertex-colored undirected finite graphs were recently classified [12].

Results. In this paper we classify ultrahomogeneous vertex-colored oriented finite graphs. This includes vertex-colored tournaments. In the language of relational structures this means we classify structures with a binary asymmetric relational structure and an arbitrary number of unary relations.

The classification essentially says the following. Up to certain forms of equivalence (namely color change and bichromatic symmetrization, see Definition 4) the graphs are color disjoint unions (i.e., disjoint unions of graphs with disjoint vertex-colors) of specific types of blow-ups of the following graphs:

1. the graph H_0,
2. a disjoint union of finitely many isomorphic copies of a discretely colored tournament,
3. a graph in which each color class forms a directed triangle and each pair of color classes is joined by a directed 6-Cycle $\overrightarrow{C_6}$.

Figure 1 shows an example containing each of the three possible building blocks but without blow-ups. See Definition 5 for a formal definition of the specific blow-ups used in our classification. Also see Theorem 6 for the formal theorem describing our classification. The monochromatic graphs that may appear as subgraphs induced by the color classes are described in Theorem 1 and shown in Fig. 2.

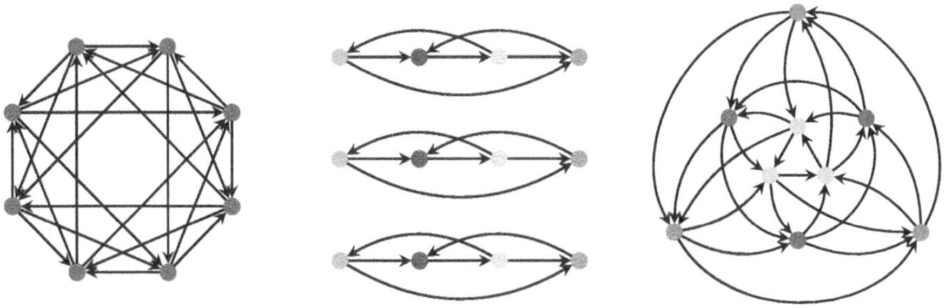

Fig. 1. Example of an ultrahomogeneous oriented vertex colored graph (no blow-ups).

Techniques. We develop techniques to analyze under what conditions and how two monochromatic ultrahomogeneous graphs with different vertex colors can be non-trivially connected to form a new ultrahomogeneous graph. The general extension theorem (Theorem 2) describes five necessary and sufficient conditions for this. The crucial insight is that the automorphism group of the one side must be compatible with what we call an ultrahomogeneous system of partitions on the other side. This severely restricts pairs of ultrahomogeneous graphs that can be connected to create new ultrahomogeneous graphs.

A second concept we introduce is a certain kind of a blow-up. Our concept here is more general than other blow-ups that have been previously used in the context of ultrahomogeneous structures. It allows for nontrivial connections between the replaced blocks. It thus allows us to organically recover some of the exceptional graphs as blow-ups of smaller graphs.

Overall these two techniques provide us with a clean and systematic way of analyzing highly symmetric graphs, dramatically reducing the number of cases that need to be considered. In fact the classification follows using fairly easy counting arguments. We first classify bichromatic oriented ultrahomogeneous graphs (Theorem 5). We then extend the bichromatic case to the general case (Theorem 6). We highlight that our general extension theorem and the more general blow-ups are not particular to the oriented case and apply to vertex colored binary relational structures in general.

Due to space limitations most proofs are omitted.

2 Preliminaries

For $n, n' \in \mathbb{N}$ we set $[n] := \{1, 2, \dots, n\}$ and $[n', n] := \{n', \dots, n\}$. For $i \in \mathbb{N}_{\geq 0}$ we denote the projection on the i-th coordinate by π_i, where the underlying set will always be clear from context. An *ordered partition* A of a set V is a tuple $(P_1, P_2, \dots, P_k)$ of disjoint non-empty subsets of V such that $\bigcup_{i \in [k]} P_i = V$. If $k = 1$, then A is called the *trivial* partition of V. A partition $(P_1, P_2, \dots, P_k)$ is *discrete* if $|P_i| = 1$ for all $i \in [k]$. We use the symbol $\dot\cup$ for the disjoint union.

Digraphs. A *directed graph* (short: *digraph*) G is a pair (V, E) where V is a non-empty set and $E \subseteq V^2$. In this paper, all considered digraphs are *finite* and *loopless*. An element of V is a *vertex* of G and an element of E is an *edge* of G. With $V(G)$ and $E(G)$ we refer to the vertices and edges of G, respectively. The *complement* $\overline{G}$ of G is the digraph with vertex set $V(G)$ and edge set $V^2 \setminus (E(G) \cup \{(v, v) \colon v \in V(G)\})$. A digraph G is an *oriented graph* if $(u, v) \in E(D)$ implies $(v, u) \notin E(D)$ for all vertices u and v of G. If $E(G) = V^2 \setminus \{(v, v) \colon v \in V(G)\}$, then G is *complete*.

(Di-)Graph families. Fix $n \in \mathbb{N}_{\geq 1}$. The *edgeless* (di-)graph E_n has vertex set $[n]$ and an empty edge set. The *directed cycle* $\overrightarrow{C_n}$ is the oriented graph on the vertex set $[n]$ with $E(\overrightarrow{C_n}) = \{(i, i+1) \colon i \in [n-1]\} \cup \{(n, 1)\}$. For simplicity, we call $\overrightarrow{C_3}$ the *directed triangle*.

Complete Colored Digraphs. A complete digraph G with a vertex coloring χ_G and an edge coloring ζ_G is a *complete colored digraph (CCD)*. Note that every directed graph D with a vertex coloring χ_D can be regarded as a CCD G with $V(G) = V(D)$, $\chi_G = \chi_D$, and $\zeta_G(u, v) = 1$ if $(u, v) \in E(D)$ and $\zeta_G(u, v) = 0$ otherwise. Note that this translation from CCDs to vertex-colored directed graphs preserves (partial) isomorphisms. Most of our techniques are formulated as statements on CCDs but in this manner they can easily be transferred to statements on directed graphs. An inclusion-wise maximal subset U of $V(G)$ with $|\chi_G(U)| = 1$ is a *vertex color class* of G. An *edge color class* of G is defined analogously. If G is a CCD and $U \subseteq V(G)$, then the CCD $G[U]$ with vertex set U, vertex coloring $\chi|_U$, and edge coloring $\psi|_{U \times U}$ is the *induced subgraph of G by U*. Observe that every (colored directed) graph H can be expressed as a CCD: all pairs of distinct vertices not present in $E(H)$ are interpreted in the CCD as edges with the same previously unused color.

Connectivity Types. Let R and B be two disjoint vertex subsets of a CCD G. We say that R and B are *homogeneously connected* if both $\zeta_G|_{R \times B}$ and $\zeta_G|_{B \times R}$ are constant. If R and B are not homogeneously connected and there exists a bijection $\alpha \colon R \to B$ where $\zeta_G|_{B \times R \setminus \{(\alpha(r), r) \colon r \in R\}}$, $\zeta_G|_{\{(r, \alpha(r)) \colon r \in R\}}$, $\zeta_G|_{\{(\alpha(r), r) \colon r \in R\}}$, and $\zeta_G|_{R \times B \setminus \{(r, \alpha(r)) \colon r \in R\}}$ are constant, then R and B are *matching-connected*. If $|\chi_G(V(G))| = |V(G)|$, then χ_G is a *discrete coloring*.

Isomorphisms and Ultrahomogeneity. Let G and G' be two CCDs. A bijection $\varphi\colon V(G) \to V(G')$ which satisfies for all vertices u and v in $V(G)$ that $\chi_{G'}(\varphi(v)) = \chi_G(v)$ and $\zeta_{G'}((\varphi(u), \varphi(v))) = \zeta_G((u,v))$ is an *isomorphism*. If additionally $G = G'$, then φ is an *automorphism* of G. The set of all automorphisms of G forms a group under composition, which we denote by $\mathrm{Aut}(G)$. An isomorphism of two induced subgraphs of G is a *partial isomorphism*. A partial isomorphism $\varphi'\colon U \to W$ of G *extends to an automorphism* of G if there exists $\psi \in \mathrm{Aut}(G)$ such that $\psi|_U = \varphi'$. A CCD G is *ultrahomogeneous* if every partial isomorphism of G extends to an automorphism of G. Observe that if G is ultrahomogeneous, then:

$$\forall (u,v), (u',v') \in E(G)\colon \zeta_G(u,v) = \zeta_G(u',v') \implies \zeta_G(v,u) = \zeta_G(v',u'). \quad (1)$$

Hence, also every induced subgraph of an ultrahomogeneous CCD satisfies (1). Throughout this article we assume without loss of generality that every considered CCD satisfies (1).

The Wreath Product of Graphs. Let D and D' be two CCDs. The *wreath product*[1] $D \cdot D'$ is a CCD with vertex set $V(D) \times V(D')$, vertex colors $\chi_{D\cdot D'}(u,u') := (\chi_D(u), \chi_{D'}(u'))$ for all $(u,u') \in V(D) \times V(D')$, and edge colors

$$\zeta_{D\cdot D'}((u,u'),(v,v')) := \begin{cases} \zeta_{D'}((u',v')) & \text{if } u = v, \\ \zeta_D(u,v) & \text{if } u \neq v. \end{cases}$$

Groups. Let V be a non-empty set. We denote the *symmetric group* of all permutations of V by $\mathrm{Sym}(V)$. A *permutation group* Γ on V is a subgroup of $\mathrm{Sym}(V)$. For $v \in V$ and $\gamma \in \Gamma$ we set $v^\gamma := \gamma(v)$. An *action* of Γ on V is a homomorphism ϕ from Γ to $\mathrm{Sym}(V)$. The image of an action of Γ on V is a subgroup of $\mathrm{Sym}(V)$ called the permutation group *induced* by Γ on V, which we denote by Γ^V. The *orbit* of an element x in V is the set $x^\Gamma := \{x^\gamma \colon \gamma \in \Gamma\}$. We say that Γ is *transitive* on V if $x^\Gamma = V$ for all $x \in V$. The *stabilizer* of an element x in V is the set $\mathrm{Stab}_\Gamma(x) := \{\gamma \in \Gamma \colon x^\gamma = x\}$. The *setwise stabilizer of a subset* X of V is the set of elements $\gamma \in \Gamma$ such that $X^\gamma = X$, that is, $x^\gamma \in X$ for every $\gamma \in \Gamma$ and $x \in X$. The *pointwise stabilizer of* X is the set $\mathrm{pw\,Stab}(X) = \bigcap_{x\in X} \mathrm{Stab}_\Gamma(x)$. We may also say *set stabilizer* and *point stabilizer*, respectively.

A *permutational isomorphism* between two permutation groups $\Gamma \leq \mathrm{Sym}(V)$ and $\Gamma' \leq \mathrm{Sym}(V')$ is a bijection $\rho\colon V \to V'$ such that $\Gamma' = \{\rho\gamma\rho^{-1} \colon \gamma \in \Gamma\}$.

Block Systems. Let $\Gamma \leq \mathrm{Sym}(V)$ be transitive. A *block* is a subset X of V such that $X^\gamma = X$ or $X^\gamma \cap X = \emptyset$ for all $\gamma \in \Gamma$. If $X = \{x\}$ for some $x \in V$ or $X = V$, then the block X is *trivial*, and otherwise it is *non-trivial*. If X is a block, then

[1] Also called the *lexicographic product*.

the set $\{X^\gamma \colon \gamma \in \Gamma\}$ is an unordered partition of V which is invariant under the action of Γ. In this case we call $\{X^\gamma \colon \gamma \in \Gamma\}$ a *block system* of V. Note that every permutation in Γ naturally induces a permutation of the blocks in a block system $\mathcal{B}$ of V. We denote the subgroup of $\mathrm{Sym}(\mathcal{B})$ which contains all permutations induced by permuations in Γ by $\Gamma^{\mathcal{B}}$.

The Wreath Product of Groups. Let $\Gamma \leq \mathrm{Sym}(V)$ and $\Gamma' \leq \mathrm{Sym}(Y)$. The *wreath product* of Γ with Γ', denoted $\Gamma \wr \Gamma'$ is the group of all permutations δ acting on $V \times Y$ for which there exist $\gamma \in \Gamma$ and an element γ'_x of Γ' for each $x \in V$, such that $\delta(x, y) = (\gamma(x), \gamma'_x(y))$ for every $(x, y) \in V \times Y$.

Remark 1. If G and G' are two edge-color disjoint CCDs with respective automorphism groups Γ and Γ', then the automorphism group of $G \cdot G'$ is permutationally equivalent to $\Gamma' \wr \Gamma$.

Families of Groups. Fix $n \in \mathbb{N}_{\geq 1}$. Set $\mathrm{Sym}(n) := \mathrm{Sym}([n])$. We denote the cyclic group of order n by $\mathbb{Z}_n$ and the alternating group on n vertices by $\mathrm{Alt}(n)$.

Monochromatic Graphs. Lachlan's [15] classic results classify the monochromatic ultrahomogeneous graphs. The oriented graphs among them, which are relevant to our classification, are depicted in Fig. 2 and are as follows.

Theorem 1 ([15]). *An oriented graph is ultrahomogeneous if and only if it is isomorphic to one of* $\overrightarrow{C_4}$, E_n, $E_n \cdot \overrightarrow{C_3}$, $\overrightarrow{C_3} \cdot E_n$, *or* H_0 *for some* $n \in \mathbb{N} \setminus \{0\}$.

3 Extension Theorems

Let G be a CCD, A an ordered partition of $V(G)$, and $\varphi \in \mathrm{Aut}(G)$. By $\varphi(A)$ we denote the ordered partition of $V(G)$ with $\pi_i(\varphi(A)) = \varphi(\pi_i(A))$ for all $i \in [|A|]$. Let $A_1, A_2, \ldots, A_k$ be a finite sequence of partitions of $V(G)$. We set

$$\chi_G(A_1, A_2, \ldots, A_k) \colon V \to \chi_G(V(G)) \times \mathbb{N}^k \quad v \mapsto (\chi_G(v), i_1, \ldots, i_k),$$

where $\forall j \in [k]$ $v \in \pi_{i_j}(A_j)$. That is, two vertices u and v of G have the same color with respect to the coloring $\chi_G(A_1, A_2, \ldots, A_k)$ precisely if they have the same color with respect to χ_G and for each $j \in [k]$ the two vertices u and v lie in the same part of the partition A_j.

Lemma 1. *Let G be a CCD and let $A_1, A_2, \ldots, A_k$ and $A'_1, A'_2, \ldots, A'_k$ be two finite sequences of partitions of $V(G)$. If there exist ψ and $\varphi_1, \varphi_2, \ldots, \varphi_{k-1}$ in $\mathrm{Aut}(G)$ such that $\psi(A_1) = A'_1$ and for all $i \in [k-1]$ it holds that $\varphi_i(A_i) = A_{i+1}$ and $\varphi_i(A'_i) = A'_{i+1}$, then $(G, \chi(A_1, A_2, \ldots, A_k))$ and $(G, \chi(A'_1, A'_2, \ldots, A'_k))$ are isomorphic via ψ.*

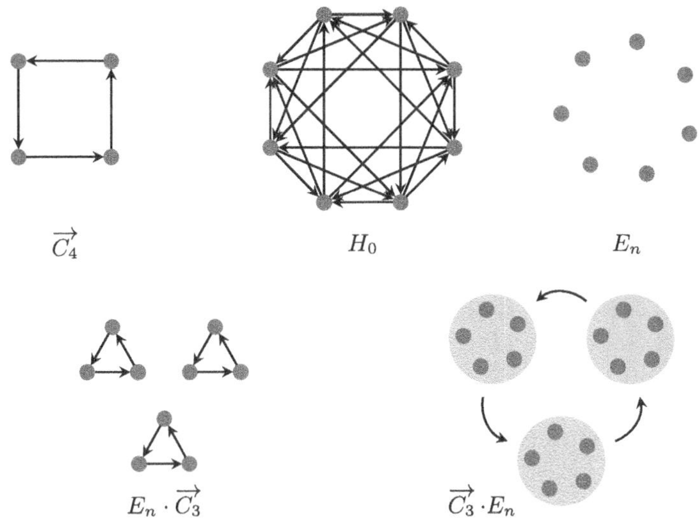

Fig. 2. The ultrahomogeneous oriented (monochromatic) graphs. For each of the three infinite families $\{E_n : n \in \mathbb{N}_{\geq 1}\}$, $\{E_n \cdot \overrightarrow{C_3} : n \in \mathbb{N}_{\geq 1}\}$, and $\{\overrightarrow{C_3} \cdot E_n : n \in \mathbb{N}_{\geq 1}\}$ we show one example. An arrow from one gray circle to another indicates all arcs from vertices in the first circle to vertices in the second are present.

Definition 1 (Ultrahomogeneous system of partitions). *Let G be a CCD. An ordered partition A of $V(G)$ is* ultrahomogeneous *if G with the vertex coloring $\chi_G(A)$ is ultrahomogeneous. We set $\mathcal{A}(A) := \{\varphi(A) : \varphi \in \mathrm{Aut}(G)\}$. If A is an ordered partition of $V(G)$ such that for every finite sequence $A_1, A_2, \ldots, A_\ell$ of partitions in $\mathcal{A}(A)$ the CCD G with the vertex coloring $\chi_G(A_1, A_2, \ldots, A_\ell)$ is ultrahomogeneous, then we call $\mathcal{A}(A)$ an* ultrahomogeneous system of partitions. *If $|\mathcal{A}(A)| \neq 1$, then $\mathcal{A}(A)$ is called* non-trivial.

Definition 2 (Easygoing CCD with respect to a block system). *Let H be a vertex-monochromatic CCD with a block system $\mathcal{B}$. We call H* easygoing *with respect to $\mathcal{B}$ if each subset $\mathcal{B}' \subseteq \mathcal{B}$ satisfies*

$$\left(\mathrm{pw\,Stab}_{\mathrm{Aut}(H)}\left(\bigcup \mathcal{B}'\right)\right)^{\mathcal{B}} = \mathrm{pw\,Stab}_{\mathrm{Aut}(H)^{\mathcal{B}}}\left(\mathcal{B}'\right),$$

i.e., the point stabilizer of the set $\bigcup \mathcal{B}' := \bigcup_{B \in \mathcal{B}'} B$ in the group $\mathrm{Aut}(H)$ induces the same group on $\mathcal{B}$ as the point stabilizer of $\mathcal{B}'$ in the induced group $\mathrm{Aut}(H)^{\mathcal{B}}$.

Note that every block system $\mathcal{B}$ of a vertex-monochromatic CCD H satisfies $\left(\mathrm{pw\,Stab}_{\mathrm{Aut}(H)}\left(\bigcup \mathcal{B}'\right)\right)^{\mathcal{B}} \subseteq \mathrm{Stab}_{\mathrm{Aut}(H)^{\mathcal{B}}}\left(\mathcal{B}'\right)$.

The property of being easygoing will be crucial for us to be able to specify how the vertices inside $\bigcup \mathcal{B}'$ are mapped while still being able to choose how blocks outside $\mathcal{B}'$ are mapped. This is done as follows.

Lemma 2. *Suppose H is easygoing with respect to a block system $\mathcal{B}$. If two automorphisms $\psi, \varphi \in \mathrm{Aut}(H)$ induce the same map on a set of blocks $\mathcal{B}' \subseteq \mathcal{B}$ (i.e. $\psi^{\mathcal{B}}|_{\mathcal{B}'} = \varphi^{\mathcal{B}}|_{\mathcal{B}'}$), then there is a $\tau \in \mathrm{Aut}(H)$ such that $\tau^{\mathcal{B}} = \psi^{\mathcal{B}}$ and $\tau(x) = \varphi(x)$ for all $x \in \bigcup \mathcal{B}'$.*

Theorem 2 (General extension theorem). *Let G be a CCD on precisely two vertex color classes R and B such that the edge colors on $B \times R$ are $1, 2, \ldots, \ell$. The graph G is ultrahomogeneous if and only if all of the following conditions are satisfied:*

1. *Both graphs $G[R]$ and $G[B]$ are ultrahomogeneous.*
2. *For $b \in B$ let $A(b) := (P_1, \ldots, P_\ell)$ where $P_i = \{r \in R \colon \zeta_G((b,r)) = i\}$. The set $\mathcal{A} := \{A(b) \colon b \in B\}$ is an ultrahomogeneous system of partitions.*
3. *For $A \in \mathcal{A}$ set $X(A) := \{b \in B \colon A(b) = A\}$. The set $\mathcal{B} := \{X(A) \colon A \in \mathcal{A}\}$ is a block system of $\mathrm{Aut}(G[B])$.*
4. *$\mathrm{Aut}(G[B])^{\mathcal{B}} = \{\hat{\varphi} \colon \varphi \in \mathrm{Aut}(G[R])^{\mathcal{A}}\}$ where $\hat{\varphi} \colon \mathcal{B} \to \mathcal{B}$, $X(A) \mapsto X(\varphi(A))$ for all $A \in \mathcal{A}$.*
5. *$G[B]$ is easygoing with respect to $\mathcal{B}$.*

We now investigate the case when the partition $A(b)$ is a block system of R for some (and thus for every) $b \in B$. For this case we are interested in minimal extensions, that is, extensions in which the sets $X(A)$ from the theorem contain only a single element $b \in B$.

Theorem 3 (Minimal extension theorem). *Let G be a CCD on precisely one vertex color class R. Suppose A is an ordered partition such that $\mathcal{A} := \mathcal{A}(A)$ is an ultrahomogeneous system of partitions. Suppose that as an unordered partition A is also a block system of $\mathrm{Aut}(G[R])$. Then the CCD $\widehat{G}$ on vertex set $B \cup R$ with $\widehat{G}[R] = G$ and with*

$$
\begin{aligned}
B &= \{b_A \colon A \in \mathcal{A}\}, \\
\zeta_G((r, b_A)) &= i \text{ if } r \in \pi_i(A) \text{ for all } r \in R \text{ and } A \in \mathcal{A}, \text{ and} \\
\zeta_G((b_A, b_{A'})) &= \textit{Iso-Type}((G, \chi(A, A'))),
\end{aligned}
$$

where Iso-Type$((G, \chi(A, A')))$ is the class of colored graphs that are isomorphic to $(G, \chi(A, A'))$, is ultrahomogeneous. In this case we call G the minimal ultrahomogeneous extension of $G[R]$ with respect to $\mathcal{A}$.

Example 1. Let $G \cong \overrightarrow{C_4}$ and label its vertices r_1, r_2, r_3 and r_4 in a cyclic fashion. Let $A = (P_1, P_2) := (\{r_1, r_3\}, \{r_2, r_4\})$. Then $\mathcal{A}(A) = \{A, A'\}$ with $A' := (P_2, P_1)$ is an ultrahomogeneous system of partitions. Note that A is a block system for $\mathrm{Aut}(G)$. The CCD $\widehat{G}$ with $\widehat{G}[R] = G$ as in Theorem 3 has the following properties. First, $B = \{b_A, b_{A'}\}$. Second, for $i \in \{1, 2\}$, $\zeta((r_i, b_A)) = \zeta((r_{i+2}, b_A)) = i$ since $r_i, r_{i+2} \in P_i$. It follows that $\zeta((r_i, b_{A'})) = 1$ if $\zeta((r_i, b_A)) = 2$ and $\zeta((r_i, b_{A'})) = 2$ if $\zeta((r_i, b_A)) = 1$ for $i \in [4]$. Lastly, $(G, \chi(A, A'))$ assigns the color $(\text{red}, 1, 2)$ to r_1 and r_3 and the color $(\text{red}, 2, 1)$ to r_2 and r_4. So $\zeta(b_A, b_{A'}) = \zeta(b_{A'}, b_A)$. Altogether we obtain that $\widehat{G}$ is indeed the minimal ultrahomogeneous extension of a red $\overrightarrow{C_4}$ with respect to $\mathcal{A}(A)$ (see Fig. 3, left).

4 Graph Operations Which Preserve Ultrahomogeneity

Two CCDs D and D' are *vertex-color disjoint* if $\chi_G(V(G)) \cap \chi_H(V(H)) = \emptyset$.

Definition 3 (Color disjoint union). *Let D and D' be two vertex-color disjoint CCDs and c a new edge color, i.e., the edge color c is neither present in D nor in D'. The* color disjoint union, *denoted $D \square D'$, is a CCD with vertex set $V(D)\dot{\cup}V(D')$ such that the vertices in $D \square D'$ inherit their original vertex colors, and edge colors*

$$\zeta_{D\square D'}(u,u') := \begin{cases} \zeta_D(u,u') & \text{if } u,u' \in D, \\ \zeta_{D'}(u,u') & \text{if } u,u' \in D', \\ c & \text{otherwise.} \end{cases}$$

Lemma 3. *The color disjoint union of two CCDs D and D' is ultrahomogeneous if and only if D and D' are ultrahomogeneous.*

If D is a CCD and $f\colon \zeta_D(E(D)) \to S$ is a bijection, then the CCD G with the same underlying graph and the same vertex coloring and with the edge coloring $f(\zeta_D)$ is an *edge color change* of D. A *vertex color change* is defined analogously using $f(\chi_D)$ for some bijection $f\colon \chi_D(V(D)) \to S$.

A *bichromatic symmetrization* for a CCD D is the operation that for some edge color $c \in \zeta_D(E(D))$ and distinct vertex colors $r, b \in \chi_D(V(G))$ replaces ζ_D by ζ_D^c where

$$\zeta_D^c(u,v) := \begin{cases} \zeta_D(v,u) & \text{if } \chi_D(u) = b \text{ and } \chi_D(v) = r, \\ \zeta_D(u,v) & \text{if } \chi_D(u) = r \text{ and } \chi_D(v) = b, \\ \zeta_D(u,v) & \text{otherwise.} \end{cases}$$

A coloring ζ_D^c is obtained from ζ_D by inverse bichromatic symmetrization, if ζ_D is a bichromatic symmetrization of ζ_D^c. We remind the reader that all our colorings satisfy $\zeta_G(u,v) = \zeta_G(u',v') \implies \zeta_G(v,u) = \zeta_G(v',u')$. Note that this property is preserved by all operations.

Definition 4 (equivalence up to color changes and bichromatic symmetrization). *We call two CCDs G and H* equivalent up to color changes and bichromatic symmetrization, *denoted $G \sim_c H$, if G can be obtained from H by a sequence of vertex color changes, edge color changes, bichromatic symmetrizations, or inverse bichromatic symmetrizations.*

We use the terms such as *equivalent up to vertex color changes*, *equivalent up to edge color changes*, and *equivalent up to (inverse) bichromatic symmetrization*

Lemma 4. *If G and H are equivalent up to color changes and bichromatic symmetrization, then both graphs are ultrahomogeneous or neither of them is.*

Lemma 5. *Suppose G is an ultrahomogeneous CCD and $\mathcal{B}$ is a block system of* $\mathrm{Aut}(G)$. *Then $G[X]$ is ultrahomogeneous for every block $X \in \mathcal{B}$.*

Definition 5. *Let G be a CCD with vertex color class R and $\Gamma := \mathrm{Aut}(G[R])$. Let H be a monochromatic CCD with a block system $\mathcal{B}$. Suppose τ is a permutational isomorphism from $\mathrm{Aut}(H)^{\mathcal{B}}$ to Γ.*

We define the blow-up $G[H \to_\tau R]$ of R in G by H via τ *to be the CCD on vertex set $(V(G) \setminus R) \dot\cup V(H)$ with colorings defined as follows. Let $\widehat{\tau}$ be the function that sends $v \in B \in \mathcal{B}$ to $\tau(B)$ if $v \in V(H)$ and to v itself otherwise (so if $v \in V(G)$).*

1. *The vertex v has color $\chi_G(\widehat{\tau}(v))$.*
2. *The edge (v, v') has edge color $\zeta_H(v, v')$ if $v, v' \in V(H)$ and $\zeta_G(\widehat{\tau}(v), (\widehat{\tau}(v')))$ otherwise.*

The blow-up is easygoing *if H is easygoing with respect to $\mathcal{B}$.*

Abusing terminology, we sometimes also talk about a blow-up of R by X rather than a blow-up of R to H in case each graph induced by a block of H is isomorphic to X. Note, however, that our notion of blow-up is more general than similar notions of homogeneous blow-up defined [12,19]. In particular the blocks of the graph H do not have to be homogeneously connected.

Lemma 6. *Suppose $Y_1 = G[H_1 \to_{\tau_1} R]$ and $Y_2 = G[H_2 \to_{\tau_2} R]$ are easygoing blow-ups of the same color class R in a graph G by ultrahomogeneous graphs H_1 and H_2, respectively. Then Y_1 is ultrahomogeneous if and only if Y_2 is ultrahomogeneous.*

If $G[R]$ is ultrahomogeneous this in particular holds with $H_1 = R$, meaning Y_2 is ultrahomogeneous if and only if G is ultrahomogeneous.

By analyzing the list of ultrahomogeneous oriented monochromatic graphs (Theorem 1) we obtain the following list of combinations where one such graphs is a blow-up of another.

Theorem 4. *Suppose G and H are monochromatic ultrahomogeneous oriented graphs such that G is a blow-up of H and $|G| > |H| > 1$. Then (G, H) is one of $(\overrightarrow{C_3} \cdot E_n, \overrightarrow{C_3})$, $(E_n \cdot \overrightarrow{C_3}, E_n)$, $(\overrightarrow{C_4}, E_2)$. In particular, a monochromatic ultrahomogeneous oriented graph arises in at most one way as a blow-up.*

Note that the oriented graph $\overrightarrow{C_4}$ as the blow-up of E_2 is the only example among the oriented graphs where the connections between the blocks are not homogeneous.

5 Classification of Ultrahomogeneous Oriented Graphs with Two Vertex Colors

In this section we focus on oriented graphs G with two vertex color classes red R and blue B.

Outline of the Proof Strategy. Without loss of generality we assume $|R| \geq |B|$. The graph $G[R]$ is monochromatic and ultrahomogeneous. It is thus one of the graphs $\overrightarrow{C_4}$, E_n, $E_n \cdot \overrightarrow{C_3}$, $\overrightarrow{C_3} \cdot E_n$, or H_0 (Theorem 1 and Fig. 2). For each choice of $G[R]$ we investigate how B can be connected to R making heavy use of the general extension theorem (Theorem 2). Specifically we set $N_+^R(b) := \{r \in R\colon (b,r) \in E(G)\}$, $N_-^R(b) := \{r \in R\colon (r,b) \in E(G)\}$, and $N^R(b) := N_+^R(b) \cup N_-^R(b)$. For each $b \in B$ it holds that $N_-^R(b) \cap N_+^R(b) = \emptyset$ and, hence, we set $A(b)$ to be the ordered partition obtained from $(N_+^R(b), N_-^R(b), R \setminus N^R(b))$ by deleting all empty parts. Conversely, for a partition A of R we set $X(A) := \{b \in B\colon A(b) = A\}$ and $\mathcal{B} := \{X(A)\colon A \in \mathcal{A}\}$. Since G is an oriented graph, we know that partitions $A(b)$ have at most three parts, so $|A(b)| \leq 3$ for every $b \in B$.

We now investigate partitions of $G[R]$ with at most 3 parts and rule out the ones that are not ultrahomogeneous. For the ones that remain, $|B|$ has size at least $|\mathcal{A}(A(b))|$ and this will rule out most of the remaining cases (since $|B| \leq |R|$ by assumption). For the cases that nevertheless remain, we determine the permutation group induced by $\mathrm{Aut}(G[R])$ on $\mathcal{A}(A(b))$. This permutation group must be the group $\mathrm{Aut}(G)^{\mathcal{B}}$ induced by $\mathrm{Aut}(G[B])$ on one of its block systems. However, $G[B]$ is ultrahomogeneous which limits the possible situations one final time. Table 1 shows all the possible groups $\mathrm{Aut}(G[B])^{\mathcal{B}}$ that can arise and the monochromatic graphs that admit them. What finally remains in the end are connections between R and B that indeed lead to ultrahomogeneous graphs.

Table 1. The oriented ultrahomogeneous digraphs G according to Lachlan's classification (Theorem 1). In the column *blocks* we list for each possible non-trivial block system of $\mathrm{Aut}(G)$ the isomorphism type of the subgraph induced by one (and thus by each) block. In the last column for each block system $\mathcal{B}$ we list the induced action $\mathrm{Aut}(G)^{\mathcal{B}}$ on $\mathcal{B}$.

G	$\lvert V(G)\rvert$	$\mathrm{Aut}(G)$	block induced graph	$\mathrm{Aut}(G)^{\mathcal{B}}$
E_n	n	$\mathrm{Sym}(n)$	E_1	$\mathrm{Sym}(n)$
$\overrightarrow{C_4}$	4	$\mathbb{Z}_4$	E_1	$\mathbb{Z}_4$
			E_2	$\mathbb{Z}_2$
H_0	8	$\mathrm{SL}(2,3)$	E_1	$\mathrm{SL}(2,3)$
			E_2	$\mathrm{Alt}(4)$
$E_n \cdot \overrightarrow{C_3}$	$3n$	$\mathbb{Z}_3 \wr \mathrm{Sym}(n)$	E_1	$\mathbb{Z}_3 \wr \mathrm{Sym}(n)$
			$\overrightarrow{C_3}$	$\mathrm{Sym}(n)$
$\overrightarrow{C_3} \cdot E_n$	$3n$	$\mathrm{Sym}(n) \wr \mathbb{Z}_3$	E_1	$\mathrm{Sym}(n) \wr \mathbb{Z}_3$
			E_n	$\mathbb{Z}_3$

Lemma 7. *Let A be a partition of $V(\overrightarrow{C_4})$ with $|A| \leq 3$. The set $\mathcal{A}(A)$ is a non-trivial ultrahomogeneous system of partitions with $|\mathcal{A}(A)| \leq 4$ if and only*

if A is a partition of $\overrightarrow{C_4}$ into two independent sets of order 2. The corresponding permutation group induced by $\mathrm{Aut}(G)$ *on* $\mathcal{A}(A)$ *is* $\mathbb{Z}_2$.

Corollary 1. *Let G be an ultrahomogeneous oriented graph on two vertex color classes R and B where $|R| \geq |B|$ and $G[R] \cong \overrightarrow{C_4}$. If R and B are not homogeneously connected, then G is equivalent (up to edge color changes) to one of the graphs of Fig. 3.*

Fig. 3. Up to color changes and bichromatic symmetrization, these are the ultrahomogeneous bichromatic oriented graphs containing $\overrightarrow{C_4}$ as a color class.

Lemma 8. *If A is an ordered partition of $V(E_n)$, then $\mathcal{A}(A)$ is an ultrahomogeneous system of partitions of E_n. Moreover, if $\mathcal{A}(A)$ is non-trivial, $|\mathcal{A}(A)| \leq n$, and $2 \leq |A| \leq 3$, then A has precisely two parts one of which is a singleton. The corresponding permutation group induced by* $\mathrm{Aut}(G)$ *on* $\mathcal{A}(A)$ *is* $\mathrm{Sym}(n)$.

Corollary 2. *Let G be an ultrahomogeneous oriented graph on two vertex color classes R and B where $|R| \geq |B|$ and $G[R] \cong E_n$ for some $n \in \mathbb{N}_{\geq 1}$. If R and B are not homogeneously connected, then $G[B] \cong E_n$ and R and B are matching-connected.*

Lemma 9. *If A is a partition of $\overrightarrow{C_3} \cdot E_n$, then $\mathcal{A}(A)$ is a non-trivial ultrahomogeneous system of partitions with $|\mathcal{A}(A)| \leq 3n$ if and only if $A = (P_1, P_2, P_3)$ such that $G[P_i] \cong E_n$ for each $i \in [3]$. The corresponding permutation group induced by* $\mathrm{Aut}(G)$ *on* $\mathcal{A}(A)$ *is* $\mathbb{Z}_3$.

Lemma 10. *If A is a partition of $G := E_n \cdot \overrightarrow{C_3}$ with $|A| \leq 3$, then $\mathcal{A}(A)$ is a non-trivial ultrahomogeneous system of partitions if and only if each part P of A satisfies $G[P] \cong E_{n'} \cdot \overrightarrow{C_3}$ for some $n' \in [n-1]$.*

Lemma 11. *The only ultrahomogeneous partition of H_0 into at most three parts is the trivial partition.*

Corollary 3. *If G is an ultrahomogeneous oriented graph on two vertex color classes R and B such that $G[R] \cong H_0$, then R and B are homogeneously connected.*

$E_n^R \rightrightarrows E_n^B$ $\overrightarrow{C_3}^R \rightleftarrows \overrightarrow{C_3}^B$

Fig. 4. Up to equivalence all ultrahomogeneous oriented bichromatic graphs can be obtained from blow-ups of graphs of the above form.

Denote by $E_n^R \rightrightarrows E_n^B$ the bichromatic graph whose color classes induce a red and a blue E_n and with a perfect matching directed from red to blue (Fig. 4 left). Also denote by $\overrightarrow{C_3}^R \rightleftarrows \overrightarrow{C_3}^B$ the ultrahomogeneous oriented graph depicted in Fig. 4 on the right non-trivially joining two directed triangles.

For blow-ups of bichromatic graphs $E_n^R \rightrightarrows E_n^B$ we introduce a specialized notation. We use the notation $E_n^R \rightrightarrows E_n^B \uparrow \mathcal{C}$ for a set of CCDs $\mathcal{C}$ to indicate that one or possibly two of the sides can be blown up to one of the graphs in $\mathcal{C}$. Similarly we use the notation $\overrightarrow{C_3}^R \rightleftarrows \overrightarrow{C_3}^B \uparrow \mathcal{C}$.

Theorem 5. *Let G be an oriented bichromatic graph. Then G is ultrahomogeneous if and only if it is equivalent (up to color changes and (inverse) bichromatic symmetrizations) to one of following graphs (see Figs. 3 and 4):*

1. $G_1 \square G_2$ *with* $G_1, G_2 \in \{\overrightarrow{C_4}, H_0, E_n, E_n \cdot \overrightarrow{C_3}, \overrightarrow{C_3} \cdot E_n\}$ *(disjoint union),*
2. $E_n^R \rightrightarrows E_n^B \uparrow \{E_n \cdot \overrightarrow{C_3} : n \in \mathbb{N}_{>1}\}$ *(matching, possibly blown up to* $E_n \cdot \overrightarrow{C_3}$*),*
3. $E_2^R \rightrightarrows E_2^B \uparrow \{E_2 \cdot \overrightarrow{C_3}, \overrightarrow{C_4}\}$ *(matching, possibly blown up to* $E_2 \cdot \overrightarrow{C_3}$ *or to* $\overrightarrow{C_4}$*), or*
4. $\overrightarrow{C_3}^R \rightleftarrows \overrightarrow{C_3}^B \uparrow \{\overrightarrow{C_3} \cdot E_n : n \in \mathbb{N}_{>1}\}$ *(special triangle connection, possibly blown up to* $\overrightarrow{C_3} \cdot E_n$*).*

6 Ultrahomogeneous Vertex-Colored Oriented Graphs

In this section we consider more than two vertex colors and finish the classification. Let us first analyze blow-ups.

Lemma 12. *If a color class R in an ultrahomogeneous graph G induces a graph $G[R]$ that is a non-trivial blow-up, then G itself is a non-trivial blow-up to R.*

Denote by $E_n^{R_1} \rightrightarrows E_n^{R_2} \rightrightarrows \cdots \rightrightarrows E_n^{R_t}$ the graph that consists of k isomorphic copies of a transitive tournament on t vertices with a discrete vertex coloring. Denote by $\overrightarrow{C_3}^{R_1} \rightleftarrows \overrightarrow{C_3}^{R_2} \rightleftarrows \cdots \rightleftarrows \overrightarrow{C_3}^{R_t}$ the CCD graph with vertex colors $R_1, R_2, \ldots, R_t$ for which each pair R_i, R_j induces $\overrightarrow{C_3} \rightleftarrows \overrightarrow{C_3}$. (Up to color changes and (inverse) bichromatic symmetrizations this graph is unique for each t).

Consistent with our previous notation, by $G \uparrow \{H_1, \cdots, H_t\}$ we denote all graphs obtained from G by blowing up an arbitrary number of color classes of G to one of the graphs in $\{H_1, \ldots, H_t\}$.

Theorem 6. *If G is an ultrahomogeneous vertex-colored oriented graph, then G is equivalent (up to color changes and (inverse) bichromatic symmetrization) to the color-disjoint union of the following graphs:*

1. H_0 *(monochromatic),*
2. $E_n^{R_1} \rightrightarrows E_n^{R_2} \rightrightarrows \cdots \rightrightarrows E_n^{R_t} \uparrow \{E_n \cdot \overrightarrow{C_3} : n \in \mathbb{N}_{>1}\}$
3. $E_2^{R_1} \rightrightarrows E_2^{R_2} \rightrightarrows \cdots \rightrightarrows E_2^{R_t} \uparrow \{E_2 \cdot \overrightarrow{C_3}, \overrightarrow{C_4}\}$
4. $\overrightarrow{C_3}^{R_1} \rightleftarrows \overrightarrow{C_3}^{R_2} \rightleftarrows \cdots \rightleftarrows \overrightarrow{C_3}^{R_t} \uparrow \{\overrightarrow{C_3} \cdot E_n : n \in \mathbb{N}_{>1}\}$

7 Further Research

One of the most pressing questions is whether our approach can be used to classify the ultrahomogeneous countably infinite vertex-colored oriented graphs. While certain concepts of our proof strategy are easily transferable to the countable infinite context (e.g., color classes induce monochromatic ultrahomogeneous structures, which are classified by Cherlin [6]) other proof concepts (e.g., most of our counting techniques) do not seem to transfer in a direct manner.

Acknowledgement. The research leading to these results has received funding from the European Research Council (ERC) under the European Union's Horizon 2020 research and innovation programme (EngageS: grant agreement No. 820148).

Disclosure of Interests. The authors have no competing interests to declare that are relevant to the content of this article.

References

1. Ahlman, O.: Limit Laws, Homogenizable Structures and Their Connections. Dissertation, Uppsala University, Uppsala (2018)
2. Babai, L.: Graph isomorphism in quasipolynomial time [extended abstract]. In: Wichs, D., Mansour, Y. (eds.) Proceedings of the 48th Annual ACM SIGACT Symposium on Theory of Computing, STOC 2016, Cambridge, MA, USA, 18–21 June 2016, pp. 684–697. ACM (2016). https://doi.org/10.1145/2897518.2897542
3. Böttcher, J., Foniok, J.: Ramsey properties of permutations. Electron. J. Comb. **20**(1), 2 (2013). https://doi.org/10.37236/2978
4. Cherlin, G.: Two problems on homogeneous structures, revisited. In: Model Theoretic Methods in Finite Combinatorics. Contemporary Mathematics, vol. 558, pp. 319–415. American Mathematical Society, Providence (2011). https://doi.org/10.1090/conm/558/11055

5. Cherlin, G., Felgner, U.: Homogeneous finite groups. J. London Math. Soc. **62**(3), 784–794 (2000). https://doi.org/10.1112/S0024610700001484, https://londmathsoc.onlinelibrary.wiley.com/doi/abs/10.1112/S0024610700001484
6. Cherlin, G.L.: The classification of countable homogeneous directed graphs and countable homogeneous n-tournaments. Mem. Amer. Math. Soc. **131**(621), xiv+161 (1998). https://doi.org/10.1090/memo/0621
7. Cherlin, G.L., Felgner, U.: Homogeneous solvable groups. J. London Math. Soc. **s2-44**(1), 102–120 (1991). https://doi.org/10.1112/jlms/s2-44.1.102, https://londmathsoc.onlinelibrary.wiley.com/doi/abs/10.1112/jlms/s2-44.1.102
8. Fraïssé, R.: Sur certaines relations qui généralisent l'ordre des nombres rationnels. Comptes Rendus l'Acad. Sci. Paris **237**, 540–542 (1953)
9. Gardiner, A.: Homogeneous graphs. J. Combin. Theory Ser. B **20**(1), 94–102 (1976). https://doi.org/10.1016/0095-8956(76)90072-1
10. Gill, N., Liebeck, M.W., Spiga, P.: Cherlin's Conjecture for Finite Primitive Binary Permutation Groups. Lecture Notes in Mathematics, vol. 2302. Springer, Cham (2022)
11. Gol'fand, Y., Klin, M.: On k-homogeneous graphs. Algorithmic Stud. Combin. (Russ.) **186**, 76–85 (1978)
12. Heinrich, I., Schneider, T., Schweitzer, P.: Classification of finite highly regular vertex-coloured graphs. arXiv preprint arXiv:2012.01058 [math.CO] (2020). https://doi.org/10.48550/arXiv.2012.01058
13. Jenkinson, T., Truss, J.K., Seidel, D.: Countable homogeneous multipartite graphs. Eur. J. Combin. **33**(1), 82–109 (2012). https://doi.org/10.1016/j.ejc.2011.04.004
14. Lachlan, A.H., Woodrow, R.E.: Countable ultrahomogeneous undirected graphs. Trans. Amer. Math. Soc. **262**(1), 51–94 (1980). https://doi.org/10.2307/1999974
15. Lachlan, A.H.: Finite homogeneous simple digraphs. In: Stern, J. (ed.) Proceedings of the Herbrand Symposium Logic Colloquium 1981, pp. 189–208. North-Holland Publishing Company (1982)
16. Lachlan, A.H.: Stable finitely homogeneous structures: a survey. In: Hart, B.T., Lachlan, A.H., Valeriote, M.A. (eds.) Algebraic Model Theory. NATO ASI Series, vol. 496, pp. 145–159. Springer, Dordrecht (1997). https://doi.org/10.1007/978-94-015-8923-9_6
17. Lachlan, A.H., Tripp, A.: Finite homogeneous 3-graphs. Math. Logic Quart. **41**(3), 287–306 (1995). https://doi.org/10.1002/malq.19950410302
18. Li, C.H.: A complete classification of finite homogeneous groups. Bull. Aust. Math. Soc. **60**(2), 331–334 (1999). https://doi.org/10.1017/S0004972700036431
19. Lockett, D.C., Truss, J.K.: Homogeneous coloured multipartite graphs. Eur. J. Combin. **42**, 217–242 (2014). https://doi.org/10.1016/j.ejc.2014.06.006
20. Macpherson, D.: A survey of homogeneous structures. Discret. Math. **311**(15), 1599–1634 (2011). https://doi.org/10.1016/J.DISC.2011.01.024
21. Mekler, A.H.: Homogeneous Partially Ordered Sets. In: Sauer, N.W., et at. (eds.) Finite and Infinite Combinatorics in Sets and Logic, pp. 279–288. Springer, Dordrecht (1993). https://doi.org/10.1007/978-94-011-2080-7_19
22. Sheehan, J.: Smoothly embeddable subgraphs. J. London Math. Soc. (2) **9**, 212–218 (1974/75). https://doi.org/10.1112/jlms/s2-9.2.212

Recognition of Unit Segment and Polyline Graphs is ∃ℝ-Complete

Michael Hoffmann[1], Tillmann Miltzow[2], Simon Weber[1(✉)], and Lasse Wulf[3]

[1] Department of Computer Science, ETH Zürich, 8001 Zürich, Switzerland
{hoffmann,simon.weber}@inf.ethz.ch

[2] Department of Information and Computing Sciences, Utrecht University, 3584 CS Utrecht, The Netherlands
t.miltzow@uu.nl

[3] Institute of Discrete Mathematics, Graz University of Technology, 8010 Graz, Austria
wulf@math.tugraz.at

Abstract. Given a set of objects O in the plane, the corresponding intersection graph is defined as follows. A vertex is created for each object and an edge joins two vertices whenever the corresponding objects intersect. We study here the case of unit segments and polylines with exactly k bends. In the recognition problem, we are given a graph and want to decide whether the graph can be represented as the intersection graph of certain geometric objects. In previous work it was shown that various recognition problems are ∃ℝ-complete, leaving unit segments and polylines as few remaining natural cases. We show that recognition for both families of objects is ∃ℝ-complete.

Keywords: Intersection graphs · Unit segment · Polyline · Recognition

1 Introduction

Many real-life problems can be mathematically described in the language of graphs. For instance, consider all the cell towers in Switzerland. We want to assign each tower a frequency such that no two towers that overlap in coverage use the same frequency. This can be seen as a graph coloring problem: Every cell tower becomes a vertex, overlap indicates an edge and a frequency assignment corresponds to a proper coloring of the vertices, see Fig. 1.

In many contexts, we have additional structure on the graph that may or may not help us to solve the underlying algorithmic problem. For instance, it might be that the graph arises as the intersection graph of unit disks in the plane such as in the example above. In that case, the coloring problem can be solved more efficiently [10], and there are better approximation algorithms for the clique problem [11]. This motivates a systematic study of geometric intersection

D. Král' and M. Milanič (Eds.): WG 2024, LNCS 14760, pp. 266–281, 2025.
https://doi.org/10.1007/978-3-031-75409-8_19

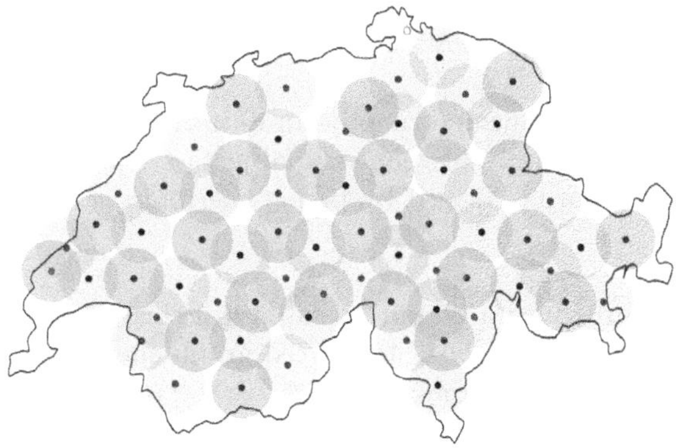

Fig. 1. A fictional illustration of mobile coverage of Switzerland using cell towers.

graphs. Another motivation is mathematical curiosity. Simple geometric shapes are easily visualized and arguably very natural mathematical objects. Studying the structural properties of intersection graphs gives insight into those geometric shapes and their possible intersection patterns.

It is known for a host of geometric shapes that it is $\exists\mathbb{R}$-complete to recognize their intersection graphs [14,26,28,29]. The class $\exists\mathbb{R}$ consists of all of those problems that are polynomial-time equivalent to deciding whether a given system of polynomial inequalities and equations with integer coefficients has a real solution. We will introduce $\exists\mathbb{R}$ in more detail below.

In this work, we focus on two geometric objects; unit segments and polylines with exactly k bends. Although we consider both types of geometric objects natural and well studied, to the best of our knowledge the complexity of their recognition problem was left open. We close this gap by showing that both recognition problems are $\exists\mathbb{R}$-complete.

1.1 Definition and Results

We define geometric intersection graphs and the corresponding recognition problem.

Intersection Graphs. Given a finite set of geometric objects O, we denote by $G(O) = (V, E)$, the corresponding *intersection graph*. The set of vertices is the set of objects ($V = O$) and two objects are adjacent ($uv \in E$) if they intersect ($u \cap v \neq \emptyset$). We are interested in intersection graphs that come from different families of geometric objects.

Families of Geometric Objects. Examples for a family of geometric objects are segments, disks, unit disks, unit segments, rays, and convex sets, to name a few of the most common ones. In general, given a geometric body $O \subset \mathbb{R}^2$ we denote by O^{+} the family of all translates of O. Similarly, we denote by $O^{\circledast}$ the family of all translates and rotations of O. For example, the family of all unit segments can be denoted as $u^{\circledast}$, where u is a unit segment.

Graph Classes. Classes of geometric objects $\mathcal{O}$ naturally give rise to classes of graphs $C(\mathcal{O})$: Given a family of geometric objects $\mathcal{O}$, we denote by $C(\mathcal{O})$ the class of graphs that can be formed by taking the intersection graph of a finite subset from $\mathcal{O}$.

Recognition. If we are given a graph, we can ask if this graph belongs to a geometric graph class. Formally, let C be a fixed graph class, then the recognition problem for C is defined as follows. As input, we receive a graph G and we have to decide whether $G \in C$. We denote the corresponding algorithmic problem by RECOGNITION(C). For example the problem of recognizing unit segment graphs can be denoted by RECOGNITION($C(u^{\circledast})$). We will use the term UNIT RECOGNITION for this problem. Furthermore, we define POLYLINE RECOGNITION as the recognition problem of intersection graphs of polylines with k bends.

Realizations. We can also say that RECOGNITION($C(\mathcal{O})$) asks about the existence of a *representation* of a given graph. A representation or *realization* of a graph G using a family of objects $\mathcal{O}$ is a function $r : V \mapsto \mathcal{O}$ such that $r(v) \cap r(w) \neq \emptyset \iff vw \in E$. For simplicity, for a set $V' \subseteq V$, we define $r(V') = \bigcup_{v \in V'} r(v)$.

Results. We show $\exists\mathbb{R}$-completeness of the recognition problems of two very natural geometric graph classes.

Theorem 1. UNIT RECOGNITION *is* $\exists\mathbb{R}$*-complete.*

Theorem 2. POLYLINE RECOGNITION *is* $\exists\mathbb{R}$*-complete, for any fixed* $k \geq 1$.

1.2 Discussion

In this section, we discuss strengths and limitations of our results from different perspectives. To supply the appropriate context, we give a comprehensive list of important geometric graph classes and the current knowledge about the complexity of their recognition problems in Table 1.

Refining the Hierarchy. We see our main contribution in refining the hierarchy of geometric graph classes for which recognition complexity is known. Both unit segments as well as polylines with k bends are natural objects that are well studied in the literature. However, the recognition of the corresponding graph classes was not studied previously. Polylines with an unbounded number of bends are equivalent to strings (it is possible to show that polylines with an unbounded number of bends are as versatile as strings with respect to the types of graphs that they can represent, since the number of intersections of any two strings can always be bounded from above [37,38]), while polylines with 0 bends are

Table 1. Classes of Geometric Intersection Graphs and their algorithmic complexity.

Intersection graphs of	Complexity	Source
circle chords	polynomial	Spinrad [44]
(unit) interval	polynomial	Booth and Lueker [12]
string	NP-complete	Schaefer and Sedgwick [24,37]
outerstring	NP-complete	Kratochvíl [23] (see also Rok and Walczak [33])
C^+, C convex polygon	NP-complete	Müller et al. [31], Kratochvíl, Matoušek [25]
(unit) disks	∃ℝ-complete	McDiarmid and Müller [29]
convex sets	∃ℝ-complete	Schaefer [34]
(downwards) rays	∃ℝ-complete	Cardinal et al. [14]
outer segments	∃ℝ-complete	Cardinal et al. [14]
segments	∃ℝ-complete	Kratochvíl and Matoušek [26] (see also [28,34])
unit balls	∃ℝ-complete	Kang and Müller [22]

just segments. Polylines with k bends thus naturally slot in between strings and segments, and their corresponding graph class is thus also an intermediate class between the class of segment intersection graphs and string graphs, as can be seen in Fig. 2. By showing that recognition for polylines with k bends is ∃ℝ-complete for all constant k, we see that the switch from ∃ℝ-completeness (segment intersection graphs) to NP-membership (string graphs) really only happens once k is infinite. Similarly, unit segment intersection graphs slot in between segment and ray intersection graphs. Intuitively, recognition of a class intermediate to two classes that are ∃ℝ-hard to recognize should also be ∃ℝ-hard, and our Theorem 1 confirms this intuition in this case.

Large Coordinates. One of the consequences of ∃ℝ-completeness is that there are no short representations of solutions known. Some representable graphs may only be representable by objects with irrational coordinates, or by rational coordinates with nominator and denominator of size at least $2^{2^{n^c}}$, for some fixed $c > 0$. In other words, the numbers to describe the position might need to be doubly exponentially large [28] for some graphs. For "flexible" objects like polylines, rational solutions can always be obtained by slightly perturbing the representation. For more "sturdy" objects like unit segments this may not be possible, however it is known that for example unit disks admit rational solutions despite their inflexibility [29].

Unraveling the Broader Story. Given the picture of Fig. 2, we wish to start unraveling a deeper story. Namely, we aim to get a better understanding of when geometric recognition problems are ∃ℝ-complete and when they are contained in NP. Figure 2 indicates that ∃ℝ-hardness comes from objects that are complicated enough to avoid a complete combinatorial characterization. Unit interval graphs, interval graphs and circle chord graphs admit such a characterization that in turn

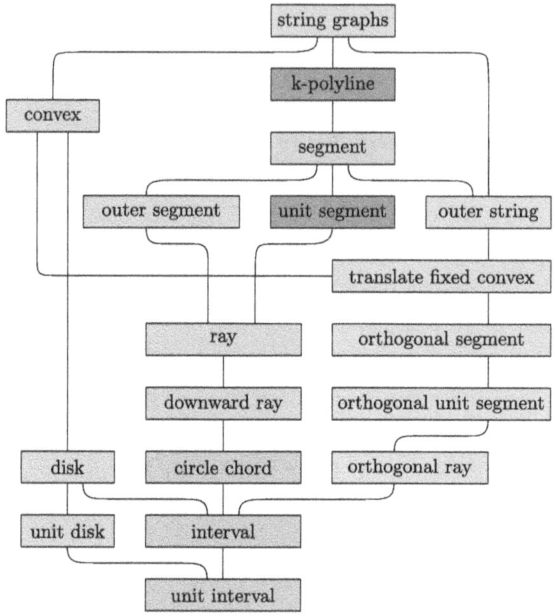

Fig. 2. Each box represents a different geometric intersection graph class. Those marked in green can be recognized in polynomial time. Those in blue are known to be ∃ℝ-complete. The ones in gray are NP-complete, and the orange ones are the new results presented in this paper. (Color figure online)

can be used to develop a polynomial-time algorithm for their recognition. On the other hand, if the geometric objects are too broad, the recognition problem is in NP. The prime example is string graphs. For string graphs, it is sufficient to know the planar graph given by the intersection pattern of the strings, which is purely combinatorial information that does not care anymore about precise coordinates. Despite the fact that there are graphs that need an exponential number of intersections, it is possible to find a polynomial-size witness [37] and thus we do not have ∃ℝ-hardness. We want to summarize this as: recognition problems are ∃ℝ-complete if the underlying family of geometric objects is at a sweet spot of neither being too simplistic nor too flexible.

When we consider Table 1 we observe two different types of ∃ℝ-complete families. The first type of family encapsulates all rotations $O^{\oplus}$ of a given object O (i.e., segments, rays, unit segments etc.). The second type of family contains translates and possibly homothets of geometric objects that have some curvature themselves (i.e., disks and unit disks). However in case we fix a specific object without curvature, i.e., a polygon, and consider all translations of it then the recognition problem also lies in NP [31]. Therefore, broadly speaking, curvature or rotation seem to be properties needed for ∃ℝ-completeness and the lack of it seems to imply NP-membership. We wish to capture one part of this intuition in the following conjecture:

Conjecture 1. Let O be a convex body in the plane with at least two distinct points. Then RECOGNITION($O^{\oplus}$) is ∃ℝ-complete.

We wonder if our intuition on curvature of geometric objects could be generalized as well. It seems plausible that RECOGNITION(O^{+}) is ∃ℝ-complete if and only if O has curvature.

Restricted Geometric Graph Classes. A classic way to make an algorithmic problem easier is to consider it only on restricted input. In the context of ∃ℝ and geometric graph recognition, we would like to mention the work by Schaefer [36], where it is shown that the recognition problem of segments is already ∃ℝ-complete for graphs of bounded degree. We conjecture that the same is true for unit segments and polylines. However, this does not follow from our reductions since we create graphs of unbounded degree.

Conjecture 2. UNIT RECOGNITION is ∃ℝ-complete already for graphs of bounded degree.

Once this would be established it would be interesting to find the exact degree threshold when the problem switches from being difficult to being easy.

1.3 Existential Theory of the Reals

The class of the existential theory of the reals ∃ℝ (pronounced as 'ER') is a complexity class defined through its canonical problem ETR, which also stands for Existential Theory of the Reals. In this problem we are given a sentence of the form $\exists x_1, \ldots, x_n \in \mathbb{R} : \Phi(x_1, \ldots, x_n)$, where Φ is a well-formed quantifier-free formula consisting of the symbols $\{0, 1, x_1, \ldots, x_n, +, \cdot, \geq, >, \wedge, \vee, \neg\}$, and the goal is to check whether this sentence is true.

The class ∃ℝ is the family of all problems that admit a polynomial-time many-one reduction to ETR. It is known that NP ⊆ ∃ℝ ⊆ PSPACE [13]. The reason that ∃ℝ is an important complexity class is that a number of common problems, mainly in computational geometry, have been shown to be complete for this class. Schaefer established the current name and pointed out first that several known NP-hardness reductions actually imply ∃ℝ-completeness [34]. Early examples are related to recognition of geometric structures: points in the plane [30,43], geometric linkages [1,35], segment graphs [26,28], unit disk graphs [22,29], ray intersection graphs [14], and point visibility graphs [14]. In general, the complexity class is more established in the graph drawing community [16,18,27,36]. Yet, it is also relevant for studying polytopes [17,32], Nash-Equilibria [6,8,9,20,39], and matrix factorization problems [15,40–42]. Other ∃ℝ-complete problems are the Art Gallery Problem [3,45], covering polygons with convex polygons [2], geometric packing [5] and training neural networks [4,7].

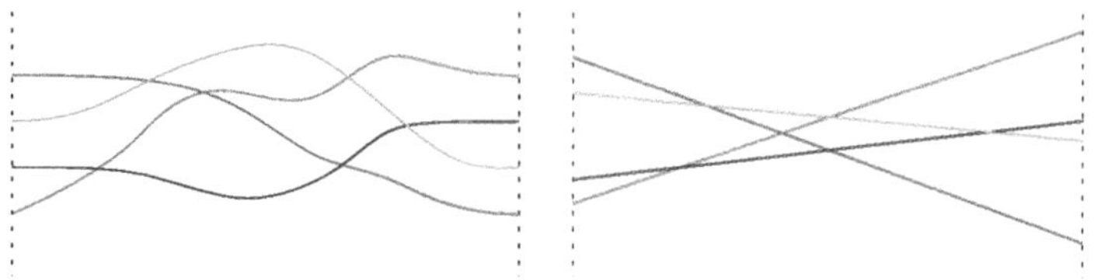

Fig. 3. The pseudoline arrangement on the left is combinatorially equivalent to the (truncated) line arrangement on the right; hence, it is stretchable.

1.4 Proof Techniques

The techniques used in this paper are similar to previous work. $\exists\mathbb{R}$-membership can be established straightforwardly by constructing concrete formulae or invoking a characterization of $\exists\mathbb{R}$ using *real* verification algorithms [19], similar to the characterization of NP. This proof can be found in the full version of this paper [21].

For $\exists\mathbb{R}$-hardness, we are in essence reducing from the SIMPLESTRETCHABILITY problem. In this problem, we are given a simple pseudoline arrangement as an input, and the question is whether this arrangement is stretchable. A pseudoline arrangement $\mathcal{A}$ is a set of n curves that are x-monotone. Furthermore, any two curves intersect exactly once and no three curves meet in a single point. We assume that there exist two vertical lines on which each curve starts and ends. The problem is to determine whether there exists a combinatorially equivalent (truncated) line arrangement. See Fig. 3 for an example.

For the reduction establishing $\exists\mathbb{R}$-hardness of UNIT RECOGNITION, we are given a pseudoline arrangement $\mathcal{A}$, and we construct a graph that is representable by unit segments (or polylines with k bends) iff $\mathcal{A}$ is stretchable. This graph is created by enhancing $\mathcal{A}$ with more curves and taking their intersection graph. By the way these additional curves are constructed, they can always be drawn as unit segments in any line arrangement combinatorially equivalent to $\mathcal{A}$, if this line arrangement is first squeezed into a canonical form as shown in Fig. 4.

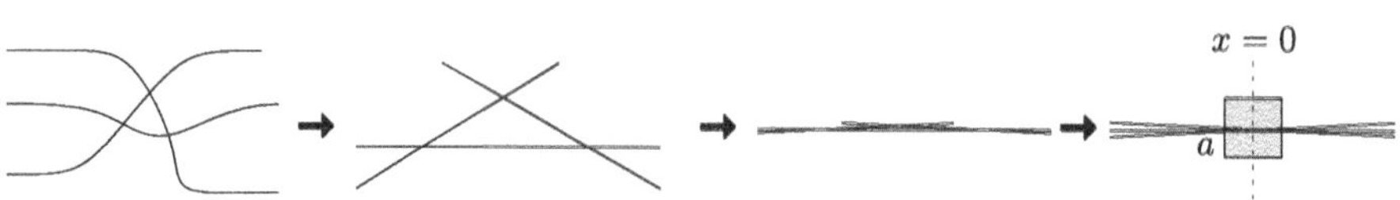

Fig. 4. Left to Right: pseudoline arrangement $\mathcal{A}$; stretched line arrangement $\mathcal{L}$; $\mathcal{L}$ squeezed to have small slopes; $\mathcal{L}$ squeezed to have all intersections in a box of side length a.

For the other direction of the equivalence we need to show that if the graph is representable by unit segments, then $\mathcal{A}$ is stretchable. To help with this proof we establish helper lemmas for both unit segments and polylines stating that cycles enforce a certain order of intersections of objects with the cycle. The details of these helper lemmas can be found in the full version [21] and a summary of the core statements can be found at the end of this section.

Using these additional lemmas we then show that the unit segments corresponding to the vertices originally representing our pseudolines must have the same combinatorial structure as our original arrangement $\mathcal{A}$. To ensure this we add so-called *probes* during the construction of our graph G. The ideas of order-enforcing cycles and probes have already been used in different contexts [14].

The main idea to show hardness of PolyLine Recognition is to enhance the construction for Unit Recognition by an additional area where the polylines representing our pseudolines have to make k bends. This then ensures that the polylines have no bends in the crucial part representing the pseudoline arrangement $\mathcal{A}$, and the same arguments as for unit segments can be used to show that this witnesses the stretchability of the original pseudoline arrangement.

Cycles and Orders. Given an induced cycle C in a graph G and a realization of G by Jordan arcs, we define two *boundary curves* wrapping around the realization infinitesimally closely, see Fig. 5.

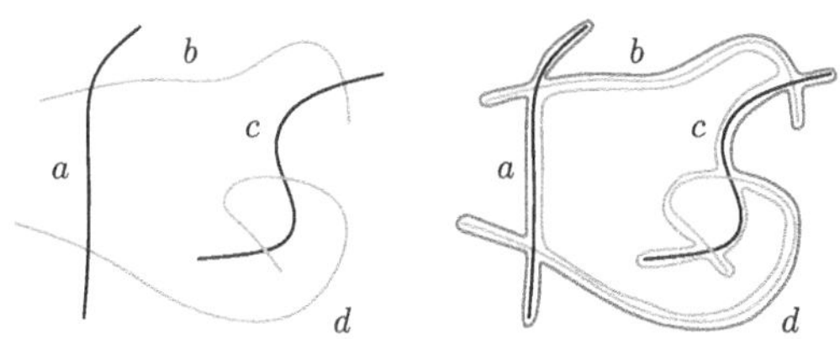

Fig. 5. Left: A representation $r(C)$ by strings. Right: The same representation with two boundary curves added.

We then show two main statements: Firstly, for certain sets of vertices $D \subseteq V \setminus C$ called *connectors*, the realization of $V \setminus (C \cup D)$ must lie completely within the inner or completely outside of the outer boundary curve. Secondly, for a more general type of sets $D \subseteq V \setminus C$ called *intersectors*, the cyclic order of intersections of realizations of $d_i \in D$ with the inner (outer) boundary curve must be the same as the cyclic order of occurence of edges between C and D when walking along C.

Examples of connectors and intersectors are shown in Fig. 6. Crucial properties of any such vertex set D are that each vertex of the cycle is incident to at most one vertex of D, and among any two neighboring vertices of C, at least one is not incident to any vertex of D.

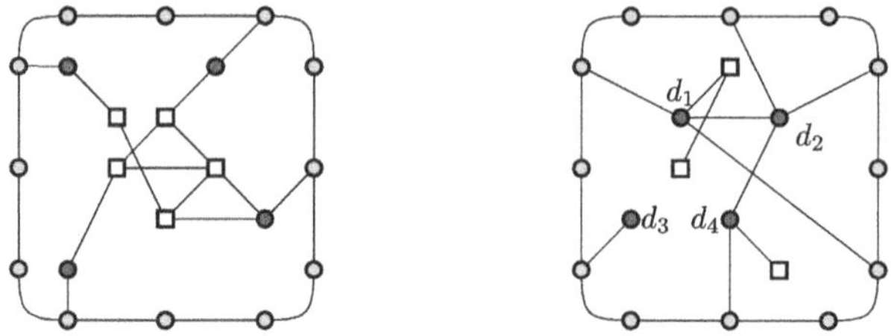

Fig. 6. Left: The green vertices are connectors of the cycle induced by the yellow vertices, and the remaining white squares form the rest of the graph. Right: The green dots are intersectors of the cycle induced by the yellow vertices. The cyclic order of intersectors along this cycle is $d_2, d_2, d_1, d_4, d_3, d_1$, when starting at the top left of the cycle and going clockwise. (Color figure online)

2 Proofs of $\exists\mathbb{R}$-Hardness

2.1 $\exists\mathbb{R}$-Hardness for Unit Segments

This section is dedicated to show $\exists\mathbb{R}$-hardness of Unit Recognition by a reduction from SimpleStretchability.

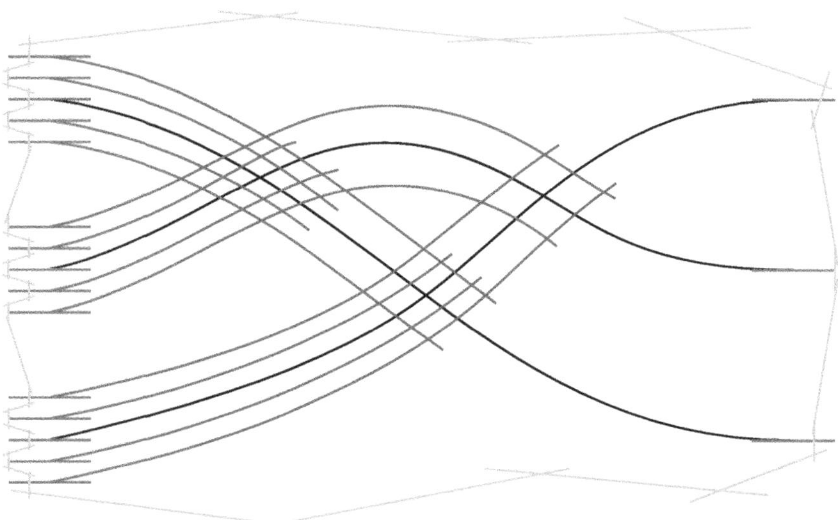

Fig. 7. A pseudoline arrangement (black), enhanced with probes (red), connectors (green) and a cycle (yellow). (Color figure online)

Construction. Given a pseudoline arrangement $\mathcal{A}$ of n pseudolines, we construct a graph G by enhancing the arrangement $\mathcal{A}$ with additional Jordan arcs. Then we define G to be the intersection graph of the pseudolines (which are also Jordan arcs) and all our added arcs. See Fig. 7 for an illustration.

First, we add so-called *probes* to our pseudolines. A probe of *depth* k of pseudoline ℓ is a Jordan arc which starts on the left vertical line, and follows ℓ closely through the arrangement until it has intersected k other pseudolines (and all probes of that other pseudoline which reach that intersection). For each pseudoline ℓ, we add $2(n-1)$ probes: Above and below ℓ we add one probe each for each depth $1 \leq k \leq n-1$. This gives a total of $p := 2n(n-1)$ probes in the arrangement. The probes are sorted according to their depth, with the probes of smallest depth situated closest to ℓ. Note that so far we have added $n+p$ arcs. We now create *connectors* for each probe and pseudoline. A connector is a short Jordan arc added to the left and/or right end of another arc. For probes, we only add connectors at the left end, while pseudolines get connectors at both ends. Note that we thus add $d := 2n + p$ connectors. Finally, we add Jordan arcs forming a cycle of length $2d+6$ around the current arrangement. At the left and right end, this cycle is placed in such a way that every second arc of the cycle intersects a connector, in the correct order. The eight additional arcs of the cycle are used to connect the left and right side of the cycle, using four arcs on the top and four arcs on the bottom.

We denote the collection of all those arcs (including the pseudolines) as the *enhanced pseudoline arrangement.* The graph G is given by the intersection graph of this arrangement. Note that G could also be described purely combinatorially, and it can be constructed in polynomial time.

Completeness. Due to space constraints we show that if $\mathcal{A}$ is stretchable then G is realizable by unit segments in the full version of this paper [21].

Soundness. This paragraph is dedicated to sketch the proof of the fact that G being realizable using unit segments implies that $\mathcal{A}$ is stretchable. The details are found in the full version [21]. We thus assume there exists a realization r of G by unit segments. We denote the vertices representing the pseudolines, probes, connectors, and the cycle by *important vertices* I, *probe vertices* P, *connector vertices* D, and *cycle vertices* C.

Note that C forms an induced cycle in G. The connector vertices D are connectors (motivating the name) as well as intersectors. Thus, the helper lemmas from the full version [21] allow us to conclude that for all realizations r of G by unit segments, $r(C)$ splits the plane into two cells, and $r(I \cup P)$ is contained completely in one of these two cells, without loss of generality the inner (bounded) one. Furthermore, we get that $r(D)$ is ordered along the interior boundary curve of C in the same way (up to cyclic shift and reflection) as it is in Fig. 7. From this, we know that also the important segments and probe segments are ordered as in Fig. 7.

We now claim that the arrangement of the important segments $r(I)$ is combinatorially equivalent to $\mathcal{A}$. Thus, extending these segments to lines yields a combinatorially equivalent line arrangement. To see this, we pick three pseudolines ℓ, a, and b, such that ℓ intersects a to the left of b in $\mathcal{A}$.

Let us consider the outermost probe segments of ℓ, their connector segments, a part of the interior boundary curve of C, and the important segment b. Since these segments form an induced cycle in the graph G, these segments bound a cell E. In this cycle, ℓ and its connector segment d_1 (the one attaching to the cycle between the probes) form a chord, thus their representations split E into two parts E_1, E_2. We furthermore consider the two probe segments p, q of ℓ, which correspond to the intersection with a. These segments are attached to the cycle between the outermost probe segments of ℓ, and ℓ itself. Note that p and q are not intersecting any segment bounding E_1 or E_2, in particular not $r(b)$. Thus the probe segments p and q are both completely contained in the interior of E_1 and E_2, respectively. See Fig. 8 for an illustration.

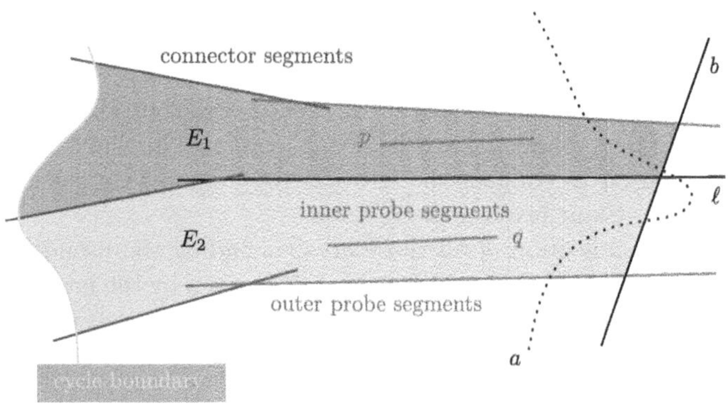

Fig. 8. If a is a line segment that intersects both p and q, it must intersect ℓ to the left of the intersection with b.

Since the line segment $r(a)$ must intersect both p and q and thus also E_1 and E_2, it is impossible for $r(a)$ to intersect ℓ outside of E. We thus conclude that $r(a)$ and $r(b)$ must intersect $r(\ell)$ in the correct order. Since this holds for all triples of pseudolines $\mathcal{A}$ must be stretchable, finishing the proof of Theorem 1.

2.2 $\exists\mathbb{R}$-Hardness for Polylines

In this section, we study polylines with k bends for a fixed constant $k > 0$. Specifically, we show that PolyLine Recognition is $\exists\mathbb{R}$-complete.

The proof works very similarly to the proof for unit segments. We only show the construction, and defer the proofs of completeness and soundness to the full version [21].

Construction. As in the construction for unit segments, we enhance our arrangement $\mathcal{A}$ of n pseudolines using additional Jordan arcs, and let our graph G be the intersection graph of the enhanced arrangement.

To create the enhanced arrangement, we first create the *frame*: We create $2k+2$ vertical chains of $c \cdot n^2$ line segments, for some sufficiently large c. We call the set of segments forming the i-th such vertical chain C_i, for $i = 1 \ldots, 2k+2$, numbered from left to right. We then connect these vertical chains by two horizontal chains, such that each pair of neighboring vertical chains forms an induced cycle with at least three horizontal segments on the top and three on the bottom. All of the segments forming these chains are called *frame segments*, and they together bound $2k+1$ bounded cells. We call the leftmost such cell the *canvas*.

We now place our pseudoline arrangement in the canvas. For every pseudoline ℓ, we first introduce a parallel *twin* pseudoline ℓ' closely below ℓ. Since $\mathcal{A}$ is assumed to be simple, every pseudoline has the same intersection pattern with all the other pseudolines as its twin.

Next, we introduce probes from the left of our pseudoline arrangement, as we did in the proof of Theorem 1. Note that every pair of twin pseudolines shares one set of probes. We attach the probes as well as the pseudolines and their twins to C_1 using connector arcs, again making sure that each connector intersects a unique arc of C_1, and that none of these intersected arcs of C_1 are intersecting each other.

We now want to weave the pseudolines and their twins through $C_2, \ldots, C_{2k+1}$. To do this, we split each set C_i into n *lanes* $L_{i,1}, \ldots, L_{i,n}$: A lane is a set of three consecutive arcs of C_i. The lanes $L_{i,j}$ are pairwise disjoint, and no arc of any lane $L_{i,j}$ may intersect an arc of a lane $L_{i,j'}$. Note that the lanes are ordered along C_i with $L_{i,1}$ being the topmost lane and $L_{i,n}$ the bottommost. We also number our pseudolines from ℓ_1 to ℓ_n by their order at the right end of the arrangement, from top to bottom. For every pseudoline ℓ_j, we now extend ℓ_j and its twin ℓ'_j as follows: For each even i, ℓ_j goes through the top arc of the lane $L_{i,j}$, and for each odd i, it goes through the bottom arc of $L_{i,j}$. The twin ℓ'_j is extended in the opposite way, going through the top arc of $L_{i,j}$ if i is odd, and through the bottom arc for i even.

At the very end, at C_{2k+2}, we use a connector arc to attach each pseudoline ℓ_j to the top element of $L_{2k+2,j}$, and its twin ℓ'_j to the bottom element. This whole construction is illustrated in Fig. 9.

The weaving of the pseudolines with their twins ensures that of each twinned pair, at least one of the two polylines has no bend within the canvas. With this property, the argument for soundness (as in the proof of Theorem 1) will carry over straightforwardly.

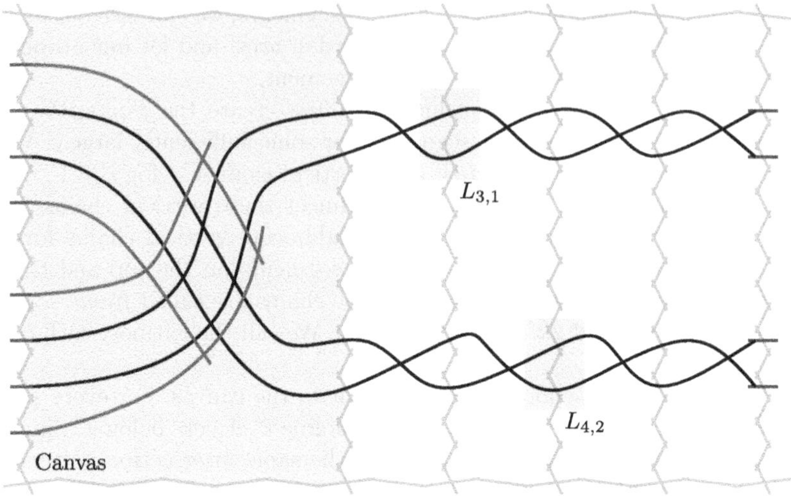

Fig. 9. A doubled pseudoline arrangement (black) enhanced for $k = 2$ with probes (red), connectors (green) and the frame (yellow). The left shaded region is the canvas, the right shaded regions are lanes. (Color figure online)

Acknowledgments. This research started at the *19th Gremo Workshop on Open Problems* in Binn VS, Switzerland, in June 2022. We thank the organizers for the invitation and for providing a very pleasant and inspiring working atmosphere. Tillmann Miltzow is generously supported by the Netherlands Organisation for Scientific Research (NWO) under project no. VI.Vidi.213.150. Lasse Wulf is generously supported by the Austrian Science Fund (FWF): W1230.

References

1. Abel, Z., Demaine, E., Demaine, M., Eisenstat, S., Lynch, J., Schardl, T.: Who needs crossings? Hardness of plane graph rigidity. In: 32nd International Symposium on Computational Geometry (SoCG 2016), pp. 3:1–3:15. Schloss Dagstuhl - Leibniz-Zentrum für Informatik (2016). https://doi.org/10.4230/LIPIcs.SoCG.2016.3
2. Abrahamsen, M.: Covering polygons is even harder. In: Vishnoi, N.K. (ed.) 2021 IEEE 62nd Annual Symposium on Foundations of Computer Science (FOCS), pp. 375–386 (2022). https://doi.org/10.1109/FOCS52979.2021.00045
3. Abrahamsen, M., Adamaszek, A., Miltzow, T.: The art gallery problem is ∃ℝ-complete. J. ACM **69**(1), 4:1–4:70 (2022). https://doi.org/10.1145/3486220
4. Abrahamsen, M., Kleist, L., Miltzow, T.: Training neural networks is ∃ℝ-complete. In: NeurIPS 2021, pp. 18293–18306 (2021). https://proceedings.neurips.cc/paper/2021/hash/9813b270ed0288e7c0388f0fd4ec68f5-Abstract.html
5. Abrahamsen, M., Miltzow, T., Seiferth, N.: Framework for ∃ℝ-completeness of two-dimensional packing problems. In: 2020 IEEE 61st Annual Symposium on Foundations of Computer Science (FOCS), pp. 1014–1021 (2020). https://doi.org/10.1109/FOCS46700.2020.00098

6. Berthelsen, M.L.T., Hansen, K.A.: On the computational complexity of decision problems about multi-player Nash equilibria. Theory Comput. Syst. **66**(3), 519–545 (2022). https://doi.org/10.1007/s00224-022-10080-1
7. Bertschinger, D., Hertrich, C., Jungeblut, P., Miltzow, T., Weber, S.: Training fully connected neural networks is ∃ℝ-complete. CoRR abs/2204.01368 (2022). To appear at NeurIPS 2024
8. Bilò, V., Mavronicolas, M.: A catalog of ∃ℝ-complete decision problems about Nash equilibria in multi-player games. In: 33rd Symposium on Theoretical Aspects of Computer Science (STACS 2016). Schloss Dagstuhl - Leibniz-Zentrum für Informatik (2016). https://doi.org/10.4230/LIPIcs.STACS.2016.17
9. Bilò, V., Mavronicolas, M.: ∃ℝ-complete decision problems about symmetric Nash equilibria in symmetric multi-player games. In: 34th Symposium on Theoretical Aspects of Computer Science (STACS 2017). Schloss Dagstuhl - Leibniz-Zentrum für Informatik (2017). https://doi.org/10.4230/LIPIcs.STACS.2017.13
10. Biró, C., Bonnet, É., Marx, D., Miltzow, T., Rzazewski, P.: Fine-grained complexity of coloring unit disks and balls. J. Comput. Geom. **9**(2), 47–80 (2018). https://doi.org/10.20382/jocg.v9i2a4
11. Bonamy, M., et al.: EPTAS and subexponential algorithm for maximum clique on disk and unit ball graphs. J. ACM **68**(2), 9:1–9:38 (2021). https://doi.org/10.1145/3433160
12. Booth, K.S., Lueker, G.S.: Testing for the consecutive ones property, interval graphs, and graph planarity using *PQ*-tree algorithms. J. Comput. Syst. Sci. **13**(3), 335–379 (1976). https://doi.org/10.1016/S0022-0000(76)80045-1
13. Canny, J.: Some algebraic and geometric computations in PSPACE. In: Proceedings of the Twentieth Annual ACM Symposium on Theory of Computing, STOC 1988, pp. 460–467. Association for Computing Machinery, New York (1988). https://doi.org/10.1145/62212.62257
14. Cardinal, J., Felsner, S., Miltzow, T., Tompkins, C., Vogtenhuber, B.: Intersection Graphs of Rays and Grounded Segments. J. Graph Algorithms Appl. **22**(2), 273–294 (2018). https://doi.org/10.7155/jgaa.00470
15. Chistikov, D., Kiefer, S., Marusic, I., Shirmohammadi, M., Worrell, J.: On Restricted nonnegative matrix factorization. In: Chatzigiannakis, I., Mitzenmacher, M., Rabani, Y., Sangiorgi, D. (eds.) 43rd International Colloquium on Automata, Languages, and Programming (ICALP 2016). Leibniz International Proceedings in Informatics (LIPIcs), vol. 55, pp. 103:1–103:14. Schloss Dagstuhl – Leibniz-Zentrum für Informatik, Dagstuhl (2016). https://doi.org/10.4230/LIPIcs.ICALP.2016.103
16. Dobbins, M.G., Kleist, L., Miltzow, T., Rzążewski, P.: ∀∃ℝ-completeness and area-universality. In: Brandstädt, A., Köhler, E., Meer, K. (eds.) WG 2018. LNCS, vol. 11159, pp. 164–175. Springer, Cham (2018). https://doi.org/10.1007/978-3-030-00256-5_14
17. Dobbins, M.G., Holmsen, A., Miltzow, T.: A universality theorem for nested polytopes. arXiv:1908.02213 (2019)
18. Erickson, J.: Optimal curve straightening is ∃ℝ-complete. arXiv:1908.09400 (2019)
19. Erickson, J., van der Hoog, I., Miltzow, T.: Smoothing the gap between NP and ER. In: 2020 IEEE 61st Annual Symposium on Foundations of Computer Science (FOCS), pp. 1022–1033 (2020). https://doi.org/10.1109/FOCS46700.2020.00099
20. Garg, J., Mehta, R., Vazirani, V.V., Yazdanbod, S.: ∃ℝ-completeness for decision versions of multi-player (symmetric) Nash equilibria. ACM Trans. Econ. Comput. **6**(1) (2018). https://doi.org/10.1145/3175494

21. Hoffmann, M., Miltzow, T., Weber, S., Wulf, L.: Recognition of unit segment and polyline graphs is ∃ℝ-complete (2024). https://doi.org/10.48550/arXiv.2401.02172
22. Kang, R., Müller, T.: Sphere and dot product representations of graphs. Discrete Comput. Geom. **47**(3), 548–569 (2012). https://doi.org/10.1007/s00454-012-9394-8
23. Kratochvíl, J.: String graphs. I. The number of critical nonstring graphs is infinite. J. Combin. Theory Ser. B **52**(1), 53–66 (1991). https://doi.org/10.1016/0095-8956(91)90090-7
24. Kratochvíl, J.: String graphs. II. Recognizing string graphs is NP-hard. J. Combin. Theory Ser. B **52**(1), 67–78 (1991). https://doi.org/10.1016/0095-8956(91)90091-W
25. Kratochvíl, J., Matoušek, J.: NP-hardness results for intersection graphs. Comment. Math. Univ. Carol. **30**(4), 761–773 (1989)
26. Kratochvíl, J., Matoušek, J.: Intersection graphs of segments. J. Combin. Theory Ser. B **62**(2), 289–315 (1994). https://doi.org/10.1006/jctb.1994.1071
27. Lubiw, A., Miltzow, T., Mondal, D.: The complexity of drawing a graph in a polygonal region. J. Graph Algorithms Appl. **26**(4), 421–446 (2022). https://doi.org/10.7155/jgaa.00602
28. Matoušek, J.: Intersection graphs of segments and ∃ℝ. ArXiv:1406.2636 (2014)
29. McDiarmid, C., Müller, T.: Integer realizations of disk and segment graphs. J. Combin. Theory Ser. B **103**(1), 114–143 (2013). https://doi.org/10.1016/j.jctb.2012.09.004
30. Mnëv, N.: The universality theorems on the classification problem of configuration varieties and convex polytopes varieties. In: Topology and Geometry – Rohlin Seminar, pp. 527–543 (1988)
31. Müller, T., van Leeuwen, E.J., van Leeuwen, J.: Integer representations of convex polygon intersection graphs. SIAM J. Discret. Math. **27**(1), 205–231 (2013). https://doi.org/10.1137/110825224
32. Richter-Gebert, J., Ziegler, G.M.: Realization spaces of 4-polytopes are universal. Bull. Am. Math. Soc. **32**(4), 403–412 (1995)
33. Rok, A., Walczak, B.: Outerstring graphs are χ-bounded. SIAM J. Discret. Math. **33**(4), 2181–2199 (2019). https://doi.org/10.1137/17M1157374
34. Schaefer, M.: Complexity of some geometric and topological problems. In: Eppstein, D., Gansner, E.R. (eds.) GD 2009. LNCS, vol. 5849, pp. 334–344. Springer, Heidelberg (2010). https://doi.org/10.1007/978-3-642-11805-0_32
35. Schaefer, M.: Realizability of Graphs and Linkages. In: Pach, J. (eds) Thirty Essays on Geometric Graph Theory, pp. 461–482. Springer, New York (2013). https://doi.org/10.1007/978-1-4614-0110-0_24
36. Schaefer, M.: Complexity of geometric k-planarity for fixed k. J. Graph Algorithms Appl. **25**(1), 29–41 (2021). https://doi.org/10.7155/jgaa.00548
37. Schaefer, M., Sedgwick, E., Štefankovič, D.: Recognizing string graphs in NP. J. Comput. Syst. Sci. **67**(2), 365–380 (2003). https://doi.org/10.1016/S0022-0000(03)00045-X. Special Issue on STOC 2002
38. Schaefer, M., Štefankovič, D.: Decidability of string graphs. J. Comput. Syst. Sci. **68**(2), 319–334 (2004). https://doi.org/10.1016/j.jcss.2003.07.002. Special Issue on STOC 2001
39. Schaefer, M., Štefankovič, D.: Fixed points, nash equilibria, and the existential theory of the reals. Theory Comput. Syst. **60**(2), 172–193 (2017). https://doi.org/10.1007/s00224-015-9662-0
40. Schaefer, M., Štefankovič, D.: The Complexity of Tensor Rank. Theory Comput. Syst. **62**(5), 1161–1174 (2017). https://doi.org/10.1007/s00224-017-9800-y

41. Shitov, Y.: A universality theorem for nonnegative matrix factorizations. ArXiv:1606.09068 (2016)
42. Shitov, Y.: The complexity of positive semidefinite matrix factorization. SIAM J. Optim. **27**(3), 1898–1909 (2017). https://doi.org/10.1137/16M1080616
43. Shor, P.W.: Stretchability of pseudolines is NP-hard. In: Gritzmann, P., Sturmfels, B. (eds.) Applied Geometry and Discrete Mathematics. DIMACS Series in Discrete Mathematics and Theoretical Computer Science, vol. 4, pp. 531–554 (1991). https://doi.org/10.1090/dimacs/004/41
44. Spinrad, J.: Recognition of circle graphs. J. Algorithms. Cogn. Inform. Log. **16**(2), 264–282 (1994).https://doi.org/10.1006/jagm.1994.1012
45. Stade, J.: Complexity of the boundary-guarding art gallery problem. CoRR abs/2210.12817 (2022). https://doi.org/10.48550/arXiv.2210.12817

Roman Cycle Hitting Set

Satyabrata Jana[1], Sounak Modak[2(✉)], Saket Saurabh[2,3], and Kushal Singanporia[2]

[1] University of Warwick, Coventry, UK
[2] The Institute of Mathematical Sciences, HBNI, Chennai, India
{sounakm,saket,kushalps}@imsc.res.in
[3] University of Bergen, Bergen, Norway

Abstract. The variant of Dominating Set known as Roman Domination has been extensively studied in the fields of graph theory and graph algorithms. Fernau and Mann [arXiv:2302.11417] introduced the concept of Roman Vertex Cover and developed a fixed-parameter tractable (FPT) algorithm for the problem with running time of $\mathcal{O}^*(2^k)$, where k is the solution size. Building on this work, we investigate the parameterized algorithms for Roman Feedback Vertex Set (RFVS) and Roman Odd Cycle Transversal (ROCT), and demonstrate that they are also FPT. Our approach to RFVS is similar to the best-known algorithm for the non-Roman variant. However, the algorithm for ROCT is significantly different and utilizes the technique of recursive understanding and structural insights on unbreakable graphs.

Keywords: feedback vertex set · odd cycle transversal · Roman hitting set · fixed-parameter tractability · kernelization

1 Introduction

The problem of finding a Hitting Set is a well-studied problem in graph theory with various applications. Given a ground set $\mathcal{U}$ of elements, a collection $\mathcal{C}$ of subsets in $\mathcal{U}$, and an integer k, the goal is to find a subset of $\mathcal{U}$ of size at most k that hits each set in $\mathcal{U}$. One specific example of this problem is the Vertex Cover problem, where the input is a graph G, the set $\mathcal{U}$ is defined as $V(G)$ (the vertices of G), and $\mathcal{C}$ consists of all the edges in G. Another one being the Dominating Set problem where we seek to find a subset S of at most k vertices such that every vertex not in S is adjacent to at least one vertex of S. Note that Hitting Set generalizes the problem Dominating Set, where the set of vertex of the given graph is considered as the ground set $\mathcal{U}$, and the set of all closed neighborhoods serves as the collection $\mathcal{C}$ of subsets in $\mathcal{U}$.

The concept of *Roman domination* is a well-explored variation of graph domination, extensively discussed in a dedicated chapter of the recent textbook by

S. Jana—Supported by the Engineering and Physical Sciences Research Council (EPSRC) via the project MULTIPROCESS (grant no. EP/V044621/1).

D. Kráľ and M. Milanič (Eds.): WG 2024, LNCS 14760, pp. 282–296, 2025.
https://doi.org/10.1007/978-3-031-75409-8_20

Chellali et al. [13]. Consider a scenario in which the armies are strategically placed in various regions. And a region is considered secure if it contains at least one army or has an adjacent region with two armies. Mathematically, the map is represented as a graph, with regions corresponding to vertices, and two vertices share an edge if the corresponding regions have a common boundary. Furthermore, the deployment of armies is described by a function that assigns a value from $\{0, 1, 2\}$ to each vertex. Such a function is known as a *Roman dominating function* if every vertex with a value 0 has a neighbor with a value 2. A natural follow-up question would be to find a Roman dominating function that minimizes the number of armies used. The decision version of this problem is called the ROMAN DOMINATING SET problem. For a comprehensive overview of the problems related to Roman domination, we refer to [10].

Fernau and Mann [11] introduced the notion of *Roman hitting set*, a generalization of ROMAN DOMINATING SET. Let $(X, \hat{S} = (s_i)_{i \in I})$ be a hypergraph. Any pair $(R_1, R_2) \in 2^I \times 2^X$ is said to be a Roman hitting set if for all $i \in I$, $s_i \cap R_2 \neq \emptyset$, or $i \in R_1$. In their study, the authors also examined Roman variants of classical VERTEX COVER, denoted by ROMAN VERTEX COVER (RVC). In the RVC problem, we are given a graph G and an integer k, and our goal is to find two sets $R_1 \subseteq E(G)$ and $R_2 \subseteq V(G)$ such that each edge in $E(G) \setminus R_1$ has an end point in R_2 and $|R_1| + 2|R_2| \leq k$. Intuitively in Roman hitting-set problems, the goal is to find a *special* kind of hitting set that looks to hits all sets except a few (defined by a given parameter) where the number of such sets (edges for RVC) may exponentially large and may not be explicitly given. Whereas in the case of RVC the number of objects being in edges is polynomial in the input size.

In the field of algorithmic graph theory, there is a well-studied problem called the cycle hitting set. It serves as an intermediate problem, which is a generalization of the VERTEX COVER problem and a restriction of the HITTING SET problem. There are two extensively researched variants of this problem known as FEEDBACK VERTEX SET (FVS) and ODD CYCLE TRANSVERSAL (OCT). The FVS (respectively, OCT) problem involves deleting a maximum of k vertices in order to transform a given graph into a forest (resp, bipartite graph). Both problems have been extensively studied in the field of parameterized complexity.

The earliest known Fixed-Parameter Tractable (FPT) algorithms for FVS date back to the late 1980s and early 1990s [9]. These algorithms relied on the groundbreaking Graph Minor Theory by Robertson and Seymour. Over time, there have been multiple advancements and refinements, leading to the current best deterministic FPT algorithm for FVS, which runs in $\mathcal{O}^*(3.460^k)$ time [14]. Additionally, Li and Nederlof gave the fastest known randomized algorithm for this problem running in time $\mathcal{O}^*(2.7^k)$ [16]. In the past decade, various variants of FVS have been extensively studied in undirected graphs, including CONNECTED FVS [6,22], CONFLICT-FREE FVS [2], SUBSET FVS [19], GROUP FVS [7], SIMULTANEOUS FVS [3], and INDEPENDENT FVS [1,17,21].

The parameterized complexity of OCT was a long-standing open problem in the area, and only in 2003 Reed et al. [23] developed an algorithm for the problem running in time $\mathcal{O}^*(3^k)$. However, there was no further improvement on this algorithm in the following 11 years until Lokshtanov et al. [18] gave the fastest known algorithm running in time $\mathcal{O}^*(2.3146^k)$.

Drawing inspiration from the Roman Hitting Set, Roman Dominating Set, and particularly the Roman Vertex Cover problem, we embark on an investigation of the Roman versions of the FVS and OCT problems. One key finding from our research is that these variants do not necessarily make problems easier than FVS and OCT. Furthermore, our algorithm for Roman Odd Cycle Transversal differs significantly. We focus on exploring these variations and uncovering their complexities, guided by the following theme.

Roman variant of **Cycle Hitting Set** problems: FVS &OCT

Let $G = (V, E)$ be an undirected graph. The $\mathcal{C}(G)$ (resp, $\mathcal{C}_0(G)$) denotes the set of all cycles (resp, odd cycles) of graph G. A pair $(R_1, R_2) \subseteq (\mathcal{C}(G) \times V)$ is a Roman feedback vertex set (in short, Roman fvs), if $\mathcal{C}(G - R_2) \subseteq R_1$, i.e. R_2 hits every cycle in $\mathcal{C}(G) \setminus R_1$. Analogously, $(R_1, R_2) \subseteq (\mathcal{C}_0(G) \times V)$ is a Roman odd cycle transversal (in short, Roman oct), if $\mathcal{C}_o(G - R_2) \subseteq R_1$, i.e. R_2 hits every cycle in $\mathcal{C}_o(G) \setminus R_1$. The size of a Roman fvs (resp, Roman oct) (R_1, R_2) of G is defined as $w(R_1, R_2) = |R_1| + 2|R_2|$. An *optimal* Roman fvs (resp, Roman oct) of G minimizes the size. We call R_1 the solution cycles, and R_2 the solution vertices.

Roman Feedback Vertex Set (RFVS)	**Parameter:** k
Input: A graph $G = (V, E)$ and $k \in \mathbb{N}$.	
Question: Does G has a Roman fvs (R_1, R_2) of size at most k?	

Roman Odd Cycle Transversal (ROCT)	**Parameter:** k
Input: A graph $G = (V, E)$ and $k \in \mathbb{N}$.	
Question: Does G has a Roman oct (R_1, R_2) of size at most k?	

It should be noted that for any graph G with n vertices, the pair $(\emptyset, V(G))$ is a Roman fvs as well as Roman oct of size $2n$. However, it is unlikely to find polynomial-time algorithm for RFVS (resp, ROCT) that aimed to minimize the solution size, which we showed in the full version of the paper. Observe that, if a graph has a Roman fvs (resp, Roman oct) (R_1, R_2) of size at most k, then there exists a subset $W \subseteq V(G)$ of size $i \leq k/2$ such that $G - W$ has at most $k - 2i$ cycles (resp, odd cycles), which are vertex-disjoint (by Lemma 8). This immediately leads to an algorithm with a runtime of $\mathcal{O}(n^{\frac{k}{2}})$ (by guessing each possible subset $|W| \leq k/2$ and checking whether $G - W$ has no pair of intersecting cycles (resp, odd cycles) and the number of vertex disjoint cycles (resp, odd cycles) in $G - W$ is at most $k - 2|W|$). In the context of parameterized complexity, this algorithm is considered an XP algorithm. Therefore, the natural question arises whether there is an algorithm with a runtime of $f(k) \cdot n^{\mathcal{O}(1)}$ exists, that is, whether the problem admits a fixed-parameter tractable (FPT) algorithm. Additionally, if in polynomial time $(n^{\mathcal{O}(1)})$ it is possible to obtain an equivalent instance (kernel) of polynomial size ($k^{\mathcal{O}(1)}$) then the problem is said to have a polynomial kernel. The existence of a kernel for a problem implies the existence of an FPT algorithm, and converse also true; however, obtaining a polynomial kernel is more challenging.

1.1 Our Results and Methods

Our main result is the following FPT algorithm for ROCT (Sect. 3).

Theorem 1. ROCT *can be solved in time* $f(k) \cdot n^{\mathcal{O}(1)}$*, for some computable function* $f : \mathbb{Z}^+ \rightarrow \mathbb{Z}^+$*, where* k *is the solution size.*

We utilizes the recursive understanding [5,12,15,20] together with the notion of unbreakable graphs to obtain a non-uniform FPT algorithm for the ROCT problem. We first show that ROCT can be expressed in MIN-CMSO (Lemma 16) which when combined with the Proposition 2 brings down the problem of finding a Roman oct from general graphs to unbreakable graphs. To get an FPT algorithm for ROCT in unbreakable graphs, we first observe the structure of instance based on a hypothetical solution. We show that for a Roman oct (R_1, R_2) of size at most c, there cannot be more than one odd cycle of length more than s in R_1 for a (s, c)-unbreakable graph G (Lemma 13). In the case when R_1 contains exactly one odd cycle of length more than s in G, we show that the treewidth of the graph G is bounded (Lemma 15) which when combined with Proposition 1 and MIN-CMSO expressibility of ROCT gives us an FPT algorithm. In the case when R_1 does not contain any odd cycle of length more than s, we utilizes the unbreakable separation and branch over the cycles of length at most $s + 1$ in G. At the end of the branching, we are left over with the odd cycles in G that has length more than $s + 1$ and we construct Parity-Torso based upon the partial solution that we constructed after the branching. By utilizing the parity preserving properties of the Parity-Torso that we developed in Sect. 3.1, we show that it is enough afterwards to solve OCT optimally in the Parity-Torso to get a solution for ROCT in G.

We obtain the following two results in RFVS[1].

Theorem 2 (♠). RFVS *admits a kernel with* $\mathcal{O}(k^2)$ *vertices and* $\mathcal{O}(k^2)$ *edges.*

Theorem 3 (♠). RFVS *can be solved in* $\mathcal{O}^*(3.618^k)$ *time.*

The idea behind our algorithm of Theorem 2 is quite similar to obtaining quadratic kernel for FVS [24]. Our kernelization procedure makes use of the *Expansion Lemma*, a combinatorial tool that played a crucial role to bound the maximum degree of the graph. Following which by employing various combinatorial arguments, we then bound the number of vertices in order to achieve our desired kernel. The FPT algorithm in Theorem 3 uses a branching subroutine following the finding of a feedback vertex set of size at most k.

[1] Missing proofs (marked with ♠) are in the full version of the paper.

2 Preliminaries

We use the standard graph-theoretic notation (see Diestel [8] for any undefined terminology). Given an undirected graph $G = (V, E)$, we let $V(G)$ denote its vertex set and $E(G)$ its edge set. When G is clear from the context, we use V and E to denote the vertex set and the edge set of G. We say a vertex u is reachable from a vertex v in G, if there is a path which contains both the vertices u and v in G. By Reach(x, G), we denote the set of all vertices (including x) that are reachable from x in G. We call a graph G *connected* if for any vertex $x \in V(G)$, we have Reach$(x, G) = V(G)$, otherwise we say that G is *disconnected.* We use the $\mathcal{O}^\star()$ notation to hides the polynomial factors in the running time of an algorithm.

Definition 1 (Block and Cut Vertex)**.** A vertex that separates two other vertices of the same component is called a *cut vertex.* A *block* in a graph is a maximal connected subgraph without a cut vertex. Thus, blocks in a graph G are either an isolated vertex, an edge, or a maximally-2-connected component.

Lemma 1 ([8])**.** *The cycles of a graph G are precisely the cycles of its blocks.*

Treewidth. It is a structural parameter that indicates how much a given graph resembles a tree. For example, a tree has treewidth 1 and an n-vertex clique has treewidth $n - 1$. Formally, the treewidth of a graph is defined as follows.

Definition 2 (Tree-decomposition)**.** A *tree decomposition* of a graph G is a pair (T, β) of a tree T and $\beta : V(T) \to 2^{V(G)}$, such that

1. $\bigcup_{t \in V(T)} \beta(t) = V(G)$, and
2. for any edge $e \in E(G)$, there exists a node $t \in V(T)$ such that both endpoints of e belong to $\beta(t)$, and
3. $\beta(t_1) \cap \beta(t_3) \subseteq \beta(t_2)$ whenever $t_1, t_2, t_3 \in T$ satisfy $t_2 \in t_1Tt_3$ i.e. t_2 lies in the unique path from t_1 to t_3 in T, and
4. for any vertex $v \in V(G)$, the subgraph of T induced by the set $T_v = \{t \in V(T) : v \in \beta(t)\}$ is a tree.

Note that Property 4 is implied by Property 3 and vice versa. The *width* of (T, β) is $\max_{v \in V(T)}\{|\beta(v)|\} - 1$. The *treewidth* of G, denoted by $\mathtt{tw}(G)$ is the minimum width of a tree-decomposition of G. A tree decomposition (T, β) is *trivial* if $|V(T)| = 1$. The width of a trivial tree-decomposition of G is $|V(G)| - 1$.

The following well-known lemma shows the relation between the treewidth of a graph G and its subgraph H.

Lemma 2. *If H is a subgraph of G, then $tw(H) \leq tw(G) \leq tw(H) + tw(G - H)$.*

The most important feature of a tree-decomposition is that it transfers the separation properties of its tree to the graph decomposed. The following lemma depicts the separation property of a tree-decomposition.

Lemma 3 ([8]). *Let t_1t_2 be any edge of T and let T_1, T_2 be the components of $T - t_1t_2$ with $t_1 \in T_1$ and $t_2 \in T_2$. Then $\beta(t_1) \cap \beta(t_2)$ separates $U_1 := \bigcup_{t \in V(T_1)} \beta(t)$ from $U_2 := \bigcup_{t \in V(T_2)} \beta(t)$ in G.*

$\beta(t_1) \cap \beta(t_2)$ is called the *adhesion set* of $\beta(t_1)$ and $\beta(t_2)$. The *adhesion* of a tree decomposition is the maximum size of its adhesion sets.

Unbreakability. To formally introduce the notion of unbreakability, we rely on the definition of a separation:

Definition 3 (Separation). A pair (X, Y) where $X \cup Y = V(G)$ is a *separation* if $E(X \setminus Y, Y \setminus X) = \emptyset$. The order of (X, Y) is $|X \cap Y|$.

Intuitively, the graph is breakable if it is possible to *break* it into two large parts by removing only a small number of vertices. Formally,

Definition 4 ((s, c)-Unbreakable graph). Let G be a graph. If there exists a separation (X, Y) of order at most c such that $|X \setminus Y| > s$ and $|Y \setminus X| > s$, called an (s, c)-*witnessing separation*, then G is (s, c)-*breakable*. Otherwise, G is (s, c)-*unbreakable*.

Definition 5 (Counting Monadic Second Order Logic). Counting Monadic Second Order Logic (CMSO) extends MSO by including atomic sentences that test whether the cardinality of a set is equal to q modulo r, where q and r are integers such that $0 \leq q < r$ and $r \geq 2$. That is, CMSO is an MSO with the following atomic sentence: $\textsc{card}_{q,r}(S) = \texttt{True}$ if and only if $|S| \equiv q \pmod{r}$, where S is a set.

A parameterized problem Π is a language $L \subseteq \Sigma^* \times \mathbb{N}$, where Σ is a finite fixed alphabet. For an instance $(x, k) \in \Sigma^* \times \mathbb{N}$, k is called the parameter. An instance $(x, k) \in \Sigma^* \times \mathbb{N}$ is called a `Yes`-instance of the problem Π if $(x, k) \in L$. We say a parameterized graph problem Π is x-CMSO expressible, if we can express the problem Π using a CMSO formula ψ such that given an instance (x, k) of the problem Π, $(x, k) \in \Pi$ if and only if ψ evaluates to `True`.

3 FPT Algorithm for ROCT

In this section, we present an FPT algorithm for ROCT problem. We start by outlining the essential tools that aid in the design of our algorithm.

3.1 Tools

We begin by defining the notion of Parity Preserving Torsos, which we will use to construct an equivalent instance of the problem.

Definition 6 (Parity Torso). Let $G = (V, E)$ be a graph and let $X \subseteq V$ be a vertex set. The graph Parity-Torso(G, X) is constructed as follows. Consider the induced subgraph $G[X]$ and perform the following operations on it. For every pair $u, v \in X$, if there is an odd path in G from u to v whose internal vertices are not in X, then add an edge (u, v). Furthermore, for every pair $u, v \in X$, if there is an even path in G from u to v whose internal vertices are not in X, then add a subdivided edge between u and v. The resulting graph is referred to as Parity-Torso(G, X) or $\texttt{PT}(G, X)$ and we call a newly added vertex which is part of a subdivided edge, a *subdivision* vertex.

Lemma 4 (♠). *Given a graph $G = (V, E)$ and a vertex set $X \subseteq V$, the graph $\texttt{PT}(G, X)$ can be constructed in polynomial time.*

Definition 7 (Dangling Subpath). Let $G = (V, E)$ be a graph, and let $X \subseteq V$ be a vertex set. Let P be a $s-t$ path where $s, t \in X$. We call a subpath $P' := aPb$ a *dangling subpath* of P, if $a, b \in X$ and no internal vertices of P' are from X. By $\#\texttt{DS}(P)$, we denote the number of dangling subpaths of P.

Lemma 5 (♠). *Given a graph $G = (V, E)$, a vertex set $X \subseteq V$ and an $s-t$ path P in G, there exist a path containing all vertices of $V(P) \cap X$ in $\texttt{PT}(G, X)$ which has same parity as of P and it is does not contain any vertex from $X \setminus \{V(P) \cap X\}$. Note that the path can contain subdivision vertices which are between pair of vertices of $V(P) \cap X$.*

Now the lemmas below depict the fact that Parity-Torso is parity preserving.

Lemma 6 (♠). *Given a graph $G = (V, E)$, a vertex set $X \subseteq V$ and $s, t \in X$. If a set $S \subset X$ intersects all the paths from s to t of some fixed parity in $\texttt{PT}(G, X)$, then S also intersects all paths from s to t in G of the same parity.*

Lemma 7. *Given a graph $G = (V, E)$, a vertex set $X \subseteq V$ and $s, t \in X$. If a set $S \subset X$ intersects all cycles containing s and t of some fixed parity in $\texttt{PT}(G, X)$, then S also intersects all cycles containing s and t in G of the same parity.*

Proof. Assume for the sake of contradiction that the statement is not true, i.e., S intersects all cycles containing s and t of some fixed parity in $\texttt{PT}(G, X)$ but S does not intersect all cycles of the same parity containing s and t in G and let C be such a cycle containing s and t in $G \setminus S$ which witnesses the fact. Let a and b be the any two vertices in $V(C) \cap X$ such that the subpath $\widehat{P} := aCb$ which is contained in C is a dangling subpath. Let P' be the path from a to b which is contained in C and is disjoint from $\widehat{P}$ other than the end vertices a and b, i.e., $C := a\widehat{P}bP'a$. Note that the existence of a and b is implied by the fact that $s, t \in X$. Now, depending on $\widehat{P}$'s parity, we have either an edge (odd parity) between a and b in $\texttt{PT}(G, X)$ or a subdivided edge (even parity) between a and b in $\texttt{PT}(G, X)$. Let P_1 be the path (edge or subdivided edge) in $\texttt{PT}(G, X)$ between a and b that has the same parity as $\widehat{P}$. Also by Lemma 5, we know there is a path P_2 containing vertices $V(P') \cap X = V(C) \cap X$ from a to b in $\texttt{PT}(G, X)$ and is disjoint from $X \setminus \{V(P') \cap X\}$. Note that $s, t \in V(P_2)$ as

$s, t \in V(P') \cap X$. Now, let us consider the cycle $C' := aP_1bP_2a$ containing vertices of $V(C) \cap X$, has same parity as C. By construction of P_1 and P_2, we know that C' is disjoint from $X \setminus \{V(P') \cap X\}$ thus it is also disjoint from $X \setminus \{V(C) \cap X\}$ as $V(P') \cap X = V(C) \cap X$. Observe that, $s, t \in V(C')$ as $s, t \in V(C) \cap X$. As S does not intersect C, therefore $S \subseteq X \setminus \{V(C) \cap X\}$, which also implies that C' is also disjoint from S. Therefore, we have a contradiction to our assumption that S intersects all cycles containing s and t of that parity (same as C) in $\mathtt{PT}(G, X)$. □

3.2 Structures

In this section, we prove some structural results that will be used in several proofs later. We fix a hypothetical Roman oct and, on the basis of that, examine the structure of the instance with respect to this solution. We start by examining the structure of a general instance based on a hypothetical Roman oct.

Lemma 8 (♠). *Given a graph $G = (V, E)$ there exists an optimal Roman oct (R_1, R_2) of G, such that the cycles in R_1 are pairwise vertex-disjoint.*

Lemma 9 (♠). *Given a graph $G = (V, E)$ and an Roman oct (R_1, R_2) of G, we can construct an Roman oct (R'_1, R'_2) of G, where the cycles in R'_1 are pairwise vertex-disjoint and $w(R_1, R_2) = w(R'_1, R'_2)$ in polynomial time.*

Now onwards we can focus on finding solution for ROCT problem which satisfy the criterion of Lemma 8. We call an Roman oct (R_1, R_2) *special* if it satisfies Lemma 8. For an special Roman oct (R_1, R_2) of G, the following lemma gives us the information on the structure of cycles in R_1 on the graph $G[V \setminus R_2]$.

Lemma 10 (♠). *Given a graph $G = (V, E)$ and an special Roman oct (R_1, R_2) of G, the graph $G[V \setminus R_2]$ does not contain any C-path for any $C \in R_1$.*

Lemma 11 (♠). *Given an special Roman oct (R_1, R_2) of a graph $G = (V, E)$, the cycles of R_1 are the non-intersecting blocks of the graph $G[V \setminus R_2]$.*

Lemma 12 (♠). *Let $G = (V, E)$ be a connected graph which contains exactly one odd cycle C then the following properties hold:*

1. *G does not contain any C-path.*
2. *If G contains a cut vertex then G also contains a cut vertex $v \in V(C)$.*
3. *If $G \neq C$, then G contains a cut vertex $v \in V(C)$.*
4. *Let $C_{cut} \subseteq V(C)$ be a non-empty set of cut vertices of the graph G, then $tw(G) \leq q$ and there exists such a tree-decomposition of G with adhesion 2 which we can find in polynomial time, where $q = max_{v \in C_{cut}}\{|Reach(v, G - \{x, y\})| \mid x, y \in N(v) \cap V(C)\} + 1$.*

Unbreakable Graphs. In this part of the section, we derive some structural findings when the input graph is an unbreakable graph.

Lemma 13. *Given a (s, c)-unbreakable graph $G = (V, E)$ and an special Roman oct (R_1, R_2) of size at most c of G, there does not exist $C_1, C_2 \in R_1$ such that $|V(C_1)| > s$ and $|V(C_2)| > s$.*

Proof. Since (R_1, R_2) has size at most c, thus $|R_2| \leq c/2$. Now, assume for the sake of contradiction that there exist $C_1, C_2 \in R_1$ such that $|C_1| > s$ and $|C_2| > s$. Observe that C_1 and C_2 are in same connected component of $G[V \setminus R_2]$, otherwise we have a witnessing (s, c)-breakable separation $(X, Y) := (\text{Reach}(x, G \setminus R_2) \cup R_2, G \setminus \text{Reach}(x, G \setminus R_2))$ where $x \in V(C_1)$, of order $c/2$ (as $X \cap Y = R_2$) having $|X \setminus Y| > s$ (as $V(C_1) \subseteq X \setminus Y$) and $|Y \setminus X| > s$ (as $V(C_2) \subseteq Y \setminus X$), which will be a contradiction. Now by Lemma 10, we know that there is no C_1-path and C_2 path in $G[V \setminus R_2]$ and also C_1 and C_2 are disjoint because (R_1, R_2) is an special Roman oct. Thus, there exist a cut vertex $v \in G[V \setminus R_2]$ where $v \notin V(C_1)$ & $v \notin V(C_2)$.

Now Consider the following separation; $(X', Y') := (\text{Reach}(x, G \setminus \{R_2 \cup \{v\}\}) \cup \{R_2 \cup \{v\}\}, G \setminus \text{Reach}(x, G \setminus \{R_2 \cup \{v\}\}))$ where $x \in V(C_1)$. The separation (X', Y') has order $c/2 + 1$ as $X' \cap Y' = R_2 \cup \{v\}$ where $|X' \setminus Y'| > s$ (as $V(C_1) \subseteq X' \setminus Y'$) and $|Y' \setminus X'| > s$ (as $V(C_2) \subseteq Y' \setminus X'$). Thus (X', Y') is an (s, c)-breakable separation which is a contradiction to the fact that G is a (s, c)-unbreakable graph. □

Lemma 14 (♠). *Given a (s, c)-unbreakable graph $G = (V, E)$ and a special Roman oct (R_1, R_2) of size at most c of G, where R_1 contains exactly one odd cycle C. If $|V(C)| > s + 1$ and $G[V \setminus R_2]$ contains a cut vertex $v \in V(C)$, then $|Reach(v, G - \{R_2 \cup \{V(C) \setminus \{v\}\}\})| \leq s$.*

Now combining both Lemma 14 and Lemma 12, we have the following result.

Lemma 15. *Given a (s, c)-unbreakable graph $G = (V, E)$ and an special Roman oct (R_1, R_2) of size at most c of G. If R_1 contains exactly one odd cycle C where $|V(C)| > s + 1$, then $tw(G) \leq c/2 + 2 \cdot s + 1$.*

Proof. Since (R_1, R_2) has size at most c, thus $|R_2| \leq c/2$. Now we have the following cases.

- **Case 1:** If $G[V \setminus R_2]$ is connected and does not contain a cut vertex, then by Property 3 of Lemma 12 we know that $V \setminus R_2 = V(C)$. As $\texttt{tw}(C) = 2$ (by Lemma 12), thus in this case we can get a tree-decomposition of G, such that $\texttt{tw}(G) \leq c/2 + 2$ by combining (using Lemma 2) the trivial tree-decomposition of $G[R_2]$ and the tree-decomposition of $G[V \setminus R_2]$.
- **Case 2:** If $G[V \setminus R_2]$ is connected and contains a cut vertex, then by Property 2 of Lemma 12 we know that G also contains a cut vertex $v \in V(C)$. Thus using Property 4 of Lemma 12 and Lemma 14, we can get a tree-decomposition of G, such that $\texttt{tw}(G) \leq c/2 + s + 1$ by combining (using Lemma 2) the trivial tree-decomposition of $G[R_2]$ and the tree-decomposition of $G[V \setminus R_2]$.

- **Case 3:** If $G[V \setminus R_2]$ is disconnected, then we know all the components of $G[V \setminus R_2]$ are bipartite except one component which contains the cycle C. As G is (s, c)-unbreakable and $|V(C)| > s + 1$, thus the size of all the bipartite components of $G[V \setminus R_2]$ is all together less than s.
  Now if $G[V \setminus R_2]$ contains a cut vertex $v \in V(C)$ then using property 4 of Lemma 12 and Lemma 14, we can get a tree-decomposition of G, such that $\mathtt{tw}(G) \leq c/2 + 2.s + 1$ by combining (using Lemma 2) trivial tree-decompositions of $G[R_2]$ and the bipartite components of $G[V \setminus R_2]$ together with the tree-decomposition of the component in $G[V \setminus R_2]$ which contains the cycle C.
  If $G[V \setminus R_2]$ does not contain a cut vertex $v \in V(C)$ then using Property 3 of Lemma 12, we know that the non-bipartite component of $G[V \setminus R_2]$ is precisely the cycle C. As $\mathtt{tw}(C) = 2$ (by Lemma 12), thus we can get a tree-decomposition of G, such that $\mathtt{tw}(G) \leq c/2 + s + 2$ by combining (using Lemma 2) trivial tree-decompositions of $G[R_2]$ and the bipartite components of $G[V \setminus R_2]$ together with the tree-decomposition of the cycle C.

Thus if $|V(C)| > s + 1$, then $\mathtt{tw}(G) \leq c/2 + 2.s + 1$ (assuming $s \geq 1$). □

3.3 FPT Algorithm for ROCT on Unbreakable Graphs

In this section we will design an FPT algorithm for ROCT on (s, c)-unbreakable graphs. We will use the following result of Bodlaender et al. [4] that allows one to obtain an FPT algorithm for X-CMSO expressible parameterized graph problems when parameterized by treewidth and the length of the X-CMSO formula.

Proposition 1 ([4]). *Let ψ be a* CMSO *sentence, then there exist a computable function $f : \mathbb{Z}^+ \times \mathbb{Z}^+ \to \mathbb{Z}^+$ and an algorithm that solves* MIN/MAX/EQ-CMSO *on* STRUCTURES *in $f(tw(G_\alpha), |\psi|) \cdot |V(G_\alpha)|$ steps.*

In order to prove the central result of this section, we begin by showing that the ROCT problem can be expressed using MIN-CMSO.

Lemma 16. ROCT *is* MIN-CMSO *expressible.*

Proof. First, we will go on by proving CMSO expressibility of some simple properties of a graph.

Consider the following formula $\mathbf{conn}(X)$ which verifies that a subset $X \subseteq V$ of a graph $G = (V, E)$ induces a connected subgraph.

$$\mathbf{conn}(X) \equiv \forall_{Y \subseteq V} [(\exists_{u \in X} u \in Y \wedge \exists_{v \in X} v \notin Y) \implies$$
$$(\exists_{e \in E} \exists_{u \in X} \exists_{v \in X} \mathbf{inc}(u, e) \wedge \mathbf{inc}(v, e) \wedge u \in Y \wedge v \notin Y)]$$

The CMSO formula $\mathbf{bip}(G)$ to verify that the given graph $G = (V, E)$ is bipartite.

$$\mathbf{bip}(G) \equiv \exists_{X_1, X_2 \subseteq V} \mathbf{partition}(V, X_1, X_2) \wedge \mathbf{indp}(X_1) \wedge \mathbf{indp}(X_2)$$

where, the auxiliary sub-formulas **partition**(V, X_1, X_2) which given two subsets $X_1, X_2 \subseteq V$ verifies that X_1 & X_2 are partition of V and **indp**(X) which verifies that a given subset of vertices X is independent, are defined as follows,

$$\mathbf{partition}(V, X_1, X_2) \equiv \forall_{v\in V}[(v \in X_1 \wedge v \notin X_2) \vee (v \notin X_1 \wedge v \in X_2)]$$

&

$$\mathbf{indp}(X) \equiv \forall_{u,v\in X} \neg \mathbf{adj}(u, v)$$

Now for the ROCT, the MIN-CMSO formula $\varphi_{\text{ROCT}}(G)$ verifies that graph $G = (V, E)$ has an Roman oct (R_1, R_2) of size at most k, where $|R_1| = k_2$ and $|R_2| = k_1$.

$$\varphi_{\text{ROCT}}(G) \equiv \exists_{W_1,\ldots,W_{k_2}\subseteq V} \exists_{S\subseteq V}\left[\bigwedge_{\ell=1}^{6} \texttt{Condition}(\ell)\right]$$

where,

- $\texttt{Condition}(1) := \bigwedge_{i=1}^{k_2}\Big(\mathbf{conn}(W_i) \wedge \forall_{v\in W_i}\mathbf{deg2}(v, W_i) \wedge \mathbf{card}_{1,2}(W_i)\Big)$
- $\texttt{Condition}(2) := \forall_{F\subseteq V}\Big[\Big(\mathbf{conn}(F) \wedge \forall_{v\in F}\mathbf{deg2}(v, F) \wedge \mathbf{card}_{1,2}(F)\Big) \Rightarrow \Big(\bigvee_{i=1}^{k_2}(F = W_i) \vee (\exists_{v\in S} v \in F)\Big)\Big]$
- $\texttt{Condition}(3) := \exists_{v_1,\ldots,v_{k_1}\in V}\forall_{u\in V}\Big((u \in S) \Rightarrow (u = v_1 \vee \cdots \vee u = v_k)\Big)$
- $\texttt{Condition}(4) := \forall_{v\in S}\Big(v \notin W_1 \wedge v \notin W_2 \wedge \ldots \wedge v \notin W_{k_2}\Big)$
- $\texttt{Condition}(5) := \bigwedge_{i=1}^{k_2}\bigwedge_{j=1}^{k_2}\Big(\mathbf{disjoint}(W_i, W_j) \vee (W_i = W_j)\Big)$
- $\texttt{Condition}(6) := \mathbf{bip}(G[V \setminus \{S \cup W_1 \cup W_2 \cup \ldots \cup W_{k_2}\}])$

The auxiliary sub-formulas **deg2**(v, X) which verifies vertex v has exactly two adjacent edges belonging to the given subset of vertices X and **disjoint**(X, Y) verifies that the given two subset of vertices X and Y are disjoint, are defined as follows,

$$\mathbf{deg2}(v, X) \equiv \exists_{e_1,e_2\in E(X)}[(e_1 \neq e_2) \wedge \mathbf{inc}(v, e_1) \wedge \mathbf{inc}(v, e_2) \wedge (\forall_{e_3\in E(X)}\mathbf{inc}(v, e_3) \implies (e_1 = e_3 \vee e_2 = e_3))]$$

&

$$\mathbf{disjoint}(X, Y) \equiv \forall_{v\in X\cup Y}(v \in X \Leftrightarrow v \notin Y)$$

Note that, the length of the formula $\varphi_{\text{ROCT}}(G)$ which is the number of existential variables, depends linearly on $k_1 + k_2$. □

Now, we will show that ROMAN OCT admits an FPT algorithm on (s, c)-unbreakable graphs when parameterized by $s + c$, where c is the solution size.

Lemma 17. ROCT *parameterized by* $s + c$ *has an* FPT *algorithm on* (s, c)*-unbreakable graphs running in time* $f(s, c) \cdot n^{\mathcal{O}(1)}$*, for some computable function* $f : \mathbb{Z}^+ \times \mathbb{Z}^+ \to \mathbb{Z}^+$*, where* c *is the solution size.*

Proof. Let $G = (V, E)$ be an (s, c)-unbreakable graph which will be our input and $s + c$ is the parameter. Our aim is to find a special Roman oct (R_1, R_2) in G of size at most c. Now, we consider a separation of (X, Y) of G, where order of the separation is at most c, i.e., $|X \cap Y| \leq c$. Now we have the following two cases based on the cardinalities of $|X \setminus Y|$ and $|Y \setminus X|$.

Case 1 ($|X \setminus Y| < s$ **and** $|Y \setminus X| < s$)**:** Observe that in this case $|V| = |X \setminus Y| + |Y \setminus X| + |X \cap Y| < 2.s + c$, thus we can get a trivial FPT algorithm to find an special Roman oct (R_1, R_2) of size at most c in G running in time $2^{\mathcal{O}(s+c)} \cdot n^{\mathcal{O}(1)}$ by guessing R_2 of size at most $c/2$ from V and checking whether or not the blocks (other than isolated vertex and edges) of $G[V \setminus R_2]$ are odd cycles. The number of odd cycles in $G[V \setminus R_2]$ can be at most $c - 2|R_2|$. The fact that upon a correct guessing of R_2, the blocks which are not isolated vertex and edges of $G[V \setminus R_2]$ will be odd cycles is guaranteed by Lemma 11 and checking of that can also be done in polynomial time (Algorithm 4.1.23 [25]).

Case 2 ($|X \setminus Y| > s$ **and** $|Y \setminus X| < s$)**:** Our objective is to find an special Roman oct (R_1, R_2) of size at most c in G. We initiate by setting $R_1' = \emptyset$ and $R_2' = \emptyset$. We incrementally construct the sets R_2' and R_1' until $R_2' = R_2$ and $R_1' = R_1$. We go ahead by finding induced odd cycles of length at most $s + 1$ in G which we can do using breadth-first-search (bfs) in polynomial time.

Branching Step: We branch over the induced odd cycles of length at most $s + 1$ for guessing their intersection with R_2 and incrementally construct the set R_2'. If for some cycle C the branch guesses $C \cap R_2 = \emptyset$ then we put C into the set R_1'. Note that while branching over a cycle C_a of length at most $s + 1$, the branching step can not guess $C_a \cap R_2 = \emptyset$ (by Lemma 8), if $V(C_a) \cap V(C_b) \neq \emptyset$ for some $C_b \in R_1'$.

Now we have Observation 1 which follows from Lemma 8, saying; For an odd cycle, if all its vertices are part of some cycle in R_1 except one vertex, then that one vertex has to be part of R_2.

Observation 1. *Consider the partial solution* $(R_1', R_2') \subseteq (R_1, R_2)$. *If* $x \in V \setminus \{R_2' \cup V(R_1')\}$ *is part of an odd cycle* C, *where* $V(C) \setminus \{x\} \subseteq V(R_1')$, *then* $x \in R_2 \setminus R_2'$.

Now, the above Observation 1 provides us with the way to extend the partial solution (R_1', R_2') that we got after the branching step, as follows:

Solution Extension Step: After the branching step, if there is an odd cycle C that contains a vertex $x \in V \setminus \{R_2' \cup V(R_1')\}$ where $V(C) \setminus \{x\} \subseteq V(R_1')$, then we extend the partial solution (R_1', R_2') by adding the vertex x to the set R_2', i.e., $(R_1', R_2') := (R_1', R_2' \cup \{x\})$. Note that we can find the vertex x in polynomial time by doing a bfs from x in the graph $G[V(R_1') \cup \{x\}]$ and looking for cross-edges that connects two vertices of $V(R_1')$ which are in the same layer of the bfs tree. After the end of the branching and solution extension step, we know that there is no cycle of length at most $s + 1$ to branch over. Let (R_1', R_2') be the partial solution that we currently have after the branching and solution extension step.

Now we construct $\texttt{PT}(G[V \setminus R'_2], V \setminus \{R'_2 \cup V(R'_1)\})$ using Lemma 4 in polynomial time. We also know by Lemma 13 that for an special Roman oct (R_1, R_2) of size at most c in G, there can not exist two odd cycles $C_1, C_2 \in G[V \setminus R_2]$ such that $|V(C_1)| > s$ and $|V(C_2)| > s$. So now we devise the following two subcases based on whether or not $G[V \setminus R_2]$ contains an odd cycle of length more than $s+1$ based on the hypothetical special Roman oct (R_1, R_2). Note that we already branched over all the odd cycles of length at most $s + 1$.

Subcase 1: If $G[V \setminus R_2]$ does not contain any odd cycle of length more than s, then we run the FPT algorithm parameterized by solution size $k = c - 2|R'_2| - |R'_1|$ for OCT (due to Reed, Smith and Vetta [23]) on $\texttt{PT}(G[V \setminus R'_2], V \setminus \{R'_2 \cup V(R'_1)\})$, running in time $3^k \cdot n^{\mathcal{O}(1)}$. If we get an affirmative answer for OCT in $\texttt{PT}(G[V \setminus R'_2], V \setminus \{R'_2 \cup V(R'_1)\})$, then we can also give an affirmative answer to ROCT in G. We can get an Roman oct (R_1, R_2) by combining the partial solution (R'_1, R'_2) in G, together with the solution returned by the OCT in $\texttt{PT}(G[V \setminus R'_2], V \setminus \{R'_2 \cup V(R'_1)\})$. Let D be the solution returned by the OCT in $\texttt{PT}(G[V \setminus R'_2], V \setminus \{R'_2 \cup V(R'_1)\})$. If D contains any subdivision vertices then we swap those vertices with one of its neighbour vertices. The swapping is safe as the subdivision vertices have degree two and thus all the odd cycles that contains the subdivision vertices, also contains the neighbours of those subdivision vertices. Therefore, we construct an Roman oct $(R_1, R_2) := (R'_1, R'_2 \cup D)$ of size at most c in G. The correctness of this subcase is assured by Observation 1 which tells that any odd cycle which does not contain any vertices of R'_2 must intersect $V \setminus \{R'_2 \cup V(R'_1)\}$ on at least two vertices and by Lemma 7 which tells that hitting all odd cycles in $\texttt{PT}(G[V \setminus R'_2], V \setminus \{R'_2 \cup V(R'_1)\})$ that contains two specific vertices of $V \setminus \{R'_2 \cup V(R'_1)\}$, is sufficient to hit all odd cycles in $G[V \setminus R'_2]$ containing those two specific vertices.

Subcase 2: If $G[V \setminus R_2]$ contains an odd cycle of length more than $s + 1$, then by Lemma 15, we know $\texttt{tw}(G) \leq c/2 + 2.s + 1$. Thus we invoke Proposition 1 as a black-box to get an FPT algorithm for ROCT parameterized by $s + c$ in G running in time $f(s, c) \cdot n^{\mathcal{O}(1)}$, for some computable function $f : \mathbb{Z}^+ \times \mathbb{Z}^+ \to \mathbb{Z}^+$.

This completes the proof. □

3.4 FPT Algorithm for ROCT on General Graphs

In this section we will design an FPT algorithm for ROCT on general graphs. We will use the following results of Lokshtanov et al. [20] that allows one to obtain a (non-uniform) FPT algorithm for X-CMSO expressible parameterized graph problems by designing an FPT algorithm for the problem on unbreakable graphs.

Proposition 2. ([20]). *Let* X$\in$ {MIN,MAX,EQ,MIN-EDGE,MAX-EDGE, EQ-EDGE}, *and let* $\widehat{\psi}$ *be a* CMSO *sentence. For all* $\widehat{c} : \mathbb{N}_0 \to \mathbb{N}_0$, *there exists* $\widehat{s} : \mathbb{N}_0 \to \mathbb{N}_0$ *such that if* X-CMSO$[\widehat{\psi}]$ *parameterized by* k *is* FPT *on* $(\widehat{s}(k), \widehat{c}(k))$-*unbreakable structures, then* X-CMSO$[\widehat{\psi}]$ *parameterized by* k *is* FPT *on general structures.*

Proposition 2 allows us to use the technique of recursive understanding in a black-box manner, leading to a streamlined proof at the expense of obtaining non-uniform algorithms. Now, we rely on Proposition 2, which requires the MIN-CMSO expressibility of the ROCT problem which we already have due to Lemma 16. Therefore, we have our main result showing that ROMAN OCT admits an FPT algorithm on general graphs when parameterized by solution size k, by invoking Proposition 2 with $\widehat{c}(k) = k$ and apply Lemma 17 on $(\widehat{s}(k), k)$-unbreakable graphs, where the function $\widehat{s}$ obtained by the invocation of Proposition 2. Hence we have our main theorem.

Theorem 1. ROCT *can be solved in time* $f(k) \cdot n^{\mathcal{O}(1)}$*, for some computable function* $f : \mathbb{Z}^+ \to \mathbb{Z}^+$*, where* k *is the solution size.*

References

1. Agrawal, A., Gupta, S., Saurabh, S., Sharma, R.: Improved algorithms and combinatorial bounds for independent feedback vertex set. In: Guo, J., Hermelin, D. (eds.) 11th International Symposium on Parameterized and Exact Computation, IPEC 2016, 24–26 August 2016, Aarhus, Denmark. LIPIcs, vol. 63, pp. 2:1–2:14. Schloss Dagstuhl - Leibniz-Zentrum für Informatik (2016). https://doi.org/10.4230/LIPIcs.IPEC.2016.2
2. Agrawal, A., Jain, P., Kanesh, L., Lokshtanov, D., Saurabh, S.: Conflict free feedback vertex set: a parameterized dichotomy. In: Potapov, I., Spirakis, P.G., Worrell, J. (eds.) 43rd International Symposium on Mathematical Foundations of Computer Science, MFCS 2018, 27–31 August 2018, Liverpool, UK. LIPIcs, vol. 117, pp. 53:1–53:15. Schloss Dagstuhl - Leibniz-Zentrum für Informatik (2018). https://doi.org/10.4230/LIPIcs.MFCS.2018.53
3. Agrawal, A., Lokshtanov, D., Mouawad, A.E., Saurabh, S.: Simultaneous feedback vertex set: a parameterized perspective. ACM Trans. Comput. Theory **10**(4):18:1–18:25 (2018). https://doi.org/10.1145/3265027
4. Bodlaender, H.L., Fomin, F.V., Lokshtanov, D., Penninkx, E., Saurabh, S., Thilikos, D.M.: (Meta) kernelization. J. ACM (JACM) **63**(5), 1–69 (2016)
5. Chitnis, R., Cygan, M., Hajiaghayi, M.T., Pilipczuk, M., Pilipczuk, M.: Designing FPT algorithms for cut problems using randomized contractions. SIAM J. Comput. **45**(4), 1171–1229 (2016). https://doi.org/10.1137/15M1032077
6. Cygan, M., Nederlof, J., Pilipczuk, M., Pilipczuk, M., van Rooij, J.M.M., Wojtaszczyk, J.O.: Solving connectivity problems parameterized by treewidth in single exponential time. ACM Trans. Algorithms **18**(2):17:1–17:31 (2022). https://doi.org/10.1145/3506707
7. Cygan, M., Pilipczuk, M., Pilipczuk, M.: On group feedback vertex set parameterized by the size of the cutset. Algorithmica **74**(2), 630–642 (2016). https://doi.org/10.1007/s00453-014-9966-5
8. Diestel, R.: Graph Theory, vol. 173. Springer, Heidelberg (2017). https://doi.org/10.1007/978-3-662-53622-3
9. Downey, R.G., Fellows, M.R.: Fixed parameter tractability and completeness. In: Ambos-Spies, K., Homer, S., Schöning, U. (eds.) Complexity Theory: Current Research, Dagstuhl Workshop, 2–8 February 1992, pp. 191–225. Cambridge University Press (1992)

10. Fernau, H.: ROMAN DOMINATION: a parameterized perspective. Int. J. Comput. Math. **85**(1), 25–38 (2008). https://doi.org/10.1080/00207160701374376
11. Fernau, H., Mann, K.: Hitting the romans. CoRR, abs/2302.11417 (2023). https://doi.org/10.48550/arXiv.2302.11417, http://arxiv.org/abs/2302.11417
12. Grohe, M., Kawarabayashi, K., Marx, D., Wollan, P.: Finding topological subgraphs is fixed-parameter tractable. In: Fortnow, L., Vadhan, S.P. (eds.) Proceedings of the 43rd ACM Symposium on Theory of Computing, STOC 2011, San Jose, CA, USA, 6–8 June 2011, pp. 479–488. ACM (2011). https://doi.org/10.1145/1993636.1993700
13. Haynes, T.W., Hedetniemi, S.T., Henning, M.A.: Topics in Domination in Graphs, vol. 64. Springer (2020)
14. Iwata, Y., Kobayashi, Y.: Improved analysis of highest-degree branching for feedback vertex set. Algorithmica **83**(8), 2503–2520 (2021). https://doi.org/10.1007/s00453-021-00815-w
15. Kawarabayashi, K., Thorup, M.: The minimum k-way cut of bounded size is fixed-parameter tractable. In: Ostrovsky, R. (ed.) IEEE 52nd Annual Symposium on Foundations of Computer Science, FOCS 2011, Palm Springs, CA, USA, 22–25 October 2011, pp. 160–169. IEEE Computer Society (2011). https://doi.org/10.1109/FOCS.2011.53
16. Li, J., Nederlof, J.: Detecting feedback vertex sets of size k in $O^\star$ $(2.7k)$ time. ACM Trans. Algorithms **18**(4), 34:1–34:26 (2022). https://doi.org/10.1145/3504027
17. Li, S., Pilipczuk, M.: An improved FPT algorithm for independent feedback vertex set. Theory Comput. Syst. **64**(8), 1317–1330 (2020). https://doi.org/10.1007/s00224-020-09973-w
18. Lokshtanov, D., Narayanaswamy, N.S., Raman, V., Ramanujan, M.S., Saurabh, S.: Faster parameterized algorithms using linear programming. ACM Trans. Algorithms **11**(2), 15:1–15:31 (2014). https://doi.org/10.1145/2566616
19. Lokshtanov, D., Ramanujan, M.S., Saurabh, S.: Linear time parameterized algorithms for subset feedback vertex set. ACM Trans. Algorithms **14**(1), 7:1–7:37 (2018). https://doi.org/10.1145/3155299
20. Lokshtanov, D., Ramanujan, M.S., Saurabh, S., Zehavi, M.: Reducing CMSO model checking to highly connected graphs. In: 45th International Colloquium on Automata, Languages, and Programming (ICALP 2018), Dagstuhl, Germany. Leibniz International Proceedings in Informatics (LIPIcs), vol. 107, pp. 135:1–135:14 (2018). https://drops.dagstuhl.de/entities/document/10.4230/LIPIcs.ICALP.2018.135
21. Misra, N., Philip, G., Raman, V., Saurabh, S.: On parameterized independent feedback vertex set. Theor. Comput. Sci. **461**, 65–75 (2012). https://doi.org/10.1016/j.tcs.2012.02.012
22. Misra, N., Philip, G., Raman, V., Saurabh, S., Sikdar, S.: FPT algorithms for connected feedback vertex set. J. Comb. Optim. **24**(2), 131–146 (2012). https://doi.org/10.1007/s10878-011-9394-2
23. Reed, B., Smith, K., Vetta, A.: Finding odd cycle transversals. Oper. Res. Lett. **32**(4), 299–301 (2004)
24. Thomassé, S.: A quadratic kernel for feedback vertex set. In: Mathieu, C. (ed.) Proceedings of the Twentieth Annual ACM-SIAM Symposium on Discrete Algorithms, SODA 2009, New York, NY, USA, 4–6 January 2009, pp. 115–119. SIAM (2009). https://doi.org/10.1137/1.9781611973068.13
25. West, D.B., et al.: Introduction to Graph Theory, vol. 2. Prentice Hall, Upper Saddle River (2001)

Popular Solutions for Optimal Matchings

Telikepalli Kavitha(✉)

Tata Institute of Fundamental Research, Mumbai, India
kavitha@tifr.res.in

Abstract. Let G be a bipartite graph where every vertex has a strict preference order over its neighbors. The preferences of a vertex over its neighbors extend naturally to preferences over matchings. A matching M is popular in G if there is no matching N such that vertices that prefer N outnumber those that prefer M. Every stable matching is popular. We consider the following variant: edges in G have utilities and it is only *max-utility* matchings that are relevant for us. We show there always exists a max-utility matching that is popular within the set of all max-utility matchings; moreover, such a matching can be efficiently computed. We focus on *largest* max-utility matchings and show a compact extended formulation for the polytope of largest max-utility matchings that are popular within the set of all largest max-utility matchings.

Keywords: Bipartite graphs · Stable matchings · Dual certificates

1 Introduction

We consider a matching problem in a bipartite graph $G = (A \cup B, E)$ where A is a set of agents and B is a set of jobs. Every vertex in $A \cup B$ has a strict preference order over its neighbors. This is a well-studied model—such a graph G is called a *marriage instance*. In several applications, along with vertex preferences, there might be other attributes that we seek to optimize, e.g., in the assignment of residents to hospitals [2] and in the assignment of sailors to billets [17,21], it is only A-perfect matchings that are admissible where A is the set of residents/sailors. More generally, let us assume there is a utility function $f : E \to \mathbb{Q}$ and it is only max-utility matchings that are admissible. Such applications include allocation problems in humanitarian organizations [1,19] where we seek an allocation of resources to beneficiaries that has maximum impact.

This motivates us to consider the setting where *edge utilities* are more important than vertex preferences. The usual solution to a two-sided matching market problem is an appropriate *stable* matching in G. A matching M is stable if there is no *blocking edge* (a, b), i.e., one where both a and b prefer each other to their respective assignments in M. Stable matchings always exist in G and can be efficiently computed by the Gale-Shapley algorithm [7]. All stable matchings have the same size [8] which may be only half the size of a maximum matching. Thus stable matchings need not be max-size matchings or max-utility matchings. So we need to go beyond stability in order to find a max-utility matching that is optimal with respect to vertex preferences.

D. Král' and M. Milanič (Eds.): WG 2024, LNCS 14760, pp. 297–311, 2025.
https://doi.org/10.1007/978-3-031-75409-8_21

Popularity. The preferences that any vertex v has over its neighbors in G extend naturally to preferences over matchings. Given any pair of matchings M and N, we say v prefers M to N if v prefers its assignment in M to its assignment in N. Note that being left unassigned is the worst choice for any vertex. Let $\phi(M, N)$ be the number of vertices that prefer M to N. Matching N is *more popular* than matching M if $\phi(N, M) > \phi(M, N)$.

Definition 1. *M is a popular matching in G if $\phi(M, N) \geq \phi(N, M)$ for all matchings N, i.e., G has no matching more popular than M.*

A popular matching is "stable" in the relaxed sense that no majority vote can force a migration from a popular matching. Popular matchings always exist in G since every stable matching is popular [9]. Recall that utility is more important in our setting than vertex preferences. We seek a max-utility matching M such that there is no *better* max-utility matching with respect to vertex preferences.

So rather than a popular matching, what we seek is a max-utility matching that is popular *within* the set of max-utility matchings, i.e., a max-utility matching M such that $\phi(M, N) \geq \phi(N, M)$ for all max-utility matchings N. Such a matching M will be called a *popular max-utility matching* or more concisely, a *popular* opt-*matching*. The "more popular than" relation is not transitive, so it is not obvious if a popular opt-matching always exists. We show the following.

Theorem 1. *Let $G = (A \cup B, E)$ be any marriage instance where there is a utility function $f : E \to \mathbb{Q}$. A popular* opt-*matching always exists in G and can be computed in polynomial time.*

The next question is whether the entire set of popular opt-matchings can be compactly described, say, as the intersection of half-spaces in $\mathbb{R}^{|E|}$. The motivation for such a description is to efficiently solve *special* popular opt-matching problems such as finding a popular opt-matching with forced/forbidden edges or more generally, one with minimum cost when every edge has an associated cost. Observe that the popular opt-matching problem with forced/forbidden edges generalizes the problem of finding a popular matching with forced/forbidden edges, which is NP-hard [6]. Thus it is NP-hard to optimize over all popular opt-matchings and solve special popular opt-matching problems.

A Useful Subset. The above hardness motivates us to focus on a natural and interesting subset of the entire set of opt-matchings. Let us restrict our attention to *largest* opt-matchings—so subject to the constraint that matchings have maximum utility, we focus on the largest ones. Indeed, it is these matchings that would be most useful in applications such as allocation problems in humanitarian organizations where subject to the constraint that maximum utility is ensured, we would like to maximize the number of beneficiaries. We will refer to largest max-utility matchings as max-opt-matchings. Hence the matchings we study now are *popular max*-opt-*matchings*, i.e., largest max-utility matchings that are popular within the set of all largest max-utility matchings.

The main idea used to prove Theorem 1 can be used to show that popular max-opt-matchings always exist and can be computed in polynomial time. Let the popular max-opt-matching polytope be the convex hull of the edge incidence vectors of all popular max-opt-matchings. We show the following structural result that the entire set of popular max-opt-matchings admits a compact description. Thus we can optimize over the set of popular max-opt-matchings in polynomial time via linear programming on this polytope.

Theorem 2. *Let $G = (A \cup B, E)$ be a marriage instance with a utility function $f : E \to \mathbb{Q}$. A compact extension of the popular max-opt-matching polytope of G exists and can be formulated in polynomial time.*

1.1 Background and Related Results

Gärdenfors [9] introduced the notion of popularity in 1975 where he observed that any stable matching in a marriage instance is popular. Popular matchings have been well-studied during the last 15–20 years and we refer to [3] for a survey.

It was shown in [11] that any marriage instance has a *popular maximum matching*, i.e., a maximum matching that is popular within the set of all maximum matchings and a polynomial time algorithm was given to find such a matching. The problem of computing a *min-cost* popular maximum matching was considered in [13] and a compact extended formulation of the popular maximum matching polytope was given. Thus a min-cost popular maximum matching can be computed in polynomial time. In contrast to this, as mentioned earlier, it is NP-hard to compute a min-cost popular matching [6]. Furthermore, the extension complexity of the popular matching polytope is near-exponential [5].

The following definitions will be useful to us. For any subset $C \subseteq A \cup B$, call a matching *C-critical* if it matches as many vertices in C as possible. A C-critical matching M is a *popular C-critical matching* if $\Delta(M, N) \geq 0$ for all C-critical matchings N in G. When $C = \emptyset$ (resp., $C = A \cup B$), popular C-critical matchings are the same as popular matchings (resp., popular maximum matchings).

It was shown in [12] that for any subset $C \subseteq A \cup B$, a popular C-critical matching always exists and can be computed in polynomial time. A related problem in the hospitals-residents setting was independently considered in [15] where every hospital had upper and lower bounds on the number of residents that it could be matched to in any feasible matching and certain special residents were to be matched in any feasible matching. It was shown that a popular feasible matching always exists and can be computed in polynomial time.

1.2 Our Techniques

A crucial observation in our result on popular opt-matchings (Theorem 1) is that for any edge utility function, the convex hull of all max-utility matchings is a face of the matching polytope of G (see Sect. 2). So max-utility matchings can be characterized as *C-perfect* matchings in a subgraph G' for some $C \subseteq A \cup B$, where a matching is C-perfect if it matches all vertices in C. Thus once we determine the set C and the edge set of G', the popular C-critical matching algorithm in G' finds a popular max-utility matching in G.

Popular opt-matchings generalize popular matchings, thus their polytope inherits the near-exponential extension complexity of the popular matching polytope. The key idea in formulating a compact extension for the popular max-opt-matching polytope is that the critical set for max-opt-matchings has rich structure. We will use the well-known Dulmage-Mendelsohn decomposition [4] to determine the critical set for popular max-opt-matchings (see Sect. 3).

We construct another bipartite graph $H = (L \cup R, E_H)$ and max-opt-matchings in G can be characterized as $(L \cup K)$-perfect matchings in H for a certain subset $K \subseteq R$. Observe that the min-cost popular maximum matching algorithm [13] in H does not solve the min-cost popular max-opt-matching problem since such a matching may leave some vertices in the set K unmatched—this would make the resulting matching a non-opt-matching in the original instance G.

So the problem we need to solve is to find a min-cost popular $(L \cup K)$-perfect matching in H. As done in [13], we use LP duality to solve the above problem. Our key technical lemma (Lemma 3) shows that every popular $(L \cup K)$-perfect matching has a very useful dual certificate which allows us to realize every such matching as a stable matching in a certain multigraph derived from H (see Sect. 4). So the stable matching polytope [18,20] of this multigraph yields a compact extended formulation for the popular max-opt-matching polytope of G.

2 Popular Opt-Matchings

We prove Theorem 1 in this section. We will show that the popular opt-matching problem in G can be solved as the popular C-critical matching problem for an appropriate set C in a subgraph $G' = (A \cup B, E')$ of $G = (A \cup B, E)$.

There is a utility function $f : E \to \mathbb{Q}$ and $f(M) = \sum_{e \in M} f(e)$ for any matching M. Let $\lambda = \max_M f(M)$, where the max is over all matchings in G. Recall that M is an opt-matching if and only if $f(M) = \lambda$. The opt-matching polytope (call it $\mathcal{X}$) is the convex hull of edge incidence vectors of all opt-matchings in G.

A face of a polytope $\mathcal{P}$ is the set of points $\boldsymbol{x}$ in $\mathcal{P}$ such that $\sum_j \alpha_j x_j = \beta$ where $\sum_j \alpha_j x_j \le \beta$ is a valid inequality for $\mathcal{P}$. Since $\sum_{e \in E} f(e) \le \lambda$ is a valid inequality for the matching polytope $\mathcal{M}$ of G, the polytope $\mathcal{X}$ is a face of $\mathcal{M}$. Suppose a polytope $\mathcal{P}$ is defined by the constraints $\sum_j \alpha_{ij} x_j \le \beta_i$ for $1 \le i \le k$. Then any face of $\mathcal{P}$ corresponds to the set of solutions to $\sum_j \alpha_{ij} x_j = \beta_i$ for all $i \in I$ and $\sum_j \alpha_{ij} x_j \le \beta_i$ for all $i \notin I$ for some set $I \subseteq \{1, \ldots, k\}$. See [10, Theorem 3.5] for a proof.

Thus the opt-matching polytope $\mathcal{X}$, which is a face of the matching polytope $\mathcal{M}$ of G, has the following formulation for some $A' \subseteq A$, $B' \subseteq B$, and $\tilde{E} \subseteq E$:

$$\sum_{e \in \delta(v)} x_e \le 1 \quad \forall v \in A \cup B \quad \text{and} \quad \sum_{e \in \delta(v)} x_e = 1 \quad \forall v \in A' \cup B'$$

$$x_e \ge 0 \quad \forall e \in E \quad \text{and} \quad x_e = 0 \quad \forall e \in \tilde{E}.$$

Here $\delta(v)$ is the set of edges incident to v in G. So a matching M is an opt-matching if and only if M matches all vertices in $A' \cup B'$ and $M \subseteq E'$, where $E' = E \setminus \tilde{E}$. The sets A', B', and E' can be computed by solving LP2. This linear program is dual to the max-utility matching LP, which is LP1.

$$\max \sum_{e \in E} f(e) \cdot x_e \qquad \text{(LP1)}$$
$$\text{s.t.} \quad \sum_{e \in \delta(v)} x_e \leq 1 \quad \forall v \in A \cup B$$
$$x_e \geq 0 \quad \forall e \in E.$$

$$\min \sum_{v \in A \cup B} y_v \qquad \text{(LP2)}$$
$$\text{s.t.} \quad y_a + y_b \geq f(a,b) \quad \forall (a,b) \in E$$
$$y_v \geq 0 \quad \forall v \in A \cup B.$$

Let $\boldsymbol{y}$ be an optimal solution to LP2. We have $A' = \{a \in A : y_a > 0\}$, $B' = \{b \in B : y_b > 0\}$, and $E' = \{(a,b) \in E : y_a + y_b = f(a,b)\}$. It follows from complementary slackness conditions that for any $\boldsymbol{x} \in \mathcal{X}$, we have $\sum_{e \in \delta(v)} x_e = 1$ for all $v \in A' \cup B'$ and $x_e = 0$ for all $e \in E \setminus E'$. Thus we can conclude the following proposition. The converse in Proposition 1 also follows from LP-duality.

Proposition 1. *M is an* opt*-matching in G if and only if M is a C-perfect matching in $G' = (A \cup B, E')$ where $C = A' \cup B'$.*

Proposition 1 implies that the popular opt-matching problem in $G = (A \cup B, E)$ reduces to the popular C-critical matching problem—more precisely, the popular C-*perfect* matching problem—in $G' = (A \cup B, E')$, where $C = A' \cup B'$.

Finding a Popular C-Perfect Matching. We now describe a multigraph G'_C such that any stable matching in G'_C projects to a popular C-perfect matching in G'. The graph G'_C is inspired by the graph used in [12] for solving the popular C-critical matching problem. Our graph has fewer vertices and edges than the graph used in [12]. Recall that $C = A' \cup B'$.

Let $|A'| = s$ and $|B'| = t$. Below we describe the edge set E'_C of G'_C.

- Initialize $E'_C = \{e_0 : e \in E'\}$.
- For every $e = (a,b) \in E'$, we will consider the following parallel edges where each e_i has a and b as its endpoints.
  1. If $a \in A'$: then add the s edges $e_1, \ldots, e_s$ to E'_C.
  2. If $b \in B'$: then add the t edges $e_{-1}, \ldots, e_{-t}$ to E'_C.

So if $a \in A'$ and $b \in B'$ then there are $s+t+1$ edges $e_{-t}, \ldots, e_s$ in E'_C. We need to specify vertex preferences in G'_C. In order to compare two edges e_i and e'_j incident to it, any vertex v first compares the subscripts i and j. Suppose $v \in A$.

- If $i < j$ then v prefers e_i to e'_j and if $i > j$ then v prefers e'_j to e_i.
- If $i = j$ then v prefers e_i if it prefers e to e'; else it prefers e'_j.

Thus any vertex in A prefers lower subscript edges to higher subscript edges. Among edges with the same subscript, it is as per its original preference order. On the other hand, any vertex in B prefers *higher* subscript edges to *lower* subscript edges. Hence for any $v \in B$ and any two edges e_i and e'_j incident to v:

- If $i < j$ then v prefers e'_j to e_i and if $i > j$ then v prefers e_i to e'_j.
- If $i = j$ then v prefers e_i if it prefers e to e'; else it prefers e'_j.

A matching M' in the multigraph G'_C is a subset of E'_C such that at most one edge of M' is incident to any vertex. Matching M' is *stable* in G'_C if there is no edge in E'_C that blocks M'. For any matching M' in G'_C, define M to be the projection of M' in G', i.e., replace every edge $e_i \in M'$ with the original edge e (so subscripts are dropped from the edges in M').

Lemma 1. *If M' is a stable matching in G'_C then M is a popular C-perfect matching in G'.*

The proof of the above lemma is similar to the proof of correctness of the popular maximum matching algorithm [11,13] and the popular critical matching algorithm [12]. This lemma can be proved in two parts. The first claim is that M is a C-perfect matching in G' where $C = A' \cup B'$. This claim follows by showing that there are no *forbidden* alternating or augmenting paths with respect to M in G'. A proof of this claim will be given in the full version of our paper.

We will now assume that M is a C-perfect matching in G' where $C = A' \cup B'$. We prove below that M is a *popular* C-perfect matching in G'. That is, for any C-perfect matching N in G', we will show that $\Delta(N, M) \le 0$. To show this, we will use the following two linear programs: LP3 and LP4.

Let $\bar{E}'$ be the edge set E' augmented with self-loops (v, v) for each vertex v where $v \in (A \cup B) \setminus (A' \cup B')$. Adding self-loops will allow us to work with perfect matchings in the augmented G' whose vertex set is $A \cup B$ and edge set is $\bar{E}'$. We will use the following edge weight function wt_M. For any $(a, b) \in E'$:

$$\text{let } \mathsf{wt}_M(a,b) = \begin{cases} 2 & \text{if } (a,b) \text{ blocks } M; \\ -2 & \text{if } a \text{ and } b \text{ prefer their assignments in } M \text{ to each other;} \\ 0 & \text{otherwise.} \end{cases}$$

For any edge e, $\mathsf{wt}_M(e)$ is the sum of *votes* of the endpoints of e for each other versus their respective assignments in M, where the vote of a vertex for one neighbor versus another is ± 1 depending on which one it likes better and v's vote for neighbor u versus neighbor u' is 0 if and only if $u = u'$. So $\mathsf{wt}_M(e) = 0$ for every $e \in M$. Let $\mathsf{wt}_M(v, v) = 0$ if the self-loop (v, v) is in the augmented M (i.e., v was originally left unmatched in M), else $\mathsf{wt}_M(v, v) = -1$. It follows from the definition of wt_M that for any perfect matching N in the augmented graph G', we have $\mathsf{wt}_M(N) = \Delta(N, M)$.

Our goal is to show that $\mathsf{wt}_M(N) \le 0$ for all perfect matchings N in the augmented graph G'. Consider the linear programs LP3 and LP4 given below.

$$\max \sum_{e\in E'} \mathsf{wt}_M(e)\cdot x_e \qquad \text{(LP3)}$$

$$\text{s.t.} \quad \sum_{e\in\delta'(v)} x_e = 1 \quad \forall v \in A\cup B$$

$$x_e \ge 0 \quad \forall e \in \tilde{E}'.$$

$$\min \sum_{v\in A\cup B} y_v \qquad \text{(LP4)}$$

$$\text{s.t.} \quad y_a + y_b \ge \mathsf{wt}_M(a,b) \quad \forall (a,b)\in E'$$

$$y_v \ge \mathsf{wt}_M(v,v) \quad \forall v \notin C.$$

LP3 is the max-weight perfect matching LP in the augmented G' with the edge weight function wt_M and LP4 is the dual LP. We will show the optimal value of LP3 is at most 0 by showing a feasible solution $\boldsymbol{y}$ to LP4 such that $\sum_{v\in A\cup B} y_v = 0$. We delete self-loops from M and set y-values as follows.

For each unmatched vertex v do: set $y_v = 0$. Let $e = (a,b) \in M$. We know that e_i is in M' for some $-t \le i \le s$. If $e_i \in M'$ then set $y_a = -2i$ and $y_b = 2i$. So for any edge $(a,b)\in M$, we have $y_a + y_b = 0$. Since $y_v = 0$ for any unmatched vertex v, we have $\sum_{v\in A\cup B} y_v = 0$.

Claim. $\boldsymbol{y}$ is a feasible solution to LP4.

Proof. It is easy to see that the constraint $y_v \ge \mathsf{wt}_M(v,v)$ holds for all $v \notin C$. For any unmatched vertex v, we have $y_v = 0 = \mathsf{wt}_M(v,v)$. Consider any $a \notin A'$. Only non-positive subscript edges are incident to a, thus $y_a = -2i$ for some $i \le 0$, so $y_a \ge 0$. Consider any $b \notin B'$. Only non-negative subscript edges are incident to b, thus $y_b = 2j$ for some $j \ge 0$, so $y_b \ge 0$.

We now need to show that $y_a + y_b \ge \mathsf{wt}_M(a,b)$ for all $(a,b) \in E'$. Let us partition the sets $A = A_{-t} \cup A_{-t+1} \cup \cdots \cup A_s$ and $B = B_{-t} \cup B_{-t+1} \cup \cdots \cup B_s$ where for each edge $e = (a,b) \in M$ (so $e_i \in M'$ for some $i \in \{-t,\ldots,s\}$), we include a in A_i and b in B_i. Since M is C-perfect, every unmatched vertex is outside $A' \cup B'$. Unmatched vertices in A (resp., B) are added to A_0 (resp., B_0).

We claim the following properties hold for this partition of A and B.

1. There is no edge in $A_i \times B_j$ for $i \ge j+2$.
2. For any edge $e \in A_i \times B_{i-1}$, we have $\mathsf{wt}_M(e) = -2$.
3. For any edge $e \in A_i \times B_i$, we have $\mathsf{wt}_M(e) \le 0$.

- Suppose property 1 does not hold for some edge e. Then the edge e_{i+1} blocks M', a contradiction to the stability of M' in G'_C.
- Suppose property 2 does not hold for some edge e. Then either the edge e_i or the edge e_{i-1} blocks M', a contradiction.[1]
- Suppose property 3 does not hold for some edge e. Then the edge e_i blocks M', a contradiction.

[1] If $\mathsf{wt}_M(e) \ne -2$, then $\mathsf{wt}_M(e) \ge 0$. Since $e \notin M$, this means that at least one of a, b prefers the other to its partner in M. If a prefers b to its partner in M then e_i blocks M'; if b prefers a to its assignment in M then e_{i-1} blocks M'.

Consider any edge (a, b) in G'. Let $a \in A_i$ and $b \in B_j$. Observe that $y_a = -2i$ and $y_b = 2j$. Property 1 above tells us that $i \le j + 1$. Consider the following three cases.

(1) $i = j + 1$. So $(a, b) \in A_i \times B_{i-1}$ and property 2 tells us that $\mathsf{wt}_M(a, b) = -2$. Thus $y_a + y_b = -2i + 2j = -2 = \mathsf{wt}_M(a, b)$.
(2) $i = j$. So $(a, b) \in A_i \times B_i$ and property 3 tells us that $\mathsf{wt}_M(a, b) \le 0$. Thus $y_a + y_b = -2i + 2i = 0 \ge \mathsf{wt}_M(a, b)$.
(3) $i \le j - 1$. Then $y_a + y_b = -2i + 2j \ge 2 \ge \mathsf{wt}_M(a, b)$.

Hence for any edge (a, b) in G', we have $y_a + y_b \ge \mathsf{wt}_M(a, b)$. So $\boldsymbol{y}$ is a feasible solution to LP4. This finishes the proof of this claim. □

Thus we showed a feasible solution $\boldsymbol{y}$ to LP4 such that $\sum_{v \in A \cup B} y_v = 0$. Hence the optimal value of LP3 is at most 0, i.e., $\Delta(N, M) \le 0$ for all C-perfect matchings N. So M is a popular C-perfect matching in G'. This completes the proof of Lemma 1.

Our Algorithm. Let us run the Gale-Shapley algorithm in G'_C where vertices in A propose. For any edge $e = (a, b)$ (say, $b \in B'$): among all the parallel edges $e_{-t}, e_{-t+1}, \ldots$ between a and b, the first time a proposes along e, it is along e_{-t}; if a proposes along e again later, then it is along e_{-t+1} and so on.

This algorithm computes a stable matching M' in G'_C. By Lemma 1, dropping the subscripts of edges in M' yields a popular C-perfect matching M in G'. So our algorithm is as follows.

1. Solve LP2 and determine the sets A', B', and E'.
2. Find a stable matching M' in G'_C where $G' = (A \cup B, E')$ and $C = A' \cup B'$.
3. Return the projection M of M' in G (so edge subscripts are dropped).

Lemma 1 and Proposition 1 show that M is a popular opt-matching in G. Since steps 1–3 can be implemented in polynomial time, Theorem 1 follows.

3 Popular Max-Opt-Matchings

In this section we will see an extension of Theorem 1 that solves the popular max-opt-matching problem in G. Consider the subgraph $G' = (A \cup B, E')$ of G. The *Dulmage-Mendelsohn* decomposition [4] in G' will be very useful to us.

Let M be a matching in G'. An alternating path with respect to M is a path whose alternate edges are in M. We have $A \cup B = \mathcal{E}_M \cup \mathcal{O}_M \cup \mathcal{U}_M$, where a vertex v is in $\mathcal{E}_M$ (resp., $\mathcal{O}_M$) if there is an even (resp., odd) length alternating path with respect to M in G' from an unmatched vertex to v; a vertex v is in $\mathcal{U}_M$ if there is no alternating path in G' from an unmatched vertex to v. So all vertices left unmatched in M are in $\mathcal{E}_M$.

The sets $\mathcal{E}_M$, $\mathcal{O}_M$, and $\mathcal{U}_M$ will be called the sets of *even, odd,* and *unreachable* vertices, respectively, with respect to M. We refer to [14,16] for a proof of the following theorem.

Theorem 3. *The sets $\mathcal{E}_M$, $\mathcal{O}_M$, and $\mathcal{U}_M$ are pairwise disjoint if and only if M is a maximum matching in G'. Any maximum matching in G' partitions the vertex set into the same sets of even, odd, and unreachable vertices. Furthermore, M is a maximum matching in G' if and only if M matches all vertices in $\mathcal{O}_M$ with those in $\mathcal{E}_M$ and it matches all vertices in $\mathcal{U}_M$ among themselves.*

Let M be a maximum matching in G'. It follows from the definition of $\mathcal{U}_M$ that G' has no edge between the sets $\mathcal{E}_M$ and $\mathcal{U}_M$. Similarly, there is no edge in G' with both its endpoints in $\mathcal{E}_M$ since it would imply $\mathcal{E}_M \cap \mathcal{O}_M \neq \emptyset$. Thus there is no edge in G' with one endpoint in $\mathcal{E}_M$ and the other endpoint in $\mathcal{E}_M \cup \mathcal{U}_M$.

Let the subset $E'' \subseteq E'$ be obtained by removing all those edges with one endpoint in $\mathcal{O}_M$ and the other endpoint in $\mathcal{O}_M \cup \mathcal{U}_M$. Thus every edge in E'' has either (i) both its endpoints in $\mathcal{U}_M$ or (ii) one endpoint in $\mathcal{O}_M$ and the other in $\mathcal{E}_M$. By Theorem 3, all maximum matchings of G' are in $G'' = (A \cup B, E'')$.

Since the decomposition of $A \cup B$ into $\mathcal{E}_M, \mathcal{O}_M$, and $\mathcal{U}_M$ is independent of the matching M, we will henceforth refer to these sets as $\mathcal{E}, \mathcal{O}$, and $\mathcal{U}$, respectively.

Lemma 2. *A matching M is a max-opt-matching in G if and only if M is a C-perfect matching in $G'' = (A \cup B, E'')$ where $C = \mathcal{O} \cup \mathcal{U} \cup (\mathcal{E} \cap (A' \cup B'))$.*

Proof. Let M be a max-opt-matching in G. Since M is an opt-matching, we have $M \subseteq E'$ and M is $(A' \cup B')$-perfect (by Proposition 1). Suppose there is an augmenting path ρ with respect to M in G'. Then the matching $M \oplus \rho$ is also an opt-matching in G because $M \oplus \rho \subseteq E'$ and every vertex matched in M is also matched in $M \oplus \rho$. Hence any max-opt-matching M has to be a maximum matching in G'. Thus by Theorem 3, M belongs to the subgraph G'' and it matches all vertices in $\mathcal{O} \cup \mathcal{U}$. Moreover, M also matches all vertices in $A' \cup B' \supseteq \mathcal{E} \cap (A' \cup B')$. Thus M is a C-perfect matching in $G'' = (A \cup B, E'')$ for the given set C.

Conversely, let M be a C-perfect matching in $G'' = (A \cup B, E'')$. So $M \subseteq E''$ and M matches all vertices in $\mathcal{O} \cup \mathcal{U}$. Thus by Theorem 3, M is a maximum matching in G'. Moreover, M matches all vertices in $A' \cup B'$ since M matches all in $\mathcal{O} \cup \mathcal{U} \cup (\mathcal{E} \cap (A' \cup B'))$. Because $M' \subseteq E'' \subseteq E'$, we can conclude that M is an opt-matching in G (by Proposition 1). Recall that every opt-matching in G belongs to the subgraph G'. Since M is a maximum matching in G', there is no larger opt-matching in G; so M is a max-opt-matching in G. □

By Lemma 2, the popular C-perfect matching algorithm in G'' solves the popular max-opt-matching problem in G. Thus we have the following corollary.

Corollary 1. *A popular max-opt-matching in $G = (A \cup B, E)$ can be computed in polynomial time.*

4 The Popular Max-Opt-Matching Polytope

We saw in Sect. 2 that any stable matching in G'_C projects to a popular C-perfect matching in G. But it is not the case that *every* popular C-perfect matching in

G is realizable as a stable matching in G'_C. Suppose $C = \emptyset$ (this happens when $f(e) = 0$ for all $e \in E$). Then every matching in G is a C-perfect matching. The stable matching algorithm in $G'_\emptyset = G$ finds only stable matchings in G whereas any popular matching in G is a popular C-perfect matching here.

However, in the case of popular C-perfect matchings in G''_C studied in Sect. 3, we will now see that *every* popular C-perfect matching in G is realizable as a stable matching in G''_C. Rather than work in G''_C, it will be easier to work in another multigraph which is essentially the same as G''_C. Recall the edge set $E'' \subseteq E'$ and the vertex sets $\mathcal{O}, \mathcal{E}$, and $\mathcal{U}$ from Sect. 3. Let $L = \mathcal{O} \cup (\mathcal{U} \cap A)$ and $R = \mathcal{E} \cup (\mathcal{U} \cap B)$. Consider the bipartite graph $H = (L \cup R, E'')$ which is G'' with certain vertices swapped between the left and right sides (see Fig. 1).

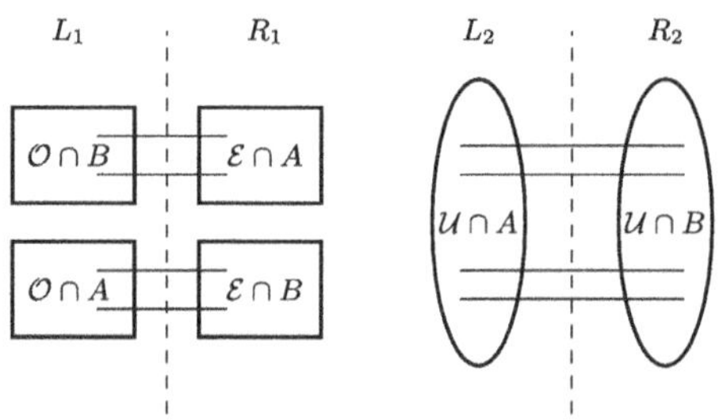

Fig. 1. $L = L_1 \cup L_2$ and $R = R_1 \cup R_2$. By moving all vertices in $\mathcal{O} \cap B$ (resp., $\mathcal{E} \cap A$) from the right of G'' to left (resp., from the left of G'' to right), we get H from G''.

It follows from Lemma 2 that a matching M is a max-opt-matching in G if and only if M is an $(\mathcal{O} \cup \mathcal{U} \cup K)$-perfect matching in $H = (L \cup R, E'')$ where $K = \mathcal{E} \cap (A' \cup B')$. In any L-perfect matching in H, note that all vertices in $\mathcal{U} \cap B$ will get matched to vertices in $\mathcal{U} \cap A$. Hence in our critical set, $\mathcal{U}$ can be replaced by $\mathcal{U} \cap A$. So M is a popular max-opt-matching in G if and only if M is a popular $(L \cup K)$-perfect matching in H where $L = \mathcal{O} \cup (\mathcal{U} \cap A)$.

Popular $(L \cup K)$-Perfect Matchings in H. It will be convenient to work with perfect matchings in H, i.e., no vertex is left unmatched. So let us add a self-loop (v, v) at each $v \in \mathcal{E} \setminus K$ in the graph H. Recall that vertices in $\mathcal{E} \setminus K$ are the only vertices that may possibly be left unmatched in a $(L \cup K)$-perfect matching. Thus every $(L \cup K)$-perfect matching becomes a perfect matching in the augmented graph H. We can now focus on perfect matchings in the augmented H.

Let M be any popular $(L \cup K)$-perfect matching in the original H. So M is a popular perfect matching in the augmented H. Consider the following edge weight function wt_M (also defined in Sect. 2). For any $(a, b) \in E''$:

$$\text{let } \mathsf{wt}_M(a,b) = \begin{cases} 2 & \text{if } (a,b) \text{ blocks } M; \\ -2 & \text{if } a \text{ and } b \text{ prefer their assignments in } M \text{ to each other;} \\ 0 & \text{otherwise.} \end{cases}$$

Furthermore, let $\mathsf{wt}_M(v,v) = 0$ if the self-loop (v,v) is in the augmented M (i.e., v was originally left unmatched in M), else $\mathsf{wt}_M(v,v) = -1$. It follows from the definition of wt_M that for any perfect matching N in the augmented graph H, we have $\mathsf{wt}_M(N) = \Delta(N,M)$.

The linear program LP5 is the max-weight perfect matching LP in the augmented H with edge weight function wt_M. Note that $\bar{E}''$ is the edge set E'' augmented with self-loops (v,v) for all $v \in \mathcal{E} \setminus K$. For any $v \in L \cup R$, the set $\delta''(v)$ is the set of edges incident to v in the augmented H, so $\delta''(v)$ includes the self-loops (v,v) for all $v \in \mathcal{E} \setminus K$.

$$\max \sum_{e \in \bar{E}''} \mathsf{wt}_M(e) \cdot x_e \qquad \text{(LP5)}$$
$$\text{s.t.} \quad \sum_{e \in \delta''(v)} x_e = 1 \quad \forall v \in L \cup R$$
$$x_e \geq 0 \quad \forall e \in \bar{E}''.$$

$$\min \sum_{v \in L \cup R} y_v \qquad \text{(LP6)}$$
$$\text{s.t.} \quad y_l + y_r \geq \mathsf{wt}_M(l,r) \quad \forall (l,r) \in E''$$
$$y_r \geq \mathsf{wt}_M(r,r) \quad \forall r \in \mathcal{E} \setminus K.$$

The optimal value of LP5 is $\max_N \Delta(N,M)$ where the max is over all perfect matchings in the augmented H, equivalently, over all $(L \cup K)$-perfect matchings N in the original H. Since M is a popular $(L \cup K)$-perfect matching, $\Delta(N,M) \leq 0$ for all $(L \cup K)$-perfect matchings N; thus the optimal value of LP5 is at most 0. Since $\mathsf{wt}_M(M) = \Delta(M,M) = 0$, it follows that the edge incidence vector of M is an optimal solution to LP5.

The linear program LP6 is the dual LP. The following is our key technical lemma. A solution to LP6 as described in this lemma will be called a *dual certificate* for the popular $(L \cup K)$-perfect matching M. Let $|K| = k$ and $|L| = \ell$.

Lemma 3. *Suppose M is a popular $(L \cup K)$-perfect matching in H. Then the linear program LP6 admits an optimal solution $\boldsymbol{y}$ that satisfies the following properties:*

- $y_l \in \{2k, 2(k-1), \ldots, 0, -2, -4, \ldots, -2(\ell-1)\}$ *for all* $l \in L$.
- $y_r \in \{-2k, -2(k-1), \ldots, 0, 2, 4, \ldots, 2(\ell-1)\}$ *for all* $r \in R$.

Proof. The constraint matrix of LP6 is totally unimodular. Thus there is an integral optimal solution $\boldsymbol{y}$ to LP6. We can also assume that all the y-values are *even* integers by updating $\boldsymbol{y}$: increase y_r by 1 for all $r \in R$ with odd y-values and decrease y_l by 1 for all $l \in L$ with odd y-values. The updated $\boldsymbol{y}$ continues to be feasible since edge weights are even; so for any edge (l,r), if $y_l + y_r$ was odd, then there was a slack of at least 1 in the constraint $y_l + y_r \geq \mathsf{wt}_M(l,r)$ and if $y_l + y_r$ was even then it remains the same.

The updated $\boldsymbol{y}$ continues to be optimal because by complementary slackness, $y_r = \mathsf{wt}_M(r,r) = 0$ for any r with $(r,r) \in M$ and moreover, for any $(l,r) \in M$, $y_l + y_r = \mathsf{wt}_M(l,r) = 0$. Hence l and its partner r have the same parity of y-values. So we have decreased y_l and increased y_r by the same amount for an equal number of vertices in L and R, respectively. Thus $\sum_{v \in L \cup R} y_v$ remains the same, i.e., $\sum_{v \in L \cup R} y_v = 0$. For $r \in \mathcal{E} \setminus K$, because the original constraint was $y_r \geq -1$ (since $y_r \geq \mathsf{wt}_M(r,r) \geq -1$), we now have $y_r \geq 0$ for all $r \in \mathcal{E} \setminus K$.

A Special Case. Suppose M matches all vertices in R, i.e., M does not use any self-loops in the augmented H. So $|L| = |R| = \ell$. Let us increase y_r for all $r \in R$ by the required amount so that $y_r \geq 0$ for all $r \in R$. Symmetrically, let us decrease y_l for all $l \in L$ by the same amount. This preserves feasibility since the sum $y_l + y_r$ is unchanged for any edge (l, r). Because M is a perfect matching, $\sum_{v \in L \cup R} y_v$ is unchanged; so we have preserved optimality. Thus $y_r \in \{0, 2, 4, \ldots\}$ for all $r \in R$ and $y_l \in \{0, -2, -4, \ldots\}$ for all $l \in L$ (recall that all the y-values are even and $y_l = -y_r$ for any $(l, r) \in M$).

Among all optimal solutions $\boldsymbol{y}$ to LP6 such that $y_r \in \{0, 2, 4, \ldots\}$ and $y_l \in \{0, -2, -4, \ldots\}$, let $\boldsymbol{y}$ be such that $\sum_{r \in R} y_r$ is the least. We claim that the y_r values are consecutive even integers. Otherwise there is a gap in the y_r values, say there is no vertex in R with a y-value of $2i$. Then we can update $y_r = y_r - 2$ for all $r \in R$ with $y_r \geq 2i + 2$ and symmetrically update $y_l = y_l + 2$ for all $l \in L$ with $y_l \leq -(2i + 2)$. Since all edge weights are at most 2, this update preserves feasibility. Any $r \in R$ has $y_r \geq 2i + 2$ if and only if $y_{M(r)} \leq -(2i + 2)$ because $y_{M(r)} + y_r = 0$; so this update preserves optimality as well.

Thus we have an optimal solution $\boldsymbol{y}$ to LP6 with a smaller value of $\sum_{r \in R} y_r$, a contradiction. Hence we can conclude that y_r values are consecutive even integers. Since $|A| = |B| = \ell$, we have $y_r \in \{0, 2, 4, \ldots, 2(\ell - 1)\}$ for all $r \in R$ and $y_l \in \{0, -2, -4, \ldots, -2(\ell - 1)\}$ for all $l \in L$. This proves the lemma for the special case when M is a perfect matching.

The General Case. Suppose M leaves some vertices in $\mathcal{E} \setminus K$ unmatched. For any $r \in \mathcal{E} \setminus K$ that is left unmatched in M, we have $y_r = \mathsf{wt}_M(r, r) = 0$ by complementary slackness. So the y-values of these vertices are fixed. Hence we cannot shift y-values as done in the special case, thus all we can say at this point is that all the y-values are even integers. Among all optimal solutions $\boldsymbol{y}$ to LP6 such that the y-values are even integers, let $\boldsymbol{y}$ be such that $\sum_{r \in R} |y_r|$ is the least. The same argument as given above shows that y-values are consecutive even integers. That is, if there is no vertex in R with a y-value of $2i$ for some $i > 0$ while there are vertices r with $y_r \geq 2i + 2$ then we update $y_r = y_r - 2$ for all $r \in R$ with $y_r \geq 2i + 2$ and symmetrically update $y_l = y_l + 2$ for all $l \in L$ with $y_l \leq -(2i + 2)$. Analogously, if there is no vertex in R with a y-value of $2i$ for some $i < 0$ while there are vertices r with $y_r \leq 2i - 2$ then we update $y_r = y_r + 2$ for all $r \in R$ with $y_r \leq 2i - 2$ and symmetrically update $y_l = y_l - 2$ for all $l \in L$ with $y_l \geq -2i + 2$. Since all edge weights are at most 2, these updates preserve feasibility. In either case, we would have an optimal solution $\boldsymbol{y}$ to LP6 with a smaller value of $\sum_{r \in R} |y_r|$, a contradiction. Note that this argument deals with matched vertices only and unmatched vertices remain untouched since their y-value is 0. Thus $y_v \in \{0, \pm 2, \pm 4, \ldots, \pm 2(\ell - 1)\}$ for all $v \in L \cup R$.

Since $\boldsymbol{y}$ is a feasible solution to LP6, we have $y_r \geq \mathsf{wt}_M(r, r) \geq -1$ for all $r \in \mathcal{E} \setminus K$. So among the vertices of R, it is only the ones in K that can take values in $\{-2, -4, \ldots, -2(\ell - 1)\}$. Since y-values are consecutive even integers and because $|K| = k$, the only possible negative y-values for vertices in K are in $\{-2, \ldots, -2k\}$. Thus we have shown that $y_r \in \{-2k, \ldots, 0, 2, 4, \ldots, 2(\ell - 1)\}$ for all $r \in R$. Hence $y_l \in \{2k, \ldots, 0, -2, -4, \ldots, -2(\ell - 1)\}$ for all $l \in L$. □

Analogous to G'_C (see Sect. 2), we define a multigraph $H_K = (L \cup R, E''_K)$ to characterize popular $(L \cup K)$-perfect matchings in H. For every $e = (l, r) \in E''$ where $l \in L$ and $r \in R$, the following parallel edges are in the edge set E''_K.

- If $r \notin K$ then there are ℓ edges $e_0, e_1, \ldots, e_{\ell-1}$ in E''_K.
- If $r \in K$ then there are $\ell + k$ edges $e_{-k}, e_{-k+1}, \ldots, e_0, \ldots, e_{\ell-1}$ in E''_K.

As before, vertices in L (resp., R) prefer lower (resp., higher) subscript edges. Among edges with the same subscript, it is as per their original preference order. We will now show that for each popular $(L \cup K)$-perfect matching M in G, there is a stable matching M' in H_K such that M is the projection of M'. We will use Lemma 3 to define M'. We know that M has a dual certificate $\boldsymbol{y}$ as described in Lemma 3. For any edge $e = (l, r) \in M$, let us set e's subscript s_e in M' to be $y_r/2$. Since $y_r \in \{-2k, \ldots, 0, 2, 4, \ldots, 2(\ell-1)\}$, we have $s_e \in \{-k, \ldots, 0, 1, 2, \ldots, \ell-1\}$.

The proof of Lemma 4 is an extension of the proof of correctness of the min-cost popular perfect matching from [13]. Our matching M need not be perfect and moreover, M may have a blocking edge incident to some unmatched vertex in $\mathcal{E} \setminus K$. However the L-endpoint of such a blocking edge would have to be matched along a *negative* subscript edge and this helps us to show the stability of M' in H_K.

Lemma 4. *For any popular $(L \cup K)$-perfect matching M in G, the matching $M' = \{e_{s_e} : e \in M\}$ is stable in H_K.*

Proof. Let $\boldsymbol{y}$ be a dual certificate of M. So $\boldsymbol{y}$ is an optimal solution to LP6 with coordinates as given in Lemma 3.

Consider any edge $e = (l, r) \in L \times R$ where r is unmatched in M. So $y_r = 0$ and $r \in \mathcal{E} \setminus K$. Since $r \notin K$, there are only ℓ copies $e_0, \ldots, e_{\ell-1}$ of e in H_K. We need to show that none of $e_0, \ldots, e_{\ell-1}$ is a blocking edge to M'. Note that $\mathsf{wt}_M(l, r) \geq 0$ since r prefers to be matched to l rather than be unmatched. Thus we have $y_l = y_l + y_r \geq \mathsf{wt}_M(l, r) \geq 0$. So $y_l \geq 0$, hence $y_l \in \{0, 2, \ldots, 2k\}$. This implies $y_{M(l)} \in \{0, -2, \ldots, -2k\}$ and l is matched in M along an edge e' with $s_{e'} \in \{0, -1, \ldots, -k\}$.

Suppose $y_l = 0$. Then $\mathsf{wt}_M(l, r) \leq y_l + y_r = 0$. Thus $\mathsf{wt}_M(l, r) = 0$. So l prefers $M(l)$ to r, i.e., l prefers e' to e, thus it prefers e'_0 to e_0; hence e_0 does not block M'. None of $e_1, \ldots, e_{\ell-1}$ blocks M' since l prefers e'_0 to any of $e_1, \ldots, e_{\ell-1}$: recall that lower subscript edges are preferred by vertices in L to higher ones.

The other case is that $y_l \in \{2, \ldots, 2k\}$. Then l is matched in M along an edge e' with $s_{e'} \in \{-1, \ldots, -k\}$. So l is matched along a negative subscript edge in M', hence none of the higher subscript edges $e_0, \ldots, e_{\ell-1}$ blocks M'.

Let us now consider the case of an edge $(l, r) \in L \times R$ where both l and r are matched in M along edges e and e', respectively. Let $y_l = -2i$ and $y_r = 2j$. So e'_i and e''_j are in M'. Since $y_l + y_r \geq \mathsf{wt}_M(l, r) \geq -2$, we have $2j = y_r \geq -y_l - 2 = 2i - 2$. So $j \geq i - 1$.

1. If $j = i - 1$ then $\mathsf{wt}_M(l, r) \leq y_l + y_r = -2i + 2i - 2 = -2$. So $\mathsf{wt}_M(l, r) = -2$, i.e., both l and r prefer their respective partners in M to each other.

- So l prefers e'_i to e_i. Also, l prefers e'_i to $e_{i+1}, \ldots, e_{\ell-1}$ since vertices in L prefers lower subscript edges to higher subscript edges in H_K.
- Also r prefers e''_{i-1} to e_{i-1} and r prefers e''_{i-1} to $e_{i-2}, \ldots$ since vertices in R prefers higher subscript edges to lower subscript edges in H_K.

Hence no copy of the edge e in H_K blocks M'.

2. If $j = i$ then $\mathsf{wt}_M(l, r) \le y_l + y_r = -2i + 2i = 0$. So (l, r) is not a blocking edge to M. Hence either $(l, r) \in M$ or one of l, r prefers its partner in M to the other, i.e., in H_K, either l prefers e'_i to e_i or r prefers e''_i to e_i. So e_i does not block M'. Furthermore, l prefers e'_i to $e_{i+1}, \ldots, e_{\ell-1}$ and r prefers e''_i to $e_{i-1}, e_{i-2}, \ldots$. Hence no copy of the edge e in H_K blocks M'.
3. Let $j \ge i+1$. Since r prefers higher subscript edges to lower subscript edges in H_K, r prefers e''_j to $e_{j-1}, e_{j-2}, \ldots$ and similarly, since l prefers lower subscript edges to higher subscript edges in H_K, l prefers e'_i to $e_{i+1}, \ldots, e_{\ell-1}$. Since $j \ge i + 1$, no copy of the edge e in H_K blocks M'.

Thus no edge in H_K blocks M'. Hence M' is a stable matching in H_K. □

The converse to Lemma 4 also holds: it is easy to check that the proof of Lemma 1 shows that any stable matching M' in H_K projects to a popular $(L \cup K)$-perfect matching M in G. Thus every stable matching in H_K maps to a popular max-opt-matching in G. Furthermore, Lemma 4 shows this mapping is surjective. A compact extended formulation of the popular max-opt-matching polytope (call it $\mathcal{Q}$) is now easy to formulate. The following constraints from [18] describe the stable matching polytope of $H_K = (A \cup B, E''_K)$.

For any vertex $v \in L \cup R$, let $\delta_{H_K}(v)$ be the set of edges incident to v in H_K. Let $\{e'_j : e'_j \succ_v e_i\} \subseteq \delta_{H_K}(v)$ be the set of all edges in E''_K that v prefers to e_i.

$$\sum_{e'_j : e'_j \succ_l e_i} x_{e'_j} + \sum_{e''_j : e''_j \succ_r e_i} x_{e''_j} + x_{e_i} \ge 1 \quad \forall e_i = (l, r)_i \in E''_K \quad (1)$$

$$x_{e_i} \ge 0 \;\; \forall e_i \in E''_K \quad \text{and} \quad \sum_{e_i : e_i \in \delta_{H_K}(v)} x_{e_i} \le 1 \quad \forall v \in L \cup R \quad (2)$$

1. Constraint (1) captures the *stability* constraint for $e_i \in E''_K$ where $e = (l, r) \in E''$ and $i \in \{-k, \ldots, 0, 1, 2, \ldots, \ell - 1\}$.
2. The constraints in (2) say that $\boldsymbol{x}$ belongs to the matching polytope of H_K.

It is easy to check that the simple proof given in [20, Theorem 1] shows that the above constraints describe the stable matching polytope in a multigraph (the original proof in [20] holds for graphs but recall that H_K is a multigraph). Thus the formulation of the stable matching polytope of H_K [18] along with the following equations is a compact extended formulation for the polytope $\mathcal{Q}$.

For each edge $e = (l, r) \in E''$:

- If $r \in K$ then include $x_e = x_{e_{-k}} + \cdots + x_{e_{\ell-1}}$ in this formulation.
- If $r \notin K$ then include $x_e = x_{e_0} + \cdots + x_{e_{\ell-1}}$ in this formulation.

The above formulation can be computed in time linear in the size of H_K. Since the graph H_K can be computed in $\mathsf{poly}(m, n)$ time where m and n are the number of edges and vertices in G, Theorem 2 stated in Sect. 1 follows. □

References

1. Ahani, N., Andersson, T., Martinello, A., Trapp, A., Teytelboym, A.: Placement optimization in refugee resettlement. Oper. Res. **69**(5), 1468–1486 (2021)
2. Biro, P., Manlove, D.F., Mittal, S.: Size versus stability in the marriage problem. Theoret. Comput. Sci. **411**, 1828–1841 (2010)
3. Cseh, A.: Popular matchings. In: Endriss, U. (ed.) Trends in Computational Social Choice (2017)
4. Dulmage, A.L., Mendelsohn, N.S.: Coverings of bipartite graphs. Can. J. Math. **10**, 517–534 (1958)
5. Faenza, Y., Kavitha, T.: Quasi-popular matchings, optimality, and extended formulations. Math. Oper. Res. **47**(1), 427–457 (2022)
6. Faenza, Y., Kavitha, T., Powers, V., Zhang, X.: Popular matchings and limits to tractability. In: Proceedings of the 30th ACM-SIAM Symposium on Discrete Algorithms, SODA, pp. 2790–2809 (2019)
7. Gale, D., Shapley, L.: College admissions and the stability of marriage. Amer. Math. Monthly **69**(1), 9–15 (1962)
8. Gale, D., Sotomayor, M.: Some remarks on the stable matching problem. Discret. Appl. Math. **11**, 223–232 (1985)
9. Gärdenfors, P.: Match making: assignments based on bilateral preferences. Behav. Sci. **20**, 166–173 (1975)
10. Goemans, M.: Combinatorial optimization (2017). https://math.mit.edu/~goemans/18453S17/18453.html
11. Kavitha, T.: A size-popularity tradeoff in the stable marriage problem. SIAM J. Comput. **43**(1), 52–71 (2014)
12. Kavitha, T.: Matchings, critical nodes, and popular solutions. In: Proceedings of the 41st IARCS Annual Conference on Foundations of Software Technology and Theoretical Computer Science, FSTTCS, pp. 25:1–25:19 (2021)
13. Kavitha, T.: Maximum matchings and popularity. SIAM J. Discr. Math. **38**(2), 1202–1221 (2024). https://doi.org/10.1137/22M1523248
14. Losász, L., Plummer, M.D.: Matching Theory. Mathematics Studies, vol. 121. North-Holland (1986)
15. Nasre, M., Nimbhorkar, P., Ranjan, K., Sarkar, A.: Popular matchings in the hospital-residents problem with two-sided lower quotas. In: Proceedings of the 41st IARCS Annual Conference on Foundations of Software Technology and Theoretical Computer Science, FSTTCS, pp. 30:1–30:21 (2021)
16. Pulleyblank, W.R.: Chapter 3, matchings and extensions. In: Graham, R.L., Grötschel, M., Lovasz, L. (ed.) The Handbook of Combinatorics (1995)
17. Robards, P.A.: Applying the two-sided matching processes to the United States Navy enlisted assignment process. Master's thesis, Naval Postgraduate School, Monterey, Canada (2001)
18. Rothblum, U.G.: Characterization of stable matchings as extreme points of a polytope. Math. Program. **54**, 57–67 (1992)
19. Soldner, M.: Optimization and measurement in humanitarian operations: addressing practical needs. Ph.D. thesis, Georgia Institute of Technology (2014)
20. Teo, C.P., Sethuraman, J.: The geometry of fractional stable matchings and its applications. Math. Oper. Res. **23**(4), 874–891 (1998)
21. Yang, W., Giampapa, J.A., Sycara, K.: Two-sided matching for the US Navy detailing process with market complication. Technical report CMU-R1-TR-03-49, Robotics Institute, Carnegie Mellon University (2003)

Degreewidth on Semi-complete Digraphs

Ryan Keeney(✉) and Daniel Lokshtanov

University of California Santa Barbara, Santa Barbara, CA, USA
{ryankeeney,daniello}@ucsb.edu

Abstract. For a digraph G and ordering of G, the *degreewidth* of the ordering is the maximum number of backward arcs incident to any vertex of G. The *degreewidth* $\Delta(G)$ of G is defined as the minimum degreewidth of an ordering of G. A digraph G is *semi-complete* if every pair of vertices is connected by at least one arc, *oriented* if every pair of vertices is connected by at most one arc, and a *tournament* if every pair of vertices is connected by exactly one arc. We give a fixed parameter tractable (FPT) algorithm, with running time $\Delta(G)^{O(\Delta(G))}n + O(n^2)$, to compute the degreewidth of semi-complete digraphs. We then show that both the Feedback Arc Set and Cutwidth problems on semi-complete digraphs admit algorithms with running time $\Delta(G)^{O(\Delta(G))}n + O(n^2)$. Our algorithms resolve in the affirmative two open problems of Davot *et al.* [WG 2023], who asked whether there exists an FPT algorithm to compute the degreewidth of a tournament, and whether Feedback Arc Set on tournaments admits an FPT algorithm when parameterized by the degreewidth of the input digraph. Additionally, extending an argument of Davot *et al.* [WG 2023], we show that sorting by in-degree is a 3-approximation algorithm for degreewidth on semi-complete digraphs. Finally we prove that it is NP-hard to determine whether a given oriented digraph has degreewidth at most 2.

Keywords: Semi-complete digraph · parameterized algorithm · degreewidth · feedback arc set · cutwidth

1 Introduction

Degreewidth is a graph parameter, recently introduced by Davot *et al.* [8]. Given a digraph G, the *degreewidth* $\Delta(G)$ of G is defined to be the minimum value k for which there exists an ordering of $V(G)$ with respect to which every vertex is incident to at most k backwards arcs. It follows from the definition that a digraph has degreewidth zero if and only if it is acyclic. Thus degreewidth joins other parameters such as minimum feedback arc set and cutwidth as a measure of the acyclicity of a directed graph. Here the *minimum feedback arc set* of a digraph G is the minimum number of backward arcs taken over all orderings of G. The *cutwidth* of a digraph G is the minimum taken over all orderings σ of G of the maximum over $i \leq n$ (where $n = |V(G)|$) of the number of arcs from the last i vertices (according to σ) to the first $n - i$ vertices of G.

D. Král' and M. Milanič (Eds.): WG 2024, LNCS 14760, pp. 312–326, 2025.
https://doi.org/10.1007/978-3-031-75409-8_22

Davot *et al.* [8] introduced the notion of degreewidth, and studied degreewidth on tournaments i.e. directed graphs in which there is exactly one arc between every pair of vertices. They show that computing the degreewidth of a tournament is NP-hard, but that sorting the vertices by their in-degree yields a 3-approximation algorithm. Tournaments of degreewidth one were previously studied under the name *sparse tournaments* [4,5,12]. Davot *et al.* [8] give an $O(n^3)$ time algorithm for determining whether a tournament is sparse. In light of these results they pose as an open problem whether there exists an FPT algorithm to compute the degreewidth of a tournament. Davot *et al.* [8] then turn their attention to degreewidth as a structural parameter. In other words, they consider the extent to which having small degreewidth is useful for obtaining efficient algorithms for other problems on tournaments. Among other results they show that FEEDBACK ARC SET, which is known to be NP-hard on tournaments [1,6], is solvable in polynomial time in sparse tournaments. They leave the existence of an FPT algorithm for FEEDBACK ARC SET in tournaments parameterized by degreewidth as an open problem.

In this paper we resolve both of the aforementioned open problems in the affirmative. In particular we give algorithms to compute the degreewidth, cutwidth, and minimum feedback arc set of a tournament with running time $\Delta(G)^{O(\Delta(G))}n + O(n^2)$. Our algorithms also work for the more general case of semi-complete digraphs i.e. digraphs in which there is *at least* one arc between every pair of vertices. If the input is provided as a degreewidth k ordering together with the set of (at most nk) backward arcs, the running time of the algorithms for CUTWIDTH and FEEDBACK ARC SET is upper bounded by $k^{O(k)}n$. It is worth noting that semi-complete digraphs are quite a bit more general than tournaments. For example, sorting by in-degree immediately produces optimal solutions to both the CUTWIDTH and OPTIMAL LINEAR ARRANGEMENT problems on tournaments [10], while both problems become NP-hard on semi-complete digraphs [2]. In fact, CUTWIDTH in semi-complete digraphs has been quite intensively studied [2,7,9,11]. We supplement our main results by showing that sorting by in-degree yields a 3-approximation for degreewidth even in semi-complete digraphs, and that determining whether an *oriented* digraph has degreewidth at most 2 is NP-complete. Here an oriented digraph is one where there is at most one arc between every pair of vertices.

Our fixed parameter tractable algorithms for DEGREEWIDTH, FEEDBACK ARC SET, and CUTWIDTH are all based on the same idea. Let π be an ordering that sorts the vertices of the input semi-complete digraph by their in-degree, breaking ties arbitrarily. We show that there exists an optimal ordering σ for the problem that we are considering such that for every vertex $v \in V(G)$ it holds that

$$|\sigma(v) - \pi(v)| \leq c\Delta(G)$$

for some constant c. In other words, for each of the considered problems we know that the optimal ordering can be obtained by first sorting the vertices by in-degree and then moving each vertex up or down at most $O(\Delta(G))$ positions. This result is shown for each problem individually. For degreewidth it is quite simple (see Lemma 1), while for feedback arc set and cutwidth it is more involved (see Lemmas 6 and 8).

Having established the "locality relative to π" property discussed above we observe that the objective function value (for each of the considered problems) of an ordering σ can be determined by considering for each $i \leq n$ the function σ restricted to the set $\{\pi^{-1}(j) \ : \ i - c\Delta(G) \leq j \leq i + c\Delta(G)\}$ for some constant c. In particular for every pair u, v of vertices of G that do not appear together in one of these "windows" we know that u and v appear very far apart in π, and that therefore their relative order according to σ will be the same as their relative order in π.

With these insights an optimal ordering σ can be computed by dynamic programming where the state space is the set of all integers $i \leq n$ and all injective functions

$$\sigma_i : \{\pi^{-1}(j) \ : \ i - c\Delta(G) \leq j \leq i + c\Delta(G)\} \rightarrow \{i - 2c\Delta(G), \ldots, i + 2c\Delta(G)\}.$$

The size of the state space, and therefore also the running time of the algorithm, is bounded by $\Delta(G)^{O(\Delta(G))}n$. The additional $O(n^2)$ term in the running time comes from reading the input and computing π.

Organization of the Paper. We begin by defining the notation and terms used throughout the paper in Sect. 2. To avoid repeating the same dynamic programming scheme for each of the three problems we express it as a path problem in an auxiliary graph whose vertices are the states of the dynamic program discussed above. The properties of the auxiliary graph that are independent of which of the three problems we are considering are stated and proved in Sect. 3. In Sects. 4, 5 and 6 we give FPT algorithms for DEGREEWIDTH, FEEDBACK ARC SET and CUTWIDTH in semi-complete digraphs, respectively, when parameterized by the degreewidth of the input digraph. In Sect. 7 we show that sorting by in-degree yields a 3-approximation for degreewidth in semi-complete digraphs, and in Sect. 8, we show that it is NP-hard to determine whether an oriented digraph has degreewidth at most 2. Proofs marked with $\star$ are omitted in the abridged version due to lack of space, and may be found in the full version.

2 Notation

We let $[n]$ denote the set $\{1, \ldots, n\}$. All digraphs are assumed to be *simple* i.e. they contain no self-loops and no more than one arc with the same head and tail. An *opposite pair* of arcs is a pair of arcs (u, v) and (v, u) for some $u, v \in V(G)$. Given a digraph G with vertex set $V(G)$ and arc set $A(G)$, we say G is *oriented* if it contains no opposite pair of arcs, that is, for all $u, v \in V(G)$ at most one of (u, v) or (v, u) is in $A(G)$. A *semi-complete* digraph is a digraph in which, for every pair of distinct vertices $u, v \in V(G)$, at least one of the arcs (u, v) or (v, u) is in $A(G)$. For $v \in V(G)$ we denote by $N^-(v) := \{u \in V(G) :$ $(u, v) \in A(G)\}$ the *in-neighborhood* of v. The *in-degree* of v is then denoted by $d^-(v) := |N^-(v)|$. Similarly, $N^+(v) := \{u \in V(G) : (v, u) \in A(G)\}$ shall denote the *out-neighborhood* of v, and $d^+(v) := |N^+(v)|$ its *out-degree*. Given a semi-complete digraph T, we let V_i denote the set of vertices of T with in-degree i, $V_{<i}$ denote the set of vertices with in-degree $< i$, and define $V_{>i}$, etc. analogously. An *ordering* of a set X of size n is a bijective function $\sigma : X \to [n]$, and we denote by $\Sigma(X)$ the set of all orderings of X. We adopt much of the following notation and terminology from [8]: Given distinct $x, y \in X$ and an ordering σ of X, we say x is *to the left of* or *before* y (with respect to σ) if $\sigma(x) < \sigma(y)$. Otherwise, we say x is *to the right of* or *after* y . Given a digraph G, an ordering σ of $V(G)$, and an arc $(u, v) \in A(G)$, we say that (u, v) is *backwards* (with respect to σ) if $\sigma(v) < \sigma(u)$; otherwise, (u, v) is *forwards*. The *length* of an arc (u, v) is defined to be $|\sigma(u) - \sigma(v)|$. For $v \in V(G)$, we denote by $d_\sigma(v)$ the number of backwards arcs incident to v with respect to σ; more precisely, $d_\sigma(v) = |\{u \in N^-(v) : \sigma(u) > \sigma(v)\}| + |\{u \in N^+(v) : \sigma(u) < \sigma(v)\}|$. The *degreewidth of* G *with respect to* σ is then defined to be $\Delta_\sigma(G) := \max_{v \in V(G)} d_\sigma(v)$. Finally, the *degreewidth* of G is defined as $\Delta(G) := \min_{\sigma \in \Sigma(V(G))} \Delta_\sigma(G)$.

3 Restricted Ordering Graph

We begin with a simple lemma, generalizing a result from [8]:

Lemma 1. *If T is a semi-complete digraph with $\Delta(T) \leq k$, σ is an optimal degreewidth ordering of $V(T)$, and $d^-(v) = i$, then $\sigma(v) \in [i - k + 1, i + k + 1]$.*

Proof. Suppose, for sake of contradiction, that $\sigma(v) \leq i - k$. Then v has at most $i - k - 1$ vertices to its left in the ordering, meaning in particular it has at most $i - k - 1$ in-neighbors to its left in the ordering. Thus it must have at least $k + 1$ in-neighbors to its right, contradicting that σ is optimal.

Now suppose instead that $\sigma(v) \geq i + k + 2$. Then v has at least $i + k + 1$ vertices to its left in the ordering, and at most i of these are in-neighbors. So, since T is semi-complete, at least $k + 1$ of them are out-neighbors of v, again contradicting that σ is optimal. □

Corollary 1. *Suppose $\Delta(T) \leq k$. Let σ be an optimal degreewidth ordering and let u, v be in $V(T)$. Then:*

(i) If $d^-(u) < d^-(v) - 2k$, then $\sigma(u) < \sigma(v)$.
(ii) If $d^-(u) > d^-(v) + 2k$, then $\sigma(u) > \sigma(v)$.

In particular, if vertices u and v have in-degrees that differ by more than $2k$, the direction of the arc between them is the same in all optimal orderings. We call these arcs *long arcs*, and note that their direction with respect to every optimal degreewidth ordering can be precomputed.

It is clear from the above lemma that, in searching for an optimal degreewidth ordering, it suffices to consider only the family of orderings in which every vertex v is in the range $[d^-(v) - k + 1, d^-(v) + k + 1]$. Later, in designing FPT algorithms for feedback arc set and cutwidth, we will prove similar lemmas (see Lemmas 6 and 8) which restrict every vertex to some small range in the optimal ordering, and so we would like a convenient way to study such families of orderings. We now define an auxiliary digraph to accomplish this. Paths through this graph will correspond exactly to the orderings we are interested in, and so, by defining weight functions on the vertices, we can frame the problems of computing degreewidth, minimum feedback arc set, and cutwidth as shortest/minimax path problems in this auxiliary graph.

Let X be a set of size n and let $X_1, \ldots, X_n$ be a family of (possibly empty) subsets of X with the property that each element of X occurs in exactly one X_i. We now define the *restricted ordering graph* $H(X, \{X_i\}_{i=1}^n, m)$ as follows: The vertex set $V(H)$ will consist of all functions of the form

$$f : \bigcup_{t=i-2m}^{i+2m} X_t \to [i - 3m, i + 3m] \cap [n]$$

where $i \in [n]$ and f is an injective function satisfying $f(X_t) \subseteq [t - m, t + m]$ for all $t \in [i - 2m, i + 2m]$. We call i the *index* of f and will often indicate it as a subscript. The only arcs in the graph will be defined as follows: Given $i < n$ and vertices f and g of index i and $i + 1$, respectively, there will be an arc from f to g if and only if

$$f(x) = g(x) \text{ for all } x \in \bigcup_{t=i+1-2m}^{i+2m} X_t$$

i.e. f and g agree on the intersection of their respective domains. We begin by proving that this property extends from arcs to paths:

Lemma 2. *Let $f_i, \ldots, f_j$ be a path in H. Then, for all $x \in \bigcup_{t=j-2m}^{i+2m} X_t$, $f_i(x) = f_j(x)$.*

Proof. ⋆

Let $m \in \mathbb{N}$, and let $\mathcal{P} = \mathcal{P}(X, \{X_i\}_{i=1}^n, m)$ denote the set of all orderings σ of X satisfying

$$\sigma(X_i) \subseteq [i - m, i + m] \cap [n]$$

for all $i \in [n]$. We let H_i denote the set of vertices of H with index i, and define an H_i-H_j-path to be a path in H whose first vertex is in H_i and whose last vertex is in H_j. We then define an explicit function F from $\mathcal{P}$ to the set of H_1-H_n-paths by

$$F(\sigma) = f_1 f_2 \ldots f_n,$$

where $f_i := \sigma|_{\bigcup_{t=i-2m}^{i+2m} X_t}$. We claim the following:

Theorem 1. *F is a one-to-one correspondence between $\mathcal{P}$ and the set of H_1-H_n-paths in H.*

Proof. ⋆

We note that, given an H_1-H_n-path, its corresponding ordering can be constructed in $O(mn)$ time, and vice versa.

We now bound the size of H when $\mathcal{P}$ is nonempty. Because, for all i, there exists an injective function from $\bigcup_{t=i-2m}^{i+2m} X_t$ to $[i-3m, i+3m]$, $|\bigcup_{t=i-2m}^{i+2m} X_t| \leq 6m+1$, meaning we have $|H_i| \leq (6m+1)^{6m+1}$, and so $|V(H)| \leq (6m+1)^{6m+1}n$.

Since the graph H is acyclic and all arcs go between H_i and H_{i+1} for some i, dynamic programming over a topological order of H immediately yields the following lemma.

Lemma 3. *Given a weight function $W : V(H) \to \mathbb{R}$, we can compute quantities of the form*

$$\min_{P \text{ an } H_1\text{-}H_n\text{-path}} \left[\max_{f_i \in P} W(f_i)\right] \quad \text{and} \quad \min_{P \text{ an } H_1\text{-}H_n\text{-path}} \left[\sum_{f_i \in P} W(f_i)\right],$$

as well as compute an optimal path, in time $O((6m+1)^{2(6m+1)}n)$.

In the following sections we will use Lemma 3 to compute optimal degreewidth, feedback arc set, and cutwidth orderings.

4 An FPT Algorithm for Degreewidth

Throughout this section and the next three, we will denote the input semi-complete digraph by T, and use H for the auxiliary restricted ordering graphs.

Lemma 4. *If $\Delta(T) = k$, then the number $b(v)$ of backwards long arcs incident to v in every optimal ordering is given by*

$$b(v) = |V_{<d^-(v)-2k} \cap N^+(v)| + |V_{>d^-(v)+2k} \cap N^-(v)|.$$

We call an ordering σ of $V(T)$ *viable* if $\sigma(v) \in [d^-(v) - k + 1, d^-(v) + k + 1]$ for all v. By Lemma 1, it suffices to check only viable orderings. Letting $X = V(T)$ and $X_i = V_{i-1}$, we then consider the restricted ordering graph $H(X, \{X_i\}_{i=1}^n, k)$. By Theorem 1, we know that H_1-H_n-paths correspond exactly with viable orderings of $V(T)$. For $f_i \in V(H)$ and $v \in X_i$, we then define $d_{f_i}(v)$ to be the number of backwards arcs incident to v with respect to the ordering f_i of $\bigcup_{t=i-2k}^{i+2k} X_t$. More precisely,

$$d_{f_i}(v) := \left|\{u : (u \in N^+(v) \wedge f_i(u) < f_i(v)) \vee (u \in N^-(v) \wedge f_i(u) > f_i(v))\}\right|$$

Finally, we define a weight function W on the vertices of H by

$$W(f_i) = \max_{v \in X_i} [b(v) + d_{f_i}(v)]$$

Lemma 5. *Let σ be a viable ordering of T and let $P = f_1 f_2 \dots f_n$ be its corresponding H_1-H_n-path. Then*

$$\Delta_\sigma(T) = \max_{i \in [n]} W(f_i).$$

Proof. ⋆

It follows immediately by Theorem 1 and Lemma 5 that

$$\Delta(T) = \min_{P \text{ an } H_1\text{-}H_n\text{-path}} \left[\max_{f_i \in P} W(f_i) \right],$$

and so by Lemma 3 we may compute $\Delta(T)$ in time $(6k+1)^{2(6k+1)} n = k^{O(k)} n$ once the restricted ordering graph H has been constructed. Computing $b(v)$ for all v incurs an $O(n^2)$ preprocessing cost which is independent of k, and so we conclude the following:

Theorem 2. *There exists an algorithm which, given as input a semi-complete digraph T on n vertices and an integer k, determines whether $\Delta(T) \leq k$ and, if this is the case, computes an optimal degreewidth ordering of $V(T)$ in time $k^{O(k)} n + O(n^2)$.*

5 FAS Parameterized by Degreewidth

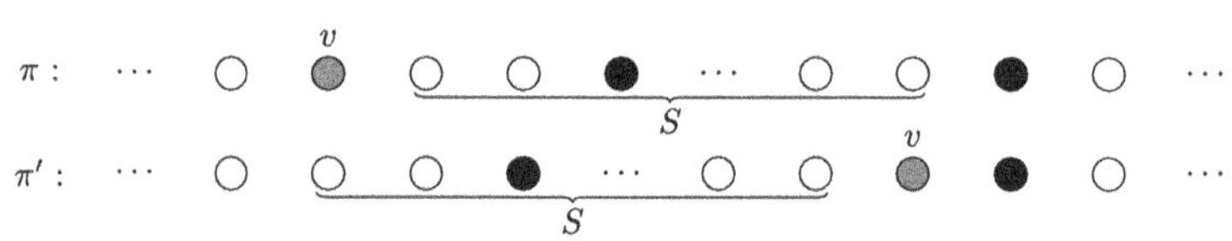

(a) When $r < \ell - 2k$, moving v to the right of S, keeping the relative order of all other vertices the same, is sufficient to reduce the total number of backwards arcs, leading to a contradiction.

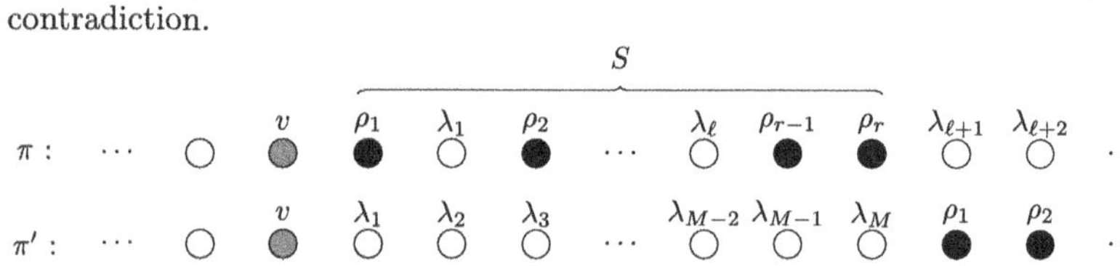

(b) When $r \geq \ell - 2k$, rearranging R_π so that all vertices in L are to the left of all vertices in R, keeping the relative orders of L and R the same, produces an ordering with strictly fewer backwards arcs, again giving a contradiction.

Fig. 1. The proposed reorderings for the two cases from the proof of Lemma 6. Vertices in L_σ are unshaded, while vertices in R_σ are shaded black. We assume $|S| = M > 10k$.

A *feedback arc set* (fas) of a digraph is a set of arcs that, when removed, renders the digraph acyclic. The problem of finding a minimum sized feedback arc set is equivalent to that of finding a minimum set of arcs that, when reversed, renders the digraph acyclic. Yet another equivalent formulation is that of finding an ordering of the vertices of the digraph which minimizes the number of backwards arcs. Given a digraph T, we let $\text{fas}(T)$ denote the size of a minimum feedback arc set. We will now employ a similar strategy as before to develop an FPT algorithm for feedback arc set on semi-complete digraphs parameterized by degreewidth, starting with the following lemma:

Lemma 6. *Suppose T is a semi-complete digraph with $\Delta(T) = k$. Let σ be an optimal degreewidth ordering and π be an optimal fas ordering of T. Then, for all $v \in V(T)$, $\pi(v) \in [\sigma(v) - 10k, \sigma(v) + 10k]$.*

Proof. We argue by contradiction. Suppose first there exists $v \in V(T)$ such that $\sigma(v) - \pi(v) > 10k$, and choose v that maximizes $M := \sigma(v) - \pi(v)$. We define $L_\sigma := \{u : \sigma(u) < \sigma(v)\}$ and $R_\sigma := \{u : \sigma(u) > \sigma(v)\}$. We also define L_π and R_π analogously, as well as let $L := L_\sigma \cap R_\pi$ and $R := R_\sigma \cap R_\pi$.

We first note that if $\sigma(u) > \sigma(v)$ then $\pi(u) > \pi(v)$ since $\sigma(v) - \pi(v)$ is maximum, and so $R = R_\sigma$. Since $|R_\pi| = |R_\sigma| + M = |L| + |R|$, we see that $|L| = M$. Let $S := \{u : \pi(v) < \pi(u) \leq \sigma(v)\}$ be the first M vertices immediately to the right of v with respect to π. Finally, let $\ell = |L \cap S|$ and $r = |R \cap S|$, so $\ell + r = |S| = M$.

We consider first the case in which $r < \ell - 2k$. We construct a new ordering π' by removing v from its current position and moving it immediately to the right of S (see Fig. 1a). Letting A denote the set of arcs with one endpoint equal to v and the other in S, it is clear that only arcs in A change direction when going from π to π'. Because σ is an optimal degreewidth ordering, at most k elements of $L \cap S$ can be in $N^+(v)$, meaning by semi-completeness that at least $\ell - k$ of them are in $N^-(v)$. Thus in π', at most $r + k < \ell - k$ of the arcs in A are backwards, and in π, at least $\ell - k$ of the arcs in A must be backwards. We see then that π' induces strictly fewer backwards arcs than π, contradicting the fact that π is an optimal fas ordering.

Suppose instead that $r \geq \ell - 2k$. Then $2r \geq \ell + r - 2k = M - 2k$, and so $r \geq (M - 2k)/2$. We again consider an alternative ordering π', this time constructed by pushing all vertices in L to the left of all vertices in R inside R_π, keeping the relative orders of L and R the same (see Fig. 1b). Letting C denote the set of arcs with one endpoint in L and the other in R, we observe that only arcs in C change direction when going from π to π'.

With respect to π', every backwards arc in C must have its tail in R and its head in L. Since every vertex in L is to the left of every vertex in R with respect to σ, each vertex in L can have at most k in-neighbors in R, meaning that at most $k|L| = kM$ arcs in C are backwards with respect to π'.

We now consider π. There are r vertices from R in S, meaning that there are exactly r vertices in $L \cap (R_\pi - S)$. Because each element of R can have at most k out-neighbors in L, by semi-completeness, each of the r vertices in $R \cap S$ has at least $r - k$ in-neighbors in $L \cap (R_\pi - S)$, and so, since every vertex in S is to the left of every vertex in $R_\pi - S$ with respect to π, we conclude that at least $r(r-k)$ arcs in C are backwards with respect to π. Of course,

$$r(r-k) \geq \frac{(M-2k)(M-4k)}{4} = \frac{M^2 - 6kM + 8k^2}{4} > \frac{4kM + 8k^2}{4} > kM,$$

where the first strict inequality follows from the fact that $M > 10k$, and so we conclude that the number of backwards arcs in C with respect to π' is strictly smaller than that with respect to π, again contradicting the fact that π is an optimal fas ordering.

This proves that $\pi(v) \geq \sigma(v) - 10k$. A symmetrical argument proves also that $\pi(v) \leq \sigma(v) + 10k$. □

Corollary 2. *Suppose $\Delta(T) = k$ and σ is an optimal degreewidth ordering of T. If $(u, v) \in A(T)$ and $\sigma(u) > \sigma(v) + 20k$, then (u, v) is contained in every minimum feedback arc set of T.*

We now use this lemma to design an FPT algorithm to compute a minimum feedback arc set on semi-complete digraphs. We label the vertices of T as $v_1, \ldots, v_n$ in optimal degreewidth ordering, which we assume is given as input. By Corollary 2, we may begin by adding every backwards arc whose length exceeds $20k$ to the solution set and then flipping it. Clearly, this does not cause the degreewidth to increase. We may then assume from here on that all arcs of length longer than $20k$ with respect to the degreewidth ordering are forwards arcs.

We now call an ordering π of $V(T)$ *viable* if $\pi(v_i) \in [i - 10k, i + 10k]$. By Lemma 6, every optimal ordering must be viable. Letting $X = V(T)$ and $X_i = \{v_i\}$, we then consider the restricted ordering graph $H(X, \{X_i\}_{i=1}^n, 10k)$. By Theorem 1, the H_1-H_n-paths in this graph correspond exactly to viable orderings. We define $d_{f_i}(v_i)$ as before to be the number of backwards arcs incident to v with respect to the ordering f_i of $\bigcup_{t=i-20k}^{i+20k} X_t$. Our weight function on $V(H)$ is then just

$$W(f_i) = d_{f_i}(v_i).$$

Given an ordering π of $V(T)$ we denote by $B(\pi)$ the number of backwards arcs induced by π. We then claim the following:

Lemma 7. *Let π be a viable ordering of $V(T)$ and let $P = f_1 f_2 \ldots f_n$ be its corresponding H_1-H_n-path. Then*

$$2B(\pi) = \sum_{f_i \in P} W(f_i)$$

Proof. ⋆

Thus by Theorem 1 we can immediately say the following:

Corollary 3.

$$2\,\mathrm{fas}(T) = \min_{P \text{ an } H_1\text{-}H_n\text{-path}} \left[\sum_{f_i \in P} W(f_i) \right]$$

By Lemma 3, once the restricted ordering graph H is constructed, we can compute a minimum feedback arc set in time $O(60k+1)^{2(60k+1)}n = k^{O(k)}n$.

Assuming we are given as input an optimal degreewidth ordering of T as well as, for each vertex, the backwards arcs incident to that vertex with respect to that ordering, the preprocessing required to construct the restricted ordering graph takes only time linear in n, meaning we can conclude the following:

Theorem 3. *There exists an algorithm which, given as input an optimal degreewidth ordering of a semi-complete digraph T on n vertices and the set of backwards arcs induced by this ordering, computes a minimum feedback arc set of T in time $\Delta(T)^{O(\Delta(T))}n$.*

6 Cutwidth Parameterized by Degreewidth

Let T be a digraph. Given a pair of disjoint subsets P, Q of $V(T)$, we define $\chi(P,Q)$ to be the number of arcs crossing from Q to P i.e.

$$\chi(P,Q) := |\{(u,v) \in A(T) : u \in Q, v \in P\}|.$$

If P or Q is a singleton, say $P = \{x\}$, then we will drop the braces and write simply $\chi(x,Q)$. Let σ be an ordering of $V(T)$. For $v \in V(T)$, we define $\sigma_{<v} := \{u \in V(T) : \sigma(u) < \sigma(v)\}$, and define $\sigma_{\geq v}$, etc. analogously. For $v \in V(T)$, we then define $c(\sigma, v) := \chi(\sigma_{<v}, \sigma_{\geq v})$. The *width* of the ordering σ is then $w(\sigma) := \max_{v \in V(T)} c(\sigma, v)$, and the *cutwidth* of T is $cw(T) := \min_{\sigma \in \Sigma(V(T))} w(\sigma)$.

We now employ a similar strategy as for feedback arc set to develop a $k^{O(k)}n$ algorithm for cutwidth parameterized by degreewidth, starting with the following lemma:

Lemma 8. *Suppose T is a semi-complete digraph with $\Delta(T) = k$, and let σ be an optimal degreewidth ordering of $V(T)$. Then there exists an optimal cutwidth ordering π of $V(T)$ such that, for all $v \in V(T)$, $\pi(v) \in [\sigma(v) - 2k, \sigma(v) + 2k]$.*

Proof. ⋆

We may now apply the same technique as before to design an FPT algorithm for cutwidth. We assume we are given as input an optimal degreewidth ordering σ of T, as well as, for each vertex, the backwards arcs incident to that vertex. We label the vertices of T as $v_1, \ldots, v_n$ in optimal degreewidth order (so $\sigma(v_i) = i$). We begin with another small lemma.

Lemma 9. *Let T be a semi-complete digraph with $\Delta(T) \leq k$ and optimal degreewidth ordering σ. Let π be an optimal cutwidth ordering satisfying $\pi(u) \in [\sigma(u) - 2k, \sigma(u) + 2k]$ for all $u \in V(T)$, and let $v \in V(T)$. Finally, suppose $x, y \in V(T)$ satisfy $\sigma(x) < \sigma(v) - 4k$ and $\sigma(y) > \sigma(v) + 4k$. Then the arc (x, y), if it exists, is forwards with respect to π while the arc (y, x), if it exists, is backwards with respect to π, and crosses the cut $(\pi_{<v}, \pi_{\geq v})$.*

Proof. $\star$

We call these arcs *long arcs*. For $v \in V(T)$, and every optimal cutwidth ordering π satisfying the above conditions, we can then precompute the number of backwards long arcs $b(v)$ crossing the cut $(\pi_{<v}, \pi_{\geq v})$ with the formula

$$b(v) := |\{(x, y) : \sigma(y) < \sigma(v) - 4k, \sigma(x) > \sigma(v) + 4k\}|.$$

Since we are given the backwards arcs induced by σ as input, $b(v)$ can be computed for all v in $O(kn)$ time using a sliding window algorithm.

As usual, letting $X = V(T)$ and $X_i = \{v_i\}$, we consider the restricted ordering graph $H(X, \{X_i\}_{i=1}^n, 2k)$. We now call an ordering π of $V(T)$ *viable* if $\pi(v_i) \in [i - 2k, i + 2k]$ for all $i \in [n]$. By Theorem 1, the H_1-H_n-paths in H correspond exactly to viable orderings, and by Lemma 8, there exists a viable optimal cutwidth ordering. Given $f_i \in V(H)$, we then define sets

$$L(f_i) := \{(x, y) \in A(T) : \sigma(y) \in [i - 4k, i + 4k], f_i(y) < f_i(v_i), \sigma(x) > i + 4k\},$$

and, similarly,

$$R(f_i) := \{(x, y) \in A(T) : \sigma(x) \in [i - 4k, i + 4k], f_i(x) \geq f_i(v_i), \sigma(y) < i - 4k\}.$$

Note that all arcs in $L(f_i)$ and $R(f_i)$ are backwards with respect to σ, meaning $L(f_i)$ and $R(f_i)$ can be computed in $O(k^2)$ time. We also define

$$I(f_i) := \{(x, y) \in A(T) : \sigma(x), \sigma(y) \in [i - 4k, i + 4k], f_i(y) < f_i(v) \leq f_i(x)\}.$$

Clearly, $I(f_i)$ can be computed in $O(k^2)$ time. Our weight function on $V(H)$ is then

$$W(f_i) = b(v_i) + |L(f_i)| + |R(f_i)| + |I(f_i)|.$$

We now claim the following:

Lemma 10. *Let π be a viable ordering of $V(T)$ and let $P = f_1 f_2 \dots f_n$ be its corresponding H_1-H_n-path. Then*

$$w(\pi) = \max_{i \in [n]} W(f_i).$$

Proof. ⋆

It follows immediately by Theorem 1 and Lemma 8 that

$$cw(T) = \min_{P \text{ an } H_1\text{-}H_n\text{-path}} \left\lceil \max_{f_i \in P} W(f_i) \right\rceil,$$

and so by Lemma 3 we may compute $cw(T)$ in time $(12k+1)^{2(12k+1)}n = k^{O(k)}n$ once the restricted ordering graph H has been constructed. As mentioned earlier, the preprocessing required to construct the restricted ordering graph takes only $O(kn)$ time, and the weights require only $O(k^2)$ time to compute, so we conclude the following:

Theorem 4. *There exists an algorithm which, given as input an optimal degreewidth ordering of a semi-complete digraph T on n vertices and the set of backwards arcs induced by this ordering, computes an optimal cutwidth ordering of T in time $\Delta(T)^{O(\Delta(T))}n$.*

7 Approximating Degreewidth on Semi-complete Graphs

We now consider the problem of approximating degreewidth on semi-complete digraphs, starting with the following observation:

Observation 1. *For all $c > 1$, a c-approximation algorithm for degreewidth on tournaments implies the existence of a $(c+1)$-approximation algorithm for degreewidth on semi-complete digraphs.*

Proof. ⋆

It is proven in [8] that sorting a tournament by in-degree is a tight 3-approximation algorithm for degreewidth, and so by the above we are immediately granted a 4-approximation on semi-complete digraphs. However, we can do better. Given a semi-complete digraph T with $\Delta(T) = k$, we again call an ordering σ of $V(T)$ *viable* if, for all $v \in V(T)$, $\sigma(v) \in [d^-(v)-k+1, d^-(v)+k+1]$, and recall by Lemma 1 that every optimal degreewidth ordering is viable. We first prove the following lemma:

Lemma 11. *Let T be a semi-complete digraph of degreewidth k. Then sorting $V(T)$ by in-degree produces a viable ordering of $V(T)$.*

Proof. ⋆

We now modify the argument in [8] and apply this lemma to prove the following:

Theorem 5. *Sorting the vertices by in-degree is a tight 3-approximation of degreewidth for semi-complete digraphs.*

Proof. ⋆

8 Degreewidth on General Digraphs

We now study degreewidth on general digraphs, proving that it is NP-hard to determine whether a digraph has degreewidth at most 2. For this section, we will now let G denote the input digraph. We let Degreewidth denote the decision problem of, given as input a digraph G and integer k, determining whether $\Delta(G) \leq k$. We then define the Generalized Degreewidth decision problem as follows: Given a directed graph G, an integer k, and, for each $v \in V(G)$, an integer t_v satisfying $t_v \leq k$, does there exist an ordering of the vertices of G such that, for all $v \in V(G)$, v is incident to at most t_v backwards arcs with respect to this ordering? We show first that this is polynomial time reducible to Degreewidth:

Lemma 12. Generalized Degreewidth *is polynomial time reducible to* Degreewidth.

Proof. ⋆

We note first that the reduction does not change the parameter k and also that, if G is an oriented graph, then so is G'. We will now prove that Generalized Degreewidth is NP-hard, even when restricted to inputs where $k = 2$ and G is an oriented digraph. To do so, we reduce from Balanced 3-Sat(4), a variant of 3-Sat in which every clause contains exactly 3 literals and every literal occurs exactly twice. This is known to be NP-hard [3]. We begin with the following construction which, given an instance of Balanced 3-Sat(4), produces an input to Generalized Degreewidth:

Construction 1. *Let ϕ be a* Balanced 3-Sat(4) *boolean formula with variables $x_1, \ldots, x_n$ and clauses $C_1, \ldots, C_m$. We construct an oriented directed graph G as follows:*

- *For each variable x_i, create three vertices y_i, u_i, and $\bar{u}_i$, as well as arcs (y_i, u_i) and $(y_i, \bar{u}_i)$. The u_i's and $\bar{u}_i$'s will be referred to as* literal vertices, *while the y_i's will be referred to as* splitters.
- *Create* clause vertices *$c_1, \ldots, c_m$ for each clause. For the three literal vertices x, y, z that satisfy c_i, add arcs (x, c_i), (y, c_i), (z, c_i).*
- *Create a* buffer *vertex b, and add arcs (c_i, b) and (b, y_j) for all $i \in [m]$, $j \in [n]$.*

We then set $t_b = 0$, $t_{y_i} = 1$ for all $i \in [n]$, and $t_v = 2$ for all other $v \in V(G)$. Finally, we set $k = 2$. The input instance to Generalized Degreewidth *is then $(G, \{t_v\}_{v \in V(G)}, k)$.*

Lemma 13. *Let ϕ be an input boolean formula to* Balanced 3-Sat(4), *and let $(G, \{t_v\}_{v \in V(G)}, k)$ be the* Generalized Degreewidth *instance produced by applying Construction 1 to ϕ. If ϕ has a satisfying assignment, then $(G, \{t_v\}_{v \in V(G)}, k)$ is a yes instance.*

Proof. ⋆

Lemma 14. *Let ϕ be an input boolean formula to* Balanced 3-Sat(4)*, and let $(G, \{t_v\}_{v \in V(G)}, k)$ be the* Generalized Degreewidth *instance produced by applying Construction 1 to ϕ. If $(G, \{t_v\}_{v \in V(G)}, k)$ is a yes instance, then ϕ has a satisfying assignment.*

Proof. ⋆

From Lemmas 12, 13, and 14, as well as the accompanying remarks, we immediately conclude the following:

Theorem 6. *The problem of determining whether an oriented digraph G has degreewidth at most 2 is NP-hard.*

9 Conclusion

In this paper, we studied the parameter degreewidth, introduced in [8] for tournaments, on the more general class of semi-complete digraphs. We gave a $k^{O(k)}n + O(n^2)$ FPT algorithm to determine whether a semi-complete digraph has degreewidth at most k and output an optimal ordering if this is the case. We also gave an algorithm which, given as input an optimal degreewidth ordering and the set of backwards arcs induced by this ordering, computes a minimum feedback arc set on semi-complete digraphs of degreewidth k in $k^{O(k)}n$ time. These results resolve two open problems posed in [8]. We then gave an algorithm which, with the same input requirements as the feedback arc set algorithm, computes an optimal cutwidth ordering for semi-complete digraphs of degreewidth k in $k^{O(k)}n$ time. The question arises as to whether this parameter is useful for solving other problems which are easy on tournaments but hard on semi-complete digraphs, such as Optimal Linear Arrangement. We next studied the problem of approximating degreewidth on semi-complete digraphs, showing that the 3-approximation algorithm described in [8] for tournaments works for semi-complete digraphs as well. Finally, we considered degreewidth on general digraphs, showing that it is NP-hard to determine whether a given digraph has degreewidth at most 2, even if the input is restricted to be oriented. This leaves open the question of whether or not there exists a polynomial time algorithm to determine whether a digraph has degreewidth 1.

References

1. Alon, N.: Ranking tournaments. SIAM J. Discret. Math. **20**(1), 137–142 (2006)
2. Barbero, F., Paul, C., Pilipczuk, M.: Exploring the complexity of layout parameters in tournaments and semicomplete digraphs. ACM Trans. Algorithms **14**(3) (2018)
3. Berman, P., Karpinski, M., Scott, A.D.: Approximation hardness of short symmetric instances of MAX-3SAT. Electron. Colloq. Comput. Complexity (49) (2003). https://eccc.weizmann.ac.il/report/2003/049/

4. Bessy, S., et al.: Packing arc-disjoint cycles in tournaments. In: MFCS 2019 - 44th International Symposium on Mathematical Foundations of Computer Science. Leibniz International Proceedings in Informatics (LIPIcs), Aachen, Germany, vol. 138, pp. 27:1–27:14 (2019). https://hal.science/hal-02277436
5. Bessy, S., Bougeret, M., Thiebaut, J.: Triangle packing in (sparse) tournaments: approximation and kernelization. In: 25th Annual European Symposium on Algorithms, ESA 2017, 4–6 September 2017, Vienna, Austria. LIPIcs, vol. 87, pp. 14:1–14:13. Schloss Dagstuhl - Leibniz-Zentrum für Informatik (2017)
6. Charbit, P., Thomassé, S., Yeo, A.: The minimum feedback arc set problem is np-hard for tournaments. Comb. Probab. Comput. **16**(1), 1–4 (2007)
7. Chudnovsky, M., Fradkin, A., Seymour, P.: Tournament immersion and cutwidth. J. Combin. Theory Ser. B **102**(1), 93–101 (2012)
8. Davot, T., Isenmann, L., Roy, S., Thiebaut, J.: DegreeWidth: a new parameter for solving problems on tournaments. In: Graph-Theoretic Concepts in Computer Science, pp. 246–260. Springer, Cham (2023)
9. Fomin, F.V., Pilipczuk, M.: On width measures and topological problems on semi-complete digraphs. J. Combin. Theory Ser. B **138**, 78–165 (2019)
10. Fradkin, A.: Forbidden structures and algorithms in graphs and digraphs. Ph.D. thesis, Princeton University (2011)
11. Pilipczuk, M.: Computing cutwidth and pathwidth of semi-complete digraphs via degree orderings. CoRR abs/1210.5363 (2012). http://arxiv.org/abs/1210.5363
12. Thiebaut, J.: Algorithmic and structural results on directed cycles in dense digraphs. Ph.D. thesis (2019)

On the Connectivity of the Flip Graph of Plane Spanning Paths

Linda Kleist[1,2(✉)], Peter Kramer[2(✉)], and Christian Rieck[2(✉)]

[1] Institute of Computer Science, Universität Potsdam, Potsdam, Germany
kleist@cs.uni-potsdam.de

[2] Department of Computer Science, TU Braunschweig, Braunschweig, Germany
{kleist,kramer,rieck}@ibr.cs.tu-bs.de

Abstract. Flip graphs of non-crossing configurations in the plane are widely studied objects, e.g., flip graph of triangulations, spanning trees, Hamiltonian cycles, and perfect matchings. Typically, it is an easy exercise to prove connectivity of a flip graph. In stark contrast, the connectivity of the flip graph of plane spanning paths on point sets in general position has been an open problem for more than 16 years.

In order to provide new insights, we investigate certain induced subgraphs. Firstly, we provide tight bounds on the diameter and the radius of the flip graph of spanning paths on points in convex position with one fixed endpoint. Secondly, we show that so-called *suffix-independent paths* induce a connected subgraph. Consequently, to answer the open problem affirmatively, it suffices to show that each path can be flipped to some suffix-independent path. Lastly, we investigate paths where one endpoint is fixed and provide tools to flip to suffix-independent paths. We show that these tools are strong enough to show connectivity of the flip graph of plane spanning paths on point sets with at most two convex layers.

Keywords: flip graph · connectivity · point set · plane spanning path · convex layer

1 Introduction

The reconfiguration of objects by small local operations is a basic concept that appears in various contexts, e.g., robot motion planning, morphing, network dynamics, discrete mathematics, and computer science [24]. A *flip graph* captures the structure of the reconfiguration space for the problem at hand: The vertices of the graph correspond to the possible configurations, i.e., states of the system, while an edge indicates that two configurations can be transformed into each other by a small local modification—a so-called *flip*. A famous example is the flip graph of triangulations of a convex n-point set where a flip operation replaces one diagonal by another one; this flip graph is the 1-skeleton of the $(n-3)$-dimensional associahedron. More generally, flip graphs of various

A full version containing all details is available at arXiv:2407.03912 [18].

D. Král' and M. Milanič (Eds.): WG 2024, LNCS 14760, pp. 327–342, 2025.
https://doi.org/10.1007/978-3-031-75409-8_23

Fig. 1. Plane spanning paths that can be transformed into each other by a flip.

non-crossing configurations on point sets are widely studied, e.g., triangulations, spanning trees, Hamiltonian cycles, or perfect matchings; see Sect. 1.2. In this paper, we investigate the flip graph of *plane spanning paths* where a flip consists of the insertion and deletion of one edge, as illustrated in Figs. 1 and 2.

Typically, one is interested in properties such as the diameter of the flip graph (the maximum length of a shortest reconfiguration sequence over all pairs), the distance between two given configurations, or whether the flip graph contains a Hamiltonian cycle, i.e., a cyclic listing of all configurations. Usually, it is a simple exercise to prove that a flip graph is connected. In stark contrast, the connectivity of the flip graph of plane spanning paths has been an open problem for more than 16 years [4,7] and has been conjectured by Akl et al. [4].

Conjecture 1 (Akl et al. [4]). *For every point set S in general position, the flip graph of plane spanning paths on S is connected.*

The conjecture has been confirmed for some special cases, namely, point sets with up to 8 points [4], sets in convex position [4], and generalized double circles [3].

In a spirit similar to Conjecture 1, Aichholzer et al. [3] conjecture the connectivity of subgraphs induced by all paths with a fixed endpoint. In fact, they show that Conjecture 2 implies Conjecture 1.

Conjecture 2 (Aichholzer et al. [3]). *For every point set S in general position and every point $s \in S$, the flip graph of plane spanning paths on S starting at s is connected.*

1.1 Our Contribution

In this work, we tackle Conjectures 1 and 2. For a point set S, we denote the set of all paths on S by $\mathcal{P}(S)$ and the subset of paths starting at a fixed $s \in S$ by $\mathcal{P}(S, s)$. Firstly, we settle Conjecture 2 for point sets in convex position. More precisely, we provide tight bounds for the diameter and the radius. Note that for $n = 1, 2$, the flip graphs are trivial, as they consist of a single vertex.

Theorem 3. *Let S be a set of n points in convex position and let $s \in S$. For $n \geq 3$, the flip graph of $\mathcal{P}(S, s)$ has diameter $2n - 5$ and radius $n - 2$. Moreover, the spirals are exactly the centers of the flip graph.*

Secondly, we show that the flip graph of so-called *suffix-independent paths* is connected. A path is *suffix-independent* if each suffix starts on the boundary of its convex hull; for precise definitions, we refer to Sect. 2. We denote the set of suffix-independent paths starting at $s \in S$ by $\mathcal{P}_{\text{sin}}(S, s)$.

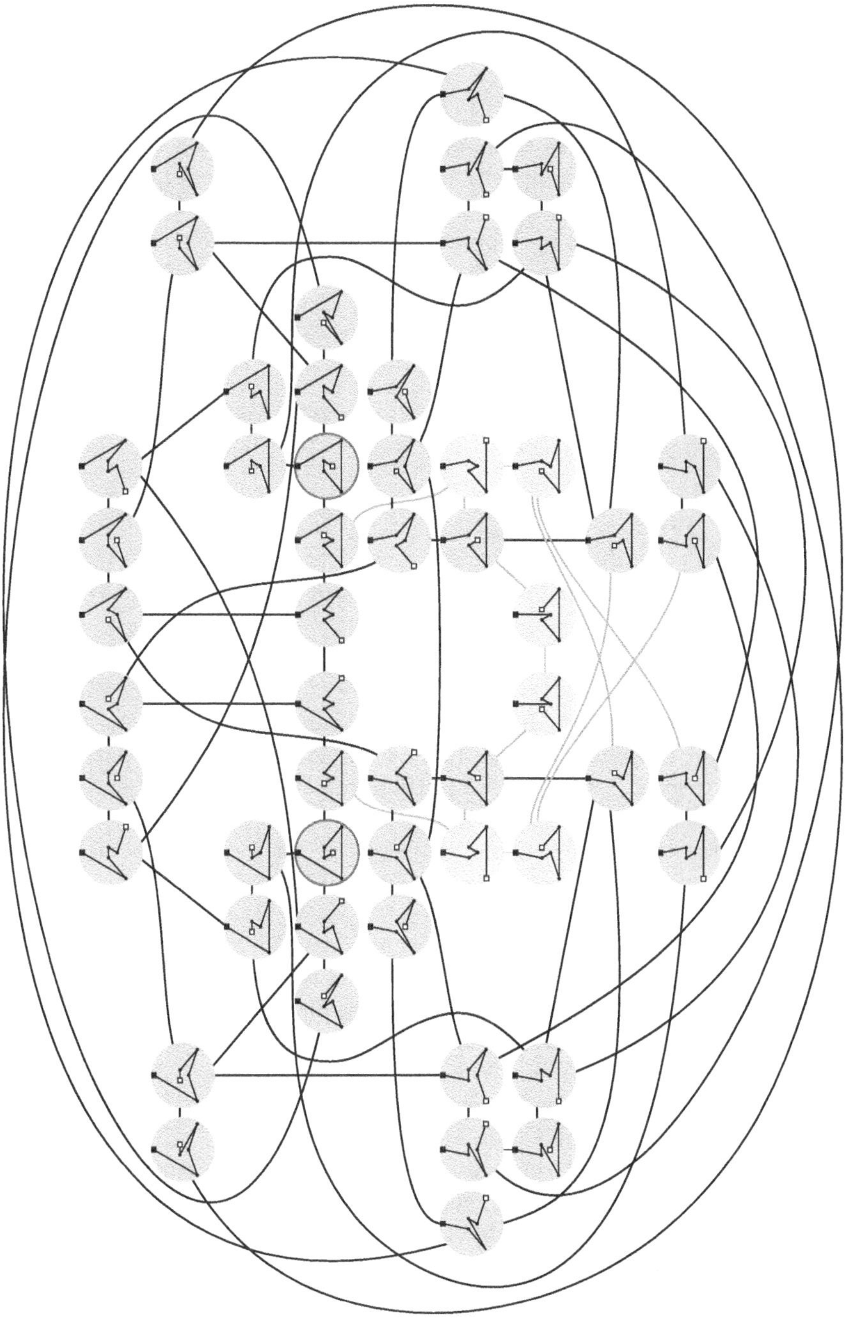

Fig. 2. The flip graph of plane spanning paths with a fixed start on a point set with two layers, where each layer contains three points. Six paths, highlighted in orange, are not suffix-independent. The two spirals, highlighted in red, have flip distance 5 and are a furthest pair. (Color figure online)

Theorem 4. *For every point set S in general position and every point $s \in S$, the flip graph of $\mathcal{P}_{\text{sin}}(S, s)$ is connected.*

Theorem 4 may serve as a crucial building block to prove Conjectures 1 and 2 as it suffices to flip each path to some suffix-independent path. We investigate a point set S based on its decomposition into $L(S)$ many *convex layers* $L_0, L_1, \dots$. Using Theorem 4 alongside other tools, we conclude connectivity of the flip graph of plane spanning paths on each point set with at most two convex layers. As a crucial tool, we consider paths that start at a fixed point on the convex hull.

Theorem 5. *Let S be a point set in general position with $L(S) \leq 2$, and $s \in L_0$. Then the flip graph of $\mathcal{P}(S, s)$ is connected.*

We then use the insights from Theorem 5 to settle Conjecture 1 for point sets with at most two convex layers.

Theorem 6 ($\star$). *Let S be a point set in general position with $L(S) \leq 2$. Then the flip graph of $\mathcal{P}(S)$ is connected.*

Proofs for all statements marked by ($\star$) can be found in the full version [18].

1.2 Related Work

Flip graphs of non-crossing configurations on point sets in the plane are widely studied, e.g., for triangulations [11,12,16,17,19,20,26], spanning paths [3,4,10, 25], spanning trees [1,2,5,6,8,9,14,23], Hamilton cycles [13], and matchings [15, 16,21].

In the following, we review the state of the art for the two most related settings, namely of plane spanning paths and plane spanning trees.

Plane Spanning Paths. While not much is known for the flip graph of plane spanning paths on general point sets, the case of convex point sets is pretty well understood. Akl et al. [4] show that the flip graph is connected and Chang and Wu [10] provide tight bounds for the diameter, i.e., for a convex n-set, its flip graph has diameter $2n - 5$ for $n = 3, 4$ and $2n - 6$ for all $n \geq 5$. Rivera-Campo and Urrutia-Galicia [25] show Hamiltonicity of the flip graph, and its chromatic number is shown to be n by Fabila-Monroy et al. [22].

For more general point sets, Akl et al. [4] show connectivity for all small point sets with up to 8 points in general position. Aichholzer et al. [3] confirm Conjecture 1 for generalized double circles; a point set is a generalized double circle if it allows for a Hamiltonian cycle that intersects no segment between any two points. While the obtained upper bound on the diameter is quadratic, they provide a linear upper bound of $2n - 4$ on the diameter for the special case of wheel sets of size n.

Plane Spanning Trees. Avis and Fukuda [5] show that the diameter of the flip graph of plane spanning trees is upper bounded by $2n-4$ for n points in general position. The best known lower bound of $1.5n - 5$ due to Hernando et al. [14] even holds for the special case of convex point sets. If one tree is a monotone path, Aichholzer et al. [2] show that it can be reconfigured into any other tree by $1.5n - 2$ flips.

For points in convex position, the upper bound on the diameter has been improved several times. Bousquet et al. [8] obtain a bound of $2n - \Omega(\sqrt{n})$ and conjecture $1.5n$ to be the tight bound. Aichholzer et al. [2] bound the distance of two given trees T_1 and T_2 on convex point sets by $2|T_1 \setminus T_2| - \Omega(\log(|T_1 \setminus T_2|))$. Bousquet et al. [9] break the barrier of 2 in the leading coefficient and show that there always exists a flip sequence between any pair of non-crossing spanning trees T_1 and T_2 of length at most $c\,|T_1 \setminus T_2|$ where $c \approx 1.95$. As a lower bound in terms of the symmetric difference, they provide a pair of trees such that a minimal flip sequence has length $5/3\,|T_1 \setminus T_2|$. Most recently, Bjerkevik et al. [6] show a lower bound of $14/9 \cdot n - \mathcal{O}(1)$ and an upper bound of $5/3 \cdot n - \mathcal{O}(1)$.

2 Fundamental Concepts and Tools

Let S be a set of n points in general position, i.e., no three points in S are collinear. A *plane spanning path* on S is a path that visits each point of S exactly once. Every edge of a path is a straight-line segment between a pair of points, and no two edges intersect except at a common endpoint. We write uv or vu for an edge between $u, v \in S$. For any path P, the points of degree 1 are the *endpoints* of P. We denote the set of all plane spanning paths of S by $\mathcal{P}(S)$. A *flip* on a path $P \in \mathcal{P}(S)$ replaces an edge e of P with another edge f on S to obtain a new path $P' \in \mathcal{P}(S)$; for an example, see Fig. 1. The flip graph of $\mathcal{P}(S)$ contains a vertex for each path in $\mathcal{P}(S)$ and an edge between any two paths that can be transformed into one another by a flip. Similarly, for each $\mathcal{P}' \subset \mathcal{P}(S)$, the flip graph of $\mathcal{P}'$ refers to the subgraph induced by $\mathcal{P}'$.

Paths with a Fixed Start. We focus in particular on paths with a fixed endpoint. For every $s \in S$, we denote the set of all paths with s as an endpoint by $\mathcal{P}(S, s)$. Figures 2, 6(a) and 6(b) illustrate examples of induced flip graphs. Typically, we consider an st-path of $\mathcal{P}(S, s)$ as oriented from s to t, and call s the *start* and t the *end*. We illustrate these as black and white squares, respectively. It is easy to see that all edges that can be introduced by a flip are incident to t, see Fig. 3.

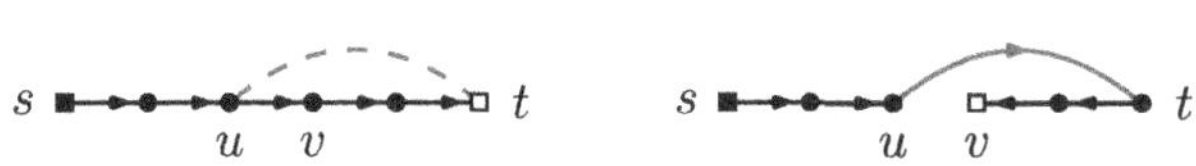

Fig. 3. A flip on a path with a fixed endpoint, illustrating Observation 7.

Observation 7. *An st-path $P \in \mathcal{P}(S, s)$ can be flipped to an st'-path $P' \in \mathcal{P}(S, s)$ if and only if some edge uv is replaced by ut, i.e., an edge incident to t.*

For two points $u, v \in S$ and a path P, we say that u *sees* v if the straight-line segment uv does not intersect any of the edges of P except at a common endpoint. Clearly, this relationship is symmetric, i.e., u sees v exactly if v sees u. Then, Lemma 7 implies that the degree of each path in the flip graph of $\mathcal{P}(S, s)$ is equal to the number of points that are visible from the endpoint of the path.

We say that a directed edge uv is *outgoing* for u and *incoming* for v. For any directed path P which contains the edge uv, we also say that u is the *predecessor* of v on P, and v the *successor* of u on P. Moreover, let $\overleftarrow{P}$ denote the *reversed* path of P, i.e., $vu \in \overleftarrow{P}$ if and only if $uv \in P$.

Convex Layers. We denote the convex hull of S by $\text{conv}(S)$ and say that a point $s \in S$ is *outer* if it lies on the boundary of the convex hull of S. A point that is not outer is *inner*. The *layers* of S are defined as follows, see Fig. 4(a): $L_0 \subseteq S$ consists of all outer points. For $i \geq 1$, L_i is the set of points that are outer in $S \setminus \bigcup_{k<i} L_k$. The number of non-empty layers in S is the *layer number* $L(S)$.

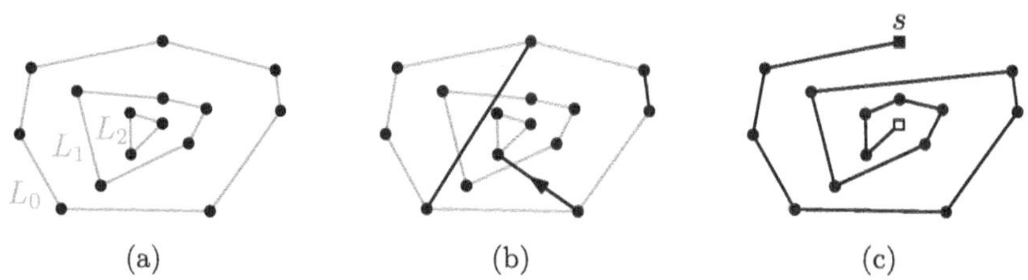

Fig. 4. A point set with (a) its layers, (b) a chord, an inward cutting edge, and a level edge (from left to right), and (c) its counterclockwise spiral from $s \in L_0$.

We speak of an L_i-*edge* if both endpoints are in L_i. For a convex point set, two points are *adjacent* if they appear consecutively on the boundary of its convex hull. We distinguish certain types of edges, as illustrated in Fig. 4(b). If u and v are adjacent in L_i, we say that uv is a *level edge* of L_i. A *chord* is then an L_0-edge that is not level. Further, an edge uv with $u \in L_i$ and $v \in L_j$ such that $i \neq j$ is called an *inter-layer edge*; we call it *inward* if $i < j$, and *outward* if $i > j$. Finally, we say that an inter-layer edge is *cutting*, if it crosses any level edge in S. For a path $P \in \mathcal{P}(S)$, we denote by $\ell_i(P)$ the number of level edges of L_i in P, and by $\psi(P)$ the number of cutting edges in P.

Suffix-independence. For a given st-path $P \in \mathcal{P}(S, s)$, let $P[a, b]$ refer to the subpath of P from a to b. We say that $P[a, b]$ is a *prefix* of P if $a = s$, and a *suffix* of P if $b = t$. A suffix $P[a, t]$ of P is called *independent*, if the convex hull of $P[a, t]$ does not contain any points of the prefix $P[s, a]$. By definition, a is then outer with respect to the points in $P[a, t]$. We say that a path $P \in \mathcal{P}(S, s)$ is *suffix-independent* if each suffix of P is independent, i.e., if every $u \in S$ is outer with respect to the points in $P[u, t]$. By definition, every suffix of a suffix-independent path is suffix-independent. We denote by $\mathcal{P}_{\text{sin}}(S, s)$ the set of all suffix-independent paths of $\mathcal{P}(S, s)$.

Note that the suffix-independent paths contain all paths that are monotone, radially monotone, or spirals. We now present a sufficient criterion to identify a family of suffix-independent paths, the *layer-monotone* paths. A path on S is layer-monotone if it visits the layers of S in increasing order, i.e., for all $i \in [0, L(S) - 1]$, all points in layer L_i are visited before any point in L_{i+1}.

Lemma 8 **($\star$).** *Let S be a point set in general position, $s \in S$, and $P \in \mathcal{P}(S, s)$.*

(1) If P is layer-monotone, then P is suffix-independent.
(2) If P is not suffix-independent, then it contains an outward edge.

A special type of layer-monotone path, the *clockwise spiral* of S from an outer point s is the path $s_1, \ldots, s_n$ such that $s_1 = s$ and s_{i+1} is an outer point of $S \setminus \{s_1, \ldots, s_{i-1}\}$ which lies clockwise adjacent to s_i. Figure 4(a) depicts a *counterclockwise spiral*, which can be constructed analogously. Finally, we define the *L_i-suffix* of a path as the maximal suffix that contains only points on layers $L_{j \geq i}$.

Useful Properties. We present two useful properties for paths starting and ending in an outer point. Two points $u, v \in S$ divide a layer $L_i \setminus \{u, v\}$ into two sets by the line through u and v; for a reference point $w \in L_i$, we denote them by $L_i^+(u, v; w)$ and $L_i^-(u, v; w)$ such that $w \in L_i^+(u, v; w)$, see Fig. 5.

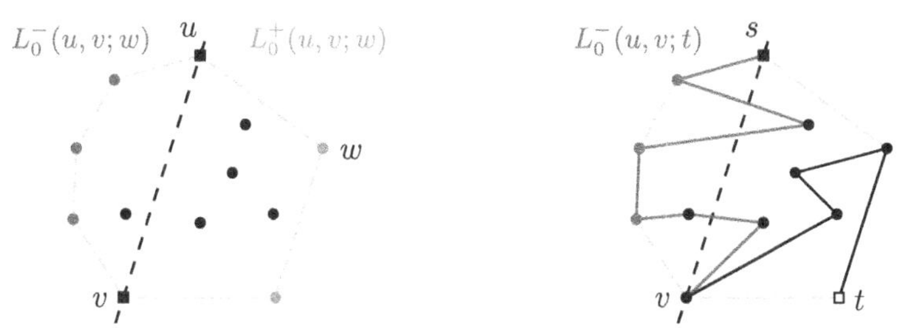

(a) A pair of points divides each layer into $L_0^+(u, v; w)$ and $L_0^-(u, v; w)$.

(b) The prefix $P[s, v]$ visits all points in $L_0^-(s, v; t)$.

Fig. 5. Illustrations for $L_0^{\pm}(u, v; w)$ in (a) and Lemma 9 in (b).

By planarity of the paths, we obtain the following statement.

Lemma 9 **($\star$).** *If an st-path P starts and ends on L_0, then the prefix $P[s, v]$ for every $v \in L_0 \setminus \{s, t\}$ visits all points in $L_0^-(s, v; t)$.*

This can be further extended to prove the following.

Lemma 10 (⋆). *If a path P starts and ends in two adjacent points on L_0, then it is chord-free.*

3 Plane Paths on Point Sets in Convex Position

In this section, we consider plane spanning paths on point sets in convex position, and present tight bounds for the radius and diameter of the flip graph. Figures 6(a) and 6(b) depict the flip graphs for $n = 4$ and $n = 5$, respectively.

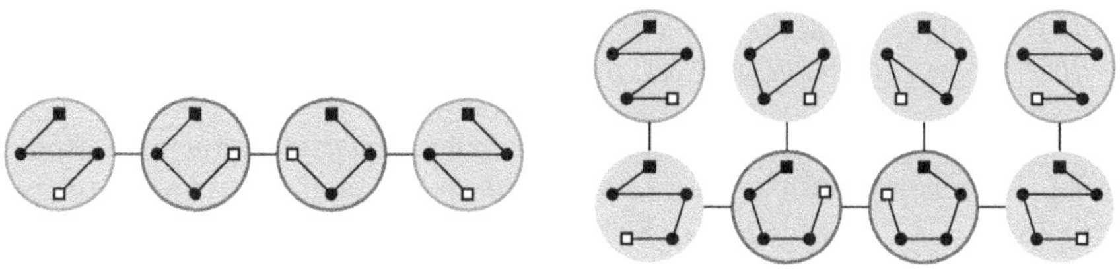

(a) Flip graph of $\mathcal{P}(S_1, s)$, for any $s \in S_1$. (b) Flip graph of $\mathcal{P}(S_2, s)$, for any $s \in S_2$.

Fig. 6. We illustrate flip graphs on two point sets in convex position, $|S_1| = 4$ and $|S_2| = 5$. The spirals are highlighted in cyan and the zigzag paths in orange. (Color figure online)

Theorem 3. *Let S be a set of n points in convex position and let $s \in S$. For $n \geq 3$, the flip graph of $\mathcal{P}(S, s)$ has diameter $2n - 5$ and radius $n - 2$. Moreover, the spirals are exactly the centers of the flip graph.*

Proof. Let C_{ccw} and C_{cw} denote the (counter-)clockwise spirals of S, i.e., the two plane spanning paths that only use convex hull edges, see Figs. 7(a) and 7(b).

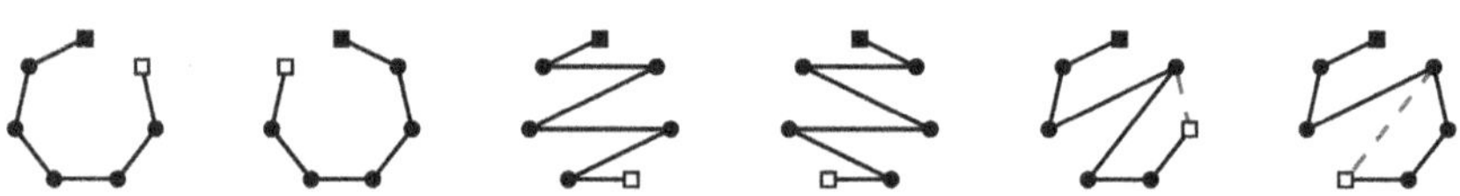

(a) C_{ccw} (b) C_{cw} (c) Z_{ccw} (d) Z_{cw} (e) (f)

Fig. 7. Illustration for the proof of Lemma 3.

For the upper bounds, we show that each path P can be transformed into C_{cw} or C_{ccw} with $n - 3$ flips. Because C_{cw} and C_{ccw} can be transformed into each other using one flip, we obtain an upper bound of $2(n-3)+1 = 2n-5$ on the diameter and of $n-2$ on the radius.

It thus remains to show that each path P can be transformed into C_{cw} or C_{ccw} with $n-3$ flips. As each path contains at least two convex hull edges, we may assume without loss of generality that P and C_{cw} share at least two edges; otherwise we consider C_{ccw}. Note that the endpoint of P is incident to a convex hull edge e which is not contained in P. We add e to P and remove the appropriate edge f, see Figs. 7(e) and 7(f). Observe that f lies on the convex hull if and only if $s = t$. Hence, the number of convex hull edges increases until we end in C_{cw} after at most $n-3$ flips.

For the lower bound on the diameter, we consider the zigzag-paths Z_{cw} and Z_{ccw} depicted in Figs. 7(c) and 7(d). We show they have distance $2n-5$; this directly implies that the radius is lower bounded by $\lceil (2n-5)/2 \rceil = n-2$. Note that the edges incident to s differ in Z_{cw} and Z_{ccw}. In order to flip this first edge, by Lemma 7, the terminal must be a neighbor of s; note that this is only possible if $P \in \{C_{\text{cw}}, C_{\text{ccw}}\}$. Because C_{cw} shares only two edges with Z_{cw} (and only one with Z_{ccw}), it takes $n-3$ flips to transform Z_{cw} into C_{cw}. Afterward, it takes at least $n-2$ flips to flip C_{cw} to Z_{ccw} because they share only one edge.

By the above constructions, it is clear that the spirals are centers. As they lie on any shortest paths from Z_{cw} to Z_{ccw}, they are the unique centers. ■

4 The Subgraph of suffix-independent Paths is Connected

In this section, we investigate the subclass of suffix-independent paths. We show that the flip graph on $\mathcal{P}_{\text{sin}}(S, s)$ is connected, i.e., Theorem 4. In order to prove this theorem, the following property is crucial. We say that a path is *strongly suffix-independent* if both of its orientations are suffix-independent.

Lemma 11. *For any two outer points $s, t \in S$, there exists an st-path P that is strongly suffix-independent.*

In fact, Theorem 4 and Lemma 11 also imply that the flip graph induced by the union of all suffix-independent paths is connected.

Corollary 12. *For every point set S in general position, the flip graph of the suffix-independent paths $\bigcup_{s \in S} \mathcal{P}_{\text{sin}}(S, s)$ is connected.*

Proof. Let P_1 and P_2 be two suffix-independent paths starting in s_1 and s_2, respectively. By definition, s_1 and s_2 are outer. If $s_1 = s_2$, Theorem 4 implies that there exists a flip sequence from P_1 to P_2. So, we consider the case that $s_1 \neq s_2$. Due to Lemma 11, there exists a strongly suffix-independent path P from s_1 to s_2, i.e., P and $\overleftarrow{P}$ are suffix-independent. So Theorem 4 ensures that P_1 can be flipped to P and the reversed path $\overleftarrow{P}$ can be flipped to P_2. ■

We now present a constructive proof of Lemma 11.

Proof (of Lemma 11). Without loss of generality, assume that s lies above t and that t is rightmost. Consider a vertical line $\mathcal{L}$ through s that splits S into two sets, the set S_1 of points contained in the left (closed) half-space of $\mathcal{L}$ and its complement $S_2 = S \setminus S_1$. We define P by two subpaths P_1 and P_2, for an illustration see Fig. 8.

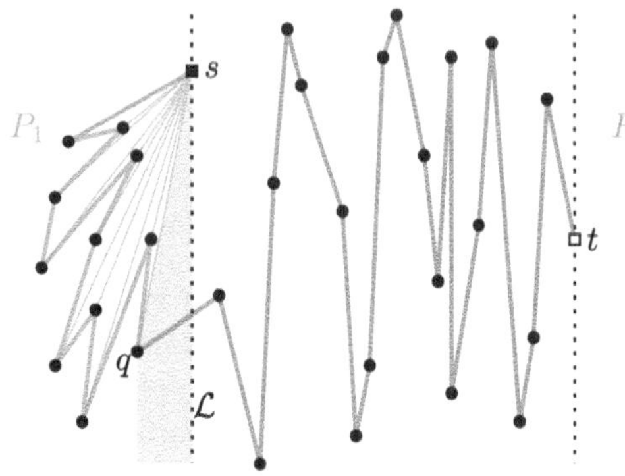

Fig. 8. Illustration for the proof of Lemma 11.

Starting from s, P_1 visits all points in S_1 in the ccw order that s sees them. In particular, if $|S_1| > 1$, we ensure that the first edge of P_1 is a convex hull edge of S; if $|S_1| = 1$, P_1 is the trivial path starting and ending in s.

Let q denote the endpoint of P_1. Observe that the cone spanned by $\mathcal{L}$ and sq contains no points; otherwise such a point would be in S_1 and visited after q. The path P_2 visits all points in $S_2 \cup \{q\}$. It starts in q and then collects the points by increasing x-coordinate. As t is rightmost, P_2 ends in t. Clearly, the concatenation of P_1 and P_2 is an st-path.

It remains to prove that P and $\overleftarrow{P}$ are suffix-independent. By construction, for each $p \in S_1$, the line through p and s separates $P[s,p]$ from $P[p,t]$. For each $p \in S_2$, the vertical line through p separates $P[s,p]$ from $P[p,t]$. Hence, each $p \in S$ lies on the boundary of the convex hull of its reversed prefix and suffix, and the specified lines separate their convex hulls. Consequently, for all $p \in S$, the reversed prefix and the suffix starting in s is independent. ■

Using Lemma 11 as an essential tool, we now prove Theorem 4.

Theorem 4. *For every point set S in general position and every point $s \in S$, the flip graph of $\mathcal{P}_{\text{sin}}(S,s)$ is connected.*

Proof. We prove this by induction on the cardinality of S. In the base case, for $|S| = 2$, there exists exactly one path from s to the other point of S. Thus, the flip graph is trivially connected.

For the induction step, consider a point set S with $|S| > 2$ and two arbitrary suffix-independent paths $P_1, P_2 \in \mathcal{P}_{\text{sin}}(S,s)$. Let s_i be the successor of s on P_i, and let P_i' be the suffix of P_i starting at s_i. We distinguish two cases.

If $s_1 = s_2$, then by induction, P'_1 can be transformed into P'_2. This directly yields a flip sequence from P_1 to P_2.

If $s_1 \neq s_2$, we consider a path H from s_1 to s_2 on $S \setminus \{s\}$ such that each prefix and suffix is independent; the existence of H is guaranteed by Lemma 11. By induction, P'_1 can be transformed into H, see Figs. 9(a) and 9(b). We may replace the edge ss_1 with ss_2 in the resulting path, see Figs. 9(b) and 9(c). This reverses H; however, as any prefix of H is independent, any suffix of $\overleftarrow{H}$ is independent as well. Thus, by induction, $\overleftarrow{H}$ can be transformed into P'_2, see Figs. 9(c) and 9(d). We conclude that the flip graph of $\mathcal{P}_{\text{sin}}(S, s)$ is connected. ■

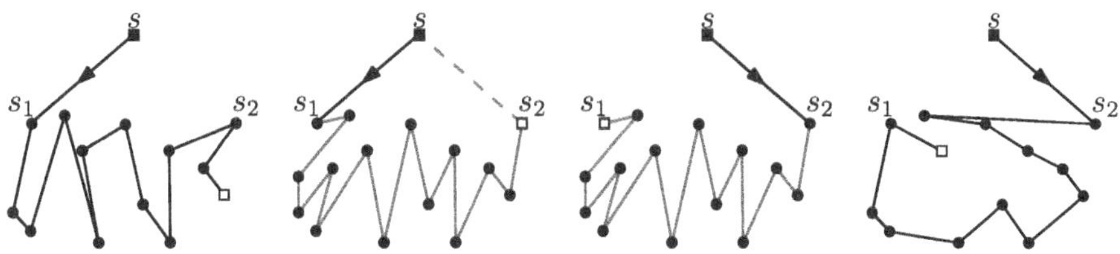

(a) P_1 (b) (c) (d) P_2

Fig. 9. Illustration for the proof of Theorem 4. By induction, we can flip P_1 and P_2 to a path that has H or $\overleftarrow{H}$ as a suffix; indicated by the cyan subpath.

5 General Tools and Point Sets on Two Convex Layers

In this section, we consider point sets with two convex layers. We start by investigating the setting with a fixed outer start point.

Theorem 5. *theorem Let S be a point set in general position with $L(S) \leq 2$, and $s \in L_0$. Then the flip graph of $\mathcal{P}(S, s)$ is connected.*

The idea for our proof of Theorem 5 is based on the insight that if each path can be flipped to some suffix-independent path, then Theorem 4 implies the connectivity of the respective flip graph. In a first step, we aim to obtain some path $P \in \mathcal{P}(S, s)$ where the (outgoing) edge of s is a *level* edge, i.e., it lies on the boundary of the convex hull. If the first edge of P is a level edge sq, then the suffix $P[q, t]$ is independent, and we may consider this suffix instead (or assume by induction that we can flip to a suffix-independent path).

To obtain a path that starts with a level edge, we prove that we can increase ℓ_0 for any given path, unless $\ell_0 = |L_0| - 1$; recall that $\ell_0(P)$ denotes the number of L_0-level edges in P. It follows that we eventually introduce a level edge that is outgoing from s. The crucial part lies in considering paths with two special properties. In particular, a path $P \in \mathcal{P}(S, s)$ with $s \in L_0$ belongs to $\mathcal{P}^*(S, s)$ if

(i) it has no chords, and
(ii) the outgoing edge of s goes to some $q \notin L_0$.

Furthermore, we say that the *k-layer property* holds for $k \in \mathbb{N}$ exactly if for every point set S in general position with $L(S) \leq k$, any st-path $P \in \mathcal{P}^*(S, s)$ with $\ell_0(P) < |L_0| - 1$ can be flipped to an st'-path $P' \in \mathcal{P}(S, s)$, such that (a) $\ell_0(P') > \ell_0(P)$ if $t \in L_0$, or (b) $\ell_0(P') = \ell_0(P)$ and $t' \in L_0$, otherwise.

Theorem 13. *If the k-layer property holds, then for every point set S in general position with $L(S) \leq k$ and $s \in L_0$, the flip graph of $\mathcal{P}(S, s)$ is connected.*

Proof. By Theorem 4, it suffices to show that each path can be flipped to some suffix-independent path. We prove this by induction on $|S|$. The induction base for $|S| = 1$ is trivial because $\mathcal{P}(S, s)$ consists of a single path.

For the induction step, consider an st-path $P \in \mathcal{P}(S, s)$ where $n := |S| > 1$ and assume that the statement holds for all point sets with less than n points.

First, we show that we can restrict our attention to chord-free paths. If the path P contains a chord, we consider the first chord uv of P. The independent suffix $P[u, t]$ visits a subset $S' \subset S$ of fewer than n points, so the induction hypothesis ensures that $P[u, t]$ can be flipped to any path in $\mathcal{P}(S', u)$. In particular, $P[u, t]$ (and by extension, P) can be flipped to a path P' ending in v (which exists by Lemma 11). For an illustration, consider Fig. 10. The path P' is chord-free by Lemma 10 and as it does not contain the edge uv (which would be a chord in S), adding the prefix $P[s, u]$ yields a chord-free path. We can therefore assume property (i) for the remainder of this proof.

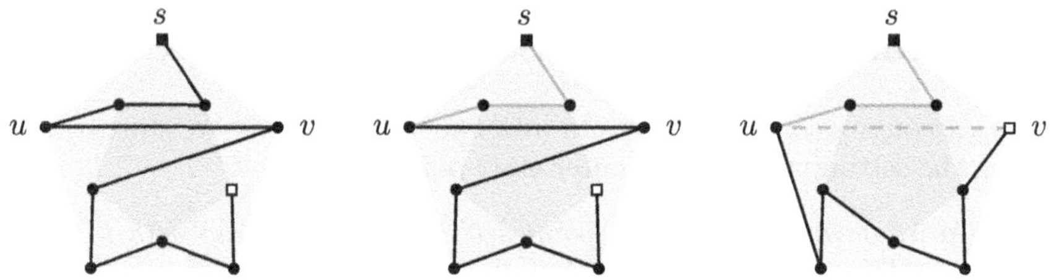

(a) The st-path P. (b) The suffix $P[u, t]$. (c) The path P'.

Fig. 10. Flipping to chord-free paths.

We now further establish property (ii). To this end, let s_1 denote a point such that ss_1 is the first edge of P. If $s_1 \in L_0$ (implying that ss_1 lies on the boundary of the convex hull), we may flip the independent suffix $P[s_1, t]$ to a suffix-independent path by the induction hypothesis. Consequently, the same applies to P. Thus, we may assume that $s_1 \notin L_0$, i.e., property (ii) holds.

It thus remains to show that each path P in $\mathcal{P}^*(S, s)$ can be flipped to a suffix-independent path. By repeatedly using the k-layer property, we flip P to a path where the first edge is a level edge in L_0. Then, as above, we may use induction to flip to a suffix-independent path. This concludes the proof. ∎

In order to prove Theorem 5, it remains to establish the 2-layer property. To this end, we introduce a crucial tool that exploits visibility between a suffix of a path and some preceding edge. The following statement allows us to argue the existence of flip sequences to paths that end at arbitrary points which, in the original path, directly precede an independent suffix, as illustrated in Fig. 11.

Lemma 14 (⋆). *Let S be a point set in general position and $P \in \mathcal{P}(S, s)$ a path with an independent suffix $X = P[a, t]$ spanning $A \subset S$, such that the flip graph of $\mathcal{P}(A, a)$ is connected. If there exists an edge $uv \in P \setminus X$ such that (i) uv is a boundary edge of* $\text{conv}(A \cup \{u, v\})$ *and (ii) the interior of* $\text{conv}(A \cup \{u, v\}) \setminus (\text{conv}(A) \cup \{uv\})$ *contains neither points nor parts of edges, then there exists a flip sequence from P to a path that ends at the predecessor of a on P. This flip sequence introduces only edges inside* $\text{conv}(A \cup \{u, v\})$.

We use Lemma 14 to prove the k-layer property for $k = 2$.

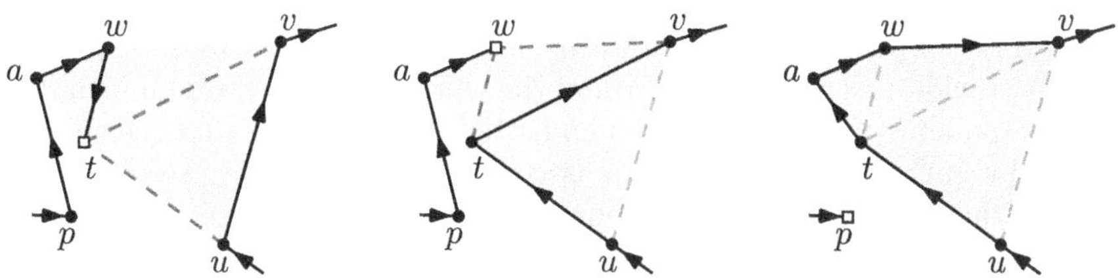

Fig. 11. Illustration for Lemma 14: We exploit the visibility between an independent suffix $P[a, t]$ and some preceding edge $uv \in P$ (some flips omitted).

Lemma 15 (⋆). *The k-layer property holds for $k \leq 2$.*

Proof (Sketch). Our approach consists of two phases. First, we show that an st_1-path with $s, t_1 \in L_0$ can be flipped to an st_2-path for some $t_2 \in L_1$ that contains an additional level edge in L_0. If applicable, we then flip to some st_3-path with $t_3 \in L_0$, without removing any level edges from L_0.

To this end, we argue three cases based on the visibility between t_2 and points with outgoing edges to L_0. Whenever such visibility is given, we show that such an st_3-path can be reached. Otherwise, we exploit Lemma 14 to iteratively remove obstructing edges, eventually obtaining the necessary visibility. ∎

Together, Theorem 13 and Lemma 15 imply Theorem 5.

Theorem 5. *Let S be a point set in general position with $L(S) \leq 2$, and $s \in L_0$. Then the flip graph of $\mathcal{P}(S, s)$ is connected.*

With similar arguments, we can deduce Corollary 6, confirming Conjecture 1 for point sets with layer number up to two. We provide a sketch of our approach.

Theorem 6 (⋆). *Let S be a point set in general position with $L(S) \leq 2$. Then the flip graph of $\mathcal{P}(S)$ is connected.*

Proof (Sketch). From Theorem 5 and Lemma 11, we can immediately conclude that the flip graph of all paths that have at least one endpoint on the convex hull, i.e., L_0, is connected. It remains to show that we can flip any st-path $P \in \mathcal{P}(S)$ with $s, t \in L_1$ to an st'-path with $t' \in L_0$.

If s and t can see one another, this is trivial. Otherwise, we interpret P as a directed path from $\mathcal{P}(S, s)$, and argue cases based on visibility between t and a point p with an outgoing edge to L_0. Whenever such visibility is given, we show that an st'-path with $t' \in L_0$ can be reached. Otherwise, we again exploit Lemma 14 to iteratively remove obstructing edges, eventually obtaining the necessary visibility. ■

6 Discussion

In this work, we present progress towards proving Conjectures 1 and 2. In particular, we provide first results concerning the connectivity of flip graphs induced by plane spanning paths with a fixed end. Moreover, we show that the subgraph induced by suffix-independent paths is connected. Lastly, we present tools to allow for the extension of known connectivity results, based on empty convex regions within spanning paths, and show how to conclude connectivity for point sets with up to two convex layers. Our insights indicate that investigating paths with a fixed end is a fruitful strategy.

We are optimistic that our tools prove to be useful to settle Conjectures 1 and 2. Specifically, we believe that Lemma 14 could enable an inductive approach to settling the conjectures, based on the layer number of point sets. Our application of Lemma 14 to extend results from convex position (Lemma 3) to two convex layers (Theorems 5 and 6) could be viewed as a first step.

References

1. Aichholzer, O., Aurenhammer, F., Hurtado, F.: Sequences of spanning trees and a fixed tree theorem. Comput. Geom. Theory Appl. **21**(1–2), 3–20 (2002). https://doi.org/10.1016/S0925-7721(01)00042-6
2. Aichholzer, O., et al.: Reconfiguration of non-crossing spanning trees (2022). https://doi.org/10.48550/arXiv.2206.03879
3. Aichholzer, O., Knorr, K., Mulzer, W., Obenaus, J., Paul, R., Vogtenhuber, B.: Flipping plane spanning paths. In: International Conference and Workshops on Algorithms and Computation (WALCOM), pp. 49–60 (2023). https://doi.org/10.1007/978-3-031-27051-2_5
4. Akl, S.G., Islam, M.K., Meijer, H.: On planar path transformation. Inf. Process. Lett. **104**(2), 59–64 (2007). https://doi.org/10.1016/j.ipl.2007.05.009
5. Avis, D., Fukuda, K.: Reverse search for enumeration. Discret. Appl. Math. **65**(1–3), 21–46 (1996). https://doi.org/10.1016/0166-218X(95)00026-N

6. Bjerkevik, H.B., Kleist, L., Ueckerdt, T., Vogtenhuber, B.: Non-crossing spanning trees. In: Proceedings of the Symposium on Discrete Algorithms (SODA 2025). https://doi.org/10.48550/arXiv.2410.23809
7. Bose, P., Hurtado, F.: Flips in planar graphs. Comput. Geom. Theory Appl. **42**(1), 60–80 (2009). https://doi.org/10.1016/j.comgeo.2008.04.001
8. Bousquet, N., Gledel, V., Narboni, J., Pierron, T.: A note on the flip distance between non-crossing spanning trees. Comput. Geom. Topol. **2**(1), 8:1–8:7 (2023). https://doi.org/10.57717/cgt.v2i1.36
9. Bousquet, N., de Meyer, L., Pierron, T., Wesolek, A.: Reconfiguration of plane trees in convex geometric graphs. In: International Symposium on Computational Geometry (SoCG), pp. 22:1–22:17 (2024). https://doi.org/10.4230/LIPIcs.SoCG.2024.22
10. Chang, J.M., Wu, R.Y.: On the diameter of geometric path graphs of points in convex position. Inf. Process. Lett. **109**(8), 409–413 (2009). https://doi.org/10.1016/j.ipl.2008.12.017
11. Eppstein, D.: Happy endings for flip graphs. J. Comput. Geom. **1**(1), 3–28 (2010). https://doi.org/10.20382/JOCG.V1I1A2
12. Felsner, S., Kleist, L., Mütze, T., Sering, L.: Rainbow cycles in flip graphs. SIAM J. Discret. Math. **34**(1), 1–39 (2020). https://doi.org/10.1137/18M1216456
13. Hernando, M.C., Houle, M.E., Hurtado, F.: On local transformation of polygons with visibility properties. Theor. Comput. Sci. **289**(2), 919–937 (2002). https://doi.org/10.1016/S0304-3975(01)00409-1
14. Hernando, M.C., Hurtado, F., Márquez, A., Mora, M., Noy, M.: Geometric tree graphs of points in convex position. Discret. Appl. Math. **93**(1), 51–66 (1999). https://doi.org/10.1016/S0166-218X(99)00006-2
15. Hernando, M.C., Hurtado, F., Noy, M.: Graphs of non-crossing perfect matchings. Graphs Comb. **18**(3), 517–532 (2002). https://doi.org/10.1007/S003730200038
16. Houle, M.E., Hurtado, F., Noy, M., Rivera-Campo, E.: Graphs of triangulations and perfect matchings. Graphs Comb. **21**(3), 325–331 (2005). https://doi.org/10.1007/S00373-005-0615-2
17. Hurtado, F., Noy, M., Urrutia, J.: Flipping edges in triangulations. Discret. Comput. Geom. **22**(3), 333–346 (1999). https://doi.org/10.1007/PL00009464
18. Kleist, L., Kramer, P., Rieck, C.: On the connectivity of the flip graph of plane spanning paths (2024). https://doi.org/10.48550/arXiv.2407.03912
19. Lawson, C.L.: Transforming triangulations. Discret. Math. **3**(4), 365–372 (1972). https://doi.org/10.1016/0012-365X(72)90093-3
20. Lubiw, A., Pathak, V.: Flip distance between two triangulations of a point set is NP-complete. Comput. Geom. Theory Appl. **49**, 17–23 (2015). https://doi.org/10.1016/J.COMGEO.2014.11.001
21. Milich, M., Mütze, T., Pergel, M.: On flips in planar matchings. Discret. Appl. Math. **289**, 427–445 (2021). https://doi.org/10.1016/J.DAM.2020.10.018
22. Monroy, R.F., Flores-Peñaloza, D., Huemer, C., Hurtado, F., Wood, D.R., Urrutia, J.: On the chromatic number of some flip graphs. Discret. Math. Theor. Comput. Sci. **11**(2), 47–56 (2009). https://doi.org/10.46298/DMTCS.460
23. Nichols, T.L., Pilz, A., Tóth, C.D., Zehmakan, A.N.: Transition operations over plane trees. Discret. Math. **343**(8), 111929 (2020). https://doi.org/10.1016/J.DISC.2020.111929
24. Nishimura, N.: Introduction to reconfiguration. Algorithms **11**(4), 52 (2018). https://doi.org/10.3390/A11040052

25. Rivera-Campo, E., Urrutia-Galicia, V.: Hamilton cycles in the path graph of a set of points in convex position. Comput. Geom. Theory Appl. **18**(2), 65–72 (2001). https://doi.org/10.1016/S0925-7721(00)00026-2
26. Wagner, U., Welzl, E.: Connectivity of triangulation flip graphs in the plane. Discret. Comput. Geom. **68**(4), 1227–1284 (2022). https://doi.org/10.1007/S00454-022-00436-2

Graph Reconstruction with Connectivity Queries

Kacper Kluk[1], Hoang La[2(✉)], and Marta Piecyk[3]

[1] Faculty of Mathematics, Informatics and Mechanics, University of Warsaw, Warsaw, Poland
kacper.kluk@mimuw.edu.pl

[2] LISN, Université Paris-Saclay, CNRS, Gif-sur-Yvette, Orsay, France
hoang.la.research@gmail.com

[3] Faculty of Mathematics and Information Science, Warsaw University of Technology, Warsaw, Poland
marta.piecyk.dokt@pw.edu.pl

Abstract. We study a problem of reconstruction of connected graphs where the input gives all subsets of size k that induce a connected subgraph. Originally introduced by Bastide *et al.* (WG 2023) for triples ($k = 3$), this problem received comprehensive attention in their work, alongside a study by Qi, who provided a complete characterization of graphs uniquely reconstructible via their connected triples, i.e. no other graphs share the same set of connected triples. Our contribution consists in output-polynomial time algorithms that enumerate every triangle-free graph (resp. every graph with bounded maximum degree) that is consistent with a specified set of connected k-sets. Notably, we prove that sufficiently large triangle-free graphs are uniquely reconstructible, while graphs with bounded maximum degree that are consistent with the same k-sets share a substantial common structure, differing only locally. We suspect that the problem is NP-hard in general and provide a NP-hardness proof for a variant where the connectivity is specified for only some k-sets (with k at least 4).

Keywords: Graph reconstruction · Triangle-free graphs · Bounded maximum degree

1 Introduction

Graph reconstruction primarily refers to the concept popularized by the Graph Reconstruction Conjecture by Kelly and Ulam [3,5]. This conjecture states that any graph with a minimum of three vertices can be uniquely reconstructed, up to isomorphism, using the multiset of subgraphs resulting from the removal of exactly one vertex. However, in this context, we delve into the reconstruction

M. Piecyk—MP was funded by Polish National Science Centre, grant no. 2022/45/N/ST6/00237.

D. Král' and M. Milanič (Eds.): WG 2024, LNCS 14760, pp. 343–357, 2025.
https://doi.org/10.1007/978-3-031-75409-8_24

of labeled graphs, where we lack explicit knowledge about the structures of subgraphs of order $n-1$, as proposed in the Graph Reconstruction Conjecture. Instead, our focus lies on reconstructing graphs based solely on the connectivity of their subgraphs. Specifically, we address the task of reconstructing graphs by querying if an induced subgraph of order k is connected, for some fixed integer $k \geq 2$. This variant of graph reconstruction was recently introduced by Bastide, Cook, Erickson, Groenland, Kreveld, Mannens, and Vermeulen [1], initially for triples, and more broadly for induced subgraphs on sets of vertices of size k, which we term k-sets for brevity.

Some problems lend themselves to straightforward solutions when access to answers for all possible queries is available, such as graph reconstruction using a distance oracle [2]. In such scenarios, our focus shifts towards optimizing the number of queries utilized. In graph reconstruction via connectivity queries on k-sets, it is evident that for $k = 2$, we are effectively querying the edges of the graphs. However, for $k \geq 3$, even identifying a graph consistent with the answers to all queries becomes a nontrivial task.

Graph reconstruction inherently suggests the existence of a target graph we aim to rebuild. Yet, within this context, we may encounter arbitrary connected and disconnected k-sets. The questions regarding the existence and uniqueness of graphs satisfying these queries become significantly more intriguing. This problem bears potential applications in computational networks where concealing its exact structure is imperative while still being able to respond to local queries about potential data transfers to a sub-network, which would require connectivity, is necessary.

Vocabularies and Notations. We recall some standard notations and introduce some vocabularies that will be useful for our graph reconstruction problem.

For a positive integer n, by $[n]$ we denote the set $\{1, \ldots, n\}$. We use standard graph notations, i.e., a graph is a pair $G = (V, E)$, where V is the set of vertices and $E \subseteq \binom{V}{2}$ is the set of edges of G. By $\deg_G(v)$ we denote the degree of v in G and by $N_G(v)$ we denote the set of neighbors of v in G. We denote the closed neighborhood of v by $N_G[v] = N_G(v) \cup \{v\}$. Similarly, for $S \subseteq V$, the closed neighborhood of S is $N_G[S] = \bigcup_{v \in S} N_G[v]$ and the open neighborhood of S is $N_G(S) = N_G[S] - S$. If G is clear from the context, we omit the subscript G. For a set of vertices S, by $G[S]$, we denote the subgraph of G induced by S.

Connected k-sets: Let $G = (V, E)$ be a graph and let k be an integer with $k \geq 2$. Let $\mathcal{S}_k(G)$ and $\overline{\mathcal{S}}_k(G)$ be the set of connected and disconnected k-sets of G respectively, i.e. $G[S]$ is connected for every $S \in \mathcal{S}_k(G)$ and $G[S']$ is disconnected for every $S' \in \overline{\mathcal{S}}_k(G)$.

Consistent Graphs: We say that *a graph G is consistent with $\mathcal{S}_k$ and $\overline{\mathcal{S}}_k$* when $\mathcal{S}_k \subseteq \mathcal{S}_k(G)$ and $\overline{\mathcal{S}}_k \subseteq \overline{\mathcal{S}}_k(G)$.

Uniquely Reconstructible Graphs: We say that a graph G is *uniquely k-reconstructible in the class of graphs $\mathcal{G}$* if it is the only graph in $\mathcal{G}$ consistent with $\mathcal{S}_k(G)$ and $\overline{\mathcal{S}}_k(G)$. We omit $\mathcal{G}$ when we consider all graphs.

Hardness of the General Case. For the case where $k = 3$, Bastide et al. showed an enumeration algorithm of graphs consistent with some given connected 3-sets applicable to general graphs, leveraging a reduction of the problem to 2-SAT. Moreover, Qi [4] provided a characterization of graphs that are uniquely 3-reconstructible from connected triples. However, extending this reduction to $k \geq 4$ merely leads to the problem being reduced to SAT. Moreover, we suspect that for $k \geq 4$, the task of identifying a graph consistent with a given set of connected k-sets may be NP-hard in general. More formally, we define the following problem.

k-RECONSTRUCTION *Input:* Set V, disjoint sets $\mathcal{S}_k, \overline{\mathcal{S}}_k \subseteq \binom{V}{k}$ such that $\mathcal{S}_k \cup \overline{\mathcal{S}}_k = \binom{V}{k}$. *Question:* Does there exists a graph G on V that is consistent with $\mathcal{S}_k$ and $\overline{\mathcal{S}}_k$?

Conjecture 1. If $k \geq 4$, then the k-RECONSTRUCTION problem is NP-hard.

To substantiate this suspicion, we show NP-hardness for a modified version of this problem, wherein not every connected and disconnected k-set is given.

k-RECONSTRUCTION WITH PARTIAL INFORMATION *Input:* Set V, disjoint sets $\mathcal{S}_k, \overline{\mathcal{S}}_k \subseteq \binom{V}{k}$. *Question:* Does there exists a graph G on V that is consistent with $\mathcal{S}_k$ and $\overline{\mathcal{S}}_k$?

Theorem 1. *If $k \geq 4$, then the k-*RECONSTRUCTION WITH PARTIAL INFORMATION *problem is NP-hard.*

We point out that such a version of the problem is still tractable for $k = 3$, as the reduction to 2-SAT can still handle the case where incomplete information is given.

Obstructions. The graphs that we aim to rebuild are all labeled, finite, and simple. It is essential to note that this form of reconstruction does not always yield a unique labeled graph: a path on vertices a, b, c, d has the same connected triples as the isomorphic path on a, c, b, d. More generally, we do not always obtain isomorphic graphs. For example, a clique and a clique lacking a matching have identical connected k-sets. Moreover, from the connected k-sets, we are able to determine the vertex sets of all connected components of size at least k, but we cannot distinguish connected components of size at most $k - 1$. Therefore, we will always assume that our graphs are connected and reasonably large.

Here, we direct your attention to two specific examples illustrating the challenge of reconstructing graphs solely through connectivity. The first example revolves around the existence of edges between twin vertices. Twin vertices share precisely the same neighbors, excluding themselves. Consequently, when examining their connected k-sets, we cannot discern between a graph where they share an edge and one where they do not.

The second example is between a star and more generally a graph with a universal vertex. Consider a graph G with a universal vertex v such that every connected component of $G - v$ has size at most $k - 1$. Then G and a star on $|V(G)|$ vertices have the same connected k-sets.

Triangle-free Graphs. In both of the previous examples, with the additional assumption of triangle-freeness, we are left with only one potential outcome. Indeed, not only do we prove that enumerating connected triangle-free graphs that are consistent with some given connected k-sets can be done efficiently, but we also show that when such a graph exists, there are no alternative connected triangle-free graphs sharing the same k-sets. More formally, we prove the following statements.

Theorem 2. *Let k be an integer with $k \geq 2$. Let V be a set of vertices. Let $\mathcal{S}_k$ and $\overline{\mathcal{S}}_k$ be disjoint sets such that $\mathcal{S}_k \cup \overline{\mathcal{S}}_k = \binom{V}{k}$. We can enumerate every connected triangle-free graph G on V that is consistent with $\mathcal{S}_k$ and $\overline{\mathcal{S}}_k$ in polynomial time in $|V|$.*

Theorem 3. *Let k be an integer with $k \geq 2$. If G is a large enough triangle-free connected graph, then G is uniquely k-reconstructible in the class of triangle-free connected graphs.*

Graphs with Bounded Maximum Degree. In the second example, the presence of the universal vertex, ensuring the graph's connectivity, opens the door to a multitude of different structures that can be concealed within its neighborhood, which is (almost) the entire graph. Intuitively, bounding the maximum degree of the graph restricts the size of such neighborhood and thus the variability to only sets of constant size, at most the maximum degree. While this constraint does not guarantee uniqueness – in fact, quite the opposite; we may still have an exponential number of possible subquartic graphs consistent with the same connected k-sets: see Fig. 1.

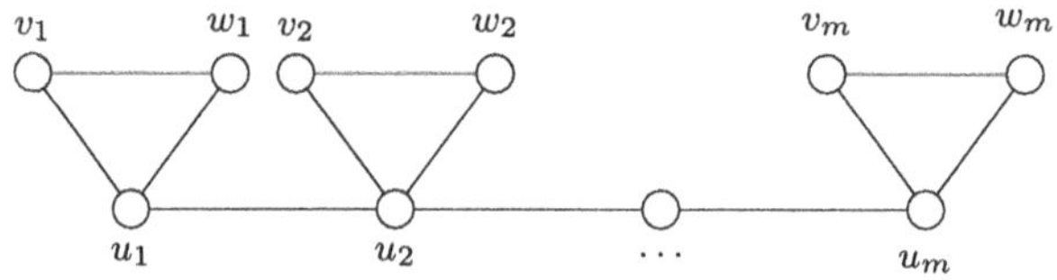

Fig. 1. The presence (or lack thereof) of a red edge does not change the connected k-sets. Therefore, there are at least 2^m connected graphs that have the same connected k-sets as the drawn graph. (Color figure online)

However, we anticipate these graphs to exhibit a substantial common structure (the path $u_1 \ldots u_m$ and the edges u_iv_i, u_iw_i) and differing only locally (the potential edges v_iw_i). To formalize these intuitions, we introduce the notion of partial graphs and skeletons.

Definition 1. *A* partial graph *is a tuple* $H = (V, E, E_N, E_U)$, *where* V *is the vertex set and* E, E_N, *and* E_U *are disjoint sets of pairs of vertices in* V *that form a partition of* $\binom{V}{2}$. *The sets* E, E_N, E_U *denote respectively (known) edges, (known) non-edges and unknown edges.*

For example, a graph $G = (V, E)$ can be seen as a partial graph $(V, E, \binom{V}{2} \setminus E, \emptyset)$. For a partial graph $H = (V, E, E_N, E_U)$, we say that a partial graph H' on V is a *partial supergraph* H' of H if $E \subseteq E(H')$, $E_N \subseteq E_N(H')$, and $E_U(H') \subseteq E_U$. We say that a graph G is a *supergraph* of H if $E \subseteq E(G)$ and $E_N \subseteq \binom{V}{2} \setminus E$. In other words we can obtain G (resp. H') from H by deciding for every unknown edge (resp. for some subset of unknown edges) if it is an edge or a non-edge. We say that H *is consistent with* $\mathcal{S}_k$ *and* $\overline{\mathcal{S}}_k$ if there exists an assignment of unknown edges (E_U) of H to obtain a graph G consistent with $\mathcal{S}_k$ and $\overline{\mathcal{S}}_k$.

A skeleton is a partial graph where the unknown edges are contained in some disjoint small sets. The idea is that using a skeleton we can describe a set of graphs in which the differences between its members only occur in small sets of constant size which do not interact with one another. Formally, a skeleton is defined as follows.

Definition 2. *Let a* skeleton *be a partial graph* H *with a family of disjoint vertex subsets* $V_1, V_2, \ldots, V_m$ *and a family of collections* $\mathcal{C}_1, \mathcal{C}_2, \ldots, \mathcal{C}_m$ *such that* $E_U(H) \subseteq \bigcup_{i \in [m]} \binom{V_i}{2}$ *and* $\mathcal{C}_i$ *is a collection of graphs on* V_i *for every* $1 \leq i \leq m$. *The* width *of a skeleton is* $\max_{i \in [m]} |V_i|$. *A graph* G *is a* completion *of a skeleton if there exists graphs* $C_1 \in \mathcal{C}_1, \ldots, C_m \in \mathcal{C}_m$ *such that* G *is a supergraph of* H *and* $G[V_i] = C_i$ *for every* $1 \leq i \leq m$.

Going back to Fig. 1, we would like to prove that every graph that is consistent with the graph drawn in Fig. 1 must be a completion of the skeleton of width 2 (less than or equal to 4, the maximum degree of the graph) defined by the partial graph $H = (V, E, E_N, E_U)$ – where $V = \{u_1, \ldots, u_m, v_1, \ldots, v_m, w_1, \ldots, w_m\}$, $E = \{u_i u_{i+1} \mid i \in [m-1]\} \cup \{u_i v_i, u_i w_i \mid i \in [m]\}$, $E_U = \{v_i w_i \mid i \in [m]\}$, and $E_N = \binom{V}{2} \setminus (E \cup E_N)$ – the family of disjoint vertex subsets $V_1, \ldots, V_m$ where $V_i = \{v_i, w_i\}$ for every $i \in [m]$, and the family of collections $\mathcal{C}_1, \ldots, \mathcal{C}_m$ where for every $i \in [m]$, $\mathcal{C}_i$ is the set of two possible graphs on v_i and w_i (one is the two isolated vertices and the other contains an edge). Indeed, we confirm these intuitions and prove the following.

Theorem 4. *Let* k *and* d *be integers such that* $k \geq 2$ *and* $d \geq 1$. *Let* V *be a set of vertices. Let* $\mathcal{S}_k$ *and* $\overline{\mathcal{S}}_k$ *be disjoint sets such that* $\mathcal{S}_k \cup \overline{\mathcal{S}}_k = \binom{V}{2}$. *In polynomial time, we can find a constant-size (depending on* k *and* d*) family* $\mathcal{F}$ *of skeletons of width at most* d *such that a connected graph of maximum degree* d *is consistent with* $\mathcal{S}_k, \overline{\mathcal{S}}_k$ *if and only if it is a completion of a skeleton of* $\mathcal{F}$.

This implies that we can enumerate every connected graph of maximum degree d that is consistent with our given k-sets in a reasonable time. Moreover, the difference between these graphs are very local since any combination of graphs in $\mathcal{C}_1, \ldots, \mathcal{C}_m$ will produce a desired consistent graph.

Corollary 1. *Let k and d be integers such that $k \geq 2$ and $d \geq 1$. Let V be a set of vertices. Let $\mathcal{S}_k$ and $\overline{\mathcal{S}}_k$ be disjoint sets such that $\mathcal{S}_k \cup \overline{\mathcal{S}}_k = \binom{V}{2}$. We can enumerate every connected graph of maximum degree d that is consistent with $\mathcal{S}_k$ and $\overline{\mathcal{S}}_k$ in polynomial time in $|V|$ and the size of the output.*

Outline. Section 2 describes our enumeration algorithms for both triangle-free and bounded maximum degree graphs. Following this, Sect. 3 presents the structural result on uniquely reconstructible triangle-free graphs. Finally, Sect. 4 shows the NP-hardness of the variant of this problem with incomplete information.

The proofs of the statements marked with (*) are not included in this version of the paper. The full version of the paper can be found at https://arxiv.org/abs/2407.07500.

2 Enumerating Consistent Graphs

The main idea behind both of our enumeration algorithm relies on a breadth-first-search (BFS for short) partition of the vertices starting from some small initial set that we can guess by brute force and then determining the edges between the layers of the partition to obtain a partial graph that is consistent with $\mathcal{S}_k$ and $\overline{\mathcal{S}}_k$. Filling in the rest will depend on the target class of graphs: triangle-free or bounded maximum degree.

Lemma 1 is the inductive step where given a big enough partial graph, we can determine the next layer in our BFS.

Lemma 1 (*). *Let $\mathcal{S}_k, \overline{\mathcal{S}}_k$ be disjoint subsets such that $\mathcal{S}_k \cup \overline{\mathcal{S}}_k = \binom{V}{2}$. Let H be a partial graph on V consistent with $\mathcal{S}_k$ and $\overline{\mathcal{S}}_k$. Let T be a subset of V such that:*

1. *$H[T]$ is connected,*
2. *$|T| \geq k - 1$,*
3. *and $N_H(T)$ is non-empty.*

We can find H' a partial supergraph of H using at most $|V| \cdot |N_H(T)|$ queries such that:

1. *for every $x \in V - N_H[T]$ and $y \in N_H(T)$, $xy \notin E_U(H')$,*
2. *and for every supergraph G of H such that $N_G(T) = N_H(T)$ and G is consistent with $\mathcal{S}_k$ and $\overline{\mathcal{S}}_k$, G is a supergraph of H'.*

Using Lemma 1, we can now determine the BFS layering of the graph, i.e. all of the edges and non-edges between the layers.

Lemma 2 (*). *Let $\mathcal{S}_k, \overline{\mathcal{S}}_k, H, T$ be defined as in Lemma 1. We can find H' a partial supergraph of H together with the partition $T = L_0, L_1, \ldots, L_\ell$ of V using at most $|V|^2$ queries such that:*

1. *for every* $0 \leq i < j \leq \ell$ *except* $i = 0, j = 1$*, and every* $x \in L_i, y \in L_j$*,* $xy \notin E_U(H')$*,*
2. L_i *are the vertices at distance exactly* i *from* T *in* H'*,*
3. *and for every supergraph* G *of* H *such that* $N_G(T) = N_H(T)$ *and* G *is consistent with* $\mathcal{S}_k, \overline{\mathcal{S}}_k$*,* G *is a supergraph of* H'*.*

We also refer to $(L_i)_{i=0}^{\ell}$ *as a layering.*

2.1 Triangle-Free Graphs

The following lemma fills in the gap that is the unknown edges inside a layer and shows an unique reconstruction of a triangle-free graph given that a starting set and its neighborhood is known.

Lemma 3 (*). *Let* $\mathcal{S}_k, \overline{\mathcal{S}}_k, H, T$ *be defined as in Lemma 1. Let all edges of* $H[N_H[T]]$ *be known. There exists at most one triangle-free supergraph* G *of* H *consistent with* $\mathcal{S}_k, \overline{\mathcal{S}}_k$ *such that* $N_G(T) = N_H(T)$*. Moreover, we can find this graph or determine that it does not exist in polynomial time.*

The only thing missing is the base case, i.e. a starting set of vertices and its neighborhood for which we can deduce the induced subgraph.

For triangle-free graphs, we will choose this set to be the closed neighborhood of a vertex u with large enough degree (when it exists) since what follows show that we can easily determine its neighborhood and vertices at distance 2 just by knowing a small part of $N(u)$.

Lemma 4 (*). *Fix* $k \geq 2$*. Let* $\mathcal{S}_k, \overline{\mathcal{S}}_k$ *be disjoint sets such that* $\mathcal{S}_k \cup \overline{\mathcal{S}}_k = \binom{V}{2}$*. Given vertices* $u, x_1, x_2, \ldots, x_{2k-4}$*, we can find subsets* $X, Y \subseteq V$ *and a partial graph* H *on* V *such that:*

1. $X = N_H(u)$ *and* $Y = N_H(\{u\} \cup X)$*,*
2. *for every* $s \in \{u\} \cup X \cup Y$ *and* $t \in V$*,* $st \notin E_U(H)$*,*
3. *and for every triangle-free graph* G *consistent with* $\mathcal{S}_k, \overline{\mathcal{S}}_k$ *such that* $x_1, \ldots, x_{2k-4} \in N_G(u)$*,* G *is a supergraph of* H*.*

We combine this base case with our previous lemmas to show that indeed, given a big vertex, we can determine an unique triangle-free graph that is consistent with our connected k-sets.

Lemma 5 (*). *Fix* $k \geq 2$*. Let* $\mathcal{S}_k, \overline{\mathcal{S}}_k$ *be disjoint sets such that* $\mathcal{S}_k \cup \overline{\mathcal{S}}_k = \binom{V}{2}$*. Given vertices* $u, x_1, x_2, \ldots, x_{2k-4}$*, there exists at most one triangle-free graph* G *consistent with* $\mathcal{S}_k, \overline{\mathcal{S}}_k$ *such that* $\{x_1, \ldots, x_{2k-4}\} \in N_G(u)$*. Moreover, we can find this graph or determine that it does not exist in polynomial time.*

To prove Theorem 2, we proceed as follows. For the base case, on the one hand, we guess a big vertex u with its closed neighborhood as our starting set and determine its neighborhood thanks to Lemma 4, on the other hand, we assume no big vertices exists and therefore it suffices to brute force a connected

k-set and its neighborhood as they have constant size. We determine the BFS layering thanks to Lemma 2. Finally, we fill in the layers with Lemma 3.

The only reason why our algorithm does not yield an unique reconstruction is due to the initial guessing in the base. However, as we will see in Sect. 3.1, for large enough connected triangle-free graphs, only one such graph can exist.

2.2 Graphs with Bounded Maximum Degree

Before proving Theorem 4, the main result of this section, we introduce a notion of importance for unknown edges of a k-set S in the current partial graph H which are edges whose presence can influence the connectivity of every supergraph of $H[S]$.

Definition 3 (Importance of an unknown edge). *Let S be a k-set on V, let H be a partial graph on V, and let $u, v \in S$ such that $uv \in E_U(H)$. We say that uv* is not important for S in H *if for every supergraph G_S of $H[S]$, $G_S + uv$ is connected if and only if $G_S - uv$ is connected. Otherwise, we say that uv* is important for S in H.

The fact that an unknown edge is not important for some k-set S in H gives us some specific structures in $H[S]$.

Proposition 1. *Let S be a k-set, let K_S be the complete graph on vertices of S, let H be a partial graph on the vertices of S, and let $u, v \in S$ such that $uv \in E_U(H)$. Then, (a) uv is not important for S in H if and only if (b) there exists a path on vertices of S and edges of $E(H)$ between u and v or (c) $K_S - E_N(H)$ is disconnected.*

Proof. First, we prove that (c) implies (a). Suppose that $K_S - E_N(H)$ is disconnected. Every supergraph G_S of $H[S]$ must be disconnected. Therefore, $G_S + uv$ and $G_S - uv$ are also disconnected. Hence, uv is not important for S in H.

Now, we prove that (b) implies (a). If there exists a path in S between u and v using known edges, i.e. edges of $E(H)$, then every supergraph G_S of $H[S]$ will contain such path. Removing or adding an edge between two vertices that are connected by another path cannot change the connectivity of a graph. Therefore, $G_S + uv$ and $G_S - uv$ must have the same connectivity. Hence, uv is not important for S in H.

Finally, we show that if (b) and (c) do not hold, then (a) also cannot hold, i.e. uv is an important edge for S in H. To show that uv is important, we will build a supergraph G_S of $H[S]$ such that $G_S + uv$ is connected and $G_S - uv$ is disconnected. First, start with $G'_S = K_S - E_N(H)$ which is a connected supergraph of $H[S]$ by assumption. Since there are no paths between u and v in S using only known edges of H, we must have a path in S between u and v using both known and unknown edges, i.e. edges of $E(H)$ and $E_U(H)$. Consider such a path P and an edge e_P on P that is unknown in H. Recall that $E(G'_S) = E(H[S]) \cup E_U(H[S])$. We can thus remove e_P from G'_S. If G'_S is still connected, then we repeat this process by choosing another path between

u and v in G'_S and removing another edge of $E_U(H[S])$ from this path until G'_S is disconnected. This process must end because at worst we remove every path between u and v in G'_S. We call G_S the resulting graph. By construction, $G_S - uv = G_S$ is disconnected. Let us show that $G_S + uv$ is connected. Let wx be the last edge removed by this process and consider two vertices y and z in G_S with no paths between them. Since $G_S + wx$ is connected, there exists a path P_{wx} from y to z going through wx. By construction, wx exists on a path in $G_S + wx$ between u and v. Therefore, there exists a path P_{uv} between w and x going through uv in $G_S + uv$. Thus, we have a path between y and z which consists of P_{wx} where wx is replaced by P_{uv} in $G_S + uv$. This shows that $G_S + uv$ is connected and concludes our proof. □

We are ready to prove Theorem 4.

Proof (Theorem 4). Let $\mathcal{S}_k, \overline{\mathcal{S}}_k$ be disjoint sets such that $\mathcal{S}_k \cup \overline{\mathcal{S}}_k = \binom{V}{2}$ and let $n = |V|$. Here, we provide a sketch of an algorithm that produces the desired skeletons of width d in time polynomial in n.

1. Let $T \in \mathcal{S}_k$. Determine the neighborhood of T. Enumerate all possible connected graphs with maximum degree d on vertices of the closed neighborhood of T to obtain a family of partial graphs. Proceed with the next steps for each partial graph.
2. For each partial graph H, obtain a layering $(L_i)_{i=0}^{\ell}$ with Lemma 2 from $\mathcal{S}_k$, $\overline{\mathcal{S}}_k$, H, and T. For every $2 \leq i \leq \ell$, let $\mathcal{V}_i$ be the family of maximal sets of vertices in layer L_i with the same neighborhood in the previous layer L_{i-1}. Let $\{V_1, \ldots, V_m\} = \bigcup_{i=2}^{\ell} \mathcal{V}_i$ be all such sets for all layers. By definition, these sets have size at most the maximum degree of G which is d. Now, determine every unknown edge that is not contained in a V_j $(1 \leq j \leq m)$.
3. For every $1 \leq j \leq m$, we determine every unknown edge uv in V_j such that u (resp. v) is contained in a connected subgraph G' (of the current partial graph) of order $k-1$, $v \notin V(G')$ (resp. $u \notin V(G')$), and v (resp. u) is not adjacent to any vertex in $G' - u$ (resp. $G' - v$), i.e. $vw \in E_N$ (resp. uw) for every $w \in V(G' - u)$ (resp. $w \in V(G' - v)$). Let the resulting partial graph be H^*.
4. For each $1 \leq j \leq m$, we enumerate every possible supergraph of $H^*[V_j]$ and keep the ones that is consistent with $\mathcal{S}_k$ and $\overline{\mathcal{S}}_k$ to obtain the family $\mathcal{C}_j$.

The key ingredient to proving the correctness of this algorithm and the reason why we can define separate sets of vertices and collections of graphs that are independent from each other is the following claim.

Claim (*). For every k-set S, there exists $1 \leq j \leq m$, such that V_j contains every important unknown edge of S in H^*.

□

3 Uniqueness

In this section, we consider a graph G with k-sets $\mathcal{S}_k(G)$ and $\overline{\mathcal{S}}_k(G)$ and we study the structure of the family of graphs consistent with $\mathcal{S}_k(G)$ and $\overline{\mathcal{S}}_k(G)$ to see when G is uniquely k-reconstructible.

Proposition 2. *Let $G = (V, E)$ be a graph and let $uv \in E$. For every $S \in \mathcal{S}_k(G)$ such that $v \in S, u \notin S$, there exists $v' \in S \setminus \{v\}$ such that $(S \setminus \{v'\}) \cup \{u\} \in \mathcal{S}_k(G)$.*

Proof. Since $G[S]$ is connected, there exists a spanning tree of $G[S]$. Since a tree always has at least two leaves, consider $v' \neq v$ to be a leaf of such a spanning tree. Observe that $G[S \setminus \{v'\}]$ also contains a spanning tree as a subgraph. Therefore, it is connected. Combining it with the fact that $uv \in E$, we conclude that $G[(S \setminus \{v'\}) \cup \{u\}]$ is connected. In other words, $(S \setminus \{v'\}) \cup \{u\} \in \mathcal{S}_k(G)$. □

This motivates the following definition.

Definition 4. *Let $G = (V, E)$ be a graph and let $uv \notin E$. We say that u and v are* clear non-neighbors *if (i) there exists $S \in \mathcal{S}_k(G)$ such that $v \in S, u \notin S$ and for every $v' \in S \setminus \{v\}$, it holds that $(S \setminus \{v'\}) \cup \{u\} \in \overline{\mathcal{S}}_k(G)$, or (ii) there exists $S \in \mathcal{S}_k(G)$ such that $u \in S, v \notin S$ and for every $u' \in S \setminus \{u\}$, it holds that $(S \setminus \{u'\}) \cup \{v\} \in \overline{\mathcal{S}}_k(G)$. Otherwise, we say that they are* fake neighbors.

Proposition 2 implies that non-edges between clear non-neighbors are present in every graph consistent with the same k-sets.

Corollary 2. *Let $G = (V, E)$ be a graph and let $uv \notin E$. If u and v are clear non-neighbors, then for every graph G' consistent with $\mathcal{S}_k(G)$ and $\overline{\mathcal{S}}_k(G)$, we have $uv \notin E(G')$.*

What follows is a sufficient condition for two vertices to be clear non-neighbors.

Proposition 3. *Let $G = (V, E)$ be a graph and let $u \in V$. If there exists a subgraph G_u such that u is isolated in G_u, $G_u - u$ is connected, and $|V(G_u - u)| \geq k$, then for every $v \in V(G_u)$, u and v are clear non-neighbors.*

Proof. Consider $v \in V(G_u)$, a connected subgraph of G_u of size k containing v, and let S be its set of vertices. We have $S \in \mathcal{S}_k(G)$, $v \in S$ and $u \notin S$. Moreover, for every vertex u' in $S - \{v\}$, $S - \{u'\} + \{u\} \in \overline{\mathcal{S}}_k(G)$ because u is an isolated vertex in G_u. Therefore, by definition, u and v are clear non-neighbors. □

Lemma 6 describes the structure of a graph in which there is some pair of fake neighbors u and v. In essence, if such a pair exists, then they have no "big private connected component" allowing us to determine their non-edge. Therefore, their common neighbors form a separator of the graph where on one side we have small components containing u and v and most of the graph is on the other side.

Lemma 6. *Let $G = (V, E)$ be a graph and let $uv \notin E$ be such that u and v are fake neighbors. Let C_u and C_v be the connected component containing u in $G - N(v)$ and v in $G - N(u)$ respectively. We have $|V(C_u)| \leq k - 1$ and $|V(C_v)| \leq k-1$. In particular, the connected component(s) containing u and v in $G-(N(u) \cap N(v))$ has size (have size in total) at most $|V(C_u) \cup V(C_v)| \leq 2k-2$. Moreover, if $|V(C_u) \cup V(C_v)| \leq k - 1$, then G is not uniquely k-reconstructible.*

Proof. First, observe that v (resp. u) is not adjacent to any vertex in C_u (resp. C_v). Therefore, v is an isolated vertex in $C_u + v$. By Proposition 3, $|V(C_u)| \leq k - 1$. Symmetrically, the same reasoning shows that $|V(C_v)| \leq k - 1$. We claim that the connected component(s) containing u and v in $G - (N(u) \cap N(v))$ is a subgraph of $G[V(C_u) \cup V(C_v)]$, so it (they) has size (have size in total) at most $|V(C_u) \cup V(C_v)| \leq 2k - 2$. Indeed, let w be a vertex in $G - (N(u) \cap N(v))$ that is in the same connected component as u. Consider a shortest path from w to u. If this path does not intersect $N(v)$, then w is in C_u. Otherwise, we cut the path short by stopping it the moment it intersects $N(v)$ and we add v to end of this shorter path. This is a path from w to v that does not intersect $N(u)$, therefore, w is in C_v. In any case, $w \in V(C_u) \cup V(C_v)$ and concludes the proof of the first case.

Now, suppose that $|V(C_u) \cup V(C_v)| \leq k - 1$. Let $G' = G + uv$, we will show that G' is consistent with $\mathcal{S}_k(G)$ and $\overline{\mathcal{S}}_k(G)$; in other words, G is not uniquely k-reconstructible. If G' is not consistent with some k-set of G, then it implies that there exists $S \in \overline{\mathcal{S}}_k(G)$ containing u and v such that $G'[S]$ is connected. Since $G[S]$ is disconnected and $G' = G + uv$, uv must be a bridge in $G'[S]$ and $G[S]$ has exactly two connected components, one containing u, the other containing v. Therefore, S cannot contain any common neighbor of u and v, and $S \subseteq V(C_u) \cup V(C_v)$. This is a contradiction because $k = |S| \leq |V(C_u) \cup V(C_v)| \leq k - 1$. This completes the proof. □

3.1 Uniqueness for Triangle-Free Graphs

We start with showing that if in a triangle-free graph we have a vertex v of large degree, then the neighborhood of v is uniquely k-reconstructible.

Lemma 7 (*). *Let k be an integer with $k \geq 3$ and $r \geq 3k - 6$. The graph $K_{1,r}$ is uniquely k-reconstructible in the class of triangle-free graphs.*

Now, we are ready to prove that triangle-free graphs are uniquely k-reconstructible. The main idea is to use Lemma 6. In triangle-free graphs, vertices with common neighbors definitely do not share an edge so if there are non-edges that we are not able to determine (fake neighbors), then the components containing them are small. Thus, we should be able to reconstruct large enough graphs uniquely.

Theorem 5. *Let k be an integer with $k \geq 3$. If G is a triangle-free connected graph on at least $(2k - 2)(3k - 7)^2 + 3k - 6$ vertices, then G is uniquely k-reconstructible in the class of connected triangle-free graphs.*

Proof. Suppose by that there exists G' that is consistent with $\mathcal{S}_k(G)$ and $\overline{\mathcal{S}}_k(G)$. If an induced subgraph $G[V']$ of G is uniquely k-reconstructible, then $G'[V'] = G[V]$. Therefore, if there is a vertex v of degree at least $3k-6$, then by Lemma 7, $G[N[v]]$ is uniquely determined. In this case, G' is uniquely k-reconstructible by Lemma 5 since $3k-6 \geq 2k-4$. In other words, $G' = G$. Now, we can assume that the maximum degree of G is at most $3k-7$.

First, we show that if $uv \notin E(G)$, then $uv \notin E(G')$. If u, v are clear non-neighbors, then by Corollary 2, $uv \notin E(G')$. So now assume that u and v are fake neighbors. By Lemma 6, the connected component(s) of $G - (N(u) \cap N(v))$ containing u and v have size at most $2k-2$. In particular, it implies that u and v have a common neighbor since $|V(G)| > 2k-2$. We will show that u and v also have a common neighbor in G', which will prove that $uv \notin E(G')$ since G' must be triangle-free.

Observe that $|N(u) \cap N(v)| \leq 3k-7$ due to the bound on the maximum degree of G. Furthermore, the number of connected components of $G - (N(u) \cap N(v))$ is at most $(3k-7)^2$, again by the bound on the maximum degree. Thus, there is at least one connected component C of size at least $2k-1$ since $|V(G) \setminus (N(u) \cap N(v))| \geq (2k-2)(3k-7)^2 + 3k - 6 - 3k + 7 = (2k-2)(3k-7)^2 + 1$. As a result, C cannot contain u or v. Moreover, for every vertex $w \in V(C)$, the u (resp. v) and w are clear non-neighbors due to Proposition 3 since u (resp. v) is isolated in $C + u$ (resp. $C + v$). By Corollary 2, the edges uw and vw are also not in $E(G')$ for every $w \in V(C)$. Since G is connected, there exists $w \in V(C)$ and $z \in N(u) \cap N(v)$ such that $wz \in E(G)$. Now, let us show that zu and zv are also present in $E(G')$.

There exists $S \in \mathcal{S}_k(G)$ such that $u, z \in S$ and $S - \{u, z\} \subseteq V(C)$. If $zu \notin E(G')$, then $G'[S]$ is disconnected since uw is a non-edge in G' for every $w \in S - \{u, z\} \subseteq V(C)$. This contradicts the fact that G' is consistent with $\mathcal{S}_k(G)$ and $\overline{\mathcal{S}}_k(G)$. Therefore, $zu \in E(G')$ and similarly, the same holds for zv. Finally, since z is a common neighbor of u and v in G' and G' is triangle-free, uv is a non-edge in G'.

By symmetry, if $uv \notin E(G')$, then $uv \notin E(G)$ as we can treat G as a graph consistent with $\mathcal{S}_k(G')$ and $\overline{\mathcal{S}}_k(G')$. Therefore $uv \in E(G)$ if and only if $uv \in E(G')$, which completes the proof. □

4 NP-Hardness

In this section, we will prove that for $k \geq 4$, if we are not given the full set of connected/disconnected k-sets, then it is NP-hard to determine, whether there exists a graph consistent with the given k-sets.

Let us point out that contrary to previous sections, we do not assume here that $\mathcal{S}_k \cup \overline{\mathcal{S}}_k = \binom{V}{k}$. Formally, we prove the following.

Theorem 6. *Let $k \geq 4$. The k-*Reconstruction with partial information* problem is NP-hard. Moreover, there is no algorithm that solves k-*Reconstruction with partial information* on instances $(V, \mathcal{S}_k, \overline{\mathcal{S}}_k)$ with*

n vertices and $t = |\mathcal{S}_k \cup \overline{\mathcal{S}}_k|$ in time $2^{o(n)} \cdot n^{O(1)}$ neither $2^{o(t)} \cdot n^{O(1)}$, unless the ETH fails.

Proof. We will reduce from 3-SAT. Let ϕ be an instance of 3-SAT with n variables $x_1, \ldots, x_n$ and m clauses. We can assume that all clauses contain exactly 3 variables.

We construct an instance $(V, \mathcal{S}_k, \overline{\mathcal{S}}_k)$ of k-RECONSTRUCTION as follows. First, we define the vertex set V. For every variable x_i, we introduce two vertices x_i, y_i and for $j \in [k-1]$ we introduce x_i^j and y_i^j. Moreover, we introduce vertex v and for all $i \in [k-3], j \in [k-1]$, we introduce vertices $u_i, u_i^j, w_i, w_i^j, v^j$. This completes the definition of the vertex set V. Note that $|V| = O(n)$.

Auxiliary Partial Graph. We will define an auxiliary partial graph H on V. First, for every $i \in [k-3]$ we add edges $u_i v$, and for every $j \in [k-1]$, we add edges $u_i u_i^j, w_i w_i^j, v v^j, x_i x_i^j, y_i y_i^j$. Moreover, we make each w_i adjacent to all x_j, y_j and we introduce all edges $x_i x_j, y_i y_j, x_i y_j$ for $i \neq j$. The edges $v x_i, v y_i$ are unknown. All the remaining pairs of vertices are non-edges (see Fig. 2).

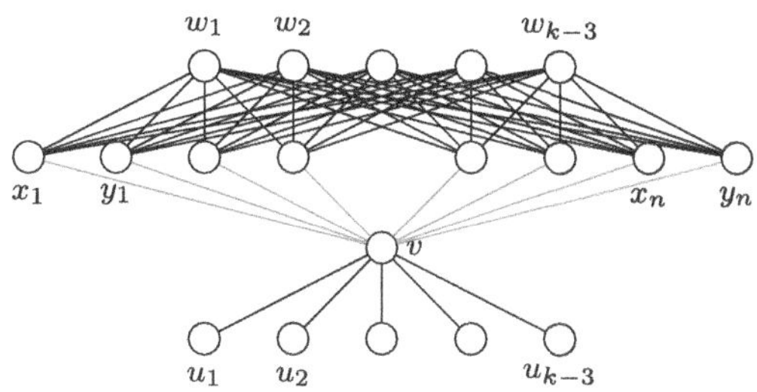

Fig. 2. The auxiliary partial graph H. Orange edges denote the unknown edges. We omit here $k-1$ private neighbors of each vertex in the figure. Furthermore, the vertices from $\{x_i, y_j \mid i, j \in [n]\}$ induce a clique with the matching $\{x_i y_i \mid i \in [n]\}$ removed. (Color figure online)

Defining Connected and Disconnected k-sets. We add the following k-sets to $\mathcal{S}_k, \overline{\mathcal{S}}_k$.

1. **Edges vu_i.** Let $G_1 := H[\{v, v^j, u_i, u_i^j \mid i \in [k-3], j \in [k-1]\}]$. Note that G_1 is a graph, since we do not have unknown edges inside the vertex set. We add all k-sets of $\mathcal{S}_k(G_1)$ to $\mathcal{S}_k$ and all k-sets of $\overline{\mathcal{S}}_k(G_1)$ to $\overline{\mathcal{S}}_k$.
2. **Non-edges vw_i.** Let $G_2 := H[\{v, v^j, w_i, w_i^j \mid i \in [k-3], j \in [k-1]\}]$, again G_2 is a graph. We add all k-sets of $\mathcal{S}_k(G_2)$ to $\mathcal{S}_k$ and all k-sets of $\overline{\mathcal{S}}_k(G_2)$ to $\overline{\mathcal{S}}_k$.
3. **Non-edges $x_i u_j, y_i u_j$.** Let $i \in [n]$. We define $G^i = H[x_i, x_i^j, y_i, y_i^j, u_\ell, u_\ell^j \mid j \in [k-1], \ell \in [k-3]\}]$, and G^i is a graph. We add all k-sets of $\mathcal{S}_k(G^i)$ to $\mathcal{S}_k$ and all k-sets of $\overline{\mathcal{S}}_k(G^i)$ to $\overline{\mathcal{S}}_k$.

4. **Variables.** For every $i \in [n]$ we add the set $\{x_i, y_i, v, u_1, \ldots, u_{k-3}\}$ to $\overline{\mathcal{S}}_k$.
5. **Clauses.** For every clause with literals ℓ_1, ℓ_2, ℓ_3, for $j \in [3]$, let $z_j = x_i$ if $\ell_i = x_i$ and $z_j = y_i$ if $\ell_j = \neg x_i$. We add the set $\{z_1, z_2, z_3, v, u_1, \ldots, u_{k-4}\}$ to $\mathcal{S}_k$.

This completes the definition of $\mathcal{S}_k$ and $\overline{\mathcal{S}}_k$. Let us verify the equivalence of the instances.

First assume that there is a graph G on V that satisfies all the k-sets from $\mathcal{S}_k, \overline{\mathcal{S}}_k$, i.e., for every $S \in \mathcal{S}_k$ the graph $G[S]$ is connected and for every $S' \in \overline{\mathcal{S}}_k$, the graph $G[S']$ is disconnected. Before we define the truth assignment for variables of ϕ, let us discuss some properties of G.

Claim (*). Let $i \in [k-3]$. Then we have that $u_i v \in E(G)$ and $w_i v \notin E(G)$ (Fig. 3).

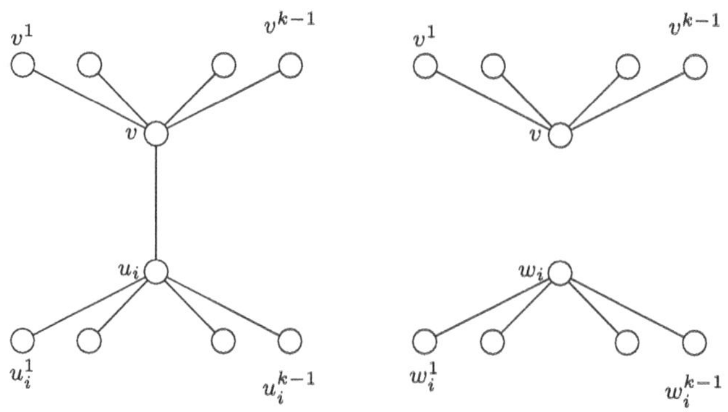

Fig. 3. The auxiliary partial graph H on vertices v, v^j, u_i, u_i^j (left) and on v, v^j, w_i, w_i^j (right).

Similarly, because of the k-sets added in 3., we can obtain the following.

Claim (*). Let $i \in [n], j \in [k-3]$. Then it holds that $x_i u_j \notin E(G)$ and $y_i u_j \notin E(G)$.

We also show that at most one of edges $x_i v, y_i v$ can be present in G.

Claim (*). Let $i \in [n]$. Either $x_i v \in E(G)$ or $y_i v \in E(G)$.

Now we are ready to define a truth assignment φ of the variables of ϕ. For $i \in [n]$, we set $\varphi(x_i) = 1$ if $x_i v \in E(G)$, and $\varphi(x_i) = 0$ if $y_i v \in E(G)$. If none of the edges is present in G, we set $\varphi(x_i)$ arbitrarily.

We leave the reader to check that such an assignment allows us to conclude this proof. $\square$

Acknowledgments. This work started at the 2023 Structural Graph Theory workshop held at the IMPAN Będlewo Conference Center in Poland in September 2023. This workshop was a part of STRUG: Stuctural Graph Theory Bootcamp, funded by the "Excellence initiative - research university (2020–2026)" of the University of Warsaw.

References

1. Bastide, P., et al.: Reconstructing graphs from connected triples. In: Paulusma, D., Ries, B. (eds.), Graph-Theoretic Concepts in Computer Science, pp. 16–29. Springer Nature, Switzerland, Cham (2023)
2. Kannan, S., Mathieu, C., Zhou, H.: Graph reconstruction and verification. ACM Trans. Algorithms **14**(4) (2018)
3. Kelly, P.J.: On isometric transformations. PhD thesis, University of Wisconsin, 1942
4. Qi, Y.: Graph reconstruction from connected triples, 2023
5. Ulam, S.M.: A Collection of Mathematical Problems. Interscience tracts in pure and applied mathematics. Interscience Publishers, 1960

On Polynomial Kernelization for Stable Cutset

Stefan Kratsch[1(✉)] and Van Bang Le[2]

[1] Institut für Informatik, Humboldt-Universität zu Berlin, Berlin, Germany
kratsch@informatik.hu-berlin.de

[2] Institut für Informatik, Universität Rostock, Rostock, Germany
van-bang.le@uni-rostock.de

Abstract. A *stable cutset* in a graph G is a set $S \subseteq V(G)$ such that vertices of S are pairwise non-adjacent and such that $G - S$ is disconnected, i.e., it is both stable (or independent) set and a cutset (or separator). Unlike general cutsets, it is NP-complete to determine whether a given graph G has any stable cutset. Recently, Rauch et al. [FCT 2023] gave a number of fixed-parameter tractable (FPT) algorithms, time $f(k) \cdot |V(G)|^c$, for Stable Cutset under a variety of parameters k such as the size of a (given) dominating set, the size of an odd cycle transversal, or the deletion distance to P_5-free graphs. Earlier works imply FPT algorithms relative to clique-width and relative to solution size.

We complement these findings by giving the first results on the existence of polynomial kernelizations for Stable Cutset, i.e., efficient preprocessing algorithms that return an equivalent instance of size polynomial in the parameter value. Under the standard assumption that $\mathsf{NP} \nsubseteq \mathsf{coNP/poly}$, we show that no polynomial kernelization is possible relative to the deletion distance to a single path, generalizing deletion distance to various graph classes, nor by the size of a (given) dominating set. We also show that under the same assumption no polynomial kernelization is possible relative to solution size, i.e., given (G, k) answering whether there is a stable cutset of size at most k. On the positive side, we show polynomial kernelizations for parameterization by modulators to a single clique, to a cluster or a co-cluster graph, and by twin cover.

Keywords: stable cutset · kernelization · parameterized complexity

1 Introduction

A *stable cutset* in a graph G is a set $S \subseteq V(G)$ such that vertices of S are pairwise non-adjacent and such that $G - S$ is disconnected, i.e., it is both stable (or independent) set and a cutset (or separator). The Stable Cutset problem asks whether or not a given graph has a stable cutset:

For Dieter Kratsch on his 65th birthday.

D. Král' and M. Milanič (Eds.): WG 2024, LNCS 14760, pp. 358–373, 2025.
https://doi.org/10.1007/978-3-031-75409-8_25

> **STABLE CUTSET**
>
> *Instance:* A graph G.
> *Question:* Does G have a stable cutset?

Unlike finding general cutsets, the STABLE CUTSET problem is NP-complete, and this remains true even on various restricted graph classes [3,7,15,18,20]. Moreover, it follows from a result by Brandstädt et al. [3] that, assuming the Exponential-Time Hypothesis, STABLE CUTSET cannot be solved in $2^{o(n)}$ time on n-vertex graphs, even when restricted to K_4-free graphs of diameter 4.

The importance of stable cutsets has been demonstrated first in connection to perfect graphs [8,25]. Tucker [25] proved that if a graph G has a stable cutset S such that no induced cycle of odd length at least five has a vertex in S, then the coloring problem on G reduces to the same problem on the smaller graphs induced by S and the connected components of $G - S$. Later, structural and algorithmic aspects of stable cutsets were studied: It is known that any n-vertex graph with at most $2n-4$ edges has a stable cutset and such one can be computed in polynomial time [6], extremal graphs with $2n - 3$ edges and without stable cutsets have been described in [19], and graphs with small stable cutsets were studied in [5]. While the complexity of STABLE CUTSET in graphs of maximum degree four is open, it is known to remain NP-complete in planar graphs of maximum degree five [18] and in 5-regular line graphs of bipartite graphs [20].

In fact, stable cutsets in *line graphs* have been studied under the notion of a matching cut. Here, a *matching cut* is an edge cut that is also a matching. It can be seen that, for graphs G with minimum degree at least two, G has a matching cut if and only if the line graph of G, $L(G)$, admits a stable cutset. Graphs having a matching cut were first discussed in [7,12]. It follows from [7,23] that STABLE CUTSET is NP-complete even on line graphs of bipartite graphs, hence on perfect graphs, and it follows from [21] that computing a *maximum* (*minimum*) stable cutset in line graphs is NP-hard. We refer to the recent papers [4,10,11,16,21] for more information on matching cuts, i.e., stable cutsets in line graphs.

The (likely) intractability of STABLE CUTSET on general graphs (and many restricted classes) as well as its relation to graph structure motivate a study of its parameterized complexity: How does structure of the input graph G influence the complexity of STABLE CUTSET? E.g., is it significantly easier to solve not just on, say, chordal graphs, but also on graphs that are close to being chordal (e.g., at most ℓ vertex deletions away)? Similarly, does having (small) treewidth at most w or asking for a stable set of (small) size at most k allow for faster algorithms? To this end, one studies STABLE CUTSET with any chosen parameter(s), say, k, and asks whether the resulting parameterized problem is *fixed-parameter tractable* (FPT), i.e., if there is an algorithm that decides inputs (x, k) in time $f(k) \cdot |x|^c$ for some computable function f (also called an FPT-algorithm). Some but (likely) not all FPT problems also admit a strong form of efficient preprocessing: A *polynomial kernelization* is an efficient algorithm that given (x, k) returns an input (x', k') with same answer but size bounded by a polynomial in k.

So far, there are only few results on the parameterized complexity of Stable Cutset, in particular, no dedicated results on kernelization. A general result of Marx et al. [22] implies that Stable Cutset[k](parameterized by solution size) is fixed-parameter tractable. Very recently, Rauch et al. [24] proved, among others, that Stable Cutset is FPT when parameterized by the domination number, by the distance to bipartite graphs, and by the distance to chordal graphs. Bodlaender observed (in private communication in 2015) that the problem can be solved in $\mathcal{O}(3^w n^c)$ when a tree decomposition of width w is given.

Our Work. We initiate the study of upper and lower bounds for kernelizations for the Stable Cutset problem. The results obtained have some similarity to what is known for 3-Coloring, in particular, results obtained by Jansen and Kratsch [14]. While the problems are different, some similarities can be observed: A set $S \subseteq V(G)$ is a stable cutset of G if and only if $G - S$ is disconnected, and hence can be partitioned into two parts $G[A]$ and $G[B]$ with no connecting edges and with $A, B \neq \emptyset$. In other words, G has a stable cutset if and only if its vertices can be partitioned into $V(G) = S \dot\cup A \dot\cup B$ such that S is a (possibly empty) stable set, A and B are both nonempty, and there is no edge with one endpoint in A and one in B. Intuitively, this leaves us with a 3-coloring-like problem (that could also be cast as H-coloring with vertices S, A, and B, suitable edges/loops, and condition that $A, B \neq \emptyset$). A key difference between Stable Cutset and 3-Coloring, however, is that a stable cutset may be rather local, in the sense of separating only a small part, say, A and having most of the graph in B. Unlike 3-Coloring, Stable Cutset is also not monotone, i.e., deletion of vertices may turn the instance from yes to no.

We obtain several several polynomial kernelizations for Stable Cutset with respect to different parameters. All but one of the results involve one or more marking-based reduction rules that preserve one vertex/component each for certain adjacencies of $G - X$ into X, similar to what was done for 3-Coloringparameterized by vertex cover [14].

Theorem 1. *Kernelizations with the following bounds exist for* Stable Cutset *parameterized by* $|X|$ *when a set* X *with the required properties is provided:*

- $\mathcal{O}(|X|)$ *vertices when* $G - X$ *is a clique.*
- $\mathcal{O}(|X|^3)$ *vertices when* X *is a vertex cover (or* $G - X$ *is a stable set).*
- $\mathcal{O}(|X|^3)$ *vertices when* X *is a twin cover.*
- $\mathcal{O}(|X|^5)$ *vertices when* $G - X$ *is a cluster graph.*
- $\mathcal{O}(|X|^3)$ *vertices when* $G - X$ *is a co-cluster graph.*

As is common, we assume that a suitable witness (in this case a suitable set X) is provided with the input. That being said, all parameterizations in the previous theorem have efficient constant-factor approximation algorithms.

Our main lower bound result is to show that Stable Cutset parameterized by the size of a deletion set (also called modulator) to a single path is hard for a class called MK[2] (introduced by Hermelin et al. [13]) under polynomial parameter transformations (a special parameterized reduction [2]). To this end, we first establish that SET SPLITTING[n]is complete for MK[2]. We then give the required reduction from SET SPLITTING[n]to STABLE CUTSET[MOD LIN FOREST], which also works for parameterization by (connected) domination number. It was proved by Hermelin et al. [13] that MK[2]-hard problems do not admit polynomial kernelizations unless NP $\subseteq$ coNP/poly. They also conjectured that such problems do not admit so-called polynomial *Turing kernelizations.*

Theorem 2. *The following parameterizations of* STABLE CUTSET *are hard for the class* MK[2] *under polynomial parameter transformations and thus do not admit polynomial kernelizations unless* NP $\subseteq$ coNP/poly *(and PH collapses):*

- *Parameterized by the size of X such that $G - X$ is a single path.*
- *Parameterized by the size of a given (connected) dominating set X.*

Furthermore, we show a simple construction that, unless NP $\subseteq$ coNP/poly, rules out polynomial kernelizations for STABLE CUTSET parameterized by solution size or by treewidth (or by vertex integrity).

Organization. Section 2 contains the preliminaries. In Sect. 3 we collect various properties of stable cutsets and develop some first reduction rules. Section 4 contains our polynomial kernelizations, while Sect. 5 contains the claimed lower bounds for kernelization. We conclude in Sect. 6. Proofs for statements marked by $*$ are deferred to the full version [17].

2 Preliminaries

Graphs. We consider only finite and simple graphs $G = (V, E)$ with vertex set $V(G) = V$ and edge set $E(G) = E$. Throughout we use $n = |V|$ and $m = |E|$. A set $S \subseteq V(G)$ is a *stable set* in G if $E(G)$ contains no edges between vertices of S, respectively a *clique* if all vertices in S are pairwise adjacent. A set $S \subseteq V(G)$ is a *cutset* of G if $G - S$ is disconnected; the empty set is a cutset of any disconnected graph G. A *consistent partition* of a disconnected graph G is a partition $V(G) = A \dot{\cup} B$ with $A, B \neq \emptyset$ and such that no edge of G has one endpoint in A and the other in B.

Proposition 1. *A set S is a cutset of G iff $G - S$ has a consistent partition.*

We use $N(v)$ to denote the (open) neighborhood of a vertex v and, $N[v] = N(v) \cup \{v\}$ for its closed neighborhood. A vertex v is *simplicial* if $N(v)$ is a clique. A set $M \subseteq V(G)$ is a *module* of G if $N(u) \setminus M = N(v) \setminus M$ holds for all $u, v \in M$. Two vertices u and v of G are *true twins* if $N[u] = N[v]$ and *false twins* if $N(u) = N(v)$, i.e., if they have the same neighbors outside of $\{u, v\}$ and they are either adjacent or non-adjacent. Say that u and v are *twins* if they are true twins or false twins.

Proposition 2. *Vertices being twins in a graph G is an equivalence relation on $V(G)$ and each equivalence class is a clique or stable set. Moreover, each equivalence class is a (clique or stable) module of G.*

Parameterized Complexity and Kernelization. A *parameterized problem* is a language $L \subseteq \Sigma^* \times \mathbb{N}$; the second component k of an instance (x, k) is called its *parameter.* A parameterized problem L is *fixed-parameter tractable* if there is an algorithm A, a computable function f, and a constant c such that A decides $(x, k) \in L$ in time $f(k)|x|^c$. A *kernelization* for a parameterized problem L is an efficient algorithm that given $(x, k) \in \Sigma^* \times \mathbb{N}$ takes time polynomial in $|x| + k$ and returns an instance (x', k') of size $|x'| + k'$ upper bounded by $h(k)$ for some computable function $h \colon \mathbb{N} \to \mathbb{N}$ and with $(x, k) \in L$ if and only if $(x', k') \in L$. The function h is also called the *size* of the kernelization and for a *polynomial kernelization* it is required that h is polynomially bounded. A *polynomial parameter transformation* from $Q \subseteq \Sigma^* \times \mathbb{N}$ to $Q' \subseteq \Gamma^* \times \mathbb{N}$ is an efficiently computable mapping $\pi \colon \Sigma^* \times \mathbb{N} \to \Gamma^* \times \mathbb{N}$ such that for each $(x, k) \in \Sigma^* \times \mathbb{N}$ and $(x', k') = \pi((x, k))$ we have that $(x, k) \in Q$ if and only if $(x', k') \in Q'$ and that k' is polynomially bounded in k.

Graph Parameters. Let $G = (V, E)$ be a graph. A *vertex cover* of G is a set $X \subseteq V$ such that $G - X$ is a stable set. A *twin cover* of G is a set $X \subseteq V$ such that $N[u] = N[v]$ holds for every edge $\{u, v\}$ of $G - X$, i.e., every two adjacent vertices of $G - X$ are twins (which means that they must be true twins). For a graph class $\mathcal{C}$, a *modulator of* G *to* $\mathcal{C}$ is a set $X \subseteq V$ such that $G - X \in \mathcal{C}$. E.g., the vertex covers are exactly the modulators to the class of stable sets, and each twin cover is also a modulator to the class of cluster graphs (but not vice versa). A *dominating set* of G is a set $X \subseteq V$ such that $V(G) = N[X]$, i.e., such that each vertex of G is contained in X or adjacent to a vertex of X.

3 Some Properties of Stable Cutsets

In this section, we observe and recall several properties of stable cutsets. This leads to a number of reduction rules that will be used in Sect. 4.

The first three rules handle several cases where G has a trivial stable cutset, such as the empty set for disconnected graphs, or no (stable) cutset, such as for cliques. These can be applied for any parameterization because they outright answer the problem rather than modifying the input. Throughout we assume that reduction rules are applied in the order that they are introduced in, so that later rules benefit from structural properties that are established by earlier ones.

Reduction Rule 1. *If G is disconnected or G has a cut-vertex then return yes.*

Reduction Rule 2. *If G is a clique then return no.*

Reduction Rule 3. *If G has any vertex whose neighborhood is a stable set then return yes.*

Lemma 1 (∗). *Reduction Rule 3 is safe, i.e., if Reduction Rules 1 and 2 do not apply to G and the neighborhood of v is a stable set then G has a stable cutset.*

Several of our reduction rules allow the deletion of a single vertex v or of a clique C from G. The following lemma eases proofs of safeness for the case of vertex deletion by showing that this cannot change the answer from yes to no.

Lemma 2 (∗). *Let G be a graph to which Reduction Rules 1, 2, and 3 do not apply and let v be any vertex of G. If G has a stable cutset then so does $G - v$.*

Something similar can be done for the case of deleting a clique C with $|C| \geq 2$ from a graph G. Crucially, any stable cutset S of G contains at most one vertex of C while the rest $C \setminus S$ must be part of a single connected component of $G - S$. For a clique C, we say that a stable cutset S is *C-simple* if $S = N(C)$ or if $S = N(C \setminus \{v\})$ for some vertex $v \in C$. For a given clique C, it can be tested efficiently if there is any C-simple stable cutset and, if so, we may return yes. (There is probably no efficient way of doing this for all cliques of G, so it has to be done "on demand" when a suitable clique is found by a reduction rule.) Deleting any clique C without C-simple stable cutset cannot change the answer from yes to no (analogous to Lemma 2).

Lemma 3 (∗). *Let G be a graph and let $C \subseteq V(G)$ be a clique in G without C-simple stable cutset. If G has a stable cutset then so does $G - C$.*

We can now present several further reduction rules that allow to delete vertices from the graph under certain conditions. They can be applied for any parameterization that is monotone under vertex deletions such as deletion distance to some hereditary graph class but not, e.g., for max leaf number.

Reduction Rule 4 (∗). *Delete any simplicial vertex from G.*

Reduction Rule 5 (∗). *If G has vertices u and v with $N(v) \subseteq N(u)$ delete v.*

Amongst others, this rule strongly restricts the size of stable modules of G.

Lemma 4 (∗). *If Reduction Rules 1, 2, 3, and 5 do not apply to G then each module of G that is a stable set has size at most one.*

For clique modules the following reduction rule can be applied.

Reduction Rule 6 (∗). *If Q is a clique module of G with at least three vertices then delete any vertex $v \in Q$ from G.*

Observation 1. If Q is a clique module of G then each $Q' \subseteq Q$ is also a clique module. In particular, if $|Q| \geq 3$ then there is a clique module $Q' \subseteq Q$ of size exactly three to which Reduction Rule 6 can be applied. Thus, exhaustively checking triplets of vertices for being a clique module (and possibly deleting one of them) suffices to efficiently implement Reduction Rule 6. Afterwards, each clique module has size at most two.

4 Polynomial Kernelizations for Stable Cutset

4.1 Vertex Cover Number and Twin Cover Number

We first give a kernelization with a cubic number of vertices for Stable Cutset parameterized by the size of a given vertex cover. The approach is similar to a kernelization for 3-Coloring parameterized by vertex cover [14]. We then generalize the result to work also for the smaller parameter twin-cover.

Stable Cutset[vc]

Instance: A graph G and a vertex cover X of G.
Parameter: $|X|$.
Question: Does G have a stable cutset?

We use Reduction Rules 1, 2, and 3 as well as the following reduction rule.

Reduction Rule 7. *If $|V(G) \setminus X| > |X|^3$, then, starting with all vertices as unmarked, mark (isolated) vertices in $G-X$ according to the following condition:*

- *For each $(x_1, x_2, x_3) \in X^3$ mark a vertex p that is adjacent to x_1, x_2, and x_3.*

Delete any unmarked vertex $v \in V(G) \setminus X$, i.e., return the instance $(G - v, X)$.

Observation 2. If $|V(G) \setminus X| > |X|^3$ then there will always be an unmarked vertex, so each application will lead to the deletion of a vertex.

Lemma 5. *Reduction Rule 7 is safe.*

Proof. Let v be the vertex that is being deleted. By Lemma 2, if G has a stable cutset then so does $G-v$. Thus, if (G, X) is yes then so is $(G-v, X)$. It remains to show the converse.

Assume that $(G - v, X)$ is yes and let S be a stable cutset of $G - v$ with consistent partition $V(G-v) \setminus S = A \dot\cup B$. If v has no neighbor in S then $S \cup \{v\}$ is a stable cutset of G with same consistent partition. Similarly, if v has no neighbor in A (respectively in B) then S is a stable cutset of G with consistent partition $V(G) \setminus S = A \dot\cup (B \cup \{v\})$ (resp. $V(G) \setminus S = (A \cup \{v\}) \dot\cup B$). It remains to consider the case that all three types of neighbors exist.

Note that $N_G(v) \subseteq X$ and let $x_1 \in N_G(v) \cap S$, let $x_2 \in N_G(v) \cap A$, and let $x_3 \in N_G(v) \cap B$. As v was unmarked, there is a marked vertex $u \in V(G-v) \setminus X$ that is adjacent to $x_1 \in S$, $x_2 \in A$, and $x_3 \in B$. Because of x_1, we must have $u \notin S$. Similarly, because of x_2 and x_3 we must have $u \notin B$ and $u \notin A$. This is a contradiction since u is present in $G-v$ and $V(G-v) \setminus S = A \dot\cup B$ is a consistent partition. It follows that this final case cannot happen, so (G, X) is yes. □

It is now easy to complete our first kernelization result.

Theorem 3. Stable Cutset[vc] *admits a polynomial kernelization to equivalent instances with at most $\mathcal{O}(|X|^3)$ vertices.*

Proof. Clearly, Reduction Rules 1, 2, 3, and 7 can be exhaustively applied in polynomial time. We also know that they are safe, so the result will be the correct answer (respectively a dummy yes- or no-instance of constant size) or an equivalent instance (G', X'), where $|X'| \le |X|$, to which none of the reduction rules applies. It remains to upper bound the size of (G', X'). Since Reduction Rule 7 does not apply to (G', X'), we have $V(G') \setminus X' \le |X'|^3 \le |X|^3$, hence $|V(G')| = \mathcal{O}(|X|^3)$. □

We now generalize this result to STABLE CUTSET parameterized by the size of a twin cover. Clearly, each vertex cover of G is also a twin cover.

STABLE CUTSET[TC]

Instance: A graph G and a twin cover X of G.
Parameter: $|X|$.
Question: Does G have a stable cutset?

By definition, if X is a twin cover of G then $N_G[u] = N_G[v]$ holds for each edge $\{u, v\}$ of $G - X$. Thus each connected component of $G - X$ is a clique module of G. Thus, adding Reduction Rule 6 reduces all connected components of $G - X$ (i.e., the clique modules of G) to at most two vertices each. It can be easily verified that Reduction Rule 7 is safe also in this context and will reduce the number of isolated vertices in $G - X$ to at most $|X|^3$ many. It remains to reduce the number of connected components of size two of $G - X$. Recall that these are clique modules, i.e., if $C = \{u, v\}$ is a connected component of $G - X$ then both vertices have the same neighborhood in X and no further neighbors.

Reduction Rule 8. *If $G - X$ has more than $|X|^2$ cliques of size two, then, starting with all cliques of $G - X$ as unmarked, mark cliques of size two in $G - X$ according to the following condition:*

- *For each $(x_1, x_2) \in X^2$ mark a clique of $G - X$ of size two that is fully adjacent to both x_1 and x_2.*

Pick any unmarked clique C of $G - X$ of size two. If there is a C-simple stable cutset S then return yes. Else delete C, i.e., return $(G - C, X)$.

Lemma 6 (∗). *Reduction Rule 8 is safe, i.e., if Reduction Rules 1, 2, 3, and 6 do not apply to G, and if C is an unmarked clique of size two in $G - X$ then*

- *if there is a C-simple stable cutset S then (G, X) is yes, and*
- *else (G, X) is yes if and only if $(G - C, X)$ is yes.*

Theorem 4 (∗). STABLE CUTSET[TC] *admits a polynomial kernelization to equivalent instances with at most $\mathcal{O}(|X|^3)$ vertices.*

4.2 Distance to Cluster

In this section we show a polynomial kernelization for Stable Cutset parameterized by the size of a given modulator X to the class of cluster graphs.

Stable Cutset[mod cluster]

Instance: A graph $G = (V, E)$ and $X \subseteq V$ s.t. $G - X$ is a cluster graph.
Parameter: $|X|$.
Question: Does G have a stable cutset?

We use Reduction Rules 1, 2, 3, and 4, as well as the following rules.

Reduction Rule 9. *If any (clique) component C of $G - X$ has a C-simple stable cutset then return yes.*

Safeness of this rule is immediate. Since we only have to test this for connected components of $G - X$ it is also efficient (unlike testing for all cliques of a graph). The following rule allows us to shrink connected (clique) components of $G - X$.

Reduction Rule 10 (∗). *Let C be a (clique) component of $G - X$ and let $v \in C$. If each neighbor $x \in X$ of v is adjacent to at least two other vertices of C then delete v, i.e., return the instance $(G - v, X)$.*

The following more involved marking-based reduction rule allows us to reduce the number of connected (clique) components of $(G - X)$.

Reduction Rule 11 (∗). *If $G - X$ has more than $|X|^3 + 3|X|^4$ (clique) components, then, starting with all vertices as unmarked, mark vertices of $G - X$ according to the following conditions:*

- *For each $(x_1, x_2, x_3) \in X^3$ mark a vertex p that is adjacent to x_1, x_2, x_3.*
- *For each $(x_1, x_2, x_3, x_4) \in X^4$ mark any two vertices p and q that are adjacent, and with p adjacent to x_1 and x_2 and q adjacent to x_3 and x_4.*
- *For each $(x_1, x_2, x_3, x_4) \in X^4$ mark any three vertices p, q, and r that are pairwise adjacent, and with p adjacent to x_1 and x_2, q adjacent to x_3, and r adjacent to x_4.*
- *For each $(x_1, x_2, x_3, x_4) \in X^4$ mark any four vertices p, q, r, and s that are pairwise adjacent, and with p adjacent to x_1, q adjacent to x_2, r adjacent to x_3, and s adjacent to x_4.*

Delete any (clique) component C of $G - X$ whose vertices are all unmarked, i.e., return the instance $(G - C, X)$.

Theorem 5 (∗). Stable Cutset[mod cluster] *admits a polynomial kernelization to equivalent instances with $\mathcal{O}(|X|^5)$ vertices.*

As a free corollary we get a vertex-linear kernelization for STABLE CUTSET parameterized by the size of a modulator to a single clique.

STABLE CUTSET[MOD CLIQUE]

Instance: A graph $G = (V, E)$ and $X \subseteq V$ such that $G - X$ is a clique.
Parameter: $|X|$.
Question: Does G have a stable cutset?

Corollary 1 (∗). STABLE CUTSET[MOD CLIQUE] *has a polynomial kernelization to equivalent instances with $\mathcal{O}(|X|)$ vertices.*

4.3 Distance to Co-cluster

In this section we show a polynomial kernelization for STABLE CUTSET parameterized by the size of a modulator to the class of co-cluster graphs (the complements of disjoint unions of cliques).

STABLE CUTSET[MOD CO-CLUSTER]

Instance: A graph $G = (V, E)$ and $X \subseteq V$ s.t. $G - X$ is a co-cluster.
Parameter: $|X|$.
Question: Does G have a stable cutset?

We use Reduction Rules 1, 2, and 3 as well as the following rules. For each instance (G, X), we have that $\overline{G - X}$ is a disjoint union of cliques, while $G - X$ is a join of stable sets. If $G - X$ is a (single) stable set, X is also a vertex cover of G and we can apply the kernelization for STABLE CUTSET[VC]. Else, the following reduction rule can be used to reduce the size of the stable sets of $G - X$.

Reduction Rule 12 (∗). *If $G - X$ is the join of at least two stable sets, and if one such stable set I has more than $|X|^2$ vertices then, starting with all vertices of I as unmarked, mark vertices of I according to the following condition:*

- *For each $(x_1, x_2) \in X^2$ mark any vertex $p \in I$ that is adjacent to x_1 and x_2.*

Delete any unmarked vertex v of I, i.e., return $(G - v, X)$.

We use a second reduction rule to reduce the *number* of stable sets whose join is $G - X$. This rule is an adaptation of Reduction Rule 10.

Reduction Rule 13 (∗). *If $G - X$ is the join of at least four stable sets and if there is a vertex $v \in V(G - X)$ such that each neighbor $x \in N_G(v) \cap X$ is adjacent to at least three different stable sets of $G - X$ (including the one of v) then delete v, i.e., return $(G - v, X)$.*

Theorem 6 (∗). STABLE CUTSET[MOD CO-CLUSTER] *admits a polynomial kernelization to equivalent instances with $\mathcal{O}(|X|^3)$ vertices.*

5 Lower Bounds for Kernelizations for Stable Cutset

Our main result in this section is a lower bound for kernelization for STABLE CUTSET parameterized by the deletion distance to a linear forest. As a free corollary, our construction also provides a lower bound for parameterization by the size of a dominating set. We also show simple lower bounds relative to parameters treewidth and the size of the stable cutset.

5.1 Parameterization by Distance to a Linear Forest

In this section, we show that STABLE CUTSET[MOD LIN FOREST] is hard under polynomial parameter transformations for the class MK[2] introduced by Hermelin et al. [13]. This implies that unless NP ⊆ coNP/poly (and the polynomial hierarchy collapses) it has no polynomial kernelization [13].

STABLE CUTSET[MOD LIN FOREST]

Instance: A graph $G = (V, E)$ and $X \subseteq V$ s.t. $G - X$ is a linear forest.
Parameter: $|X|$.
Question: Does G have a stable cutset?

The class MK[2] can be defined via its complete problem HITTING SET[n], i.e., it consists of all parameterized problems that have a PPT to HITTING SET[n]. We establish MK[2]-hardness of STABLE CUTSET[MOD LIN FOREST] via the intermediate problems MULTICOLORED HITTING SET[n]and SET SPLITTING[n], both of which we show to be complete for MK[2] under PPTs, which may be of independent interest (though not at all surprising). For proving MK[2]-hardness of SET SPLITTING[n]we adapt a reduction due to Cygan et al. [9].

SET SPLITTING[n]

Instance: A ground set U and a family $\mathcal{F}$ of subsets of U.
Parameter: $n = |U|$.
Question: Is there $S \subseteq U$ with $F \nsubseteq S$ and $F \nsubseteq U \setminus S$ for all $F \in \mathcal{F}$?

Lemma 7 (∗). *MULTICOLORED HITTING SET[n] and SET SPLITTING[n] are complete for* MK[2] *under PPTs.*

Theorem 7 (∗). *STABLE CUTSET[MOD LIN FOREST] is* MK[2]*-hard under PPTs.*

Proof (construction only). We give a PPT from SET SPLITTING[n] to STABLE CUTSET[MOD LIN FOREST]. Let $(U, \mathcal{F})$ be an instance of SET SPLITTING[n]with $n = |U|$ and $m = |\mathcal{F}|$. If any set $F \in \mathcal{F}$ has size at most one then the instance is trivially no and we return a dummy no-instance of STABLE CUTSET[MOD LIN FOREST]. Otherwise, we construct an instance (G, X) as follows:

- We begin by adding five vertices s, a_1, a_2, b_1, and b_2, called *special vertices* along with edges $\{s, a_1\}$, $\{s, a_2\}$, $\{s, b_1\}$, $\{s, b_2\}$, $\{a_1, a_2\}$, and $\{b_1, b_2\}$.
- For each element $u \in U$ we add two vertices u_a and u_b, called *element vertices*, and the edge $\{u_a, u_b\}$ to G.
- For each set $F \in \mathcal{F}$ we add the following two paths of length $2|F| - 1$ each, called *set paths* to G: A path $L_F = (\ell_{F,1}, \ell'_{F,1}, \ell_{F,2}, \ell'_{F,2}, \ldots, \ell_{F,|F|}, \ell'_{F,|F|})$ and a path $R_F = (r_{F,1}, r'_{F,1}, r_{F,2}, r'_{F,2}, \ldots, r_{F,|F|}, r'_{F,|F|})$.
- Between special vertices and element vertices we add the following edges:
  - We make a_1 and a_2 adjacent to each vertex u_a with $u \in U$.
  - We make b_1 and b_2 adjacent to each vertex u_b with $u \in U$.
- Between special vertices and set paths we add the following edges:
  - We make s adjacent to each vertex $\ell'_{F,i}$ with $F \in \mathcal{F}$ and $1 \le i \le |F|$.
  - We make s adjacent to each vertex $r'_{F,i}$ with $F \in \mathcal{F}$ and $1 \le i \le |F|$.
  - We make a_1 and a_2 adjacent to the final vertex of each set path R_F, i.e., to $r'_{F,|F|}$ for each $F \in \mathcal{F}$.
  - We make b_1 and b_2 adjacent to the final vertex of each set path L_F, i.e., to $\ell'_{F,|F|}$ for each $F \in \mathcal{F}$.
- Between element vertices and set paths we add the following edges:
  - For each set $F \in \mathcal{F}$ we arbitrarily enumerate its elements as $c_{F,1}, \ldots, c_{F,|F|}$, e.g., by following any linear ordering of U.
  - For each $u \in F$ with $u = c_{F,i}$ we make u_a adjacent to $r_{F,i}$ and $r'_{F,i}$ in set path R_F, and u_b adjacent to $\ell_{F,i}$ and $\ell'_{F,i}$ in set path L_F.
  - For the first and second element of each set F, according to the above enumeration, there are additional edges between element vertices and set paths (recall also that each set has size at least two):
    - For $u = c_{F,1}$ we make u_a adjacent to $\ell_{F,2}$ in set path L_F, and we make u_b adjacent to $r_{F,2}$ in set path R_F.
    - For $u = c_{F,2}$ we make u_a adjacent to $r_{F,1}$ in set path R_F, and we make u_b adjacent to $\ell_{F,1}$ in set path L_F.
- We let $X = \{s, a_1, a_2, b_1, b_2\} \cup \{u_a, u_b \mid u \in U\}$, i.e., X contains all special and element vertices. Clearly $G - X$ is a disjoint union of all the set paths, i.e., a linear forest.

Clearly this construction can be carried out in polynomial time and the set X is a modulator to the class of linear forests. The parameter value $|X| = 5 + 2n$ is polynomial in the input parameter n. It remains to verify correctness, i.e., that $(U, \mathcal{F})$ is yes for SET SPLITTING[n] if and only if (G, X) is yes for STABLE CUTSET[MOD LIN FOREST]. □

As a free corollary, we get the same hardness result for STABLE CUTSET parameterized by the size of a given (connected) dominating set. It suffices to observe that the set X in the reduction from SET SPLITTING[n] to STABLE CUTSET[MOD LIN FOREST] is also a (connected) dominating set of size $5 + 2n$.

Corollary 2. STABLE CUTSET *parameterized by the size of a (connected) dominating set is hard for* MK[2] *under PPTs.*

By a small modification of the reduction, one can ensure that $G - X$ is a single path, i.e., that X is a modulator to the class of all paths, a subclass of the linear forests (so we get a slightly stricter parameter). This is similar in spirit to a lower bound for 3-COLORINGparameterized by distance from a single path [14].

Corollary 3 (∗). STABLE CUTSET[MOD PATH] *is hard for* MK[2] *under PPTs.*

5.2 Lower Bounds for Other Parameterizations

We show a simple construction that given t graphs $G_1, \dots, G_t$ returns a single graph G such that G has a stable cutset if and only if at least one graph G_i has a stable cutset. This construction is somewhat similar to a construction in obtaining lower bounds for Matching Cut in [16]. Using the standard framework for kernelization lower bounds via cross-compositions [1] then allows to rule out polynomial kernelizations for STABLE CUTSET parameterized by either treewidth or solution size unless NP ⊆ coNP/poly.

STABLE CUTSET[k]

Instance: A graph G and $k \in \mathbb{N}$.
Parameter: k.
Question: Does G have a stable cutset of size at most k?

STABLE CUTSET[TW]

Instance: A graph G and a tree decomposition $\mathcal{T}$ of G.
Parameter: The width of $\mathcal{T}$.
Question: Does G have a stable cutset?

Lemma 8. *Given t graphs $G_1, \dots, G_t$, one can efficiently construct a graph G such that G has a stable cutset if and only if at least one graph G_i has a stable cutset. Moreover, the minimum size of stable cutset in G is at most the minimum size of any stable cutset of any graph G_i, and the treewidth of G is at most the maximum treewidth of any graph G_i plus an additive constant.*

Proof. W.l.o.g., each of the graphs G_i has at least one edge. (If any graph G_i is disconnected then return $G := G_i$, clearly fulfilling the lemma. We may also discard any graphs G_i that are cliques, including the empty graph and the graph with a single vertex.) Take the disjoint union of all the graphs and choose an arbitrary edge $\{u_i, v_i\}$ in each graph G_i. Add two adjacent vertices p and q, and make them fully adjacent to all vertices u_i and v_i. Return the obtained graph G. Clearly, its treewidth is at most the maximum of the treewidths of the graph G_i and three (due to the cliques of size four on p, q, and any u_i and v_i). It remains to check the existence of stable cutsets.

We show that any stable cutset $S \subseteq V(G_i)$ is also a stable cutset of G, which also proves the remaining moreover part of the lemma. (The converse of the main statement is proved below.) Let $V(G_i) \setminus S = A \dot{\cup} B$ be a consistent partition for S in G_i. Clearly, u_i and v_i cannot be one in A and one in B, as they are adjacent. By symmetry, assume that $u_i, v_i \in S \cup A$. It follows directly that $(A \cup \{p,q\} \cup \bigcup_{j \neq i} V(G_j)) \dot{\cup} B$ is a consistent partition for S in G.

For the converse, let $S \subseteq V(G)$ be a stable cutset of G. Then at least one of p and q, and for each i, at least one of u_i and v_i are not in S. Hence $S \cap V(G_i)$ is a stable cutset of G_i for some i (otherwise $G - S$ would be connected). □

Theorem 8. Stable Cutset*[k] and* Stable Cutset[tw] *have no polynomial kernelization unless* NP $\subseteq$ coNP/poly.

6 Conclusion

We have initiated the study of polynomial kernelizations for Stable Cutset. Our most general positive results are the kernelizations relative to the size of modulators to the classes of cluster respectively co-cluster graphs. Complementing this, the lower bound relative to size of a modulator to a single path rules out positive results for many (if not most) often-studied graph classes above cluster and/or co-cluster graphs. An interesting case to study may be the size of a modulator to the class of cographs. It would also be interesting to prove lower bounds on the size of kernelizations for the positive cases that we obtained, and to push towards making these bounds tight.

Acknowledgments. VBL thanks Hoang-Oanh Le for discussions on stable cutsets, and beyond.

References

1. Bodlaender, H.L., Jansen, B.M.P., Kratsch, S.: Kernelization lower bounds by cross-composition. SIAM J. Discret. Math. **28**(1), 277–305 (2014). https://doi.org/10.1137/120880240
2. Bodlaender, H.L., Thomassé, S., Yeo, A.: Kernel bounds for disjoint cycles and disjoint paths. Theor. Comput. Sci. **412**(35), 4570–4578 (2011). https://doi.org/10.1016/J.TCS.2011.04.039
3. Brandstädt, A., Dragan, F.F., Le, V.B., Szymczak, T.: On stable Cutsets in graphs. Discret. Appl. Math. **105**(1–3), 39–50 (2000). https://doi.org/10.1016/S0166-218X(00)00197-9
4. Chen, C.-Y., Hsieh, S.-Y., Le, H.-O., Le, V.B., Peng, S.-L.: Matching Cut in graphs with large minimum degree. Algorithmica **83**(5), 1238–1255 (2020). https://doi.org/10.1007/s00453-020-00782-8

5. Chen, G., Faudree, R.J., Jacobson, M.S.: Fragile graphs with small independent cuts. J. Graph Theory **41**(4), 327–341 (2002). https://doi.org/10.1002/jgt.10074
6. Chen, G., Yu, X.: A note on fragile graphs. Discret. Math. **249**(1–3), 41–43 (2002). https://doi.org/10.1016/S0012-365X(01)00226-6
7. Chvátal, V.: Recognizing decomposable graphs. J. Graph Theory **8**(1), 51–53 (1984). https://doi.org/10.1002/jgt.3190080106
8. Corneil, D.G., Fonlupt, J.: Stable set bonding in perfect graphs and parity graphs. J. Comb. Theory, Ser. B **59**(1), 1–14 (1993). https://doi.org/10.1006/jctb.1993.1049
9. Cygan, M., et al.: On problems as hard as CNF-SAT. ACM Trans. Algorithms **12**(3), 1–24 (2016). https://doi.org/10.1145/2925416
10. Feghali, C., Lucke, F., Paulusma, D., Ries, B.: Matching Cuts in graphs of high girth and H-free graphs. In: Iwata, S., Kakimura, N. (eds.) 34th International Symposium on Algorithms and Computation, ISAAC 2023, Kyoto, Japan. LIPIcs, vol. 283, pp. 1–16. Schloss Dagstuhl - Leibniz-Zentrum für Informatik (2023). https://doi.org/10.4230/LIPICS.ISAAC.2023.31
11. Golovach, P.A., Komusiewicz, C., Kratsch, D., Le, V.B.: Refined notions of parameterized enumeration kernels with applications to matching cut enumeration. STACS 2021, LIPIcs, vol. 187, pp. 1–18 (2021). also in: J. Comput. Syst. Sci. **123**, 76–102 (2022). https://doi.org/10.1016/j.jcss.2021.07.005
12. Graham, R.L.: On primitive graphs and optimal vertex assignments. Ann. N. Y. Acad. Sci. **175**(1), 170–186 (1970)
13. Hermelin, D., Kratsch, S., Soltys, K., Wahlström, M., Wu, X.: A completeness theory for polynomial (Turing) kernelization. Algorithmica **71**(3), 702–730 (2015). https://doi.org/10.1007/S00453-014-9910-8
14. Jansen, B.M.P., Kratsch, S.: Data reduction for graph coloring problems. Inf. Comput. **231**, 70–88 (2013). https://doi.org/10.1016/j.ic.2013.08.005
15. Klein, S., de Figueiredo, C.M.: The NP-completeness of multi-partite Cutset testing. Congressus Numeranium **119**, 217–222 (1996)
16. Komusiewicz, C., Kratsch, D., Le, V.B.: Matching cut: kernelization, single-exponential time FPT, and exact exponential algorithms. IPEC 2018, pp. 1- 13. also in: Discret. Appl. Math. **283**, 44–58 (2020). https://doi.org/10.1016/j.dam.2019.12.010
17. Kratsch, S., Le, V.B.: On polynomial kernelization for stable Cutset. CoRR **abs/2407.02086** (2024)
18. Le, V.B., Mosca, R., Müller, H.: On stable Cutsets in claw-free graphs and planar graphs. J. Discrete Algorithms **6**(2), 256–276 (2008). https://doi.org/10.1016/j.jda.2007.04.001
19. Le, V.B., Pfender, F.: Extremal graphs having no stable Cutset. Electr. J. Comb. **20**(1), P35 (2013). https://doi.org/10.37236/2513
20. Le, V.B., Randerath, B.: On stable Cutsets in line graphs. Theor. Comput. Sci. **1–3**(301), 463–475 (2003). https://doi.org/10.1016/S0304-3975(03)00048-3
21. Lucke, F., Paulusma, D., Ries, B.: Dichotomies for maximum Matching Cut: H-freeness, bounded diameter, bounded radius. In: 48th International Symposium on Mathematical Foundations of Computer Science, MFCS 2023. LIPIcs, vol. 272, pp. 1–15. Schloss Dagstuhl - Leibniz-Zentrum für Informatik (2023). https://doi.org/10.4230/LIPIcs.MFCS.2023.64
22. Marx, D., O'Sullivan, B., Razgon, I.: Finding small separators in linear time via treewidth reduction. ACM Trans. Algorithms **9**(4), 1–35 (2013). https://doi.org/10.1145/2500119

23. Moshi, A.M.: Matching Cutsets in graphs. J. Graph Theory **13**(5), 527–536 (1989). https://doi.org/10.1002/jgt.3190130502
24. Rauch, J., Rautenbach, D., Souza, U.S.: Exact and parameterized algorithms for the independent Cutset problem. In: Fernau, H., Jansen, K. (eds.) Fundamentals of Computation Theory - 24th International Symposium, FCT 2023, Trier, Germany, Proceedings. Lecture Notes in Computer Science, vol. 14292, pp. 378–391. Springer (2023). https://doi.org/10.1007/978-3-031-43587-4_27
25. Tucker, A.: Coloring graphs with stable Cutsets. J. Comb. Theory, Ser. B **34**(3), 258–267 (1983). https://doi.org/10.1016/0095-8956(83)90039-4

Revisiting Path Contraction and Cycle Contraction

R. Krithika[1], V. K. Kutty Malu[1(✉)], and Prafullkumar Tale[2]

[1] Indian Institute of Technology Palakkad, Palakkad, India
krithika@iitpkd.ac.in, 112104002@smail.iitpkd.ac.in
[2] Indian Institute of Science Education and Research Bhopal, Bhopal, India
prafullkumar@iiserb.ac.in

Abstract. The PATH CONTRACTION and CYCLE CONTRACTION problems take as input an undirected graph G with n vertices, m edges and an integer k and determine whether one can obtain a path or a cycle, respectively, by performing at most k edge contractions in G. We revisit these NP-complete problems and prove the following results.

- PATH CONTRACTION admits an $\mathcal{O}^*(2^k)$-time algorithm. This improves over the current algorithm known for the problem [Algorithmica 2014].
- CYCLE CONTRACTION admits an $\mathcal{O}^*((2+\epsilon_\ell)^k)$-time algorithm where $0 < \epsilon_\ell \leq 0.5509$ and ϵ_ℓ decreases as length of the resultant cycle ℓ increases.

Central to these results is an algorithm for a general variant of PATH CONTRACTION, namely, PATH CONTRACTION WITH CONSTRAINED ENDS. We also give an $\mathcal{O}^*(2.5191^n)$-time algorithm to solve the optimization version of CYCLE CONTRACTION.

Next, we turn our attention to restricted graph classes and show the following results.

- PATH CONTRACTION on planar graphs admits a polynomial-time algorithm.
- PATH CONTRACTION on chordal graphs does not admit an $\mathcal{O}(n^{2-\epsilon} \cdot 2^{o(tw)})$-time algorithm for any $\epsilon > 0$, unless the Orthogonal Vectors Conjecture fails. Here, tw is the treewidth of the input graph.

The second result complements the $\mathcal{O}(nm)$-time, i.e., $\mathcal{O}(n^2 \cdot tw)$-time, algorithm known for the problem [Discret. Appl. Math. 2014].

1 Introduction

Graph editing problems have consistently been benchmark problems against which the power of new algorithmic tools and techniques are tested. Typical editing operations are vertex deletions, edge deletions, edge additions, edge subdivisions and edge contractions. In this work, we focus on editing a simple undirected graph by only performing edge contractions. Contracting an edge results in the addition of a new vertex adjacent to the neighbors of its endpoints followed

R. Krithika—Supported by SERB MATRICS grant number MTR/2022/000306.

D. Král' and M. Milanič (Eds.): WG 2024, LNCS 14760, pp. 374–386, 2025.
https://doi.org/10.1007/978-3-031-75409-8_26

by the deletion of the endpoints. One may view an edge contraction as "merging" its endpoints into one thereby making them indistinguishable. Any loops or parallel edges created in the process are deleted so that the graph remains simple. In graph contraction problems, given a graph G and an integer k, the interest is in determining if G can be modified into a graph that belongs to a specific family of graphs by a sequence of at most k edge contractions. Most basic such graph families are paths and cycles. This brings us to the problems studied in this paper, namely, Path Contraction and Cycle Contraction.

> Path Contraction
> **Input:** A connected undirected graph G and an integer k.
> **Question:** Can one contract at most k edges in G to obtain a path?

> Cycle Contraction
> **Input:** A connected undirected graph G and an integer k.
> **Question:** Can one contract at most k edges in G to obtain a cycle?

Let P_ℓ and C_ℓ denote the path and cycle respectively on ℓ vertices. When viewed as vertex sets rather than graphs, P_ℓ and C_ℓ are referred to as induced ℓ-path and induced ℓ-cycle, respectively. Early results by Brouwer and Veldman [4] show that Path Contraction and Cycle Contraction are NP-complete even when the target graph (which is a path or a cycle) has four vertices (i.e., when $k = n - 4$). However, Path Contraction and Cycle Contraction are polynomial-time solvable when the target graph is a P_3 or a C_3 [4]. In fact, Brouwer and Veldman [4] showed that Path Contraction is NP-complete when the target path has ℓ vertices for every $\ell \geq 4$. Later, Hammack [9] showed that Cycle Contraction is NP-complete when the target cycle has ℓ vertices for every $\ell \geq 5$. These results imply that Path Contraction and Cycle Contraction are NP-complete in general.

Path Contraction and Tree Contraction were the first graph contraction problems to be studied in the parameterized complexity framework. The most natural parameter is k and Heggernes et al. [10] showed that Path Contraction admits a kernel with at most $5k + 3$ vertices and an FPT algorithm running in $2^{k+\mathcal{O}(\sqrt{k}\log k)}$ time. They also showed that Path Contraction can be solved in $\mathcal{O}^*(2^n)$ time[1] by a simple algorithm – color the graph with two colors, contract the connected components in the monochromatic subgraphs and check if the resulting graph is a path. Later, Li et al. [13] improved the size of the kernel to at most $3k + 4$ vertices. While there is no improvement to the $2^{k+\mathcal{O}(\sqrt{k}\log k)}$-time algorithm known till date, Agrawal et al. [1] gave an $\mathcal{O}^*(1.99987^n)$-time algorithm breaking the $\mathcal{O}^*(2^n)$ barrier.

Our approach to Path Contraction has a different style when compared to these results. We focus on the following generalization of Path Contraction. Refer to Sect. 2 for the definition of witness structure.

[1] $\mathcal{O}^*(.)$ suppresses polynomial factors.

> PATH CONTRACTION WITH CONSTRAINED ENDS
> **Input:** A connected graph G, two disjoint subsets $X, Y \subseteq V(G)$ and an integer k.
> **Question:** Can one contract at most k edges in G to obtain a path with witness structure $(W_1, \ldots, W_\ell)$ such that $X \subseteq W_1$ and $Y \subseteq W_\ell$?

This problem is NP-hard as it generalizes the 2-DISJOINT CONNECTED SUBGRAPHS problem which is known to be NP-hard [12]. Our first result is an algorithm for PATH CONTRACTION WITH CONSTRAINED ENDS using a dynamic programming formulation that crucially uses the notions of *potential k-witness sets* (Definition 1) and *potential k-prefix sets* (Definition 2). It is inspired by an exact exponential-time algorithm by Agrawal et al. [1] for PATH CONTRACTION and an FPT algorithm by Saurabh et al. [14] for GRID CONTRACTION.

Theorem 1. *PATH CONTRACTION WITH CONSTRAINED ENDS can be solved in $\mathcal{O}^*(2^{k-|X|-|Y|})$ time.*

Observe that PATH CONTRACTION WITH CONSTRAINED ENDS when $X = \emptyset$ and $Y = \emptyset$ is PATH CONTRACTION. Therefore, we obtain the following result.

Corollary 1. *PATH CONTRACTION can be solved in $\mathcal{O}^*(2^k)$ time.*

This is the first improvement for PATH CONTRACTION since its fixed-parameter tractability was established by Heggernes et al. [10].

Our original motivation for focusing on PATH CONTRACTION WITH CONSTRAINED ENDS was to understand its applicability in designing faster algorithms for CYCLE CONTRACTION since deleting one vertex from a cycle results in a path whose endpoints are neighbours of the deleted vertex. As cycles have treewidth 2, one may use Courcelle's theorem to show the existence of FPT algorithms for CYCLE CONTRACTION (see [5, Chapter 7] for related definitions and arguments). However, the resulting algorithm has huge exponential factors in the running time. Belmonte et al. [3] showed that CYCLE CONTRACTION admits a kernel with at most $6k+6$ vertices which was later improved to $5k+4$ by Sheng and Sun [15]. This immediately results in a single-exponential time algorithm, i.e., an $\mathcal{O}^*(c^k)$-time algorithm for some constant $c > 200$, for CYCLE CONTRACTION – simply color the graph (on at most $5k + 4$ vertices) with two or three colors, contract the connected components in the monochromatic subgraphs and check if the resultant graph is a cycle. We describe a faster algorithm using the algorithm for PATH CONTRACTION WITH CONSTRAINED ENDS as a subroutine.

Theorem 2. *CYCLE CONTRACTION can be solved in $\mathcal{O}^*((2 + \epsilon_\ell)^k)$ time where $0 < \epsilon_\ell \leq 0.5509$ and ϵ_ℓ decreases as $\ell = n - k$ increases.*

This also implies an exact exponential-time algorithm for CYCLE CONTRACTION that runs in $\mathcal{O}^*(2.5509^n)$ time. We give a faster algorithm that runs in $\mathcal{O}^*(2.5191^n)$ time (Theorem 5).

CYCLE CONTRACTION is related to an important graph parameter, namely, *cyclicity*. The cyclicity of a graph is the largest integer ℓ such that the graph is contractible to C_ℓ. Computing cyclicity is NP-hard in general [4], however, Hammack [8] showed that the cyclicity of planar graphs can be computed in polynomial time. Therefore, CYCLE CONTRACTION is polynomial-time solvable on planar graphs. To the best of our knowledge, the status of PATH CONTRACTION on planar graphs is open. We resolve this question by showing the following result.

Theorem 3. *PATH CONTRACTION on planar graphs can be solved in polynomial time.*

This result uses the CYCLE CONTRACTION algorithm on planar graphs as a subroutine to obtain a polynomial-time algorithm for PATH CONTRACTION. Note that TREE CONTRACTION is NP-hard on planar graphs [2]. Hence, planar graphs is an example graph class where PATH CONTRACTION is easier than TREE CONTRACTION.

On chordal graphs, while CYCLE CONTRACTION is trivially solvable (in linear time), the status of PATH CONTRACTION is interesting. Heggernes et al. [11] proved that PATH CONTRACTION is polynomial-time solvable on chordal graphs by giving an $\mathcal{O}(nm)$-time algorithm where m is the number of edges in the input graph. This is also an $\mathcal{O}(n^2 \cdot tw)$-time algorithm where tw is the treewidth (the size of a maximum clique minus 1) of G. We prove that this algorithm is optimal in the following sense.

Theorem 4. *Unless the Orthogonal Vectors Conjecture is false, PATH CONTRACTION on chordal graphs cannot be solved in $\mathcal{O}(n^{2-\epsilon} \cdot 2^{o(tw)})$ time for any $\epsilon > 0$.*

This result specifically shows that the $\mathcal{O}(n^2 \cdot tw)$-time algorithm given by Heggernes et al. is optimal – if one wishes to reduce the quadratic dependency on n, an exponential overhead in terms of treewidth is inevitable. Heggernes et al. [11] also showed that TREE CONTRACTION on chordal graphs admits an $\mathcal{O}(n+m)$-time algorithm. This result combined with Theorem 4 implies that chordal graphs is an example graph class where TREE CONTRACTION is easier than PATH CONTRACTION.

Due to space constraints, the proofs of most results are omitted. The proof of such results can be found in the full paper.

2 Preliminaries

For $k \in \mathbb{N}$, $[k]$ denotes the set $\{1, 2, \dots, k\}$. We refer to the book by Diestel [6] for standard graph-theoretic definitions and terminology not defined here. For details on parameterized algorithms and complexity, we refer to the book by Cygan et al. [5]. For an undirected graph G, sets $V(G)$ and $E(G)$ denote its set of vertices and edges, respectively. For a vertex v, $N_G(v)$ denotes the set of its neighbors in G. We omit the subscript in the notation for neighborhood if the graph under consideration is clear. For a set $S \subseteq V(G)$, $\overline{S}$ denotes the set

$V(G) \setminus S$. For a set $S \subseteq V(G)$, we denote the graph obtained by deleting S from G by $G - S$ and the subgraph of G induced on the set S by $G[S]$. For two subsets $S_1, S_2 \subseteq V(G)$, let $E(S_1, S_2)$ denote the set of edges with with one endpoint in S_1 and other in S_2 and we say that S_1 and S_2 are adjacent if $E(S_1, S_2) \neq \emptyset$. A set S of $V(G)$ is said to be a *connected set* if $G[S]$ is connected. We state the following two notions of special connected sets given in [1]. For integers $\alpha, \beta \in \mathbb{N}$ and a non-empty set $Q \subseteq V(G)$, a connected set X is a (Q, α, β)-connected set if $Q \subseteq X$, $|X| = \alpha$, and $|N(X)| = \beta$. For integers $\alpha, \beta \in \mathbb{N}$, a connected set X is called an (α, β)-connected set if $|X| = \alpha$ and $|N(X)| = \beta$. Next, we state a result on the enumeration of (Q, α, β)-connected sets and (α, β)-connected sets.

Proposition 1. [1,7] *Given integers $\alpha, \beta \in \mathbb{N}$, the number of (α, β)-connected sets in G is at most $2^{\alpha+\beta} \cdot n$ and the set of all (α, β)-connected sets can be enumerated in $\mathcal{O}^*(2^{\alpha+\beta})$ time. Further, given a non-empty set $Q \subseteq V(G)$, the number of (Q, α, β)-connected sets in G is at most $2^{\alpha+\beta-|Q|}$ and the set of all (Q, α, β)-connected sets can be enumerated in $\mathcal{O}^*(2^{\alpha+\beta-|Q|})$ time.*

Edge Contractions. For a graph G and an edge $e = uv \in E(G)$, G/e denote the graph obtained from G by contracting edge e. A graph G is said to be *contractible* to the graph H if there is an onto function $\psi : V(G) \to V(H)$ such that the following properties hold (1) For any vertex $h \in V(H)$, $G[W(h)]$ is connected and not empty where $W(h) = \{v \in V(G) \mid \psi(v) = h\}$ (2) For any two vertices $h, h' \in V(H)$, $W(h)$, $W(h')$ are adjacent in G if and only if $hh' \in E(H)$. For a vertex h in H, set $W(h)$ is called a *witness set* associated with/corresponding to h. We define *H-witness structure* of G, denoted by $\mathcal{W}$, as a collection of all witness sets. Note that a witness structure $\mathcal{W}$ is a partition of $V(G)$. If G is contractible to a P_ℓ or C_ℓ, we denote the corresponding witness structure as a sequence $(W_1, \ldots, W_\ell)$ to emphasize the order of the witness sets. For a cycle witness structure $(W_1, \ldots, W_\ell)$, W_1 and W_ℓ are regarded as consecutive witness sets just like the other obvious consecutive witness sets. We say that G is *k-contractible* to H if H can be obtained from G by at most k edge contractions, i.e., $\sum_{i=1}^{\ell}(|W_i| - 1) \leq k$.

3 Path Contraction

In this section, we describe an algorithm to solve Path Contraction With Constrained Ends. Consider an instance (G, X, Y, k). Let n denote the number of vertices in G. We begin with the following observation on the structure of a witness set in the required path witness structure (if it exists).

Observation 1. *If G is k-contractible to P_ℓ with witness structure $(W_1, \ldots, W_\ell)$, then $\sum_{i=1}^{\ell} |W_i| \leq k + \ell$.*

Observation 1 immediately leads to a $\mathcal{O}^*(2^{k-|X|-|Y|})$-time algorithm for the problem when ℓ is a constant.

Lemma 1. *There is an algorithm that determines if G is k-contractible to a path of at most five vertices with the first witness set containing X and the last witness set containing Y in $\mathcal{O}^*(2^{k-|X|-|Y|})$ time.*

Subsequently, we consider the case $\ell \geq 6$, i.e. $k \leq n-6$. We critically use the fact that any path witness structure of G has at least 6 witness sets to bound the cardinality of the closed neighborhood of a witness set.

Observation 2. *Suppose G is k-contractible to P_ℓ with $\ell \geq 6$ and $(W_1, W_2, \ldots, W_\ell)$ is the corresponding witness structure with $X \subseteq W_1$ and $Y \subseteq W_\ell$.*

1. *For any $2 \leq i \leq \ell - 1$, $G - W_i$ has two connected components C_X, C_Y containing X and Y, respectively.*
2. *For any $2 \leq i \leq \ell-1$, $|(N(W_i) \cap W_{i-1}) \setminus X| + |W_i| + |(N(W_i) \cap W_{i+1}) \setminus Y| \leq k + 5 - |X| - |Y|$.*
3. *For any $2 \leq i \leq \ell - 2$, $|(N(W_i) \cap W_{i-1}) \setminus X| + |W_i| + |W_{i+1}| + |(N(W_{i+1}) \setminus W_i) \setminus Y| \leq k + 6 - |X| - |Y|$.*

The first two properties are about a witness set whereas the third property is about a pair of consecutive witness sets. The first property states that a witness set W is essentially a connected set in G whose deletion results in exactly two components C_X and C_Y with $X \subseteq C_X$ and $Y \subseteq C_Y$. We use the second property to efficiently enumerate all candidate witness sets W along with the sets $C_X \cup W$ and $C_Y \cup W$. Triples consisting of a candidate witness set W, one of the sets $C_X \cup W$ or $C_Y \cup W$ and an integer in $[k] \cup \{0\}$ will describe a state of the dynamic programming table. The third property is used to bound the time needed for the algorithm to update all the entries in this table.

3.1 Potential Witness Sets and Potential Prefix Sets

The first two properties in Observation 2 motivates the notions of *potential k-witness sets* and *potential k-prefix sets.* Intuitively, if G is k-contractible to P_ℓ with the witness structure $(W_1, W_2, \ldots, W_\ell)$ satisfying the required properties, a potential k-witness set is a candidate for W_i with $2 \leq i \leq \ell-1$ and a potential k-prefix set is a candidate for the set of vertices in a 'prefix witness set'. See Fig. 1 for an illustration.

Definition 1 (Potential k-witness Set). *A connected set $W \subseteq V(G) \setminus (X \cup Y)$ is called a potential k-witness set of G if $|(N(W) \setminus (X \cup Y)| + |W| \leq k + 5 - |X| - |Y|$ and $G - W$ has two connected components C_X, C_Y containing X and Y, respectively.*

Note that every non-terminal witness set in a P_ℓ-witness structure of G is a potential k-witness set. Next, we define the notion of a *potential k-prefix set.*

Definition 2 (Potential k-prefix Set). *For a potential k-witness set $W \subseteq V(G) \setminus (X \cup Y)$ with C_X and C_Y being the two connected components of $G - W$ containing X and Y, respectively, the sets $S_X := C_X \cup W$ and $S_Y := C_Y \cup W$ are called the* potential k-prefix sets *associated with W.*

Note that a potential k-prefix set is a non-empty connected set $S \subseteq V(G)$ that contains a potential k-witness set W such that $G[S] - W$ is one of the connected components of $G - W$. The number of potential k-prefix sets is twice the number of potential k-witness sets since every potential k-witness set is associated with two potential k-prefix sets. However, we note that a potential k-prefix set S can be associated with more than one potential k-witness set W. Formally, there may exist potential k-witness sets $W \neq W'$ such that $S = C \cup W = C' \cup W'$, where C and C' are the connected components of $G - W$ and $G - W'$, respectively.

We will describe a dynamic programming algorithm in which every entry of the dynamic programming table that we subsequently construct will be indexed by a triple consisting of a potential k-witness set, a potential k-prefix set and a non-negative integer that is at most k. Therefore, our algorithm's running time will crucially depend upon an efficient enumeration of all potential k-witness sets and all potential k-prefix sets. Targeting towards this enumeration, we have the following lemma.

Lemma 2. *For given integers α, β, the number of connected sets $W \subseteq V(G) \setminus (X \cup Y)$ in G such that $|W| = \alpha$ and $|N(W) \setminus (X \cup Y)| = \beta$ is $\mathcal{O}^*(2^{\alpha+\beta})$ and these sets can be enumerated in $\mathcal{O}^*(2^{\alpha+\beta})$ time.*

As every potential k-witness set is also a connected set $W \subseteq V(G) \setminus (X \cup Y)$ such that $|W| = \alpha$, $|N(W) \setminus (X \cup Y)| = \beta$ and $\alpha + \beta \leq k + 5 - |X| - |Y|$, we have the following corollary.

Corollary 2. *The number of potential k-witness sets in G is $\mathcal{O}^*(2^{k-|X|-|Y|})$ and these sets can be enumerated in $\mathcal{O}^*(2^{k-|X|-|Y|})$ time.*

3.2 A Dynamic Programming Algorithm

Towards defining the entries in the dynamic programming table Γ, we define the notion of a *valid index*. Informally, it is a tuple consisting of a potential k-witness set W, a potential k-prefix set S associated with W, and an integer $k' \in \{0\} \cup [k]$. The value corresponding to this index is set to `True`, if $G[S]$ is k'-contractible to a path (with desired properties) and W is one of its end witness set. Note that there are two types of potential k-prefix sets, one containing X and the other containing Y. We differentiate between these two types by adding a subscript, i.e., S_X denotes a potential k-prefix set containing X and S_Y denotes a potential k-prefix set containing Y.

Definition 3 (Valid Index). *For sets $S_X, W \subseteq V(G)$ and $k' \in \{0\} \cup [k]$, the triple (S_X, W, k') is a* valid index *if*

- *W is a potential k-witness set,*
- *S_X is a potential k-prefix set associated with W, and*
- *$k' \in \{0\} \cup [k]$ and $k' + |N(S_X)| - 1 \leq k$.*

Similarly, we define valid indices corresponding to $S_Y, W \subseteq V(G)$ and $k' \in \{0\} \cup [k]$.

The table Γ has entries only corresponding to valid indices. We define

$$\Gamma[S_X, W, k'] = \texttt{True} \text{ if and only if } G[S_X] \text{ is } k'\text{-contractible to a path with witness structure } (W_1, \ldots, W_q = W) \text{ for some } q \geq 2 \text{ such that } X \subseteq W_1$$

and

$$\Gamma[S_Y, W, k''] = \texttt{True} \text{ if and only if } G[S_Y] \text{ is } k''\text{-contractible to a path with witness structure } (W'_1, \ldots, W'_p = W) \text{ for some } p \geq 2 \text{ such that } Y \subseteq W'_1$$

Note that the third condition in Definition 3 insists that k' and k'' are small enough to accommodate 'future' edge contractions which are at least $|N(S_X)| - 1$ or $|N(S_Y)| - 1$ in number, respectively. Next we describe the initialization of entries of Γ. We describe the algorithm for entries of type $\Gamma(S_X, W, k')$ and a similar process is applicable for entries of type $\Gamma(S_Y, W, k')$. First, all entries are set to `False`. Then, the algorithm uses Corollary 2 to enumerate all potential k-witness sets of a predefined size and neighbourhood size (which we specify later). In this step, the algorithm enumerates all the subsets that are potential candidates for either W_2 or $W_{\ell-1}$. For each such set W, let S_X and S_Y be the two potential k-prefix sets associated with W. The algorithm sets $\Gamma[S_X, W, k'] = \texttt{True}$ for every $|S_X| - 2 \leq k' \leq k - |N(S_X)| + 1$. Similarly, it sets $\Gamma[S_Y, W, k'] = \texttt{True}$ for every $|S_Y| - 2 \leq k' \leq k - |N(S_Y)| + 1$. The correctness of this initialization follows from the definition. Observe that in this phase we only care if $G[S_X]$ is k'-contractible to a path on two vertices with the corresponding witness structure containing W as the second witness set and set the entry $\Gamma[S_X, W, k']$ to be `True` if this is indeed possible.

We now describe how the algorithm updates the entries in Γ. Unlike most of the dynamic programming algorithms, our algorithm does not 'look back' previously set `True` entries to update an entry that is set to `False` in the initialization phase. Instead, it checks entries $\Gamma[S_X, W, k']$ that are `True` and 'looks forward' and sets certain entries $\Gamma[S_X^\circ, A, k^\circ]$ to `True` where $S_X \subseteq S_X^\circ$, $(N(W) \setminus S_X) \subseteq A$, and $k' + |A| - 1 \leq k^\circ \leq k - |N_{G-S_X}(A)| + 1$. Intuitively, A is a subset that can be appended to the path witness structure of $G[S_X]$ as a new witness set to obtain a path witness structure of $G[S_X \cup A]$. Hence, if we can obtain a path from $G[S_X]$ (with desired properties) by contracting at most r edges, then we can obtain a path from $G[S_X \cup A]$ (with desired properties) by contracting at most $r + |A| - 1$ edges. Thus, $\Gamma[S_X^\circ, A, k^\circ]$ can be set to `True`. Refer to Fig. 1 for an illustration.

This brings us to the following notion of *potential* (S_X, W)*-attachments.*

Definition 4 (Potential (S_X, W)-attachment). *For a potential k-witness set W and the potential k-prefix set S_X associated with W, the set $\mathcal{A}(S_X, W)$ of all potential (S_X, W)-attachments is defined as $\bigcup_{a=|N(S_X)|}^{k+1} \mathcal{A}_a(S_X, W)$ where*

$$\mathcal{A}_a(S_X, W) = \{A \subseteq V(G) \setminus S_X \mid G[A] \textit{ is connected, } N(S_X) \subseteq A, |A| = a, \textit{ and } |(N(W) \cap S_X) \setminus X| + |W| + |A| + |(N(A) \setminus S_X) \setminus Y| \leq k + 6 - |X| - |Y|\}$$

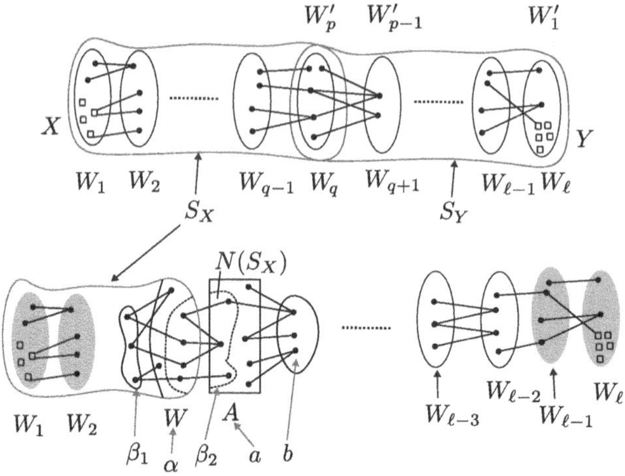

Fig. 1. (Top) W_q is one of the potential k-witness sets and S_X and S_Y are the corresponding potential k-prefix sets. (Bottom) W_1, W_2, W_ℓ, $W_{\ell-1}$ (shaded sets) are identified during the initialization phase. For a potential k-witness set W and the corresponding potential k-prefix set S_X, the algorithm enumerates all the possible sets A that can be appended to a path witness structure of $G[S_X]$.

We now justify the upper bound on the cardinality of sets mentioned in the definition. Suppose G is k-contractible to P_ℓ with witness structure $(W_1, \ldots, W_{i-1}, W_i, W_{i+1}, W_{i+2}, \ldots, W_\ell)$ where $X \subseteq W_1$ and $Y \subseteq W_\ell$. In the initialization phase, the algorithm identifies sets W_1, W_2 and $W_{\ell-1}, W_\ell$. Hence, a set A in the potential (S_X, W)-attachment $\mathcal{A}_a(S_X, W)$ corresponds to the case when W is some internal witness set W_i where $2 \leq i \leq \ell - 2$. In fact, we can improve the upper bound to $\ell - 3$ and our algorithm tries to expand S_X only till $W = W_{\ell-3}$ to find A which is a potential candidate for $W_{\ell-2}$. This implies that $W_1, W_{i-1}, W_i, W_{i+1}, W_{i+2}, W_\ell$ are six different witness sets when $2 < i \leq \ell - 3$ and are five different sets when $i = 2$ since $6 \leq \ell$.

By the definitions of the relevant sets, when $W = W_i$ we have $S_X = W_1 \cup W_2 \cup \cdots \cup W_i$. Hence, $N(W) \cap S_X$ is part of W_{i-1}, A is a potential candidate for W_{i+1} and $N(A) \setminus S_X$ is part of W_{i+2}. As X and Y are part of W_1 and W_ℓ, respectively, for $2 < i \leq \ell - 3$ we have

$$|W_1| + |W_{i-1}| + |W_i| + |W_{i+1}| + |W_{i+2}| + |W_\ell| \leq k + 6$$

$$|X| + |N(W) \cap S_X| + |W| + |A| + |N(A) \setminus S_X| + |Y| \leq k + 6$$

$$|N(W) \cap S_X| + |W| + |A| + |N(A) \setminus S_X| \leq k + 6 - |X| - |Y| \quad (1)$$

For $i = 2$, we have $W_1 = W_{i-1}$ and hence

$$|(N(W) \cap S_X) \setminus X| + |W| + |A| + |N(A) \setminus S_X| \leq k + 5 - |X| - |Y| \quad (2)$$

Note that Eq. 1 subsumes Eq. 2. Coming back to the updating phase of the algorithm, it processes each entry $\Gamma[S_X, W, k']$ in the increasing order of $|S_X|$,

$|W|$ and k'. If $\Gamma[S_X, W, k'] = \texttt{False}$, then it skips to the next entry. Otherwise, it compute $\mathcal{A}(S_X, W)$ (we specify the steps in this computation later). Then, for every set A in $\mathcal{A}(S_X, W)$ and every $k'+(|A|-1) \leq k^\circ \leq k-|N_{G-S_X}(A)|+1$, it sets $\Gamma[S_X \cup A, A, k^\circ] = \texttt{True}$. This completes the updating phase of the algorithm.

In the final phase, the algorithm iterates over all potential k-witness sets W and the corresponding valid entries. Suppose S_X and S_Y are the two potential k-prefix sets associated with W. If there is pair of integers k_1, k_2 such that $k_1 + k_2 - (|W| - 1) \leq k$, and $\Gamma[S_X, W, k_1] = \texttt{True}$ and $\Gamma[S_Y, W, k_2] = \texttt{True}$, then the algorithm concludes that G is k-contractible to a path with desired properties. This concludes the description of the algorithm. We show that this algorithm correctly computes $\Gamma[S_X, W, k']$ for every valid index (S_X, W, k') in $\mathcal{O}^*(2^{k-|X|-|Y|})$ time. This establishes Theorem 1. As Path Contraction is Path Contraction With Constrained Ends when $X = \emptyset$ and $Y = \emptyset$, we obtain Corollary 1.

4 Cycle Contraction

In this section, we give improved FPT and exponential-time algorithms for Cycle Contraction. Let n denote the number of vertices in G.

Let (G, k) be an instance of Cycle Contraction. If $k \geq n-3$, then (G, k) is a Yes-instance if and only if G is k-contractible to a triangle (which can be determined in polynomial time). Hence, we may assume that $k \leq n-4$ and so we wish to determine if G is contractible to C_ℓ where $\ell \geq 4$. We begin with two basic observations.

Observation 3. *If G is k-contractible to C_ℓ with witness structure $(W_1, \ldots, W_\ell)$, then $\sum_{i=1}^{\ell}(|W_i| - 1) \leq k$.*

Observation 4. *If G is k-contractible to C_ℓ where $\ell \geq 4$ with witness structure $(W_1, W_2, \ldots, W_\ell)$, then is a set of three consecutive witness sets W_i, W_{i+1}, W_{i+2} such that $|W_i| + |W_{i+1}| + |W_{i+2}| \leq 3 + \frac{3k}{\ell}$.*

Observation 3 immediately leads to a $\mathcal{O}^*(2^k)$-time algorithm for determining if G is k-contractible to C_4 or not.

Lemma 3. *There is an algorithm that determines if G is k-contractible to C_4 or not in $\mathcal{O}^*(2^k)$ time.*

Next, we prove the following lemma on enumerating connected sets with restrictions on its size and its neighbourhood's size.

Lemma 4. *Given integers $\alpha, \beta \in \mathbb{N}$, the number of triples (X, Z, Y) where Z is a (α, β)-connected set and $X \uplus Y = N(Z)$ is $3^{\alpha+\beta} \cdot n$ and the set of all such triples can be enumerated in $\mathcal{O}^*(3^{\alpha+\beta})$ time.*

Now, we have the following result.

Theorem 2. *Cycle Contraction can be solved in $\mathcal{O}^*((2+\epsilon_\ell)^k)$ time where $0 < \epsilon_\ell \leq 0.5509$ and ϵ_ℓ decreases as $\ell = n - k$ increases.*

Proof. (sketch) Suppose G is k-contractible to C_ℓ with witness structure $(W_1, W_2, \ldots, W_\ell)$. From Observation 4, we know that there is a set of three consecutive witness whose total size is at most $3 + \frac{3k}{\ell}$. Without loss of generality, we may assume that these sets are W_1, W_2 and W_3. Observe that W_1, W_2 and W_3 are connected sets with $N(W_2) \subseteq W_1 \cup W_3$. Let $X = W_1 \cap N(W_2)$ and $Y = W_3 \cap N(W_2)$. Once W_2, X and Y is known, the problem reduces to solving a Path Contraction With Constrained Ends instance of determining if $G - W_2$ is k'-contractible to a path with witness structure $(W'_1, W'_2, \ldots, W'_q)$ where $X \subseteq W'_1$ and $Y \subseteq W'_q$ where $k' = k - (|W_2| - 1)$. Such a path witness structure $(W'_1, W'_2, \ldots, W'_q)$ along with W_2 is a cycle witness structure of G. An important subroutine in this algorithm is an efficient enumeration of (X, W_2, Y) given by Lemma 4. □

Since $1 \leq k \leq n - 1$, by invoking this algorithm for different values of k, we can determine the largest cycle to which G can be contracted to in $\mathcal{O}^*(2.5509^n)$ time. We give a faster algorithm for this problem showing the following result.

Theorem 5. *Cycle Contraction can be solved in* $\mathcal{O}^*(2.5191^n)$ *time.*

The idea is as follows. First, we describe an algorithm that determines the longest even cycle C_ℓ to which G is contractible to in $\mathcal{O}^*(2^n)$ time. Further, this algorithm also enumerates the set of all C_ℓ-witness structures of G in $\mathcal{O}^*(2^n)$ time. The largest cycle to which G can be contracted to is either C_ℓ or $C_{\ell+1}$. Suppose G is contractible to $C_{\ell+1}$ with $(W_1, W_2, W_3, \ldots, W_\ell, W_{\ell+1})$ being the corresponding witness structure. Then, by contracting one more edge with one endpoint in W_i and the other endpoint in W_{i+1}, we get a C_ℓ-witness structure $(W_1, W_2, W_3, \ldots, W_i \cup W_{i+1}, \ldots, W_\ell, W_{\ell+1})$ of G. Therefore, we iterate over all C_ℓ-witness structures and determine which edge to "uncontract" in order to get a $C_{\ell+1}$-witness structure (if one exists).

5 Planar Graphs and Chordal Graphs

In this section, we discuss the complexity of Path Contraction and Cycle Contraction on planar graphs and chordal graphs. *Planar graphs* are graphs that can be drawn on the plane such that no pair of edges intersect except possibly at their endpoints. Such a drawing is called a *planar embedding*. Edge contractions play an important role in planar graph theory and planarity is preserved under edge contractions. Hammack [8] gave a polynomial-time algorithm that determines the largest cycle to which the given planar graph is contractible to. We use this algorithm as a subroutine to solve Path Contraction on planar graphs. The idea is as follows. Let G be a connected planar graph with a planar embedding fixed. If $|V(G)| < k + 2$, (G, k) is a trivial Yes-instance. Subsequently, assume $|V(G)| \geq k + 2$. Suppose G is k-contractible to a path with witness structure $(W_1, W_2, \ldots, W_q)$. Then, there are vertices $x \in W_1$ and $y \in W_q$ that are in the same face of the embedding. Construct a graph $G_{x,y}$ by adding a path $(z_1, \ldots, z_{k+1})$ of $k + 1$ new vertices to G such that z_1 is adjacent to x and z_{k+1} is adjacent to y. Observe that $G_{x,y}$ is also planar. We claim that G is k-contractible to a path with witness structure $(W_1, W_2, \ldots, W_q)$ such that $x \in W_1$ and $y \in W_q$ if and only if $G_{x,y}$ is k-contractible to a cycle. Now, in order to determine if G is k-contractible to a path, we go over all possible choices of x and y, construct $G_{x,y}$, use the algorithm of Hammack [8] to determing if $G_{x,y}$ is k-contractible to a cycle or not which in turn by determines if G is k-contractible to a path or not. As the number of choices for x, y is at most the number of pairs of vertices of G that are in the same face, we obtain a polynomial-time algorithm for Path Contraction on planar graphs.

Now, we move on to chordal graphs. Chordal graphs are graphs in which every induced cycle is a triangle. Chordal graphs can be recognized in linear time and are amenable to linear time algorithms for several problems that are NP-complete in general. Chordality is a hereditary property and edge contractions preserve chordality. It is easy to verify that Cycle Contraction is linear-time solvable on chordal graphs – a chordal graph is contractible to a cycle if and only if it is contractible to a triangle. Heggernes et al. [11] proved that Path Contraction is polynomial-time solvable on chordal graphs by giving an $\mathcal{O}(nm)$-time algorithm where m is the number of edges in the input graph. This algorithm is an $\mathcal{O}(n^2 \cdot tw)$-time algorithm where tw is the treewidth of the input graph[2]. We give a fine-grained reduction from Orthogonal Vectors[3] to Path Contraction to conclude that no $\mathcal{O}(n^{2-\epsilon} \cdot 2^{o(tw)})$-time algorithms for any $\epsilon > 0$ is likely under the Orthogonal Vectors Conjecture. We remark that this fine-grained complexity analysis is non-trivial only when the input graph is sparse; if $m = \mathcal{O}(n^2)$, then we need to spend $\Omega(n^2)$ time just to read the graph.

6 Concluding Remarks

Just as breaking the $\mathcal{O}^*(2^n)$ barrier proved challenging for Path Contraction, we believe that beating the $\mathcal{O}^*(2^k)$ bound would be interesting. Similarly, coming up with an $\mathcal{O}^*(2^k)$-time algorithm for Cycle Contraction is a natural future direction. As such an algorithm would result in solving Cycle Contraction in $\mathcal{O}^*(2^n)$ time, coming up with an $\mathcal{O}^*(2^n)$-time algorithm for contracting to C_5 maybe a first step towards this goal. Recall that our algorithm for Cycle Contraction has the worst case running time when the target graph is C_5.

Acknowledgement. We thank Roohani Sharma for initial discussions.

Disclosure of Interests. The authors have no competing interests to declare that are relevant to the content of this article.

[2] The treewidth of a graph measures how close the graph is to a tree. See [5] for the definitions of tree decomposition and treewidth.

[3] In Orthogonal Vectors, given two sets of n d-dimensional boolean vectors X and Y, the objective is to determine if there is a pair of vectors $x \in X$ and $y \in Y$ that are orthogonal. The problem can be solved in $\mathcal{O}(n^2 d)$ time and the Orthogonal Vectors Conjecture [16] states that it cannot be solved in $\mathcal{O}(n^{2-\epsilon})$ time for any $\epsilon > 0$ when $d = \mathcal{O}(\log n)$.

References

1. Agrawal, A., Fomin, F.V., Lokshtanov, D., Saurabh, S., Tale, P.: Path contraction faster than 2^n. SIAM J. Discret. Math. **34**(2), 1302–1325 (2020). https://doi.org/10.1137/19M1259638
2. Asano, T., Hirata, T.: Edge-contraction problems. J. Comput. Syst. Sci. **26**(2), 197–208 (1983). https://doi.org/10.1016/0022-0000(83)90012-0
3. Belmonte, R., Golovach, P.A., van 't Hof, P., Paulusma, D.: Parameterized complexity of three edge contraction problems with degree constraints. Acta Informatica **51**(7), 473–497 (2014). https://doi.org/10.1007/s00236-014-0204-z
4. Brouwer, A.E., Veldman, H.J.: Contractibility and NP-completeness. J. Graph Theory **11**(1), 71–79 (1987). https://doi.org/10.1002/jgt.3190110111
5. Cygan, M., et al.: Parameterized Algorithms. Springer, Cham (2015). https://doi.org/10.1007/978-3-319-21275-3
6. Diestel, R.: Graph Theory. Graduate texts in mathematics, vol. 173, 4th edn. Springer, Berlin (2012)
7. Fomin, F.V., Villanger, Y.: Treewidth computation and extremal combinatorics. Combinatorica **32**(3), 289–308 (2012). https://doi.org/10.1007/S00493-012-2536-Z
8. Hammack, R.H.: Cyclicity of graphs. J. Graph Theory **32**(2), 160–170 (1999). https://doi.org/10.1002/(SICI)1097-0118(199910)32:2<160::AID-JGT6>3.0.CO;2-U
9. Hammack, R.H.: A note on the complexity of computing cyclicity. Ars Comb. **63** (2002)
10. Heggernes, P., van 't Hof, P., Lévêque, B., Lokshtanov, D., Paul, C.: Contracting graphs to paths and trees. Algorithmica **68**(1), 109–132 (2014). https://doi.org/10.1007/s00453-012-9670-2
11. Heggernes, P., van 't Hof, P., Lévêque, B., Paul, C.: Contracting chordal graphs and bipartite graphs to paths and trees. Discret. Appl. Math. **164**, 444–449 (2014). https://doi.org/10.1016/j.dam.2013.02.025
12. van 't Hof, P., Paulusma, D., Woeginger, G.J.: Partitioning graphs into connected parts. Theor. Comput. Sci. **410**(47-49), 4834–4843 (2009). https://doi.org/10.1016/J.TCS.2009.06.028
13. Li, W., Feng, Q., Chen, J., Hu, S.: Improved kernel results for some fpt problems based on simple observations. Theor. Comput. Sci. **657**, 20–27 (2017). https://doi.org/10.1016/j.tcs.2016.06.012, https://www.sciencedirect.com/science/article/pii/S0304397516302481. frontiers of Algorithmics
14. Saurabh, S., dos Santos Souza, U., Tale, P.: On the parameterized complexity of grid contraction. In: 17th Scandinavian Symposium and Workshops on Algorithm Theory, SWAT. LIPIcs, vol. 162, pp. 1–17 (2020). https://doi.org/10.4230/LIPIcs.SWAT.2020.34
15. Sheng, B., Sun, Y.: An improved linear kernel for the cycle contraction problem. Inf. Process. Lett. **149**, 14–18 (2019). https://doi.org/10.1016/j.ipl.2019.05.003, https://www.sciencedirect.com/science/article/pii/S0020019019300912
16. Williams, R., Yu, H.: Finding orthogonal vectors in discrete structures. In: Chekuri, C. (ed.) Proceedings of the Twenty-Fifth Annual ACM-SIAM Symposium on Discrete Algorithms, SODA 2014, Portland, Oregon, USA, 5-7 January 2014, pp. 1867–1877. SIAM (2014). https://doi.org/10.1137/1.9781611973402.135

Many Views of Planar Point Sets

Jan Kynčl and Jan Soukup(✉)

Department of Applied Mathematics, Faculty of Mathematics and Physics, Charles University, Malostranské nám. 25, 118 00 Praha 1, Czech Republic
{kyncl,soukup}@kam.mff.cuni.cz

Abstract. Given a set P of n points in the plane and two points x and y not in P, such that their union is in general position, we say that x and y have the *same view* of P if the points of P are visible in the same cyclic order from x and y. We show that for every set P of n points in general position in the plane, there are $\Omega(n^4)$ points with mutually distinct views of P, confirming a conjecture by Díaz-Báñez, Fabila-Monroy and Pérez-Lantero and a conjecture by Bieri and Schmidt. We also provide an easier alternative proof for point sets in strong general position.

Keywords: Finite point set · view of a point set · radial ordering · strong general position

1 Introduction

Let P be a set of n points in the plane in general position; that is, no three points of P are collinear. Let x be an additional reference point not in P, such that the set $P \cup \{x\}$ is in general position. The *view* of P from x, also called the *radial ordering* of P with respect to x [8], is the clockwise cyclic order in which a rotating ray with origin at x meets the points of P. How many points with mutually distinct views of P are there?

It is easy to see that the number of mutually distinct views of P is always at most $O(n^4)$: indeed, consider the arrangement of $\binom{n}{2}$ lines drawn through every pair of points from P, and observe that two points within the same cell of this arrangement have the same view of P [3,6–8,14,15]. On the other hand, there are constructions of sets of n points with $\Omega(n^4)$ distinct views [8,15]. The problem of estimating the maximum possible number of distinct views of a set of n points in the plane also appears as Exercise 6.1.7 in Matoušek's textbook [9].

How many distinct views are guaranteed in an arbitrary point set P? Díaz-Báñez, Fabila-Monroy and Pérez-Lantero [8] proved that an arbitrary set P of n points in strong general position in the plane has $\Omega(n^3)$ points with mutually distinct views of P, and conjectured that this lower bound can be improved to $\Omega(n^4)$. Here the condition of *strong general position* means that P is in general position, and three distinct lines determined by pairs of points in P cannot cross at a common point except at a point from P. Bieri and Schmidt [5] conjectured that the number of views of any set of n points in general position is $\Theta(n^4)$, and

D. Král' and M. Milanič (Eds.): WG 2024, LNCS 14760, pp. 387–399, 2025.
https://doi.org/10.1007/978-3-031-75409-8_27

similarly for the number of "projective" cyclic orders obtained by rotating a line (rather than a ray), which we might call *projective views.*

We confirm these conjectures for point sets in strong general position and also confirm that the number of views of any set of n points in general position is $\Theta(n^4)$.

Theorem 1. *For every set P of n points in general position in the plane, there are $\Omega(n^4)$ points with mutually distinct views of P.*

We also provide an alternative proof of a restricted version of Theorem 1 for the case of strong general position, which guarantees the lower bound also for projective views.

Theorem 2. *For every set P of n points in strong general position in the plane, there is a subset $X \subseteq P$ and a set Y of $\Omega(n^4)$ points with mutually distinct views of X, such that the convex hull of Y is disjoint with the convex hull of X.*

We prove Theorem 1 in Sect. 3 and Theorem 2 in Sect. 4. The proof of Theorem 1 is a generalization of the proof of Díaz-Báñez, Fabila-Monroy and Pérez-Lantero [8], who showed a weaker lower bound of $\Omega(n^3)$ on the number of distinct views. The main idea of the proof of Theorem 2 is that every point set in strong general position contains a subset of linear size that "looks like" the special construction of a point set with $\Omega(n^4)$ views [8,15].

2 Preliminaries

We always assume that $|P| \geq 3$ and that P together with the observation points is in general position. We denote the convex hull of a set X by $\text{conv}(X)$. A *triangle* is a convex hull of three points in general position and, in particular, it contains its interior.

For every pair of distinct points a, b, we denote the line passing through them by $\overleftrightarrow{ab}$. Additionally, we denote the ray contained in $\overleftrightarrow{ab}$ with endpoint b and not containing a by $ab\!\rightarrow$, and the ray contained in $\overleftrightarrow{ab}$ with endpoint a and containing b by $\overrightarrow{ab}$. Furthermore, we call the union of rays $ab\!\rightarrow$ and $ba\!\rightarrow$ the *double-ray ab* and denote it by $\leftarrow\!ab\!\rightarrow$.

Let P be a set of n points in the plane. We call the arrangement of $\binom{n}{2}$ double-rays $\leftarrow\!ab\!\rightarrow$ formed by all pairs of points $a, b \in P$ the *double-ray arrangement of P*; see Fig. 1. In the literature, the double-ray arrangement was called an *order partition* [8], or a *CL-arrangement* [7]. The rays in the double-ray arrangement were also called *swap lines* [14].

Consider an observation point in the plane. When we move the point around, the view of P from the point changes exactly when we cross a double-ray from the double-ray arrangement of P. Hence, views of P from points in the same cell of the double-ray arrangement of P are the same, and therefore we may talk about views from cells.

The double-ray arrangement of a three-point set $\{a, b, c\}$ forming a triangle with vertices in clockwise order (a, b, c) has four cells, the views from the three

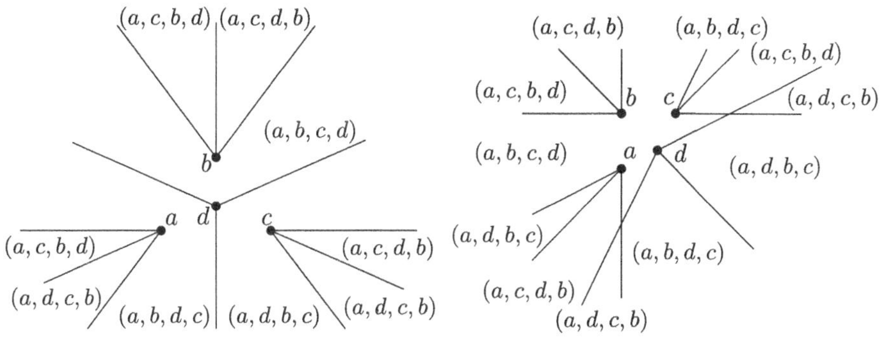

Fig. 1. Six double-rays determined by four points a, b, c, d split the plane into regions. We call this arrangement the double-ray arrangement of $\{a, b, c, d\}$. If the four points are not in convex position (left), the arrangement has nine cells. If they are in convex position and no two double-rays are parallel (right), the arrangement has eleven cells. In both cases, all six possible views of $\{a, b, c, d\}$ are realized.

convex cells are (a, c, b), and the view from the remaining cell is (a, b, c); see Fig. 2. We immediately get the following observation.

Observation 1. *The views of $\{a, b, c\}$ from two reference points p and q are distinct if and only the segment pq intersects an odd number of double-rays in the double-ray arrangement of $\{a, b, c\}$.*

Furthermore, if the views of P from two observation points p and q are distinct, then p and q have distinct views of some triple of points from P. Additionally, in order to prove that the views of P are distinct from two different reference points, it is sufficient that the views of some triple of points of P from the reference points are distinct.

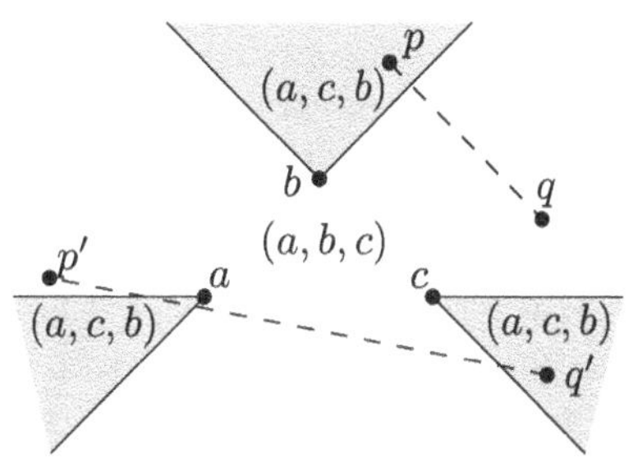

Fig. 2. An arrangement of the three double-rays determined by three points a, b, and c. The three "corner" regions have the same view (a, c, b), and the remaining region has the view (a, b, c). Two reference points have distinct views of $\{a, b, c\}$ if and only if the segment between them intersects exactly one or three rays of the arrangement, as witnessed by pairs (p, q) and (p', q'), respectively.

3 Proof of Theorem 1

Let P be a set of n points in general position in the plane. Our goal is to find $\Omega(n^4)$ reference points with mutually distinct views of P. We do this in two steps. First, we show that the double-ray arrangement of P has $\Omega(n^4)$ cells, generalizing the result of Díaz-Báñez, Fabila-Monroy and Pérez-Lantero who proved this lower bound for point sets in strong general position [8, Theorem 2.4]. Then we show that for $|P| \geq 4$, at most two cells can share the same view. This will imply that there are $\Omega(n^4)$ cells with pairwise distinct views.

3.1 Many Cells in the Double-Ray Arrangement

Theorem 3. *The number of cells in the double-ray arrangement of every set P of n points in general position in the plane is $\Omega(n^4)$.*

Proof. The plane, together with an additional line at infinity, is a representation of the projective plane $\mathbb{R}P^2$. Hence, every double-ray can be considered as a simple curve in the projective plane, intersecting the line at infinity once. The n points of P, together with the $\binom{n}{2}$ double-rays between all pairs of points from P, can then be considered as a drawing D of the complete graph K_n on the projective plane. The drawing D is *simple*, since every pair of edges intersects at most once, either at a common vertex or at a common crossing point. Since we do not assume strong general position, some crossing points can be shared by more than two edges; this is the main reason why obtaining a lower bound on the number of cells is not as straightforward as in the case of strong general position.

For our purposes, it will be beneficial to use the following transformation, which we have learned from Rom Pinchasi [11], and which is described in detail in a recent paper by Ackerman, Damásdi, Keszegh, Pinchasi and Raffay [1, Proof of Theorem 1]. Consider the universal covering map $p : S^2 \to \mathbb{R}P^2$ and let $\widetilde{D}$ be the corresponding lift of D to the sphere S^2. Geometrically, $\widetilde{D}$ can be obtained by a central projection from the plane to a tangent sphere, where every point in the plane is projected to a pair of antipodal points on the sphere. Since the map p is a double-cover, every vertex of D corresponds to a pair of vertices in $\widetilde{D}$, and every double-ray corresponds to a pair of edges in $\widetilde{D}$, each intersecting the *equator*; that is, the great circle that is projected by p to the line at infinity. Every bounded cell in the double-ray arrangement of P corresponds to a pair of cells in $\widetilde{D}$, one in the north hemisphere and one in the south hemisphere. The unbounded cells in the double-ray arrangement of P correspond to cells in $\widetilde{D}$ that intersect or touch the equator. The number of such unbounded cells is $O(n^2)$.

The *degenerate crossing number* of a graph G, denoted by $\mathrm{cr}^*(G)$, is the minimum possible number of crossing points in a simple drawing of G in the plane (or, equivalently, on the sphere), where common crossings of more than two edges are allowed, and each such crossing point is counted only once.

Pach and Tóth [10] proved that the degenerate crossing number of every graph with n vertices and e edges satisfies $\mathrm{cr}^*(G) \geq \Omega(e^4/n^4)$. This was later improved by Ackerman and Pinchasi [2] to $\mathrm{cr}^*(G) \geq \Omega(e^3/n^2)$, which is the same asymptotic lower bound as in the well-known crossing lemma for the standard crossing number. For our purposes, each of these results implies that $\widetilde{D}$ has $\Omega(n^4)$ crossing points, since the drawing $\widetilde{D}$ has $2n$ vertices and $n(n-1)$ edges. Let $\mathrm{cr}^*(\widetilde{D})$ be the number of crossing points in $\widetilde{D}$.

For every $k \geq 2$, let t_k be the number of crossing points in $\widetilde{D}$ shared by exactly k edges. Clearly, we have $\mathrm{cr}^*(\widetilde{D}) = \sum_{k\geq 2} t_k$.

Let $\widetilde{D}'$ be the *planarization* of $\widetilde{D}$; that is, the crossing-free drawing obtained from $\widetilde{D}$ by introducing a new vertex at each crossing point and subdividing the edges accordingly. Let n', e', and f' be the numbers of vertices, edges, and faces of $\widetilde{D}'$, respectively. We have

$$n' = 2n + \mathrm{cr}^*(\widetilde{D}) = 2n + \sum_{k\geq 2} t_k \quad \text{and}$$
$$e' = n(n-1) + \sum_{k\geq 2} kt_k.$$

By Euler's formula, the number of faces of $\widetilde{D}'$ satisfies

$$f' = e' + 2 - n' = n(n-3) + 2 + \sum_{k\geq 2}(k-1)t_k \geq \sum_{k\geq 2} t_k = \mathrm{cr}^*(\widetilde{D}) \geq \Omega(n^4).$$

Since at most $O(n^2)$ of the faces of $\widetilde{D}'$ intersect or touch the equator, this implies by the previous discussion that the double-ray arrangement of P has $\Omega(n^4)$ bounded cells. □

3.2 Multiplicity of Cells with the Same View

Since the number of cells is of the right order, it is sufficient to bound the maximum possible number of cells with the same view by a constant. In the case of cells inside the convex hull of P, it is known that points from different cells have distinct views of P [6, Theorem 4.2], [14, Theorem 1], [8, Lemma 2.6]. Díaz-Báñez, Fabila-Monroy and Pérez-Lantero identified a slightly more general condition that guarantees distinct views for a pair of observation points [8, Lemma 2.5], which implies distinct views also for some pair of cells outside of the convex hull of P. We state this more general result in the following lemma and include the proof for completeness.

Lemma 1. *Let p and q be two reference points in different cells of the double-ray arrangement of P such that the line $\overleftrightarrow{pq}$ intersects the convex hull of P. Then the view of P from p is distinct from the view of P from q.*

Proof. Since p and q are in different cells, there exist two points $a, b \in P$ such that the segment pq intersects the double ray $\leftarrow\!ab\!\rightarrow$. Hence, points a and b lie in the same halfplane determined by $\overleftrightarrow{pq}$. Furthermore, there exists a different point $c \in P$ in the opposite halfplane determined by $\overleftrightarrow{pq}$, since $\overleftrightarrow{pq}$ intersects the convex hull of P. It follows that the double-rays $\leftarrow ca \rightarrow$ and $\leftarrow bc \rightarrow$ do not intersect the line $\overleftrightarrow{pq}$. Hence, the segment pq intersects exactly one double-ray of the double-ray arrangement of $\{a, b, c\}$. Therefore, the views of P from p and q are distinct by Observation 1. □

It remains to consider the cells that are completely outside of conv(P). By definition, a view of P is a cyclic order of points of P. We can obtain a linear order by selecting a distinguished first point. For any reference point x outside of conv(P), this can be done in a canonical way: by the hyperplane separation theorem, x can be strictly separated from conv(P) by a line h_x, and so we can represent the view from x by the order of the intersections of the segments xp, $p \in P$, with h_x. Formally, starting from the cyclic order representing the view from x, we select the first point of the linear order as the point of P a clockwise rotating ray around x intersects when intersecting conv(P) for the first time; in other words, the first point is the "leftmost" point of P visible from x. We call this linear order the *linear view* of P from x.

In contrast to views, linear views cannot be the same in two different exterior cells.

Observation 2. *Let p and q be two reference points outside of conv(P) lying in different cells of the double-ray arrangement of P. Then their linear views are distinct.*

Proof. Since p and q lie in different cells, there exist points $a, b \in P$ such that the segment pq intersects the ray $ab\rightarrow$. It follows that exactly one of the points p and q has a before b in its linear view of P. □

Finally, we give an upper bound on the maximum possible number of cells with the same view. Combining this with the lower bound in Theorem 3, Theorem 1 immediately follows.

Proposition 1. *Assume that $|P| \geq 4$ and let p, q, and r be three reference points lying in pairwise different cells of the double-ray arrangement of P. Then some pair of points from $\{p, q, r\}$ have distinct views of P.*

Proof. We may assume that $P \cup \{p, q, r\}$ is in general position, since the cells are open subsets of the plane. Suppose, for contradiction, that the views of P from p, q, and r are the same. Hence, by Lemma 1, the points p, q, and r lie outside of conv(P). By Observation 2, their linear views are mutually distinct. Thus, the linear views of P from p, q, and r are non-trivial cyclic shifts of each other and we can partition the set P into three sets $A = \{a_1, \ldots, a_i\}$, $B = \{b_1, \ldots, b_j\}$, and $C = \{c_1, \ldots, c_k\}$, so that $i, j, k \geq 1$ and the linear view from p is $(a_1, \ldots, a_i, b_1, \ldots, b_j, c_1, \ldots, c_k)$, and the linear views from q and r are $(b_1, \ldots, b_j, c_1, \ldots, c_k, a_1, \ldots, a_i)$ and $(c_1, \ldots, c_k, a_1, \ldots, a_i, b_1, \ldots, b_j)$. Without loss of generality, we may assume the first linear view is from q and the

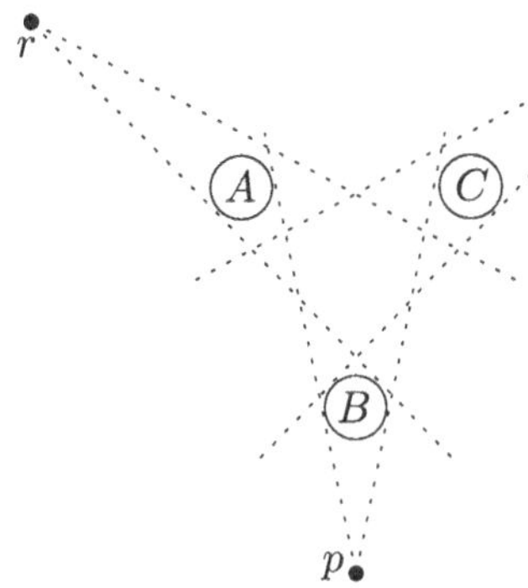

Fig. 3. If the points p, q, and r have the same view of the set of points P and lie outside of $\mathrm{conv}(P)$, then we can partition P into sets A, B, and C so that the three linear views from p, q and r are ABC, BCA and CAB.

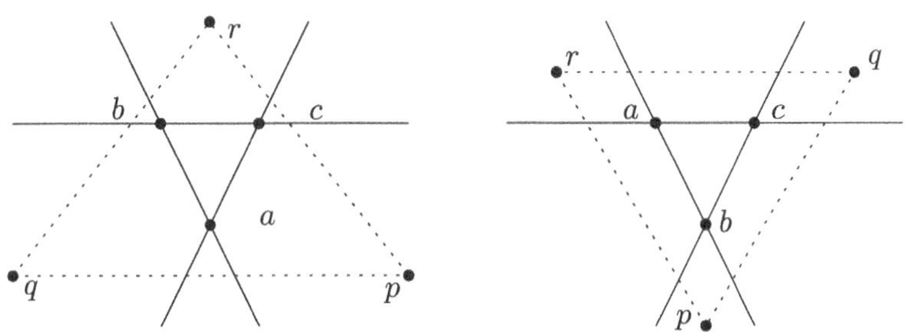

Fig. 4. Points p, q, and r with the same view of $\{a, b, c\}$ but pairwise distinct linear views of $\{a, b, c\}$. They must lie either in the three different unbounded cells sharing an edge with the triangle abc (left) or in the three different unbounded cells sharing only a vertex with the triangle abc (right). In both cases, triangles abc and pqr have a non-empty intersection.

second is from r. By the definition of the linear view, the convex hulls $\mathrm{conv}(A)$, $\mathrm{conv}(B)$, and $\mathrm{conv}(C)$ are mutually disjoint. See Fig. 3 for an illustration.

We will now show that $\mathrm{conv}(P) \cap \mathrm{conv}(\{p, q, r\}) \neq \emptyset$. Let $a \in A, b \in B$, and $c \in C$ be selected arbitrarily. Consider the arrangement of the three lines $\overleftrightarrow{ab}$, $\overleftrightarrow{bc}$, and $\overleftrightarrow{ca}$. See Fig. 4. Since the points p, q, and r have pairwise distinct linear views of $\{a, b, c\}$, they must lie in three different unbounded cells of the arrangement; either the three cells sharing an edge with the triangle abc, or the three cells sharing only a vertex with the triangle abc. In both cases, the triangles abc and pqr intersect.

By Lemma 1, no line formed by a pair of points from $\{p, q, r\}$ intersects $\mathrm{conv}(P)$ because the views of P from p, q, and r are the same. Hence, P lies entirely in the interior of the triangle pqr.

Since $|P| \geq 4$, we may without loss of generality assume that $|A| \geq 2$. Since P lies inside the triangle pqr, all double-rays formed by pairs of points of P intersect two sides of the triangle pqr. Refer to Fig. 5. Since the linear views of $\{a_1, a_2, b\}$ from the points p, q, and r are (a_1, a_2, b), (b, a_1, a_2) and (a_1, a_2, b), respectively, the double-rays $\leftarrow a_1 b \rightarrow$ and $\leftarrow a_2 b \rightarrow$ intersect the same pair of edges of the triangle pqr. Since $\leftarrow a_1 a_2 \rightarrow$ also intersects two edges of the triangle pqr, one edge of the triangle pqr is intersected by all the three double-rays $\leftarrow a_1 a_2 \rightarrow$, $\leftarrow a_1 b \rightarrow$, and $\leftarrow a_2 b \rightarrow$; we may without loss of generality assume that it is the edge pq. Thus, by Observation 1, p and q have distinct views of $\{a_1, a_2, b\}$; a contradiction. □

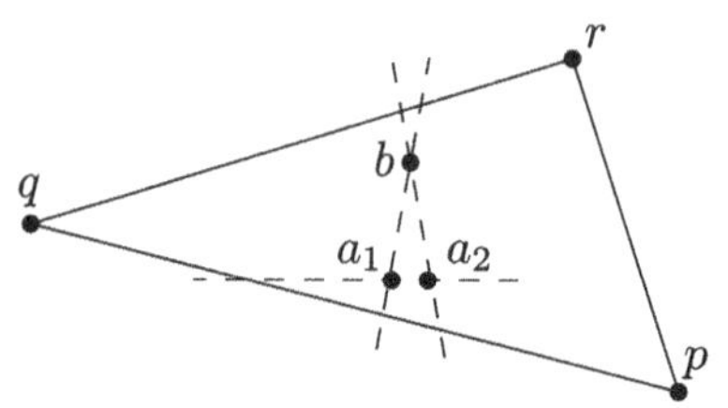

Fig. 5. The triangle $a_1 a_2 b$ inside the triangle pqr. Since the linear views of $\{a_1, a_2, b\}$ from p, q, and r are prescribed and b is never seen as the middle one, the lines $\overleftrightarrow{a_1 b}$ and $\overleftrightarrow{a_2 b}$ intersect the same pair of edges of the triangle pqr. One of these edges is then intersected by all the three lines $\overleftrightarrow{a_1 b}$, $\overleftrightarrow{a_2 b}$, and $\overleftrightarrow{a_1 a_2}$.

4 Proof of Theorem 2

Before we start with the proof, we need some additional definitions. Let $k \geq 3$ and let C be a sequence $x_1 x_2 \dots x_k$ of k points in the plane, ordered by increasing x-coordinate. We say that C is a *concave chain* (also called a k-*cap*) if for every pair of consecutive points $x_i, x_{i+1} \in C$, all the remaining points of C lie below the line $\overleftrightarrow{x_i x_{i+1}}$. In particular, the points of C form the vertices of a convex polygon, ordered clockwise along its boundary. Similarly, we say that C is a *convex chain* (also called a k-*cup*) if for every pair of consecutive points $x_i, x_{i+1} \in C$, all the remaining points of C lie above the line $\overleftrightarrow{x_i x_{i+1}}$.

In the proof of Theorem 2, we will use the Positive Fraction Erdős–Szekeres Theorem by Bárány and Valtr [4,12]. A finite planar point set is called a *convex k-clustering* if it is a disjoint union of k sets $X_1, \dots, X_k$ of equal sizes such that every k-tuple $(x_1, \dots, x_k)$, $x_1 \in X_1, \dots, x_k \in X_k$, forms a convex polygon with vertices in clockwise order $(x_1, \dots, x_k)$. The *size of a convex k-clustering* is the size of the sets X_i in the convex k-clustering.

Theorem 4 (Positive Fraction Erdős–Szekeres Theorem [4,12]). *For every $k \geq 3$ there is an $\varepsilon_k > 0$ and an integer $f(k)$ such that if X is a finite set of points in general position in the plane with $|X| \geq f(k)$, then it contains a convex k-clustering of size at least $\varepsilon_k |X|$.*

See also a more precise formulation of this theorem by Suk [13, Theorem 2.4].

We are now ready to prove Theorem 2. We have a set P of n points in the plane in strong general position and we need to prove that there is a subset $X \subseteq P$ and at least $\Omega(n^4)$ reference points with mutually distinct views of X so that the convex hull of the reference points is disjoint with the convex hull of X.

By Theorem 4 there exists a convex 20-clustering of P of size $\Omega(n)$. In other words, there exist mutually disjoint subsets $X_1, \ldots, X_{20} \subset P$, such that all X_i are of equal size $s = \Omega(n)$, and every 20-tuple $(x_1, \ldots, x_{20})$ with $x_1 \in X_1, \ldots, x_k \in X_{20}$ forms a convex 20-gon with vertices in clockwise order $(x_1, \ldots, x_{20})$. We fix one such tuple $(x_1, \ldots, x_{20})$. Eleven points from this tuple form either a convex or a concave chain; without loss of generality, let the points $x_1, \ldots, x_{11}$ form a concave chain $x_1 x_2 \ldots x_{11}$. Let X be the subset of P containing the points $x_1, x_2, x_4, x_6, x_8, x_{10}, x_{11}$ and all the points in $X_3 \cup X_5 \cup X_7 \cup X_9$; see Fig. 6 for an illustration of X. We will show that there exist $\Omega(n^4)$ distinct views of X.

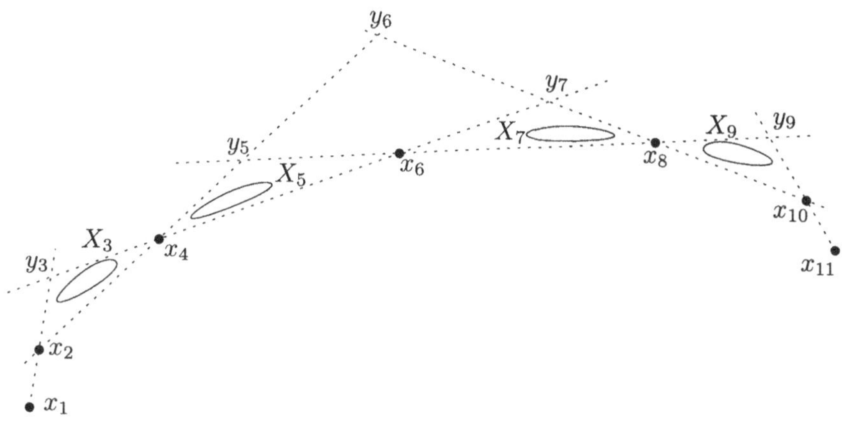

Fig. 6. A configuration of linear size that exists in any point set of size n in general position.

We define five intersection points y_3, y_5, y_7, y_9, y_6 as follows:

$$\begin{aligned}
\{y_3\} &= x_1x_2{\rightarrow} \cap x_6x_4{\rightarrow},\\
\{y_5\} &= x_2x_4{\rightarrow} \cap x_8x_6{\rightarrow},\\
\{y_7\} &= x_4x_6{\rightarrow} \cap x_{10}x_8{\rightarrow},\\
\{y_9\} &= x_6x_8{\rightarrow} \cap x_{11}x_{10}{\rightarrow},\\
\{y_6\} &= x_2x_4{\rightarrow} \cap x_{10}x_8{\rightarrow}.
\end{aligned}$$

Since $x_1x_2x_4x_6x_8x_{10}x_{11}$ is a concave chain, all these intersections are well-defined. Let L be the set of all rays $l_3l_5{\rightarrow}$ where $l_3 \in X_3$ and $l_5 \in X_5$, and let R be the set of all rays $r_9r_7{\rightarrow}$ where $r_9 \in X_9$ and $r_7 \in X_7$. The sizes of L and R are both $s^2 = \Omega(n^2)$.

Observation 3. *For every selection $l_3 \in X_3$, $l_5 \in X_5$, $r_7 \in X_7$, and $r_9 \in X_9$, the sequence $x_1x_2l_3x_4l_5x_6r_7x_8r_9x_{10}x_{11}$ is a concave chain.*

Proof. The tuple $(x_1, x_2, l_3, x_4, l_5, x_6, r_7, x_8, r_9, x_{10}, x_{11})$ forms a convex 11-gon with vertices in clockwise order by the selection of X. Since x_1, x_2, x_{10}, x_{11} are fixed and form a concave chain, it follows that $x_1x_2l_3x_4l_5x_6r_7x_8r_9x_{10}x_{11}$ is a concave chain. □

As an immediate consequence we get the following observation.

Observation 4. *For every $i \in \{3, 5, 7, 9\}$ the set X_i lies inside the triangle $x_{i-1}y_ix_{i+1}$.*
Moreover, every ray from L intersects segments x_6y_5 and y_7y_6, and every ray from R intersects segments x_6y_7 and y_5y_6.
Furthermore, every ray from L intersects with every ray from R inside the quadrilateral $x_6y_5y_6y_7$.

The rest of the proof of Theorem 2 is straightforward. The rays of $R \cup L$ form $\Omega(n^4)$ cells inside the quadrilateral $x_6y_5y_6y_7$ and the views of X from these cells are mutually distinct. We provide the remaining details in the full version.

5 Conclusion and Open Questions

By Proposition 1, in the double-ray arrangement of at least 4 points in the plane, at most two cells can have the same view. It would be interesting to determine how many times this can happen.

Question 1. Let P be a set of $n \geq 4$ points in general position in the plane. What is the maximum possible number of pairs of cells of the double-ray arrangement of P with the same view of P?

It is easy to see that if P is in convex position, then there are at least $\Omega(n)$ such pairs; see Fig. 7. Using similar techniques as in Sect. 3, we can prove that there are always at most $O(n^2)$ such pairs. We present the missing proofs in the full version.

Lemma 2. *Let P be a set of $n \geq 4$ points in general position in the plane. Let p and q be two points in distinct cells of the double-ray arrangement of P with the same view of P. Then the line $\overleftrightarrow{pq}$ does not intersect the convex hull of P and there exists a partition of P into two non-empty sets $A = \{a_1, \ldots, a_k\}$ and $B = \{b_1, \ldots, b_m\}$ such that*

a) *the linear view of P from p is $(a_1, \ldots, a_k, b_1, \ldots, b_m)$,*
b) *the linear view of P from q is $(b_1, \ldots, b_m, a_1, \ldots, a_k)$, and*
c) *the sets A and B are* mutually separated; *that is, no line determined by a pair of points from A separates B, and similarly, no line determined by a pair of points from B separates A.*

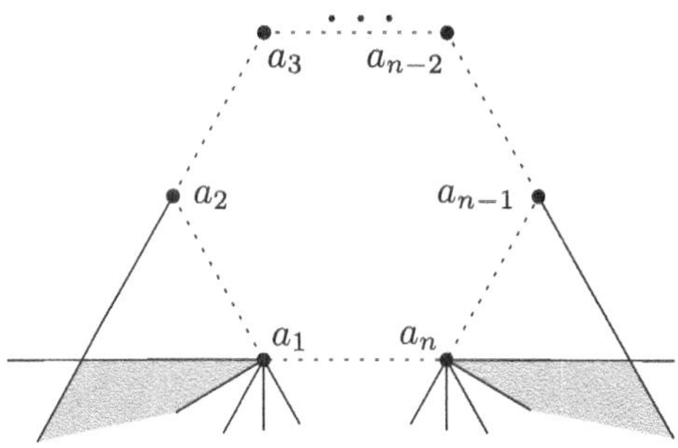

Fig. 7. If the points of a set $P = \{a_1, \ldots, a_n\}$ are the vertices of a convex polygon, there are at least $\Omega(n)$ pairs of cells with the same of view of P. One such pair is highlighted.

Lemma 3. *Let P be a set of $n \geq 4$ points in general position in the plane. Let $A \cup B$ be a partition of P into two mutually separated subsets. Then there is at most one pair (Z_p, Z_q) of distinct cells of the double-ray arrangement of P such that for every $p \in Z_p$ and every $q \in Z_q$,*

a) *the views of P from p and q are the same,*
b) *the whole set P is in the halfspace to the left of the ray $\overrightarrow{pq}$,*
c) *in the linear view of P from p all points of A precede all points of B, and*
d) *in the linear view of P from q all points of B precede all points of A.*

Corollary 1. *Let P be a set of $n \geq 4$ points in general position in the plane. The number of pairs of cells of the double-ray arrangement of P with the same view of P is at most $O(n^2)$.*

Proof. By Lemma 2 and Lemma 3, every pair of cells with the same view of P determines a unique ordered pair of mutually separated subsets forming a partition of P. See Fig. 8. The corollary follows from the fact that n points in general position in the plane can be partitioned into two sets with disjoint convex hulls in at most $O(n^2)$ different ways. □

For projective views, we have been able to prove the $\Omega(n^4)$ lower bound only for point sets in strong general position. The approach in the proof of Theorem 1 does not directly generalize to projective views. For projective views, instead of the double-ray arrangement, it is more natural to consider the arrangement of lines determined by all pairs of points of P, since the projective view changes every time the observation point crosses one of the lines. We cannot bound the multiplicity of cells with the same projective view by a constant, because for some point sets there are $\Omega(n)$ interior cells with the same projective view; see Fig. 9.

Question 2. Is it true that for every set P of n points in general position in the plane there are $\Omega(n^4)$ points with pairwise distinct projective views of P?

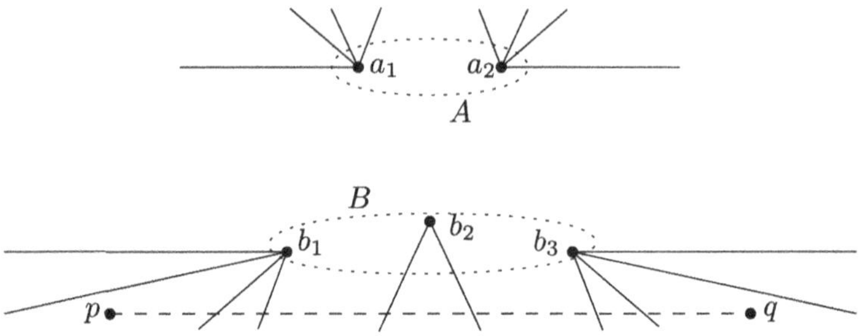

Fig. 8. Points p and q lying in cells of the double-ray arrangement of P with the same view of P but distinct linear views of P. The points p and q determine a unique ordered pair of mutually separated subsets A and B forming a partition of P. The same unordered partition is determined by at most one other pair of points from different cells (in the picture such cells are to the left of a_1 and to the right of a_2).

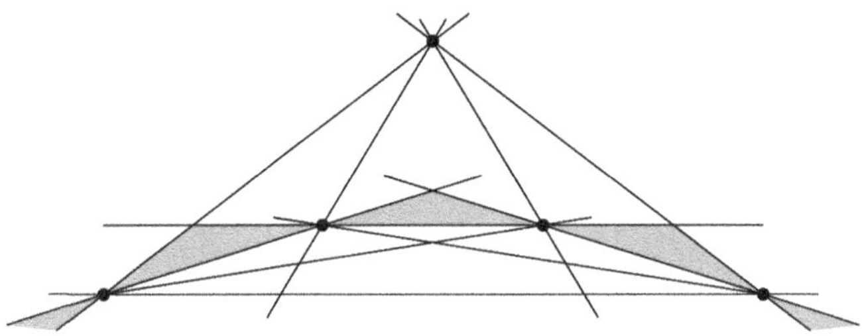

Fig. 9. A set P of n points can induce an arrangement of lines with n cells sharing the same projective view of P.

Acknowledgments. This work was supported by project 21-32817S of the Czech Science Foundation (GAČR) and by the grant SVV–2023–260699.

Disclosure of Interests. The authors have no competing interests to declare that are relevant to the content of this article.

References

1. Ackerman, E., Damásdi, G., Keszegh, B., Pinchasi, R., Raffay, R.: On the number of digons in arrangements of pairwise intersecting circles (2024). arXiv:2406.02276v1
2. Ackerman, E., Pinchasi, R.: On the degenerate crossing number. Discrete Comput. Geom. **49**(3), 695–702 (2013). https://doi.org/10.1007/s00454-013-9493-1
3. Auer, T., Held, M.: Heuristics for the generation of random polygons. In: Proceedings of the 8th Canadian Conference on Computational Geometry, Ottawa, Ontario, Canada, pp. 38–43 (1996)
4. Bárány, I., Valtr, P.: A positive fraction Erdős–Szekeres theorem. Discrete Comput. Geom. **19**(3, Special Issue), 335–342 (1998). https://doi.org/10.1007/PL00009350. Dedicated to the memory of Paul Erdős

5. Bieri, H., Schmidt, P.M.: On the permutations generated by rotational sweeps of planar point sets. In: Proceedings of the 8th Canadian Conference on Computational Geometry, Ottawa, Ontario, Canada, pp. 179–184 (1996)
6. Deneen, L., Shute, G.: Polygonizations of point sets in the plane. Discrete Comput. Geom. **3**(1), 77–87 (1988). https://doi.org/10.1007/BF02187898
7. Devillers, O., Dujmović, V., Everett, H., Hornus, S., Whitesides, S., Wismath, S.: Maintaining visibility information of planar point sets with a moving viewpoint. Int. J. Comput. Geom. Appl. **17**(4), 297–304 (2007). https://doi.org/10.1142/S0218195907002343
8. Díaz-Báñez, J.M., Fabila-Monroy, R., Pérez-Lantero, P.: On the numbers of radial orderings of planar point sets. Discrete Math. Theor. Comput. Sci. **16**(3), 291–304 (2014)
9. Matoušek, J.: Lectures on Discrete Geometry. Graduate Texts in Mathematics, vol. 212. Springer, New York (2002). https://doi.org/10.1007/978-1-4613-0039-7
10. Pach, J., Tóth, G.: Degenerate crossing numbers. Discrete Comput. Geom. **41**(3), 376–384 (2009). https://doi.org/10.1007/s00454-009-9141-y
11. Pinchasi, R.: On digons in planar arrangements of pairwise intersecting circles and pseudo-circles (2023). Graph Drawing and Combinatorial Geometry Workshop, Budapest
12. Pór, A., Valtr, P.: The partitioned version of the Erdős-Szekeres theorem. Discrete Comput. Geom. **28**(4), 625–637 (2002). https://doi.org/10.1007/s00454-002-2894-1
13. Suk, A.: On the Erdős-Szekeres convex polygon problem. J. Amer. Math. Soc. **30**(4), 1047–1053 (2017). https://doi.org/10.1090/jams/869
14. Tovar, B., Freda, L., LaValle, S.M.: Using a robot to learn geometric information from permutations of landmarks. In: Topology and Robotics, Contemporary Mathematics, vol. 438, pp. 33–45. American Mathematical Society, Providence (2007). https://doi.org/10.1090/conm/438/08443
15. Ulber, R.: On the number of star-shaped polygons and polyhedra. In: Proceedings 11th Canadian Conference on Computational Geometry, CCCG 1999, Vancouver, Canada, p. 4 (1999)

On the Complexity of Simultaneous Geometric Embedding for Edge-Disjoint Graphs

Benedikt Künzel and Jonathan Rollin(✉)

FernUniversität in Hagen, Hagen, Germany
jonathan.rollin@fernuni-hagen.de

Abstract. Simultaneous Geometric Embedding (SGE) asks whether, for a given collection of graphs on the same vertex set V, there is an embedding of V in the plane that admits a crossing-free drawing with straight-line edges for each of the given graphs. It is known that SGE is $\exists\mathbb{R}$-complete, that is, the problem is polynomially equivalent to deciding whether a system of polynomial equations and inequalities with integer coefficients has a real solution. We prove that SGE remains $\exists\mathbb{R}$-complete for edge-disjoint input graphs, that is, for collections of graphs without so-called public edges.

As an intermediate result, we prove that it is $\exists\mathbb{R}$-complete to decide whether a directional walk without repeating edges is realizable. Here, a directional walk consists of a sequence of not-necessarily distinct vertices (a walk) and a function prescribing for each inner position whether the walk shall turn left or right. A directional walk is realizable, if there is an embedding of its vertices in the plane such that the embedded walk turns according to the given directions. Previously it was known that realization is $\exists\mathbb{R}$-complete to decide for directional walks repeating each edge at most 336 times.

This answers two questions posed by Schaefer [*On the Complexity of Some Geometric Problems With Fixed Parameters*, JGAA 2021].

Keywords: Simultaneous geometric embedding · Existential theory of the reals · Partial order type realizability

1 Introduction

The complexity status of many problems from computational geometry and graph drawing has been clarified in more detail in recent years. While NP-hardness had often been established earlier, membership with NP seemed unlikely as solutions often require super-exponentially large coordinates [24]. In 2009, Schaefer [24] introduced the complexity class $\exists\mathbb{R}$ as the closure of the problem of deciding the *Existential Theory of the Reals* (ETR) under polynomial-time reductions. Here, ETR is the set of true statements over the reals that can

D. Král' and M. Milanič (Eds.): WG 2024, LNCS 14760, pp. 400–414, 2025.
https://doi.org/10.1007/978-3-031-75409-8_28

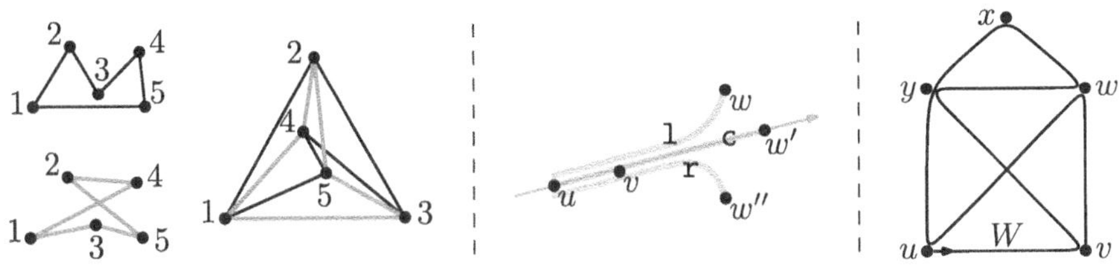

Fig. 1. Left: Two edge-disjoint 5-cycles on the same vertex set and a simultaneous geometric embedding of the cycles. Middle: Order types $\chi(u, v, w) = \mathtt{l}$, $\chi(u, v, w') = \mathtt{c}$, and $\chi(u, v, w'') = \mathtt{r}$. Right: A realization of the directional walk $uv^{\mathtt{l}}y^{\mathtt{r}}w^{\mathtt{l}}x^{\mathtt{l}}y^{\mathtt{l}}u^{\mathtt{l}}w^{\mathtt{r}}v$.

be written in the form $\exists(x_1, \dots, x_n) : \phi(x_1, \dots, x_n)$, where ϕ is a Boolean formula over polynomial equations and inequalities with integer coefficients. It is known that $\mathsf{NP} \subseteq \exists\mathbb{R} \subseteq \mathsf{PSPACE}$ [7,24] and many geometric decision problems turned out to be $\exists\mathbb{R}$-complete (for examples see [1,8,19]).

Among these problems is also the question of *Simultaneous Geometric Embedding* (SGE), asking whether several graphs with a common vertex set admit a drawing with straight-line edges on the same vertex locations in the plane such that each graph individually is drawn without crossings. Fig. 1 (right) shows an example of two simultaneously embedded graphs. Cardinal and Kusters [9] describe a reduction from the $\exists\mathbb{R}$-complete problem *Abstract Order Type Realizability* to Simultaneous Geometric Embedding. Abstract Order Type Realizability, or the dual problem of stretchability of an arrangement of pseudolines, forms the base of many $\exists\mathbb{R}$-hardness proofs of geometric decision problems [24] and we will describe this problem in more detail next.

Let u, v, and w be distinct points in $\mathbb{R}^2$. Then w is either to the left of, or to the right of, or on the directed line from u through v, see Fig. 1 (left). We call this the *order type* of the triple (u, v, w) and denote it $\chi(u, v, w) \in \{\mathtt{l}, \mathtt{c}, \mathtt{r}\}$, where $\mathtt{l}$ stands for left, $\mathtt{c}$ for collinear, and $\mathtt{r}$ for right. The *order type* χ of a set of points $P \subset \mathbb{R}^2$ is then defined as $\chi : P^3 \to \{\mathtt{l}, \mathtt{c}, \mathtt{r}\}$. Given an abstract universe of points $U = \{u_1, \dots, u_n\}$, an *abstract order type* of U is a mapping $\chi : U^3 \to \{\mathtt{l}, \mathtt{c}, \mathtt{r}\}$. We say that χ is *realizable* if the points in U can be mapped to points in $\mathbb{R}^2$ such that the order type of the resulting point set matches χ. The problem *Abstract Order Type Realizability* asks whether a given abstract order type is realizable. It follows from results of Mnëv [20] that this problem is $\exists\mathbb{R}$-complete. Knuth defined a set of five axioms and called abstract order types satisfying these axioms *CC-Systems* [16]. Knuth then showed that any order type of a set of points in $\mathbb{R}^2$ is a CC-System. It follows that all realizable abstract order types are CC-Systems. Order types and CC-systems are special cases of chirotopes and oriented matroids, where realizability is also of fundamental interest [22].

Interesting questions arise when the order type is not prescribed for all triples of points, but only for a certain subset. The corresponding realization problems are called *Partial Order Type Realizability*. A result of Schaefer [25, Theorem 1] implies that Partial Order Type Realizability remains $\exists\mathbb{R}$-complete, even for the set of instances where each vertex is contained in at most $\binom{131}{2}$ triples with prescribed order type[1]. One of the motivations for studying partial order types are reductions of Partial Order Type Realizability to restricted variants of other $\exists\mathbb{R}$-complete geometric problems. For instance, Schaefer used the result mentioned above to show $\exists\mathbb{R}$-completeness of deciding whether *bounded-degree* graphs are segment intersection graphs – a result that was known for graphs without degree constraints before [17]. In the next paragraph, we describe another variant of partial order types, which is then used to show that a restricted variant of simultaneous geometric embedding is $\exists\mathbb{R}$-complete. A discussion of further variants of partial order types is included in the full version of this paper [18].

Schaefer [25] introduced the following *Directional Walk Problem*. A *directional walk* W consists of a sequence of not-necessarily distinct vertices $u_1, \ldots, u_t$ (a walk) and a function $d\colon [t-2] \to \{\mathtt{l}, \mathtt{r}\}$ prescribing directions. The idea is that the walk turns in direction $d(i)$ after traveling the straight-line segment from vertex u_i to vertex u_{i+1}. We write $V(W)$ to denote the set (not the sequence) of vertices in W and d_W to denote the function prescribing directions in W. The pairs $\{u_i, u_{i+1}\}$ are called the *edges* of W and t is called the *length* of W (note that t might be much larger than $|V(W)|$ and much larger than the number of edges of W). We also write $u_1, u_2^{d(1)}, u_3^{d(2)}, \ldots, u_t$ to describe a directional walk.

Figure 1 (middle) shows an example of a directional walk. Note that a directional walk can be directly interpreted as a partial order type $\chi(u_i, u_{i+1}, u_{i+2}) = d_W(i)$, with $1 \leq i \leq t-2$.

Definition 1 (Directional Walk Problem [25]**).** *Given a directional walk W, decide whether W is* realizable, *that is, whether there is an embedding of $V(W)$ in the plane such that for each i, $1 \leq i \leq t-2$, we have $\chi(u_i, u_{i+1}, u_{i+2}) = d_W(i)$.*

Theorem 1 (Schaefer [25]**).** *The Directional Walk Problem is $\exists\mathbb{R}$-complete, even if each edge appears at most* 336 *times in the walk.*

Schaefer [25] asked whether the problem remains $\exists\mathbb{R}$-hard, even if the walk does not repeat any edges. We answer this question in the affirmative.

Theorem 2. *The Directional Walk Problem is $\exists\mathbb{R}$-complete for walks without repeated edges.*

Note that we consider undirected edges and hence the walks in the above theorem are not allowed to revisit an edge in either direction.

We now formally introduce *Simultaneous Geometric Embedding* (SGE) which we already mentioned as an $\exists\mathbb{R}$-complete problem above. A *geometric drawing* of a graph is a drawing where each edge is represented as a straight-line segment.

---

[1] Schaefer's notion of *thin* [25] implies that each vertex has at most 131 neighbors in the associated 3-uniform hypergraph. Schaefer's result applies to even more restricted systems of prescribed order types and a more careful analysis might give a smaller bound on the degrees.

Definition 2 (SGE [2]). *Given a finite set $\mathcal{G}$ of graphs on the same vertex set V, decide whether there is an embedding of V in the plane such that for each $G \in \mathcal{G}$ a geometric drawing of G on that embedding is crossing-free. If such an embedding exists, then we say that the graphs in $\mathcal{G}$ can be* simultaneously embedded.

$\exists\mathbb{R}$-hardness of SGE follows from works of Kyncl [19] and Gassner et al. [15] (see [9]). Using the result on partial order types mentioned above, Schaefer [25] showed that SGE is $\exists\mathbb{R}$-complete, even if each edge occurs in at most two input graphs. In the language of SGE, each edge that occurs in more than one input graph is called *public*. If all public edges occur in all graphs, we say that the graphs in $\mathcal{G}$ share a *Sunflower Intersection*. The *Sunflower Variant* of SGE restricts the input to sets of graphs that share a Sunflower Intersection [25].

We use our result from Theorem 2 above to prove that SGE remains $\exists\mathbb{R}$-hard, even if the input graphs are edge-disjoint, answering a question of Schaefer. In the language of SGE this means that there are no public edges and so it is an extreme case of the sunflower variant.

Theorem 3. *SGE is $\exists\mathbb{R}$-complete for edge-disjoint input graphs.*

Remark: After this paper was finished, Förster et al. [12] published a preprint which independently proves $\exists\mathbb{R}$-completeness for SGE on edge-disjoint input graphs using a different approach, namely by reduction from so-called *Pseudo-Segment Stretchability*. They show that SGE is $\exists\mathbb{R}$-complete for families of at most 58 edge-disjoint input graphs.

Related Work on SGE

There are many different variants of drawing graphs simultaneously. A systematic study of simultaneous embeddings was initiated by Brass et al. [6] and surveys are given by Bläsius et al. [5] and Rutter [23].

Concerning the setting studied in this paper, Brass et al. [6] showed that two paths always admit a simultaneous geometric embedding while there are three paths that do not admit such an embedding. Angelini et al. [2] described a path and an edge-disjoint tree that do not admit a simultaneous geometric embedding. SGE is $\exists\mathbb{R}$-complete for collections of 9360 paths or collections of 240 graphs [25] and the complexity status is open in case of just two input graphs.

In our setting we have a fixed common labeling of the vertices across all input graphs. If for each graph the embedded vertex set may be permuted we obtain a version called *SGE without mapping*. Recently, Steiner [26] proved the existence of a collection of $O(\log n)$ many n-vertex graphs that do not admit a simultaneous geometric embedding without fixed mapping. It is open whether

[2] In our definition, the term "SGE" refers to the problem with an arbitrary number of input graphs. In the literature the term is often used in case $|\mathcal{G}| = 2$ exclusively, while the general problem is called "k-SGE with unbounded k", where $k = |\mathcal{G}|$.

any two graphs do admit such a simultaneous embedding and the complexity appears to have not been investigated.

Another variant is called *Simultaneous Embedding with Fixed Edges* (SEFE). In this setting the vertex labeling is fixed across all input graphs but the edges are not required to be straight-line segments anymore. Instead, each edge must be drawn on the same curve in the plane for each graph it occurs in. The surveys mentioned above present many results for this variant. For example, Frati [13] showed that any tree and any planar graph do admit a SEFE while there are two outerplanar graphs that do not. There are polynomial-time algorithms checking whether two graphs with a connected intersection graph admit a SEFE [3,4,14] but it is unknown whether such an algorithm exists for arbitrary pairs of graphs. Gassner et al. [15] showed that the problem becomes NP-complete for three or more input graphs, which stays true even if the graphs share a sunflower intersection that is a tree and all the input graphs are biconnected [2]. Further polynomially tractable as well as hard instances were identified by Fink et al. [11]. Note that edge-disjoint graphs always admit a SEFE as every planar graph can be drawn in a planar way with arbitrary vertex locations [21] which is in stark contrast to our Theorem 3 above. This also shows that arbitrary graphs can be drawn simultaneously, if each graph's edges can be drawn on individual curves.

Finally, we mention geometric thickness. A graph has geometric thickness at most k if it can be partitioned into k subgraphs that admit a simultaneous geometric embedding. It is known that deciding whether a graph has geometric thickness 2 is NP-hard [10] and contained in $\exists\mathbb{R}$ [25]. Very recently, Förster et al. [12] published a preprint claiming $\exists\mathbb{R}$-hardness of deciding whether a multigraph has geometric thickness at most 57.

Notation: Let $\neg\mathtt{l} = \mathtt{r}$, $\neg\mathtt{r} = \mathtt{l}$, and $\neg\mathtt{c} = \mathtt{c}$. We will use the identity $\chi(u, v, w) = \neg\chi(w, u, v)$ several times which holds for any three points $u, v, w \in \mathbb{R}^2$. This might be immediate from the geometry of the plane or, for instance, it is the first axiom of CC-systems [16].

2 Proof of Theorem 2

We will use Theorem 1 which states that the Directional Walk Problem is $\exists\mathbb{R}$-complete. This directly implies that the restricted variant lies in $\exists\mathbb{R}$. We start with a brief outline of the $\exists\mathbb{R}$-hardness proof. Given a directional walk W with vertex sequence $u_1, \dots, u_t$ we shall construct a new directional walk W' using the vertices of W and additional dummy vertices such that W' does not repeat any of its edges and is realizable if and only if W is realizable. To achieve the latter property, W' enforces the same partial order type on $V(W)$ as prescribed by W. For each triple u_i, u_{i+1}, u_{i+2} of vertices along W we create several dummy vertices x such that in any realization of W we have $\chi(u_i, u_{i+1}, x) = d_W(i)$. As we do not want to use the edge $\{u_i, u_{i+1}\}$ in W' we use the equivalent order type $\chi(u_i, x, u_{i+1}) = \neg d_W(i)$ and walk along u_i, x, u_{i+1} in W' instead, or we use the

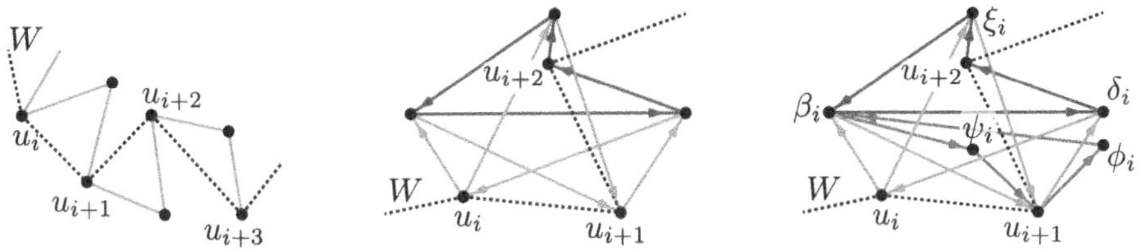

Fig. 2. Left: A walk (dotted black) with some hooks (solid orange). Middle: For each edge $\{u_i, u_{i+1}\}$ three hooks (orange) and a walk (blue) are placed so that u_{i+2} is placed in the triangle formed by the dummy vertices of the hooks. Right: The hooks and the placement of u_{i+2} are connected to a triangle gadget via two additional dummy vertices. (Color figure online)

order type $\chi(u_{i+1}, x, u_i) = d_W(i)$ and walk along u_{i+1}, x, u_i in W', see Fig. 2 (left). We call such a part of W' a *hook*. After placing three such hooks at the $d_W(i)$-side of $\{u_i, u_{i+1}\}$ we proceed with W' in such a way that u_{i+2} is placed in the triangle formed by the dummy vertices in the hooks. This will enforce the correct order type of (u_i, u_{i+1}, u_{i+2}), see Fig. 2 (middle).

We give a precise definition of W' next. For each i, $1 \leq i \leq t-2$, we use five dummy vertices β_i, δ_i, ξ_i, ϕ_i, and ψ_i. The first three are used in hooks and the latter two are used to combine the hooks and the placement of u_{i+1} to a single walk, starting at u_i and ending at u_{i+1}. The *triangle gadget* for i is the following directional walk, where $d = d_W(i)$:

$$W_i = \underbrace{u_i, \beta_i^{\neg d}, \overbrace{u_{i+1}^d}^{}}_{\text{hook (i)}}\overbrace{{}, \delta_i^d, u_i^{\neg d}}^{\text{hook (ii)}}\!\!\!\!\!\!\!\!\underbrace{\phantom{u_i^{\neg d}}, \xi_i^{\neg d}, u_{i+1}^d}_{\text{hook (iii)}}, \phi_i^d, \underbrace{\beta_i^{\neg d}, \delta_i^d, u_{i+2}^{\neg d}, \xi_i^d, \beta_i^d}_{u_{i+2} \text{ in triangle (iv)}}, \psi_i^{\neg d}, u_{i+1} \tag{1}$$

Figure 2 (right) shows an example of a triangle gadget. Now, W' is defined as the concatenation of $W_1, \ldots, W_{t-2}$ where for each i, with $1 < i \leq t-2$, the first vertex of W_i is identified with the last vertex of W_{i-1} and the first occurrence of u_i in W_i receives direction $\neg d_W(i)$. For example, assuming $d_W(1) = \mathtt{l}$ and $d_W(2) = \mathtt{r}$, the beginning of W' is

$$W' = \underbrace{u_1, \beta_1^{\mathtt{r}}, u_2^{\mathtt{l}}, \delta_1^{\mathtt{l}}, u_1^{\mathtt{r}}, \xi_1^{\mathtt{r}}, u_2^{\mathtt{l}}, \phi_1^{\mathtt{l}}, \beta_1^{\mathtt{r}}, \delta_1^{\mathtt{l}}, u_3^{\mathtt{r}}, \xi_1^{\mathtt{l}}, \beta_1^{\mathtt{l}}, \psi_1^{\mathtt{r}}, u_2^{\mathtt{l}}}_{W_1}, \beta_2^{\mathtt{l}}, u_3^{\mathtt{r}}, \delta_2^{\mathtt{r}}, u_2^{\mathtt{l}}, \xi_2^{\mathtt{l}}, u_3^{\mathtt{r}}, \ldots$$

By construction, there are no repeated edges in W'. Moreover, the walk W' can be constructed in polynomial time (with respect to the length of W) as each occurrence of an edge in W is replaced by a walk of constant length. It remains to show that W' is realizable if and only if W is realizable.

If W' is realizable, consider some realization of W'. We will show that for each i, $1 \leq i \leq t-2$, we have $\chi(u_i, u_{i+1}, u_{i+2}) = d_W(i)$ in this realization. Due to the hooks in W_i we have

$$\chi(u_i, u_{i+1}, \beta_i) = \neg\chi(u_i, \beta_i, u_{i+1}) \overset{\text{(i) in (1)}}{=} d_W(i),$$
$$\chi(u_i, u_{i+1}, \delta_i) = \chi(u_{i+1}, \delta_i, u_i) \overset{\text{(ii) in (1)}}{=} d_W(i),$$
$$\chi(u_i, u_{i+1}, \xi_i) = \neg\chi(u_i, \xi_i, u_{i+1}) \overset{\text{(iii) in (1)}}{=} d_W(i).$$

That is, the dummy vertices β_i, δ_i, and ξ_i are placed on the $d_W(i)$-side of the straight line from u_i through u_{i+1}. The latter part of the triangle gadget gives

$$\chi(\beta_i, \delta_i, u_{i+2}) \overset{\text{(iv) in (1)}}{=} d_W(i),$$
$$\chi(\delta_i, \xi_i, u_{i+2}) = \neg\chi(\delta_i, u_{i+2}, \xi_i) \overset{\text{(iv) in (1)}}{=} d_W(i),$$
$$\chi(\xi_i, \beta_i, u_{i+2}) = \chi(u_{i+2}, \xi_i, \beta_i) \overset{\text{(iv) in (1)}}{=} d_W(i).$$

The three order types above show that u_{i+2} has the same relative position to each side of the triangle formed by β_i, δ_i, and ξ_i when considering these sides in some cyclic order around the triangle. This implies that u_{i+2} is placed in the interior of the triangle formed by β_i, δ_i, and ξ_i. Hence, $\chi(u_i, u_{i+1}, u_{i+2}) = d_W(i)$, as desired.

Now assume that W is realizable and consider some realization of W. We will construct a realization of W' by positioning the dummy vertices, processing $W_1, \ldots, W_{t-2}$ one after the other in that order. For each i, $1 \leq i \leq t-2$, we have $\chi(u_i, u_{i+1}, u_{i+2}) = d_W(i)$ in the realization of W. We now place the dummy vertices β_i, δ_i, ξ_i, ϕ_i, and ψ_i. First, we apply an affine transformation of the plane to all points constructed so far such that

- if $d_W(i) = \mathtt{l}$, then u_i is mapped to the point $(4, -2)$, u_{i+1} is mapped to the point $(4, 2)$, and u_{i+2} is mapped to the point $(0, 0)$,
- if $d_W(i) = \mathtt{r}$, then u_i is mapped to the point $(-4, -2)$, u_{i+1} is mapped to the point $(-4, 2)$, and u_{i+2} is mapped to the point $(0, 0)$.

Observe that the chosen transformation preserves the order types of all triples of points constructed so far, as it preserves the order type of $\{u_i, u_{i+1}, u_{i+2}\}$ (which prevents reflection). Now we place the dummy vertices like in Fig. 3, depending on the values of $d_W(i)$ and $d_W(i+1)$:

- If $d_W(i) = \mathtt{l}$, then β_i is placed at $(1, -3)$, δ_i is placed at $(1, 3)$, ξ_i is placed at $(-1, 0)$, and ϕ_i is placed at $(2, 3)$. The point ψ_i is chosen with x-coordinate 2 and non-negative y-coordinate (possibly not an integer) such that it is in the interior of the triangle formed by u_{i+1}, u_{i+2}, and u_{i+3}.
- If $d_W(i) = \mathtt{r}$, then β_i is placed at $(-1, -3)$, δ_i is placed at $(-1, 3)$, ξ_i is placed at $(1, 0)$, and ϕ_i is placed at $(-2, 3)$. The point ψ_i is chosen with x-coordinate -2 and non-negative y-coordinate (possibly not an integer) such that it is in the interior of the triangle formed by u_{i+1}, u_{i+2}, and u_{i+3}.

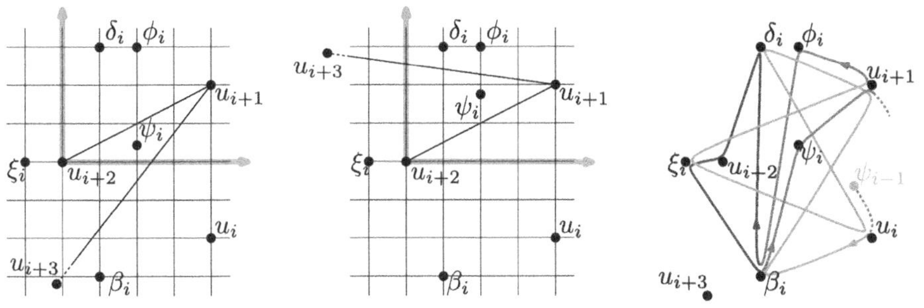

Fig. 3. Left/Middle: Placement of the dummy vertices in the cases $d_W(i) = \mathtt{l}$ and $d_W(i+1) = \mathtt{l}$ (left) and $d_W(i+1) = \mathtt{r}$ (middle). The thin lines represent the integer grid with origin at u_{i+2}. The positions of u_i, u_{i+1}, and u_{i+2} are fixed by an affine transformation. The position of u_{i+3} in the figure is just sketched for illustrating the placement of ψ_i: u_{i+3} is shown on the correct side of the line through u_{i+1} and u_{i+2} but its actual position on that side might differ. The placement in case $d_W(i) = \mathtt{r}$ is obtained by mirroring the points along the y-axis. Right: The walk W' in case $d_W(i) = \mathtt{l}$ and $d_W(i+1) = \mathtt{l}$.

Finally, we apply a small perturbation to ensure that the point set is in general position while preserving the order types of non-collinear triples of points. It can be readily checked that this placement indeed realizes W' by following the walk in Fig. 3 (right). The placement of ψ_i in the triangle formed by u_{i+1}, u_{i+2}, and u_{i+3} (and the placement of β_{i+1} below u_{i+1}) ensures that $\chi(\psi_i, u_{i+1}, \beta_{i+1}) = \neg d_W(i+1)$.

Altogether, we have a polynomial-time reduction that, for each directional walk W, gives a directional walk W' such that W' does not repeat any edges and W' is realizable if and only if W is realizable. This finishes the proof, since deciding whether a directional walk is realizable is $\exists\mathbb{R}$-complete by Theorem 1.

3 Proof of Theorem 3

We shall give a polynomial-time reduction from the Directional Walk Problem to SGE. To prove $\exists\mathbb{R}$-hardness of the restricted variant of SGE, we restrict the input of the reduction to directional walks without repeating edges and receive SGE-instances with edge-disjoint input graphs. The reduction includes the vertices of the input directional walk W in the vertex set of the SGE-instance such that, if the instance admits a simultaneous embedding, then the embedding of $V(W)$ in any such embedding virtually realizes W. We briefly describe the construction next and give precise definitions afterward. The SGE-instance contains, for each position i in the walk prescribing some order type, an individual graph G_i that is responsible for realizing that order type. Each G_i consists of two isomorphic subgraphs (distinguished by primed and non-primed vertex labels) which are glued together at specific vertices x, y, z. An embedding of G_i realizes the order type correctly, if one of these subgraphs is completely embedded within the

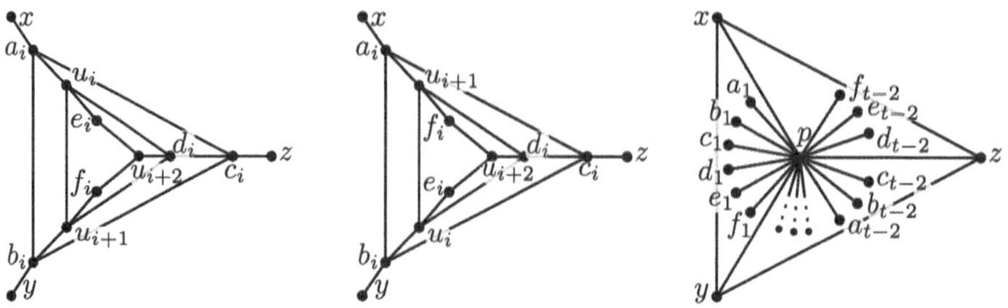

Fig. 4. Left: The graph $\widetilde{G}_i$ in case $d_W(i) = \mathtt{l}$. Middle: The graph $\widetilde{G}_i$ in case $d_W(i) = \mathtt{r}$ (only the vertices u_i, u_{i+1}, e_i, and f_i switched locations). Right: The graph G_{frame} (primed vertices omitted).

triangle formed by x, y, and z. An additional graph G_{frame} in the SGE-instance ensures that this happens for all graphs G_i.

Given a directional walk W with vertex sequence $u_1, \ldots, u_t$ the reduction is precisely defined as follows. The vertex set of the SGE-instance is the disjoint union $V = \cup_{i=1}^{t-2} V_i \cup V_{\text{frame}} \cup V(W) \cup \{u_j' : u_j \in V(W)\}$ where

$$V_i = \{a_i, a_i', b_i, b_i', c_i, c_i', d_i, d_i', e_i, e_i', f_i, f_i'\} \text{ and } V_{\text{frame}} = \{p, p', x, y, z\}.$$

We define the graphs G_i in two steps. For each i, with $1 \le i \le t-2$, let $\widetilde{G}_i$ denote graph given in Fig. 4 (left or middle): its vertex set is given by $\widetilde{V}_i = \{a_i, b_i, c_i, d_i, e_i, f_i, x, y, z\}$ and its edge set is given by the disjoint union $E(\widetilde{G}_i) = \widetilde{E}^i_{\text{frame}} \cup \widetilde{E}_i$ where

$$\widetilde{E}^i_{\text{frame}} = \{xa_i, yb_i, zc_i, a_ib_i, a_ic_i, b_ic_i\}$$

and in case $d_W(i) = \mathtt{l}$

$$\widetilde{E}_i = \{a_iu_i, b_iu_{i+1}, c_id_i, u_iu_{i+1}, u_id_i, u_{i+1}d_i, u_ie_i, u_{i+1}f_i, u_{i+2}d_i, u_{i+2}e_i, u_{i+2}f_i\},$$

and in case $d_W(i) = \mathtt{r}$

$$\widetilde{E}_i = \{a_iu_{i+1}, b_iu_i, c_id_i, u_iu_{i+1}, u_id_i, u_{i+1}d_i, u_ie_i, u_{i+1}f_i, u_{i+2}d_i, u_{i+2}e_i, u_{i+2}f_i\}.$$

That is, the graph $\widetilde{G}_i$ consists of edges xa_i, yb_i, and zc_i, a triangle on $\{a_i, b_i, c_i\}$, a copy of K_4 on $\{u_i, u_{i+1}, u_{i+2}, d_i\}$ where the edge u_iu_{i+2} is subdivided by e_i and the edge $u_{i+1}u_{i+2}$ is subdivided by f_i and, in case $d_W(i) = \mathtt{l}$, further edges a_iu_i, b_iu_{i+1}, c_id_i, and, in case $d_W(i) = \mathtt{r}$, further edges a_iu_{i+1}, b_iu_i, c_id_i. Let $\widetilde{G}_i'$ denote a copy of $\widetilde{G}_i$ where each vertex label is primed, that is, $V(\widetilde{G}_i') = \{a_i', b_i', c_i', d_i', e_i', f_i', x', y', z'\}$. The graph G_i is now obtained from the disjoint union of $\widetilde{G}_i$ and $\widetilde{G}_i'$ by identifying x and x', y and y', as well as z and z' to single vertices x, y, and z. Additionally, all vertices from V that are not yet

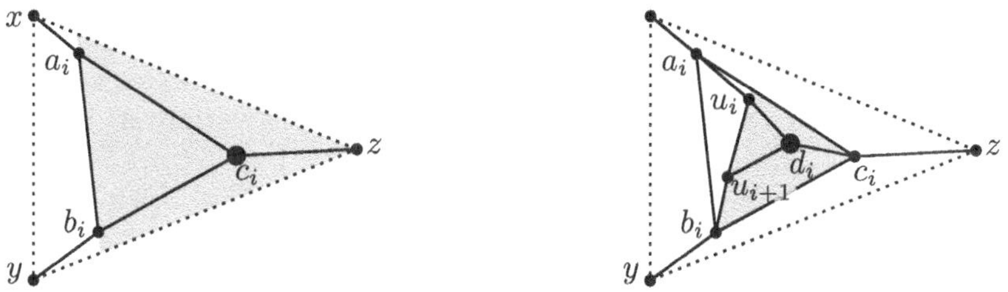

Fig. 5. The gray areas shows all possible locations of c_i (left), respectively d_i (right). (Color figure online)

present in G_i are added as isolated vertices. So the vertices p and p', all vertices in V_j, with $j \neq i$, and all vertices u_j, u'_j, with $j \notin \{i, i+1, i+2\}$, are isolated in G_i. Hence, the only edges shared by G_i and G_j, with $i \neq j$, might be $u_i u_{i+1}$ and $u'_i u'_{i+1}$. This, however, happens only if $\{u_i, u_{i+1}\} = \{u_j, u_{j+1}\}$ (as sets) which is impossible if the walk W does not repeat any edges.

Further, let G_{frame} denote the graph illustrated in Fig. 4 (right, primed vertices omitted) with vertex set V consisting of a triangle on $\{x, y, z\}$, all edges between p and $\{x, y, z\} \cup (\bigcup_{i=1}^{t-2} \widetilde{V}_i)$ (all non-primed vertices distinct from p), and all edges between p' and $\{x, y, z\} \cup (\bigcup_{i=1}^{t-2} \{a'_i, b'_i, c'_i, d'_i, e'_i, f'_i\})$ (x, y, z and all primed vertices distinct from p').

The graph G_{frame} does not share any edges with any of the graphs $G_1, \ldots, G_{t-2}$ by construction. As $|V| < 14t$, the reduction can be carried out in polynomial time in t. We claim that $G_1, \ldots, G_{t-2}$, and G_{frame} can be simultaneously embedded if and only if the directional walk W is realizable.

First assume that $G_1, \ldots, G_{t-2}$, and G_{frame} can be simultaneously embedded. That is, there is an embedding of V such that for each of the graphs a geometric drawing on that embedding is crossing-free. We will prove that we can derive an embedding of $V(W) \subseteq V$ that is a realization of W. Consider the triangle Δ on x, y, and z in G_{frame} and the vertices p and p'. Due to the edges between $\{p, p'\}$ and $\{x, y, z\}$ in the crossing-free embedding of G_{frame}, one of p and p' is embedded in the interior of Δ while the other is embedded in the exterior. If p is in the exterior, then we swap all primed vertices with their corresponding non-primed vertices (while fixing x, y, z). The swapped embedding still admits a simultaneous embedding of the graphs as this mapping is an isomorphism for each of $G_1, \ldots, G_{t-2}$ and G_{frame}. So we assume that p is embedded in the interior of Δ. Let $\widetilde{V} = \bigcup_{i=1}^{t-2} \widetilde{V}_i$ denote the set of all non-primed vertices, except x, y, z. Then all vertices from $\widetilde{V}$ are embedded in the interior of Δ as well, as they are adjacent to p in G_{frame}. We reflect the plane, if necessary, to ensure that $\chi(x, y, z) = 1$. Note that the reflected embedding of V still admits a simultaneous embedding of the graphs and still has all vertices from $\widetilde{V}$ in the interior of Δ.

We now show that for each i, with $1 \leq i \leq t-2$, we have $\chi(u_i, u_{i+1}, u_{i+2}) = d_W(i)$ in the embedding, that is, the embedding of $V(W) \subseteq \widetilde{V}$ realizes W. To prove this, consider the path x, a_i, b_i, y in G_i that lies within Δ. Since c_i is in the interior of Δ and adjacent to a_i, b_i, and z we have $\chi(a_i, b_i, c_i) = \mathtt{l}$ as $\chi(x, y, z) = \mathtt{l}$ and as G_i is drawn without crossings, see Fig. 5 (left). Since all vertices from $\widetilde{V}$ are in the interior of Δ, also the triangle T on $\{d_i, u_i, u_{i+1}\}$ in G_i is embedded within Δ. As it has an edge to each of a_i, b_i, and c_i in the crossing-free drawing of G_i, T is embedded in the interior of the triangle on $\{a_i, b_i, c_i\}$. With a similar argument as above we see that $\chi(u_i, u_{i+1}, d_i) = d_W(i)$: If $d_W(i) = \mathtt{l}$, we have a path a_i, u_i, u_{i+1}, b_i in G_i that lies within the triangle on $\{a_i, b_i, c_i\}$ and, if $d_W(i) = \mathtt{r}$, we have a path a_i, u_{i+1}, u_i, b_i in G_i, see Fig. 5 (right). The case distinction comes from the differences in the construction of G_i (respectively $\widetilde{G}_i$) depending on the value of $d_W(i)$. In both cases, d_i is adjacent to u_i, u_{i+1}, and c_i which implies that $\chi(u_i, u_{i+1}, d_i) = d_W(i)$. Finally observe that u_{i+2} lies in the interior of the triangle T due to its connections to d_i, u_i (via e_i), and u_{i+1} (via f_i). Hence $\chi(u_i, u_{i+1}, u_{i+2}) = \chi(u_i, u_{i+1}, d_i) = d_W(i)$. Altogether, this shows that the embedding of $V(W)$ realizes W.

Next assume that the directional walk W is realizable. We need to show that the graphs $G_1, \ldots, G_{t-2}$, and G_{frame} can be simultaneously embedded. To this end, we will specify the locations of the remaining vertices in V. Consider an embedding of $V(W)$ that realizes W. We apply an affine transformation of the plane, if necessary, to ensure that the embedded vertices lie within the unit square centered at the origin of the plane. The transformed embedding still realizes W. First place the vertex x at position $(-1, 2)$, the vertex y at position $(-1, -2)$, and z at position $(1, 0)$. Then $\chi(x, y, z) = \mathtt{l}$ and the embedded vertices from $V(W)$ are in the interior of the triangle Δ formed by x, y, and z. We proceed with the remaining non-primed vertices which will be placed in the interior of the triangle Δ. For each i, with $1 \leq i \leq t-2$, we place the vertices from $\widetilde{V}_i$ (the non-primed vertices of G_i) in the interior of Δ as follows. The vertices e_i and f_i are placed in the interior of the segments $u_{i+2}u_i$ and $u_{i+2}u_{i+1}$, respectively. The vertex d_i is placed close to u_{i+2} such that u_{i+2} is in the interior of the triangle T formed by u_i, u_{i+1}, and d_i. So far, all edges of G_i induced by already embedded vertices are crossing-free. Then we place the vertices a_i, b_i, and c_i such that

- they are in the interior of Δ and in the exterior of T,
- for some (arbitrary but fixed) point o in T each of the three rays emanating from o through one of the corners of T contains one of a_i, b_i, and c_i,
- $\chi(a_i, b_i, c_i) = \mathtt{l}$,
- the straight-line segments xa_i, yb_i, and zc_i do not intersect each other.

Such a placement is possible, since at least two sides of Δ are intersected by the chosen rays emanating from T, see Fig. 6 (left and middle). The placement on these rays ensures that for each of a_i, b_i, and c_i the straight-line segment to any vertex in $\{d_i, u_i, u_{i+1}\}$ does not intersect the interior of T. Moreover, the neighbor of a_i in $\{d_i, u_i, u_{i+1}\}$ comes, counterclockwise, before the neighbor of b_i in $\{d_i, u_i, u_{i+1}\}$ which in turn comes before the neighbor of c_i in $\{d_i, u_i, u_{i+1}\}$,

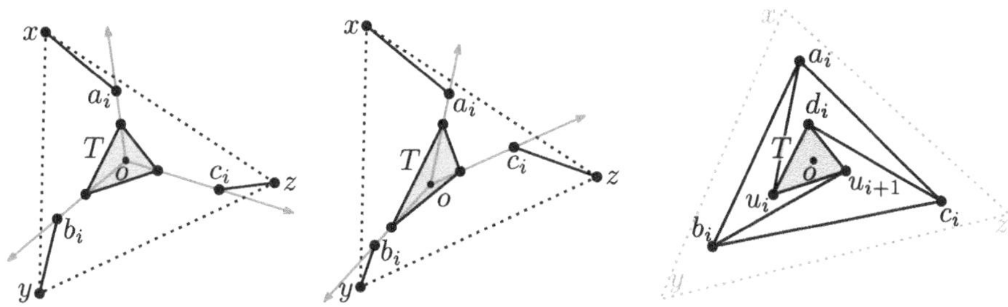

Fig. 6. Left & Middle: Placement of a_i, b_i, and c_i with $\chi(a_i, b_i, c_i) = 1$ on rays emanating from some point o in T. Right: This placement and the construction of $\widetilde{G}_i$ depending on the value of $d_W(i)$ ensure that the edges between $\{a_i, b_i, c_i\}$ and T do not cross.

which is due to the construction of $\widetilde{G}_i$ dependent on the value of $d_W(i)$. As $\chi(a_i, b_i, c_i) = 1$ is required above, the edges between $\{a_i, b_i, c_i\}$ and T do not cross, see Fig. 6 (right). Hence, the edges incident to a_i, b_i, or c_i in G_i neither intersect each other nor any other edges of G_i drawn yet. Finally, we place p arbitrarily in the interior of Δ. If necessary, we apply a small perturbation, preserving the order types of non-collinear triples of points, to ensure that the vertices are mapped to distinct points and not mapped to the interior of edges. With this placement, all graphs $\widetilde{G}_1, \ldots, \widetilde{G}_{t-2}$ and the non-primed subgraph of G_{frame} are drawn without crossings.

It remains to place the primed vertices, which we will place in the exterior of Δ. We start with the vertices from $S = \bigcup_{i=1}^{t-2}\{d_i', e_i', f_i'\} \cup \{u_j' : u_j \in V(W)\}$. We embed S such that its embedding forms a copy of the already constructed embedding of the corresponding non-primed vertices. However, the embedding of S is reflected along the y-axis, rotated so that no two points lie on a common horizontal line, and translated to the right such that each straight line through two points from S intersects the horizontal lines through x and y at x-coordinates larger than 1 (to the right of Δ). Note that S is embedded strictly between the horizontal lines through x and y. Now we place, for each i, with $1 \leq i \leq t-2$, the vertices a_i', b_i', and c_i'. Consider the triangle T' formed by d_i', u_i', and u_{i+1}'. We call the corners of T' the *first*, *second*, and *third* corner such that the topmost corner is the first and the second corner follows clockwise along the boundary of T'. The rotation above ensures that there is a unique topmost vertex. We place a_i' on the horizontal line through x such that it lies on a ray which starts in the interior of T' and runs through the first corner of T' (that is, within the gray area labeled 1 in Fig. 7). We place b_i' on the horizontal line through y such that it lies on a ray which starts in the interior of T' and runs through the second corner of T' (that is, within the gray area labeled 2 in Fig. 7). We place c_i' close to the third corner, and in the exterior of T', such that it lies on a ray starting in the interior of T' and running through the third corner (that is, within the gray area labeled 3 in Fig. 7). Finally, we place p' at position $(2, 0)$. This placement ensures that

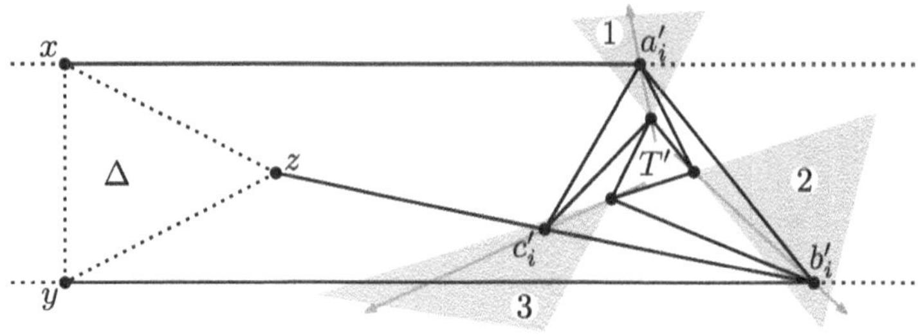

Fig. 7. The placement of a'_i, b'_i, and c'_i regarding T'.

- the triangle formed by a'_i, b'_i, and c'_i lies to the right of Δ and has T' in its interior (due to the placement within the gray areas and due to the translation of S to the right of Δ),
- $\chi(a'_i, b'_i, c'_i) = \mathtt{r}$ (due to placing the vertices clockwise around T'),
- the edges between Δ and $\{a'_i, b'_i, c'_i\}$ do not intersect each other, intersect Δ only in their respective endpoint, and intersect the triangle formed by a'_i, b'_i, and c'_i only in their respective endpoint (as zc'_i lies horizontally strictly between the other two edges),
- the edges from a'_i, b'_i, and c'_i to T' neither intersect each other nor the interior of T' (since $\chi(a'_i, b'_i, c'_i) = \mathtt{r}$ and a'_i, b'_i, and c'_i are placed into the three gray regions distinctly).

If necessary, we apply a small perturbation to ensure that the vertices are mapped to distinct points and not mapped to the interior of edges while preserving the order types of non-collinear triples of points. So we have a simultaneous crossing-free geometric drawing of $G_1, \ldots, G_{t+2}$, and of G_{frame}, as desired.

This reduction shows that SGE is $\exists\mathbb{R}$-hard for edge-disjoint input graphs, since the Directional Walk Problem is $\exists\mathbb{R}$-hard for directional walks without repeated edges by Theorem 2. As SGE is $\exists\mathbb{R}$-complete, the restricted variant is contained in $\exists\mathbb{R}$ as well and hence $\exists\mathbb{R}$-complete.

References

1. Abrahamsen, M., Adamaszek, A., Miltzow, T.: The art gallery problem is $\exists\mathbb{R}$-complete. J. ACM **69**(1), 1–70 (2022). https://doi.org/10.1145/3486220
2. Angelini, P., Geyer, M., Kaufmann, M., Neuwirth, D.: On a tree and a path with no geometric simultaneous embedding. J. Graph Algorithms Appl. **16**(1), 37–83 (2012). https://doi.org/10.7155/jgaa.00250
3. Angelini, P., Lozzo, G.D.: SEFE = c-planarity? Comput. J. **59**(12), 1831–1838 (2016). https://doi.org/10.1093/COMJNL/BXW035
4. Bläsius, T., Fink, S.D., Rutter, I.: Synchronized planarity with applications to constrained planarity problems. ACM Trans. Algorithms **19**(4), 1–23 (2023). https://doi.org/10.1145/3607474
5. Bläsius, T., Kobourov, S.G., Rutter, I.: Simultaneous embedding of planar graphs. In: Tamassia, R. (ed.) Handbook on Graph Drawing and Visualization, pp. 349–381. Chapman and Hall/CRC (2013)

6. Brass, P., et al.: On simultaneous planar graph embeddings. Comput. Geom. **36**(2), 117–130 (2007). https://doi.org/10.1016/j.comgeo.2006.05.006
7. Canny, J.F.: Some algebraic and geometric computations in PSPACE. In: Simon, J. (ed.) 20th Annual ACM Symposium on Theory of Computing, STOC 1988, pp. 460–467. ACM (1988). https://doi.org/10.1145/62212.62257
8. Cardinal, J.: Computational geometry column 62. SIGACT News **46**(4), 69–78 (2015). https://doi.org/10.1145/2852040.2852053
9. Cardinal, J., Kusters, V.: The complexity of simultaneous geometric graph embedding. J. Graph Algorithms Appl. **19**(1), 259–272 (2015). https://doi.org/10.7155/jgaa.00356
10. Durocher, S., Gethner, E., Mondal, D.: Thickness and colorability of geometric graphs. Comput. Geom. **56**, 1–18 (2016). https://doi.org/10.1016/j.comgeo.2016.03.003
11. Fink, S.D., Pfretzschner, M., Rutter, I.: Parameterized complexity of simultaneous planarity. In: Bekos, M.A., Chimani, M. (eds.) Graph Drawing and Network Visualization, 31st International Symposium, GD 2023. LNCS, vol. to appear. Springer (2023)
12. Förster, H., Kindermann, P., Miltzow, T., Parada, I., Terziadis, S., Vogtenhuber, B.: Geometric thickness of multigraphs is ∃ℝ-complete. In: Soto, J.A., Wiese, A. (eds.) LATIN 2024: Theoretical Informatics - 16th Latin American Symposium. LNCS, vol. 14578, pp. 336–349. Springer (2024). https://doi.org/10.1007/978-3-031-55598-5_22
13. Frati, F.: Embedding graphs simultaneously with fixed edges. In: Kaufmann, M., Wagner, D. (eds.) GD 2006. LNCS, vol. 4372, pp. 108–113. Springer, Heidelberg (2007). https://doi.org/10.1007/978-3-540-70904-6_12
14. Fulek, R., Tóth, C.D.: Atomic embeddability, clustered planarity, and thickenability. J. ACM **69**(2), 1–34 (2022). https://doi.org/10.1145/3502264
15. Gassner, E., Jünger, M., Percan, M., Schaefer, M., Schulz, M.: Simultaneous graph embeddings with fixed edges. In: Fomin, F.V. (ed.) WG 2006. LNCS, vol. 4271, pp. 325–335. Springer, Heidelberg (2006). https://doi.org/10.1007/11917496_29
16. Knuth, D.E. (ed.): Axioms and Hulls. LNCS, vol. 606. Springer, Heidelberg (1992). https://doi.org/10.1007/3-540-55611-7
17. Kratochvíl, J., Matoušek, J.: Intersection graphs of segments. JCTB **62**(2), 289–315 (1994). https://doi.org/10.1006/jctb.1994.1071
18. Künzel, B., Rollin, J.: On the complexity of simultaneous geometric embedding for edge-disjoint graphs (2023). preprint
19. Kynčl, J.: Simple realizability of complete abstract topological graphs in P. Discret. Comput. Geom. **45**(3), 383–399 (2011). https://doi.org/10.1007/s00454-010-9320-x
20. Mnev, N.E.: The universality theorems on the classification problem of configuration varieties and convex polytopes varieties. In: Viro, O.Y., Vershik, A.M. (eds.) Topology and Geometry — Rohlin Seminar. LNM, vol. 1346, pp. 527–543. Springer, Heidelberg (1988). https://doi.org/10.1007/BFb0082792
21. Pach, J., Wenger, R.: Embedding planar graphs at fixed vertex locations. Graphs Comb. **17**(4), 717–728 (2001). https://doi.org/10.1007/PL00007258
22. Richter-Gebert, J., Ziegler, G.M.: Oriented matroids. In: Goodman, J.E., O'Rourke, J. (eds.) Handbook of Discrete and Computational Geometry, Second Edition, pp. 129–151. Chapman and Hall/CRC (2004).https://doi.org/10.1201/9781420035315.CH6

23. Rutter, I.: Simultaneous embedding. In: Hong, S.-H., Tokuyama, T. (eds.) Beyond Planar Graphs, pp. 237–265. Springer, Singapore (2020). https://doi.org/10.1007/978-981-15-6533-5_13
24. Schaefer, M.: Complexity of some geometric and topological problems. In: Eppstein, D., Gansner, E.R. (eds.) GD 2009. LNCS, vol. 5849, pp. 334–344. Springer, Heidelberg (2010). https://doi.org/10.1007/978-3-642-11805-0_32
25. Schaefer, M.: On the complexity of some geometric problems with fixed parameters. J. Graph Algorithms Appl. **25**(1), 195–218 (2021). https://doi.org/10.7155/jgaa.00557
26. Steiner, R.: A logarithmic bound for simultaneous embeddings of planar graphs. In: Bekos, M.A., Chimani, M. (eds.) Graph Drawing and Network Visualization, 31st International Symposium, GD 2023. LNCS, vol. 14466, pp. 133–140. Springer (2023).https://doi.org/10.1007/978-3-031-49275-4_9

Finding d-Cuts in Graphs of Bounded Diameter, Graphs of Bounded Radius and H-Free Graphs

Felicia Lucke[1(✉)], Ali Momeni[2], Daniël Paulusma[3], and Siani Smith[4]

[1] Department of Informatics, University of Fribourg, Fribourg, Switzerland
felicia.lucke@unifr.ch

[2] Faculty of Computer Science, UniVie Doctoral School Computer Science DoCS, University of Vienna, Vienna, Austria
ali.momeni@univie.ac.at

[3] Department of Computer Science, Durham University, Durham, UK
daniel.paulusma@durham.ac.uk

[4] Heilbronn Institute for Mathematical Research, University of Bristol, Bristol, UK
siani.smith@bristol.ac.uk

Abstract. The d-Cut problem is to decide if a graph has an edge cut such that each vertex has at most d neighbours at the opposite side of the cut. If $d = 1$, we obtain the intensively studied Matching Cut problem. The d-Cut problem has been studied as well, but a systematic study for special graph classes was lacking. We initiate such a study and consider classes of bounded diameter, bounded radius and H-free graphs. We prove that for all $d \geq 2$, d-Cut is polynomial-time solvable for graphs of diameter 2, $(P_3 + P_4)$-free graphs and P_5-free graphs. These results extend known results for $d = 1$. However, we also prove several NP-hardness results for d-Cut that contrast known polynomial-time results for $d = 1$. Our results lead to full dichotomies for bounded diameter and bounded radius and to partial dichotomies for H-free graphs; for $d \geq 3$, our classification of d-Cut for H-free graphs only has three open cases.

Keywords: matching cut · d-cut · diameter · H-free graph

1 Introduction

We consider the generalization d-Cut of a classic graph problem Matching Cut (1-Cut). First, we explain the original graph problem. Consider a connected graph $G = (V, E)$, and let $M \subseteq E$ be a subset of edges of G. The set M is an *edge cut* of G if it is possible to partition V into two non-empty sets B (set of *blue* vertices) and R (set of *red* vertices) in such a way that M is the set of all edges with one end-vertex in B and the other one in R. Now, suppose that M is in addition also a *matching*, that is, no two edges in M have a common end-vertex. Then M is said to be a *matching cut*. See Fig. 1 for an example.

D. Král' and M. Milanič (Eds.): WG 2024, LNCS 14760, pp. 415–429, 2025.
https://doi.org/10.1007/978-3-031-75409-8_29

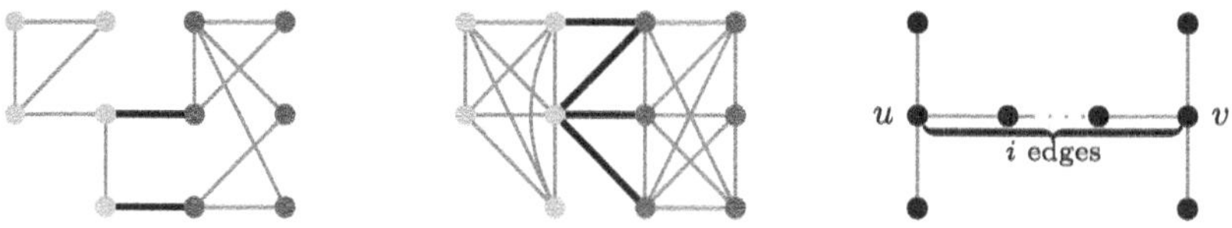

Fig. 1. Left: a graph with a matching cut (1-cut). Middle: a graph with a 3-cut but no d-cut for $d \leq 2$. Right: the graph H_i^*.

Graphs with matching cuts were introduced in the context of number theory [14] and have various other applications [1,8,11,26]. The MATCHING CUT problem is to decide if a connected graph has a matching cut. This problem was shown to be NP-complete by Chvátal [7]. Several variants and generalizations of matching cuts are known. In particular, a *perfect matching cut* is a matching cut that is a perfect matching, whereas a *disconnected perfect matching* is a perfect matching containing a matching cut. The corresponding decision problems PERFECT MATCHING CUT [15] and DISCONNECTED PERFECT MATCHING [6] are also NP-complete; see [3,10,18,20,24] for more complexity results for these two problems. The optimization versions MAXIMUM MATCHING CUT and MINIMUM MATCHING CUT are to find a matching cut of maximum and minimum size in a connected graph, respectively; see [19,23] for more details.

Our Focus. Matching cuts have also been generalized as follows. For an integer $d \geq 1$ and a connected graph $G = (V, E)$, a set $M \subseteq E$ is a *d-cut* of G if it is possible to partition V into two non-empty sets B and R, such that: (i) the set M is the set of all edges with one end-vertex in B and the other one in R; and (ii) every vertex in B has at most d neighbours in R, and vice versa (see also Fig. 1). Note that a 1-cut is a matching cut. We consider the d-CUT problem: does a connected graph have a d-cut? Here, $d \geq 1$ is a fixed integer, so not part of the input. Note that 1-CUT is MATCHING CUT. The d-CUT problem was introduced by Gomes and Sau [13] who proved its NP-completeness for all $d \geq 1$.

Our Goal. To get a better understanding of the hardness of an NP-complete graph problem, it is natural to restrict the input to belong to some special graph classes. We will first give a brief survey of the known complexity results for MATCHING CUT and d-CUT for $d \geq 2$ under input restrictions. As we will see, for MATCHING CUT many more results are known than for d-CUT with $d \geq 2$. Our goal is to obtain the same level of understanding of the d-CUT problem for $d \geq 2$. This requires a currently lacking systematic study into the complexity of this problem. We therefore consider the following research question:

*For which graph classes $\mathcal{G}$ does the complexity of d-*CUT*, restricted to graphs from $\mathcal{G}$, change if $d \geq 2$instead of $d = 1$?*

As *testbeds* we take classes of graphs of bounded diameter, graphs of bounded radius and H-free graphs. The *distance* between two vertices u and v in a connected graph G is the *length* (number of edges) of a shortest path between u and v in G. The *eccentricity* of a vertex u is the maximum distance between u

and any other vertex of G. The *diameter* of G is the maximum eccentricity over all vertices of G, whereas the *radius* of G is the minimum eccentricity over all vertices of G. A graph G is H*-free* if G does not contain a graph H as an *induced subgraph*, that is, G cannot be modified into H by vertex deletions.

Existing Results. We focus on classical complexity results; see [2,13] for exact and parameterized complexity results for d-Cut. Let $K_{1,r}$ denote the $(r+1)$-vertex star, which has vertex set $\{u, v_1, \ldots, v_r\}$ and edges uv_i for $i \in \{1, \ldots, r\}$. Chvátal [7] showed that Matching Cut is NP-complete even for $K_{1,4}$-free graphs of maximum degree 4, but polynomial-time solvable for graphs of maximum degree at most 3. Gomes and Sau [13] extended these results by proving that for every $d \geq 2$, the d-Cut problem is NP-complete for graphs in which every vertex has degree $2d+2$, but polynomial-time solvable for graphs of maximum degree at most $d+2$. Feghali et al. [10] proved that for every $d \geq 1$ and every $g \geq 3$, there is a function $f(d)$, such that d-Cut is NP-complete for bipartite graphs of girth at least g and maximum degree at most $f(d)$. The *girth* of a graph G that is not a forest is the length of a shortest induced cycle in G. It is also known that Matching Cut is polynomial-time solvable for graphs of diameter at most 2 [5,17], and even radius at most 2 [22], while being NP-complete for graphs of diameter 3 [17], and thus radius at most 3. Hence, we obtain:

Theorem 1 ([17,22])**.** *For $r \geq 1$, Matching Cut is polynomial-time solvable for graphs of diameter r and graphs of radius r if $r \leq 2$ and NP-complete if $r \geq 3$.*

To study a problem in a systematic way on graph classes that can be characterized by forbidden induced subgraphs, an often used approach is to first focus on the classes of H-free graphs. As Matching Cut is NP-complete for graphs of girth g for every $g \geq 3$ [10] and for $K_{1,4}$-free graphs [7], Matching Cut is NP-complete for H-free graphs whenever H has a cycle or is a forest with a vertex of degree at least 4. What about when H is a forest of maximum degree 3?

We let P_t be the path on t vertices. We denote the disjoint union of two vertex-disjoint graphs $G_1 + G_2$ by $G_1 + G_2 = (V(G_1) \cup V(G_2), E(G_1) \cup E(G_2))$. We let sG be the disjoint union of s copies of G. Feghali [9] proved the existence of an integer t such that Matching Cut is NP-complete for P_t-free graphs, which was narrowed down to $(3P_5, P_{15})$-free graphs in [24] and to $(3P_6, 2P_7, P_{14})$-free graphs in [18]. Let H_1^* be the "H"-graph, which has vertices u, v, w_1, w_2, x_1, x_2 and edges $uv, uw_1, uw_2, vx_1, vx_2$. For $i \geq 2$, let H_i^* be the graph obtained from H_1^* by subdividing uv exactly $i-1$ times; see Fig. 1. It is known that Matching Cut is NP-complete for $(H_1^*, H_3^*, H_5^*, \ldots)$-free bipartite graphs [25] and for $(H_1^*, \ldots, H_i^*)$-free graphs for every $i \geq 1$ [10].

On the positive side, Matching Cut is polynomial-time solvable for *claw-free* graphs ($K_{1,3}$-free graphs) and for P_6-free graphs [24]. Moreover, if Matching Cut is polynomial-time solvable for H-free graphs for some graph H, then it is so for $(H + P_3)$-free graphs [24].

For two graphs H and H', we write $H \subseteq_i H'$ if H is an induced subgraph of H'. Combining the above yields a partial classification (see also [10,23]):

Theorem 2 ([4,7,10,18,22,24,25]). *For a graph H,* Matching Cut *on H-free graphs is*

- *polynomial-time solvable if $H \subseteq_i sP_3 + K_{1,3}$ or $sP_3 + P_6$ for some $s \geq 0$;*
- NP*-complete if $H \supseteq_i K_{1,4}$, P_{14}, $2P_7$, $3P_5$, C_r for some $r \geq 3$, or H_i^* for some $i \geq 1$.*

Our Results. We first note that d-Cut is straightforward to solve for graphs of radius 1 (i.e., graphs with a dominating vertex) and extend Theorem 1:

Theorem 3. *Let $d \geq 2$. For $r \geq 1$, d-*Cut *is polynomial-time solvable for graphs of diameter r if $r \leq 2$ and* NP*-complete if $r \geq 3$.*

Theorem 4. *Let $d \geq 2$. For $r \geq 1$, d-*Cut *is polynomial-time solvable for graphs of radius r if $r \leq 1$ and* NP*-complete if $r \geq 2$.*

Comparing Theorem 1 with Theorems 3 and 4 shows no difference in complexity for diameter but a complexity jump from $d = 1$ to $d = 2$ for radius.

For every $d \geq 2$, we also give polynomial-time algorithms for d-Cut for $(P_3 + P_4)$-free graphs and P_5-free graphs. Our proof techniques use novel arguments, as we can no longer rely on a polynomial-time algorithm for radius 2 or a reduction to 2-SAT as for $d = 1$ [17,22]. Moreover, we show that for $d \geq 2$, d-Cut is polynomial-time solvable for $(H + P_1)$-free graphs whenever d-Cut is so for H-free graphs, thus the cases $\{H + sP_1 \mid s \geq 0\}$ are all *(polynomially) equivalent.* All these results extend the known results for $d = 1$, as can be seen from Theorem 2.

As negative results, we prove that d-Cut is NP-complete for $3P_3$-free graphs for $d = 2$, which we can strengthen to $3P_2$-free graphs for $d \geq 3$. We also show that for $d \geq 2$, d-Cut is NP-complete for $(H_1^*, H_2^*, \ldots, H_i^*)$-free graphs for every $i \geq 1$. Finally, we prove that d-Cut is NP-complete for $K_{1,4}$-free graphs for $d = 2$, which we can strengthen to line graphs for $d \geq 3$. The NP-completeness for graphs of large girth from [10] implies that for $d \geq 2$, d-Cut is NP-complete for H-free graphs if H has a cycle. Hence, by combining the above results, we obtain the following two partial complexity classifications for $d = 2$ and $d \geq 3$:

Theorem 5. *For a graph H, 2-*Cut *on H-free graphs is*

- *polynomial-time solvable if $H \subseteq_i sP_1 + P_3 + P_4$ or $sP_1 + P_5$ for some $s \geq 0$;*
- NP*-complete if $H \supseteq_i K_{1,4}$, $3P_3$, C_r for some $r \geq 3$, or H_i^* for some $i \geq 1$.*

Theorem 6. *Let $d \geq 3$. For a graph H, d-*Cut *on H-free graphs is*

- *polynomial-time solvable if $H \subseteq_i sP_1 + P_3 + P_4$ or $sP_1 + P_5$ for some $s \geq 0$;*
- NP*-complete if $H \supseteq_i K_{1,3}$, $3P_2$, C_r for some $r \geq 3$, or H_i^* for some $i \geq 1$.*

Theorem 5 leaves an infinite number of non-equivalent open cases for $d = 2$. In contrast, Theorem 6 leaves only three non-equivalent open cases for $d \geq 3$, namely when $H = 2P_4$, $H = P_6$ and $H = P_7$.

From Theorems 2–6 we can make the following observations. First, the case where $H = K_{1,3}$ is still open for $d = 2$. We could only prove NP-completeness of d-Cut for $K_{1,4}$-free graphs. However, for $K_{1,3}$-free graphs, there is still a complexity jump from $d = 1$ to $d = 3$ (which might possibly occur even at $d = 2$). Second, there are complexity jumps from $d = 1$ to $d = 2$ for sP_3-free graphs when $s = 3$, and from $d = 1$ to $d = 3$ for sP_2-free graphs when $s = 3$; we do not know if the latter jump even occurs at $d = 2$.

We prove our polynomial-time results in Sect. 3 and our NP-completeness results in Sect. 4. We finish our paper with several open problems in Sect. 5. We start with providing some basic results in Sect. 2.

2 Preliminaries

The *line graph* $L(G)$ of a graph $G = (V, E)$ has the edges of G as its vertices, with an edge between two vertices in $L(G)$ if and only if the corresponding edges in G share an end-vertex. Let $u \in V$. The set $N(u) = \{v \in V \mid uv \in E\}$ is the *neighbourhood* of u, and $|N(u)|$ is the *degree* of u. Let $S \subseteq V$. The *neighbourhood* of S is the set $N(S) = \bigcup_{u \in S} N(u) \setminus S$, and $G[S]$ denotes the subgraph of G *induced* by S. If every vertex of $V \setminus S$ has a neighbour in S, then S is a *dominating* set of G. We also say that $G[S]$ *dominates* G. The *domination number* of G is the size of a smallest dominating set of G. Let $T \subseteq V \setminus S$. The sets S and T are *complete* to each other if every vertex of S is adjacent to every vertex of T.

A *red-blue colouring* of a graph G assigns every vertex of G either the colour red or blue (see, e.g., [9,22]). For $d \geq 1$, a red-blue colouring is a *red-blue d-colouring* if every blue vertex has at most d red neighbours; every red vertex has at most d blue neighbours; and both colours red and blue are used at least once. See Fig. 1 for examples of a red-blue 1-colouring and a red-blue 3-colouring.

Observation 7 *For every $d \geq 1$, a connected graph G has a d-cut if and only if it has a red-blue d-colouring.*

If every vertex of a set $S \subseteq V$ has the same colour (either red or blue) in a red-blue colouring, then S, and also $G[S]$, are *monochromatic*. An edge with a blue and a red end-vertex is *bichromatic*. Note that for every $d \geq 1$, the graph K_{2d+1} is monochromatic in every red-blue d-colouring, and that every connected graph with a red-blue d-colouring contains a bichromatic edge.

We now generalize a known lemma for Matching Cut (see, e.g., [22])).

Lemma 8. *For $d, g \geq 1$, it is possible to find in $O(2^g n^{dg+2})$-time a red-blue d-colouring (if it exists) of a graph G with n vertices and domination number g.*

Proof. Let $d, g \geq 1$ and G be a graph on n vertices with domination number g. Let D be a dominating set D of G that has size at most g.

We consider all $2^{|D|} \leq 2^g$ options of giving the uncoloured vertices of D either colour red or blue. For each red-blue colouring of D we do as follows. For every red vertex of D, we consider all $O(n^d)$ options of colouring at most d of its uncoloured neighbours blue, and we colour all of its other uncoloured neighbours red. Similarly, for every blue vertex of D, we consider all $O(n^d)$

options of colouring at most d of its uncoloured neighbours red, and we colour all of its other uncoloured neighbours blue. As D dominates G, we obtained a red-blue colouring c of the whole graph G. We discard the option if c is not a red-blue d-colouring of G.

We note that any red-blue d-colouring of G, if it exists, will be found by the above algorithm. As the total number of options is $O(2^g n^{dg})$ and checking if a red-blue colouring is a red-blue d-colouring takes $O(n^2)$ time, our algorithm has total running time $O(2^g n^{dg+2})$. □

Let $X = \{x_1, x_2, ..., x_n\}$ be a set of variables and $\mathcal{C} = \{C_1, C_2, ..., C_m\}$ be a set of clauses over X. The problem Not-All-Equal Satisfiability asks if $(X, \mathcal{C})$ has a *satisfying not-all-equal truth assignment* ϕ that is, ϕ sets, in each C_i, at least one literal true and at least one literal false. The next theorem is well-known (where the second part follows from a folklore trick; see e.g. [12]).

Theorem 9 ([27]). Not-All-Equal Satisfiability *is* NP*-complete even for*

- *instances, in which each literal occurs only positively, and in which each clause contains exactly three literals;*
- *instances, in which each literal occurs only positively and in two or three different clauses, and in which each clause contains either two or three literals.*

3 Polynomial-Time Results

We now show our polynomial-time results for d-Cut for $d \geq 2$, which complement corresponding known polynomial-time results for $d = 1$ (see Sect. 1) and their proofs yield alternative proofs for $d = 1$. We omit the proof of our first result.

Theorem 10. *For $d \geq 2$, d-*Cut *is polynomial-time solvable for graphs of diameter at most 2.*

Theorem 11. *For $d \geq 2$, d-*Cut *is polynomial-time solvable for P_5-free graphs.*

Proof. Let $d \geq 2$. Let $G = (V, E)$ be a connected P_5-free graph on n vertices. As G is P_5-free and connected, G has a dominating set D, such that either D induces a cycle on five vertices or D is a clique [21]. Moreover, we find such a dominating set D in $O(n^3)$ time [16]. If $|D| \leq 3d$, then we apply Lemma 8. Now assume that $|D| \geq 3d + 1 \geq 7$, and thus D is a clique. If $D = V$, then G has no d-cut, as $|D| \geq 3d + 1$ and D is a clique. Assume that $D \subsetneq V$, so $G - D$ has at least one connected component. Below we explain how to find in polynomial time a d-cut of G, or to conclude that G does not have a d-cut. By Observation 7, we need to decide in polynomial time if G has a red-blue d-colouring.

We first enter the *blue phase* of our algorithm. As $|D| \geq 3d + 1$ and D is a clique, D is monochromatic in any red-blue d-colouring of G. Hence, we may colour, without loss of generality, all the vertices of D blue, and we may colour all vertices of every connected component of $G - D$ except one connected component blue as well. We branch over all $O(n)$ options of choosing the connected component L_1 of $G - D$ that will contain a red vertex.[1] For each option, we are going to repeat the process of colouring vertices blue until we colour at least one vertex red. We first explain how this process works if we do not do this last step.

As L_1 is connected and P_5-free, we find in $O(n^3)$ time (using the algorithm of [16]) a dominating set D_1 of L_1 that is either a cycle on five vertices or a clique. We colour the vertices of D_1 blue. As we have not used the colour red yet, we colour all vertices of every connected component of $L_1 - D_1$ except one connected component blue. So, we branch over all $O(n)$ options of choosing the connected component L_2 of $L_1 - D_1$ that will contain a red vertex. Note that every uncoloured vertex of G, which belongs to L_2, has both a neighbour in D (as D dominates G) and a neighbour in D_1 (as D_1 dominates L_1). We now find a dominating set D_2 of L_2 and a connected component L_3 of $G - D_2$ with a dominating set D_3, and so on. See also Fig. 2.

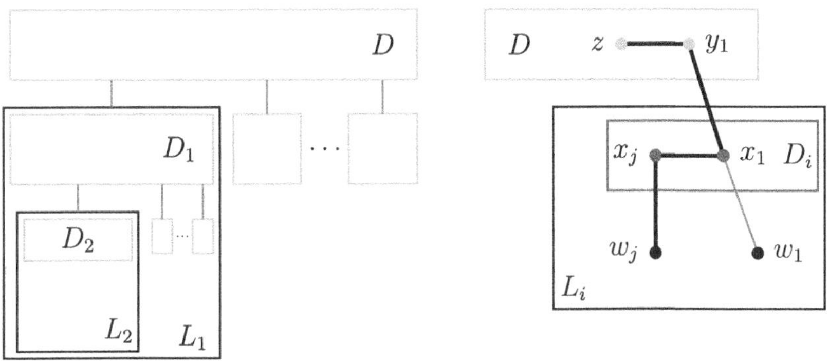

Fig. 2. The graph G in the blue phase (left) and in the red phase: Case 2 (right). (Color figure online)

If we repeat the above process more than d times, we have either coloured every vertex of G blue, or we found $d + 1$ pairwise disjoint, blue sets $D, D_1, \ldots, D_d$, such that every uncoloured vertex u has a neighbour in each of them. The latter implies that u has $d + 1$ blue neighbours, so u must be coloured blue as well. Hence, in each branch, we would eventually end up with the situation where all vertices of G will be coloured blue. To prevent this from happening, we must colour, in each branch, at least one vertex of D_i red, for some $1 \leq i \leq d-1$. As soon as we do this, we end the blue phase for the branch under consideration, and our algorithm enters the *red phase*.

[1] For $d = 1$, up to now, the same approach is used for P_5-free graphs [9]. But the difference is that for $d = 1$, the algorithm and analysis is much shorter: one only has to check, in this stage, if there is a component of $G - D$, whose vertices can all be safely coloured red. Then one either finds a matching cut, or G has no matching cut.

Each time we have $O(n)$ options to select a connected component L_i and, as argued above, we do this at most d times. Hence, we enter the red phase for $O(n^d)$ branches in total. From now on, we call these branches the *main branches* of our algorithm. For a main branch, we say that we *quit the blue phase at level i* if we colour at least one vertex of D_i red. If we quit the blue phase for a main branch at level i, for some $1 \leq i \leq d-1$, then we have constructed, in polynomial time, pairwise disjoint sets $D, D_1, \ldots, D_i$ and graphs $L_1, \ldots, L_i$, such that:

- for every $h \in \{1, \ldots, i-1\}$, L_{h+1} is a connected component of $L_h - D_h$;
- every vertex of G that does not belong to L_i has been coloured blue;
- for every $h \in \{1, \ldots, i\}$, D_h induces a cycle on five vertices or is a clique;
- D dominates G, so, in particular, D dominates L_i; and
- for every $h \in \{1, \ldots, i\}$, D_h dominates L_h.

We now prove the following claim, which shows that we branched correctly.

Claim 11.1 *The graph G has a red-blue d-colouring that colours every vertex of D blue if and only if we have quit a main branch at level i for some $1 \leq i \leq d-1$, such that G has a red-blue d-colouring that colours at least one vertex of D_i red and all vertices not in L_i blue.*

Proof of the Claim. Suppose G has a red-blue d-colouring c that colours every vertex of D blue (the reverse implication is immediate). By definition, c has coloured at least one vertex u in $G - D$ red. As we branched in every possible way, there is a main branch that quits the blue phase at level i, such that u belongs to L_i for some $1 \leq i \leq d-1$. We pick a main branch with largest possible i, so u is not in L_{i+1}. Hence, $u \in D_i$. We also assume that no vertex u' that belongs to some D_h with $h < i$ is coloured red by c, as else we could take u' instead of u. Hence, c has coloured every vertex in $D_1 \cup \ldots \cup D_{i-1}$ (if $i \geq 2$) blue. Therefore, we may assume that every vertex $v \notin D \cup D_1 \cup \ldots \cup D_{i-1} \cup V(L_i)$ has been coloured blue by c. If not, may just recolour all such vertices v blue for the following reasons: u is still red and no neighbour of v is red, as after a possible recolouring all red vertices belong to L_i, while v belongs to a different component of $G - (D \cup D_1 \cup \cdots \cup D_{i-1})$ than L_i. So, we proved the claim. ◁

Claim 11.1 allows us to do some specific branching once we quit the blue phase for a certain main branch at level i. Namely, all we have to do is to consider all options to colour at least one vertex of D_i red. We call these additional branches *side branches*. We distinguish between the following two cases:

Case 1. $|D_i| \leq 2d+1$.
We consider each of the at most 2^{2d+1} options to colour the vertices of D_i either red or blue, such that at least one vertex of D_i is coloured red. Next, for each vertex $u \in D_i$, we consider all $O(n^d)$ options to colour at most d of its uncoloured neighbours blue if u is red, or red if u is blue. Note that the total number of side branches is $O(2^{2d+1} n^{d(2d+1)})$. As D_i dominates L_i and the only uncoloured vertices were in L_i, we obtained a red-blue colouring c of G. We

check in polynomial time if c is a red-blue d-colouring of G. If so, we stop and return c. If none of the side branches yields a red-blue d-colouring of G, then by Claim 11.1 we can safely discard the main branch under consideration.

Case 2. $|D_i| \geq 2d+2$.
Recall that D_i either induces a cycle on five vertices or is a clique. As $|D_i| \geq 2d+2 \geq 6$, we find that D_i is a clique. As $|D_i| \geq 2d+2$, this means that D_i must be monochromatic, and thus every vertex of D_i must be coloured red. We check in polynomial time if D_i contains a vertex with more than d neighbours in D (which are all coloured blue), or if D contains a vertex with more than d neighbours in D_i (which are all coloured red). If so, we may safely discard the main branch under consideration due to Claim 11.1.

From now on, assume that every vertex in D_i has at most d neighbours in D, and vice versa. By construction, every uncoloured vertex belongs to $L_i - D_i$. Hence, if $V(L_i) = D_i$, we have obtained a red-blue colouring c of G. We check, in polynomial time, if c is also a red-blue d-colouring of G. If so, we stop and return c. Otherwise, we may safely discard the main branch due to Claim 11.1.

Now assume $V(L_i) \supsetneq D_i$. We colour all vertices in $L_i - D_i$ that are adjacent to at least $d+1$ vertices in D blue; we have no choice as the vertices in D are all coloured blue. If there are no uncoloured vertices left, we check in polynomial time if the obtained red-blue colouring is a red-blue d-colouring of G. If so, we stop and return it; else we may safely discard the main branch due to Claim 11.1.

Assume that we still have uncoloured vertices left. We recall that these vertices belong to $L_i - D_i$, and that by construction they have at most d neighbours in D. Consider an uncoloured vertex w_1. As D_i dominates L_i, we find that w_1 has a neighbour x_1 in D_i. As D dominates G, we find that x_1 is adjacent to some vertex $y_1 \in D$. We consider all $O(n^{2d})$ possible ways to colour the uncoloured neighbours of x_1 and y_1, such that x_1 (which is red) has at most d blue neighbours, and y_1 (which is blue) has at most d red neighbours. If afterwards there is still an uncoloured vertex w_2, then we repeat this process: we choose a neighbour x_2 of w_2 in D_i (so x_2 is coloured red). We now branch again by colouring the uncoloured neighbours of x_2. If we find an uncoloured vertex w_3, then we find a neighbour x_3 of w_3 in D_i and so on. So, we repeat this process until there are no more uncoloured vertices. This gives us $2p$ distinct vertices $w_1, \ldots, w_p$, $x_1, \ldots, x_p$ for some integer $p \geq 1$, together with vertex y_1. The total number of side branches for the main branch is $O(n^{2pd})$.

We claim that $p \leq d$. For a contradiction, assume that $p \geq d+1 \geq 3$. Let $2 \leq j \leq p$. As D_i is a clique that contains x_1 and x_j, we find that x_1 and x_j are adjacent. Hence, G contains the 4-vertex path $w_j x_j x_1 y_1$. By construction, w_j was uncoloured after colouring the neighbours of x_1 and y_1, so w_j is neither adjacent to x_1 nor to y_1. This means that $w_j x_j x_1 y_1$ is an induced P_4 if and only if x_j is not adjacent to y_1. Recall that w_j and every vertex of D_i, so including x_1 and x_j, has at most d neighbours in D. As $|D| \geq 3d+1$, this means that D contains a vertex z that is not adjacent to any of w_j, x_j, x_1. As D is a clique, z is adjacent to y_1. Hence, x_j must be adjacent to y_1, as otherwise $w_j x_j x_1 y_1 z$ is an induced P_5; see also Fig. 2. The latter is not possible as G is P_5-free. We now find that y_1 is adjacent to $x_1, \ldots, x_p$, which all belong to D_i. As we assumed that $p \geq d+1$ and every vertex of D, including y_1, is adjacent to at most d vertices of D_i, this yields a contradiction. We conclude that $p \leq d$.

As $p \leq d$, the total number of side branches is $O(n^{2d^2})$. Each side branch yields a red-blue colouring of G. We check in polynomial time if it is a red-blue d-colouring of G. If so, then we stop and return it; else we can safely discard the side branch, and eventually the associated main branch, due to Claim 11.1.

The correctness of our algorithm follows from its description. We now analyze its running time. We started the algorithm with searching for a set D. If D has size at most $3d$, then we applied Lemma 8, which takes polynomial time. Otherwise we argue as follows. The total number of main branches is $O(n^d)$. For each main branch we have either $O(2^{2d+1}n^{d(2d+1)})$ side branches (Case 1) or $O(n^{2d^2})$ side branches (Case 2). Hence, the total number of branches is $O(n^d 2^{2d+1} n^{d(2d+1)})$. As d is fixed, this number is polynomial. As processing each branch takes polynomial time (including the construction of the D_i-sets and L_i-graphs), we conclude that our algorithm runs in polynomial time. This completes the proof. □

We omit the proofs of our final two polynomial-time results.

Theorem 12. *For every $d \geq 2$, d-*Cut *is polynomial-time solvable for (P_3+P_4)-free graphs.*

Theorem 13. *For every graph H and every $d \geq 2$, if d-*Cut *is polynomial-time solvable for H-free graphs, then it is so for $(H + P_1)$-free graphs.*

4 NP-Completeness Results

In this section we show our NP-completeness results for d-Cut for $d \geq 2$ (some of these results only hold for $d \geq 3$). As d-Cut is readily seen to be in NP for each $d \geq 1$, we only show NP-hardness in our proofs. The proof of our first result, which we omit, generalizes the arguments of the corresponding proof for Matching Cut [7].

Theorem 14. *For $d = 2$, d-*Cut *is* NP*-complete for $K_{1,4}$-free graphs.*

For $d \geq 3$, we prove a stronger result than Theorem 14. An edge colouring of a graph $G = (V, E)$ with colours red and blue is called a *red-blue edge colouring* of G, which is a *red-blue edge d-colouring* of G if every edge of G is adjacent to at most d edges of the other colour and both colours are used at least once. Now, G has a red-blue edge d-colouring if and only if $L(G)$ has a red-blue d-colouring. A set $S \subseteq V$ is *monochromatic* if all edges of $G[S]$ are coloured alike.

Theorem 15. *For $d \geq 3$, d-*Cut *is* NP*-complete for line graphs.*

Proof. We reduce from Not-All-Equal Satisfiability with positive literals only and in which each clause contains exactly three literals. This problem is NP-complete by Theorem 9. Let $(X, \mathcal{C})$ be such an instance, where $X = \{x_1, x_2, \ldots, x_n\}$ and $\mathcal{C} = \{C_1, C_2, \ldots, C_m\}$.

We construct, in polynomial time, a graph G; see also Fig. 3:

- Build a clique $S = \{v^S_{x_1}, \ldots, v^S_{x_n}\} \cup \{v^{c_1}_1, \ldots, v^{c_1}_{d-2}\} \cup \cdots \cup \{v^{c_m}_1, \ldots, v^{c_m}_{d-2}\}$.
- Build a clique $\overline{S} = \{v^{\overline{S}}_{x_1}, \ldots, v^{\overline{S}}_{x_n}\} \cup \{u^{c_1}_1, \ldots, u^{c_1}_{d-2}\} \cup \cdots \cup \{u^{c_m}_1, \ldots, u^{c_m}_{d-2}\}$.
- For every $x \in X$, add cliques $V_x = \{v^x_1, \ldots, v^x_{d-1}\}$ and $V_{\overline{x}} = \{v^{\overline{x}}_1, \ldots, v^{\overline{x}}_{d-1}\}$.
- For every $x \in X$, add a vertex v_x with edges $v_x v^S_x$, $v_x v^{\overline{S}}_x$, $v_x v^x_1, \ldots, v_x v^x_{d-1}$, $v_x v^{\overline{x}}_1, \ldots, v_x v^{\overline{x}}_{d-1}$.
- For every $C \in \mathcal{C}$, add a clause vertex v_c with edges $v_c v^c_1, \ldots, v_c v^c_{d-2}$, $v_c u^c_1, \ldots, v_c u^c_{d-2}$, and if $C = \{x_i, x_j, x_k\}$, also add a vertex $v^{x_i}_c$ to V_{x_i}, a vertex $v^{x_j}_c$ to V_{x_j} and a vertex $v^{x_k}_c$ to V_{x_k}, and add the edges $v_c v^{x_i}_c$, $v_c v^{x_j}_c$, $v_c v^{x_k}_c$.
- Add, if needed, some auxiliary vertices to $S, \overline{S}, V_{x_1}, \ldots, V_{x_n}, V_{\overline{x_1}}, \ldots, V_{\overline{x_n}}$, such that in the end all these sets are cliques of size at least $2d + 2$.

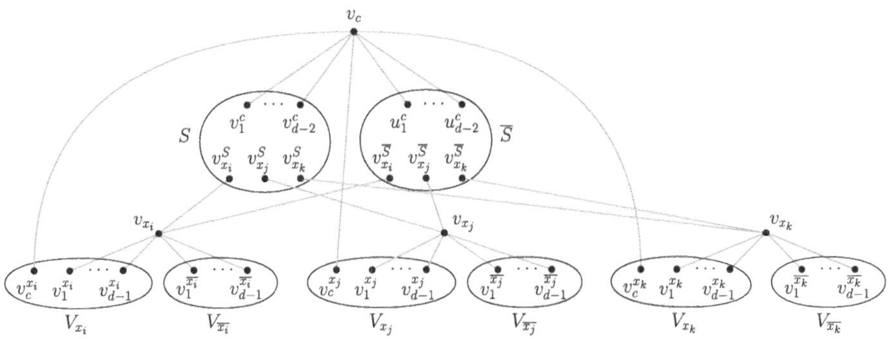

Fig. 3. An example of vertices in the reduction related to clause $C = \{x_i, x_j, x_k\}$.

We claim that $(X, \mathcal{C})$ has a satisfying not-all-equal truth assignment if and only if the line graph $L(G)$ has a d-cut. Recall that, by Observation 7, $L(G)$ has a d-cut if and only if $L(G)$ has a red-blue d-colouring. Furthermore, $L(G)$ has a red-blue d-colouring if and only if G has a red-blue edge d-colouring. Hence, we will show that $(X, \mathcal{C})$ has a satisfying not-all-equal truth assignment if and only if G has a red-blue edge d-colouring.

First suppose $(X, \mathcal{C})$ has a satisfying not all-equal truth assignment. We colour all edges in S red and in $\overline{S}$ blue. For every $x \in X$ set to true, we colour the edges in V_x red and those in $V_{\overline{x}}$ blue. For every $x \in X$ set to false, we colour the edges in V_x blue and those in $V_{\overline{x}}$ red. Consider an edge uv, with $v \in \{v_x, v_c \mid x \in X, c \in C\}$. Then u is contained in a clique $D \in \{S, \overline{S}, V_{x_1}, V_{\overline{x_1}}, \ldots, V_{x_n}, V_{\overline{x_n}}\}$. Colour uv with the same colour as the edges of D.

Now, let $D \in \{S, \overline{S}, V_{x_1}, V_{\overline{x_1}}, \ldots, V_{x_n}, V_{\overline{x_n}}\}$. Every $uu' \in E(D)$ is adjacent to only edges of the same colour. For $u \in V(D)$ and $v \in \{v_x, v_c \mid x \in X, c \in C\}$, the edge uv has the same colour as all edges in D. Since S and $\overline{S}$ have different colours and V_x and $V_{\overline{x}}$ have different colours for every $x \in X$, uv has at most d adjacent edges of each colour. Hence, we obtained a red-blue edge d-colouring of G.

Now suppose that G has a red-blue edge d-colouring. We prove a series of claims:

Claim 15.1 *Every clique* $D \in \{S, \overline{S}, V_{x_1}, V_{\overline{x_1}}, \ldots, V_{x_n}, V_{\overline{x_n}}\}$ *is monochromatic.*

Proof of the Claim. First assume $G[D]$ has a red edge uv and a blue edge uw. As $|D| \geq 2d+2$, we know that u is incident to at least $2d+1$ edges. Hence, we may assume without loss of generality that u is incident to at least $d+1$ red edges. However, now the blue edge uw is adjacent to $d+1$ red edges, a contradiction. As every $u \in D$ is incident to only edges of the same colour and D is a clique, it follows that D is monochromatic. ◁

By Claim 15.1, we can speak about the *colour* (either red or blue) of a clique D if D belongs to $\{S, \overline{S}, V_{x_1}, V_{\overline{x_1}}, \ldots, V_{x_n}, V_{\overline{x_n}}\}$.

Claim 15.2 *For each* $x \in X$ *and* $c \in C$*, each edge from* v_x *or* v_c *to a vertex in a clique* $D \in \{S, \overline{S}, V_{x_1}, V_{\overline{x_1}}, \ldots, V_{x_n}, V_{\overline{x_n}}\}$ *has the same colour as* D.

Proof of the Claim. This follows directly from the fact that $|D| \geq 2d+1$. ◁

Claim 15.3 *The cliques* S *and* $\overline{S}$ *have different colours if and only if for every variable* $x \in X$*, it holds that* V_x *and* $V_{\overline{x}}$ *have different colours.*

Proof of the Claim. First suppose S and $\overline{S}$ have different colours, say S is red and $\overline{S}$ is blue. For a contradiction, assume there exists a variable $x \in X$, such that V_x and $V_{\overline{x}}$ have the same colour, say blue. By Claim 15.2, we have that the $2d-2$ edges between v_x and $V_x \cup V_{\overline{x}}$ and the edge $v_x v_x^{\overline{S}}$ are all blue, while $v_x v_x^S$ is red. Hence, the red edge $v_x v_x^S$ is adjacent to at least $2d-1 \geq d+1$ blue edges, a contradiction.

Now suppose that for all $x \in X$, V_x and $V_{\overline{x}}$ have different colours, say V_x is red and $V_{\overline{x}}$ is blue. For a contradiction, assume that S and $\overline{S}$ have the same colour, say blue. Let $x \in X$. By Claim 15.2, we have that the edges between v_x and V_x are red, while all other edges incident to v_x are blue. Now every (red) edge between v_x and V_x is incident to $d+1$ blue edges, a contradiction. ◁

Claim 15.4 *The cliques* S *and* $\overline{S}$ *have different colours.*

Proof of the Claim. For a contradiction, assume S and $\overline{S}$ have the same colour, say blue. By Claim 15.2, we have that $v_x v_x^S$ is blue. By Claim 15.3, we find that for every $x \in X$, V_x and $V_{\overline{x}}$ have the same colour. If V_x and $V_{\overline{x}}$ are both red, then the $2d-2$ edges between v_x and $V_x \cup V_{\overline{x}}$ are red due to Claim 15.2. Consequently, the blue edge $v_x v_x^S$ is adjacent to $2d-2 \geq d+1$ red edges. This is not possible. Hence, V_x and $V_{\overline{x}}$ are blue, and by Claim 15.2, all edges between v_x and $V_x \cup V_{\overline{x}}$ are blue as well. This means that every edge of G is blue, a contradiction. ◁

Claim 15.5 *For every clause $C = \{x_i, x_j, x_k\}$ in $\mathcal{C}$, the cliques V_{x_i}, V_{x_j} and V_{x_k} do not all have the same colour.*

Proof of the Claim. For a contradiction, assume V_{x_i}, V_{x_j} and V_{x_k} have the same colour, say blue. By Claim 15.4, S and $\overline{S}$ are coloured differently, say S is red and $\overline{S}$ is blue. By Claim 15.2, we have that the three edges $v_c v_c^{x_i}$, $v_c v_c^{x_j}$ and $v_c v_c^{x_k}$ are all blue, just like the $d-2$ edges between v_c and $\overline{S}$, while every edge between v_c and S is red. Consider such a red edge e. We find that e is adjacent to $d+1$ blue edges, a contradiction. $\lhd$

For each variable x, if the clique V_x is coloured red, then set x to true, and else to false. By Claim 15.5, this yields a satisfying not-all-equal truth assignment. □
We now show a result for $d = 2$ and strengthen it for $d \geq 3$. The gadgets in the two proofs, which we omit, are similar to each other. They are not $2P_4$-free, P_6-free or P_7-free.

Theorem 16. *For $d = 2$, d-*Cut *is* NP*-complete for $3P_3$-free graphs of radius 2 and diameter 3.*

Theorem 17. *For $d \geq 3$, d-*Cut *is* NP*-complete for $3P_2$-free graphs of radius 2 and diameter 3.*

Our final theorem is a straightforward generalization of the case $d = 1$ from [24], and we omit its proof.

Theorem 18. *For $d \geq 2$ and $i \geq 1$, d-*Cut *is* NP*-complete for $(H_1^*, \ldots, H_i^*)$-free graphs.*

5 Conclusions

We considered the natural generalization of Matching Cut to d-Cut [13] and proved dichotomies for graphs of bounded diameter and graphs of bounded radius. We also started a systematic study on the complexity of d-Cut for H-free graphs. For $d = 2$, our results for H-free graphs still left an infinite number of non-equivalent open cases, just like for $d = 1$. However, for $d = 3$ we were able to obtain an almost-complete complexity classification of d-Cut for H-free graphs, with only three non-equivalent open cases left. We finish our paper with some open problems on H-free graphs resulting from our systematic study.

We recall that 1-Cut is polynomial-time solvable for claw-free graphs [4], while we showed that d-Cut is NP-complete even for line graphs if $d \geq 3$. What is the complexity of 2-Cut for line graphs and for claw-free graphs? We also ask whether 2-Cut is polynomial-time solvable for $(H + P_2)$-free graphs if it is for H-free graphs. A positive answer would reduce the number of non-equivalent open cases to a finite number for $d = 2$. In particular, we ask for the complexity of 2-Cut for $3P_2$-free graphs. Finally, we recall the only three non-equivalent open cases $H = 2P_4$, $H = P_6$, $H = P_7$ for d-Cut on H-free graphs for $d \geq 3$. The first and last case are also open for $d = 1$; for $d = 2$ all three cases are open.

Acknowledgments. We thank Carl Feghali and Édouard Bonnet for fruitful discussions.

References

1. Araújo, J., Cohen, N., Giroire, F., Havet, F.: Good edge-labelling of graphs. Discret. Appl. Math. **160**, 2502–2513 (2012)
2. Aravind, N.R., Saxena, R.: An FPT algorithm for Matching Cut and d-Cut. In: Proceedings of IWOCA 2021, LNCS, vol. 12757, pp. 531–543 (2021)
3. Bonnet, E., Chakraborty, D., Duron, J.: Cutting Barnette graphs perfectly is hard. In: Proceedings of WG 2023, LNCS, vol. 14093, pp. 116–129 (2023)
4. Bonsma, P.S.: The complexity of the Matching-Cut problem for planar graphs and other graph classes. J. Graph Theory **62**, 109–126 (2009)
5. Borowiecki, M., Jesse-Józefczyk, K.: Matching Cutsets in graphs of diameter 2. Theoret. Comput. Sci. **407**, 574–582 (2008)
6. Bouquet, V., Picouleau, C.: The complexity of the perfect Matching-Cut problem. CoRR **abs/2011.03318** (2020)
7. Chvátal, V.: Recognizing decomposable graphs. J. Graph Theory **8**, 51–53 (1984)
8. Farley, A.M., Proskurowski, A.: Networks immune to isolated line failures. Networks **12**, 393–403 (1982)
9. Feghali, C.: A note on Matching-Cut in P_t-free graphs. Inf. Process. Lett. **179**, 106294 (2023)
10. Feghali, C., Lucke, F., Paulusma, D., Ries, B.: Matching Cuts in graphs of high girth and H-free graphs. In: Proceedings of ISAAC 2023, LIPIcs, vol. 283, pp. 1–16 (2023)
11. Golovach, P.A., Paulusma, D., Song, J.: Computing vertex-surjective homomorphisms to partially reflexive trees. Theoret. Comput. Sci. **457**, 86–100 (2012)
12. Golovach, P.A., Paulusma, D., Song, J.: Closing complexity gaps for coloring problems on H-free graphs. Inf. Comput. **237**, 204–214 (2014)
13. Gomes, G., Sau, I.: Finding cuts of bounded degree: complexity, FPT and exact algorithms, and kernelization. Algorithmica **83**, 1677–1706 (2021)
14. Graham, R.L.: On primitive graphs and optimal vertex assignments. Ann. N. Y. Acad. Sci. **175**, 170–186 (1970)
15. Heggernes, P., Telle, J.A.: Partitioning graphs into generalized dominating sets. Nord. J. Comput. **5**, 128–142 (1998)
16. van't Hof, P., Paulusma, D.: A new characterization of P_6-free graphs. Discrete Appl. Math. **158**, 731–740 (2010)
17. Le, H.O., Le, V.B.: A complexity dichotomy for Matching Cut in (bipartite) graphs of fixed diameter. Theoret. Comput. Sci. **770**, 69–78 (2019)
18. Le, H., Le, V.B.: Complexity results for matching cut problems in graphs without long induced paths. In: Proceedings of WG 2023, LNCS, vol. 14093, pp. 417–431 (2023)
19. Le, V.B., Lucke, F., Paulusma, D., Ries, B.: Maximizing Matching Cuts. CoRR **abs/2312.12960** (2023)
20. Le, V.B., Telle, J.A.: The perfect Matching Cut problem revisited. Theoret. Comput. Sci. **931**, 117–130 (2022)
21. Liu, J., Zhou, H.: Dominating subgraphs in graphs with some forbidden structures. Discret. Math. **135**, 163–168 (1994)
22. Lucke, F., Paulusma, D., Ries, B.: On the complexity of Matching Cut for graphs of bounded radius and H-free graphs. Theoret. Comput. Sci. **936** (2022)

23. Lucke, F., Paulusma, D., Ries, B.: Dichotomies for Maximum Matching Cut: H-freeness, bounded diameter, bounded radius. In: Proceedings of MFCS 2023, LIPIcs, vol. 272, pp. 1–15 (2023)
24. Lucke, F., Paulusma, D., Ries, B.: Finding matching cuts in H-free graphs. Algorithmica **85**, 3290–3322 (2023)
25. Moshi, A.M.: Matching Cutsets in graphs. J. Graph Theory **13**, 527–536 (1989)
26. Patrignani, M., Pizzonia, M.: The complexity of the Matching-Cut problem. In: Brandstädt, A., Le, V.B. (eds.) WG 2001. LNCS, vol. 2204, pp. 284–295. Springer, Heidelberg (2001). https://doi.org/10.1007/3-540-45477-2_26
27. Schaefer, T.J.: The complexity of satisfiability problems. Proc. STOC **1978**, 216–226 (1978)

Exact and Approximate k-planarity Testing for Maximal Graphs of Small Pathwidth

Miriam Münch[1(✉)], Maximilian Pfister[2], and Ignaz Rutter[1]

[1] Fakultät für Informatik und Mathematik, Universität Passau, Passau, Germany
{muenchm,rutter}@fim.uni-passau.de

[2] Institut für Informatik, Eberhard Karls Universität Tübingen, Tübingen, Germany
maximilian.pfister@uni-tuebingen.de

Abstract. A graph is k-planar, if it admits a drawing with at most k crossings per edge. Testing whether a given graph is k-planar is known to be NP-complete. For $k = 1$ the problem remains NP-complete even for graphs of bounded pathwidth [3]. In this paper we give linear-time algorithms for efficiently testing 1-planarity of w-paths, where a w-path is a maximal graph with pathwidth w. Closely related to the concept of k-planarity is the local crossing number of a graph G; i.e. the minimum number k such that G is k-planar. For general graphs of pathwidth 3 we give a 7-approximation for the local crossing number. Finally we employ a technique used by Biedl et al. to derive an $O(w)$-approximation of the local crossing number for maximal pathwidth-w graphs.

Keywords: Approximation · 1-planarity · Pathwidth · Local crossing number · k-planarity

1 Introduction

The recognition of planar graphs is one of the most researched topics in graph theory. Almost a century ago, Kuratowski [11] and independently Wagner [19] characterized a planar graph in terms of forbidden subdivisions/minors of the graphs K_5 and $K_{3,3}$. Later research focused on making planarity testing efficient and culminated in several linear-time algorithms, e.g., the algorithm by Hopcroft and Tarjan [8] or the one by Boyer and Myrvold [5]. An empirical study [12] showed that, when edge-crossings cannot be avoided (i.e., the given graph is non-planar), the quality of the drawing not only depends on the number of crossings, but also on the topological properties of the crossings. This observation gave rise to the field of *beyond-planarity*, where graph classes are defined by a set of forbidden crossing configurations. One of the most researched beyond-planar graph classes is the class of k-planar graphs, which are graphs that admit a drawing where every edge is crossed at most k times.

The class of 1-planar graphs was first considered by Ringel [14] in the context of a simultaneous coloring of the vertices and faces of a planar graph. Since

D. Kráľ and M. Milanič (Eds.): WG 2024, LNCS 14760, pp. 430–443, 2025.
https://doi.org/10.1007/978-3-031-75409-8_30

then, numerous results have appeared in the literature, we refer the reader to the recent survey of Suzuki [16]. Regarding the recognition problem of k-planar graphs, we first observe that by subdividing every edge of our graph sufficiently many times, we always obtain a 1-planar drawing. This implies that the class of k-planar graphs is not minor-closed for any $k > 1$ and hence we cannot leverage results from the graph minor theory by Robertson and Seymour [15]. In fact, it was shown that the recognition of 1-planar graphs is NP-complete [7,10]. Urschel and Wellens [18] extended this result to any k and also showed that the corresponding minimization problem is APX-Hard (in particular, k-planarity cannot be approximated by a factor smaller than 2). For the special case of 1-planarity, the hardness results hold even for graphs of bounded pathwidth [3].

To overcome these hardness-results we employ an *approximation* framework. For a fixed k the goal is to derive a so called *c-approximation*; i.e. an algorithm that either yields a drawing Γ in which every edge is crossed at most $c \cdot k$ times, or correctly recognizes that the input graph is not k-planar.

Our Contribution. In Sect. 3 we provide exact algorithms for testing whether 3- or 4-paths are 1-planar in linear time. Section 4 deals with the problem of k-planarity testing. In Sect. 4.1, we consider pathwidth 3 and provide an algorithm that yields a 6-approximation for $k \leq 4$ and a 7-approximation for $k > 4$. Finally, in Sect. 4.2, we consider w-paths and show that a construction of Biedl et al. [17] yields an $\mathcal{O}(w)$-approximation.

2 Preliminaries

A *drawing* Γ of a graph $G = (V, E)$ is a function that maps every vertex $v \in V$ to a point $\Gamma(v)$ in the plane and every edge $uv \in E$ to a Jordan arc with endpoints $\Gamma(u)$ and $\Gamma(v)$ that does not pass through any $\Gamma(w)$ for $w \in V \setminus \{u, v\}$. For simplicity we usually identify edges and vertices with their images in a drawing. A *crossing* is then a common interior point of two edges. A drawing Γ of a graph G is called *k-planar*, if every edge of G is crossed at most k-times in Γ. A graph G is *k-planar* if it admits a k-planar drawing. The *local crossing number* $\text{lcr}(G)$ is the smallest k such that G is k-planar.

A *path decomposition* P of G is a sequence $X_1, \ldots, X_l$ of subsets of vertices of G, called *bags* such that (i) $\bigcup_{i=1}^{l} X_i = V$, (ii) for every edge uv in G there is a bag X_i in P with $u, v \in X_i$ and (iii) every vertex appears in a contiguous subsequence of bags. The *width* of a path decomposition $P = X_1, \ldots, X_l$ is $\max_{1 \leq i \leq l} |X_i| - 1$. The *pathwidth* $pw(G)$ of a graph G is the smallest width among all path decompositions of G. A pathwidth-w graph G is *maximal* if adding any edge to G increases its pathwidth. A maximal pathwidth-w graph is called *w-path*. Note that in a width-w path decomposition of a w-path G each bag forms a clique. For a subset $S \subseteq V$, we write $G[S]$ to denote the subgraph induced by S. For a drawing Γ of G, we use $\Gamma[S]$ to denote the drawing of $G[S]$ induced by Γ. For a path decomposition $P = X_1, \ldots, X_l$ of G, it is further convenient to define $G[i] := G[X_i]$ for $1 \le i \le l$ and $G[i : j] := G[\bigcup_{k=i}^{j} X_k]$. We also write $G[: i]$ and $G[i :]$ to denote $G[1 : i]$ and $G[i : l]$, respectively. For a drawing Γ of G, $\Gamma[i]$ as well as $\Gamma[i : j]$, $\Gamma[: i]$ and $\Gamma[i :]$ are defined similarly.

In this paper we use so-called *alternating* path decompositions introduced by Biedl et al. [17]. A path decomposition P of G of width w is alternating if P contains exactly $l = 2|V(G)| - 2w - 1$ bags such that $|X_i| = w + 1$ if i is odd and $|X_i| = w$ if i is even with $X_{i-1} \supsetneq X_i \subsetneq X_{i+1}$ for every even $1 < i < l$; see Fig. 1. Let G be a graph of pathwidth w and let $P = X_1, \ldots, X_l$ be an

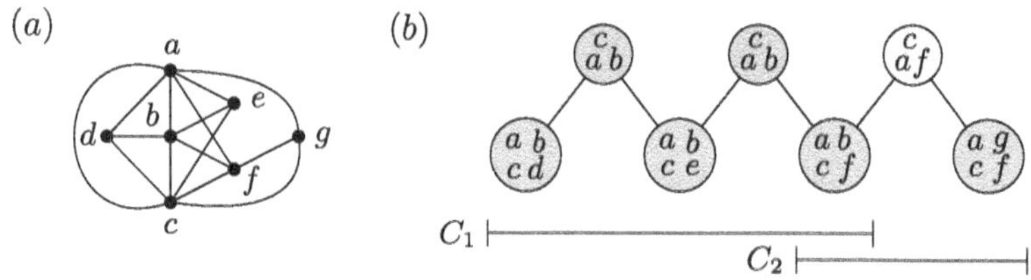

Fig. 1. (*a*) A 3-path G with (*b*) alternating path decomposition P of width 3 with two clusters C_1, C_2 and anchor triplets $T_1 = \{a, b, c\}$ and $T_2 = \{a, c, f\}$.

alternating path decomposition of G. Let Y be a set of w vertices such that there exists a bag $X_i = Y$. Note that i is even and that all bags containing Y are consecutive in P. For any such Y, the maximal consecutive set C of bags, that contain Y, i.e. $C = \{X_j \mid Y \subseteq X_j\}$, is called *cluster*; see Fig. 1. The set Y is called the *anchor set* of C. Note that two consecutive clusters overlap in exactly one bag. In the special case of $w = 3$ we call the anchor sets *anchor triplets.* Biedl et al. [17] showed that given a graph G and a path decomposition P of G, an alternating path decomposition P' of G can be found in $O(w \cdot |P|)$ time.

Let $G = (V, E)$ be a w-path with an alternating path decomposition of width w, let X be a bag of size $w + 1$ in the decomposition, and let Y be the anchor set of X. If $d \in V \setminus Y$ is connected to $1 \le k \le w$ vertices of Y, then d is called a *k-stack* of the set $N(d) \cap Y$. In Fig. 1 vertex f is a 3-stack of $\{a, b, c\}$. We call a k-stack d *proper*, if it does not belong to any anchor set; i.e. it is contained in exactly one bag and this bag has an odd index.

For $n, m \in \mathbb{N}$, let $K_{n,m}$ denote the complete bipartite graph on $n+m$ vertices and let K_n denote the complete graph on n vertices. We denote the graph that we obtain from K_n by deleting one edge by K_n^-. The graph that is obtained from K_n by deleting three edges that form a triangle is called $K_n^{-\triangle}$. The following proposition summarizes the tools we use to conclude that a graph is not k-planar.

Proposition 1. *The following graphs are not k-planar: (i) $K_{3,7k+1}$ for any k, (ii) $K_{3,4k+3}$ for $k \le 4$, (iii) any w-path with $w + 2 > 7.62\sqrt{k}$. Moreover, $K_7^{-\triangle}$ is not 1-planar.*

Proof. For (i), the crossing number of $K_{3,a}$ is $\lfloor \frac{a}{2} \rfloor \lfloor \frac{a-1}{2} \rfloor$ [13]. For $a \ge 7k + 1$, averaging over the $3a$ edges of $K_{3,a}$ yields crossing number that exceeds k.

Statement (ii) is shown by Angelini et al. [2]. For (iii) observe that a w-path contains K_{w+1}, whose density exceeds the maximum of $3.81\sqrt{k}n$ for k-planar graphs [1]. Korzhik [9] showed that $K_7^{-\Delta}$ is not 1-planar.

3 Exact Algorithms for 1-Planarity of w-Paths

In this section we show that the problem of testing 1-planarity for w-paths can be solved efficiently for all values of w. First, note that 2-paths are series-parallel and hence planar, i.e., the problem is trivial for $w \leq 2$. On the other hand, every w-path with $w \geq 6$ contains K_7, which has more than $4n-8$ edges and hence is not 1-planar. Further observe that a 5-path on $n > 6$ vertices contains K_7^-, which is not 1-planar [4]. Hence a 5-path is 1-planar if and only if it is isomorphic to K_6. The remaining cases are therefore $w = 3$ and $w = 4$. In the following sections we give efficient algorithms for testing 1-planarity for 3-paths and 4-paths, respectively. Together with the previous discussion, this proves the main result of this section.

Theorem 1. *There is a linear-time algorithm for testing 1-planarity of w-paths.*

3.1 3-Paths

For the rest of this section let G be a 3-path and let $P = X_1, \ldots, X_l$ be an alternating path decomposition of G. Let $T_1, \ldots, T_s$ and $C_1, \ldots, C_s$ be the corresponding anchor triplets and clusters, respectively, in the order given by P. Let X be the set of vertices of G that are contained in at least one anchor triplet. We call the subgraph $G[X]$ of G induced by X the *skeleton* of G.

Our first goal is to show that if G is 1-planar, then it has a 1-planar drawing whose restriction to the skeleton is planar. Hence, in what follows, we assume that G is 1-planar. First consider a 1-planar drawing Γ of G containing a non-planar $K_4 = \{u, v, w, z\}$ such that uw crosses vz and $\deg(z) = 3$. Then we can resolve this crossing by moving z to a point on the former Jordan arc vz close to uw but inside the triangular region bounded by u, v, w; see Fig. 2. Observe that it is possible to draw the edges incident to z in such a way that we do not introduce new crossings; i.e., uz, wz are drawn close to uw. This observation allows us to transform an arbitrary 1-planar drawing Γ of G into a 1-planar drawing Γ' of G such that $\Gamma'[i]$ is planar for every $1 \leq i \leq l$ and any two crossing edges of Γ' cross also in Γ.

Lemma 1. *Let G be a 1-planar 3-path with an alternating width-3 path decomposition P. Let X be the set of vertices of G that are contained in at least one anchor triplet. Then there exists a 1-planar drawing Γ of G such that the sub-drawing $\Gamma[X]$ is planar.*

Proof (Sketch). Let Γ be a 1-planar drawing of G. By the above observation we may assume that each bag of P induces a plane drawing in Γ. Assume that edges of two distinct anchor triplets $T_i = \{a_i, b_i, c_i\}, T_j = \{a_j, b_j, c_j\}$ cross in Γ.

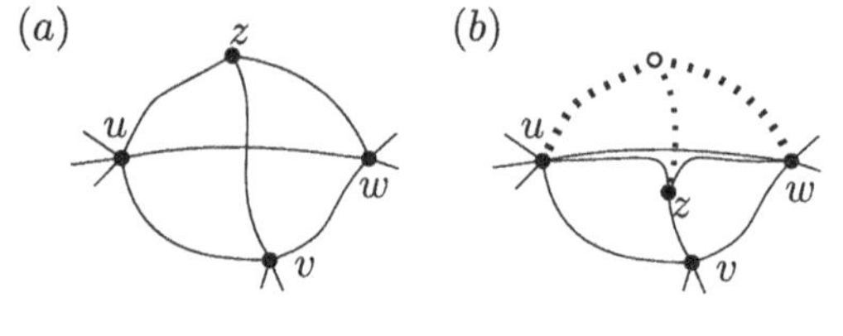

Fig. 2. (*a*) A non-planar subdrawing containing a K_4 with degree-3 vertex z. (*b*) We can move z inside the region bounded by u, v, w without causing new crossings.

Without loss of generality assume $i < j$ and that there is no index $i < k < j$ such that T_k crosses T_i or T_j. Note that $i \neq j - 1$, since then P would contain a bag $X = T_i \cup T_j$ but by assumption all bags are planar in Γ. Further observe that for every anchor triplet T_k in G there exists a vertex $x_k \in G[: k]$ with $T_k \subseteq N(x_k)$ and there is a vertex $y_k \in G[k :]$ with $T_k \subseteq N(y)$ and $x_k \neq y_k$. Note that $x_i \notin T_j$ and $y_j \notin T_i$. We distinguish cases based on the number of vertices shared by T_i and T_j. Here we only consider the case $T_i \cap T_j = \emptyset$; the remaining cases follow a similar strategy.

Assume that T_i and T_j are disjoint. After renaming the vertices, by 1-planarity $\Gamma[T_i \cup T_j]$ has to be embedded as in Fig. 3(*a*). Since both triangles T_i and T_j have only one uncrossed edge, there is only one possible way to place x_i and y_j, respectively; see Fig. 3(*b*). Then no further 3-stack to T_i or to T_j can be added without loosing 1-planarity, hence $y_i \in T_j$, $x_j \in T_i$. The only vertex of T_j from which all vertices in T_i can be reached while maintaining 1-planarity is a_j. Since by assumption there is no anchor triplet between T_i and T_j that crosses T_i or T_j, we have $T_{i+1} = \{a_j, b_i, c_i\}$. No vertex adjacent to all vertices in T_{i+1} can be added without loosing 1-planarity, hence $y_{i+1} \in T_j$. But none of the vertices in T_{i+1} can be reached from both b_j and c_j - a contradiction to the 1-planarity of Γ.

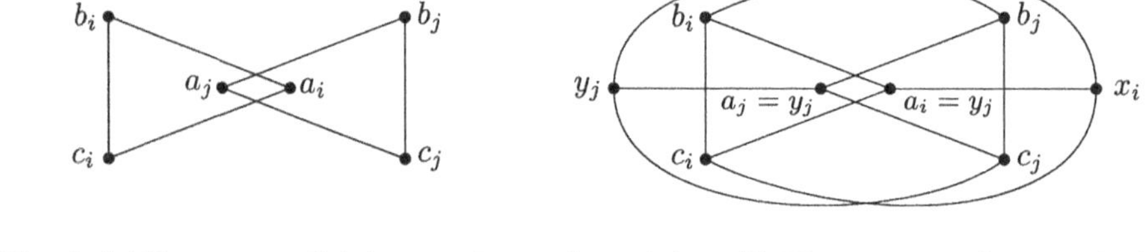

Fig. 3. (*a*) Two vertex-disjoint crossing anchor triplets. (*b*) The unique 1-planar embedding of the graph induced by T_i, T_j, x_i and y_j.

Let Γ be a drawing of G whose restriction to the skeleton $\Gamma[X]$ is planar and where every bag is drawn planar, which exists by Lemma 1. Note that the skeleton is triconnected and thus has a unique planar embedding. Hence we can change the outer face of Γ such that $\Gamma[X]$ remains planar and for every $1 \leq i < s$ the triangle $\Gamma[T_{i+1}]$ is drawn in the interior (or on the boundary) of $\Gamma[T_i]$. We call a drawing with these properties *nice.*

For every $1 \leq i \leq s$ let i' be the smallest index such that $X_{i'} = T_i$. A proper 3-stack v of an anchor triplet T_i is *inside* T_i if it is in the interior of $\Gamma[T_i]$ in $\Gamma[i']$ and it is *outside* T_i if it is in the exterior of $\Gamma[T_i]$ in $\Gamma[i']$. Note that the fact that $\Gamma[i']$ is planar, guarantees that no edge incident to a 3-stack of T_i crosses an edge of the triangle formed by the vertices in T_i.

Lemma 2. *Let G be a 1-planar 3-path and let Γ be a nice drawing of G that minimizes the number of crossings among all such drawings. Then for any two consecutive anchor triplets T_i and T_{i+1} there is either no proper 3-stack of T_i inside T_i or no proper 3-stack of T_{i+1} outside T_{i+1}.*

Proof. Assume for the sake of contradiction that u is a 3-stack inside T_i and v is a 3-stack outside T_{i+1}. If u is drawn inside T_{i+1}, then v must be drawn outside T_i; see Fig. 4(a). We can then swap the positions of u and v (and reroute some edges as indicated in Fig. 4(b)) to remove some crossings while maintaining 1-planarity; a contradiction to our choice of Γ. Hence u must be drawn in the region between T_i and T_{i+1} in Γ. A symmetric argument applies to v. However, there is no way to 1-planarly draw u and v in this region simultaneously.

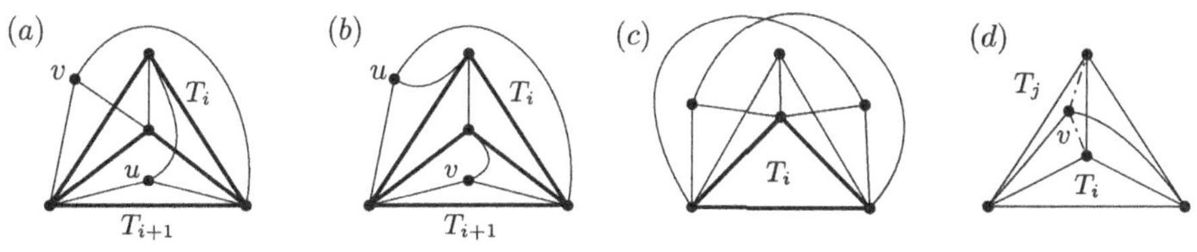

Fig. 4. Illustrations for the proofs of Lemmas 2–4.

Lemma 3. *Let Γ be a nice drawing of a 1-planar 3-path G. Then*

(i) no anchor triplet T_i with $i \geq 2$ has two proper 3-stacks outside T_i and
(ii) no anchor triplet T_j with $j < s$ has two proper 3-stacks inside T_j.

Proof. We show only (i) as (ii) follows from (i) by changing the outer face and reversing the path decomposition. Assume for the sake of contradiction that there is an anchor triplet T_i, $i \geq 2$ with two proper 3-stacks u, v outside T_i. Then u, v and the vertex $x \in T_{i-1} \setminus T_i$ are all 3-stacks of T_i that lie outside T_i, and they are hence drawn as shown in Fig. 4(c) (up to the choice of x). However, since T_{i-1} is an anchor triplet, there needs to be an additional 3-stack on T_{i-1}, which is impossible to add as all candidates for x are already enclosed by a cycle that is formed from edges that already have a crossing.

Let v be a proper 3-stack. We say that v is *between* T_i and T_{i+1} if it is either a 3-stack of T_i inside T_i or a 3-stack of T_{i+1} outside T_{i+1}. Lemma 2–3 together imply that in a crossing minimal nice drawing of G there is at most one proper 3-stack between any two consecutive anchor triplets. Observe that in a nice drawing Γ of G every crossing involves at most one edge incident to a proper 3-stack. The following lemma shows that for a 1-planar 3-path there exists a 1-planar drawing such that only edges incident to proper 3-stacks cross.

Lemma 4. *Let G be a 1-planar 3-path, then there exists a nice 1-planar drawing Γ of G in which the anchor triplets are crossing free.*

Proof. By Lemma 1 there exists a nice 1-planar drawing of G whose restriction to the skeleton is planar and where every bag is drawn planar. Let Γ be a nice drawing that minimizes the number of crossings among all such drawings. We now modify Γ in such a way that all anchor triplets become crossing-free.

First we remove from Γ all proper 3-stacks that are between two consecutive anchor triplets. This leaves only proper 3-stacks that are outside T_1 or inside T_l and therefore removes all crossings on anchor triplets. We then reinsert the removed proper 3-stacks and put each 3-stack v between T_i and T_{i+1} into the region that is bounded by T_i and T_{i+1}; see Fig. 4(d). For a single proper 3-stack, this preserves 1-planarity. Since there is at most one proper 3-stack between any two consecutive anchor triplets and since the regions between consecutive anchor triplets in the drawing are pairwise disjoint, we can add all removed proper 3-stacks simultaneously without violating 1-planarity.

Theorem 2. *There is a linear-time algorithm for testing whether a given 3-path is 1-planar. In the affirmative case, the algorithm computes a 1-planar drawing.*

Proof. If G is 1-planar, then there exists a 1-planar drawing satisfying the conditions guaranteed by Lemma 4. We associate such a drawing with a vector $(x_i)_{i=1}^{s}$ where x_i is the number of proper 3-stacks of T_i that are outside T_i. The properties from Lemma 4 give that $x_i \in \{0,1\}$ for $i \geq 2$. Moreover, clearly $x_1 \leq 3$. Given two drawings Γ and Γ' with these properties and associated vectors $x = (x_i)_{i=1}^{s}$ and $x' = (x'_i)_{i=1}^{s}$, if x is lexicographically larger than x', let j be the first position such that $x_j > x'_j$. If $j = 1$, we can simply move an additional proper 3-stack of T_1 outside of T_1 in Γ' to obtain a drawing whose vector is lexicographically larger. We therefore may assume $j > 1$. Then $x'_j = 0$ and $x_j = 1$. Moreover, x_{j-1} equals the number p_{j-1} of proper 3-stacks of T_{j-1} as by Lemma 2 and 3 there can be at most one proper 3-stack between T_j and T_{j-1}. By the choice of j, we find $x'_{j-1} = x_{j-1} = p_{j-1}$ and hence no proper 3-stack of T_{j-1} is between T_{j-1} and T_j in Γ'. We can hence take a proper 3-stack of T_j and move it between T_j and T_{j-1}. This gives a 1-planar drawing with a lexicographically larger vector.

This implies that a drawing satisfying the conditions from Lemma 4 that has the lexicographically largest vector can be found by a simple greedy algorithm that starts with the unique planar drawing Γ' of the skeleton with T_1 on the outer face and then iteratively adds the proper 3-stacks of T_i for $i = 1, \ldots, s$. Every time, we embed as many proper 3-stacks outside T_i as possible without violating 1-planarity and the remaining ones between T_i and T_{i+1}. If this process fails to produce a 1-planar drawing, the graph is not 1-planar.

3.2 4-Paths

For this section let G be a 4-path. Let $P = X_1, \ldots, X_l$ be an alternating path decomposition of G with anchor sets $T_1, \ldots, T_s$. Since K_5 and K_6 are 1-planar [6] we assume for the rest of this section that $G \neq K_5$, $G \neq K_6^-$, i.e., $s > 1$. We begin by observing that if G contains three 4-stacks to a single anchor set in P, then G contains the graph $K_7^{-\triangle}$, which is not 1-planar by Lemma 1.

For every $1 \leq i \leq s$ let i' be the smallest index such that $X_{i'} = T_i$. If G is 1-planar, then by the above observation we have that for every anchor set T_i, G contains exactly two 4-stacks x_i and y_i of T_i with $x_i \in G[{:}\ i' - 1]$ and $y_i \in G[i' + 1 :]$. Note that $x_i \in T_{i-1}$ if $i > 1$ and $y_i \in T_{i+1}$ if $i < s$. Since $x_i \neq y_i$ are non-adjacent, for every $1 \leq i \leq s$ the subgraph $G[T_i \cup \{x_i, y_i\}]$ induces a K_6^-, which has three 1-planar embeddings up to isomorphism [9]; see Fig. 5.

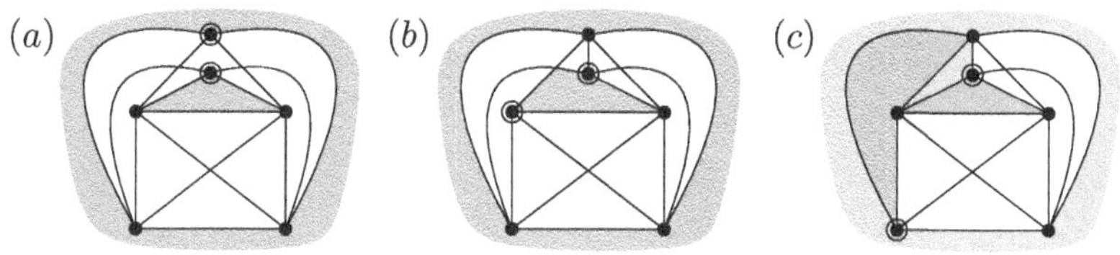

Fig. 5. The three non-isomorphic 1-planar embeddings of K_6^- with the two non-adjacent vertices encircled.

Lemma 5. *If G is 1-planar, then in every 1-planar drawing Γ of G, $\Gamma[T_i]$ is planar for every $1 \leq i < s$.*

Proof. Let $1 \leq i < s$ and let Γ be a 1-planar drawing of $G[{:}\ i + 1]$. For the sake of contradiction assume that the K_4 induced by T_i is not drawn planarly. This implies that $\Gamma[T_i \cup \{x_i, y_i\}]$ is embedded as in Fig. 5(a) such that x_i and y_i are the two encircled vertices. Now we consider possible positions for y_{i+1}, which is adjacent to one of the encircled vertices and three of the non-encircled vertices. Since y_{i+1} has degree 4 it cannot be placed inside a triangular face bounded by two vertices and a crossing. Thus only the shaded faces remain. But from none of these two faces a third non-encircled vertex can be reached while maintaining 1-planarity – a contradiction.

Let $1 \leq i < s$ and let Γ be a 1-planar drawing of G. Observe that $\Gamma[T_i \cup \{x_i, y_i\}]$ cannot be embedded as in Fig. 5(b), since y_{i+1} has degree at least 4 and thus can only be placed in one of the shaded faces, but from none of them one encircled and three non-encircled vertices can be reached. Thus we may assume that in any 1-planar drawing of G for every i, the subdrawing $\Gamma[T_i \cup \{x_i, y_i\}]$ is embedded as in Fig. 5(c) and y_{i+1} is placed in one of the four shaded faces.

Lemma 6. *Let G be a 1-planar 4-path with alternating path decomposition P. For every 1-planar drawing Γ' of $G[T_1 \cup \{x_1, y_1\}]$ there is at most one 1-planar drawing Γ of $G - y_s$ such that Γ contains Γ'. If Γ exists, it can be extended to a 1-planar drawing of G.*

Proof. If $\Gamma'[T_1]$ or $\Gamma'[T_2]$ is not planar, then Γ' cannot be extended to a 1-planar drawing of $G - y_s$ by Lemma 5. Hence assume that $\Gamma'[T_1]$ and $\Gamma'[T_2]$ are planar. We construct Γ iteratively and claim that for every $2 \le i < s$ either y_i cannot be added while maintaining 1-planarity or the following invariant holds.

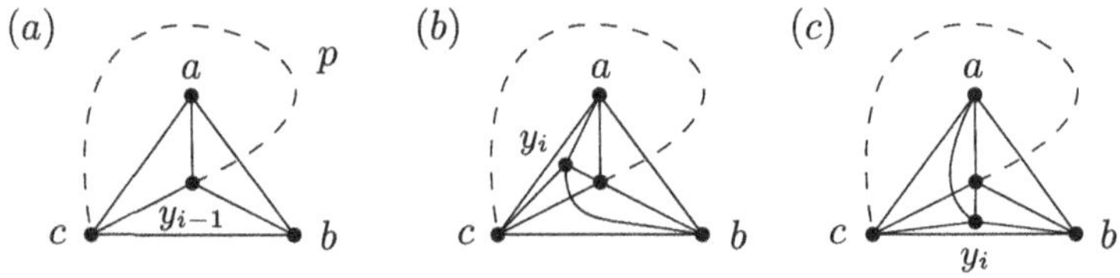

Fig. 6. Illustrations for the proof of Lemma 6.

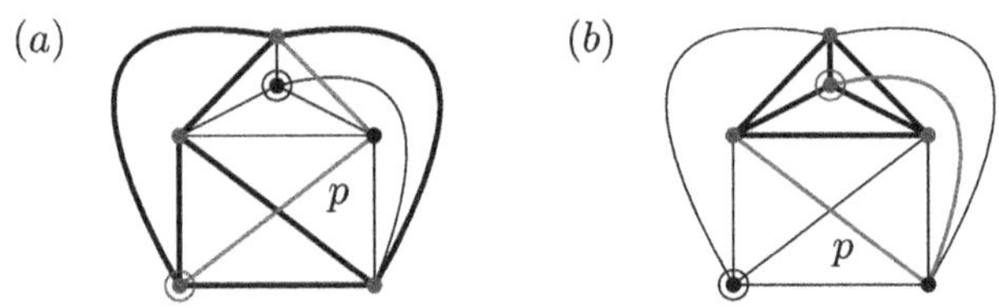

Fig. 7. The two planar K_4's containing exactly three non-encircled vertices.

Invariant 1 *The K_4 induced by T_i is crossing-free except for a path p of crossed edges from y_{i-1} to another vertex in T_i that crosses exactly one edge of the boundary of the K_4. Moreover, the inner faces of this K_4 are empty, except for, possibly, the path p; see Fig. 6(a).*

Recall that $\Gamma[T_1 \cup \{x_1, y_1\}]$ is embedded as in Fig. 5(c). Therefore the invariant holds for $i = 2$, since for both planar K_4 consisting of three non-encircled and one encircled vertex in Fig. 5(c), there is a path of crossed edges from the encircled to one of the non-encircled vertices; see Fig. 7. Hence assume that the invariant holds before inserting y_i. We show that there is at most one way to place y_i. Note that y_i cannot be placed into a face incident to p without violating 1-planarity. Let $T_i = \{a, b, c, y_{i-1}\}$ be such that ab is crossed by p. We distinguish cases based on which of the vertices are contained in T_{i+1}.

First assume that $T_{i+1} = \{y_i, y_{i-1}, a, c\}$. Then we have to place y_i inside the common face of y_{i-1}, a, c; see Fig. 6(b); since otherwise $y_i a$ would cross $y_{i-1}c$ and then $\Gamma[T_{i+1}]$ would not be crossing-free. Observe that after inserting y_i the invariant still holds with $p = y_i b, ba$. The case $T_{i+1} = \{y_i, y_{i-1}, b, c\}$ is symmetric.

It remains to consider the case $T_{i+1} = \{y_i, y_{i-1}, a, b\}$. Then y_i can be inserted either into the face shared by a, c, y_{i-1} or into the face shared by b, c, y_{i-1}; see Fig. 6(c). In both cases the vertices in T_{i+1} induce a planar K_4 but it is impossible to insert the vertex y_{i+1} adjacent to all vertices in T_{i+1} without violating 1-planarity; thus no 1-planar drawing of $G - y_s$ that contains Γ' exists. Note that if Γ exists, the invariant still holds after inserting y_{s-1}, thus the last vertex y_s can be drawn 1-planarly in any case.

Theorem 3. *There is a linear-time algorithm for testing whether a given 4-path is* 1*-planar. In the affirmative case, the algorithm yields a* 1*-planar drawing.*

Proof. Recall that $G[T_1 \cup \{x_1, y_1\}]$ is a K_6^- and thus has only three non-isomorphic 1-planar embeddings. There are $2 \cdot 4! = 48$ ways to assign the vertices of $G[T_1 \cup \{x_1, y_1\}]$ to such an embedding. For each of those 48 non-isomorphic 1-planar embeddings, we test whether it can be extended to a 1-planar drawing Γ of $G - y_s$ as follows. We insert the vertices in the order in which they appear in P. Given a 1-planar drawing Γ_i of $G[: i]$ for $4 \leq i < l$, there is at most one face in which the vertex introduced in bag X_i can be placed. By Lemma 6, given a 1-planar drawing Γ' of $G[T_1 \cup \{x_1, y_1\}]$, a 1-planar drawing Γ of $G - y_s$ extending Γ' is either unique or does not exist. Moreover, if Γ exists, it is always possible to add y_s without violating 1-planarity. Thus the test only takes linear time.

4 Approximation Algorithms

4.1 Graphs with Pathwidth-3

In this section we provide an approximation algorithm for testing whether a pathwidth-3 graph is k-planar. More precisely, given a graph G of pathwidth at most 3, our algorithm either produces a drawing with at most $7k$ crossings per edge or determines that G is not k-planar.

Theorem 4. *Given a graph G of pathwidth* 3 *and an integer k there is a linear-time algorithm that either outputs a $7k$-drawing of G or correctly determines that G is not k-planar.*

Proof. Without loss of generality, we assume that G is connected and contains no degree-1 vertices. Let $P = X_1, X_2, \ldots, X_l$ be an alternating path decomposition of G with width 3, clusters $C_1, \ldots, C_s$ and anchor triplets $T_1, \ldots, T_s$. Let G_i be the subgraph of G induced by $T_i \cup \bigcup_{j=1}^{i-1} C_j$. Observe that $G_1 = G[T_1]$ and that G_s is the graph obtained from G after removing all proper stacks of T_s. Let Δ_i be the triangle consisting of the vertices in T_i. We claim that if G is k-planar, then $G_i + \Delta_i$ for $i = 1, \ldots, s$ has a $7k$-planar drawing Γ_i where Δ_i is crossing-free and bounds a face.

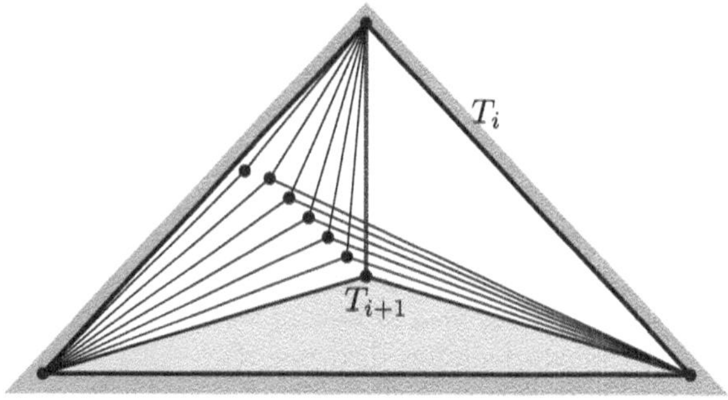

Fig. 8. One 2-stack and k 3-stacks of an anchor triplet T_i. The blue edge has the most crossings, namely $k-1$. The subdrawing $\Gamma[: i'-1]$ completely lies in the dark shaded region, while $\Gamma[i'+1 :]$ completely lies in the light shaded region. (Color figure online)

Clearly Γ_1 can be chosen as an arbitrary planar drawing of the triangle $G_1 + \Delta_1$. We proceed inductively from i to $i+1$. Observe that $V(G_{i+1}) \setminus V(G_i)$ consists of a set Y of proper stacks of T_i and the vertex $x \in T_{i+1} \setminus T_i$. Let $Y_2, Y_3 \subseteq Y$ be the vertices of degree 2 and 3, respectively. The degree-2 vertices can be inserted without crossings into Γ_i since Δ_i is planar. Moreover, $|Y_3 \cup \{x\}| \leq 7k$ as otherwise G contains $K_{3,7k+1}$, which is not k-planar by Lemma 1. We now place the vertices in $Y_3 \cup \{x\}$ inside Δ_i as indicated in Fig. 8 with at most $7k-1$ crossings per edge in such a way that the triangle Δ_{i+1} can be added without crossings. This yields Γ_{i+1}. Finally, given Γ_s, we can add the proper stacks of T_s to complete the drawing similarly to the inductive step.

For $k \leq 4$, Lemma 1 yields $|Y_3| \leq 6$, which gives a better approximation.

Corollary 1. *Given an input graph G of pathwidth 3, the algorithm computes a $6k$-approximation to k-planarity for $k \leq 4$ in linear time.*

4.2 w-Paths

Let G be a w-path on n vertices and let $P = X_1, \ldots, X_l$ be a path decomposition of G. Without loss of generality assume $n > w+1$. In the following we show that a construction of Biedl et al. [17], which was used to obtain an approximation to the crossing number of w-paths (which is quadratic in the pathwidth), in fact yields a linear approximation (in the pathwidth) to the local crossing number.

We briefly sketch the construction, see also Fig. 9; refer to [17] for details. We order the vertices as $v_1, \ldots, v_n$, where the first $w+1$ vertices from X_1 are in the order in which they disappear from P and $v_{w+2}, \ldots, v_n$ are in the order in which they appear in P. The construction is incremental and places v_i at coordinate $(i, 0)$. We start with a drawing of $G[X_1]$, which is a complete graph K_{w+1}, where every edge is drawn as a half-circle above the x-axis. Let now $j > 1$ and assume that bag X_j introduces vertex v_i with $i \geq w+2$. By construction, v_i is placed to the right of all other vertices that are placed so far. Denote by $p_1^i, \ldots, p_w^i$ the neighbors of v_i in $G[:j]$ according to the previously defined order. Observe that $\deg(p_j^i) \geq \deg(p_{j+1}^i)$ for every $1 \leq j < w$. The edge $v_i p_1^i$ is routed counterclockwise around the drawing $\Gamma[:j]$, see the red edge in Fig. 9. The other edges are routed clockwise as a bundle, see the blue, gold and green edge in Fig. 9, each wrapping around the other vertices of X_j so that they remain visible from the bottom. Clearly the drawing can be computed in linear time.

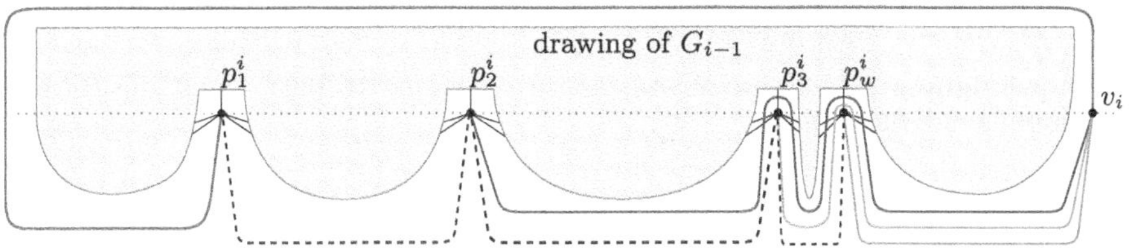

Fig. 9. The construction of [17].

Theorem 5. *Let G be a w-path on n vertices. The construction by Biedl. et al. [17] yields an $O(w)$-approximation to the local crossing number of G in linear time.*

Proof. Let Γ be the drawing of G obtained by the construction. Let $e = v_k v_i$ with $k < i$ be the edge of Γ that is involved in the most crossings. Let X_j be the bag that introduces v_i. Let y' be the number of edges in $\Gamma[:j]$ that cross e and let y be the number of the remaining edges that cross e. First assume that $y' \geq y$. Then e is crossed $y' + y \leq 2y'$ many times. If $j = 1$, i.e., v_i belongs to X_1, then e is part of the initial K_{w+1} and we have $y' \leq \binom{w}{2}$ by construction. Since G contains a K_{w+1}, we have $\mathrm{lcr}(G) \geq (w+2)^2/7.62^2$ by Proposition 1. Then there is a constant c such that $\mathrm{lcr}(\Gamma) = \mathrm{cr}(e) \leq 2y' \leq 2 \cdot \binom{w}{2} \leq c \cdot w \cdot ((w+2)^2/7.62^2) \leq c \cdot w \cdot \mathrm{lcr}(G)$. Otherwise, if $j > 1$, then by construction e crosses the incident edges of $w - 2$ many neighbors of v_i. Let u be the neighbor of v_i with the highest degree x. Then $y' \leq (w-2) \cdot x \leq w \cdot x$. Since $\deg(p_a^i) \geq \deg(p_b^i)$ for any $1 \leq a < b \leq w$ and since no edge wraps around p_1^i or p_2^i by construction, it follows that $u = p_3^i$. Since G is a w-path, every neighbor of p_3^i is also a neighbor of p_2^i and p_1^i – hence, we have a $K_{3,x-2}$ in G, and thus $\mathrm{lcr}(G) > (x-3)/7$ by Proposition 1. Then there is a constant c such that $\mathrm{lcr}(\Gamma) = \mathrm{cr}(e) \leq 2y' \leq 2 \cdot w \cdot x \leq c \cdot w \cdot (x-3)/7 \leq c \cdot w \cdot \mathrm{lcr}(G)$.

It remains to consider the case $y' < y$. Then e is crossed $y' + y \leq 2y$ many times. We call a vertex v_r a *witness vertex*, if $r > i$ and $v_i = p_l^r$ with $l \geq 3$. Since every witness vertex contributes at most $2w$ crossings to e, there are at least $\frac{y}{2w}$ witness vertices. Consider the last witness vertex v_s of e. Since $v_i = p_l^s$ for $l \geq 3$, p_1^s and p_2^s as well as v_i are adjacent to all $\frac{y}{w}$ many witness vertices. Hence we have a $K_{3,\frac{y}{2w}}$, and thus $\mathrm{lcr}(G) > (\frac{y}{2w} - 1)/7$ by Proposition 1. Since $\mathrm{lcr}(G)$ is a natural number we have $\mathrm{lcr}(G) \geq (y/14w)$. Then there is a constant c such that $\mathrm{lcr}(\Gamma) = \mathrm{cr}(e) \leq 2y \leq c \cdot w \cdot (y/14w) < c \cdot w \cdot \mathrm{lcr}(G)$. In fact all inequalities are true for $c = 7.62^2$ and $w \geq 5$, since $x \geq w$.

5 Conclusions and Open Problems

We exploited the alternating path decomposition of a graph to efficiently test 1-planarity for restricted graph classes and to bound the local crossing number of a given graph. It remains open whether exact algorithms for testing 1-planarity of general pathwidth-3 or pathwidth-4 graphs can be given. Another interesting question is whether considering the tree decomposition instead of the path decomposition is helpful for testing k-planarity for small values of k.

Acknowledgments. This work was funded by grant RU 1903/4-1 of the German Research Foundation (DFG).

References

1. Ackerman, E.: On topological graphs with at most four crossings per edge. Comput. Geom. 85 (2019). https://doi.org/10.1016/j.comgeo.2019.101574
2. Angelini, P., Bekos, M.A., Kaufmann, M., Schneck, T.: Efficient generation of different topological representations of graphs beyond-planarity. J. Graph Algorithms Appl. **24**(4), 573–601 (2020). https://doi.org/10.7155/jgaa.00531
3. Bannister, M.J., Cabello, S., Eppstein, D.: Parameterized complexity of 1-planarity. J. Graph Algorithms Appl. **22**(1), 23–49 (2018). https://doi.org/10.7155/jgaa.00457
4. Bodendiek, R., Schumacher, H., Wagner, K.: Über 1-optimale Graphen. Math. Nachr. **117**, 323–339 (1984)
5. Boyer, J.M., Myrvold, W.J.: On the cutting edge: simplified O(n) planarity by edge addition. J. Graph Algorithms Appl. **8**(3), 241–273 (2004). https://doi.org/10.7155/jgaa.00091
6. Czap, J., Hudák, D.: 1-planarity of complete multipartite graphs. Discret. Appl. Math. **160**(4–5), 505–512 (2012). https://doi.org/10.1016/j.dam.2011.11.014
7. Grigoriev, A., Bodlaender, H.L.: Algorithms for graphs embeddable with few crossings per edge. Algorithmica **49**(1), 1–11 (2007). https://doi.org/10.1007/s00453-007-0010-x
8. Hopcroft, J.E., Tarjan, R.E.: Efficient planarity testing. J. ACM 21(4) (1974). https://doi.org/10.1145/321850.321852
9. Korzhik, V.P.: Minimal non-1-planar graphs. Discrete Math. **308**(7), 1319–1327 (2008). https://www.sciencedirect.com/science/article/pii/S0012365X07002087
10. Korzhik, V.P., Mohar, B.: Minimal obstructions for 1-immersions and hardness of 1-planarity testing. J. Graph Theory **72**(1), 30–71 (2013). https://doi.org/10.1002/jgt.21630
11. Kuratowski, K.: Sur le problème des courbes gauches en topologie. Fund. Math. **15**, 217–283 (1930)
12. Mutzel, P.: An alternative method to crossing minimization on hierarchical graphs. In: GD. Lecture Notes in Computer Science, vol. 1190, pp. 318–333. Springer (1996)
13. Richter, R.B., Širáň, J.: The crossing number of $K_{3,n}$ in a surface. J. Graph Theory **21**(1), 51–54 (1996)
14. Ringel, G.: Ein Sechsfarbenproblem auf der Kugel. Abh. Math. Sem. Univ. Hamb. **29**, 107–117 (1965)
15. Robertson, N., Seymour, P.D.: Graph minors. XX. Wagner's conjecture. J. Comb. Theory, Ser. B **92**, 325–357 (2004)

16. Suzuki, Y.: 1-Planar graphs. In: Hong, S.-H., Tokuyama, T. (eds.) Beyond Planar Graphs, pp. 47–68. Springer, Singapore (2020). https://doi.org/10.1007/978-981-15-6533-5_4
17. Biedl, T., Chimani, M., Derka, M., Mutzel, P.: Crossing number for graphs with bounded pathwidth. Algorithmica **82**(2), 355–384 (2020). https://doi.org/10.1007/s00453-019-00653-x
18. Urschel, J.C., Wellens, J.: Testing gap k-planarity is NP-complete. Inf. Process. Lett. **169**, 106083 (2021). https://doi.org/10.1016/j.ipl.2020.106083
19. Wagner, K.: Über eine Eigenschaft der ebenen Komplexe. Math. Ann. **114**, 570–590 (1937)

The Complexity of Diameter on H-free Graphs

Jelle J. Oostveen[1(✉)], Daniël Paulusma[2], and Erik Jan van Leeuwen[1]

[1] Department of Information and Computing Sciences, Utrecht University, Utrecht, The Netherlands
{j.j.oostveen,e.j.vanleeuwen}@uu.nl

[2] Department of Computer Science, Durham University, Durham, UK
daniel.paulusma@durham.ac.uk

Abstract. The intensively studied DIAMETER problem is to find the diameter of a given connected graph. We investigate, for the first time in a structured manner, the complexity of DIAMETER for H-free graphs, that is, graphs that do not contain a fixed graph H as an induced subgraph. We first show that if H is not a linear forest with small components, then DIAMETER cannot be solved in subquadratic time for H-free graphs under SETH. For some small linear forests, we do show linear-time algorithms for solving DIAMETER. For other linear forests H, we make progress towards linear-time algorithms by considering specific diameter values. If H is a linear forest, the maximum value of the diameter of any graph in a connected H-free graph class is some constant $d_{\max}$ dependent only on H. We give linear-time algorithms for deciding if a connected H-free graph has diameter $d_{\max}$, for several linear forests H. In contrast, for one such linear forest H, DIAMETER cannot be solved in subquadratic time for H-free graphs under SETH. Moreover, we even show that, for several other linear forests H, one cannot decide in subquadratic time if a connected H-free graph has diameter $d_{\max}$ under SETH.

1 Introduction

The DIAMETER problem asks to find the diameter of an undirected, unweighted graph $G = (V, E)$, that is, the longest of the shortest paths between all pairs of nodes; formally, $\text{diam}(G) = \max_{u,v \in V} d(u, v)$. We shall denote $|V| = n$ and $|E| = m$. A trivial algorithm executes a Breadth First Search (BFS) from every node in the graph, and has a running time of $O(nm)$. The best known matrix multiplication-based algorithms achieve a running time of $\tilde{O}(n^{\omega})$ [16,43,45] to find the diameter of a graph, where $\tilde{O}$ hides logarithmic factors and ω is the matrix multiplication constant, with current known value $\omega < 2.371866$ [26]. A search for improvement led to a hardness result: under the Strong Exponential Time Hypothesis (SETH), one cannot decide between diameter 2 or 3 on (sparse) split graphs in $O(n^{2-\epsilon})$ time, for any $\epsilon > 0$ [42]. SETH is a hypothesis that

J.J. Oostveen is supported by the NWO grant OCENW.KLEIN.114 (PACAN).

D. Král' and M. Milanič (Eds.): WG 2024, LNCS 14760, pp. 444–459, 2025.
https://doi.org/10.1007/978-3-031-75409-8_31

states that SATISFIABILITY cannot be solved in $2^{(1-\epsilon)n}$ time, for any $\epsilon > 0$, where n is the number of variables [37,38]. On other simple graph classes like constant degree graphs, truly subquadratic time algorithms for DIAMETER are also ruled out under SETH [34]. Directed versions of the DIAMETER problem admit similar barriers under SETH [4]. No clear bound is known for DIAMETER on dense graphs, and no such lower bound can be derived when $\omega = 2$, but we do know that there is a subcubic equivalence between DIAMETER and computing the reach centrality of a graph, that is, a truly subcubic algorithm for one implies such an algorithm for the other and vice versa [1].

Given that the hardness results are based on long-standing conjectures, it is natural to approach diameter computation and other similar problems on restricted graph classes. Related literature also concerns computation of eccentricities, as computing the diameter of a graph is equivalent to computing the largest eccentricity over all vertices. A conceptually simple algorithm called LexBFS can solve DIAMETER in $O(n + m)$ time for distance-hereditary chordal graphs and interval graphs [25]. Distance-hereditary graphs have been studied separately, and admit linear-time algorithms of all eccentricities [19,22,24]. Interval graphs admit computation of the eccentricity of the center of the graph in linear time, next to linear-time diameter computation [40]. Subquadratic algorithms for DIAMETER and computing eccentricities have been studied for more graph classes, including asteroidal triple (AT)-free graphs [28], directed path graphs [13], strongly chordal graphs [17], dually chordal graphs [6,18], Helly graphs and graphs of bounded Helly number [21,30,31], α_i-metric graphs [20], retracts [27], δ-hyperbolic graphs [11,12,22,23], planar graphs [2,8,36], and outerplanar graphs [35]. DIAMETER was also studied from the parameterized perspective, see e.g. [4,7,15,29,32], and a large body of work exists on approximation algorithms, see e.g. [4,5,10,13,14,20,42,44].

For graph classes with forbidden patterns, Ducoffe et al. [32] show subquadratic-time algorithms of DIAMETER for $\mathcal{H}$-minor-free graphs, where the precise exponent depends on $\mathcal{H}$, improved upon by Duraj et al. [33]. Johnson et al. [39] showed for $\mathcal{H}$-subgraph-free graphs that there is a dichotomy for DIAMETER between almost-linear time solvability and quadratic-time conditional lower bounds depending on the family $\mathcal{H}$. Also note that many of the studied graph classes listed above can be characterized as $\mathcal{H}$-free graphs for a family of graphs $\mathcal{H}$, but for each one, $\mathcal{H}$ has size 2 or larger. As far as we are aware, a structured study into forbidden (monogenic) induced patterns is absent in the literature.

Our Contributions. We initiate a structured study into diameter computation on H-free graphs, where H is a single graph. Recall that a graph is H-free if it does not contain H as an induced subgraph. The question we consider is:

For which H-free graph classes $\mathcal{G}$ can the diameter of an n-vertex graph $G \in \mathcal{G}$ be computed in time $O(n^{2-\epsilon})$?

Our first result analyses existing lower bounds to find hardness for H-free graph classes. Here, P_t denotes the path on t vertices, and sP_t denotes the disjoint union of s copies of a P_t. For graphs G, H let $G + H$ denote their disjoint union. Recall that a *linear forest* is a disjoint union of one or more paths.

Theorem 1 (♠[1]). *Let H be a graph that contains an induced $2P_2$ or is not a linear forest. Under SETH,* Diameter *on H-free graphs cannot be solved in $O(n^{2-\epsilon})$ time for any $\epsilon > 0$.*

Theorem 1 shows the most prominent gap to be for graph classes which exclude a small linear forest. Ducoffe [28] proved a hardness result for Diameter on AT-free graphs that holds under the hypothesis that the currently asymptotically fastest (combinatorial) algorithms for finding simplicial vertices (vertices with a complete neighbourhood) are optimal, which we refer to as the *Simplicial Vertex Hypothesis.* Because the vertex set of the graph in Ducoffe's hardness construction can be partitioned into four cliques, the result of [28] can be formulated as follows.

Theorem 2 ([28]). *For any fixed $k \geq 5$, under the Simplicial Vertex Hypothesis, there does not exist a combinatorial algorithm for* Diameter *on kP_1-free graphs that runs in $O(m^{3/2-\epsilon})$ time for any $\epsilon > 0$.*

To complement the hardness results, we show linear-time algorithms for several classes of H-free graphs for which H is a small linear forest.

Theorem 3 (♠). *Let H be a graph. If H is an induced subgraph of $P_2 + 2P_1$, $P_3 + P_1$, or P_4, then* Diameter *on H-free graphs can be solved in $O(n + m)$ time.*

We achieve Theorem 3 by careful structural analysis of the graph class and then show that a constant number of Breadth First Searches suffice algorithmically. Note that a running time of $O(n + m)$ clearly beats the naive algorithm of $O(nm)$ time and the matrix multiplication algorithms of $\tilde{O}(n^{\omega})$ time, but also rules out any quadratic lower bound in n, as the classes of graphs contain abitrarily large families of sparse graphs, e.g. stars.

Combining Theorems 1, 2 and 3, the only open cases for the complexity of Diameter on H-free graphs are $H = 4P_1$, $H = P_2 + 3P_1$, $H = P_3 + 2P_1$, $H = P_4 + 2P_1$, and $H = P_4 + P_1$. The smallest graph H that is an open case is that of $H = 4P_1$. As a 'hardness' result for $4P_1$-free graphs, one could try to take the split graph construction of Roditty and Williams [42], and add edges to make the graph consist of three cliques (as in [13]). Conceptually, this would seem to work: the diameter distinction is still 2 or 3 and translates to a SAT positive or negative answer. However, this approach fails due to the quantity of edges one adds to the graph. The lower bound shows that no $O(n^{2-\epsilon})$ time algorithm may exist for this new instance, which is now a relatively empty lower bound: the graph has a quadratic number of edges, so this lower bound does not even rule out an $O(n + m)$ time algorithm. The density of graphs matters

[1] Proofs of results marked with ♠ are deferred to the full version, see [41].

in relation to lower bounds, and seems to provide a barrier to finding a lower bound that rules out a linear-time algorithm.

However, if we adopt the perspective from the other side, a linear-time algorithm for $4P_1$-free graphs would still be surprising. Indeed, such an algorithm that can decide between diameter 2 or 3 on the three-clique instance described earlier implies an algorithm for ORTHOGONAL VECTORS in time $O(n^2 + d^2)$, where d is the dimension of the vectors and n the size of the vector sets (♠). Although lower bounds do not rule out this possibility, such a result would be highly non-trivial, as the best known algorithms for ORTHOGONAL VECTORS do not achieve this running time for all d [3,9]. Any linear-time algorithm would even beat the best known matrix-multiplication algorithms of $\tilde{O}(n^\omega)$ time, even if $\omega = 2$. It thus seems we are at an impasse to find or exclude a linear-time algorithm for computing the diameter of $4P_1$-free graphs.

However, as it turns out, we *can* decide in linear time whether the diameter of a $4P_1$-free graph is exactly 5. Our approach avoids the above barriers by focusing on specific diameter values instead of deciding on the diameter of a graph completely.

In general, for a graph class $\mathcal{G}$, we call $d_{\max}(\mathcal{G})$ the maximum diameter that any graph in $\mathcal{G}$ can have; formally $d_{\max}(\mathcal{G}) = \sup_{G\in\mathcal{G}} \operatorname{diam}(G)$. We omit $\mathcal{G}$ when it is clear from context. In particular, for $4P_1$-free graphs, $d_{\max}$ is equal to 5. We define the $d_{\max}(\mathcal{G})$-DIAMETER problem as deciding for a graph $G \in \mathcal{G}$ whether it holds that $\operatorname{diam}(G) = d_{\max}(\mathcal{G})$. The research question we investigate is:

*For which H-free graph classes $\mathcal{G}$ can we solve $d_{\max}(\mathcal{G})$-*DIAMETER *in linear time?*

For some classes $\mathcal{G}$, it is easy to see $d_{\max}(\mathcal{G})$ is bounded. For instance, the class of cliques has $d_{\max} = 1$. Any graph class that contains paths of arbitrary length has $d_{\max} = \infty$. For deciding whether the diameter of a graph is equal to $d_{\max}$, only classes with bounded $d_{\max}$ value are interesting to consider. It turns out that for classes of connected H-free graphs, $d_{\max}$ is bounded exactly when H is a linear forest (♠).

Our contributions with respect to the $d_{\max}(\mathcal{G})$-DIAMETER problem are twofold. Firstly, we find several examples of H-free classes $\mathcal{G}$ where H is a linear forest of more than one path where we solve $d_{\max}(\mathcal{G})$-DIAMETER in linear time. Note that $d_{\max}$ can differ vastly for classes where H is a linear forest, depending on H. Here, $H \subseteq_i H'$ denotes that H is an induced subgraph of H'.

Theorem 4 (♠). *Let $H = 2P_2 + P_1$ or $H \subseteq_i P_2 + 3P_1$, $P_3 + 2P_1$, or $P_4 + P_1$, and let $\mathcal{G}$ be the class of H-free graphs. Then $d_{\max}(\mathcal{G})$-*DIAMETER *can be solved in $O(n + m)$ time.*

Note that in particular, Theorem 4 shows that we can decide whether $\operatorname{diam}(G) = d_{\max}(\mathcal{G})$ for all previously stated open cases for DIAMETER computation on H-free graphs, except for the case of $H = P_4 + 2P_1$.

Secondly, we extend known hardness constructions to hold for the $d_{\max}(\mathcal{G})$-DIAMETER problem for certain H-free graph classes $\mathcal{G}$. Note that one needs different hardness proofs for different H, H', even if $H \subseteq_i H'$, because $d_{\max}$ can differ for both classes.

Theorem 5 (♠). *Let $H = 2P_2$ or $H = P_t$ for some odd $t \geq 5$, and let $\mathcal{G}$ be the class of H-free graphs. Under SETH, it is not possible to solve $d_{\max}(\mathcal{G})$-*Diameter *in time $O(n^{2-\epsilon})$ for any $\epsilon > 0$.*

Theorems 4 and 5 together cover almost all cases where $d_{\max} \leq 4$: $H = 2P_2$ is hard by Theorem 5, $H = P_3+P_2$ is open, and all other cases with $d_{\max} \leq 4$ are linear-time solvable by Theorem 4. Theorem 4 also gives linear-time algorithms for some cases where $d_{\max} > 4$; $H = P_3+2P_1$ and $H = 2P_2+P_1$ have $d_{\max} = 5$, and $H = P_2 + 3P_1$ has $d_{\max} = 6$. It appears that, algorithmically, the presence of a P_1 in H helps out in structural analysis, which may explain the inability to attain a result for $H = P_3 + P_2$. We further discuss particular cases and possible generalizations of our theorems in the conclusion.

Our algorithmic results are attained through careful analysis of the structure of the graph with respect to the forbidden pattern. This limits the ways in which a shortest path that realizes the diameter can appear in the graph. However, even for small patterns H, such analysis quickly becomes highly technical.

In this extended abstract, we give several sketches of our algorithmic technical proofs. We sketch a linear-time algorithm for Diameter on $(P_2 + 2P_1)$-free graphs, which also contributes to most of the linear-time algorithm for Diameter on $(P_3 + P_1)$-free graphs (♠). These two results together prove Theorem 3. We also sketch a part of the proof that we can solve $d_{\max}(\mathcal{G})$-Diameter on $(P_4 + P_1)$-free graphs in linear time. This result is the most technical result in our work, and contributes to Theorem 4.

Preliminaries. Graphs are denoted $G = (V, E)$ and are connected, undirected, and unweighted. For any $v \in V$, denote $N(v)$ as the neighbourhood of v, and $N[v] = N(v) \cup \{v\}$. For a set of vertices $S \subseteq V$, let $G[S]$ denote the induced subgraph on the vertices of S. A vertex $v \in V$ is *complete* to a set $S \subseteq V$ when $S \subseteq N(v)$, and *anti-complete* to a set S when $S \cap N(v) = \emptyset$. A set is $A \subseteq V$ is complete to a set $S \subseteq V$ when every vertex in A is complete to S, and A is anti-complete to S when every vertex in A is anti-complete to S. For vertices $v_1, \ldots, v_k \in V$, $\langle v_1, \ldots, v_k \rangle$ denotes an induced path from v_1 to v_k.

Two vertices $u, v \in V$ are *twins* when $N[u] = N[v]$, also called *true twins*. Two vertices $u, v \in V$ are *false twins* when $N(u) = N(v)$. We shall always be explicit when we talk about false twins; in general, twins will refer to true twins. A *twin class* is a set of vertices all of which are pairwise twins.

Removing twins can be done in linear time and keeps most distance properties in a graph (see e.g. Coudert et al. [15]).

Proposition 1. *Given a graph $G = (V, E)$, in $O(n + m)$ time we can detect true twins and remove all-but-one vertex from each twin class resulting in a graph $G' = (V', E')$ with $V' \subseteq V$, $E' \subseteq E$. The following hold:*

(i) the distance between two vertices u, v in G' is equal to the distance between u, v in G,

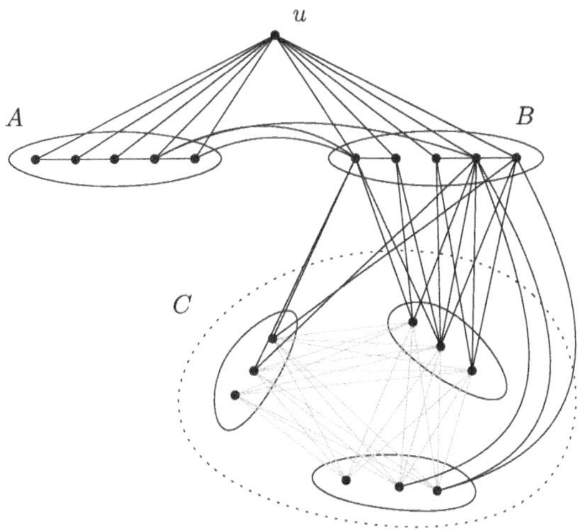

Fig. 1. A sketch of a $(P_2 + 2P_1)$-free graph as seen from some vertex u.

(ii) the diameter of G' is equal to the diameter of G, unless G is a clique,
(iii) if G is H-free for some graph H then G' is also H-free.

There are similar results with respect to false twins. The following was proved by Ducoffe [29] (formulated in terms of modules):

Corollary 1. *Given a graph G and a vertex set $B \subseteq V(G)$, we can partition B into classes of false twins with respect to their neighbourhoods towards $V(G) \setminus B$ in time $O(n + m)$.*

2 Algorithmic Results

2.1 $(P_2 + 2P_1)$-free Graphs

We first prove that the diameter of a $(P_2 + 2P_1)$-free graph can be computed in linear time. The statement of the theorem is slightly stronger however, as we use this algorithm as a subroutine in another proof (♠).

Theorem 6 (♠). *Given a graph G, there is an algorithm that in $O(n+m)$ time either (a) correctly decides that G is not (P_2+2P_1)-free; or (b) outputs a shortest path, which is diametral if G is $(P_2 + 2P_1)$-free.*

Proof (Sketch). Let $G = (V, E)$ be a graph. The diameter of any (P_2+2P_1)-free graph is at most 4. If the diameter of G is 1, then the graph is a clique, which we can check in $O(n + m)$ time, and return any arbitrary pair of vertices, or a single vertex if $|V| = 1$.

Remove twins from the graph in $O(n+m)$ time. By Proposition 1, distances and the diameter are not affected, and the graph is H-free if it was H-free for some graph H. By abuse of notation, we still call this graph $G = (V, E)$.

Let u be a vertex in G with lowest degree, which can be found in $O(n+m)$ time. Now execute a BFS from u. We distinguish the structure of the graph as seen from u; see Fig. 1. Let $C = V \setminus N[u]$ and let $A \subseteq N(u)$ be the subset of vertices of $N(u)$ with no edges to C. Let $B = N(u) \setminus A$. Note that A, B, C can be identified by the BFS from u. If $C = \emptyset$, then the diameter of G is at most 2, and we are done. Hence, $C \neq \emptyset$ and $B \neq \emptyset$. We note that $A = \emptyset$, which can be seen by the following. For any $a \in A$ it holds that $\deg(a) \leq \deg(u)$ by definition of A. u was picked to be a vertex of lowest degree in G, so for any $a \in A$ we have $\deg(a) = \deg(u)$. But as $N(a) \subseteq \{u\} \cup A \cup B$ for all $a \in A$ and $N(u) = A \cup B$, it follows that every $a \in A$ is a twin of u. But then $A = \emptyset$ as we removed twins.

If there is a vertex at distance 5 or more from u, then return that the graph is not $(P_2 + 2P_1)$-free. If there is a vertex at distance 4 from u, then return this shortest path; it is diametral if G is (P_2+2P_1)-free. Both of these cases are identified by the BFS from u. Now observe that any shortest path of length 3 or 4 must have at least one endpoint in C, as the distances between vertices in $A \cup B \cup \{u\}$ are at most 2 by u.

We prove properties of G under the assumption that it is $(P_2 + 2P_1)$-free.

Claim 1 (♠). *If G is $(P_2 + 2P_1)$-free, then (a) $G[C]$ is a complete r-partite graph for some $r \geq 1$; and (b) every $b \in B$ has at most one non-neighbour in every part of $G[C]$.*

Claim 2 (♠). *In $O(n + m)$ time, we can decide whether $G[C]$ is a complete r-partite graph for some $r \geq 1$ and, if so, return its parts.*

Claim 3 (♠). *In $O(n+m)$ time, either (a) we find a length-4 shortest path with both endpoints in C; (b) we find a length-3 shortest path with both endpoints in C and conclude no such length-4 shortest path exists in G; (c) we conclude no length-3 or length-4 shortest path with both endpoints in C exists in G; or (d) we conclude G is not $(P_2 + 2P_1)$-free.*

Run the algorithm of Claim 3. If it returns option (d), then output that G is not (P_2+2P_1)-free. If it returns option (a), then output the length-4 shortest path the algorithm gives; it is diametral if G is $(P_2 + 2P_1)$-free. In both other cases, we argue no length-4 diametral path can exist if G is $(P_2 + 2P_1)$-free. If G is (P_2+2P_1)-free, distances from vertices in B to vertices in C are at most 3, because $G[C]$ is r-partite and every $b \in B$ has at most one non-adjacent vertex per part of $G[C]$ by Claim 1. We already found a length-4 shortest path with u as an endpoint, if it exists. But then any length-4 shortest path has both endpoints in C, if it exists, as we already knew at least one endpoint was in C. In both options (b) and (c) we can conclude that no length-4 shortest path with both endpoints in C exists.

We continue as follows. If u has a vertex at distance 3, then we would already know this by the BFS from u. Otherwise, every $c \in C$ is at distance 2 from u.

If the algorithm of Claim 3 returned option (b), we can output a shortest path of length 3 with both endpoints in C; it is diametral if G is $(P_2 + 2P_1)$-free. If the algorithm of Claim 3 returned option (c), no length-3 shortest path with both endpoints in C exists in G. Hence, the only remaining case for a length-3 shortest path is that there is a vertex in B with distance 3 to a vertex in C.

Claim 4 (♠). *If G is $(P_2 + 2P_1)$-free and a length-3 shortest path exists from some $b \in B$ to some $c \in C$, then (a) $G[C]$ has exactly one part with more than one vertex, and (b) all vertices in B with a vertex in C at distance 3 are adjacent to only that multi-vertex part and have exactly one non-neighbour in that part.*

By Claim 4, we can look for $G[C]$ to have simple structure. In particular, only one part may have multiple vertices. By Claim 2, we can detect if $G[C]$ is complete r-partite in $O(n + m)$ time, and, given the parts, check whether only one part has multiple vertices in $O(n+m)$ time. If this is not the case, then there is no length-3 shortest path from a vertex in B to a vertex in C by Claim 4, if G is $(P_2 + 2P_1)$-free, and we may return a length-2 shortest path with as witness some non-adjacent pair of vertices and a common neighbour.

Otherwise, the structure is as Claim 4(a) and (b) suggest. Find all $b \in B$ only adjacent to the multi-vertex part with one non-neighbour in that part in $O(n+m)$ time, by iterating over the adjacency lists of the vertices in B. Let this be the set of vertices B'. Then look for each vertex in B' whether all its neighbours in B have the same non-adjacency in the multi-vertex part in $O(n + m)$ time. If there is a vertex $b \in B'$ that meets this requirement, and G is $(P_2 + 2P_1)$-free, then this is a witness for diameter-3 shortest path: b is non-adjacent to one vertex $c \in C$ in a multi-vertex part of $G[C]$, which is the only part it is adjacent to, and $N(b) \cap N(c) = \emptyset$. So the distance from b to c is at least (and at most) 3. To verify, execute a BFS from any single one of these vertices. If a shortest path is found of length 4 or longer, then return that G is not $(P_2 + 2P_1)$-free. If a length-3 shortest path is found, then return it. Otherwise, there is no length-3 shortest path in G from B to C, if G is $(P_2 + 2P_1)$-free.

If in none of the above cases a confirmation for a shortest path of length 3 or 4 was found, and the graph is (P_2+2P_1)-free and not a clique, then the diameter of G must be 2. Return some non-adjacent pair with some common neighbour as a shortest path in $O(n + m)$ time. □

Note that any $3P_1$-free graph is $(P_2 + 2P_1)$-free.

Corollary 2. *Given a $3P_1$-free graph G, we can compute the diameter of G in $O(n + m)$ time.*

2.2 $(P_4 + P_1)$-free Graphs

We show that we can decide whether the diameter of a $(P_4 + P_1)$-free graph is equal to $d_{\max}$ in $O(n + m)$ time. The proof will identify all possibilities of a diameter-4 path occurring in relation to a BFS from an arbitrary vertex. Luckily, most cases reduce to some other case in the proof, and algorithmically speaking, only a few cases require algorithmic computation.

Theorem 7 (♠). *Given a $(P_4 + P_1)$-free graph G, we can decide whether the diameter of G is equal to $d_{\max} = 4$ in $O(n + m)$ time.*

Proof (Sketch). Let $G = (V, E)$ be a (connected) $(P_4 + P_1)$-free graph. Indeed, $d_{\max} = 4$. We view the structure of G from a BFS from an arbitrary vertex u. Let $C = V \setminus N[u]$ and denote $B = N(u)$. If $C = \emptyset$ or $B = \emptyset$, then the diameter of G is at most 2, so assume this is not the case. Note that both sets can be identified during the BFS from u with no overhead. Moreover, $G[C]$ is P_4-free. We use the convention that $b_i \in B$ and $c_i \in C$.

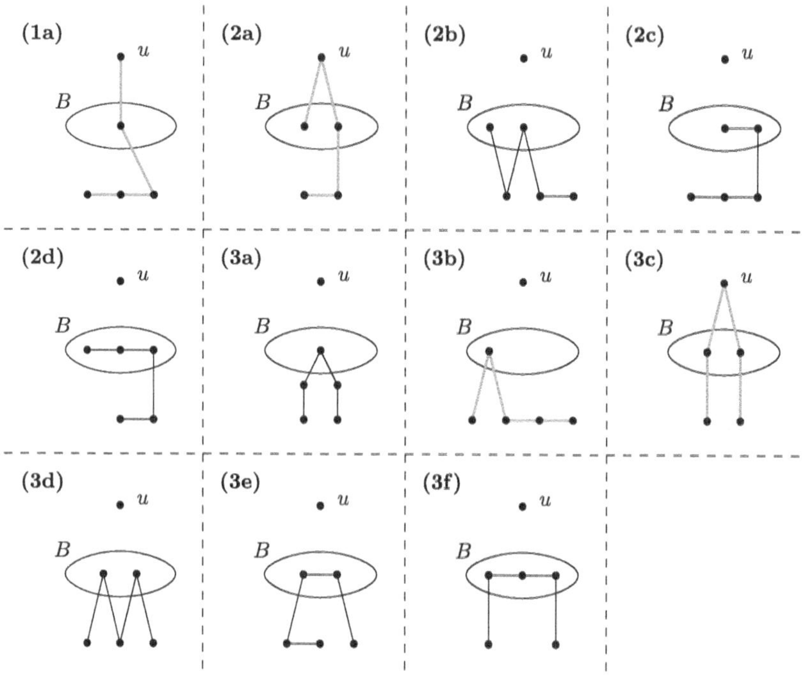

Fig. 2. An illustration of the types of a diameter-4 shortest path appearing in a $(P_4 + P_1)$-free graph, with respect to the vertex u and its neighbourhood $N(u) = B$. Only the highlighted types **(1a)**, **(2a)**, **(3b)**, **(3c)** require algorithmic computation.

We first list all possibilities for a diameter-4 shortest path to exist with respect to its structure; see Fig. 2 for an illustration and ♠ for the full case distinction. We will call each such possibility a 'type'. We show that we can decide whether the diameter of G is 4 in time $O(n + m)$ by resolving each type. Algorithmically speaking, only types **(1a)**, **(2a)**, **(3b)**, and **(3c)** will require computation to find diametral paths corresponding to them (highlighted in the figure). We will show that all other types are either covered by these computations, or are non-existent in G. As illustrations, we show **(1a)**, **(2a)**, and **(2b)** all with proof of correctness, and we give the algorithm for **(3c.1)**, a part of the proof for type **(3c)**; all remaining types and full proofs are discussed in ♠.

(1a) $\langle u, b, c_1, c_2, c_3 \rangle$ This type is identified by the initial BFS from u if and only if it occurs in G.

(2a) $\langle b_2, u, b_1, c_1, c_2 \rangle$ For this type, let us further partition C into sets C' and D, where D consists of the vertices with no neighbours in B (i.e. the vertices at distance 3 from u), and $C' = C \setminus D$. It must be that $c_2 \in D$ for a diametral path of type **(2a)** to exist, otherwise, $\langle c_2, b_i, u, b_2 \rangle$ is a shorter path for some $b_i \in B$. So rename the vertex c_2 as $d = c_2 \in D$ and look for a path $\langle b_2, u, b_1, c_1, d \rangle$. Further partition C' into C_1 and C_2, where C_1 are the vertices with edges towards D and $C_2 = C' \setminus C_1$. This partitioning can be done during the BFS from u, or alternatively using another linear pass over all vertices and edges. Note that every vertex in D has at least one neighbour in C_1; otherwise, a diametral path of type **(1a)** exists in G, and we are done. Let us first describe the algorithmic steps necessary for this type.

(2a.algorithm) Find a vertex $d \in D$ with the smallest degree with respect to C_1, and execute a BFS from d. If a distance-4 vertex is found, return that the diameter of G is 4. If we do not find a vertex at distance 4, no distance-4 diametral path of type **(2a)** exists in G.

We next prove correctness of **(2a.algorithm)**. We prove correctness when $G[D]$ is not connected in **(2a.1)** and correctness when $G[D]$ is connected in **(2a.2)**. To do this, we analyse the structure of G under the assumption that a diametral path of type **(2a)** exists, to conclude the structure of G must then be 'simple' in some way, to the extent that the above algorithmic steps suffice.

(2a.1) Assume $G[D]$ is not connected. We first prove that every vertex in C_1 is complete to D. Assume for sake of contradiction that there is a $c \in C_1$ which is not complete to D. Let $d' \in D$ be a non-neighbour of c. Let $d \in D$ be some neighbour of c, which exists because $c \in C_1$. Now $\langle d, c, b, u \rangle$ is an induced P_4 for some $b \in B$ which exists by definition of C_1. But then d' must be a neighbour of d; otherwise, it would induce a $P_4 + P_1$. We see that every neighbour of c is adjacent to every non-neighbour of c in D. We get a contradiction with the assumption that $G[D]$ is not connected.

So, every vertex in C_1 must be complete to D. But then, from the viewpoint of shortest paths, for any $b \in B$, the distances from all $d \in D$ to b must be equal, as the shortest path to b must go through some vertex of C_1 in the first step, and C_1 is complete to D. Hence, if the BFS that **(2a.algorithm)** executes does not find a length-4 diametral path, no diametral path of type **(2a)** exists in G.

(2a.2) Assume $G[D]$ is connected. Then $G[C_1 \cup D]$ is a connected cograph, so it has diameter at most 2. Let $B_1 \subseteq B$ be the vertices of B with neighbours in C_1. Vertices in $B_2 = B \setminus B_1$ have no neighbours in C_1. Every vertex in B_1 has distance at most 3 to any $d \in D$, as the diameter of $G[C_1 \cup D]$ is at most 2. So, for a shortest path of type **(2a)** $\langle b_2, u, b_1, c_1, d \rangle$ we get $b_2 \in B_2$ and $b_1 \in B_1$.

Let us call a pair (d, b_2) with $d \in D$, $b_2 \in B_2$ 'good' if d has distance 4 to b_2, and 'bad' when d has distance at most 3 to b_2. Assuming a diametral path of form $\langle b_2, u, b_1, c_1, d\rangle$ exists in G, with $b_2 \in B_2$, $b_1 \in B_1$, $c_1 \in C_1$, $d \in D$, it is clear that (d, b_2) is good. Assume that we also have that (d', b_2) is bad for some $d \neq d' \in D$. Then d' has distance exactly 3 to b_2, as b_2 has no neighbour in C_1. Let c_1' be the neighbour of d' on some distance-3 path from d' to b_2. Then $c_1 \neq c_1'$; otherwise, d has distance 3 to b_2. Then the shortest path from d' to b_2 is of the form $\langle d', c_1', c_2', b_2\rangle$ with $c_2' \in C_2$ (case **(2a.2.1)**), or $\langle d', c_1', b_1', b_2\rangle$ with $b_1' \in B_1$, $b_1' \neq b_1$ (case **(2a.2.2)**). See Fig. 3 for an illustration of both scenarios.

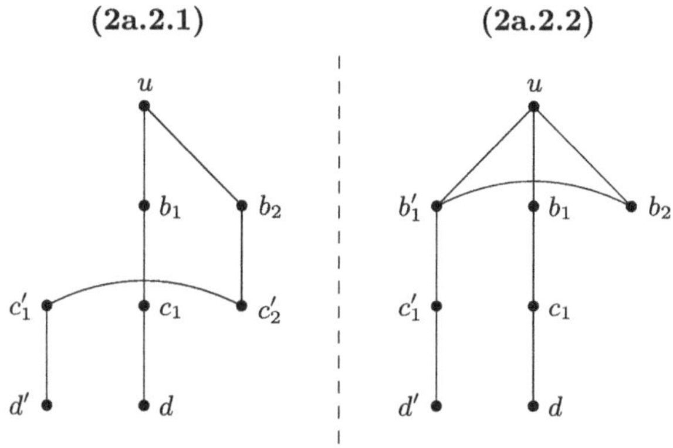

Fig. 3. Structure for type **(2a)** with respect to a $d \in D$ and a $d' \in D$ for which (d, b_2) is good and (d', b_2) is bad for $b_2 \in B_2$. The path from d' to b_2 goes through either some $c_2' \in C_2$ (left, **(2a.2.1)**) or some $b_1' \in B_1$ (right, **(2a.2.2)**).

(2a.2.1) The shortest path from d' to b_2 is of the form $\langle d', c_1', c_2', b_2\rangle$ with $c_2' \in C_2$. Note that $(c_1, u), (c_1', u) \notin E$ by definition of C, and $(c_1, b_2), (c_1', b_2) \notin E$ by definition of B_2, and $(b_1, b_2), (d, c_1'), (c_1, c_2'), (d, c_2') \notin E$ as (d, b_2) is good. But then $(c_1', c_1) \in E$ or $(c_1', b_1) \in E$; otherwise, $\langle c_1, b_1, u, b_2\rangle + c_1'$ is an induced $P_4 + P_1$ in the graph.

- If $(c_1', c_1) \in E$, then $\langle c_2', c_1', c_1, d\rangle + u$ is an induced $P_4 + P_1$ in the graph; u is non-adjacent to all of c_2', c_1', c_1, d by definition of C and D.
- If $(c_1', b_1) \in E$ then $\langle b_2, u, b_1, c_1'\rangle + d$ is an induced $P_4 + P_1$ in the graph; $(c_1', u) \notin E$ by definition of C, and (d, b_1), (d, b_2), $(d, u) \notin E$ by definition of D.

Hence, we get a contradiction, and this case cannot occur in G.

(2a.2.2) The shortest path from d' to b_2 is of the form $\langle d', c_1', b_1', b_2\rangle$ with $b_1' \in B_1$, $b_1' \neq b_1$. Note that $(c_1, u) \notin E$ by definition of C, $(d', b_1), (d', b_2), (d', u) \notin E$ by definition of D, and $(b_1, b_2), (c_1, b_2) \notin E$, as (d, b_2) is good. But then $(d', c_1) \in E$: otherwise, $\langle c_1, b_1, u, b_2\rangle + d'$ is an induced $P_4 + P_1$ in the graph.

Say that d has some other neighbour $c_1'' \in C_1$, so $(c_1'', b_1'') \in E$ for some $b_1'' \in B_1$ (possibly $b_1'' = b_1$), but $(b_1'', b_2) \notin E$ because (d, b_2) is good. Then, c_1'' can fulfil the role of c_1 in the above analysis, so it must be that $(d', c_1'') \in E$. From this analysis we can conclude that, for any $b_2 \in B_2$, if $d \in D$ is such that (d, b_2) is good, and $d' \in D$ is such that (d', b_2) is bad, then it must be that $N(d) \cap C_1 \subset N(d') \cap C_1$. Hence, if some $d \in D$ has minimum degree to C, it may be good.

We now have the tool to prove correctness of **(2a.algorithm)** for this case. If the BFS from the picked $d \in D$ finds a vertex at distance 4, then clearly a length-4 diametral path exists in G. The only risk is that we conclude there is no diametral path corresponding to this case even though it does exist. To this end, let $d \in D$ be the vertex picked by the algorithm, and assume (d, b) is bad for all $b \in B_2$. Assume to the contrary that there exist $d' \in D$, $b_2 \in B_2$ which are at distance 4 in a path of type **(2a)**. Then (d', b_2) is good. So, by the analysis above, it must be that $N(d') \cap C_1 \subset N(d) \cap C_1$, which contradicts the assumption that d was picked to have the minimal size neighbourhood with respect to C_1. So all pairs (d, b_2) with $d \in D$, $b_2 \in B_2$ must be bad, and we are correct to conclude that a diametral path of type **(2a)** does not exist.

(2b) $\langle b_1, c_1, b_2, c_2, c_3 \rangle$ As b_1 and b_2 are both adjacent to u, if type **(2b)** exists in G, $\langle b_1, u, b_2, c_2, c_3 \rangle$ is also a shortest path of distance 4 from b_1 to c_3, which is a distance-4 path of type **(2a)**. The algorithm for type **(2a)** finds a diametral path of length 4 or concludes that no diametral path of type **(2a)** exists in G, which also rules out that a diametral path of type **(2b)** exists in G.

(3c) $\langle c_1, b_1, u, b_2, c_2 \rangle$ We only treat one subcase: **(3c.1)** Every $b \in B$ is complete or anti-complete to every component of $G[C]$. It can be checked in $O(n+m)$ time whether we are in this case. We give the algorithm to solve this case, and the proof of correctness is deferred (♠).

(3c.1.algorithm) For every component of $G[C]$, delete all-but-one vertex in $O(n+m)$ time, each component is a twin class with respect to B. Let the resulting set of vertices be C'. Identify twins within B with respect to their neighbourhood in C', which can be done in linear time by Corollary 1. We identify every 'class' of vertices with the same neighbourhood in C' with single vertices. Let B' be the set of vertices corresponding to classes. The neighbourhoods of the vertices in B' are the union of all neighbourhoods of vertices in the class, which can be computed in $O(n+m)$ time total (♠). Call the resulting graph $G' = (V', E')$ with $B', C' \subseteq V'$. In G', check that $G'[B']$ consists of two non-empty disjoint cliques B_1, B_2, where B_1 and B_2 are anti-complete, with possibly a third clique X complete to $B' \setminus X$. This check takes $O(n+m)$ time by inspecting the neighbourhoods of all vertices in B'. If $G'[B']$ does not have this structure, then return that there is no diametral path of type **(3c)**. If it does, then return that a diametral path of type **(3c)** exists if there are vertices $c_1, c_2 \in C'$ with $N(c_1) = B_1$ and $N(c_2) = B_2$. This can be checked in linear time by inspecting the edge lists of vertices $c \in C'$.

The correctness of **(3c.1.algorithm)** follows from structural analysis on the graph G'. However, G' must be treated with care; it is not necessarily a (P_4+P_1)-free graph. However, the fact that G is (P_4+P_1)-free leads to a strong structural property: for every non-edge $(b'_1, b'_2) \notin E'$ in $G'[B']$, it holds that b'_1, b'_2 together dominate C'. Such a non-edge must exist if a path of type **(3c)** exists in G.

The remaining cases and correctness are treated in the full version (♠). □

3 Conclusion and Discussion

We analysed the complexity of computing the diameter of H-free graphs. For several H-free graph classes $\mathcal{G}$ where H is a linear forest, we found linear-time algorithms for solving $d_{\max}(\mathcal{G})$-DIAMETER. We conjecture that this generalizes to a broader set of H-free graphs. In particular, this conjecture emphasizes that H should include some set of isolated vertices, as it seems that in our algorithms, the presence of a P_1 in H helps out in structural analysis.

Conjecture 1. Let $\mathcal{G}$ be the class of $rP_t + sP_1$-free graphs, where $r, s, t \geq 1$. Then $d_{\max}(\mathcal{G})$-DIAMETER can be solved in $O(n+m)$ time.

Note that all our results support Conjecture 1. A linear-time algorithm for the only open case with $d_{\max} = 4$, which is for $H = P_3 + P_2$, would suggest Conjecture 1 is not the full truth, if it holds true at all. In the full discussion and conclusion (♠), we give two more conjectures in an attempt to reveal the underlying patterns in our results.

The main open problem posed by our work is whether our conjectures hold true. The smallest open cases for solving $d_{\max}(\mathcal{G})$-DIAMETER are algorithms for the classes of H-free graphs with $H = 5P_1$ ($d_{\max} = 7$), $H = 2P_2 + 2P_1$ ($d_{\max} = 7$), $H = P_3 + P_2$ ($d_{\max} = 4$), $H = P_4 + 2P_1$ ($d_{\max} = 6$), or $H = P_5 + P_1$ ($d_{\max} = 5$), and a hardness result for the class of P_6-free graphs. The specific case of $H = P_4 + 2P_1$ is interesting because it is the only graph H for which we have no result.

Settling the following open question would form a complete dichotomy for $(2P_2 + P_1)$-free graphs:

Given a $(2P_2 + P_1)$-free graph G, can we decide in $O(n+m)$ time whether the diameter of G is equal to 4?

We make progress for DIAMETER computation on H-free graphs, but do not settle its complexity completely, so we ask:

When $H = 4P_1$, $H = P_2 + 3P_1$, $H = P_3 + 2P_1$, $H = P_4 + 2P_1$, or $H = P_4 + P_1$, can we solve the DIAMETER *problem on connected H-free graphs in $O(n+m)$ time?*

Acknowledgments. We thank the reviewers for their helpful comments and suggestions.

References

1. Abboud, A., Grandoni, F., Williams, V.V.: Subcubic equivalences between graph centrality problems, APSP, and diameter. ACM Trans. Algorithms **19**(1), 1–30 (2023)
2. Abboud, A., Mozes, S., Weimann, O.: What else can Voronoi diagrams do for diameter in planar graphs? In: Proceedings of ESA 2023. LIPIcs, vol. 274, pp. 1–20. Schloss Dagstuhl - Leibniz-Zentrum für Informatik (2023)
3. Abboud, A., Williams, R.R., Yu, H.: More applications of the polynomial method to algorithm design. In: Proceedings of SODA 2015, pp. 218–230. SIAM (2015)
4. Abboud, A., Williams, V.V., Wang, J.R.: Approximation and fixed parameter subquadratic algorithms for radius and diameter in sparse graphs. In: Proceedings of SODA 2016, pp. 377–391. SIAM (2016)
5. Backurs, A., Roditty, L., Segal, G., Williams, V.V., Wein, N.: Towards tight approximation bounds for graph diameter and eccentricities. In: Proceedings of STOC 2018, pp. 267–280. ACM (2018)
6. Brandstädt, A., Chepoi, V., Dragan, F.F.: The algorithmic use of hypertree structure and maximum neighbourhood orderings. Discret. Appl. Math. **82**(1–3), 43–77 (1998)
7. Bringmann, K., Husfeldt, T., Magnusson, M.: Multivariate analysis of orthogonal range searching and graph distances. Algorithmica **82**(8), 2292–2315 (2020)
8. Cabello, S.: Subquadratic algorithms for the diameter and the sum of pairwise distances in planar graphs. ACM Trans. Algorithms **15**(2), 1–38 (2019)
9. Chan, T.M., Williams, R.R.: Deterministic APSP, orthogonal vectors, and more: quickly derandomizing Razborov-Smolensky. ACM Trans. Algorithms **17**(1), 1–14 (2021)
10. Chechik, S., Larkin, D.H., Roditty, L., Schoenebeck, G., Tarjan, R.E., Williams, V.V.: Better approximation algorithms for the graph diameter. In: Proceedings of SODA 2014, pp. 1041–1052. SIAM (2014)
11. Chepoi, V., Dragan, F.F., Estellon, B., Habib, M., Vaxès, Y.: Notes on diameters, centers, and approximating trees of delta-hyperbolic geodesic spaces and graphs. Electron. Notes Discret. Math. **31**, 231–234 (2008)
12. Chepoi, V., Dragan, F.F., Habib, M., Vaxès, Y., Alrasheed, H.: Fast approximation of eccentricities and distances in hyperbolic graphs. JGAA **23**(2), 393–433 (2019)
13. Corneil, D.G., Dragan, F.F., Habib, M., Paul, C.: Diameter determination on restricted graph families. Discret. Appl. Math. **113**(2–3), 143–166 (2001)
14. Corneil, D.G., Dragan, F.F., Köhler, E.: On the power of BFS to determine a graph's diameter. Networks **42**(4), 209–222 (2003)
15. Coudert, D., Ducoffe, G., Popa, A.: Fully polynomial FPT algorithms for some classes of bounded clique-width graphs. ACM Trans. Algorithms **15**(3), 1–57 (2019)
16. Cygan, M., Gabow, H.N., Sankowski, P.: Algorithmic applications of Baur-Strassen's theorem: shortest cycles, diameter, and matchings. J. ACM **62**(4), 1–30 (2015)
17. Dragan, F.F.: Centers of graphs and the Helly property. Ph. D. Thesis, Moldova State University (1989)
18. Dragan, F.F.: HT-graphs: centers, connected r-domination and Steiner trees. Comput. Sci. J. Moldova **1**(2), 64–83 (1993)
19. Dragan, F.F.: Dominating cliques in distance-hereditary graphs. In: Schmidt, E.M., Skyum, S. (eds.) SWAT 1994. LNCS, vol. 824, pp. 370–381. Springer, Heidelberg (1994). https://doi.org/10.1007/3-540-58218-5_34

20. Dragan, F.F., Ducoffe, G.: α_i-metric graphs: radius, diameter and all eccentricities. In: Proceedings of WG 2023. LNCS, vol. 14093, pp. 276–290. Springer (2023)
21. Dragan, F.F., Ducoffe, G., Guarnera, H.M.: Fast deterministic algorithms for computing all eccentricities in (Hyperbolic) Helly graphs. In: Lubiw, A., Salavatipour, M. (eds.) WADS 2021. LNCS, vol. 12808, pp. 300–314. Springer, Cham (2021). https://doi.org/10.1007/978-3-030-83508-8_22
22. Dragan, F.F., Guarnera, H.M.: Eccentricity function in distance-hereditary graphs. Theor. Comput. Sci. **833**, 26–40 (2020)
23. Dragan, F.F., Habib, M., Viennot, L.: Revisiting radius, diameter, and all eccentricity computation in graphs through certificates. CoRR **abs/1803.04660** (2018)
24. Dragan, F.F., Nicolai, F.: LexBFS-orderings of distance-hereditary graphs with application to the diametral pair problem. Discret. Appl. Math. **98**(3), 191–207 (2000)
25. Dragan, F.F., Nicolai, F., Brandstädt, A.: LexBFS-orderings and powers of graphs. In: Proceedings of WG 1996, pp. 166–180. Springer (1997)
26. Duan, R., Wu, H., Zhou, R.: Faster matrix multiplication via asymmetric hashing. In: Proceedings of FOCS 2023, pp. 2129–2138. IEEE (2023)
27. Ducoffe, G.: Beyond Helly graphs: the diameter problem on absolute retracts. In: Kowalik, Ł, Pilipczuk, M., Rzążewski, P. (eds.) WG 2021. LNCS, vol. 12911, pp. 321–335. Springer, Cham (2021). https://doi.org/10.1007/978-3-030-86838-3_25
28. Ducoffe, G.: The diameter of AT-free graphs. J. Graph Theory **99**(4), 594–614 (2022)
29. Ducoffe, G.: Optimal centrality computations within bounded clique-width graphs. Algorithmica **84**(11), 3192–3222 (2022)
30. Ducoffe, G.: Distance problems within Helly graphs and k-Helly graphs. Theor. Comput. Sci. **946**, 113690 (2023)
31. Ducoffe, G., Dragan, F.F.: A story of diameter, radius, and (almost) Helly property. Networks **77**(3), 435–453 (2021)
32. Ducoffe, G., Habib, M., Viennot, L.: Diameter computation on H-minor free graphs and graphs of bounded (distance) VC-dimension. In: Proceedings of SODA 2020, pp. 1905–1922. SIAM (2020)
33. Duraj, L., Konieczny, F., Potepa, K.: Better diameter algorithms for bounded VC-dimension graphs and geometric intersection graphs. CoRR **abs/2307.08162** (2023)
34. Evald, J., Dahlgaard, S.: Tight hardness results for distance and centrality problems in constant degree graphs. CoRR **abs/1609.08403** (2016)
35. Farley, A.M., Proskurowski, A.: Computation of the center and diameter of outerplanar graphs. Discret. Appl. Math. **2**(3), 185–191 (1980)
36. Gawrychowski, P., Kaplan, H., Mozes, S., Sharir, M., Weimann, O.: Voronoi diagrams on planar graphs, and computing the diameter in deterministic $\tilde{o}(n^{5/3})$ time. SIAM J. Comput. **50**(2), 509–554 (2021)
37. Impagliazzo, R., Paturi, R.: On the complexity of k-SAT. J. Comput. Syst. Sci. **62**(2), 367–375 (2001)
38. Impagliazzo, R., Paturi, R., Zane, F.: Which problems have strongly exponential complexity? J. Comput. Syst. Sci. **63**(4), 512–530 (2001)
39. Johnson, M., et al.: Complexity framework for forbidden subgraphs. CoRR **abs/2211.12887** (2022)
40. Olariu, S.: A simple linear-time algorithm for computing the center of an interval graph. Int. J. Comput. Math. **34**(3–4), 121–128 (1990)
41. Oostveen, J.J., Paulusma, D., van Leeuwen, E.J.: The complexity of diameter on H-free graphs. CoRR **abs/2402.16678** (2024)

42. Roditty, L., Williams, V.V.: Fast approximation algorithms for the diameter and radius of sparse graphs. In: Proceedings of STOC 2013, pp. 515–524. ACM (2013)
43. Shoshan, A., Zwick, U.: All pairs shortest paths in undirected graphs with integer weights. In: Proceedings of FOCS 1999, pp. 605–615. IEEE (1999)
44. Weimann, O., Yuster, R.: Approximating the diameter of planar graphs in near linear time. ACM Trans. Algorithms **12**(1), 1–13 (2016)
45. Zwick, U.: All pairs shortest paths using bridging sets and rectangular matrix multiplication. J. ACM **49**(3), 289–317 (2002)

Approximating Branchwidth on Parametric Extensions of Planarity

Dimitrios M. Thilikos[1(✉)] and Sebastian Wiederrecht[2]

[1] LIRMM, Univ Montpellier, CNRS, Montpellier, France
sedthilk@thilikos.info

[2] Institute for Basic Science, Daejeon, Korea

Abstract. The *branchwidth* of a graph has been introduced by Roberson and Seymour as a measure of the tree-decomposability of a graph, alternative to treewidth. Branchwidth is polynomially computable on planar graphs by the celebrated "Ratcatcher"-algorithm of Seymour and Thomas. We investigate an extension of this algorithm to minor-closed graph classes, further than planar graphs, as follows: Let H_1 be a graph embeddable in the torus and H_2 be a graph embeddable in the projective plane. We prove that every $\{H_1, H_2\}$-minor free graph G contains a subgraph G' where the difference between the branchwidth of G and the branchwidth of G' is bounded by some constant, depending only on H_1 and H_2. Moreover, the graph G' admits a tree decomposition where all torsos are planar. This decomposition can be used for deriving a constant-additive approximation for branchwidth: For $\{H_1, H_2\}$-minor free graphs, there is a constant c (depending on H_1 and H_2) and an $\mathcal{O}(|V(G)|^3)$-time algorithm that, given a graph G, outputs a value b such that the branchwidth of G is between b and $b + c$.

Keywords: Branchwidth · Tree decomposition · Tangle · Slope · Approximation algorithm

1 Introduction

Graph parameters are important in graph algorithm design as they abstract away certain structural characteristics of graphs that appear to be useful in certain algorithmic applications. Perhaps the most eminent among all known graph parameters is *treewidth*. Graphs with small treewidth are those admitting a *tree decomposition* of small width, i.e. a tree-like arrangement into bounded size pieces (see Sect. 2.1). To have such a decomposition at hand is important as, for many problems on graphs, it enables the design of dynamic programming algorithms [8].

The first author was supported by French-German Collaboration ANR/DFG Project UTMA (ANR-20-CE92-0027) and by the Franco-Norwegian project PHC AURORA 2024 (Projet n° 51260WL). The second author was supported by the Institute for Basic Science (IBS-R029-C1 and IBS-R029-Y3).

D. Kráľ and M. Milanič (Eds.): WG 2024, LNCS 14760, pp. 460–474, 2025.
https://doi.org/10.1007/978-3-031-75409-8_32

While a considerable amount of research has been concentrated to treewidth, one may not neglect the attention that has been given to alternative tree-decomposability measures. The most famous "sibling" of treewidth is *branchwidth*, defined in terms of *branch decompositions*, that provide an alternative way to see graphs as tree-like structures. Branchwidth was introduced by Robertson & Seymour (and was adopted henceforth) in the tenth volume of their Graph Minors series [28]. One of the reasons for showing preference to branchwidth is that it enjoys combinatorial properties that makes it easier to handle and generalize than treewidth. Fundamental attempts to generalize the theory of graph minors to matroids used branch decompositions [18,19]. Also, several algorithmic and meta-algorithmic results related to dynamic programming procedures in matroids and hypergraphs utilize branch decompositions [22–25].

For more practical purposes, branchwidth appeares to be useful for particular implementations. Cook and Seymour used branch decompositions to solve the ring routing problem, related to the design of reliable cost effective SONET networks and for solving TSP [7]. Alekhnovich & Razborov [1] used branchwidth of hypergraphs to design algorithms for SAT. Finally, Dorn and Telle introduced decompositions that combined algorithm design on tree and branch decompositions to a single dynamic programming scheme [15].

Computing Branchwidth. Treewidth and branchwidth, as observed by Robertson and Seymour [28], are linearly equivalent in the sense that the one is a constant factor approximation of the other. Computing branchwidth is NP-complete, even on graph classes where treewidth can be computed exactly, [2,26,27,30] but fixed parameter tractable [3,5,6].

Perhaps the most important algorithmic result on branchwidth is the *ratcatcher algorithm* of Seymour and Thomas [30], stating that it can be computed in time $\mathcal{O}(|V(G)|^3)$ on planar graphs (see also [21]). A first proper extension of the ratcatcher method to H-minor-free graphs where H is a single-crossing minor[1] appears implicitly in the work of Demaine, Hajiaghayi, Nishimura, Ragde, and Thilikos [9] who used the method for the design of a polynomial-time constant factor approximation algorithm for treewidth on these graphs. To our knowledge, no polynomial algorithm for branchwidth is known for minor-closed graph classes beyond single-crossing minor-free graphs. However, Feige, Hajiaghayi, and Lee [16] found a polynomial-time algorithm that approximates branchwidth within a factor of $\mathcal{O}(|V(H)|^2)$ on H-minor-free graphs.This last result implies that branchwidth admits a polynomial time constant-factor approximation in every minor-closed graph class. The emerging question is:

> *Are there minor-closed graph classes where a "better than constant-factor" approximation can be achieved?*

This question is the main algorithmic motivation of this paper.

[1] A graph H is a *minor* of a graph G if H can be obtained from some subgraph of G by contracting edges. A graph is *singly-crossing* if it can be embedded in the plane with at most one crossing. Moreover, a graph is a *single-crossing minor* if it is a minor of some singly-crossing graph.

1.1 Our Contribution

Let $\mathcal{E}_1$, and $\mathcal{E}_2$ be the classes of graphs that are embeddable in the torus and the projective plane respectively. Our main result is that whenever we exclude some graph in $\mathcal{E}_1$ and some graph in $\mathcal{E}_2$ as minors, a *constant-additive* approximation of branchwidth is possible. This extends the result of [9] on single-crossing minor free graphs.

Theorem 1. *Branchwidth admits a polynomial-time additive approximation algorithm for every graph class that excludes as a minor a graph $H_1 \in \mathcal{E}_1$ and a graph $H_2 \in \mathcal{E}_2$. In particular, there is a function $f_1\colon \mathbb{N} \to \mathbb{N}$ and an algorithm that, given as input two graphs $H_1 \in \mathcal{E}_1$ and $H_2 \in \mathcal{E}_2$, and an $\{H_1, H_2\}$-minor free graph G, outputs a value b where $b \leq \mathsf{bw}(G) \leq b + f_1(k)$, where $k = |V(H_1)| + |V(H_2)|$. Moreover, $f_1(k) = 2^{\mathsf{poly}(k)}$ and the algorithm runs in time $\mathcal{O}(2^{2^{\mathsf{poly}(k)}}|V(G)|^3)$.*

In the above, by the term $\{H_1, H_2\}$*-minor free graph* we mean a graph that does neither contain H_1 nor H_2 as a minor. We also use the term $\mathsf{poly}(x)$ as a shortcut of $x^{O(1)}$. We next remark that Theorem 1 cannot be extended to general graphs, assuming $\mathsf{P} \neq \mathsf{NP}$.

Theorem 2. *Branchwidth does not admit a polynomial time additive approximation algorithm, unless $\mathsf{P} = \mathsf{NP}$.*

An easy consequence of Theorem 1 is the following EPTAS.

Theorem 3. *Branchwidth admits an EPTAS for every graph class that excludes as a minor a graph $H_1 \in \mathcal{E}_1$ and a graph $H_2 \in \mathcal{E}_2$. In particular, there is a function $f_3\colon \mathbb{N} \to \mathbb{N}$ and an algorithm that, given as input two graphs $H_1 \in \mathcal{E}_1$ and $H_2 \in \mathcal{E}_2$, and an $\{H_1, H_2\}$-minor free graph G, outputs a value b where $b \leq \mathsf{bw}(G) \leq (1+\epsilon)b$. Moreover, the algorithm runs in time $\mathcal{O}(2^{2^{\mathsf{poly}(k)}}|V(G)|^3 + f_3(\epsilon, k)|V(G)|)$, where $k = |V(H_1)| + |V(H_2)|$ and $f_3(\epsilon, k) = 2^{\mathsf{poly}(\frac{1}{\epsilon}\cdot 2^{\mathsf{poly}(k)})}$.*

We stress that, in Theorems 1 and 3, the graphs H_1 and H_2 might be the same graph, i.e., a graph that can be embedded both in the projective plane and the torus. For instance, this implies that the algorithms of Theorem 1 and Theorem 3 apply for K_6-minor free graphs or, alternatively, for graphs excluding the Petersen graph (or any other graph in the Petersen family) as a minor.

1.2 Main Combinatorial Result

We define the *elementary annulus wall* as the parametric graph $\mathscr{W}^{\mathtt{A}} = \langle \mathscr{W}^{\mathtt{A}}_k \rangle_{k \in \mathbb{N}}$, where $\mathscr{W}^{\mathtt{A}}_k$ is the graph obtained from a $(k \times 8k)$-cylindrical grid[2] $\mathscr{G}_k$ by removing some perfect matching that does not intersect the "cycles" of $\mathscr{G}_k$. We consider two more parametric graphs, namely the *elementary handle wall* $\mathscr{W}^{\mathtt{H}} = \langle \mathscr{W}^{\mathtt{H}}_k \rangle_{k \in \mathbb{N}}$ and

[2] Given $k, k' \in \mathbb{N}$ we denote by $\mathscr{G}_k$, the $(k \times k')$-cylindical grid, that is a Cartesian product of a cycle on k vertices and a path on k' vertices.

the *elementary crosscap wall* $\mathscr{W}^{\mathtt{C}} = \langle \mathscr{W}_k^{\mathtt{C}} \rangle_{k \in \mathbb{N}}$, where $\mathscr{W}_k^{\mathtt{H}}$ and $\mathscr{W}_k^{\mathtt{C}}$ are obtained by adding edges inside the "internal disk" of the elementary annulus wall $\mathscr{W}_k^{\mathtt{A}}$, as indicated in Fig. 1.

We may see graphs in $\mathcal{E}_1$ and $\mathcal{E}_2$ as minors of the *elementary handle wall* $\mathscr{W}^{\mathtt{H}} = \langle \mathscr{W}_k^{\mathtt{H}} \rangle_{k \in \mathbb{N}}$ and the *elementary crosscap wall* $\mathscr{W}^{\mathtt{C}} = \langle \mathscr{W}_k^{\mathtt{C}} \rangle_{k \in \mathbb{N}}$. Indeed, as proved by Gavoille and Hilaire in [17], there is a constant c such that every graph H in $\mathcal{E}_1$ (resp. $\mathcal{E}_2$) is a minor of $\mathscr{W}_k^{\mathtt{C}}$ (resp. $\mathscr{W}_k^{\mathtt{H}}$), for every $k \geq c \cdot |V(H)|^2$.

A *handle wall* (resp. *crosscap wall*) of *order* k is some subdivision of $\mathscr{W}_k^{\mathtt{H}}$ (resp. $\mathscr{W}_k^{\mathtt{C}}$). A *handle wall* (resp. *crosscap wall*) of a graph G is a subgraph of G that is a *handle wall* (resp. *crosscap wall*). It is easy to observe that the exclusion of $\mathscr{W}_k^{\mathtt{H}}$ (resp. $\mathscr{W}_k^{\mathtt{C}}$) as minor is equivalent to the exclusion of a *handle wall* (resp. *crosscap wall*) of order k as a subgraph. This is because the maximum degree of $\mathscr{W}_k^{\mathtt{H}}$ and $\mathscr{W}_k^{\mathtt{C}}$ is at most three for all k. This, combined with the aforementioned results of [17], implies the following:

Observation 1. *There is some constant c such that if a graph excludes some graph H in $\mathcal{E}_1$ (resp. $\mathcal{E}_2$) as a minor, it also excludes the handle (resp. crosscap) wall of order $c \cdot |V(H)|^2$ as a subgraph.*

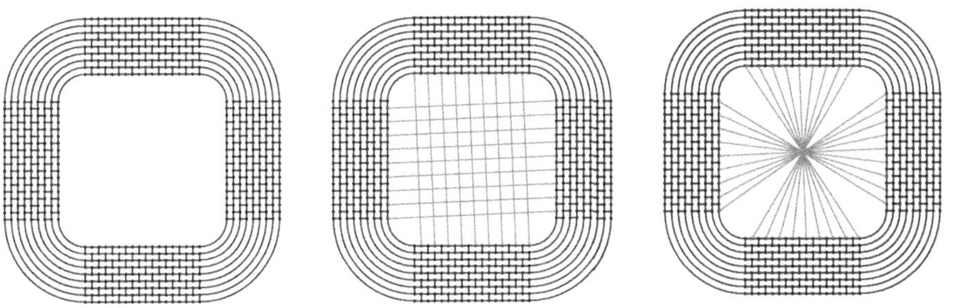

Fig. 1. The elementary annulus wall $\mathscr{W}_9^{\mathtt{A}}$, the elementary handle wall $\mathscr{W}_9^{\mathtt{H}}$, and the elementary crosscap wall $\mathscr{W}_9^{\mathtt{C}}$.

Our algorithm is based on a special case of the main result of [32] that reveals the structure of graphs that do not contain a handle wall or a crosscap wall of order k as subgraphs. Each such graph admits a tree decomposition in pieces that are "bw-almost planar", in the sense that their branchwidth is not much bigger than the branchwidth of some of their planar subgraphs (Proposition 4). Moreover, such a decomposition (for fixed values of k) can be constructed in polynomial time.

Theorem 4. *There is a function $f_4 \colon \mathbb{N} \to \mathbb{N}$ and an algorithm that, given a graph G and a non-negative integer k, either outputs a handle wall or a crosscap wall of G of order k or outputs a subgraph G' of G where $\mathsf{bw}(G) - \mathsf{bw}(G') \leq f_4(k)$ and a tree decomposition (T', β') of G' such that, for every $t \in V(T')$,*

(a) the torso G'_t is planar,
(b) the torso G'_t is a minor of G', and
(c) every adhesion set of t has size at most three.

Moreover, $f_4(k) = 2^{\mathsf{poly}(k)}$ *and the above algorithm runs in time* $\mathcal{O}(2^{2^{\mathsf{poly}(k)}}|V(G)|^3)$.

For the formal definitions of tree decompositions, torsos, and adhesion sets, see Sect. 2.1. Notice that the tree decomposition (T', β') of G' in Theorem 4 implies that G' enjoys the structure of K_5-minor free graphs given by Wagner's theorem [33]. This permits the following restatement of Theorem 4.

Corollary 1. *For every k, there exists some r such that every graph that does not contain a handle wall nor a crosscap wall of order k as a subgraph, contains a K_5-minor free subgraph G' where $\mathsf{bw}(G) - \mathsf{bw}(G') \leq r$.*

The decomposition of Theorem 4 permits us to call the polynomial algorithm of [30] in each G'_t and suitably combine the obtained values in order to design the constant-additive approximation for branchwidth claimed by Theorem 1. The technical details of how to prove Theorem 4, and how to make algorithmic use of it, are given in Sect. 2. In Sect. 3 we give a brief outline on the combinatorial results supporting our algorithm. We conclude with some open problems and research directions in Sect. 4.

2 Decompositions and Algorithms

Given two non-negative integers $a, b \in \mathbb{N}$ we denote the set $\{z \in \mathbb{N} \mid a \leq z \leq b\}$ by $[a, b]$. In case $a > b$ the set $[a, b]$ is empty. For an integer $p \geq 1$, we set $[p] := [1, p]$ and $\mathbb{N}_{\geq p} := \mathbb{N} \setminus [0, p-1]$. The graphs considered in this paper are undirected, finite, without loops, and may have multiple edges, i.e., the edge set is a multi-set. In fact, multiple edges will be relevant in Sect. 3 where we deal with spherical embeddings. We use standard graph-theoretic notation, and we refer the reader to [10] for any undefined terminology.

2.1 Decompositions and Tangles

Many "width parameters" for graphs are defined by "gluing" together simpler graphs of some particular property in a tree-like fashion. A possible way to formalize this notion of tree-likeness is given by tree decompositions.

Let G be a graph. A *tree decomposition* of G is a tuple (T, β) where T is a tree and $\beta\colon V(T) \to 2^{V(G)}$ is a function such that

1. $\bigcup_{t \in V(T)} \beta(t) = V(G)$,
2. for every $e \in E(G)$ there exists $t \in V(T)$ with $e \subseteq \beta(t)$, and
3. for every $v \in V(G)$ the set $\{t \in V(T) \mid v \in \beta(t)\}$ induces a subtree of T.

The *torso* of (T, β) on a node t is the graph, denoted by G_t, that is obtained by $G[\beta(t)]$ by adding all edges between the vertices of $\beta(t) \cap \beta(t')$ for every neighbor t' of t in T. For each $t \in V(T)$, we define the *adhesion sets of* t as the sets in $\{\beta(t) \cap \beta(d) \mid dt \in E(T)\}$ and the maximum size of an adhesion set of t is called the *adhesion of* t.

The *adhesion* of (T, β) is the maximum adhesion of a node of (T, β). The *treewidth* of G, denoted by $\mathsf{tw}(G)$, is the minimum k for which G has a tree decomposition where all torsos have at most $k+1$ vertices.

Observation 2. *Let G be a graph and let (T, β) be a tree decomposition of G. Then* $\mathsf{tw}(G) \leq \max\{\mathsf{tw}(G_t) \mid t \in V(T)\}$.

Branchwidth. A *branch decomposition* of a graph G is a pair (T, δ), where T is a tree with vertices of degree one or three and δ is a bijection from $E(G)$ to the set of leaves of T. The *order function* $\omega \colon E(T) \to 2^{V(G)}$ of a branch decomposition maps every edge e of T to a subset of vertices $\omega(e) \subseteq V(G)$ as follows. The set $\omega(e)$ consists of all vertices $v \in V(G)$ such that there exist edges $f_1, f_2 \in E(G)$ with $v \in f_1 \cap f_2$, and such that the leaves $\delta(f_1)$, $\delta(f_2)$ are in different components of $T - e$.

The *width* of (T, δ) is equal to $\max_{e \in E(T)} |\omega(e)|$ and is zero, in case $|V(T)| = 1$. The *branchwidth* of G, denoted by $\mathsf{bw}(G)$, is the minimum width over all branch decompositions of G. If $E(G) = \emptyset$, then $\mathsf{bw}(G) = 0$.

The next two results reveal important algorithmic properties on branchwidth.

Proposition 1 (Ratcatcher algorithm [30]). *There exists an algorithm that, given a planar graph G, outputs* $\mathsf{bw}(G)$ *in time* $\mathcal{O}(|V(G)|^3)$.

Proposition 2 ([6]). *There exists a function $f_2 : \mathbb{N} \to \mathbb{N}$ and an algorithm that, given a graph G, outputs* $\mathsf{bw}(G)$ *in time* $\mathcal{O}(f_2(\mathsf{bw}(G)) \cdot |V(G)|)$, *where* $f_2(k) = 2^{\mathsf{poly}(k)}$

According to [28], for every graph G with at least one edge it holds that $\max\{\mathsf{bw}(G), 2\} \leq \mathsf{tw}(G) + 1 \leq \max\{\lfloor \frac{3}{2}\mathsf{bw}(G) \rfloor, 2\}$. For our purposes we rewrite the inequality in a slightly more relaxed form as follows.

Lemma 1. *For every graph G,* $\mathsf{bw}(G) \leq \mathsf{tw}(G) + 1 \leq \lfloor \frac{3}{2}\mathsf{bw}(G) \rfloor + 2$.

Tangles. In [28], Robertson and Seymour introduced tangles as the max-min counterpart of branchwidth. Tangles have been important in the proofs of their seminal series of papers on graph minors. Moreover, Tangles also play an important role in an abstract theory of connectivity, see for example [11–14,20].

Tangles. A *separation* in a graph G is a pair (A, B) of vertex sets such that $A \cup B = V(G)$ and there is no edge in G with one endpoint in $A \setminus B$ and the other in $B \setminus A$. The *order* of (A, B) is $|A \cap B|$. Let G be a graph and k be a positive integer. We denote by $\mathcal{S}_k(G)$ the collection of all tuples (A, B) where $A, B \subseteq V(G)$ and (A, B) is a separation of order $< k$ in G. An *orientation* of

$\mathcal{S}_k(G)$ is a set $\mathcal{O}$ such that for all $(A, B) \in \mathcal{S}_k(G)$ exactly one of (A, B) and (B, A) belongs to $\mathcal{O}$. A *tangle* of order k in G is an orientation $\mathcal{T}$ of $\mathcal{S}_k(G)$ such that

$$\text{for all } (A_1, B_1), (A_2, B_2), (A_3, B_3) \in \mathcal{T}, \text{ it holds } A_1 \cup A_2 \cup A_3 \neq V(G).$$

Proposition 3 ([28]). *Let G be a graph where $\mathsf{bw}(G) \geq 2$. Then the maximum order of a tangle of G is equal to $\mathsf{bw}(G)$.*

The next observation is a direct consequence of Proposition 3 and the definition of a tangle.

Observation 3 *Let G be a graph and let $X \subseteq V(G)$ such that $\mathsf{bw}(G - X) \geq 2$. Then $\mathsf{bw}(G - X) \geq \mathsf{bw}(G) - |X|$.*

In this paper we repetitively make use of Proposition 3 in order to work with tangles instead of branch decompositions. Notice that we only make use of the concept of tangles to prove the correctness of our algorithm. Next we need to observe that, in order to find an upper bound on the branchwidth of a graph G with a given tree decomposition, it suffices to consider the torsos.

Lemma 2 ($\star$[3]). *Let $r \in \mathbb{N}_{\geq 1}$. Let G be a graph where $\mathsf{bw}(G) \geq r + 1$ that has a tree decomposition (T, β) of adhesion at most r. Then the branchwidth of G is at most the maximum branchwidth of the torsos of (T, β).*

Given a graph G and a positive integer k we define the graph $G^{(k)}$ by replacing each vertex v of G with a k-clique K_v and each edge $e = uv$ with all k^2 edges between the vertices of K_v and the vertices of K_u.

Lemma 3. *Let G be a graph where $\mathsf{bw}(G) \geq 2$ and $k \in \mathbb{N}_{\geq 1}$. Then $\mathsf{bw}(G^{(k)}) = k \cdot \mathsf{bw}(G)$.*

Proof. By Proposition 3, we may prove that every tangle $\mathcal{T}$ in G of order t corresponds to a tangle $\mathcal{T}'$ in $G^{(k)}$ of order kt.

Suppose that $\mathcal{T}$ is a tangle in G of order t. Let now $(A', B') \in \mathcal{S}_{kt}(G^{(k)})$. Notice that, for every $v \in V(G)$, the vertices of K_v should be either all in A' or all in B'. This means that we may consider a separation (A, B) of G where $A = \{v \in V(G) \mid V(K_v) \subseteq A\}$ and $B = \{v \in V(G) \mid V(K_v) \subseteq B\}$ that has order k. By orienting each separation (A', B') of $G^{(k)}$ in the same way as the corresponding (A, B) in $\mathcal{T}$ we obtain a tangle of $G^{(k)}$ of order kt. Suppose now that $\mathcal{T}'$ is a tangle in $G^{(k)}$ of order kt and let (A, B) be a separation of G of order t. Observe that $(A', B') = (\bigcup_{v \in B} V(K_v), \bigcup_{v \in A} V(K_v)) \in \mathcal{S}_{kt}(G^{(k)})$. As before, we may use the orientation of $\mathcal{S}_{kt}(G^{(k)})$ defined by $\mathcal{T}'$, in order to define an orientation $\mathcal{T}$ of $\mathcal{S}_t(G)$ that is a tangle of G of order t. □

[3] The proofs of the results marked with "$\star$" can be found in the full version of the paper in [31].

Lemma 3 permits us to prove that one may not expect, in general, any constant-additive approximation for branchwidth that runs in polynomial time. The following proof of Theorem 2 uses a reduction employed by [4] for deriving the analogous result for treewidth.

Proof (of Theorem 2). Suppose that there is some integer $k \geq 0$ and an algorithm that, for every graph G, outputs a number b such that $\mathsf{bw}(G) \leq b \leq \mathsf{bw}(G)+k$, for some constant $k \geq 0$. We use this algorithm as a subroutine to compute $\mathsf{bw}(G)$. If $\mathsf{bw}(G) < 2$, then we may compute $\mathsf{bw}(G)$ using Proposition 2. If $\mathsf{bw}(G) \geq 2$, we apply the additive approximation algorithm on $G^{(k+1)}$ and, by calling upon Lemma 3, obtain a value b where $\mathsf{bw}(G^{(k+1)}) \leq b \leq \mathsf{bw}(G^{(k+1)}) + k \Rightarrow \mathsf{bw}(G) \leq b/(k+1) \leq \mathsf{bw}(G)+1-\frac{1}{k+1} < \mathsf{bw}(G)+1$, which implies that $\mathsf{bw}(G) = \lfloor b/(k+1) \rfloor$. This would make possible to decide in polynomial time whether, given a graph G and an integer r, $\mathsf{bw}(G) \leq r$, which, as proven in [30], is an NP-complete problem. □

Notice that the blow-up operation used to define $G^{(k)}$ introduces large cliques to the graph. This means that the operation does not respect the exclusion of a minor which means that Lemma 3 and Theorem 2 do not preclude additive approximation algorithms for the branchwidth of graphs excluding a minor.

2.2 Approximation Algorithms

The combinatorial engine of our approximation algorithms is the following result.

Theorem 5. *There exists a function $f_5 \colon \mathbb{N} \to \mathbb{N}$ and an algorithm that, given a graph G and an integer k, either outputs a subgraph of G that is a handle wall or a crosscap wall of order k or outputs a tree decomposition (T, β), where every torso G_t, $t \in V(T)$, contains a set X_t such that*

1. *$G_t - X_t$ is planar,*
2. *$G_t - X_t$ is a minor of G,*
3. *every adhesion of t contains at most $f_5(k)$ vertices and, among them, at most three do not belong to X_t, and*
4. *$\mathsf{bw}(G_t) - \mathsf{bw}(G_t - X_t) \leq f_5(k)$.*

Moreover, $f_5(k) = 2^{\mathsf{poly}(k)}$ and the above algorithm runs in time $\mathcal{O}(2^{2^{\mathsf{poly}(k)}}|V(G)|^3)$.

We postpone the proof of Theorem 5 to Sect. 3.

Theorem 5, is used for the proof of Theorem 4, as follows.

Proof (of Theorem 4). Consider a tree decomposition (T, β) of G where every torso G_t contains a set X_t such that conditions 1, 2, 3, and 4 of Theorem 5 are satisfied. By Theorem 5, there exists an algorithm that computes such a decomposition in time $\mathcal{O}(2^{2^{\mathsf{poly}(k)}}|V(G)|^3)$. For each $t \in V(T)$, we set $G'_t := G_t - X_t$. Let $X := \bigcup_{t \in V(T)} X_t$ and $G' := G - X$. We consider the tree decomposition

(T, β') of G' where, for each $t \in V(T)$, $\beta'(t) = \beta(t) \setminus X_i$ (clearly $\beta'(t) = V(G'_t)$). The algorithm outputs G' and (T, β'). For every $t \in V(T)$, conditions (a), (b), and (c) hold for (T, β') because conditions 1, 2, and 3 of Theorem 5 respectively hold for (T, β'). It remains to prove that

$$\mathsf{bw}(G) - \mathsf{bw}(G') \leq f_5(k). \tag{1}$$

If $\mathsf{bw}(G) \leq f_5(k)$, we directly have that $\mathsf{bw}(G) - \mathsf{bw}(G') \leq \mathsf{bw}(G) \leq f_5(k)$ and (1) follows.

If $\mathsf{bw}(G) > f_5(k)$, then, using Theorem 5.(3), we may apply Lemma 2 for G and $r = f_5(k)$ and obtain that $\mathsf{bw}(G) \leq \max\{\mathsf{bw}(G_t) \mid t \in V(T)\} \leq^{\text{Theorem5.(4)}} \max\{\mathsf{bw}(G'_t) \mid t \in V(T)\} + f_5(k) \leq^{(\text{b})} \mathsf{bw}(G') + f_5(k)$. This yields (1) and the result follows by setting $f_4(k) = f_5(k) = 2^{\mathsf{poly}(k)}$. □

Using now Theorem 4, we can prove Theorem 1.

Proof (of Theorem 1). By Observation 1, G does not contain a handle wall or a crosscap wall of order $k = c|V(H)|^2$ as a subgraph. Thus, using Theorem 4, we can find, in time $\mathcal{O}(2^{2^{\mathsf{poly}(|V(H)|)}}|V(G)|^3)$, a subgraph G' of G where

$$\mathsf{bw}(G) \leq \mathsf{bw}(G') + f_4(k) \tag{2}$$

and a tree decomposition (T', β') of G' such that, for every $t \in V(T')$, conditions (a), (b), and (c) of Theorem 4 hold. From condition (a) it follows that each torso $G'_t, t \in V(T)$, is a planar graph. Therefore we can use the algorithm of Proposition 1 in order to compute the branchwidth of G'_t in time $\mathcal{O}(|G'_t|^3)$. Our algorithm returns the value $b = \max\{\mathsf{bw}(G'_t) \mid t \in V(T')\}$, that can be computed in time $\mathcal{O}(|V(G')|)$. We next claim that

$$\mathsf{bw}(G') \leq b + 3 \tag{3}$$

In order to prove (3), we distinguish two cases.

Case 1: $b > 3$: In this case we may use property (c) in order to apply Lemma 2 for G' and $r = 3$ and obtain that $\mathsf{bw}(G') = b \leq b + 3$.

Case 2: $b \leq 3$: Here, because of Lemma 1 and Observation 2, we have that $\mathsf{bw}(G') \leq \mathsf{tw}(G') + 1 \leq \max\{\mathsf{tw}(G'_t) \mid t \in V(T)\} + 1 \leq \max\{\lfloor \frac{3}{2}\mathsf{bw}(G'_t) \rfloor + 2 \mid t \in V(T)\} = \lfloor \frac{3}{2}b \rfloor + 2 \leq b + 3$.

We conclude that $b \leq^{\text{Theorem4.b)}} \mathsf{bw}(G') \leq \mathsf{bw}(G) \leq^{(2)} \mathsf{bw}(G') + f_4(k) \leq^{(3)} b + 3 + f_4(k)$ and the theorem holds if we set $f_1 = 3 + f_4(k) = 2^{\mathsf{poly}(k)}$. □

We now use Theorem 4 for the EPTAS of Theorem 3.

Proof (of Theorem 3). Let $\epsilon > 0$. We give an $(1 + \epsilon)$-approximation algorithm as follows. First, we run the algorithm of Theorem 1 and obtain some value b such that $b \leq \mathsf{bw}(G) \leq b + f_1(k)$ where $k = |V(H_1)| + |V(H_2)|$. This algorithm runs in $\mathcal{O}(2^{2^{\mathsf{poly}(k)}}|V(G)|^3)$ steps.

If $f_1(k)/b < \epsilon$, then $\mathsf{bw}(G) \leq b + f_1(k) < (1 + \epsilon)b$. The algorithm outputs b.

If $f_1(k)/b \geq \epsilon$, this means that $\mathsf{bw}(G) \leq b + f_1(k) \leq f_1(k)/\epsilon + f_1(k) = (1/\epsilon + 1) \cdot f_1(k)$, and we use Proposition 2 in order to output the exact value of $\mathsf{bw}(G)$ in time $\mathcal{O}(f_2((1/\epsilon + 1) \cdot f_1(k)) \cdot |V(G)|)$.

In both cases, we output a value b' where $b' \leq \mathsf{bw}(G) \leq (1 + \epsilon)b'$, i.e., b' is an $(1 + \epsilon)$-factor approximation of $\mathsf{bw}(G)$. The result follows if we set $f_3(\epsilon, k) = f_2((1/\epsilon + 1) \cdot f_1(k)) = 2^{\mathsf{poly}(\frac{1}{\epsilon} \cdot 2^{\mathsf{poly}(k)})}$. □

3 A Sketch for the Proof for Theorem 5

In this section we give a rough sketch of the technical proof of Theorem 5.

We make use of some powerful tools from the theory of graph minors, specifically, we use a refinement of the *Graph Minors Structure Theorem* by Robertson and Seymour [29] as it appeared in [32]. A core concept in these types of structure theorems is the idea of *almost embeddings* in surfaces. For our purposes it suffices to define such embeddings for the sphere only.

We use $\mathbb{S}^2$ to denote the *sphere*, i.e., the points of $\{(x, y, z) \mid x^2 + y^2 + z^2 = 1\}$. A *spherical embedding* is a pair (Γ, U) where $U \subseteq \Gamma \subseteq \mathbb{S}_0$, Γ is a closed set, U is a finite set of points of $\mathbb{S}^2$, and each connected component of $\Gamma \setminus U$ is homeomorphic to the open set $(0, 1)$. A *noose* of (Γ, U) is a closed curve (i.e., a set that is homeomorphic to the cycle $\{(x, y) \mid x^2 + y^2 = 1\}$) that intersects Γ only in points of U.

We refer to the points in U as the *points* of (Γ, U) and to the connected components of $\Gamma \setminus U$ as the *arcs* of (Γ, U). Given an arc a of (Γ, U), we define its endpoints as the points of $\overline{a} \setminus a$ (we use the notation $\overline{a}$ for the closure of a set a). We say that (Γ, U) is a *spherical embedding* of a graph G if the graph $(U, \{xy \mid x \text{ and } y \text{ are the endpoints of some arc in } (\Gamma, U)\})$ is isomorphic to G.

An *almost spherical embedding* of a graph G is a quadruple $(G', \Gamma, U, \mathcal{B}, \mathcal{P})$ where

1. G' is a subgraph of G and $\mathcal{B} = \{B_1, \ldots, B_b\}$ are subgraphs of G, such that $G = G' \cup \bigcup \mathcal{B}$.
2. (Γ, U) is a spherical embedding of G'.
3. for each $i \in [b]$, there exists a noose N_i of (Γ, U) where $V(G') \cap V(B_i) = V_{G'}(N_i) = \{v_1^i, \ldots, v_{n_i}^i\}$, ordered according to the appearance of the vertices in $V_{G'}(N_i)$ while traversing along N_i and such that one, say Δ_i, of the two open disks bounded by N_i does not contain any arc or point of (Γ, U). Moreover, it holds that for $1 \leq i < j \leq b$, $\Delta_i \cap \Delta_j \subseteq U$.
4. $\mathcal{P} = \{(P_i, \beta_i) \mid i \in [b]\}$, where each (P_i, β_i) is a path decomposition[4] of B_i such that
   (a) P_i is a path on the vertices $p_1^i, \ldots, p_{n_i}^i$, appearing in this order, and
   (b) for each $j \in [n_i]$ we have $v_j^i \in \beta_i(p_j^i)$.

[4] A *path decomposition* of a graph G is a tree decomposition (P, β) for G where P is a path.

We call the graphs B_i, $i \in [b]$, the *vortices* of $(G', \Gamma, U, \mathcal{B}, \mathcal{P})$ and we call the noose N_i (resp. open disk Δ_i) the *boundary noose* (resp. *disk*) of the vortex $B_i, i \in [b]$. We call the vertices in $B_i \setminus V_{G'}(N_i)$ (resp. $V_{G'}(N_i)$) *internal* (resp. *boundary*) vertices of the vortex B_i. The *width* of an almost spherical embedding $(G', \Gamma, U, \mathcal{B}, \mathcal{P})$ of G is the maximum width over the path decompositions in $\mathcal{P}$ and its *breadth* is b, i.e., is the number of vortices.

We now present the following result that can be derived from [32, Lemma 5.5] for the case where the surface is a sphere.

Proposition 4. *There exists a function $\zeta : \mathbb{N} \to \mathbb{N}$ and an algorithm that, given a graph G and integer k, either finds a subgraph of G that is either a handle wall or a crosscap wall of order k or outputs a tree decomposition (T, β) of G of adhesion less than $\zeta(k)$ such that, for every $t \in V(T)$, assuming that G_t is the torso of G at $t \in V(T)$, it holds that either*

- $|V(G_t)| \leq \zeta(k)$ *or*
- *there exist a set $A_t \subseteq \beta(t)$, called* apex set, *where $|A_t| \leq \zeta(k)$ such that the graph $G'_t := G_t - A_t$ has an almost spherical embedding $(G'_t, \Gamma_t, U_t, \mathcal{B}_t, \mathcal{P}_t)$ of width at most $\zeta(k)$ and breadth at most $\zeta(k)$, such that*

  B1 *for every neighbor $\hat{t}$ of t in T, $G'_t \cap G'_{\hat{t}}$ is a complete graph on at most three vertices and*

  B2 *G'_t is a minor of G.*

Moreover, $\zeta(k) = 2^{\mathsf{poly}(k)}$ and the above algorithm runs in time $\mathcal{O}(2^{2^{\mathsf{poly}(k)}}|V(G)|^3)$.

We stress that Proposition 4 is a special case of the main result in [32] that provides tree decompositions of spherical embeddings instead of similar decompositions for general surfaces. Proposition 4 is tight in the sense that no such tree decomposition is possible for graphs containing a large handle wall or a crosscap wall.

Remark 1. In Proposition 4 we may also assume that, in the second outcome, we also have

B3 $\mathsf{bw}(G'_t) \geq 2$.

The above remark is supported by the following observation that follows easily from known results.

Lemma 4 (⋆). *There exists a function $f_4 : \mathbb{N}^2 \to \mathbb{N}$ such that if $(G', \Gamma, U, \mathcal{B}, \mathcal{P})$ is an almost spherical embedding of a graph G of width at most w and breadth b and where $\mathsf{bw}(G') \leq 1$, then $\mathsf{tw}(G) \leq f_{mp}(w, b)$.*

Indeed, for the t's where $\mathsf{bw}(G'_t) \leq 1$, we may use Lemma 4, in order to construct a tree decomposition of G_t of adhesion $\leq f_4(q)$ where each torso has at most $f_4(q) + 1$ vertices, assuming that $q = \zeta(k)$. Then we use these decompositions in order to transform the tree decomposition of Proposition 4, where we may update $\zeta(k)$ to be $f_4(q)$. This is possible because each adhesion set X of $\beta(t)$

is a clique in G_t and thus, the tree decomposition constructed above must have some bag which fully contains X.

A graph is 2-*edge connected* if it is connected and *bridgeless*, i.e., does not contain edges whose removal increase the number of connected components. The ratcatcher algorithm relies heavily on planar embeddings of graphs and their duals. As a consequence, several arguments rely on a gentle connectivity assumption that can easily be achieved by adding a small amount of additional edges.

Lemma 5 (⋆). *Let* $(G', \Gamma, U, \mathcal{B}, \mathcal{P})$ *be an almost spherical embedding of a graph* G*, with* $\mathsf{bw}(G') \geq 2$*. Then there is a set* $E \subseteq \binom{V(G')}{2}$ *such that, if* G^{new} *(resp.* $G'^{,\text{new}}$*) is the graph obtained from* G *(resp.* G'*) by adding* E *to the existing edges, agreeing that if an edge in* E *already exists then its multiplicity becomes double, then* $G'^{,\text{new}}$ *is* 2*-edge-connected and* $\mathsf{bw}(G'^{\text{new}}) = \mathsf{bw}(G')$*. Moreover,* G^{new} *has a an almost spherical embedding* $(G'^{,\text{new}}, \Gamma^{\text{new}}, U, \mathcal{B}, \mathcal{P})$*.*

We are now ready to state the key tool that will facilitate the proof of Theorem 5.

Theorem 6 (⋆). *Let* G *be a graph and let* $(G', \Gamma, U, \mathcal{B}, \mathcal{P})$ *be an almost spherical embedding of* G *of breadth* b *and width at most* w*. Assume also that* G' *is* 2*-edge connected. Then* $\mathsf{bw}(G) - \mathsf{bw}(G') \leq 2b^2 \cdot w$*.*

The proof of Theorem 6 is technical and amounts to adapting the notion of *slopes* to the setting of graphs with almost spherical embeddings. Slopes where defined by Seymour and Thomas in [30] and can be seen as the topological analogue of tangles designed to understand the interaction between vertex separators and nooses in plane graphs. The result in [30] roughly says that, for plane graphs, in the duality of Proposition 3 between tangles and branchwidth, we may restrict our tangles to separations defined by noses. The proof of Theorem 6, makes strong use of this result.

Proof (*of Theorem* 5). We consider the decomposition (T, β) of Proposition 4 and let $q := \zeta(k)$.

Let $t \in V(T)$. If $|V(G_t)| \leq q$, we set $X_t = V(G_t)$ and conditions 1–4 hold trivially as $G_t - X_t$ is the empty graph.

Let now $|V(G_t) > q$ and let $(G'_t, \Gamma_t, U_t, \mathcal{B}_t, \mathcal{P}_t)$ be an almost spherical embedding of $\overline{G}_t := G_t - A_t$ of width $\leq q$ and breadth $\leq q$ and for some $A_t \subseteq V(G_t)$ where $|A_t| \leq q$.

Let $Z_t = (\bigcup \mathcal{B}_t) \setminus V(G'_t)$ and $X_t = A_t \cup Z_t$, i.e., X_t contains all apices and all the internal vertices of the vortices in $\mathcal{B}_t$. Clearly, $G_t - X_t = \overline{G}_t - Z_i = G'_t$. Condition 1 follows as G'_t has a spherical embedding. Conditions 2 and 3 follow from B1 and B2 respectively.

For Condition 4, recall first that **B3** permits us to assume that $\mathsf{bw}(G'_t) \geq 2$ (Remark 1). Therefore we may apply Lemma 5 and obtain a 2-edge-connected graph $G'^{,\text{new}}_t$ where $\mathsf{bw}(G'_t) = \mathsf{bw}(G'^{,\text{new}}_t)$ and an almost spherical embedding

$(G_t^{\prime,\text{new}}, \Gamma_t^{\text{new}}, U_t, \mathcal{B}_t, \mathcal{P}_t)$ of $\overline{G}_t^{\text{new}}$. As $\overline{G}_t$ is a spanning subgraph of $\overline{G}_t^{\text{new}}$, we also have that $\mathsf{bw}(\overline{G}_t) \leq \mathsf{bw}(\overline{G}_t^{\text{new}})$.

We may now apply Theorem 6 on $(G_t^{\prime,\text{new}}, \Gamma_t^{\text{new}}, U_t, \mathcal{B}_t, \mathcal{P}_t)$ and derive that $\mathsf{bw}(\overline{G}_t^{\text{new}}) \leq 2q^3 + \mathsf{bw}(G_t^{\prime,\text{new}})$. This implies that $\mathsf{bw}(\overline{G}_t) \leq 2q^3 + \mathsf{bw}(G'_t)$. Therefore, we obtain $\mathsf{bw}(G_t) \leq |A_t| + \mathsf{bw}(\overline{G}_t) \leq q + 2q^3 + \mathsf{bw}(G'_t)$, from Observation 3, and thus, Condition 4 also holds by setting $f_5(k) := q + 2q^3$. □

4 Conclusion

In this paper we show that for classes of graphs not containing as a subgraph a handle wall or a crosscap wall, there exists an constant-additive approximation algorithm for branchwidth (Theorem 3).

As we already mentioned, the exact polynomial computation of branchwidth is known to be possible for graph classes excluding a single-crossing graph as a minor [9]. As single-crossing graphs are minors of both the handle wall and the crosscap wall, our result can be seen as an "approximation" extension of the results in [9]. However the question on whether (and when) an exact polynomial algorithm exists remains open. Our guess is the following.

Conjecture 1. If H is a graph that can be embedded both in the torus and in the projective plane, then branchwidth can be computed in polynomial time in H-minor free graphs.

Notice that single-crossing graphs are minors of both handle walls or a crosscap walls. However, the inverse is not the case. For instance K_6 can be embedded both in the torus and in the projective plane, but it is not a single-crossing graph. So a first step towards pursuing a proof of the above conjecture would be to look for a polynomial time algorithm for branchwidth on K_6-minor free graphs or proving that this is not possible subject to some complexity theory assumption.

Acknowledgements. The authors wish to thank the anonymous reviewers for their remarks and suggestions on earlier versions of this paper.

References

1. Alekhnovich, M., Razborov, A.: Satisfiability, branch-width and Tseitin tautologies. Comput. Complexity **20**(4), 649–678 (2011). https://doi.org/10.1007/s00037-011-0033-1
2. Arnborg, S., Corneil, D.G., Proskurowski, A.: Complexity of finding embeddings in a k-tree. SIAM J. Algebraic Discrete Methods **8**(2), 277–284 (1987). https://doi.org/10.1137/0608024
3. Bodlaender, H.L.: A linear-time algorithm for finding tree-decompositions of small treewidth. SIAM J. Comput. **25**(6), 1305–1317 (1996). https://doi.org/10.1137/S0097539793251219
4. Bodlaender, H.L., Gilbert, J.R., Hafsteinsson, H., Kloks, T.: Approximating treewidth, pathwidth, frontsize, and shortest elimination tree. J. Algorithms **18**(2), 238–255 (1995). https://doi.org/10.1006/jagm.1995.1009

5. Bodlaender, H.L., Kloks, T.: Efficient and constructive algorithms for the pathwidth and treewidth of graphs. J. Algorithms **21**(2), 358–402 (1996). https://doi.org/10.1006/jagm.1996.0049
6. Bodlaender, H.L., Thilikos, D.M.: Constructive linear time algorithms for branchwidth. In: Degano, P., Gorrieri, R., Marchetti-Spaccamela, A. (eds.) ICALP 1997. LNCS, vol. 1256, pp. 627–637. Springer, Heidelberg (1997). https://doi.org/10.1007/3-540-63165-8_217
7. Cook, W., Seymour, P.D.: An algorithm for the ring-routing problem. Bellcore technical memorandum, Bellcore (1993)
8. Courcelle, B.: The monadic second-order logic of graphs. I. Recognizable sets of finite graphs. Inform. and Comput. **85**(1), 12–75 (1990). https://doi.org/10.1016/0890-5401(90)90043-H
9. Demaine, E.D., Hajiaghayi, M.T., Nishimura, N., Ragde, P., Thilikos, D.M.: Approximation algorithms for classes of graphs excluding single-crossing graphs as minors. J. Comput. System Sci. **69**(2), 166–195 (2004). https://doi.org/10.1016/j.jcss.2003.12.001
10. Diestel, R.: Graph theory, Graduate Texts in Mathematics, vol. 173, 5th edn. Springer, Berlin (2017). https://doi.org/10.1007/978-3-662-53622-3
11. Diestel, R.: Abstract separation systems. Order **35**(1), 157–170 (2018). https://doi.org/10.1007/s11083-017-9424-5
12. Diestel, R., Erde, J., Weiß auer, D.: Structural submodularity and tangles in abstract separation systems. J. Combin. Theory Ser. A **167**, 155–180 (2019). https://doi.org/10.1016/j.jcta.2019.05.001
13. Diestel, R., Kneip, J.: Profinite separation systems. Order **37**(1), 179–205 (2020). https://doi.org/10.1007/s11083-019-09499-y
14. Diestel, R., Oum, S.I.: Tangle-tree duality: in graphs, matroids and beyond. Combinatorica **39**(4), 879–910 (2019). https://doi.org/10.1007/s00493-019-3798-5
15. Dorn, F., Telle, J.A.: Semi-nice tree-decompositions: the best of branchwidth, treewidth and pathwidth with one algorithm. Discrete Appl. Math. **157**(12), 2737–2746 (2009). https://doi.org/10.1016/j.dam.2008.08.023
16. Feige, U., Hajiaghayi, M., Lee, J.R.: Improved approximation algorithms for minimum weight vertex separators. SIAM J. Comput. **38**(2), 629–657 (2008). https://doi.org/10.1137/05064299X
17. Gavoille, C., Hilaire, C.: Minor-universal graph for graphs on surfaces. arXiv preprint: arXiv:2305.06673 (2023)
18. Geelen, J., Whittle, G.: Branch-width and Rota's conjecture. J. Combin. Theory Ser. B **86**(2), 315–330 (2002). https://doi.org/10.1006/jctb.2002.2130
19. Geelen, J.F., Gerards, A.M.H., Whittle, G.: Branch-width and well-quasi-ordering in matroids and graphs. J. Combin. Theory Ser. B **84**(2), 270–290 (2002). https://doi.org/10.1006/jctb.2001.2082
20. Grohe, M.: Tangled up in blue (A survey on connectivity, decompositions, and tangles). CoRR **abs/1605.06704** (2016). http://arxiv.org/abs/1605.06704
21. Gu, Q.P., Tamaki, H.: Optimal branch-decomposition of planar graphs in $O(n^3)$ time. ACM Trans. Algorithms **4**(3), Art. 30, 13 (2008). https://doi.org/10.1145/1367064.1367070
22. Hliněný, P.: A parametrized algorithm for matroid branch-width. SIAM J. Comput. **35**(2), 259–277, loose erratum (2005). https://doi.org/10.1137/S0097539702418589
23. Hliněný, P.: Branch-width, parse trees, and monadic second-order logic for matroids. J. Combin. Theory Ser. B **96**(3), 325–351 (2006). https://doi.org/10.1016/j.jctb.2005.08.005

24. Hliněný, P., Oum, S.I.: Finding branch-decompositions and rank-decompositions. SIAM J. Comput. **38**(3), 1012–1032 (2008). https://doi.org/10.1137/070685920
25. Jeong, J., Kim, E.J., Oum, S.I.: Finding branch-decompositions of matroids, hypergraphs, and more. SIAM J. Discrete Math. **35**(4), 2544–2617 (2021). https://doi.org/10.1137/19M1285895
26. Kloks, T., Kratochvíl, J., Müller, H.: Computing the branchwidth of interval graphs. Discrete Appl. Math. **145**(2), 266–275 (2005). https://doi.org/10.1016/j.dam.2004.01.015
27. Paul, C., Telle, J.A.: Branchwidth of chordal graphs. Discrete Appl. Math. **157**(12), 2718–2725 (2009). https://doi.org/10.1016/j.dam.2008.08.006
28. Robertson, N., Seymour, P.D.: Graph minors. X. Obstructions to tree-decomposition. J. Combin. Theory Ser. B **52**(2), 153–190 (1991). https://doi.org/10.1016/0095-8956(91)90061-N
29. Robertson, N., Seymour, P.D.: Graph minors. XVI. Excluding a non-planar graph. J. Combin. Theory Ser. B **89**(1), 43–76 (2003). https://doi.org/10.1016/S0095-8956(03)00042-X
30. Seymour, P.D., Thomas, R.: Call routing and the ratcatcher. Combinatorica **14**(2), 217–241 (1994). https://doi.org/10.1007/BF01215352
31. Thilikos, D.M., Wiederrecht, S.: Approximating branchwidth on parametric extensions of planarity (2023). https://arxiv.org/abs/2304.04517
32. Thilikos, D.M., Wiederrecht, S.: Excluding surfaces as minors in graphs (2024). https://arxiv.org/abs/2306.01724
33. Wagner, K.: Über eine Eigenschaft der ebenen Komplexe. Math. Ann. **114**(1), 570–590 (1937). https://doi.org/10.1007/BF01594196

Author Index

D. Král′ and M. Milanič (Eds.): WG 2024, LNCS 14760, pp. 475–476, 2025.
https://doi.org/10.1007/978-3-031-75409-8

The manufacturer's authorised representative in the EU is Springer Nature Customer Service Centre GmbH, Europaplatz 3, 69115 Heidelberg, Germany. If you have any concerns regarding our products, please contact ProductSafety@springernature.com

Printed and bound by CPI Group (UK) Ltd, Croydon, CR0 4YY

29/07/2026

02180491-0004